Signals and Images

Advances and Results in Speech, Estimation, Compression, Recognition, Filtering, and Processing

Signals and Images

Advances and Results in Speech, Estimation, Compression, Recognition, Filtering, and Processing

EDITED BY

Rosângela Fernandes Coelho • Vítor Heloiz Nascimento
Ricardo Lopes de Queiroz
João Marcos Travassos Romano • Charles Casimiro Cavalcante

CRC Press
Taylor & Francis Group
Boca Raton London New York

CRC Press is an imprint of the
Taylor & Francis Group, an **informa** business

CRC Press
Taylor & Francis Group
6000 Broken Sound Parkway NW, Suite 300
Boca Raton, FL 33487-2742

First issued in paperback 2017

© 2016 by Taylor & Francis Group, LLC
CRC Press is an imprint of Taylor & Francis Group, an Informa business

No claim to original U.S. Government works

ISBN-13: 978-1-4987-2236-0 (hbk)
ISBN-13: 978-1-138-89301-6 (pbk)

Visit the Taylor & Francis Web site at
http://www.taylorandfrancis.com

and the CRC Press Web site at
http://www.crcpress.com

To our families

Contents

Acoustic Signal Processing

Editors: Rosângela Fernandes Coelho
Vítor Heloiz Nascimento

Image Processing

Editor: Ricardo Lopes de Queiroz

Signal Processing in Communications

Editors: João Marcos Travassos Romano
Charles Casimiro Cavalcante

Selected Topics in Signal Processing

Editors: João Marcos Travassos Romano
Charles Casimiro Cavalcante

Preface

Signal processing is a dynamic and fast growing research area with application in all modern technologies and sectors. Signal processing techniques underlie smartphones, video-cameras, biomedical diagnosis, credit card services and finance market applications.

Since the early 70's, the fundamentals of digital signal processing are being continuously revisited to cope with new challenging scenarios. The primary aim of the book is to present a comprehensive, yet cohesive, account of some of the current advances and results with the objective of serving as a future perspective of the area. It provides a means of increasing the interest of students in the field.

The book was designed for advanced undergraduate or graduate students, showing how signal processing theory is applied in current and emerging applications. It can also be used as supplementary textbook in signal processing, stochastic processes, detection and estimation, and information theory courses. It was conceived to be an invitation to dive in this very interesting field of research by addressing a variety of key problems. For example, how to enhance speech in time-domain, how to improve audio quality, how to meet the desired electrical consumption target for controlling carbon emissions. The book is designated to serve as a guide to the next generation of signal processing solutions for speech and video coding, hearing aid devices, big data processing, smartphones, smart digital communications, acoustic sensors. It is hoped that it will be an invaluable tool for undergraduate and graduate students, researchers, and practitioners working in signal processing.

The contributors are researchers from universities and the industry that are active in their particular research fields. These experts present recent advances and results in signal processing including: audio and speech enhancement, acoustic field estimation, video compression, biometric recognition, hyperspectral image analysis, tensor decomposition with applications in communications, adaptive sparse-interpolated filtering, signal processing for power line communications, bio-inspired signal processing, seismic data processing, arithmetic transforms for spectrum computation, particle filtering in cooperative networks, and three-dimensional television. The textbook is a collection of nineteen chapters that brings the latest relevant research results in signal processing. The chapters are organized into five main parts: Theory and Methods, Acoustic Signal Processing, Image Processing, Signal Processing in Communications and Selected Topics in Signal Processing.

Part I (Theory and Methods): Chapter 1 provides an introduction to the Blind Source Separation (BSS) problem and to classical solutions based on Independent Component Analysis. A more recent technique based on sparse component analysis is also reviewed. Chapter 2 introduces Kernel-based nonlinear signal processing and its application to non-parametric modeling of nonlinear functions. Chapter 3 employs a discrete-time formalism to derive a unifying mathematical treatment for arithmetic transform methods for trigonometric discrete transforms. Chapter 4 describes new distributed particle filtering algorithms for cooperative estimation in networks of collaborative agents. The performance of the proposed algorithms is illustrated for target tracking and blind equalization applications.

Part II (Acoustic Signal Processing): Chapter 5 is concerned with empirical mode decomposition theory focused on speech enhancement solutions under nonstationary acoustic noise, in time domain. Chapter 6 presents the problem of acoustic imaging methods for reducing the number of computations in the estimation of acoustic images. Chapter 7 addresses full- and no-reference automatic assessment of the quality of acoustically degraded full-band speech in modern communication systems. Chapter 8 treats temporal, spectral and time-frequency representations and models for speech signals, enabling the understanding of the basic pillars of current speech technology future improvement.

Part III (Image Processing): Chapter 9 presents a dynamic rate-distortion-energy-optimized framework for software-based realtime video compression, meeting the desired electrical consumption target while operating at the best compression performance. Chapter 10 describes classic template matching and presents two techniques to obtain rotation and scale-invariant template matching. Chapter 11 provides an overview of stereo and multi-view coding techniques, including the current MPEG, ISO/IEC and ITU-T standards for 3D television.

Part IV (Signal Processing in Communications): Chapter 12 contains an extensive overview of tensor-based models and methods for several applications in communications, where the authors discuss their characteristics, advantages and constraints. In Chapter 13, detection and parameter estimation strategies for massive (large number of elements) MIMO systems future generations of wireless communication systems are discussed. Chapter 14 covers the sparse-interpolated approach in adaptive filters, with reduced complexity, both in linear models as well as an nonlinear ones, using Volterra models. Aiming for more reliable and faster communication systems, Chapter 15 is devoted to smart reuse of the spectrum on the power electric system for performing data communication over power lines while sharing resources among both systems.

Part V (Selected Topics in Signal Processing): Chapter 16 focuses on information geometry modeling for statistical distributions, where the curvature of signal spaces is taken into account, in order to provide new models for problems with non-Gaussian assumptions. Several aspects of information-theoretical learning and biological-inspired models are covered in Chapter 17, where the authors depict the advantages of such models in linear and nonlinear problems in applications such as independent component analysis and probability density estimation. Chapter 18 is devoted to signal processing methods applied to seismic signals, in order to find the structure on the subsurface in a region of interest, which is a key stage for the gas and oil industry since it provides the information about the potential of reserves in a given location. Chapter 19 discusses methods based on imaging on synthetic aperture, for non-destructive testing, in order to allow inspection of objects without producing changes. These methods are of high interest in areas such as aerospace and power generation.

As co-editors, we would like to express our deepest gratitude and sincere thanks to all the contributors. Their commitment and dedication during the development of this special project enabled the productive and interesting experience that made this book possible. We are also thankful to the professional team of CRC Press Taylor & Francis Group for their support during the production of the book. We are particularly grateful to the publisher, Nora Konopka, for generosity, certainty of purpose, and encouragement to the completion of the book.

The Editors

Authors

ANDRÉ LIMA FERRER DE ALMEIDA
Department of Teleinformatics Engineering
Federal University of Ceará (UFC)
Fortaleza, Brazil

LARYSSA RAMOS AMADO
Faculty of Technology
Federal University of Juiz de Fora (UFJF)
Juiz de Fora, Brazil

ROMIS RIBEIRO DE FAISSOL ATTUX
Department of Computer Engineering and Industrial Automation (DCA)
School of Electrical and Computer Engineering (FEEC)
University of Campinas (UNICAMP)
Campinas, Brazil

TIAGO TAVARES LEITE BARROS
School of Electrical and Computer Engineering (FEEC)
University of Campinas (UNICAMP)
Campinas, Brazil

YVES-MARIE BATANY
MINES ParisTech
Paris, France

EDUARDO LUIZ ORTIZ BATISTA
Circuits and Signal Processing Laboratory (LINSE)
Department of Electrical and Electronics Engineering (EEL)
Federal University of Santa Catarina (UFSC)
Florianópolis, Brazil

JOSÉ CARLOS MOREIRA BERMUDEZ
Digital Signal Processing Laboratory (LPDS)
Department of Electrical and Electronic Engineering (EEL)
Federal University of Santa Catarina (UFSC)
Florianópolis, Brazil

LUIZ WAGNER PEREIRA BISCAINHO
Federal University of Rio de Janeiro (UFRJ)
Rio de Janeiro, Brazil

LEVY BOCCATO
Department of Computer Engineering and Industrial Automation (DCA)
School of Electrical and Computer Engineering (FEEC)
University of Campinas (UNICAMP)
Campinas, Brazil

MARCELO GOMES DA SILVA BRUNO
Aeronautics Institute of Technology (ITA)
São José dos Campos, Brazil

CHARLES CASIMIRO CAVALCANTE
Department of Teleinformatics Engineering
Federal University of Ceará (UFC)
Fortaleza, Brazil

RENATO J. CINTRA
Signal Processing Group
Federal University of Pernambuco (UFPE)
Pernambuco, Brazil

ROSÂNGELA FERNANDES COELHO
Military Institute of Engineering (IME)
Rio de Janeiro, Brazil

JOÃO PAULO CARVALHO LUSTOSA DA COSTA
Department of Electrical Engineering
University of Brasília (UnB)
Brasília, Brazil

MARCOS RICARDO COVRE
School of Electrical and Computer Engineering (FEEC)
University of Campinas (UNICAMP)
Campinas, Brazil

STIVEN SCHWANZ DIAS
Embraer Defense and Security
São José dos Campos, Brazil

VASSIL S. DIMITROV
Department of Electrical and Computer Engineering
University of Calgary
and
Computer Modelling Group, Ltd
Calgary, Canada

LEONARDO TOMAZELI DUARTE
School of Applied Sciences (FCA)
University of Campinas (UNICAMP)
Campinas, Brazil

DENIS GUSTAVO FANTINATO
Department of Computer Engineering and Industrial Automation (DCA)
School of Electrical and Computer Engineering (FEEC)
University of Campinas (UNICAMP)
Campinas, Brazil

GÉRARD FAVIER
I3S Laboratory
University of Nice Sophia Antipolis (UNS)
National Center for Scientific Research (CNRS)
Sophia Antipolis, France

JUGURTA ROSA MONTALVÃO FILHO
Department of Electrical Engineering
Federal University of Sergipe (UFS)
Aracaju, Brazil

WEILER ALVES FINAMORE
Faculty of Technology
Federal University of Juiz de Fora (UFJF)
Juiz de Fora, Brazil

TIAGO ALVES DA FONSECA
Gama Faculty
University of Brasília (UnB)
Brasília, Brazil

SILVIO CESAR GARCIA GRANJA
UNEMAT
Sinop, Brazil

RICARDO TOKIO HIGUTI
UNESP
Ilha Solteira, Brazil

PAUL HONEINE
Laboratory of Systems Modeling and Dependability (LM2S)
Charles Delaunay Institute (ICD), CNRS UMR 6281
University of Technology of Troyes (UTT)
Troyes, France

CLAUDIO JOSÉ BORDIN JÚNIOR
Federal University of ABC (UFABC)
Santo André, Brazil

HAE YONG KIM
University of São Paulo (USP)
São Paulo, Brazil

CLÁUDIO KITANO
UNESP
Ilha Solteira, Brazil

RAFAEL KRUMMENAUER
DSPGeo
Campinas, Brazil

RODRIGO CAIADO DE LAMARE
Centre for Telecommunications Studies (CETUC)
Pontifical Catholic University of Rio de Janeiro (PUC-Rio)
Rio de Janeiro, Brazil
and
Communications Research Group
Department of Electronics
University of York
York, United Kingdom

RENATO DA ROCHA LOPES
School of Electrical and Computer Engineering (FEEC)
University of Campinas (UNICAMP)
Campinas, Brazil

OSCAR MARTÍNEZ-GRAULLERA
Instituto de Tecnologias Fisicas y de la Informacion Leonardo Torres Quevedo
(ITEFI), Consejo Superior de Investigaciones Científicas (CSIC)
Madrid, Spain

BRUNO SANCHES MASIERO
University of Campinas (UNICAMP)
Campinas, Brazil

ALAM SILVA MENEZES
Petrobras Transporte S.A.
Rio de Janeiro, Brazil

MÁRIO MINAMI
Federal University of ABC (UFABC)
Santo André, Brazil

JOÃO CESAR MOURA MOTA
Department of Teleinformatics Engineering
Federal University of Ceará (UFC)
Fortaleza, Brazil

VÍTOR HELOIZ NASCIMENTO
University of São Paulo (USP)
São Paulo, Brazil

KENJI NOSE-FILHO
Department of Communications (DECOM)
School of Electrical and Computer Engineering (FEEC)
University of Campinas (UNICAMP)
Campinas, Brazil

LEONARDO DE OLIVEIRA NUNES
Microsoft
Rio de Janeiro, Brazil

HÉLIO M. DE OLIVEIRA
Signal Processing Group
Federal University of Pernambuco (UFPE)
Pernambuco, Brazil

CARLA LIBERAL PAGLIARI
Military Institute of Engineering (IME)
Rio de Janeiro, Brazil

ALINE DE OLIVEIRA NEVES PANAZIO
Engineering, Modeling and Applied Social Sciences Center
Federal University of ABC (UFABC)
Santo André, Brazil

VANDER TEIXEIRA PRADO
Universidade Tecnológica Federal do Paraná (UTFPR)
Departamento Acadêmico da Elétrica (DAELE)
Cornélio Procópio, Brazil

RICARDO LOPES DE QUEIROZ
Computer Science Department
University of Brasília (UnB)
Brasília, Brazil

MIGUEL ARJONA RAMÍREZ
University of São Paulo (USP)
São Paulo, Brazil

FLÁVIO PROTASIO RIBEIRO
Applied Sciences Group, Microsoft Corporation
Redmond, USA

MOISÉS VIDAL RIBEIRO
Faculty of Technology
Federal University of Juiz de Fora (UFJF)
Juiz de Fora, Brazil

CÉDRIC RICHARD
Lagrange Laboratory
Côte d'Azur Observatory
University of Nice Sophia Antipolis (UNS)
Nice, France

JOÃO MARCOS TRAVASSOS ROMANO
School of Electrical and Computing Engineering (FEEC)
University of Campinas (UNICAMP)
Campinas, Brazil

DAVID ROMERO-LAORDEN
Instituto de Tecnologias Fisicas y de la Informacion Leonardo Torres Quevedo
(ITEFI), Consejo Superior de Investigaciones Científicas (CSIC)
Madrid, Spain

RAIMUNDO SAMPAIO-NETO
Centre for Telecommunications Studies (CETUC)
Pontifical Catholic University of Rio de Janeiro (PUC-Rio)
Rio de Janeiro, Brazil

Rui Seara
Circuits and Signal Processing Laboratory (LINSE)
Department of Electrical and Electronics Engineering (EEL)
Federal University of Santa Catarina (UFSC)
Florianópolis, Brazil

Daniel Guerreiro e Silva
Department of Electrical Engineering (ENE)
University of Brasilia (UnB)
Brasília, Brazil

Eduardo Antônio Barros da Silva
Federal University of Rio de Janeiro (UFRJ)
Rio de Janeiro, Brazil

Thippur Venkatanarasaiah Sreenivas
Indian Institute of Science (IISc)
Bangalore, India

Ricardo Suyama
Engineering, Modeling and Applied Social Sciences Center
Federal University of ABC (UFABC)
Santo André, Brazil

André Kazuo Takahata
School of Electrical and Computer Engineering (FEEC)
University of Campinas (UNICAMP)
Campinas, Brazil

Jean-Yves Tourneret
Toulouse Institute of Computer Science Research (IRIT)
National Polytechnic Institute of Toulouse (INPT)
University of Toulouse
Toulouse, France

Rui Facundo Vigelis
Computer Engineering, Campus Sobral
Federal University of Ceará (UFC)
Fortaleza, Brazil

Leonardo Augusto Zão
Military Institute of Engineering (IME)
Rio de Janeiro, Brazil

Benedito ...
Circuits and Signal Processing Laboratory (LCS)
Department of Electrical and Electronics Engineering (EEL)
Federal University of Santa Catarina (UFSC)
Florianópolis, Brazil

Daniel Guerreiro e Silva
Department of Electrical Engineering (ENE)
University of Brasília (UnB)
Brasília, Brazil

Eduardo Antônio Barros da Silva
Federal University of Rio de Janeiro (UFRJ)
Rio de Janeiro, Brazil

Thippur Venkatanarayana Srinivas
Indian Institute of Science (IISc)
Bangalore, India

Ricardo Suyama
Engineering, Modeling and Applied Social Sciences Center
Federal University of ABC (UFABC)
Santo André, Brazil

Anand Kumar Takhar
School of Electrical and Computer Engineering (FEEC)
University of Campinas (UNICAMP)
Campinas, Brazil

Jean-Yves Tourneret
Toulouse Institute of Computer Science Research (IRIT)
National Polytechnic Institute of Toulouse (INPT)
University of Toulouse
Toulouse, France

Rui Facundo Vigelis
Computer Engineering, Campus Sobral
Federal University of Ceará (UFC)
Sobral, Brazil

Leonardo Aguayo Vas
Military department of Engineering (IME)
Rio de Janeiro, Brazil

Theory and Methods

Rosângela Fernandes Coelho
Vítor Heloiz Nascimento

1 Blind Source Separation: Principles of Independent and Sparse Component Analysis

LEONARDO TOMAZELI DUARTE
School of Applied Sciences (FCA), University of Campinas (UNICAMP)

YVES-MARIE BATANY
MINES ParisTech

JOÃO MARCOS TRAVASSOS ROMANO
School of Electrical and Computing Engineering (FEEC), University of Campinas (UNICAMP)

1.1 INTRODUCTION

Signal separation is one of the most important problems in the theory of signal processing and finds application in audio [46] and image processing [37, 60], and biomedical [63] and chemical [6, 26] analysis, to cite a few examples. Signal separation is also relevant in areas other than signal processing, such as machine learning [7] and statistical analysis [39]. It is also worth mentioning that several data analysis methods that are employed to deal with a large amount of data (big data) were initially developed or refined by researchers working on the problem of signal separation [7].

Solutions to the signal separation problem arise from the general theory of filtering and can be traced back to the Fourier transform [45]. For instance, let us consider the three signals shown in Figure 1.1 and their additive mixtures. If the Fourier transforms of theses signals do not overlap, i.e., if they are disjoint orthogonal in the Fourier domain, it becomes possible to design a set of filters in order to retrieve the original signals based on the observed mixture. Such an approach is the most straightforward one to separate signals and is still intensely applied in practice.

The solution illustrated in Figure 1.1 strongly relies on the knowledge of the spectral contents of each desired signal. Indeed, such information is used to set the parameters of each filter, which remains fixed during the separation process. This framework can be extended by considering tools from the adaptive filter theory [32, 24]. For instance, the seminal work of Widrow et al. [64] introduced the adaptive noise canceller (ANC), an adaptive filter that can be used to extract a desired signal from a mixture. As illustrated in Figure 1.2, the ANC works with two input signals: 1) a

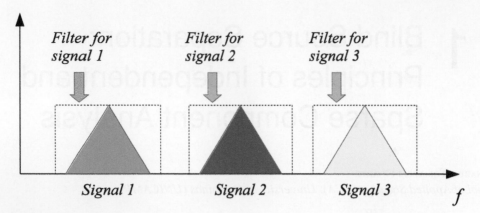

Figure 1.1   The use of Fourier transform-based filtering to separate an additive mixture of three signals.

signal $x(n)$ that corresponds to an additive mixture of the desired signal, $s(n)$, and an undesired signal (noise), $r(n)$, that is

$$x(n) = s(n) + r(n),$$

and 2) a reference signal $z(n)$ that must be as correlated as possible to $r(n)$. The adaptive filter, represented by the vector $\mathbf{w}$, acts on $z(n)$ providing a filtered signal represented by

$$y(n) = (\mathbf{w} * z)(n),$$

where $*$ stands for the convolution operator.

Given these signals, the ANC searches for a filter by solving the following optimization problem

$$\min_{\mathbf{w}} E\{(x(n) - y(n))^2\} = E\{(s(n) + r(n) - (\mathbf{w} * z)(n))^2\}. \tag{1.1}$$

In other words, the filter is adjusted so that the power of the signal $x(n) - y(n)$ be as low as possible. The rationale behind such a principle is that, if $z(n)$ is close enough[1] to $r(n)$, and if $r(n)$ and $s(n)$ are uncorrelated, the problem expressed in (1.1) is solved when $(\mathbf{w} * z)(n)$ approaches $r(n)$ — in this case the residual signal $x(n) - y(n)$ provides an estimate of the desired signal $s(n)$.

A remarkable difference between the ANC separation principle and classical separation based on transforms is that the later method considers two sensors. Such a feature incorporates additional diversity into the solution, thus easing the requirement of having signals with spectral contents that do not overlap. Notwithstanding,

---

[1]Here, close enough means that $z(n)$ can be modeled as a filtered version of $r(n)$.

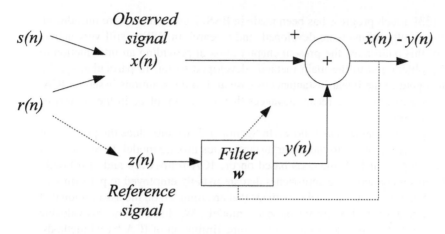

Figure 1.2   Separation by means of an adaptive noise canceller (ANC).

there is still a strong limitation in using the ANC to perform signal separation: the performance of the ANC is strongly affected when the reference signal $z(n)$ contains a certain level of the desired signal $s(n)$ [64]. Unfortunately, in many practical cases, it is difficult to obtain a reference signal that is correlated to the undesired signal but not correlated to the desired signal.

Important extensions to the separation framework based on classical and adaptive filtering took place in the 1980s. In their seminal work, the French researchers Hérault, Jutten, and Ans [33], motivated by an application in neurophysiology, introduced a novel formulation in which signal separation was described in terms of a multiple-input multiple-output (MIMO) model.[2] In other words, the problem became to separate a set of desired signals (sources) by processing a set of mixtures of these sources. This task had to be carried out without relying on training (or calibration points), and, for that reason, the problem was named *blind source separation* (BSS) — the terms *blind* and *unsupervised* were already employed in the context of blind adaptive equalization [54], which is a slightly older problem with respect to BSS.

Another major contribution of [33] was to introduce a solution that made use of higher-order statistics [44]. This idea has eventually led to a powerful data processing methodology known as independent component analysis (ICA) [37], which, among other applications in machine learning and multivariate analysis, is the basis of the main BSS algorithms. It is worth mentioning that a source separation method based on some elements present in ICA was proposed in 1984 by researchers working on the separation of geophysical sources [31].

---

[2]An interesting historical aspect is that Joseph Fourier developed his theory on heat propagation, which eventually led to the Fourier series, at the city of Grenoble (France). This is the same city where Hérault, Jutten, and Ans pioneered the field of blind source separation.

Since [33], much progress has been made in BSS. New criteria were introduced, more efficient algorithms were developed, and research in BSS is still very active. In view of this panorama, the present chapter aims at providing an introduction to the BSS problem and to the main methods developed so far. In particular, special attention is paid to the linear instantaneous case and to the solutions based on ICA. Moreover, we discuss more recent advances that are taking place in the context of sparse separation.

The chapter is organized as follows. In Section 1.2, we introduce the problem of BSS with special attention to the linear instantaneous mixing model. In Section 1.3, we discuss important BSS methods based on the ICA framework and also briefly comment on some alternative approaches that are usually employed to perform separation. In Section 1.4, we describe methods that are founded on the assumption that sources can be described by means of sparse models. We discuss how this valuable prior information can be used to overcome some limitation of ICA-based methods. The chapter is closed with a concluding section, where we provide a discussion on some perspectives for BSS research.

This chapter has no pretension to be a thorough survey of the BSS problem. Rather, our goal here is to provide the reader with a flavor of this relevant problem in signal processing, and, in view of this orientation, some details are omitted in this document. For readers that are interested in a detailed and complete description of the BSS problem and their solutions can refer to several textbooks. For instance, the books [16, 21, 54, 20] provide a detailed description of the problem. Another very interesting reference is the textbook [38], which focuses on ICA methods, but also discusses other important topics related to BSS. There are also several survey papers and tutorials on BSS that can serve as an introduction to the field. Examples include references [12, 35, 36]. A more recent paper addresses the problem of BSS and its extension to multiple datasets [1].

1.2   THE BLIND SOURCE SEPARATION PROBLEM

In order to introduce the problem of blind source separation (BSS), let us consider a set of Q source signals that are represented by the vector.[3]:

$$\mathbf{s}(n) = \begin{bmatrix} s_1(n) \\ s_2(n) \\ \vdots \\ s_Q(n) \end{bmatrix}.$$

These signals correspond to the desired ones, but in BSS the observations comprise a set of P signals that are given by

---

[3]In our notation, n represents the time index or, in image applications, the pixel index.

$$\mathbf{x}(n) = \begin{bmatrix} x_1(n) \\ x_2(n) \\ \vdots \\ x_P(n) \end{bmatrix} = \begin{bmatrix} \mathcal{A}_1(\mathbf{s}(n), \mathbf{s}(n-1), \dots, \mathbf{s}(0)) \\ \mathcal{A}_2(\mathbf{s}(n), \mathbf{s}(n-1), \dots, \mathbf{s}(0)) \\ \vdots \\ \mathcal{A}_P(\mathbf{s}(n), \mathbf{s}(n-1), \dots, \mathbf{s}(0)) \end{bmatrix}, \tag{1.2}$$

where $\mathcal{A}_i(\cdot)$ is a function that models how the sources are mixed within the observation $x_i(n)$. The goal of a BSS method is therefore *to provide an estimate of* $\mathbf{s}(n)$ *based on the observation of* $\mathbf{x}(n), \mathbf{x}(n-1), \dots, \mathbf{x}(0)$ *and, possibly, on some mild information on* $\mathcal{A}_i(\cdot)$.

Usually, the BSS problems are classified according to the nature of the mixing process. In particular, there are three characteristics related to (1.2) that are considered:

1. If $\mathcal{A}_i(\cdot)$s are linear functions of the sources, then the resulting BSS problem is said to be linear. Otherwise, one ends up with a nonlinear BSS problem [40].
2. An instantaneous (or memoryless) BSS problem arises when all the $\mathcal{A}_i(\cdot)$s depend only on $\mathbf{s}(n)$. If $\mathcal{A}_i(\cdot)$s represent dynamical systems, then the problem is said to be dynamical. The linear dynamical BSS problem is often called convolutive BSS [52].
3. BSS problems can be classified according to the numbers of available mixtures and sources. If $Q = P$, $Q > P$, $Q < P$ then the problem is said to be determined, underdetermined, and overdetermined, respectively. An important remark here is that the overdetermined case can be dealt with by methods tailored to the determined case — this can be done by applying dimensional reduction methods previously to BSS.

Closely related to BSS is the problem of blind source extraction (BSE). In this case, one is interested in recovering only one source from the mixtures. An interesting aspect here is that BSS can be solved by performing Q (or more) successive executions of a BSE method. Of course, this approach requires the introduction of a mechanism that assures that different executions of the BSE method lead to different estimates of the sources. A possible solution can be achieved by attenuating, at each execution of the considered BSE method, the contribution of the previously estimated source to the mixtures, in a procedure that bears resemblance to the ANC.

In this chapter, we shall discuss both the BSS and BSE problems. Moreover, the focus of our work will be on the most classic separation setup, which is the case of linear, determined and instantaneous mixtures. The reader interested in convolutive BSS is referred to [21, 52]. More details on nonlinear BSS can be found in [21, 40].

1.2.1 HOW TO SOLVE THE LINEAR INSTANTANEOUS CASE?

When the mixing model is determined, linear, and instantaneous, the mixing process described by Equation (1.2) can be simplified as follows

$$\mathbf{x}(n) = \mathbf{A}\mathbf{s}(n), \tag{1.3}$$

where the mixing matrix $\mathbf{A}$ has dimension $Q \times Q$. In order to counterbalance the action of $\mathbf{A}$, a natural strategy is to set up a separating matrix $\mathbf{W}$, also of dimension $Q \times Q$, so that

$$\mathbf{y}(n) = \mathbf{W}\mathbf{x}(n) \tag{1.4}$$

provides estimates of the sources. In the case of BSE, there is no need to consider the whole separating matrix $\mathbf{W}$. Instead, since one is interested in a single source, the estimation is done by means of the extracting vector $\mathbf{w}$, as follows:

$$y(n) = \mathbf{w}^T \mathbf{x}(n). \tag{1.5}$$

Ideally, the BSS problem is solved when $\mathbf{W} = \mathbf{A}^{-1}$, assuming that $\mathbf{A}^{-1}$ exists. However, the best one can expect in an unsupervised scenario is to estimate the sources up to permutation and scaling ambiguities. In other words, there is no guarantee that the retrieved sources are arranged according to the order of the original sources, if such an order makes sense. Moreover, their scales are not necessarily the same as that of the sources. In view of these ambiguities, the BSS problem expressed in (1.3) is said to be solved when:

$$\mathbf{W} = \mathbf{P}\mathbf{D}\mathbf{A}^{-1}, \tag{1.6}$$

where $\mathbf{P}$ and $\mathbf{D}$ correspond to a permutation and a diagonal matrix, respectively.

A crucial point in BSS is how to adjust the separating matrix $\mathbf{W}$. While in a supervised context this problem is simple and can be solved by defining a cost function that quantifies the distance between the estimated sources and a set of training points, finding $\mathbf{W}$ in an unsupervised fashion requires the knowledge of prior information on a given feature of the sources. Therefore, there are basically three steps when setting up a BSS method:

1. Definition of the original property of the sources that will be considered. Very often, one considers a property that is lost after the mixing process;
2. Definition of a separation criterion that aims at recovering the original property of the sources. A crucial point here is to prove that the optimization of the adopted criterion indeed leads to source separation;
3. Definition of a method to optimize the adopted separation criterion.

In order to illustrate the application of these steps, let us consider that the sources respect a given mathematical property, say "A," which is lost after the mixing process. According to step 2, the idea to perform source separation is to adjust the separating matrix $\mathbf{W}$ so that the estimated sources $\mathbf{y}(n)$ respect again property "A." This is done by setting up a cost function $J(\cdot)$ that attains its optimum value if, and only if, $\mathbf{y}(n)$ respect property "A." In order words, resolution of BSS can be formulated as the following minimization problem:[4]

$$\min_{\mathbf{W}} \left\{ J(\mathbf{y}(n)) = J(\mathbf{W}\mathbf{x}(n)) \right\}. \tag{1.7}$$

---

[4]The problem here is stated as a minimization problem only for convenience. Depending on the context, people may prefer to refer to it as a maximization problem.

In the case of BSE, the corresponding optimization problem is given by:

$$\min_{\mathbf{w}} \left\{ J(y(n)) = J(\mathbf{w}^T \mathbf{x}(n)) \right\}. \tag{1.8}$$

Very often, the optimization problems expressed in (1.7) and (1.8) cannot be solved in an analytical fashion, thus requiring the implementation of iterative methods. For instance, if a gradient-based optimization method is considered, then a BSS method can be implemented by means of the following learning rule:

$$\mathbf{W}^{(k+1)} = \mathbf{W}^{(k)} - \mu \frac{\partial J(\mathbf{y}(n))}{\partial \mathbf{W}}, \tag{1.9}$$

where μ is the learning rate and $\frac{\partial J(\mathbf{y}(n))}{\partial \mathbf{W}}$ corresponds to the gradient of $J(\cdot)$ with respect to $\mathbf{W}$. The superscript index here stands for the learning algorithm iteration. A similar learning rule can be derived for a BSE cost function $J(\cdot)$, as follows:

$$\mathbf{w}^{(k+1)} = \mathbf{w}^{(k)} - \mu \frac{\partial J(y(n))}{\partial \mathbf{w}}. \tag{1.10}$$

In BSS, it is of paramount importance the concept of contrast function (or simply contrast), i.e., a cost function $J(\cdot)$ that attains its minimum if, and only if, $\mathbf{W} = \mathbf{PDA}^{-1}$ [18] (ideal separation). Therefore, according to the scheme described above, the contrast property ensures that retrieving the property "A" leads to source separation. Of course, it is expected that a cost function $J(\cdot)$ acts as a contrast only when some conditions are verified. For example, if the sources do not respect property "A," a cost function tailored to measure property "A" is unlikely to act as a contrast. Therefore, a central question is: *What are the conditions (on the sources and possibly on the mixing process) to a given cost function $J(\cdot)$ act as a contrast?* In the main BSS approach, known as independent component analysis (ICA), these conditions rely on the assumption that the sources are statistically independent, as will be discussed in the next section.

1.3 BSS METHODS BASED ON INDEPENDENT COMPONENT ANALYSIS

In the 1980s, research on statistical signal processing was already very active [42] and was mainly focused on Gaussian models and, as a consequence, on second-order statistics such as variance and covariance. In that context, a first natural attempt to solve the BSS problem was to tackle it for the case in which the sources can be modeled as uncorrelated random variables. Since the mixing process makes the observed signals $\mathbf{x}(n)$ correlated, it would be possible to perform BSS by recovering uncorrelated signals $\mathbf{y}(n)$.

Unfortunately, as illustrated in Figure 1.3, the recovery of uncorrelated variables, which is known as *whitening*,[5] does not lead to source separation. There is a very

---

[5]This nomenclature was borrowed from deconvolution where it means to render a signal temporarily uncorrelated, thus presenting a spectrum that is flat or, equivalently, white.

intuitive explanation for such a limitation. Actually, if one takes uncorrelated random variables and applies a rotation matrix on them, the resulting variables will be still uncorrelated. Therefore, when performing whitening, there arises a sort of rotation ambiguity, which, unlike scaling and permutation ambiguities, is still associated with a mixing process.

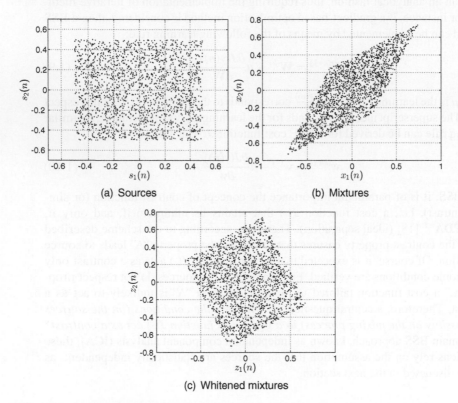

(a) Sources

(b) Mixtures

(c) Whitened mixtures

Figure 1.3 Limitations of whitening to perform source separation. After whitening, the signals are still mixtures of the sources.

On the other hand, although whitening is not enough to achieve source separation, it provides a halfway solution [12]. Indeed, after performing whitening on the mixtures $\mathbf{x}(n)$, the resulting signals $\mathbf{z}(n)$ become a rotated version of the sources $\mathbf{s}(n)$. For instance, in the case of two sources and two mixtures, the whitened mixtures $\mathbf{z}(n)$ can be expressed as follows:

$$\begin{bmatrix} z_1(n) \\ z_2(n) \end{bmatrix} = \begin{bmatrix} \cos(\theta) & -\sin(\theta) \\ \sin(\theta) & \cos(\theta) \end{bmatrix} \begin{bmatrix} s_1(n) \\ s_2(n) \end{bmatrix}, \tag{1.11}$$

where θ is the angle that parametrizes the rotation between $\mathbf{s}(n)$ and $\mathbf{z}(n)$. Therefore, there is only a single mixing parameter θ instead of 4 mixing coefficients that arise when whitening is not performed.

A possible way to perform whitening is to apply a classic multivariate statistical procedure known as principal component analysis (PCA) [39]. Bearing in mind that the whitening/PCA approach leads to the rotation ambiguity discussed above, Hérault, Jutten, and Ans [33] came up with a novel idea: instead of considering only the standard covariance measure, they went further by introducing a measure of nonlinear decorrelation. Such an approach allowed the method to take into account higher-order statistics and eventually led to independent component analysis (ICA), which was formalized by Pierre Comon [18]. In ICA, instead of assuming that the sources are uncorrelated, one assumes that the sources can be modeled as independent random variables. Since statistical independence is lost after the mixing process, the idea in ICA is to retrieve estimates that are again independent.

Differently from PCA, ICA does not suffer from the rotation ambiguity and leads to source separation, up to permutation and scale ambiguities, if the following conditions are held:[6]

1. The sources are statistically independent;
2. There is, at most, one Gaussian source;
3. The mixing matrix $\mathbf{A}$ is invertible.

An interesting aspect here is condition 2. The use of higher-order statistics, which is implicitly taken into account in ICA, does not bring relevant information about a Gaussian random variable, since such a variable is fully characterized by second-order statistics. As a consequence, ICA cannot separate mixtures of Gaussian sources. This is only possible by exploiting temporal features of the sources, as will be briefly discussed in Section 1.3.4.

Concerning the practical aspects related to the implementation of ICA, the pioneer solution of [33], as already mentioned, made use of nonlinear correlation measurements. However, the Hérault-Jutten-Ans ICA algorithm presented several shortcomings, such as convergence issues [38]. Following the work of [33], researchers developed alternative ICA methods. In the sequel, we shall review three relevant approaches, namely: methods based on mutual information minimization, the maximum likelihood approach, and methods based on non-gaussianity maximization. There are other ICA strategies that will not be addressed here, such as nonlinear PCA and geometrical methods. More details on these approaches can be found in [38].

1.3.1   MUTUAL INFORMATION MINIMIZATION

A first natural approach to perform ICA is to define a measure of statistical independence. This is actually a tricky question because statistical independence is a binary feature, in the sense that a set of random variables is either joint independent or dependent. Notwithstanding, it is possible to build smooth cost functions that attain their optima for independent variables. The most straightforward example relies on

---

[6]These conditions were established by Pierre Comon [18] based on the Darmois-Skitovich theorem [41].

the definition of statistical independence. A set of Q random variables represented by $\mathbf{y}(n)$ is mutually independent if and only if

$$p(\mathbf{y}(n)) = \prod_{i=1}^{Q} p(y_i(n)), \tag{1.12}$$

where $p(\mathbf{y}(n))$ is the joint probability distribution function (PDF) of $\mathbf{y}(n)$ and $p(y_i(n))$ corresponds to the marginal PDF of the i-th variable $y_i(n)$ — in our notation, we are keeping the index n but the reader must keep in mind that the values observed for different n correspond to realizations of a random variable when a probabilistic model is considered.

From Equation (1.12), statistical independence is achieve if, and only if, the joint PDF is given by the products between the marginals PDFs. Therefore, by considering a function that somehow measures the distance between the joint PDF and the marginal ones, one ends up with a cost function that attains its minimum value for independent variables. A possible way to implement this idea comes from the Kullback–Leibler (KL) divergence between the joint and the marginal PDFs, which is given by

$$D\left(p(\mathbf{y}(n)), \prod_{i=1}^{Q} p(y_i(n)) \right) = \int p(\mathbf{y}(n)) \log\left(\frac{p(\mathbf{y}(n))}{\prod_{i=1}^{Q} p(y_i(n))} \right) d\mathbf{y}, \tag{1.13}$$

The KL divergence is always non-negative and takes 0 only when the variables $\mathbf{y}(n)$ are independent.

Equation (1.13) is also the definition of the mutual information $\mathbf{y}(n)$. This concept, which is originated from information theory [22], can also be defined as follows:

$$I(\mathbf{y}(n)) = \sum_{i=1}^{Q} H(y_i(n)) - H(\mathbf{y}(n)), \tag{1.14}$$

where $H(\cdot)$ corresponds to Shannon's differential entropy [22] and is given by:

$$H(\mathbf{y}(n)) = -\int p(\mathbf{y}(n) \log p(\mathbf{y}(n) d\mathbf{y}. \tag{1.15}$$

In the context of BSS, since $\mathbf{y}(n) = \mathbf{W}\mathbf{x}(n)$, the joint entropy $H(\mathbf{y}(n))$ can be written as [22, 38]:

$$H(\mathbf{y}(n)) = H(\mathbf{x}(n)) + \log |\det \mathbf{W}|. \tag{1.16}$$

The term $H(\mathbf{x}(n))$ does not depend on $\mathbf{W}$. As a consequence, from Equations (1.14) and (1.16), the task of finding the separating matrix $\mathbf{W}$ that provides independent components can be written as the following optimization problem:

$$\min_{\mathbf{W}} \left\{ J(\mathbf{y}(n)) = \sum_{i=1}^{Q} H(y_i(n)) + \log |\det \mathbf{W}| \right\}. \tag{1.17}$$

In order to solve this optimization problem, one may apply the gradient descent rule expressed in Equation (1.9), which leads to the following learning rule

$$\mathbf{W}^{(k+1)} = \mathbf{W}^{(k)} - \mu \left(E\{\Psi_\mathbf{y}(\mathbf{y}(n))\mathbf{x}(n)^T\} - \mathbf{W}^{-T} \right), \tag{1.18}$$

where $E\{\cdot\}$ corresponds to the expected value operator and $\Psi_\mathbf{y}(\mathbf{y}(n))$ is the vector of score functions — its i-th element is given by

$$\psi_{y_i}(y_i(n)) = -\frac{d \log p(y_i(n))}{dy_i}.$$

The learning rule expressed in (1.18) suffers from an important inconvenience: it requires the estimation of the vector of score functions for each iteration. Although there exist several methods that can provide these estimations [53], this task is rather cumbersome, since its complexity is equivalent to the problem of estimating PDFs from a set of samples. Fortunately, in the case of linear BSS,[7] researchers noticed that it is possible to achieve separation even in the case where these score functions are roughly estimated. Curiously enough, this result was not found by researchers working on the minimal mutual information principle, but rather it came to light by the works of Cardoso on the maximum likelihood BSS estimator [12] and of Bell and Sejnowski on the Infomax principle [3]. We shall briefly review these two approaches in the sequel.

1.3.2 MAXIMUM LIKELIHOOD AND THE INFOMAX PRINCIPLE

The maximum likelihood estimation (MLE) principle was formalized by Sir Ronald Fisher [56]. Since then, it has been intensively applied in a number of areas, including, of course, signal processing [42]. The MLE principle is a tool employed in the context of probabilistic models. More precisely, given a probabilistic model for a given event, the MLE approach is used to estimate the parameters of this model from the observation of data generated by the event under analysis.

In order to introduce the mathematical aspects of the MLE principle, let us consider a simple example in which one is interested in modeling a given parameter x, say the length of products that are being manufactured in a given industry. Our model, which is of probabilistic nature, is based on the Gaussian distribution so the PDF associated with x is given by:

$$p(x) = \frac{1}{\sqrt{2\pi\sigma^2}} \exp\left(-\frac{(x-\mu)^2}{2\sigma^2}\right). \tag{1.19}$$

We shall assume that the standard deviation, σ, of the model is known. Therefore, in order to have a complete model, it is necessary to estimate μ. How can this be done from a set of N samples $x_1, x_2, \ldots, x_N$?

---

[7]In some nonlinear BSS models, such as the post-nonlinear (PNL) [59], it is not possible to accomplish the task of separation when the score functions are roughly estimated.

The MLE approach relies on the likelihood function $L(\mu)$, which, in our example, is given by the following condition distribution:

$$L(\mu; x_1, x_2, \ldots, x_N) = p(x_1, x_2, \ldots, x_N | \mu). \tag{1.20}$$

The likelihood function is a function of μ — the rationale here is to assume that $x_1, x_2, \ldots, x_N$ are fixed. A central assumption in the MLE approach is that the samples $x_1, x_2, \ldots, x_N$ correspond to independent and identically distributed (i.i.d.) observations. Therefore, Equation (1.20) can be simplified as follows

$$L(\mu; x_1, x_2, \ldots, x_N) = \prod_{i=1}^{N} p(x_i | \mu). \tag{1.21}$$

By considering (1.19), the likelihood becomes

$$L(\mu; x_1, x_2, \ldots, x_N) = \prod_{i=1}^{N} \frac{1}{\sqrt{2\pi\sigma^2}} \exp\left(-\frac{(x_i - \mu)^2}{2\sigma^2}\right), \tag{1.22}$$

which, after straightforward calculation, can be written as follows:

$$L(\mu; x_1, x_2, \ldots, x_N) = \frac{1}{(2\pi\sigma^2)^{N/2}} \exp\left(-\sum_{i=1}^{N} \frac{(x_i - \mu)^2}{2\sigma^2}\right). \tag{1.23}$$

Given the likelihood function (1.23), the idea in the MLE approach is to find the parameter μ that maximizes $L(\mu; x_1, x_2, \ldots, x_N)$, or, equivalently, the logarithm of $L(\mu; x_1, x_2, \ldots, x_N)$, that is:

$$\max_{\mu} \log L(\mu; x_1, x_2, \ldots, x_N). \tag{1.24}$$

In our example, the solution of this problem is straightforward and can be obtained analytically by imposing the optimality condition

$$\frac{\partial \log L(\mu; x_1, x_2, \ldots, x_N)}{\partial \mu} = 0. \tag{1.25}$$

By proceeding this way, one readily obtains the maximum likelihood estimator for μ, which is given by

$$\frac{1}{N} \sum_{i=1}^{N} x_i, \tag{1.26}$$

and corresponds to the sample mean of the observed data.

Let us now discuss the MLE approach for BSS. First of all, it is important to stress that we are interested in estimating the mixing matrix $\mathbf{A}$ — note that in the determined case, estimating $\mathbf{A}$ is equivalent to estimate the separating matrix $\mathbf{W}$, since, ideally, $\mathbf{W} = \mathbf{A}^{-1}$. Another important point here is that the probabilistic nature of the model will be related to the PDF of each source, which will be represented by

$p_{s_1}(s_1), p_{s_2}(s_2), \ldots, p_{s_Q}(s_Q)$. These PDFs are assumed to be known in the derivation of the maximum likelihood estimator. Finally, the N samples of the observations (mixtures in our case) will be represented by the vectors $\mathbf{x}(1), \ldots, \mathbf{x}(N)$, which are assumed to be i.i.d. observations.

Under the conditions described above, the likelihood function related to the BSS model is given by:

$$L(\mathbf{A}; \mathbf{x}(1), \ldots, \mathbf{x}(N)) = \prod_{n=1}^{N} p(\mathbf{x}(n)|\mathbf{A}). \qquad (1.27)$$

Since $\mathbf{x}(n) = \mathbf{A}\mathbf{s}(n)$ and $\mathbf{A} = \mathbf{W}^{-1}$, it asserts that [12]

$$p(\mathbf{x}(n)|\mathbf{A}) = \frac{1}{|\det(\mathbf{A})|} p_{\mathbf{s}}(\mathbf{s}(n)) = |\det(\mathbf{W})| p_{\mathbf{s}}(\mathbf{W}\mathbf{x}(n)). \qquad (1.28)$$

Moreover, since the sources are mutually independent by assumption, then

$$p(\mathbf{x}(n)|\mathbf{W}) = |\det(\mathbf{W})| \prod_{j=1}^{Q} p_{s_j}(s_j(n)) = |\det(\mathbf{W})| \prod_{j=1}^{Q} p_{s_j}(\mathbf{w}_j \mathbf{x}(n)), \qquad (1.29)$$

where $\mathbf{w}_j$ denotes the j-row of $\mathbf{W}$. Therefore, the likelihood function (1.27) becomes:

$$L(\mathbf{W}; \mathbf{x}(1), \ldots, \mathbf{x}(N)) = \prod_{n=1}^{N} |\det(\mathbf{W})| \prod_{j=1}^{Q} p_{s_j}(\mathbf{w}_j \mathbf{x}(n)). \qquad (1.30)$$

Finally, a BSS learning scheme can be defined by considering the maximization of the logarithm of $L(\mathbf{W}; \mathbf{x}(1), \ldots, \mathbf{x}(N))$, according to the following optimization problem:

$$\max_{\mathbf{W}} = \log L(\mathbf{W}; \mathbf{x}(1), \ldots, \mathbf{x}(N)). \qquad (1.31)$$

It can be shown [38] that a gradient-based learning rule for (1.31) is given by:

$$\mathbf{W}^{(k+1)} = \mathbf{W}^{(k)} - \mu \left(E\{\Psi_{\mathbf{s}}(\mathbf{y}(n))\mathbf{x}(n)^T\} - \mathbf{W}^{-T} \right), \qquad (1.32)$$

where $\Psi_{\mathbf{s}}(\cdot)$ is the vector of score functions related to the sources.

It is clear the similarity between the learning rules based on mutual information (Equation (1.18)) and on MLE (Equation (1.32)). The only difference is related to the score functions. In the mutual information approach the score functions are associated with the outputs, and, thus, should be estimated, whereas in the MLE approach the score functions are associated with the actual sources, which means that one assumes that the true PDFs of the sources are known beforehand. Of course, in a blind context, it would be unrealistic to count on such information about the sources.

There is a third paradigm that provides a more practical learning rule than Equation (1.32): the information maximization (Infomax) principle. The main idea behind Infomax is to maximize the joint entropy of the outputs of a neural network [3]. By applying this principle, it is possible to show [3] that one ends up with the following BSS learning rule

$$\mathbf{W}^{(t+1)} = \mathbf{W}^{(t)} - \mu \left(E\{\mathbf{f}(\mathbf{y}(n))\mathbf{x}(n)^T\} - \mathbf{W}^{-T} \right), \qquad (1.33)$$

where $\mathbf{f}(\mathbf{y}(n))$ represents a set of nonlinear functions given by

$$\mathbf{f}(\mathbf{y}(n)) = [f_1(y_1(n)), f_2(y_2(n)), \ldots, f_Q(y_Q(n))].$$

Again, the learning rule (1.33) is very close to that of the MLE approach — this similarity was first shown in [11]. The difference now is that the nonlinear functions $f_i(y_i(n))$ are fixed and defined beforehand and should not necessarily be the score functions of the sources. Actually, there are some general conditions on choosing $f_i(y_i(n))$ [13]. A practical rule of thumb is to consider [38]

$$f_i(y_i(n)) = -\tanh(y_i(n))$$

for super-Gaussian[8] sources and

$$f_i(y_i(n)) = -y_i(n)^3$$

for sub-Gaussian sources.

To illustrate the application of the learning rule (1.33), we here consider the separation of the three sources from the linear mixtures shown in Figure 1.4. The first two sources are sub-Gaussian (uniform distribution) whereas the third one is super-Gaussian (exponential distribution). Therefore, we adopted the following nonlinear functions

$$\mathbf{f}(\mathbf{y}(n)) = [-y_1(n)^3, -y_2(n)^3, \ldots, -\tanh(y_3(n))].$$

The learning rate was set to $\mu = 0.03$. As can be seen in Figure 1.4, the recovered sources are indeed very close to the actual ones, thus confirming that the task of source separation was fulfilled. In Figure 1.5, we plot the evolution of the matrix $\mathbf{W}$ during the learning process.

1.3.3 NON-GAUSSIANITY MAXIMIZATION

Another very popular route to perform ICA can be explained by the principle of non-gaussianity maximization. This approach is motivated by the central limit theorem (CLT), which states that the sum of independent random variables tends to a Gaussian variable no matter the distribution of each original random variable [51]. This result can be applied to BSS by observing that a given mixture is a linear combination of the sources. Therefore, it is expected that the mixtures be closer to Gaussian variables than the original sources. This fact is illustrated in Figure 1.6, which shows the histograms of three sources and their linear mixture.

---

[8] Super-Gaussian variables have a positive kurtosis, whereas sub-Gaussian variables have a negative kurtosis.

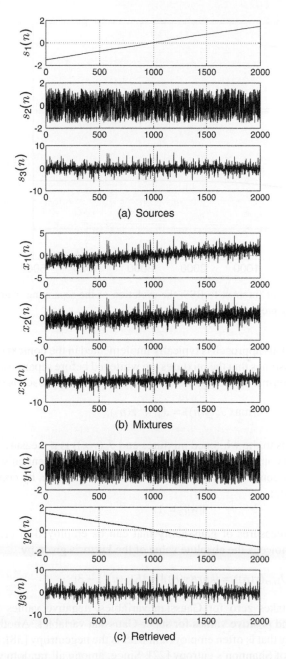

Figure 1.4   Application of Infomax to BSS considering $\mu = 0.03$ and 5000 iterations.

Given that the mixtures are more Gaussian with respect to the sources, a natural idea to perform source separation is to search for components that are as non-

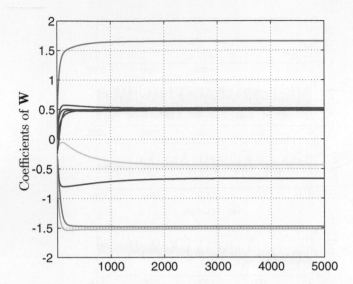

Figure 1.5 Application of Infomax to BSS: the evolution of the separating matrix coefficients during the learning process.

Gaussian as possible. This approach is typically implemented in the context of source extraction and often assumes that the mixtures are whitened before separation. Therefore, in mathematical terms, the non-gaussianity maximization approach is given by:

$$\max_{\mathbf{w}} J_g(y(n)) = J_g(\mathbf{w}^T \mathbf{z}(n)), \tag{1.34}$$

where $\mathbf{z}(n)$ corresponds to the whitened mixtures and $J_g(y(n))$ is a measure of gaussianity. Since the most common measures of gaussianity are dependent on scale, it is necessary to somehow constrain $\mathbf{w}$, which can be done, for instance, by considering the following restriction:

$$\|\mathbf{w}\|_2 = 1.$$

There are several measures of gaussianity that can be employed in (1.34). For example, a possible choice is the absolute value of the kurtosis given by

$$J_{kurt}(y(n)) = |kurt(y(n))| = |E\{y(n)^4\} - 3|. \tag{1.35}$$

Indeed, the kurtosis takes zero for Gaussian variables, negative values for sub-Gaussian variables, and positive values for super-Gaussian variables. Another measure of non-gaussianity that is often employed in BSS is the negentropy [38], which is a normalized version of Shannon's entropy [22]. Since, among all random variables of equal variance, the Gaussian variable is the most entropic one [22], the entropy can be used as a measure of gaussianity.

The optimization problem expressed in (1.34) can be solved by a gradient-based algorithm. There also other alternatives to solve (1.34) such as the application of an

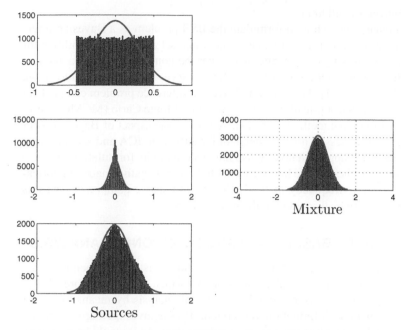

Figure 1.6 Illustration of the *gaussianization* phenomenon that takes place in a linear mixture. The distribution of the sources are, respectively, uniform, Laplace, and triangle. For each distribution, we show the best fit given by a Gaussian distribution. Note that the mixture is very close to a Gaussian distribution.

approximation of the standard Newton algorithm [2]. This idea leads to the FastICA algorithm [34], which is among the most popular ICA algorithms.

A last point to be discussed here is that, although the non-gaussianity principle is usually applied for source extraction, it can be equally used to perform source separation. This can be done by running the source extraction algorithm several times in a sequential fashion. However, one must adopt a scheme to avoid different executions of the method that provide the same source. This can be done by a deflation procedure, which was firstly proposed in [23]. The idea behind the deflation is simple: after a given execution, the contribution of the extracted source is subtracted from the mixtures in a procedure that bears strong resemblance to the adaptive noise canceler.

1.3.4 OTHER BSS METHODS

In addition to ICA, there are several other approaches that can be applied to perform source separation. For instance, a very popular approach is to exploit the temporal structure of the sources by taking into account the correlation matrices of the mixtures for different delays. This approach, which is known as the second-order approach and is the basis of the well-known methods AMUSE [61] and SOBI [4], allows one to separate non-white Gaussian sources under the condition that the spectra of these sources are different.

Another popular approach is to formulate the BSS problem as a Bayesian inference problem [21, 43, 55, 47]. In this case, one must set *a priori* probabilities for the sources and the mixing coefficients, which can be done by taking into account additional information about the problem — for instance, in some applications one expects smooth sources [21]. The resulting Bayesian inference problem can be dealt with by, for instance, the application of Markov Chain Monte Carlo (MCMC) methods [48, 26] or variational methods [65, 14]. A remarkable aspect of Bayesian BSS methods is their robustness to noise. Indeed, differently from ICA and second-order methods, the Bayesian approach takes noise into account in its formulation [47].

Other BSS approaches for the linear case include: non-negative matrix factorization (NMF) [17], time-frequency methods [5, 21] and sparse component analysis (SCA) [30], which will be discussed in the next section.

1.4   BSS METHODS BASED ON SPARSE COMPONENT ANALYSIS

The main drawback related to ICA is the need for the statistical independence assumption. Indeed, there are several applications in which the sources are clearly dependent — an example may be found in chemical analysis by means of sensor arrays [26]. In view of such limitation, research on BSS is investigating for alternatives that may work even when the sources are dependent. As it would be expected, those new strategies avoid a probabilistic framework in order to not rely on the independence assumption. A powerful approach exploits a property known as *sparsity*.

For more than ten years, signal models based on sparsity have been at the center of signal processing theory. Indeed, we are still living what some researchers call the sparsity revolution. Curiously enough, sparseness is an old concept and its application to signal processing dates back to 1979 in the context of geophysical signal processing [25]. More recently, theoretical works and effective applications to compressive sensing [10], which is a tool that allows sampling beyond the Shannon-Nyquist rate [62], have renewed the interest of the signal processing community.

Although sparsity is now a ubiquitous concept in signal processing, there is no precise and satisfactory definition of a sparse signal yet. Nevertheless, in practical terms, a sparse signal presents many coefficients that are zeros (or close to zero) and a few coefficients of large magnitude. Sparsity can show up in the time-domain but also in other transformed domains, such as Fourier and wavelets representations [28] — this is illustrated in Figure 1.7, where we show a signal that is sparse in the frequency domain.

In sparsity framework, researchers usually introduce the L_0 pseudo-norm counting the number of non-null elements of a signal in a predefined domain. From this definition, some interesting properties emerge such as invariance to a scale factor and a relative easiness to be numerically manipulated. In sparse signal representation, in a specified dictionary, it has been shown that the L_1 norm can be employed as a convex relaxation of a L_0 minimization problem. This result allows the use of efficient linear program solvers to tackle sparse representation problems [28, 57].

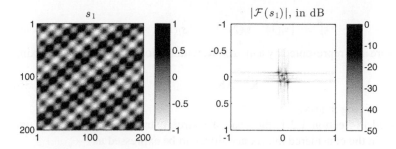

Figure 1.7   An example of a 2D signal that is sparse in the Fourier domain.

Sparse models have also been considered in BSS, particularly for solving under-determined problems. The first application in that respect was found in [9] where the authors discerned the possibility to exploit the sparse nature of the signals for estimating the mixing matrix. A lot of works have been done in that sense, especially via clustering-based approaches [9, 29, 66, 58]. Then, given an estimate of the mixing matrix, the problem is shortened to a sparse representation problem. Moreover, sparsity is at the core of morphological component analysis (MCA), a framework that allows one to separate signals having different morphological aspects [8].

More recently, sparse models have been considered to deal with the linear determined mixing model. Those approaches are interesting because they eschew the need for two steps (estimating first the mixing matrix and then the sparse signals). In [49], for instance, the authors developed a framework based on the L_1 norm, whereas in [27] the application of the L_0 pseudo-norm to source extraction was studied. There are also other works that investigated the application of sparsity to BSS (see for instance [50]). In the next section, we shall focus on [27], since this work provides a good illustration of the benefits that sparse models can bring to BSS.

1.4.1   SPARSE SOURCE EXTRACTION

The L_0 pseudo-norm of a discrete signal is given by the number of non-null elements of the signal. Therefore, sparse signals tend to present a small L_0 pseudo-norm.[9]In [27], the authors investigated the conditions for which the L_0 pseudo-norm is a contrast for source extraction. In order to discuss these conditions, let us consider the case of two sources and two mixtures. Moreover, we shall consider a matrix notation in which the sources are represented by

$$\mathbf{S} = \begin{bmatrix} \mathbf{s}_1^T \\ \vdots \\ \mathbf{s}_Q^T \end{bmatrix}$$

---

[9]The number of non-null elements of a vector is called pseudo-norm L_0 because, although it satisfies the triangle inequality, it is scale-invariant, and hence is not a norm.

where

$$\mathbf{s}_i^T = [s_i(1) \quad s_i(2) \quad \ldots \quad s_i(N)].$$

The mixtures are also represented by a matrix, $\mathbf{X}$, which allows us to write the mixing process as

$$\mathbf{X} = \mathbf{AS}, \tag{1.36}$$

where $\mathbf{A}$ is the mixing matrix.

As discussed in Section 1.2.1, the goal of source extraction is to retrieve one source, which, in the considered matrix notation, can be expressed as

$$\mathbf{y}^T = \mathbf{w}^T \mathbf{X}. \tag{1.37}$$

Assuming, without loss of generality, that $\mathbf{s}_1$ is sparser than $\mathbf{s}_2$ in the sense of the L_0 pseudo-norm, that is

$$\|\mathbf{s}_1\|_0 < \|\mathbf{s}_2\|_0,$$

where $\|\mathbf{s}_i\|_0$ denotes the L_0 pseudo-norm of $\mathbf{s}_i$, recovering the sparsest signal can be mathematically formulated as the following optimization problem

$$\min_{\mathbf{w}} J(\mathbf{w}) = \|\mathbf{y}\|_0 = \|\mathbf{w}^T \mathbf{X}\|_0, \tag{1.38}$$

with $\mathbf{w} \neq \mathbf{0}$. In a practical situation, when other norms are used, other restrictions must be considered such that, for instance, $\|\mathbf{w}\|_2 = 1$.

The very relevant question addressed in [27] is related to the conditions for which solving (1.38) leads to source extraction. It can be shown that a sufficient condition for the L_0 pseudo-norm to be a contrast is given by

$$\|\mathbf{s}_1\|_0 < \frac{\|\mathbf{s}_2\|_0}{2}. \tag{1.39}$$

There are several interesting points related to this condition. The first one is that it does not rely on statistical independence. Indeed, this condition is a deterministic one and does not impose any probabilistic constraint. As a consequence, the approach based on the L_0 pseudo-norm allows one to separate classes of dependent sources. This feature is illustrated in Figure 1.8, where we show the extraction of a source from a mixture of dependent signals.

Another interesting aspect related to (1.39) comes to light if we analyze it in the Fourier domain — this condition is valid for linear transforms such as the Fourier one. As it can be observed, if the signals are sparse in the frequency domain, then the sparsest one, $\mathbf{s}_1$, is allowed to have at most half of the non-null frequency elements of $\mathbf{s}_2$. Therefore, if signal $\mathbf{s}_2$ has only non-null frequency elements, then condition (1.39) tells us that separation is possible when the overlap between the two signals does not exceed half of $\|\mathbf{s}_2\|_0$. This result is very illustrative since it shows that, in the analyzed case, the L_0 pseudo-norm framework is somehow between a filtering approach based on Fourier and an ICA approach. In other words, the L_0 pseudo-norm minimization allows one to separate a class of dependent sources but, on the other hand, requires that the spectra of the sources present a minimum of overlapping.

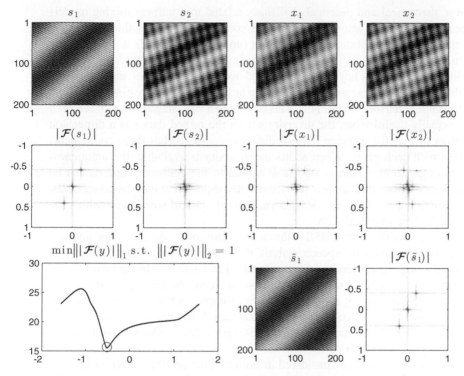

Figure 1.8 Extraction of a sparse source s_1 from two mixtures via L_1 minimization in the Fourier domain. Interestingly, the two sources s_1 and s_2 are statically dependent in the time domain due to a common low frequency component. The two mixtures x_1 and x_2 are linear combinations of s_1 and s_2. The objective function used is the normalized L_1 norm as presented in [49] but applied on the amplitude of the Fourier transform.

A last remark on (1.39) is that it provides a sufficient, but not necessary, condition for the L_0 pseudo-norm to be a contrast. Therefore, there is still room for obtaining less restrictive conditions.

1.5 CONCLUSIONS

In this chapter, we provided an introduction to the problem of blind source separation (BSS). Our focus was on linear, instantaneous, and determined mixing models, and on learning algorithms based on independent component analysis (ICA). Three important ICA paradigms were discussed. Moreover, we also briefly introduced the basics of sparse component analysis (SCA) with special attention to its application to the separation of dependent sources.

There are important topics in the field of BSS that were not covered in this chapter. For instance, the problem of BSS in convolutive mixing models — and notably its application to audio separation — is a recurrent topic in the current BSS literature.

Moreover, theoretical and practical challenges related to nonlinear mixing models have been attracting the attention of the BSS community. Finally, the situation in which there are fewer mixtures than sources (undetermined case) is still a tricky problem.

Besides the extensions beyond the linear and instantaneous model, there are some topics that might guide research on BSS for the next years. An interesting example is the application of tensors to BSS [19]. This is already a very active theme of research and is specially useful when the diversity within the mixed data has a dimension higher than two — for instance, a RGB digital color image has three components associated with each pixel. When additional diversity is available, the uniqueness of the decomposition, which is somehow related to the contrast property, can be assured for conditions that are less restrictive with respect to matrix-based methods. This feature may pave the way for the separation of class of sources for which the current methods cannot deal with.

Another trend in the area of BSS is the case of multimodal data. In some applications, e.g., brain imaging and speech analysis, there are different types of sensors that can be used to register the desired source signals. The basic idea behind multimodal separation is to integrate all this information so one can take advantage of the benefits of each type of sensing mechanism — for example, in brain imaging it is well known that functional magnetic resonance imaging (fMRI) has a very good spatial resolution whereas electroencephalogram (EEG) provides a more precise temporal resolution.

A last topic that might be considered in future works on BSS concerns its relationship with data representation methods. Indeed, with the advent of sparse representations and, more recently, structured models [15], a natural question that arises is whether BSS can be seen as a particular case of these data representation methods. The main interest here is to make use of the efficient data representation methods (usually based on convex optmization algorithms) in BSS. The difficulty, however, is that, when expressed as a data representation problem, even the canonical linear BSS model results in a nonlinear representation problem.

REFERENCES

1. T. Adali, M. Anderson, and F. Geng-Shen. Diversity in independent component and vector analyses: Identifiability, algorithms, and applications in medical imaging. *IEEE Signal Processing Magazine*, 31(3):18–33, May 2014.

2. A. Antoniou and W.-S. Lu. *Practical optimization*. Springer, 2007.

3. A. J. Bell and T. J. Sejnowski. An information-maximization approach to blind separation and blind deconvolution. *Neural Computation*, 7(6):1129–1159, 1995.

4. A. Belouchrani, K. Abed-Meraim, J.-F. Cardoso, and E. Moulines. A blind source separation technique using second-order statistics. *IEEE Transactions on Signal Processing*, 45(2):434–444, Feb. 1997.

5. A. Belouchrani and M. G. Amin. Blind source separation based on time-frequency signal representations. *IEEE Transactions on Signal Processing*, 46(11):2888–2897, Nov. 1998.

6. S. Bermejo, C. Jutten, and J. Cabestany. ISFET source separation: foundations and techniques. *Sensors and Actuators B*, 113:222–233, 2006.
7. C. M. Bishop. *Pattern Recognition and Machine Learning*. Springer, 2006.
8. J. Bobin, J.-L. Starck, Y. Moudden, and M. J. Fadili. Blind source separation: the sparsity revolution. *Advances in Imaging and Electron Physics*, 152:221–302, 2008.
9. P. Bofill and M. Zibulevsky. Underdetermined blind source separation using sparse representations. *Signal processing*, 81(11):2353–2362, 2001.
10. E. J. Candès and M. B. Wakin. An introduction to compressive sampling. *IEEE Signal Processing Magazine*, 25(2):21–30, 2008.
11. J.-F. Cardoso. Infomax and maximum likelihood for blind source separation. *IEEE Signal Processing Letters*, 4:112–114, 1997.
12. J.-F. Cardoso. Blind signal separation: statistical principles. *Proceedings of the IEEE*, 86(10):2009–2025, Oct. 1998.
13. J.-F. Cardoso. On the stability of source separation algorithms. *Journal of VLSI signal processing systems for signal, image and video technology*, 26(1-2):7–14, 2000.
14. A. T. Cemgil, C. Févotte, and S. J. Godsill. Variational and stochastic inference for bayesian source separation. *Digital Signal Processing*, 17:891–913, 2007.
15. V. Chandrasekaran, B. Recht, P. A. Parrilo, and A. S. Willsky. The convex geometry of linear inverse problems. *Foundations of Computational Mathematics*, 12(6):805–849, 2012.
16. A. Cichocki and S. Amari. *Adaptive blind signal and image processing*. John Wiley & Sons, 2002.
17. A. Cichocki, R. Zdunek, A. H. Phan, and S. Amari. *Nonnegative matrix and tensor factorizations : applications to exploratory multiway data analysis and blind source separation*. John Wiley & Sons, 2009.
18. P. Comon. Independent component analysis, a new concept? *Signal Processing*, 36:287–314, 1994.
19. P. Comon. Tensors: a brief introduction. *IEEE Signal Processing Magazine*, 31(3):44–53, 2014.
20. P. Comon and C. Jutten, editors. *Séparation de sources 1 : concepts de base et analyse en composantes indépendantes*. Hermes Science Publications, 2007.
21. P. Comon and C. Jutten, editors. *Handbook of blind source separation: independent component analysis and applications*. Academic Press, 2010.
22. T. M. Cover and J. A. Thomas. *Elements of information theory*. Wiley-Interscience, 1991.
23. N. Delfosse and P. Loubaton. Adaptive blind separation of independent sources: A deflation approach. *Signal Processing*, 45:59–83, 1995.
24. P. S. R. Diniz. *Adaptive filtering: algorithms and practical implementation*. Springer, 2008.
25. D. Donoho. Scanning the technology. *Proceedings of the IEEE*, 98(6):910–912, 2010.
26. L. T. Duarte, C. Jutten, and S. Moussaoui. A bayesian nonlinear source separation method for smart ion-selective electrode arrays. *IEEE Sensors Journal*, 9(12):1763–1771, 2009.
27. L. T. Duarte, R. Suyama, R. Attux, J. M. T. Romano, and C. Jutten. Blind extraction of sparse components based on l0-norm minimization. In *Proc. IEEE Statistical Signal Processing Workshop (SSP)*, pages 617–620, 2011.
28. M. Elad. *Sparse and redundant representations from theory to applications in signal and image processing*. Springer, 2010.

29. P. Georgiev, F. J. Theis, and A. Cichocki. Sparse component analysis and blind source separation of underdetermined mixtures. *IEEE Transactions on Neural Networks*, 16(4):992–996, 2005.

30. R. Gribonval and S. Lesage. A survey of sparse component analysis for blind source separation: principles, perspectives, and new challenges. In *Proceedings of the European Symposium on Artificial Neural Networks (ESANN)*, 2006.

31. W. S. Harlan, J. F. Claerbout, and F. Rocca. Signal/noise separation and velocity estimation. *Geophysics*, 49(11):1869–1880, 1984.

32. S. Haykin. *Adaptive filter theory*. Prentice-Hall, 1995.

33. J. Hérault, C. Jutten, and B. Ans. Détection de grandeurs primitives dans un message composite par une architecture de calcul neuromimétique en apprentissage non supervisé. In *Proceedings of the GRETSI*, 1985. In Portuguese.

34. A. Hyvärinen. Fast and robust fixed-point algorithms for independent component analysis. *IEEE Transactions on Neural Networks*, 10(3):626–634, 1999.

35. A. Hyvärinen. Survey on independent component analysis. *Neural computing surveys*, 2(4):94–128, 1999.

36. A. Hyvärinen. Independent component analysis: recent advances. *Philosophical Transactions of the Royal Society A: Mathematical, Physical and Engineering Sciences*, 371(1984), 2013.

37. A. Hyvärinen, J. Hurri, and P. O. Hoyer. *Natural Image Statistics: A probabilistic approach to early computational vision*. Springer, 2008.

38. A. Hyvärinen, J. Karhunen, and E. Oja. *Independent component analysis*. John Wiley & Sons, 2001.

39. A. J. Izenman. *Modern multivariate statistical techniques*. Springer, 2008.

40. C. Jutten and J. Karhunen. Advances in blind source separation (BSS) and independent component analysis (ICA) for nonlinear mixtures. *International Journal of Neural Systems*, 14:267–292, 2004.

41. A. M. Kagan, C. R. Rao, and Y. V. Linnik. Characterization problems in mathematical statistics. 1973.

42. S. M. Kay. *Fundamentals of statistical signal processing: estimation theory*. Prentice-Hall, 1993.

43. K. H. Knuth. Informed source separation: A bayesian tutorial. In *Proceedings of the 13th European Signal Processing Conference (EUSIPCO)*, 2005.

44. J.-L. Lacoume, P.-O. Amblard, and P. Comon. *Statistique d'ordre superior pour le traitment de signal*. Masson, 2003.

45. B. P. Lathi. *Signal processing and linear system*. Berkeley-Cambridge, 1998.

46. S. Makino, T. W. Lee, and H. Sawada, editors. *Blind speech separation*. Springer, 2007.

47. A. Mohammad-Djafari. *Approche bayésienne*, chapter 12, pages 483–514. Hermes Science Publications, 2007.

48. S. Moussaoui, D. Brie, A. Mohammad-Djafari, and C. Carteret. Separation of non-negative mixture of non-negative sources using a Bayesian approach and MCMC sampling. *IEEE Transactions on Signal Processing*, 54:4133–4145, 2006.

49. E. Z. Nadalin, A. K. Takahata, L. T. Duarte, R. Suyama, and R. Attux. Blind extraction of the sparsest component. *Lectures Notes on Computer Science*, 6365:263–270, 2010.

50. P. D. O'Grady, B. A. Pearlmutter, and S. T. Rickard. Survey of sparse and non-sparse methods in source separation. *International Journal of Imaging Systems and Technology*, 15(1):18–33, 2005.

51. A. Papoulis and U. Pillai. *Probability, random variables, and stochastic processes*.

McGraw-Hill, 2004.

52. M. S. Pedersen, J. Larsen, U. Kjems, and L. C. Parra. A survey of convolutive blind source separation methods. *Multichannel Speech Processing Handbook*, pages 1065–1084, 2007.

53. D.-T. Pham. Fast algorithms for mutual information based independent component analysis. *IEEE Transactions on Signal Processing*, 52(10):2690–2700, Oct. 2004.

54. J. M. T. Romano, R. R. F. Attux, C. C. Cavalcante, and R. Suyama. *Unsupervised signal processing: channel equalization and source separation*. CRC Press, 2011.

55. D. B. Rowe. A bayesian approach to blind source separation. *Journal of Interdisciplinary Mathematics*, 5(1):49–76, 2002.

56. D. Salsburg. *The lady tasting tea: How statistics revolutionized science in the twentieth century*. Macmillan, 2001.

57. J.-L. Starck, F. Murtagh, and J. M. Fadili. *Sparse image and signal processing: wavelets, curvelets, morphological diversity*. Cambridge University Press, 2010.

58. Y. Sun and J. Xin. Underdetermined sparse blind source separation of nonnegative and partially overlapped data. *SIAM Journal on Scientific Computing*, 33(4):2063–2094, 2011.

59. A. Taleb and C. Jutten. Source separation in post-nonlinear mixtures. *IEEE Transactions on Signal Processing*, 47(10):2807–2820, Oct. 1999.

60. A. Tonazzini, I. Gerace, and F. Martinelli. Multichannel blind separation and deconvolution of images for document analysis. *IEEE Transactions on Image Processing*, 19(4):912–925, 2010.

61. L. Tong, R.-W. Liu, V. C. Soon, and Y.-F. Huang. Indeterminacy and identifiability of blind identification. *IEEE Transactions on Circuits and Systems*, 38(5):499–509, 1991.

62. J.A. Tropp, J.N. Laska, M.F. Duarte, J.K. Romberg, and R.G. Baraniuk. Beyond nyquist: Efficient sampling of sparse bandlimited signals. *Information Theory, IEEE Transactions on*, 56(1):520–544, 2010.

63. R. Vigario and E. Oja. BSS and ICA in neuroinformatics: from current practices to open challenges. *IEEE Reviews in Biomedical Engineering*, 1:50–61, 2008.

64. B. Widrow, J. R. Glover, J. M. McCool, J. Kaunitz, C. S. Williams, R. H. Hearn, J. R. Zeidler, E. Dong, and R. C. Goodlin. Adaptive noise cancelling: Principles and applications. *Proceedings of the IEEE*, 63(12):1692–1716, 1975.

65. O. Winther and K. B. Petersen. Bayesian independent component analysis: variational methods and non-negative decompositions. *Digital Signal Processing*, 2007.

66. M. Zibulevsky and B. Pearlmutter. Blind source separation by sparse decomposition in a signal dictionary. *Neural computation*, 13(4):863–882, 2001.

2 Kernel-Based Nonlinear Signal Processing

José Carlos Moreira Bermudez
Federal University of Santa Catarina (UFSC)

Paul Honeine
University of Technology of Troyes (UTT)

Jean-Yves Tourneret
University of Toulouse

Cédric Richard
University of Nice Sophia Antipolis (UNS)

2.1 CHAPTER SUMMARY

In this chapter we discuss kernel-based nonlinear signal processing. Sections 2.3 and 2.4 present a brief introduction to the theory of reproducing kernel Hilbert spaces (RKHS) and their application to nonparametric modeling of nonlinear functions. We give special emphasis to nonlinear function estimation and regression, and illustrate their relevance on several nonlinear signal processing applications. In Section 2.5 we explore online kernel-based function approximation. We focus on finite order modeling, sparsification techniques, and their application to online learning. In Section 2.6 we discuss kernel-based online system identification, with emphasis on the popular kernel least-mean squares algorithm (KLMS) and its properties. In Section 2.7 we present an introduction to Gaussian processes as applied to nonlinear regression, and we describe its application to spectral unmixing of hyperspectral images.

2.2 INTRODUCTION

An effective way to extend the scope of linear models to nonlinear processing is to map the input data u_i into a high-dimensional space using a nonlinear function $\varphi(\cdot)$, and apply linear modeling techniques to the transformed data $\varphi(u_i)$. However, this strategy may fail when the image of $\varphi(\cdot)$ lies in a very high, or even infinite, dimensional space. More recently, kernel-based methods have been proposed for applications in classification and regression problems. These methods exploit the central idea of this research area, known as the *kernel trick*, to evaluate inner products in the high dimensional space without the knowledge of the function $\varphi(\cdot)$. Well-known examples can be found in [36, 46]. This chapter discusses the application of kernel-based methods to solve nonlinear function estimation and regression problems.

2.3 REPRODUCING KERNEL HILBERT SPACES (RKHS)

This section briefly reviews the main definitions and properties related to reproducing kernel Hilbert spaces [3] and Mercer kernels [24].

Let $\mathcal{H}$ denote a Hilbert space of real-valued functions $\psi(\cdot)$ on a compact $\mathcal{U} \subset \mathbb{R}^\ell$, and let $\langle \cdot, \cdot \rangle_{\mathcal{H}}$ be the inner product in $\mathcal{H}$. Suppose that the evaluation functional L_u defined by $L_u[\psi] \triangleq \psi(u)$ is linear with respect to $\psi(\cdot)$ and bounded, for all u in $\mathcal{U}$. By virtue of the Riesz representation theorem, there exists a unique positive definite function $u_i \mapsto \kappa(u_i, u_j)$ in $\mathcal{H}$, denoted by $\kappa(\cdot, u_j)$ and called *representer of evaluation* at u_j, that satisfies [3]

$$\psi(u_j) = \langle \psi(\cdot), \kappa(\cdot, u_j) \rangle_{\mathcal{H}}, \quad \forall \psi \in \mathcal{H} \tag{2.1}$$

for every fixed $u_j \in \mathcal{U}$. A proof of this may be found in [3]. Replacing $\psi(\cdot)$ by $\kappa(\cdot, u_i)$ in (2.1) yields

$$\kappa(u_j, u_i) = \langle \kappa(\cdot, u_i), \kappa(\cdot, u_j) \rangle_{\mathcal{H}} \tag{2.2}$$

for all $u_i, u_j \in \mathcal{U}$. Equation (2.2) is the origin of the now generic term *reproducing kernel* to refer to $\kappa(\cdot, \cdot)$, which is also known as a Mercer kernel. Note that $\mathcal{H}$ can be restricted to the span of $\{\kappa(\cdot, u) : u \in \mathcal{U}\}$ because, according to (2.1), nothing outside this set affects $\psi(\cdot)$ evaluated at any point of $\mathcal{U}$. Denoting by $\varphi(\cdot)$ the map that assigns to each input u the kernel function $\kappa(\cdot, u)$, (2.2) implies that $\kappa(u_i, u_j) = \langle \varphi(u_i), \varphi(u_j) \rangle_{\mathcal{H}}$. The kernel then evaluates the inner product of any pair of elements of $\mathcal{U}$ mapped to $\mathcal{H}$ without any explicit knowledge of either $\varphi(\cdot)$ or $\mathcal{H}$. This key idea is known as the *kernel trick*. The kernel trick has been widely used to transform linear algorithms expressed only in terms of inner products into nonlinear ones. Examples are the nonlinear extensions to the principal component analysis [25] and the Fisher discriminant analysis [1, 39]. Recent work has also been focused on kernel-based online prediction of time series [11, 12, 13, 21, 34].

Classic examples of kernels are the radially Gaussian kernel $\kappa(u_i, u_j) = \exp\left(-\|u_i - u_j\|^2 / 2\beta_0^2\right)$, with $\beta_0 \geq 0$ the kernel bandwidth, and the Laplacian kernel $\kappa(u_i, u_j) = \exp\left(-\|u_i - u_j\| / \beta_0\right)$. Another example which deserves attention in signal processing is the q-th degree polynomial kernel defined as $\kappa(u_i, u_j) = (\eta_0 + u_i^t u_j)^q$, with $\eta_0 \geq 0$ and $q \in \mathbb{N}^*$. The nonlinear function $\varphi(\cdot)$ related to the latter transforms every observation u_i into a vector $\varphi(u_i)$, in which each component is proportional to a monomial of the form $(u_{i,1})^{k_1}(u_{i,2})^{k_2} \ldots (u_{i,p})^{k_p}$ for every set of exponents $(k_1, ..., k_p) \in \mathbb{N}^p$ satisfying $0 \leq \sum_{r=1}^{p} k_r \leq q$. For details, see [10, 47] and references therein. The models of interest then correspond to q-th degree Volterra series representations.

2.4 NONLINEAR REGRESSION IN AN RKHS

The regression problem is an example of learning in which we want to predict the values of one or more continuous variables (*outputs*) as a function of a set of *inputs* that are measured or preset. Problems in which a collection of known input-output pairs is available for training the regression model are known as supervised learning problems. The regression is nonlinear when the function to be modeled is nonlinear.

A nonlinear regression problem with a scalar output variable is characterized by the expression

$$d_i = \psi(u_i) + \eta_i, \qquad i = 1, \ldots, n \qquad (2.3)$$

where d_i is the i-th output, u_i is the i-th input vector, and η_i is the so-called noise sample, which in general represents the part of d_i that cannot be modeled by $\psi(\cdot)$. Noise η_i is frequently modeled as a sample of a white Gaussian process.

Sometimes the mathematical form of $\psi(\cdot)$ is known or can be reasonably inferred from practical or theoretical considerations, except for some parameters. In these cases, the objective is to estimate the unknown parameter values as precisely as possible [20, 44, 45]. However, in several practical problems the underlying process is complex and not well understood. In such cases, one possible approach is to use a family of functions that is flexible enough to approximate a sufficiently large variety of functional forms. The objective is then to estimate the parameters of the family of functions. Examples of popular families of functions are polynomial models and kernel expansions in RKHS. Another approach that has recently gained popularity in the signal processing community is to avoid the choice of specific functional models and imagine that the output samples are samples from a multivariate Gaussian distribution. The sequence of samples is modeled as a Gaussian process characterized by a covariance function in RKHS [32]. This chapter reviews recent solutions to the nonlinear regression problem in the RKHS.

To solve the nonlinear regression problem in the RKHS, let $\kappa : \mathcal{U} \times \mathcal{U} \to \mathbb{R}$ be a kernel, and let $\mathcal{H}$ be the RKHS associated with it. The problem is to determine a function $\psi(\cdot)$ of $\mathcal{H}$ that minimizes a cost function $\sum_{i=1}^{n} \mathcal{L}(d_i, \psi(u_i))$, where $\mathcal{L}(d_i, \psi(u_i))$ is an appropriate loss. Considering the least-squares approach for instance, the cost function is the sum of the squared errors between samples d_i of the desired response and the corresponding model output samples $\psi(u_i) = \langle \psi(\cdot), \kappa(\cdot, u_i) \rangle_{\mathcal{H}}$ for $i = 1, \ldots, n$, namely,

$$\min_{\psi \in \mathcal{H}} \sum_{i=1}^{n} |d_i - \psi(u_i)|^2. \qquad (2.4)$$

While the solution to this problem is not unique, one often considers a regularized cost function

$$\min_{\psi \in \mathcal{H}} \sum_{i=1}^{n} |d_i - \psi(u_i)|^2 + \epsilon \, \| \psi(\cdot) \|_{\mathcal{H}}^2 \qquad (2.5)$$

where a regularization term is introduced in order to impose the uniqueness and smoothness of the solution. The tradeoff between training errors and smoothness of the solution is controlled by the tunable non-negative parameter ϵ. Taking the functional derivative of the above cost function with respect to $\psi(\cdot)$ [19] and nullifying it, leads to

$$-\sum_{i=1}^{n} (d_i - \psi(u_i)) \kappa(\cdot, u_i) + \epsilon \psi(\cdot) = 0$$

where we have used the linearity with respect to $\psi(\cdot)$ in $\psi(u_i) = \langle \psi(\cdot), \kappa(\cdot, u_i) \rangle_{\mathcal{H}}$, as well as the fact that $\psi(\cdot)$ is the derivative of the quadratic form $\frac{1}{2} \| \psi(\cdot) \|_{\mathcal{H}}^2$. Therefore, the function $\psi(\cdot)$ of $\mathcal{H}$ minimizing (2.4) can be written as a kernel expansion in terms of available data, namely

$$\psi(\cdot) = \sum_{i=j}^{n} \alpha_j \kappa(\cdot, u_j). \tag{2.6}$$

By injecting this expression in the problem (2.4), we obtain the optimization problem $\min_{\alpha} \| d - K\alpha \|^2 + \epsilon \alpha^\top K \alpha$, where K is the Gram matrix whose (i, j)-th entry is $\kappa(u_i, u_j)$ and d is the vector of desired output values whose i-th entry is d_i. By taking the derivative of this cost function with respect to α and nullifying it, we get the $n \times n$ linear system of equations $(K^\top K + \epsilon K) \alpha = K^\top d$.

It is worth noting that the regularization term given in (2.5) is often dropped in adaptive learning. In this case, the uniqueness and regularity of the solution of (2.4) are controlled by other principles, such as the minimal disturbance principle (see Section 2.6).

2.5 ONLINE KERNEL-BASED FUNCTION APPROXIMATION

2.5.1 INTRODUCTION

Online solution of the nonlinear regression problem formulated in the RHKS raises the question of how to process an increasing amount of observations and update the model (2.6) as new data is collected. Because the order of the model (2.6) is equal to the number n of available data u_i, this approach cannot be considered for online applications. Hence, one needs to look for sparse solutions. One way to obtain a sparse solution is to consider fixed-size models of the form

$$\psi(\cdot) = \sum_{j=1}^{m} \alpha_j \kappa(\cdot, u_{\omega_j}) \tag{2.7}$$

where the $u_{\omega_1}, \ldots, u_{\omega_m}$ form an m-element subset of the available input vectors.

Online applications, however, impose time-varying input signals. Hence, it is convenient to adapt the notation of (2.7) by denoting the input vector at time n by u_n, and by defining ω_j, $j = 1, \ldots, m$ to form a subset of $\mathcal{J}_n = \{1, 2, \ldots, n\}$ corresponding to the time indexes of the $m < n$ input vectors chosen to build the m-th order model. Under this notation, (2.7) becomes

$$\psi_n(\cdot) = \sum_{j=1}^{m} \alpha_{j,n} \kappa(\cdot, u_{\omega_j}). \tag{2.8}$$

Once the fixed-order model (2.8) is adopted, the next problem is to choose an adequate set of m input vectors $\boldsymbol{u}_{\omega_j}$ to build the model. The m kernel functions $\kappa(\cdot, \boldsymbol{u}_{\omega_j})$ form the *dictionary* $\mathcal{D}$, and the selection of the elements to compose $\mathcal{D}$ must follow some sparsification strategy. In the following we briefly review existing sparsification rules that can be used to this end.

2.5.1.1 Sparsification Rules

Discarding a kernel function from the model expansion (2.8) may degrade its performance. Sparsification rules aim at identifying kernel functions whose removal is expected to have a negligible effect on the quality of the model. An extensive literature addressing this issue in batch and online modes exists, see, e.g., [15] and references therein. In particular, much attention has been recently focused on least-squares support vector machines since they suffer from the loss of sparsity due to the use of a quadratic loss function [42]. In batch mode, this problem was addressed using pruning [8, 41] and fixed-size approaches [42, 6, 16]. Truncation and approximation processes were considered for online scenarios [15].

The most informative sparsification criteria use approximate linear dependence conditions to evaluate whether the contribution of a candidate kernel function can be distributed over the elements of $\mathcal{D}$ by adjusting their multipliers. In [37], determination of the kernel function that is best approximated by the others is carried out by an eigendecomposition of the Gram matrix. This process is not appropriate for online applications, as its complexity at each time step is cubic in the size m of $\mathcal{D}$. In [13], the kernel function $\kappa(\cdot, \boldsymbol{u}_n)$ is inserted at time step n into $\mathcal{D}$ if the following condition is satisfied

$$\min_{\gamma} \left\| \kappa(\cdot, \boldsymbol{u}_n) - \sum_{\omega_j \in \mathcal{J}_{n-1}} \gamma_j \kappa(\cdot, \boldsymbol{u}_{\omega_j}) \right\|_{\mathcal{H}}^2 \geq \nu \qquad (2.9)$$

where κ is a unit-norm kernel,[1] that is, $\kappa(\boldsymbol{u}_k, \boldsymbol{u}_k) = 1$ for all $\boldsymbol{u}_k$. The threshold ν determines the level of sparsity of the model. Note that condition (2.9) ensures the linear independence of the elements of the dictionary. A similar criterion is used in [11, 12], but in a different form. After updating the model parameters, a complementary pruning process is executed in [12] to limit the increase in the model order. It estimates the error induced in $\psi(\cdot)$ at time n by the removal of each kernel and discards those kernels found to have the smallest contribution. A major criticism that can be made of rule (2.9) is that it leads to elaborate and costly operations with quadratic complexity in the cardinality m of $\mathcal{D}$. In [11, 12], the model reduction step is computationally more expensive than the parameter update step, the latter being a stochastic gradient descent with linear complexity in m. In [13], the authors focus their study on a parameter update step of the RLS type with quadratic complexity in m. To reduce the overall computational effort, the parameter update and the model reduction steps share intermediate calculation results. This excludes very useful and popular online

---

[1]Replace $\kappa(\cdot, \boldsymbol{u}_k)$ with $\kappa(\cdot, \boldsymbol{u}_k)/\sqrt{\kappa(\boldsymbol{u}_k, \boldsymbol{u}_k)}$ in (2.9) if $\kappa(\cdot, \boldsymbol{u}_k)$ is not unit-norm.

regression techniques. In [17, 34] a coherence-based sparsification rule has been proposed which requires much less computational complexity than the rules discussed so far. According to this rule, kernel $\kappa(\cdot, \boldsymbol{u}_n)$ is inserted into the dictionary if

$$\max_{\omega_j} |\kappa(\boldsymbol{u}_n, \boldsymbol{u}_{\omega_j})| \leq \varepsilon_0 \qquad (2.10)$$

with ε_0 a parameter determining the dictionary coherence. It was shown in [34] that the dictionary dimension determined under rule (2.10) is always finite.

The next section presents a set of kernel-based adaptive algorithms and discusses their application to online time-series prediction.

2.5.2   ALGORITHMS FOR ONLINE TIME-SERIES PREDICTION

Consider the m-th order model at time step n, as given in (2.8), namely

$$\psi_n(\cdot) = \sum_{j=1}^{m} \alpha_{j,n} \, \kappa(\cdot, \boldsymbol{u}_{\omega_j})$$

where the m kernel functions $\kappa(\cdot, \boldsymbol{u}_{\omega_j})$ form a ε_0-coherent dictionary, obtained for instance by the coherence criterion (2.10). By injecting the above model in the problem (2.4), we get the optimization problem $\min_{\alpha_n} \|\boldsymbol{d} - \boldsymbol{H}\alpha_n\|^2$, where $\boldsymbol{H}$ is the $n \times m$ matrix whose (i, j)-th entry is $\kappa(\boldsymbol{u}_i, \boldsymbol{u}_{\omega_j})$. By taking the derivative of this cost function with respect to α_n and nullifying it, we get the optimal solution

$$\alpha_n = \left(\boldsymbol{H}^\top \boldsymbol{H}\right)^{-1} \boldsymbol{H}^\top \boldsymbol{d}$$

assuming $\boldsymbol{H}^\top \boldsymbol{H}$ nonsingular and where $\alpha_n = [\alpha_{1,n}, \ldots, \alpha_{m,n}]^\top$. In the following, we provide adaptive techniques to estimate α_n from its previous estimate, α_{n-1}.

We describe in details the Kernel Affine Projection (KAP) algorithm, and then we present several variants. Under the minimal disturbance principle, the estimate is determined at each time step by projecting the previous estimate, subject to solving an underdetermined least-squares problem. The latter is defined only on the p most recent inputs and outputs at each time step n, respectively $\{\boldsymbol{u}_n, \boldsymbol{u}_{n-1}, \ldots, \boldsymbol{u}_{n-p+1}\}$ and $\{d_n, d_{n-1}, \ldots, d_{n-p+1}\}$. In the following, $\boldsymbol{H}_n$ denotes the $p \times m$ matrix whose (i, j)-th entry is $\kappa(\boldsymbol{u}_{n-i+1}, \boldsymbol{u}_{\omega_j})$, namely,

$$\boldsymbol{H}_n = \begin{bmatrix} \boldsymbol{h}_n & \boldsymbol{h}_{n-1} & \ldots & \boldsymbol{h}_{n-p+1} \end{bmatrix}^\top$$

where

$$\boldsymbol{h}_{n-i+1} = [\kappa(\boldsymbol{u}_{n-i+1}, \boldsymbol{u}_{\omega_1}) \quad \ldots \quad \kappa(\boldsymbol{u}_{n-i+1}, \boldsymbol{u}_{\omega_m})]^\top, \quad i = 1, \ldots, p$$

and $\boldsymbol{d}_n$ is the column vector whose i-th entry is d_{n-i+1} for any $i = 1, \ldots, p$.

The affine projection problem is defined at time step n with the following constrained optimization problem:

$$\min_{\alpha} \|\alpha - \alpha_{n-1}\|^2 \qquad \text{subject to} \qquad \boldsymbol{d}_n = \boldsymbol{H}_n \alpha. \qquad (2.11)$$

In other words, α_n is obtained by projecting α_{n-1} onto the intersection of the p manifolds defined, for $i = 1, 2, \ldots, p$, by

$$\left\{\alpha : h_{n-i+1}^\top \alpha - d_{n-i+1} = 0\right\}, \qquad i = 1, \ldots, p.$$

At time step n, upon arrival of a new sample u_n, one of the following alternatives holds. If $\kappa(\cdot, u_n)$ violates the coherence-based condition (2.10), the dictionary remains unaltered. On the other hand, if the condition (2.10) is satisfied, $\kappa(\cdot, u_n)$ is inserted into the dictionary where it is denoted by $\kappa(\cdot, u_{\omega_{m+1}})$. The matrix H_n is augmented by appending to H_{n-1} the column vector $[\kappa(u_n, u_{\omega_{m+1}}) \ \kappa(u_{n-1}, u_{\omega_{m+1}}) \ \ldots \ \kappa(u_{n-p+1}, u_{\omega_{m+1}})]^\top$. One more entry is also added to the vector α_n. The corresponding updating rules, depending on the coherence-based sparsification rule, are detailed in the following.

First case: $\max\limits_{j=1,\ldots,m} |\kappa(u_n, u_{\omega_j})| > \varepsilon_0$

In this case, $\kappa(\cdot, u_n)$ is not appended to the dictionary, since it can be reasonably well represented by the kernel functions already in the dictionary. The solution to the constrained optimization problem (2.11) is determined by minimizing the Lagrangian function

$$\|\alpha - \alpha_{n-1}\|^2 + \lambda^\top (d_n - H_n \alpha) \tag{2.12}$$

where λ denotes the vector of Lagrange multipliers. Differentiating this expression with respect to α, setting it to zero and solving for $\alpha = \alpha_n$, and making $d_n - H_n \alpha_n = 0$ to determine λ, we get the following equations:

$$2(\alpha_n - \alpha_{n-1}) = H_n^\top \lambda \quad \text{and} \quad \lambda = 2(H_n H_n^\top)^{-1}(d_n - H_n \alpha_{n-1}). \tag{2.13}$$

In order to ensure the nonsingularity of the matrix to be inverted in (2.13), one may use the regularized version, with $(H_n H_n^\top + \epsilon I)$. Solving (2.13) for α_n, we obtain the following recursive update rule:

$$\alpha_n = \alpha_{n-1} + \eta H_n^\top (H_n H_n^\top + \epsilon I)^{-1}(d_n - H_n \alpha_{n-1}) \tag{2.14}$$

where we have introduced the step-size control parameter η. This update rule requires inverting the usually small $p \times p$ matrix $(H_n H_n^\top + \epsilon I)$.

Second case: $\max\limits_{j=1,\ldots,m} |\kappa(u_n, u_{\omega_j})| \leq \varepsilon_0$

In this case, the kernel function $\kappa(\cdot, u_n)$ is included in the dictionary since it cannot be efficiently represented by the elements already in the dictionary. Hence, it is denoted by $\kappa(\cdot, u_{\omega_{m+1}})$. The model order m in (2.7) is increased by one, and H_n is updated to a $p \times (m + 1)$ matrix. To accommodate the new entry α_{m+1} in α_n, the optimization problem (2.11) is rewritten as

$$\min_\alpha \|\alpha_{1,\ldots,m} - \alpha_{n-1}\|^2 + \alpha_{m+1}^2 \qquad \text{subject to} \quad d_n = H_n \alpha \tag{2.15}$$

where $\alpha_{1,\ldots,m}$ denotes the first m entries of the vector α and H_n has been increased by one column as described before. Note that the $(m + 1)$-th entry of α, namely α_{m+1}, acts as a regularizing term in the objective function. By following the same steps as in the derivation of (2.14), we get the following update rule

$$\alpha_n = \begin{bmatrix} \alpha_{n-1} \\ 0 \end{bmatrix} + \eta\, H_n^\top (H_n H_n^\top + \epsilon I)^{-1} \left(d_n - H_n \begin{bmatrix} \alpha_{n-1} \\ 0 \end{bmatrix} \right). \tag{2.16}$$

Next, we investigate the instantaneous approximations for the gradient vectors, in order to derive the KNLMS and the KLMS algorithms.

2.5.2.1 Instantaneous Approximations — The KNLMS Algorithm

By considering an instantaneous approximation with $p = 1$, the affine projection algorithm described above leads to the kernel normalized least-mean square (KNLMS) algorithm. At each time step n, the algorithm described above enforces a null *a posteriori* error, namely $d_n = h_n^\top \alpha_n$, where h_n is the column vector whose i-th entry is $\kappa(u_n, u_{\omega_i})$. The update rules (2.14) and (2.16) reduce to

1. If $\max_j |\kappa(u_n, u_{\omega_j})| > \varepsilon_0$: Let $h_n = [\kappa(u_n, u_{\omega_1}) \ldots \kappa(u_n, u_{\omega_m})]^\top$, then

$$\alpha_n = \alpha_{n-1} + \frac{\eta}{\epsilon + \|h_n\|^2} (d_n - h_n^\top \alpha_{n-1})\, h_n. \tag{2.17}$$

2. If $\max_j |\kappa(u_n, u_{\omega_j})| \leq \varepsilon_0$: Let $h_n = [\kappa(u_n, u_{\omega_1}) \ldots \kappa(u_n, u_{\omega_{m+1}})]^\top$, then

$$\alpha_n = \begin{bmatrix} \alpha_{n-1} \\ 0 \end{bmatrix} + \frac{\eta}{\epsilon + \|h_n\|^2} \left(d_n - h_n^\top \begin{bmatrix} \alpha_{n-1} \\ 0 \end{bmatrix} \right) h_n. \tag{2.18}$$

2.5.2.2 Instantaneous Approximations — The KLMS Algorithm

One of the easiest ways to derive an adaptive rule is to consider the stochastic gradient of (2.4), which leads to the so-called kernel least-mean square (KLMS) algorithm. Since the *a priori* error at time step n is $d_n = h_n^\top \alpha_{n-1}$, we get the same steps as in the KNLMS algorithm without the step size normalization, and the update rules (2.17) and (2.18) become, respectively,

$$\alpha_n = \alpha_{n-1} + \eta\, (d_n - h_n^\top \alpha_{n-1})\, h_n$$

and

$$\alpha_n = \begin{bmatrix} \alpha_{n-1} \\ 0 \end{bmatrix} + \eta \left(d_n - h_n^\top \begin{bmatrix} \alpha_{n-1} \\ 0 \end{bmatrix} \right) h_n.$$

2.5.3 EXAMPLE 1

Consider the "Henon map" defined by the nonlinear system

$$\begin{cases} d_n = 1 - \gamma_1\, d_{n-1}^2 + \gamma_2\, d_{n-2}; \\ d_0 = -0.3; \\ d_1 = 0. \end{cases}$$

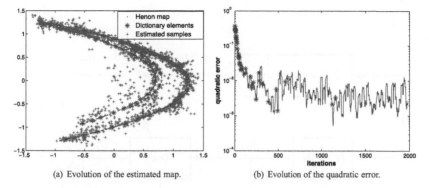

(a) Evolution of the estimated map.   (b) Evolution of the quadratic error.

Figure 2.1   Illustration of the Henon map in two dimensions (d_n, d_{n-1}), and the evolution of the quadratic error. Elements of the dictionary are illustrated by red stars.

For $\gamma_1 = 1.4$ and $\gamma_2 = 0.3$, the corresponding time-series has chaotic behavior. The investigated model takes the form $d_n = \psi(d_{n-1}, d_{n-2})$. The length of the time series was 2000 samples. We considered the Gaussian kernel, and the value of its bandwidth was set to $\beta_0 = 0.35$. The choice of the threshold $\varepsilon_0 = 0.6$ led to a dictionary with 52 entries. The relevance of the proposed online kernel-based function approximation is illustrated in Figure 2.1. Figure 2.1(a) shows the evolution of the pairs (d_n, d_{n-1}) with n corresponding to the actual Henon map, and those predicted using the kernel-based nonlinear model with coefficients α_n adapted using the KAP algorithm with $\eta = 1$. Figure 2.1(b) shows the evolution of the quadratic error.

2.6   ONLINE NONLINEAR SYSTEM IDENTIFICATION

The adaptive algorithms presented above can be employed in any application requiring the estimation of parameters associated with a time series. One particularly important application is nonlinear system identification. The block diagram of a kernel-based adaptive system identification problem is shown in Figure 2.2. Here, $\mathcal{U}$ is a compact subspace of $\mathbb{R}^\ell$, $\kappa\colon \mathcal{U}\times\mathcal{U} \to \mathbb{R}$ is a reproducing kernel, $(\mathcal{H}, \langle\cdot,\cdot\rangle_{\mathcal{H}})$ is the induced RKHS with its inner product, and z_n is a zero-mean additive noise uncorrelated with any other signal.

It is well known that a nonlinear adaptive filtering problem with input signal in $\mathcal{U}$ can be solved using a linear adaptive filter [22]. The linear adaptive filter input is a nonlinear mapping of $\mathcal{U}$ to an Hilbert space $\mathcal{H}$ possessing a reproducing kernel. As discussed in Section 2.5.1, the order of the linear adaptive filter can be finite if a proper input sparsification rule is employed [34], even if the dimensionality of the transformed input in $\mathcal{H}$ is infinite as in the case of a Gaussian kernel.

A proper design of an adaptive algorithm to solve the nonlinear system identification problem depicted in Figure 2.2, however, requires a good understanding of the

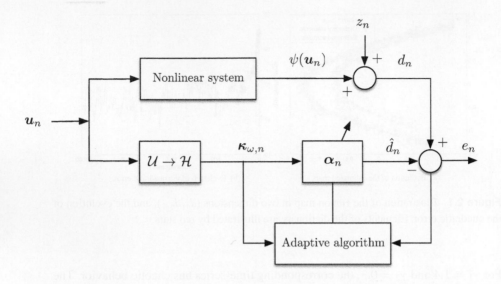

Figure 2.2 Kernel-based adaptive system identification.

algorithm properties in such an application. However, very few analyses related to the behavior of kernel-based adaptive algorithms are available in the literature. A statistical analysis of the behavior of the KLMS adaptive algorithm using a fixed-order dictionary $\mathcal{D}$ and a Gaussian kernel for system identification has been presented in [31]. Other aspects of the KLMS behavior have been studied in [29, 30, 31, 33]. In the analysis, the environment is assumed stationary, meaning that $\psi(\boldsymbol{u}_n)$ is stationary for $\boldsymbol{u}_n$ stationary. This assumption is satisfied by several nonlinear systems used to model practical situations, such as memoryless, Wiener, and Hammerstein systems. System inputs are assumed to be zero-mean, independent, and identically distributed Gaussian ($\ell \times 1$) vectors $\boldsymbol{u}_n$ so that $E\{\boldsymbol{u}_{n-i}\boldsymbol{u}_{n-j}^\top\} = \boldsymbol{0}$ for $i \neq j$. The components of the input vector $\boldsymbol{u}_n$ can, however, be correlated. Let $\boldsymbol{R}_{uu} = E\{\boldsymbol{u}_n\boldsymbol{u}_n^\top\}$ denote their autocorrelation matrix.

For a dictionary of size m, let $\boldsymbol{\kappa}_{\omega,n}$ be the vector of kernels at time $n > m$,[2] that is,

$$\boldsymbol{\kappa}_{\omega,n} = [\kappa(\boldsymbol{u}_n, \boldsymbol{u}_{\omega_1}), \ldots, \kappa(\boldsymbol{u}_n, \boldsymbol{u}_{\omega_m})]^\top \tag{2.19}$$

where $\boldsymbol{u}_{\omega_i}$ is the i-th element of the dictionary, with $\boldsymbol{u}_{\omega_i} \neq \boldsymbol{u}_n$ for $i = 1, \ldots, m$. Here we consider that the vectors $\boldsymbol{u}_{\omega_i}$, $i = 1, \ldots, m$ may change at each iteration following some dictionary updating schedule. The only limitation imposed in the following analysis is that $\boldsymbol{u}_{\omega_i,n} \neq \boldsymbol{u}_{\omega_j,n}$ for $i \neq j$ so that the dictionary vectors which are arguments of different entries of $\boldsymbol{\kappa}_{\omega,n}$ are statistically independent. Note that this

---

[2]If the dictionary size m is adapted online, we assume that n is sufficiently large so that the size m does not increase anymore.

framework does not allow the user to pre-tune the dictionary to improve performance. In [7] we proposed a theoretical analysis of the KLMS algorithm with a Gaussian kernel that considers the dictionary as part of the filter parameters to be set.

To keep the notation simple, however, we will not show explicitly the dependence of ω_i on n and represent $u_{\omega_i,n}$ as u_{ω_i} for all i.

From Figure 2.2 and model (2.8), the estimated system output is

$$\hat{d}(n) = \alpha_n^\top \kappa_{\omega,n}. \tag{2.20}$$

The corresponding estimation error is defined as

$$e_n = d_n - \hat{d}_n. \tag{2.21}$$

Squaring both sides of (2.21) and taking the expected value leads to the MSE

$$J_{\mathrm{mse},n} = E\{e_n^2\} = E\{d_n^2\} - 2p_{\kappa d}^\top \alpha_n + \alpha_n^\top R_{\kappa\kappa} \alpha_n \tag{2.22}$$

where $R_{\kappa\kappa} = E\{\kappa_{\omega,n} \kappa_{\omega,n}^\top\}$ is the correlation matrix of the kernelized input, and $p_{\kappa d} = E\{d_n \kappa_{\omega,n}\}$ is the cross-correlation vector between $\kappa_{\omega,n}$ and d_n. It is shown in [31] that $R_{\kappa\kappa}$ is positive definite. Thus, the optimum weight vector is given by

$$\alpha_{\mathrm{opt}} = R_{\kappa\kappa}^{-1} p_{\kappa d}. \tag{2.23}$$

The optimal estimation error is

$$e_{0,n} = d_n - \kappa_{\omega,n}^\top \alpha_{\mathrm{opt}} \tag{2.24}$$

and the corresponding minimum MSE is

$$J_{\min} = E\{d_n^2\} - p_{\kappa d}^\top R_{\kappa\kappa}^{-1} p_{\kappa d}. \tag{2.25}$$

These are the well-known expressions of the Wiener solution and minimum MSE, where the input signal vector has been replaced by the kernelized input vector. Determining the optimum α_{opt} requires the determination of the covariance matrix $R_{\kappa\kappa}$, given the statistical properties of u_n and the reproducing kernel.

To evaluate $R_{\kappa\kappa}$, we note that its entries are given by

$$[R_{\kappa\kappa}]_{ij} = \begin{cases} E\{\kappa^2(u_n, u_{\omega_i})\}, & i = j \\ E\{\kappa(u_n, u_{\omega_i}) \kappa(u_n, u_{\omega_j})\}, & i \neq j \end{cases} \tag{2.26}$$

with $1 \leq i, j \leq m$. Note that $R_{\kappa\kappa}$ remains time-invariant even if the dictionary is updated at each iteration, as u_n is stationary and u_{ω_i} and u_{ω_j} are statistically independent for $i \neq j$. Also, we assume, in the following, the use of the Gaussian kernel $\kappa(u_i, u_j) = \exp(-\|u_i - u_j\|^2 / 2\beta_0^2)$.

Let us introduce the following notations

$$\|u_n - u_{\omega_i}\|^2 = y_2^\top Q_2 y_2$$
$$\|u_n - u_{\omega_i}\|^2 + \|u_n - u_{\omega_j}\|^2 = y_3^\top Q_3 y_3, \quad i \neq j \tag{2.27}$$

where $\| \cdot \|$ is the ℓ_2 norm and

$$y_2 = \left(u_n^\top u_{\omega_i}^\top \right)^\top$$

$$y_3 = \left(u_n^\top u_{\omega_i}^\top u_{\omega_j}^\top \right)^\top \tag{2.28}$$

and

$$Q_2 = \begin{pmatrix} I & -I \\ -I & I \end{pmatrix} \qquad Q_3 = \begin{pmatrix} 2I & -I & -I \\ -I & I & O \\ -I & O & I \end{pmatrix} \tag{2.29}$$

where I is the $(\ell \times \ell)$ identity matrix and O is the $(\ell \times \ell)$ null matrix. From [28, p. 100], we know that the moment-generating function of a quadratic form $\xi = y^\top Q y$, where y is a zero-mean Gaussian vector with covariance matrix R_y, is given by

$$\psi_\xi(s) = E\{e^{s\xi}\} = \det\{I - 2 s Q R_y\}^{-1/2}. \tag{2.30}$$

Making $s = -1/(2\beta_0^2)$ in (2.30), we find that the (i, j)-th element of $R_{\kappa\kappa}$ is given by

$$[R_{\kappa\kappa}]_{ij} = \begin{cases} r_{\mathrm{md}} = \det\left\{ I_2 + 2 Q_2 R_2/\beta_0^2 \right\}^{-1/2}, & i = j \\ r_{\mathrm{od}} = \det\left\{ I_3 + Q_3 R_3/\beta_0^2 \right\}^{-1/2}, & i \neq j \end{cases} \tag{2.31}$$

with $1 \leq i, j \leq M$. The main diagonal entries $[R_{\kappa\kappa}]_{ii}$ are all equal to r_{md} and the off-diagonal entries $[R_{\kappa\kappa}]_{ij}$ are all equal to r_{od} because u_{ω_i} and u_{ω_j} are i.i.d. In (2.31), R_q is the $(q\ell \times q\ell)$ correlation matrix of vector y_q, I_q is the $(q\ell \times q\ell)$ identity matrix, and $\det\{\cdot\}$ denotes the determinant of a matrix. Finally, note that matrix R_q is block-diagonal with R_{uu} along its diagonal.

The analysis in [31] uses the following statistical assumptions for feasibility:

A1: $\kappa_{\omega,n}\kappa_{\omega,n}^\top$ is statistically independent of v_n. This assumption is justified in detail in [26] and has been successfully employed in several adaptive filter analyses. It is called here for further reference "modified independence assumption" (MIA). This assumption has been shown in [26] to be less restrictive than the classical independence assumption [35].

A2: The finite-order model provides a close enough approximation to the infinite-order model with minimum MSE, so that $E[e_{0,n}] \approx 0$.

A3: $e_{0,n}$ and $\kappa_{\omega,n}\kappa_{\omega,n}^\top$ are uncorrelated. This assumption is also supported by the arguments supporting the MIA (**A1**) [26].

The following are the main results of the analysis presented in [31]. The reader is directed to [31] for more details. Defining the weight error vector $v_n = \alpha_n - \alpha_{\mathrm{opt}}$ leads to the KLMS weight-error vector update equation

$$v_{n+1} = v_n + \eta e_n \kappa_{\omega,n} \tag{2.32}$$

and the estimation error can be written as

$$e_n = d_n - \kappa_{\omega,n}^\top v_n - \kappa_{\omega,n}^\top \alpha_{\mathrm{opt}}. \tag{2.33}$$

The analysis in [31] shows that the mean behavior of the adaptive weights is given by

$$E\{v_{n+1}\} = (I - \eta R_{\kappa\kappa}) E\{v_n\}. \tag{2.34}$$

For the analysis of the MSE behavior, we note that using (2.33) and the MIA (**A1**), the second-order moments of the weights are related to the MSE through [35]

$$J_{\mathrm{ms}}(n) = J_{\min} + \mathrm{trace}\{R_{\kappa\kappa} C_{v,n}\} \tag{2.35}$$

where $C_{v,n} = E\{v_n v_n^\top\}$ is the autocorrelation matrix of v_n and $J_{\min} = E\{e_{0,n}^2\}$ is the minimum MSE. The study of the MSE behavior (2.35) requires a model for $C_{v,n}$. This model is highly affected by the transformation imposed on the input signal u_n by the kernel. An analytical model for the behavior of $C_{v,n}$ is derived in [31] and is given by

$$C_{v,n+1} \approx C_{v,n} - \eta (R_{\kappa\kappa} C_{v,n} + C_{v,n} R_{\kappa\kappa}) + \eta^2 T_n + \eta^2 R_{\kappa\kappa} J_{\min} \tag{2.36a}$$

with

$$T_n = E\{\kappa_n \kappa_n^\top v_n v_n^\top \kappa_n \kappa_n^\top\}. \tag{2.36b}$$

The moments in (2.36b) are evaluated in [31], yielding the following recursive expressions for the entries of the autocorrelation matrix $C_{v,n}$:

$$
\begin{aligned}
[C_{v,n+1}]_{ii} =&(1 - 2\eta r_{\mathrm{md}} + \eta^2 \mu_1) [C_{v,n}]_{ii} + \eta^2 \mu_3 \sum_{\substack{\ell=1 \\ \ell \neq i}}^{M} [C_{v,n}]_{\ell\ell} \\
&+ (2\eta^2 \mu_2 - 2\eta r_{\mathrm{od}}) \sum_{\substack{\ell=1 \\ \ell \neq i}}^{M} [C_{v,n}]_{i\ell} + \eta^2 \mu_4 \sum_{\substack{\ell=1 \\ \ell \neq i}}^{M} \sum_{\substack{p=1 \\ p \neq \{i,\ell\}}}^{M} [C_{v,n}]_{\ell p} \\
&+ \eta^2 r_{\mathrm{md}} J_{\min}
\end{aligned}
\tag{2.37}
$$

and, for $j \neq i$,

$$
\begin{aligned}
[C_{v,n+1}]_{ij} =&(1 - 2\eta r_{\mathrm{md}} + 2\eta^2 \mu_3) [C_{v,n}]_{ij} + \eta^2 \mu_4 \sum_{\substack{\ell=1 \\ \ell \neq \{i,j\}}}^{m} [C_{v,n}]_{\ell\ell} \\
&+ (\eta^2 \mu_2 - \eta r_{\mathrm{od}})([C_{v,n}]_{ii} + [C_{v,n}]_{jj}) \\
&+ (2\eta^2 \mu_4 - \eta r_{\mathrm{od}}) \sum_{\substack{\ell=1 \\ \ell \neq \{i,j\}}}^{m} ([C_{v,n}]_{i\ell} + [C_{v,n}]_{j\ell}) \\
&+ \eta^2 \mu_5 \sum_{\substack{\ell=1 \\ \ell \neq \{i,j\}}}^{m} \sum_{\substack{p=1 \\ p \neq \{i,j,\ell\}}}^{m} [C_{v,n}]_{\ell p} + \eta^2 r_{\mathrm{od}} J_{\min}
\end{aligned}
\tag{2.38}
$$

where $r_{\mathrm{md}} = [R_{\kappa\kappa}]_{ii}$ and $r_{\mathrm{od}} = [R_{\kappa\kappa}]_{ij}$ with $j \neq i$, as defined in (2.31), $\mu_1 = \left[\det\{I_2 + 4Q_2 R_2/\beta_0^2\}\right]^{-1/2}$, $\mu_2 = [\det\{I_3 + Q_3, R_3/\beta_0^2\}]^{-1/2}$, $\mu_3 = [\det\{I_3 +$

$2\,Q_3\,R_3/\beta_0^2\}]^{-1/2}$, $\mu_4 = [\det\{I_4 + Q_4\,R_4/\beta_0^2\}]^{-1/2}$ and $\mu_5 = [\det\{I_5 + Q_5\,R_5/\beta_0^2\}]^{-1/2}$ with Q_2 and Q_3 defined in (2.29) and

$$Q_{3'} = \begin{pmatrix} 4I & -3I & -I \\ -3I & 3I & O \\ -I & O & I \end{pmatrix}. \tag{2.39a}$$

$$Q_4 = \begin{pmatrix} 4I & -2I & -I & -I \\ -2I & 2I & O & O \\ -I & O & I & O \\ -I & O & O & I \end{pmatrix}. \tag{2.39b}$$

$$Q_5 = \begin{pmatrix} 4I & -I & -I & -I & -I \\ -I & I & O & O & O \\ -I & O & I & O & O \\ -I & O & O & I & O \\ -I & O & O & O & I \end{pmatrix}. \tag{2.39c}$$

The interested reader is referred to [31] for stability and steady-state analyses of the KLMS algorithm. In the following we illustrate the accuracy of the analytical model through a simulation example.

2.6.1 EXAMPLE 2

Consider the problem studied in [27, 23], for which

$$\begin{cases} y_n = \dfrac{y_{n-1}}{1 + y_{n-1}^2} + u_{n-1}^3 \\ d_n = y_n + z_n \end{cases} \tag{2.40}$$

where the output signal y_n was corrupted by a zero-mean white Gaussian noise z_n with variance $\sigma_z^2 = 10^{-4}$. The input sequence $u(n)$ is zero-mean i.i.d. Gaussian with standard deviation $\sigma_u = 0.15$.

The proposed method was tested with a maximum MSE $J_{\max} = -21.5$ dB, a coherence level $\varepsilon_0 = 10^{-5}$, and a set of kernel bandwidths $\beta_0 \in \{0.0075, 0.01, 0.025, 0.05\}$.[3] For each value of β_0, 500 dictionary dimensions m_i, $i = 1,\dots,500$, were determined using 500 realizations of the input process. The length of each realization was 500 samples. Each m_i was determined as the minimum dictionary length required to achieve the coherence level ε_0. The value $m(\beta_0)$ was determined as the average of all m_i, rounded to the nearest integer. The values of $J_{\min}(\beta_0)$ were calculated from (2.25) for each pair (β_0, M). To this end, second-order moments $p_{\kappa d}$ and $E\{d_n^2\}$ were estimated by averaging over 500 runs.

The step size value for each values of β_0 was chosen as $\eta = \eta_{\max}/10$ with $\eta_{\max}$ determined using the stability analysis in [31]. Table 2.1 shows the values of β_0 and η

---

[3]These values of β_0 are samples within a range of values experimentally verified to be adequate for the application.

Table 2.1

Summary of simulation results for Example 1.

β_0	m	η	$J_{\min}$ [dB]	$J_{\mathrm{mse}}(\infty)$ [dB]	$J_{\mathrm{ex}}(\infty)$ [dB]	n_ϵ
0.0075	17	0.143	-22.19	-22.04	-36.85	1271
0.01	13	0.152	-21.84	-21.69	-36.27	914
0.025	6	0.007	-21.53	-21.52	-49.15	7746
0.05	3	0.011	-21.52	-21.51	-47.00	2648

used in the simulations. For each simulation, the order m of the dictionary remained fixed. It was initialized for each realization by generating input vectors in $\mathcal{U}$ and filling the m positions with vectors that satisfy the desired coherence level. Thus, the initial dictionary is different for each realization. During each realization, the dictionary elements were updated at each iteration n so that the least recently added element is replaced with $\boldsymbol{u}_{n-1}$.

Figures 2.3 illustrates the accuracy of the analytical model for the four cases presented in Table 2.1. Figure 2.3 shows an excellent agreement between Monte Carlo simulations, averaged over 500 runs, and the theoretical predictions made by using (2.35).

2.7 BAYESIAN APPROACHES TO KERNEL-BASED NONLINEAR REGRESSION

2.7.1 GAUSSIAN PROCESSES FOR REGRESSION

The more classical solutions of the nonlinear regression problem (2.3) using Bayesian techniques assume the knowledge of function $\psi(\cdot)$ except for a set of parameters that can be included in a parameter vector $\boldsymbol{\theta}$. This includes the cases in which the nonlinear function $\psi(\cdot)$ is actually known from physical considerations and cases in which $\psi(\cdot)$ represents a family of functions that can approximate a wide range of functional forms. In the former case, the entries of $\boldsymbol{\theta}$ tend to have physical interpretations. In the latter, they are usually parameters of an expansion to be fitted to the unknown function $\psi(\cdot)$. Kernel-based classical Bayesian solutions use the form (2.7) for $\psi(\cdot)$. Then, using the fixed-size model (2.6) in (2.3), the nonlinear regression problem becomes

$$d_i = \sum_{j=1}^{m} \alpha_j \kappa(\boldsymbol{u}_i, \boldsymbol{u}_{\omega_j}) + \eta_i, \qquad i = 1, \ldots, m \qquad (2.41)$$

with $\boldsymbol{\theta} = [\alpha_1, \alpha_2, \ldots, \alpha_m, \sigma_\eta^2]^\top$, where $\eta_i \sim \mathcal{N}(0, \sigma_\eta^2)$ is assumed. The solution of the problem includes defining a prior distribution $p(\boldsymbol{\theta})$ for the unknown parameters and determining the posterior distribution $p(\boldsymbol{\theta}|d_1, \ldots, d_m)$. Once the posterior distribution

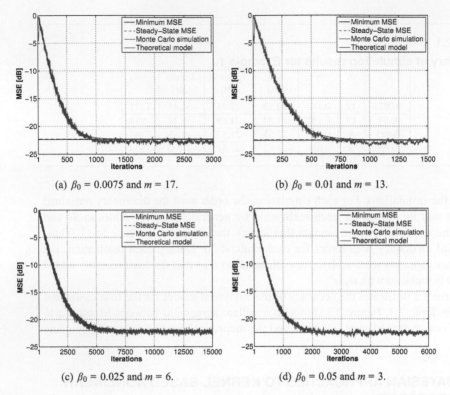

Figure 2.3 Theoretical model and Monte Carlo simulation of KLMS for different kernel bandwidths. Ragged curves (blue): simulation results averaged over 500 runs. Continuous curves (red): theory using (2.35).

is determined, an estimation criterion such as the minimum mean-square (MSE) error or the maximum *a posteriori* (MAP) can be used to estimate θ. Such classical Bayesian techniques are well documented and have given rise to popular estimators such as the least absolute shrinkage and selection operator (LASSO) proposed in [43] or the ridge regression [14]. The reader is referred, for instance, to [40, 5] for more details.

An alternative Bayesian approach that has gained popularity recently is to assume that the sequence of samples d_i, $i = 1, \ldots, m$ is characterized by a zero-mean Gaussian process ($\mathcal{GP}$) with covariance function in an RKHS [32] given by $\kappa(\boldsymbol{u}_i, \boldsymbol{u}_j)$. A common choice is the Gaussian kernel $\kappa(\boldsymbol{u}_i, \boldsymbol{u}_j) = \exp\left(-\|\boldsymbol{u}_i - \boldsymbol{u}_j\|^2 / 2\beta_0^2\right)$, where β_0 is the kernel bandwidth.

Defining the matrix $\boldsymbol{U} = [\boldsymbol{u}_1, \ldots, \boldsymbol{u}_m]$ and $\boldsymbol{d} = [d_1, \ldots, d_m]^\top$ we define the prior distribution for $\boldsymbol{d}$ as

$$\boldsymbol{d} \sim \mathcal{N}\left(\boldsymbol{0}, \boldsymbol{K} + \sigma_\eta^2 \boldsymbol{I}\right) \tag{2.42}$$

with $\boldsymbol{K}$ the Gram matrix whose entries $\boldsymbol{K}_{ij} = \kappa(\boldsymbol{u}_i, \boldsymbol{u}_j)$ are the kernel (covariance) functions [38] of the inputs u_i and u_j, and I is the $m \times m$ identity matrix.

Gaussian process regression aims at inferring the latent function distribution of f_* for a new (or test) input u_*. Using the marginalization property [32], (2.42) can be obtained by integrating out f_* in the following joint distribution of $(d, f_*)^\top$

$$\begin{bmatrix} d \\ f_* \end{bmatrix} \sim \mathcal{N}\left(0, \begin{bmatrix} K + \sigma_\eta^2 I & \kappa_* \\ \kappa_*^\top & \kappa_{**} \end{bmatrix}\right) \tag{2.43}$$

with $\kappa_*^\top = [\kappa(u_*, u_1), \ldots, \kappa(u_*, u_m)]$ and $\kappa_{**} = \kappa(u_*, u_*)$. The predictive distribution of f_*, or posterior of f_*, can be obtained by conditioning (2.43) on the data as

$$f_* | d, U, u_* \sim \mathcal{N}\left(\kappa_*^\top \left[K + \sigma_\eta^2 I\right]^{-1} d, \kappa_{**} - \kappa_*^\top \left[K + \sigma_\eta^2 I\right]^{-1} \kappa_*\right). \tag{2.44}$$

Using the Gaussian kernel to model the Gaussian process, the function estimation is done in an RKHS with universal approximating capability [22, p. 35] due to the smoothness and non-informativeness of this kernel.

2.7.2 SPECTRAL UNMIXING OF HYPERSPECTRAL IMAGES

The problem of unmixing hyperspectral images has been receiving an increasing interest in the recent years for various remote sensing applications (see for instance review papers [4, 9] and references therein). Spectral unmixing (SU) consists of identifying the spectral signatures of the pure materials (referred to as "endmembers") contained in a hyperspectral image and estimating the proportions of these materials (referred to as "abundances") in each pixel of the image. A classical model used for this unmixing procedure is the linear mixing model (LMM), which expresses a given pixel of the image $y \in \mathbb{R}^L$ (acquired in L spectral bands) as a linear combination of a given number R of endmembers m_r:

$$y = \sum_{r=1}^{R} a_r m_r + n = Ma + n \tag{2.45}$$

where a_r is the abundance of the material m_r in the pixel y, $a = (a_1, ..., a_R)^\top$, $M = [m_1, ..., m_R]$ is a matrix built with the different endmembers contained in the image and $n \in \mathbb{R}^L$ is an additive noise vector. Under this model, the abundances are required to satisfy the so-called positivity and sum-to-one constraints

$$a_r \geq 0, \quad \sum_{r=1}^{R} a_r = 1 \tag{2.46}$$

when the vectors m_r for $r = 1, ..., R$ form set of all pure materials contained in the image. The LMM (2.45) has shown interesting properties for the SU of hyperspectral images. However, as explained in [9], this model has some severe limitations when the image contains intimate mixtures or when the light scattered by a given material reflects on other materials before reaching the sensor (leading to multipath effects). In these cases, nonlinear mixing models should be preferred to identify the endmembers contained in the image and to estimate the abundances of the pure materials in each pixel of the image. A generic model for nonlinear spectral unmixing can be written as

$$y = f(M, a) + n \tag{2.47}$$

where f is a nonlinear function whose shape can be difficult to be known a priori. In this context, Gaussian processes can be investigated to infer this nonlinear function and thus to perform nonlinear unmixing of hyperspectral images.

To apply Gaussian processes to the SU problem, we first define s_i as the transpose of the i-th row of matrix M. Hence, $s_i^\top$ contains the contributions of all endmembers to the spectral component y_i of y in the LMM, and $y_i = a^\top s_i + n_i$ in (2.45).

Using the joint distribution (2.43) with inputs $u_i = s_i$, it is straightforward to show that the posterior distribution (also referred to as predictive distribution) of f_* for a new input s_* can be written as

$$f_*|y, M, s_* \sim \mathcal{N}\left(\kappa_*^\top \left[K + \sigma_n^2 I\right]^{-1} y, \kappa_{**} - \kappa_*^\top \left[K + \sigma_n^2 I\right]^{-1} \kappa_*\right) \tag{2.48}$$

where $\kappa_*^\top = [\kappa(s_*, s_1), \ldots, \kappa(s_*, s_L)]$ and $\kappa_{**} = \kappa(s_*, s_*)$.

The extension to a multivariate predictive distribution with test data $M_* = [s_{*1}, \ldots, s_{*L}]^\top$ is straightforward and yields

$$f_*|y, M, M_* \sim \mathcal{N}\left(K_* \left[K + \sigma_n^2 I\right]^{-1} y, K_{**} - K_* \left[K + \sigma_n^2 I\right]^{-1} K_*^\top\right) \tag{2.49}$$

where $[K_*]_{ij} = \kappa(s_{*i}, s_j)$ and $[K_{**}]_{ij} = \kappa(s_{*i}, s_{*j})$.

The predicted function f_* can finally be estimated by computing the mean of (2.49) following the MMSE principle. Then,

$$f_*^{\text{MMSE}} = K_* \left[K + \sigma_n^2 I\right]^{-1} y. \tag{2.50}$$

Assuming the choice of the Gaussian kernel $\kappa(s_i, s_j) = \exp\left(-\|s_i - s_j\|^2/2\beta_0^2\right)$, the practical evaluation of f_*^{MMSE} requires the values of the model parameters $\theta = [\beta_0, \sigma_n^2]$ to be estimated. Following a Bayesian approach, a prior has to be defined for θ and the posterior $p(\theta|M, y)$ has to be maximized with respect to θ. Using the type II maximum likelihood approximation [32], another solution for estimating θ is to maximize the marginal likelihood $\log p(y|M, y)$ with respect to θ, where

$$\log p(y|M, y) = -\frac{1}{2} y^\top \left[K + \sigma_n^2 I\right]^{-1} y - \frac{1}{2} \log \left|K + \sigma_n^2 I\right| - \frac{L}{2} \log(2\pi). \tag{2.51}$$

2.7.3 EXAMPLE 3

Consider a nonlinearly mixed hyperspectral image pixel generated using the simplified generalized bilinear model (GBM) used in [2] with a new scaling that permits

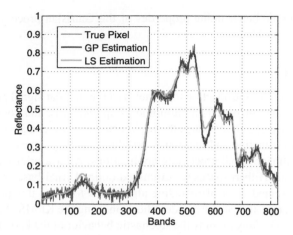

Figure 2.4 Gaussian process-based nonlinear function estimation. Actual pixel: Blue ragged curve. Least squares prediction assuming a linear model: Green curve. Gaussian process based nonlinear prediction: Magenta curve.

the control of the degree of nonlinearity for each nonlinear pixel generated. More precisely, the nonlinearly mixed pixels are generated using the following model

$$y = k\,Ma + \mu + n \tag{2.52}$$

where $0 \le k \le 1$, $\mu = \gamma \sum_{i=1}^{R-1} \sum_{j=i+1}^{R} a_i a_j m_i \odot m_j$ is the nonlinear term, γ is the parameter that governs the amount of nonlinear contribution, $\odot$ is the Hadamard product, and n is an additive white Gaussian noise with variance σ_n^2. Given the parameters M, a, γ, and σ_n^2, this model generates samples with the same energy and SNR as the LMM if $k = \left[-2E_{\ell\mu} + \sqrt{4E_{\ell\mu}^2 - 4E_\ell(E_\mu - E_\ell)}\right]/2E_\ell$, where $E_\ell = \|y_\ell\|^2$ is the energy of a noiseless linear pixel (i.e., $a^\top M^\top Ma$), $E_{\ell\mu} = y_\ell^\top \mu$ is the "cross-energy" of the linear and nonlinear parts, and $E_\mu = \|\mu\|^2$ is the energy of the nonlinear contribution. The degree of nonlinearity of a pixel is then defined as the ratio of the nonlinear portion to the total pixel energy

$$\eta_d = \frac{2kE_{\ell\mu} + E_\mu}{k^2 E_\ell + 2kE_{\ell\mu} + E_\mu} \tag{2.53}$$

so that $0 \le \eta_d \le 1$. For the simulations presented here, the endmember matrix M was composed of $R = 3$ materials (green grass, olive green paint, and galvanized steel metal) extracted from the spectral library of the software ENVI^{TM} [18]. Each endmember m_i, $i = 1, 2, 3$, has $L = 826$ bands. The model parameters were $a = [0.3, 0.6, 0, 1]^\top$, $k = 0.3162$, and $\gamma = 6.1464$ so that $\eta_d = 0.9$ and $\sigma_n^2 = 4.074 \times 10^{-4}$ was chosen to produce an SNR of 25dB.

Figure 2.4 shows the actual values of a test observation vector y_* (in blue) and the estimations using both a least-squares fitting assuming a linear model (2.45) (in green) and a nonlinear Gaussian process based nonlinear fitting (in magenta). These results clearly illustrate the better fitting properties of the Gaussian process model for nonlinear regression problems.

2.8 CONCLUSION

This chapter presented a brief introduction to kernel-based nonlinear signal processing. This is a vast area that encounters a multitude of applications, and constitutes a very active area of research. To introduce the basic principles we concentrated on the nonlinear function estimation and regression problems, as they appear in a significant amount of nonlinear signal processing applications. After a brief introduction to reproducing kernel Hilbert spaces we discussed online kernel-based function approximation and presented a set of recursive algorithms to solve this problem. We then discussed online nonlinear system identification using the KLMS algorithm with emphasis on a methodology to study the stochastic behavior of the kernel-based adaptive estimator. Finally, we returned to the nonlinear function estimation problem, this time using a Bayesian approach based on Gaussian processes, and discussed its application to the unmixing of hyperspectral images. We presented illustrative examples for each of the applications discussed. We hope that this brief introduction will provide the reader with an appreciation of the potential of kernel-based methods for solving nonlinear signal processing problems.

REFERENCES

1. F. Abdallah, C. Richard, and R. Lengellé. An improved training algorithm for nonlinear kernel discriminants. *IEEE Transactions on Signal Processing*, 52(10):2798–2806, 2004.
2. Y. Altmann, N. Dobigeon, J.-Y. Tourneret, and J. C. M. Bermudez. A robust test for nonlinear mixture detection in hyperspectral images. In *IEEE International Conference on Acoustics, Speech and Signal Processing*, pages 2149–2153, Vancouver, Canada, 2013. IEEE Signal Processing Society.
3. N. Aronszajn. Theory of reproducing kernels. *Transactions of the American Mathematical Society*, 68(3):337–404, May 1950.
4. J. M. Bioucas-Dias, A. Plaza, N. Dobigeon, M. Parente, Q. Du, P. Gader, and J. Chanussot. Hyperspectral unmixing overview: Geometrical, statistical, and sparse regression-based approaches. *Journal of Selected Topics in Applied Earth Observations and Remote Sensing*, 5(2):354–379, 2012.
5. C. M. Bishop, ed. *Pattern Recognition and Machine Learning*. Information Science and Statistics. Springer, New York, 2006.
6. G. C. Cawley and N. L. C. Talbot. Improved sparse least-squares support vector machines. *Neurocomputing*, 48:1025–1031, 2002.
7. J. Chen, W. Gao, C. Richard, and J. C. M. Bermudez. Convergence analysis of kernel LMS algorithm pre-tuned dictionary. In *2014 IEEE International Conference on Acoustics, Speech and Signal Processing (ICASSP)*, pages 7293–7297, Florence, Italy, May 2014. IEEE.
8. B. J. de Kruif and T. J. A. de Vries. Pruning error minimization in least squares support vector machines. *IEEE Transactions on Neural Networks*, 14(3):696–702, 2003.

9. N. Dobigeon, J.-Y. Tourneret, C. Richard, J. C. M. Bermudez, S. McLaughlin, and A. O Hero. Nonlinear unmixing of hyperspectral images: Models and algorithms. *IEEE Signal Processing Magazine*, 31(1):82–94, Jan. 2014.

10. T. J. Dodd and R. F. Harrison. Estimating Volterra filters in Hilbert space. In *Proceedings of the IFAC Conference on Intelligent Control Systems and Signal Processing*, pages 538–543, Faro, Portugal, 2003.

11. T. J. Dodd, V. Kadirkamanathan, and R. F. Harrison. Function estimation in Hilbert space using sequential projections. In *Proceedings of the IFAC Conference on Intelligent Control Systems and Signal Processing*, pages 113–118, Faro, Portugal, 2003.

12. T. J. Dodd, B. Mitchinson, and R. F. Harrison. Sparse stochastic gradient descent learning in kernel models. In *Proceedings of the Second International Conference on Computational Intelligence, Robotics and Autonomous Systems*, Singapore, 2003.

13. Y. Engel, S. Mannor, and R. Meir. Kernel recursive least squares. *IEEE Transactions on Signal Processing*, 52(8):2275–2285, 2004.

14. I. E. Frank and J. H. Frieman. A statistical view of some chemometrics regression tools. *Technometrics*, 35(2):109–135, 1993.

15. L. Hoegaerts. *Eigenspace methods and subset selection in kernel based learning*. PhD thesis, Katholieke Universiteit Leuven, 2005.

16. L. Hoegaerts, J. A. K. Suykens, J. Vandewalle, and B. de Moor. Subset based least squares subspace regression in RKHS. *Neurocomputing*, 63:293–323, 2005.

17. P. Honeine, C. Richard, and J. C. M. Bermudez. On-line nonlinear sparse approximation of functions. In *Proceedings of the IEEE ISIT '07*, pages 956–960, Nice, France, Jun. 2007.

18. RSI (Research Systems Inc.). ENVI User's guide Version 4.0, Sep. 2013.

19. J. Jahn. *Introduction to the Theory of Nonlinear Optimization*. Springer, 3rd ed., 2007.

20. S. M. Kay. *Fundamentals of Statistical Signal Processing: Estimation Theory*. Prentice Hall, Upper Saddle River, 1993.

21. J. Kivinen, A.J. Smola, and R.C. Williamson. Online learning with kernels. *IEEE Transactions on Signal Processing*, 52(8):2165–2176, 2004.

22. W. Liu, J. C. Principe, and S. Haykin. *Kernel Adaptive Filtering*. John Wiley & Sons, Inc., 2010.

23. D. P. Mandic. A generalized normalized gradient descent algorithm. *IEEE Signal Processing Letters*, 2:115–118, Feb. 2004.

24. J. Mercer. Functions of positive and negative type and their connection with the theory of integral equations. *Philos. Trans. Roy. Soc. London Ser. A*, 209:415–446, 1909.

25. S. Mika, G. Rätsch, J. Weston, B. Schölkopf, and K. R. Müller. Fisher discriminant analysis with kernels. In Y. H. Hu and J. Larsen and E. Wilson and S. Douglas, eds., *Proceedings of the Advances in neural networks for signal processing*, pages 41–48, San Mateo, CA, 1999. Morgan Kaufmann.

26. J. Minkoff. Comment: On the unnecessary assumption of statistical independence between reference signal and filter weights in feedforward adaptive systems. *IEEE Transactions on Signal Processing*, 49(5):1109, May 2001.

27. K. S. Narendra and K. Parthasarathy. Identification and control of dynamical systems using neural networks. *IEEE Transactions on Neural Networks*, 1(1):3–27, March 1990.

28. J. Omura and T. Kailath. Some useful probability distributions. Technical Report 7050-6, Stanford Electronics Laboratories, Stanford University, Stanford, California, USA, 1965.

29. W. D. Parreira, J. C. M. Bermudez, C. Richard, and J.-Y. Tourneret. Steady-state behav-

ior and design of the Gaussian KLMS algorithm. In *19th European Signal Processing Conference (EUSIPCO 2011)*, pages 121–125, Barcelona, Spain, 2011. EURASIP.

30. W. D. Parreira, J. C. M. Bermudez, C. Richard, and J.-Y. Tourneret. Stochastic behavior analysis of the Gaussian kernel least mean square algorithm. In *2011 IEEE International Conference on Acoustics, Speech and Signal Processing*, pages 4116–4119, Prague, 2011.

31. W. D. Parreira, J. C. M. Bermudez, C. Richard, and J.-Y. Tourneret. Stochastic behavior analysis of the Gaussian kernel least-mean-square algorithm. *IEEE Transactions on Signal Processing*, 60(5):2208–2222, May 2012.

32. C. E. Rasmussen and C. K. I. Williams. *Gaussian Processes for Machine Learning*. The MIT Press, 2006.

33. C. Richard and J. C. M. Bermudez. Closed-form conditions for convergence of the Gaussian kernel-least-mean-square algorithm. In *2012 Conf. Rec. Forty Sixth Asilomar Conf. Signals, Syst. Comput.*, pages 1797–1801, Pacific Grove, CA, Nov. 2012. IEEE.

34. C. Richard, J. C. M. Bermudez, and P. Honeine. Online prediction of time series data with kernels. *IEEE Transactions on Signal Processing*, 57(3):1058 –1067, Mar. 2009.

35. A. H. Sayed. *Fundamentals of adaptive filtering*. John Wiley & Sons, Hoboken, NJ, 2003.

36. B. Schölkopf, J. C. Burges, and A. J. Smola. *Advances in kernel methods*. MIT Press, Cambridge, MA, 1999.

37. B. Schölkopf, S. Mika, C. J. C. Burges, P. Knirsch, K. R. Müller, G. Rätsch, and A. J. Smola. Input space versus feature space in kernel-based methods. *IEEE Transactions On Neural Networks*, 10(5):1000–1017, 1999.

38. B. Schölkopf and A. J. Smola. *Learning with Kernels*. MIT Press, Cambridge, MA, 2002.

39. B. Schölkopf, A. J. Smola, and K. R. Müller. Nonlinear component analysis as a kernel eigenvalue problem. *Neural Computation*, 10(5):1299–1319, 1998.

40. G. A. F. Seber and C. J. Wild. *Nonlinear Regression*. Wiley-Interscience, 2003.

41. J. A. K. Suykens, J. de Brabanter, L. Lukas, and J. Vandewalle. Weighted least squares support vector machines: robustness and sparse approximation. *Neurocomputing*, 48:85–105, 2002.

42. J. A. K. Suykens, T. van Gestel, J. de Brabanter, B. de Moor, and J. Vandewalle. *Least Squares Support Vector Machines*. World Scientific, Singapore, 2002.

43. R. Tibshirani. Regression shrinkage and selection via the lasso. *Journal of the Royal Statistical Society. Series B (Methodological)*, 58(1):267–288, 1996.

44. H. L. Van Trees. *Detection, Estimation, and Modulation Theory - Part I*. Wiley Interscience, New York, 2001.

45. Adriaan van den Bos. *Parameter Estimation for Scientists and Engineers*. John Wiley & Sons, 2007.

46. V. N. Vapnik. *The nature of statistical learning theory*. Springer, New York, NY, 1995.

47. Y. Wan, C. X. Wong, T. J. Dodd, and R. F. Harrison. Application of a kernel method in modeling friction dynamics. In *Proceedings of the IFAC World Congress*, Czech Republic, 2005.

3 Arithmetic Transforms: Theory, Advances, and Challenges

RENATO J. CINTRA
Federal University of Pernambuco (UFPE)

HÉLIO M. DE OLIVEIRA
Federal University of Pernambuco (UFPE)

VASSIL S. DIMITROV
University of Calgary
Computer Modelling Group, Ltd

3.1 INTRODUCTION

Trigonometric transforms have a central role as signal processing tools. In particular, the discrete Fourier transform (DFT), the discrete Hartley transform (DHT), and the discrete cosine transform (DCT) are the most employed transformations in both theoretical and practical contexts [36]. Computed according to their definitions, these transforms exhibit quadratic complexity, which is enough to prevent their application in several contexts. Nevertheless, fast algorithms are capable of computing them with a computational complexity in $O(N \log N)$, where N is the transform blocklength. Considering the design of fast algorithms for discrete transforms, it is a well-known fact that the number of multiplications is an important figure of merit that significantly affects the overall computational complexity of a given algorithm. Thus, as a rule of thumb, fast algorithm designers often aim at the minimization of the multiplicative complexity.

Arithmetic transforms are algorithms based on number-theoretical tools aiming at transform computation with minimal arithmetic complexity. In principle, the ultimate goal of an arithmetic transform method is to allow transform computation by means of addition operations only. Nevertheless, arithmetic methods require input data to be sampled nonuniformly. This can be a significant hindrance since input signals are usually uniformly sampled. Usual approaches employed to obtain the necessary nonuniform samples include: (i) signal oversampling and (ii) sample interpolation. On the one hand, oversampling often requires high sampling rates, which surpass the Nyquist sampling rate. Thus, oversampling can be prohibitive in several practical systems and it is often avoided [48]. On the other hand, obtaining nonuniform samples from uniformly sampled data can be attained through interpolation methods.

However, most of arithmetic transform literature lacks a precise description of an exact interpolation method to obtain the required nonuniform samples. In a possibly too simple approach, zero-order approximations and linear interpolation methods have been considered [27]. Although not furnishing exact computations, these crude interpolation methods could attain acceptable approximations, when large blocklengths were considered [49]. However, for small blocklengths, the implied interpolation errors could be large enough to totally preclude a meaningful computation. Overall, this is one of the main reasons that prevent a larger degree of adoption of arithmetic transform methods.

By far the most popular arithmetic transform is the arithmetic Fourier transform (AFT) and several applications for the AFT have been proposed. In [33], the AFT is considered as an alternative to the Goertzel algorithm for the single spectral component evaluation. The AFT could also provide frequency domain testing units for built-in self-test routines with reduced hardware overhead [58]. Additionally, hardware considerations have been directed to enhance the associated nonuniform sampling of the AFT [45]. Finally, the AFT has been considered as a tool for DCT evaluation [53]. In all above-mentioned applications, the AFT procedure as described in [56, 48] was considered. In particular, even when the DCT was required as in [53], it was obtained by means of the AFT [55]. Indeed, the DFT spectrum can be mapped into the DCT spectrum at the cost of extra computations.

This chapter has four major goals. First, we aim at revisiting the AFT theory, which was originally devoted to Fourier series computation based on continuous functions. Not only updating the almost 30-years-old notation, a new presentation for the AFT is sought embracing the discrete-time formalism, which is more adequate to practical considerations. Our approach seems to be suitable, since ultimately a discrete transform relates two set of points, not continuous functions. Our derivations aim at employing the generalized Möbius inversion formula instead of the more restrictive Möbius inversion formula which is the common approach to arithmetic transforms. Second, we derive the mathematical foundations of the arithmetic Hartley transform (AHT) and the arithmetic cosine transform (ACT) for the computation of the DHT and DCT, respectively. We aim at a unified treatment for the arithmetic transforms based on trigonometric functions. We emphasize the similarities among different arithmetic transforms aiming at identifying the core principles that govern the arithmetic transform theory. Third, we pursue a precise description of the arithmetic transform underlying mechanism: the interpolation process. In fact, an arithmetic transform method is capable of exact computation only if a precise interpolation scheme is available. To the best of our knowledge, a mathematical analysis of the AFT interpolation capable of furnishing error-free computation is lacking in literature. Such analysis for the AHT and ACT was disseminated in a previous work in [13, 15]. This chapter aims at offering a comprehensive approach for the AFT, AHT, and ACT interpolation processes. Finally, we also plan to propose approximate expressions for the arithmetic transform interpolation. For each discussed transform, we provide approximations, which can be submitted to actual transform computation.

This chapter unfolds as follows. Section 3.2 conveys a brief historical background tracing the origins of the arithmetic transform from the early 20th century until the present. In Section 3.3 we supply fundamental theorems on which the arithmetic transforms are based. Such results stem from analytic number theory and trigonometric identities. Section 3.4 furnishes an updated presentation for the AFT. Not only the notation was renovated, but the derivation of key results was sought to be simpler and more direct. A detailed analysis of the AFT scheme is supplied as well as the complete characterization for the 8-point AFT. In a similar way, we advance the arithmetic Hartley transform using derivations and arguments comparable with the AFT case. In Section 3.5, we show that both the AFT and AHT share the same mathematical structure. Section 3.6 is devoted to the arithmetic cosine transform, which is the most recently proposed arithmetic transform. The ACT is mathematically comparable to the AFT and AHT, but its structure is not entirely identical. Section 3.7 examines the interpolation schemes required by the AFT, AHT, and ACT. We introduce a precise mathematical characterization of the arithmetic transform interpolation process. Our derivations are sought to be sufficiently general to be applied to other transformations not addressed in this work. Not only exact closed-form expressions are sought, but also approximate interpolation formulas. In Section 3.8, we further discuss the role of interpolation and its connection to signal sampling and offer concluding remarks and open problems.

3.2 HISTORICAL CONTEXT

The research on arithmetic transforms dates back to 1903 when the German mathematician Ernest Heinrich Bruns, a former student of Weierstrass and Kummer, published the *"Grundlinien des wissenschaftlichnen Rechnens"* [10], which addressed the computation of numerical differentiation, integration, and interpolation by means of arithmetical operations based on simple summations. During the next 42 years after the publication of Bruns' fundamental manuscript, the arithmetic transform technique remained largely unnoticed in the scientific community. In 1945, Aurel Freidrich Wintner, an Austro-Hungarian mathematician emigrated to the United States, privately published a monograph entitled *"An Arithmetical Approach to Ordinary Fourier Series"* [61], where an arithmetic method for computing the Fourier series of even periodic functions by means of the Möbius function was presented. Wintner coined the term "arithmetic Fourier transform" [56].

The monograph by Wintner could only find its way to the engineering community after a wait of 43 years. Circa 1988, Tufts and Sadasiv, of the University of Rhode Island, independently reinvented the arithmetical procedure by Wintner; thus rekindling the the theory. In [56] they acknowledge the role of Oved Shisha also from the University of Rhode Island and Charles Rader of MIT Lincoln Laboratories for putting them in contact with Wintner's monograph. In 1988 *"The Arithmetic Fourier Transform"* by Tufts and Sadasiv was published [56]. This particular work represents the inception of the arithmetic transform into the realms of engineering. The AFT by Tufts-Sadasiv requires band-limited, even-symmetric input signals. As a result, only the real part of the Fourier series could be evaluated [56, 55]. Wintner approach also possessed this latter characteristic.

In 1990, Irving Reed—known for his contributions in coding theory—proposed with Tufts and others a variation of the AFT capable of processing asymmetric signals. The even-symmetry restriction was removed. In 1992, another significant improvement was introduced by Reed and collaborators in *"A VLSI Architecture for Simplified Arithmetic Fourier Transform Algorithm"* [48]. Re-designed to be more computationally efficient, this AFT version inherited all improvements of previous AFT versions. Notably this version of the AFT was demonstrated to be identical to Bruns' original method [48].

However, the technological environment found by the rediscovered AFT was dramatically different from the atmosphere found by Bruns and Wintner. Computational capabilities and digital signal processing chips made possible AFT to leave theoretical frameworks and achieve practical implementations. Implementations were proposed in [21, 59, 60, 40, 38, 27, 41, 47, 48, 5, 20, 37, 39, 17]. Besides spectral estimation [18], early applications of the AFT were in the field of pattern matching [1], measurement and instrumentation [3, 2], z-transform computation [26, 25], and imaging [55]. Two-dimensional versions of the AFT were proposed following the mathematical foundation described in [46, 12, 5, 22, 11]. Alternative approaches for the AFT computation were proposed in [34, 54, 32]. However, the formalism by Tufts-Sadasiv and Reed remained the most popular approach to the arithmetic transform.

Although the main and original motivation of the arithmetic transform was the computation of the Fourier series, further generalizations allowed the calculation of other transforms. Knockaert proposed generalizations on the Bruns' procedure, defining the generalized Möbius transform [28, 29]. In the early 2000s, the DHT was subject to the arithmetic transform formalism and the AHT was introduced in [16, 14, 13]. In 2010, the arithmetic cosine transform was introduced in [15].

3.3 MATHEMATICAL BACKGROUND

In this section, we describe the necessary mathematical building blocks for the development of arithmetic transforms. In the following, $k_1|k_2$ denotes that k_1 is a divisor of k_2 and $\lfloor \cdot \rfloor$ is the floor function.

Theorem 3.1: Generalized Möbius Inversion Formula for Finite Series

Let $f[n]$ be a sequence (e.g., signal samples) such that it is non-null for $n = 1, 2, \ldots, N$ and null for $n > N$. Admit another sequence $g[n]$ defined as

$$g[n] = \sum_{s=1}^{\lfloor N/n \rfloor} a[s] \cdot f[sn],$$

where $a[s]$ is a sequence of real numbers. Then,

$$f[n] = \sum_{l=1}^{\lfloor N/n \rfloor} b[l] \cdot g[ln],$$

where $b[l]$ is the Dirichlet inverse sequence of $a[s]$, given that it exists [8]. ∎

Proof: It follows from the assumptions described in the proof of Theorem 3 from [49, p. 469] into the Möbius inversion formula as shown in [23, p. 556]. □

If we consider the unitary sequence $a[n] = 1$, for $n = 1, 2, 3, \ldots$, then the associated Dirichlet inverse sequence is the Möbius sequence $b[n] = \mu(n)$, for $n = 1, 2, 3, \ldots$ The Möbius function $\mu(n)$ [4] is defined over the positive integers and is given by

$$\mu(n) = \begin{cases} 1, & \text{if } n = 1, \\ (-1)^q, & \text{if } n \text{ can be factorized into } q \text{ distinct primes}, \\ 0, & \text{if } n \text{ is divisible by a square number}. \end{cases}$$

In this particular case, the arithmetic transform literature simply refers to the above theorem as the Möbius inversion formula for finite series [49, 48].

Corollary 1 (Möbius Inversion Formula for Finite Series)**.** *Let n be a positive integer and $f[n]$ a non-null sequence for $n = 1, 2, \ldots, N$ and null for $n > N$. If*

$$g[n] = \sum_{s=1}^{\lfloor N/n \rfloor} f[sn],$$

then

$$f[n] = \sum_{l=1}^{\lfloor N/n \rfloor} \mu(l)g[ln].$$

□

The following lemmas are pivotal for subsequent developments. They link usual trigonometric functions to number-theoretic tools, a connection that is not always trivial [42, 31].

Lemma 3.1

Let $k > 0$ and $k' \geq 0$ be integers. Then,

$$\sum_{m=0}^{k-1} \cos\left(2\pi m \frac{k'}{k}\right) = \begin{cases} k, & \text{if } k|k', \\ 0, & \text{otherwise}, \end{cases}$$

and

$$\sum_{m=0}^{k-1} \sin\left(2\pi m \frac{k'}{k}\right) = 0.$$

∎

Proof: Consider the expression $\sum_{m=0}^{k-1} \left[\exp\left(2\pi j \frac{k'}{k}\right)\right]^m$. When $k|k'$, we have:

$$\sum_{m=0}^{k-1} \left[\exp\left(2\pi j \frac{k'}{k}\right)\right]^m = \sum_{m=0}^{k-1} 1 = k.$$

Otherwise, it yields the following expression:

$$\sum_{m=0}^{k-1} \left[\exp\left(2\pi j \frac{k'}{k}\right)\right]^m = \frac{1 - \exp\{j2\pi k'\}}{1 - \exp\{j2\pi \frac{k'}{k}\}} = 0.$$

Therefore, we obtain:

$$\sum_{m=0}^{k-1} \exp\left(2\pi j m \frac{k'}{k}\right) = \begin{cases} k, & \text{if } k|k', \\ 0, & \text{otherwise.} \end{cases}$$

By taking real and imaginary parts of the above expression, the proof is concluded.

□

3.4 ARITHMETIC FOURIER TRANSFORM

In this section, we present a new derivation for the AFT based on the discrete Fourier transform. Our presentation contrasts with usual AFT theory found in literature, where the Fourier series is considered. Instead of considering a continuous-time signal which is sampled, we assume a natively define discrete-time signal Thus, the starting point of the development is the discrete transform, not the series expansion, as it was done in the AFT algorithm. We understand that adopting the discrete-time formalism is more adequate, since actual devices and hardware implementations can only be realized when based on discrete tools.

Let $x[n]$, $n = 0, 1, \ldots, N - 1$, be an N-point discrete-time input signal and $X_F[k]$, $k = 0, 1, \ldots, N - 1$, be its associated discrete Fourier transform. Such signals are related according to the forward and inverse DFT, respectively, given by [36]:

$$X_F[k] = \sum_{n=0}^{N-1} x[n] \cdot \exp\left(-\frac{2\pi}{N} nk\right), \quad k = 0, 1, \ldots, N - 1,$$

$$x[n] = \frac{1}{N} \sum_{k=0}^{N-1} X_F[k] \cdot \exp\left(\frac{2\pi}{N} kn\right), \quad n = 0, 1, \ldots, N - 1. \tag{3.1}$$

The AFT theory prescribes time averages of the input signal at non-integer sampling points. Notice that a discrete-time signal is sampled at integer samples. One way of obtaining fractional index samples is by means of interpolation based on the available integer index samples. Thus, we denote the interpolated signal based on $x[n]$ as $x_F[\cdot]$. Of course, for integer values m, $x_F[m] = x[m]$. We assume that the interpolated signal $x_F[\cdot]$ is available. The details of the interpolation process is discussed in Section 3.7. Thus, we introduce the following AFT average.

Definition 1 (AFT Average). *The kth AFT average is defined according to the following expression:*

$$S_F[k] = \frac{1}{k} \sum_{m=0}^{k-1} x_F\left[\frac{m+\beta}{k}N\right], \quad k = 1, 2, \ldots, N-1, \tag{3.2}$$

where $\beta \in [0, 1]$.

The above definition stems from the AFT average proposed in [56]. Previous versions of the AFT were based on continuous-time signals and employed similar averages. The formulation introduced in this work aims at a discrete-time formalism.

Now our goal is to derive an expression for the DFT coefficients. Therefore, we apply (3.1) in (3.2) as described below:

$$
\begin{aligned}
S_F[k] &= \frac{1}{k} \sum_{m=0}^{k-1} \left[\frac{1}{N} \sum_{k'=0}^{N-1} X_F[k'] \cdot \exp\left\{ \frac{2\pi}{N} jk' \left(\frac{m+\beta}{k}N \right) \right\} \right] \\
&= \frac{1}{k} \frac{1}{N} \sum_{k'=0}^{N-1} X_F[k'] \cdot \exp\left(2\pi j \frac{k'}{k}\beta \right) \cdot \sum_{m=0}^{k-1} \exp\left(2\pi jm \frac{k'}{k} \right) \\
&= \frac{1}{k} \frac{1}{N} \sum_{k'=0}^{N-1} X_F[k'] \cdot \exp\left(2\pi j \frac{k'}{k}\beta \right) \cdot \left\{ \begin{array}{ll} k, & \text{if } k|k', \\ 0, & \text{otherwise} \end{array} \right\}.
\end{aligned}
$$

Letting $k' = k \cdot s$, for integer s, we obtain:

$$S_F[k] = \frac{1}{N} \sum_{s=0}^{\lfloor \frac{N-1}{k} \rfloor} X_F[ks] \cdot \exp\{2\pi js\beta\}.$$

Considering $S'[k] = N \cdot S_F[k] - X_F[0]$, we have that:

$$S'_F[k] = \sum_{s=1}^{\lfloor \frac{N-1}{k} \rfloor} X_F[ks] \cdot \exp\{2\pi js\beta\}.$$

Observe that this expression is suitable for the application of the generalized Möbius inversion formula for finite series. Considering the notation of Theorem 3.1, we recognize $a[s] = \exp\{2\pi js\beta\}$, for $s = 1, 2, 3, \ldots$

Incidentally, not always is the Dirichlet inverse of $a[s]$ well-defined. Only when $a[1] \neq 0$, the existence of the Dirichlet inverse can be considered [8]. Thus, we must impose $\exp(2\pi j\beta) \neq 0$ as a necessary condition for the derivation of the AFT. However, finding the Dirichlet inverse of $a[n]$, say $b[n]$, for arbitrary values of β may not be a trivial task. Nevertheless, we separated two particular useful cases: (i) $\beta = 0$ and (ii) $\beta = 1/2$.

For $\beta = 0$, we have $a[s] = 1$ and $b[l] = \mu(l)$, for $s, l = 1, 2, 3, \ldots$ This is usually the situation addressed in standard AFT analysis [56, 49, 48]. On the other hand, setting $\beta = 1/2$ yields $a[s] = (-1)^s$, $s = 1, 2, 3, \ldots$ In this case, the Dirichlet inverse is not immediately recognized, but it can be obtained analytically. In the Appendix, we derive the sought Dirichlet inverse, which is given by

$$b[l] = \begin{cases} -\mu(l), & \text{if } l \text{ is odd,} \\ -2^{m-1}\mu(2^{-m}l), & \text{if } l = 2^m s, \text{ where } s \text{ is odd.} \end{cases} \tag{3.3}$$

The first 32 terms of $b[l]$ are listed below:

$$\begin{array}{cccccccc}
-1, & -1, & 1, & -2, & 1, & 1, & 1, & -4, \\
0, & 1, & 1, & 2, & 1, & 1, & -1, & -8, \\
1, & 0, & 1, & 2, & -1, & 1, & 1, & 4, \\
0, & 1, & 0, & 2, & 1, & -1, & 1, & -16.
\end{array}$$

In the framework of digital signal processing, this is a potentially useful sequence, since multiplying a given number by a power of two can be implemented by simple bit-shifting operations, which possess a low computational complexity.

For simplicity, let us adopt $\beta = 0$. Thus, we have that $\exp\{2\pi j s\beta\}$ furnishes the unitary sequence. Therefore, we obtain:

$$S'_F[k] = \sum_{s=1}^{\lfloor \frac{N-1}{k} \rfloor} X_F[ks]. \tag{3.4}$$

Consequently, applying the Möbius inversion formula (Corollary 1) to (3.4), we obtain the following result:

$$X_F[k] = \sum_{l=1}^{\lfloor \frac{N-1}{k} \rfloor} \mu(l) \cdot S'_F[kl], \quad k = 1, 2, \ldots, N-1. \tag{3.5}$$

Now, considering again that $S'_F[k] = N \cdot S_F[k] - X_F[0]$, we obtain:

$$X_F[k] = \sum_{l=1}^{\lfloor \frac{N-1}{k} \rfloor} \mu(l) \cdot (N \cdot S_F[k] - X_F[0])$$

$$= N \left(\sum_{l=1}^{\lfloor \frac{N-1}{k} \rfloor} \mu(l) S_F[kl] \right) - X_F[0] \cdot \sum_{l=1}^{\lfloor \frac{N-1}{k} \rfloor} \mu(l)$$

$$= N \left(\sum_{l=1}^{\lfloor \frac{N-1}{k} \rfloor} \mu(l) S_F[kl] \right) - X_F[0] \cdot M \left(\left\lfloor \frac{N-1}{k} \right\rfloor \right), \quad k = 1, 2, \ldots, N-1, \quad (3.6)$$

where $M(n) = \sum_{k=1}^{n} \mu(k)$ is the Mertens function [52, p. 272]. For $\beta = 1/2$, the derivation is similar. The above expression (3.6) is of paramount importance, since it relates the spectral coefficients $X_F[k]$ to the AFT averages, which depend only on the time domain data at specific non-uniform sampling points. As far as the computational complexity of the Möbius inversion formulae is concerned, we can provide the following probabilistic reasoning. The probability that a randomly chosen integer is not divisible by a perfect square is $6/\pi^2 \approx 0.61$ [8, p. 4]. Therefore, 61% of the values of the Möbius function are zeros; meaning that the computation of $X_F[k]$ requires $(1 - 6/\pi^2) \lfloor (N-1)/k \rfloor$ additions/subtractions on average.

Above AFT derivation differs from the usual presentations found in literature. In particular, we notice the following points:

- The proposed derivation fully embraces the discrete-time formalism;
- Derivations are directly based on the discrete Fourier transform instead of continuous-time functions and Fourier series (possibly with infinite terms) as commonly found in literature;
- In [56] only even-symmetric signals could be processed. The proposed AFT has no restriction in terms of signal symmetry;
- In [56, 49, 48] only null mean signals could be processed. The proposed AFT removes this limitation.

3.4.1 8-POINT AFT

In this section, as an illustrative example, we fully derive the details of the AFT computation for $N = 8$. For simplicity, we selected $\beta = 0$. By means of (3.2), we can compute the AFT averages as follows:

$$S_F[1] = x[0],$$
$$2 \cdot S_F[2] = x[0] + x[4],$$
$$3 \cdot S_F[3] = x[0] + x_F\left[\frac{8}{3}\right] + x_F\left[\frac{16}{3}\right],$$
$$4 \cdot S_F[4] = x[0] + x[2] + x[4] + x[6]$$
$$= 2 \cdot S_F[2] + x[2] + x[6],$$
$$5 \cdot S_F[5] = x[0] + x_F\left[\frac{8}{5}\right] + x_F\left[\frac{16}{5}\right] + x_F\left[\frac{24}{5}\right] + x_F\left[\frac{32}{5}\right],$$
$$6 \cdot S_F[6] = x[0] + x_F\left[\frac{4}{3}\right] + x_F\left[\frac{8}{3}\right] + x[4] + x_F\left[\frac{16}{3}\right] + x_F\left[\frac{20}{3}\right]$$
$$= 3 \cdot S_F[3] + x_F\left[\frac{4}{3}\right] + x[4] + x_F\left[\frac{20}{3}\right],$$
$$7 \cdot S_F[7] = x[0] + x_F\left[\frac{8}{7}\right] + x_F\left[\frac{16}{7}\right] + x_F\left[\frac{24}{7}\right] + x_F\left[\frac{32}{7}\right] + x_F\left[\frac{40}{7}\right] + x_F\left[\frac{48}{7}\right].$$
$$(3.7)$$

The computation of the AFT averages requires 18 additions. Figure 3.1 depicts the signal flow graph for this computation.

Subsequently, considering (3.5), we have that:

$$X_F[1]/8 = S_F[1] - S_F[2] - S_F[3] - S_F[5] + S_F[6] - S_F[7],$$
$$X_F[2]/8 = S_F[2] - S_F[4] - S_F[6],$$
$$X_F[3]/8 = S_F[3] - S_F[6],$$
$$X_F[4]/8 = S_F[4],$$
$$X_F[5]/8 = S_F[5],$$ $$(3.8)$$
$$X_F[6]/8 = S_F[6],$$
$$X_F[7]/8 = S_F[7].$$

This is the Möbius inversion procedure, which demands 8 additions. Figure 3.2 shows the signal flow diagram for the Möbius inversion stage.

By scaling each of the above equations, respectively, by 210, 12, 6, 4, 5, 6, 7, we

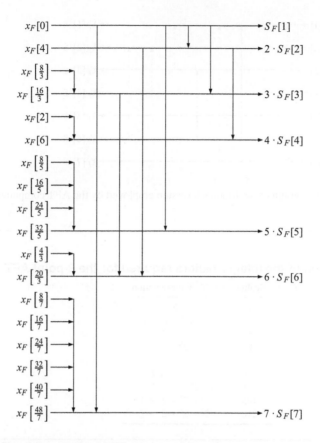

Figure 3.1 Signal flow graph of AFT averages computation.

can re-write the above set of equations as follows:

$$210 \cdot X_F[1]/8 = 210 \cdot S_F[1] - 105 \cdot (2 \cdot S_F[2]) - 70 \cdot (3 \cdot S_F[3])$$
$$- 42 \cdot (5 \cdot S_F[5]) + 35 \cdot (6 \cdot S_F[6]) - 30 \cdot (7 \cdot S_F[7]),$$
$$12 \cdot X_F[2]/8 = 6 \cdot (2 \cdot S_F[2]) - 3 \cdot (4 \cdot S_F[4]) - 2 \cdot (6 \cdot S_F[6]),$$
$$6 \cdot X_F[3]/8 = 2 \cdot (3 \cdot S_F[3]) - (6 \cdot S_F[6]),$$
$$4 \cdot X_F[4]/8 = (4 \cdot S_F[4]), \tag{3.9}$$
$$5 \cdot X_F[5]/8 = (5 \cdot S_F[5]),$$
$$6 \cdot X_F[6]/8 = (6 \cdot S_F[6]),$$
$$7 \cdot X_F[7]/8 = (7 \cdot S_F[7]).$$

The parenthetical terms were already computed from the set of equations (3.7). Notice that we require extra scaling factors, namely 2, 3, 6, 30, 35, 42, 70, 105, and 210. However, these factors are integer multiplications which can be precisely implemented in fixed-point arithmetic. Considering the canonical sign digit (CSD) rep-

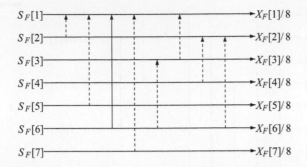

Figure 3.2 Signal flow graph of the Möbius inversion employed by the AFT computation.

Table 3.1

CSD representation of the integer factors required for the 8-point AFT

Scaling	CSD Representation
2	2
3	$2 + 1$
6	$2^8 - 2$
30	$2^5 - 2$
35	$2^5 + 2^2 - 1$
42	$2^5 + 2^3 + 2$
70	$2^6 + 2^3 - 2$
105	$2^7 - 2^5 + 2^3 + 1$
210	$2^8 - 2^6 + 2^4 + 2$

---

resentation, one may convert such multiplications in additions and bit-shifting operations. Table 3.1 displays the CSD representation for the discussed scaling factors. This particular representation results in 16 extra additions. Above procedure results in a scaled DFT computation with scaling factors of $105/4$, $3/2$, $3/4$, $1/2$, $5/8$, $3/4$, $7/8$ for coefficients $X_F[1], X_F[2], \ldots, X_F[7]$, respectively.

3.5 ARITHMETIC HARTLEY TRANSFORM

In this section, we condense the main results of the arithmetic Hartley transform [16]. The derivation of the AHT made some aspects of the arithmetic transform clearer, such as the role of interpolation.

Let $x[n]$, $n = 0, 1, \ldots, N - 1$, be an N-point discrete-time input signal and $X_H[k]$, $k = 0, 1, \ldots, N - 1$, be the associated transformed data according to the DHT. The

DHT establishes the following forward and inverse relationships, respectively:

$$X_H[k] = \sum_{n=0}^{N-1} x[n] \cdot \text{cas}\left(\frac{2\pi nk}{N}\right), \quad k = 0, 1, \ldots, N-1,$$

$$x[n] = \frac{1}{N} \cdot \sum_{k=0}^{N-1} X_H[k] \cdot \text{cas}\left(\frac{2\pi kn}{N}\right), \quad n = 0, 1, \ldots, N-1.$$

where $\text{cas}(x) = \cos(x) + \sin(x)$ is the cosine-and-sine function [43].

In order to design a arithmetic algorithm for the DHT evaluation, let us introduce the AHT averages $S_H[k]$ as follows.

Definition 2 (AHT Average). *Let the kth AHT average be defined according to the following expression:*

$$S_H[k] = \frac{1}{k} \cdot \sum_{m=0}^{k-1} x_H\left[\frac{m+\beta}{k}N\right], \quad k = 1, 2, \ldots, N-1,$$

where $x_H[\cdot]$ returns the input signal at non-integer sampling instants.

Considering (i) an application of inverse DHT on $x_H\left[\frac{m+\beta}{k}N\right]$, (ii) the cas function properties [43, Chapter 14], and (iii) Lemma 3.1, the following algebraic manipulation holds true:

$$\begin{aligned}
S_H[k] &= \frac{1}{k}\frac{1}{N} \sum_{k'=0}^{N-1} X_H[k'] \sum_{m=0}^{k-1} \text{cas}\left(\frac{2\pi}{N}\frac{m+\beta}{k}Nk'\right) \\
&= \frac{1}{k}\frac{1}{N} \sum_{k'=0}^{N-1} X_H[k']\left[\sum_{m=0}^{k-1} \cos\left(2\pi m\frac{k'}{k}\right)\text{cas}\left(2\pi\beta\frac{k'}{k}\right)\right. \\
&\quad \left. + \sum_{m=0}^{k-1} \sin\left(2\pi m\frac{k'}{k}\right)\text{cas}\left(-2\pi\beta\frac{k'}{k}\right)\right] \\
&= \frac{1}{k}\frac{1}{N} \sum_{k'=0}^{N-1} X_H[k']\left[\text{cas}\left(2\pi\beta\frac{k'}{k}\right)\sum_{m=0}^{k-1} \cos\left(2\pi m\frac{k'}{k}\right)\right. \\
&\quad \left. + \text{cas}\left(-2\pi\beta\frac{k'}{k}\right)\cdot\sum_{m=0}^{k-1} \sin\left(2\pi m\frac{k'}{k}\right)\right] \\
&= \frac{1}{N} \sum_{k'=0}^{N-1} X_H[k']\left[\text{cas}\left(2\pi\beta\frac{k'}{k}\right)\left\{\begin{array}{ll} 1, & \text{if } k|k', \\ 0, & \text{otherwise} \end{array}\right\} + 0\right].
\end{aligned}$$

As previously done for the AFT derivation, $k' = k \cdot s$, for integer s. Thus, we have that:

$$S_H[k] = \frac{1}{N} \sum_{s=0}^{\lfloor \frac{N-1}{k} \rfloor} X_H[sk] \cdot \text{cas}(2\pi\beta s).$$

Letting $S'_H[k] = N \cdot S_H[k] - X_H[0]$, we obtain:

$$S'_H[k] = \sum_{s=1}^{\lfloor \frac{N-1}{k} \rfloor} X_H[sk] \cdot \mathrm{cas}\,(2\pi\beta s)$$

$$= \sum_{s=1}^{\lfloor \frac{N-1}{k} \rfloor} a_H[s] \cdot X_H[sk],$$

where $a_H[s] = \mathrm{cas}\,(2\pi\beta s)$. For $\beta = 0$, we obtain the $a_H[s] = 1$, whose Dirichlet inverse sequence is the Möbius sequence. For $\beta = 1/2$, similar to the AFT case, we have that $a_H[s] = (-1)^s$, $s = 1, 2, 3, \ldots$

Invoking the generalized Möbius inversion formula, we derive the following expression, for $\beta = 0$:

$$X_F[k] = \sum_{l=1}^{\lfloor \frac{N-1}{k} \rfloor} \mu(l) \cdot S'_H[kl], \quad k = 1, 2, \ldots, N-1.$$

Now, considering again that $S'_F[k] = N \cdot S_F[k] - X_F[0]$, we obtain:

$$X_H[k] = N\left(\sum_{l=1}^{\lfloor \frac{N-1}{k} \rfloor} \mu(l)S_H[kl]\right) - X_H[0] \cdot \mathrm{M}\left(\left\lfloor \frac{N-1}{k} \right\rfloor\right), \quad k = 1, 2, \ldots, N-1. \quad (3.10)$$

A careful examination of above results shows an unexpected result: both AFT and AHT share the mathematical formalism. In particular, expressions (3.6) and (3.10), for the AFT and AHT, respectively, are identical. Apart from the notation, the AFT computation example shown in Section 3.4.1 also describes the AHT computation for $N = 8$. The computational complexity of both methods is exactly the same. Therefore, there seems to be an ambiguity with respect to the evaluated spectrum. Such ambiguity can be resolved in terms of how the input data at non-integer sample points are obtained. In other words, for the spectra to be different, $x_F[\cdot]$ must be different of $x_H[\cdot]$—the interpolation must not be the same. In fact, the interpolation process is the key factor that distinguishes the AFT from the AHT. The mathematical details of the interpolation process are discussed in Section 3.7.

3.6   ARITHMETIC COSINE TRANSFORM

Among the several types of DCTs, we separated the DCT-II, which can be regarded as the most popular version among the possible DCTs [9]. This transformation relates two N-point discrete-time signals according to the following expressions [9]:

$$X_C[k] = \sqrt{\frac{2}{N}}\, \alpha_k \sum_{n=0}^{N-1} x[n] \cos\left(\frac{\pi(n + 1/2)k}{N}\right), \quad k = 0, 1, \ldots, N-1,$$

$$x[n] = \sqrt{\frac{2}{N}} \sum_{k=0}^{N-1} \alpha_k X_C[k] \cos\left(\frac{\pi k(n+1/2)}{N}\right), \quad n = 0, 1, \ldots, N-1, \quad (3.11)$$

where $\alpha_k = 1/\sqrt{2}$, if $k = 0$, and $\alpha_k = 1$, otherwise. Hereafter, we refer to this transformation simply as DCT. Based on existing definitions and concepts inherited from the AFT/AHT theory, we introduce the following average specially tailored for the DCT.

Definition 3 (ACT Average). *The kth ACT average is given by:*

$$S_C[k] = \frac{1}{k} \sum_{m=0}^{k-1} x_C\left[2(m+\beta)\frac{N}{k} - \frac{1}{2}\right], \quad k = 1, 2, \ldots, N-1, \quad (3.12)$$

where $\beta \in [0, 1]$ is a fixed real number and $x_C[\cdot]$ returns the input signal at non-integer sampling points. □

Applying the inverse discrete cosine transform formula (3.11) into (3.12), we obtain:

$$
\begin{aligned}
S_C[k] &= \frac{1}{k} \sum_{m=0}^{k-1} \left[\sqrt{\frac{2}{N}} \sum_{k'=0}^{N-1} \alpha_{k'} \cdot X_C[k'] \cdot \cos\left(\frac{\pi k'(2(m+\beta)\frac{N}{k} - \frac{1}{2} + \frac{1}{2})}{N}\right) \right] \\
&= \sqrt{\frac{2}{N}} \frac{1}{k} \sum_{m=0}^{k-1} \left[\frac{1}{\sqrt{2}} X_C[0] + \sum_{k'=1}^{N-1} X_C[k'] \cos\left(2\pi(m+\beta)\frac{k'}{k}\right) \right] \\
&= \sqrt{\frac{1}{N}} X_C[0] \\
&\quad + \sqrt{\frac{2}{N}} \frac{1}{k} \sum_{k'=1}^{N-1} X_C[k'] \sum_{m=0}^{k-1} \cos\left(2\pi(m+\beta)\frac{k'}{k}\right), \quad k = 1, 2, \ldots, N-1.
\end{aligned}
$$

The following proposition allows further simplifications for the derivation of $S_C[k]$.

Proposition 1. *Let $k > 0$ and $k' \geq 0$ be integers and α be a real quantity. Then,*

$$\sum_{m=0}^{k-1} \cos\left(2\pi\frac{k'}{k}(m+\beta)\right) = \begin{cases} k, & \text{if } k' = 0, \\ k\cos(2\pi\frac{k'}{k}\beta), & \text{if } k|k', \; k' \neq 0, \\ 0, & \text{otherwise.} \end{cases}$$

Proof: It follows from usual trigonometric manipulations and an application of Lemma 3.1. □

Thus, invoking Proposition 1 and performing the substitution $k' = ks$, it follows that:

$$S_C[k] = \sqrt{\frac{1}{N}} X_C[0] + \sqrt{\frac{2}{N}} \frac{1}{k} \sum_{k'=1}^{N-1} X_C[k'] \left\{ \begin{array}{ll} k\cos(2\pi \frac{k'}{k}\beta), & \text{if } k|k', \\ 0, & \text{otherwise,} \end{array} \right\}$$

$$= \sqrt{\frac{1}{N}} V_0 + \sqrt{\frac{2}{N}} \sum_{s=1}^{\lfloor \frac{N-1}{k} \rfloor} \cos(2\pi s\beta) \cdot X_C[sk], \quad k = 1, 2, \ldots, N-1.$$

Let us adopt the following substitution: $S'_C[k] = (\sqrt{N} \cdot S_C[k] - V_0)/\sqrt{2}$. As a consequence, we have that:

$$S'_C[k] = \sum_{s=1}^{\lfloor \frac{N-1}{k} \rfloor} \cos(2\pi s\beta) \cdot X_C[sk], \quad k = 1, 2, \ldots, N-1.$$

Similarly to the AFT and AHT cases, the above expression is adequate for the generalized Möbius inversion formula. Using the terminology of Theorem 3.1, we have $a[s] = \cos(2\pi s\beta)$, for $s = 1, 2, 3, \ldots$. For $\beta = 0$, we have:

$$S'_C[k] = \sum_{s=1}^{\lfloor \frac{N-1}{k} \rfloor} X_C[sk], \quad k = 1, 2, \ldots, N-1.$$

Consequently, it follows that:

$$X_C[k] = \sum_{s=1}^{\lfloor \frac{N-1}{k} \rfloor} \mu(l) \cdot S'_C[kl]$$

$$= \sum_{s=1}^{\lfloor \frac{N-1}{k} \rfloor} \mu(l) \cdot \left(\sqrt{\frac{N}{2}} S_C[kl] - \frac{1}{\sqrt{2}} X_C[0] \right)$$

$$= \sqrt{\frac{N}{2}} \left(\sum_{s=1}^{\lfloor \frac{N-1}{k} \rfloor} \mu(l) \cdot S[kl] \right) - \sum_{s=1}^{\lfloor \frac{N-1}{k} \rfloor} \frac{1}{\sqrt{2}} X_C[0]$$

$$= \sqrt{\frac{N}{2}} \left(\sum_{s=1}^{\lfloor \frac{N-1}{k} \rfloor} \mu(l) \cdot S[kl] \right) - \frac{1}{\sqrt{2}} X_C[0] \, \mathrm{M}\left(\left\lfloor \frac{N-1}{k} \right\rfloor \right), \quad k = 1, 2, \ldots, N-1.$$

$$(3.13)$$

Notice that $X_C[0] = X_F[0]/\sqrt{N} = X_H[0]/\sqrt{N} = \sqrt{N} \cdot \bar{x}$, where $\bar{x}$ is the mean value of $x[n]$. Thus, the final AFT expression is given by:

$$X_C[k] = \sqrt{\frac{N}{2}} \cdot \left(\sum_{s=1}^{\lfloor \frac{N-1}{k} \rfloor} \mu(l) \cdot S[kl] \right) - \sqrt{\frac{N}{2}} \cdot \bar{x} \cdot \mathrm{M}\left(\left\lfloor \frac{N-1}{k} \right\rfloor \right)$$

For $\beta = 1/2$, a similar expression can be obtained based on the alternate sequence $-1, 1, -1, 1, \ldots$ and its Dirichlet inverse as described in (3.3).

3.6.1 8-POINT ACT

In this section, we detail the ACT computation for $N = 8$. This particular block-length is widely adopted in several image and video coding standards, such as JPEG, MPEG-1, MPEG-2, H.261, and H.263 [50]. The 8-point DCT is also subject to an extensive analysis in [9]. In the following, we set $\beta = 0$.

The first step of the ACT procedure consists of the identification of the necessary interpolation points. According to Definition 3, these points are given by $2m\frac{8}{k} - \frac{1}{2}$ for $k = 1, 2, \ldots, 7$ and $m = 0, 1, 2, \ldots k - 1$. Therefore, we find the following fractional values: $r \in \left\{ -\frac{1}{2}, \frac{25}{14}, \frac{13}{6}, \frac{27}{10}, \frac{7}{2}, \frac{57}{14}, \frac{29}{6}, \frac{59}{10}, \frac{89}{14}, \frac{15}{2} \right\}$.

The input signal at fractional index sampling points are obtained by a particular interpolation process to be detailed in Section 3.7. Then, the ACT averages are computed according to the following set of equations:

$$S_C[1] = x_C\left[-\frac{1}{2}\right]$$

$$2 \cdot S_C[2] = x_C\left[-\frac{1}{2}\right] + x_C\left[\frac{15}{2}\right] = S_C[1] + x_C\left[\frac{15}{2}\right]$$

$$3 \cdot S_C[3] = x_C\left[-\frac{1}{2}\right] + 2 \cdot x_C\left[\frac{29}{6}\right] = S_C[1] + 2 \cdot x_C\left[\frac{29}{6}\right]$$

$$4 \cdot S_C[4] = x_C\left[-\frac{1}{2}\right] + x_C\left[\frac{15}{2}\right] + 2 \cdot x_C\left[\frac{7}{2}\right] = 2 \cdot S_C[2] + 2 \cdot x_C\left[\frac{7}{2}\right]$$

$$5 \cdot S_C[5] = x_C\left[-\frac{1}{2}\right] + 2 \cdot x_C\left[\frac{27}{10}\right] + 2 \cdot x_C\left[\frac{59}{10}\right] = S_C[1] + 2 \cdot x_C\left[\frac{27}{10}\right] + 2 \cdot x_C\left[\frac{59}{10}\right]$$

$$6 \cdot S_C[6] = x_C\left[-\frac{1}{2}\right] + 2 \cdot x_C\left[\frac{13}{6}\right] + 2 \cdot x_C\left[\frac{29}{6}\right] + x_C\left[\frac{15}{2}\right]$$

$$= 3 \cdot S_C[3] + 2 \cdot x_C\left[\frac{13}{6}\right] + x_C\left[\frac{15}{2}\right]$$

$$7 \cdot S_C[7] = S_C[1] + 2 \cdot x_C\left[\frac{25}{14}\right] + 2 \cdot x_C\left[\frac{57}{14}\right] + 2 \cdot x_C\left[\frac{89}{14}\right].$$

Figure 3.3 displays the signal flow graph of the ACT averages computation for $N = 8$. Finally, the ACT averages are combined with respect to the Möbius function (cf. (3.13)). Being the ACT formula (3.13) identical—expect for a scaling constant—to the AFT formula (3.6), the remaining computation follows the structure described in (3.8) and (3.9). Thus, the signal flow graph shown in Figure 3.2 is also valid for the ACT. The only minor change is the spectrum scaling factor: for the AFT, we have a scaling of $1/8$, whereas, for the ACF, the factor is $1/2$. It is important to notice that this calculation involves no approximation and furnishes the exact spectrum.

3.7 ARITHMETIC TRANSFORM INTERPOLATION

In this section, we address the computation of the non-integer samples required by the arithmetic transforms. Usual arithmetic theory simply assumes that the required

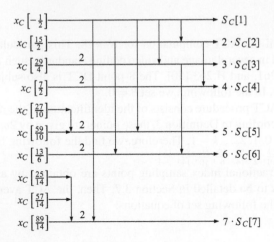

Figure 3.3 Signal flow graph of ACT averages computation.

fractional index samples can be obtained according to two main methods:

-   Apply simple zero- or first-order interpolation methods based on the samples located at the vicinity of the sought sample [49, 48, 25]. For example, $x[29/6]$ is suggested to be approximated by $x[5]$, since 5 is the closest integer to $29/6$ [49, p. 461];
-   Perform over-sampling to the original continuous signal in order to obtain all necessary samples [48].

We aim at showing that both approaches are ineffective for the computation of arithmetic transforms over the discrete-time formalism.

In the following analysis, we assume that only usual discrete-time signal is available: $x[0], x[1], \ldots, x[N-1]$. Our goal is to derive an interpolation weighting function that could furnish the input data at non-integer sampling points. Thus, the data associated with fractional sampling points can be obtained after a linear combination of the available uniformly sampled data according to:

$$x_K[r] = \sum_{n=0}^{N-1} w_K[n; r] \cdot x[n], \qquad (3.14)$$

where $x_K[r]$ is the interpolated signal at a particular non-integer sampling point r and $w_K[n; r]$ is the sought weighting function. In a general manner, let us assume that the discrete transform under analysis satisfies:

$$X[k] = \sum_{n=0}^{N-1} x[n] \cdot \text{kernel}(n, k), \qquad (3.15)$$

$$x[n] = \sum_{k=0}^{N-1} X[k] \cdot \text{kernel}^{-1}(k, n), \qquad (3.16)$$

Table 3.2

Trigonometric orthogonal kernels

Transform	kernel(n, k)	kernel$^{-1}(k, n)$
DFT	$\exp(-2\pi jnk/N)$	$\frac{1}{N}\exp(2\pi jkn/N)$
DHT	$\mathrm{cas}(2\pi nk/N)$	$\frac{1}{N}\mathrm{cas}(2\pi kn/N)$
DCT	$\sqrt{\frac{2}{N}}\,\alpha_k\cos\left[\frac{\pi}{N}(n+1/2)k\right]$	$\sqrt{\frac{2}{N}}\,\alpha_k\cos\left[\frac{\pi}{N}k(n+1/2)\right]$

where kernel(n, k) and kernel$^{-1}(k, n)$ are the forward and inverse kernel functions. Table 3.7 summarizes the kernel functions discussed in this work.

If we (i) relax the integer index assumption required in above transform expressions, (ii) invoke the inverse transformation at r, and (iii) apply the forward transformation at k, then we obtain:

$$x_K[r] = \sum_{k=0}^{N-1} X_K[k] \cdot \mathrm{kernel}^{-1}(k, r)$$

$$= \sum_{k=0}^{N-1}\left[\sum_{n=0}^{N-1} x[n] \cdot \mathrm{kernel}(n, k)\right] \cdot \mathrm{kernel}^{-1}(k, r)$$

$$= \sum_{n=0}^{N-1} x[n] \cdot \left[\sum_{k=0}^{N-1} \mathrm{kernel}(n, k) \cdot \mathrm{kernel}^{-1}(k, r)\right]. \qquad (3.17)$$

Thus, comparing (3.14) with (3.17), we maintain that the weighting function is given by:

$$w_K[n; r] = \sum_{k=0}^{N-1} \mathrm{kernel}(n, k) \cdot \mathrm{kernel}^{-1}(k, r). \qquad (3.18)$$

Notice that, being the kernel functions orthogonal, it follows that

$$w_K[n; m] = \begin{cases} 1, & \text{if } n = m, \\ 0, & \text{otherwise.} \end{cases}$$

Now with this general expression for the interpolation weighting function, in the next sections, we consider (3.18) to derive the particular weighting functions for the AFT, AHT, and ACT, denoted by $w_F[n; r]$, $w_H[n; r]$, and $w_C[n; r]$, respectively.

3.7.1 AFT INTERPOLATION

Considering (3.18) for the particular case of the DFT kernel, we obtain the following expression for the AFT interpolation weighting function:

$$w_F[n; r] = \frac{1}{N} \sum_{k=0}^{N-1} \exp\left(-\frac{2\pi}{N} jnk\right) \exp\left(\frac{2\pi}{N} jkr\right)$$

$$= \frac{1}{N} \sum_{k=0}^{N-1} \exp\left\{\frac{2\pi}{N} jk(r-n)\right\}$$

$$= \frac{1}{N} \frac{1 - \exp\{2\pi j(r-n)\}}{1 - \exp\left\{\frac{2\pi}{N} j(r-n)\right\}} \tag{3.19}$$

$$= \exp\left\{\frac{N-1}{N} \pi j(r-n)\right\} \text{Diric}_N\left\{\frac{2\pi}{N}(r-n)\right\}, \tag{3.20}$$

where $\text{Diric}_N(x) = \frac{\sin(Nx/2)}{N\sin(x/2)}$ is the Dirichlet or periodic sinc function [35]. Expression 3.19 is very close to the frequency response of the delay-line filter proposed by Tufts-Sadasiv for the AFT [56, p. 17].

By invoking Lemma 3.1, we derive the following manipulation:

$$\sum_{n=0}^{N-1} w_F[n; r] = \sum_{n=0}^{N-1} \frac{1}{N} \sum_{k=0}^{N-1} \exp\left\{\frac{2\pi}{N} jk(r-n)\right\}$$

$$= \frac{1}{N} \sum_{k=0}^{N-1} \exp\left\{\frac{2\pi}{N} jkr\right\} \sum_{n=0}^{N-1} \exp\left\{-\frac{2\pi}{N} jkn\right\}$$

$$= \frac{1}{N} \sum_{k=0}^{N-1} \exp\left\{\frac{2\pi}{N} jkr\right\} N \cdot \delta_{k,0}$$

$$= 1,$$

where $\delta_{k,0} = 1$, if $k = 0$, and $\delta_{k,0} = 0$, if $k \neq 0$. This shows that $w_F[n; r]$ is a normalized weighting function.

If N is odd, then the periodic sinc function has a period equal to 2π, otherwise its period is 4π. Because $|r - n| < N - 1$, we have that the argument of the periodic sinc function in (3.20) is always less than $\frac{N-1}{N} 2\pi < 2\pi$. Thus, in practice, only one period of $\text{Diric}_N(\cdot)$ is employed. Therefore, we can safely approximate the periodic sinc function by the discrete Fourier kernel [6, p. 179], yielding:

$$\text{Diric}_N(x) \approx \text{sinc}\left[\frac{N}{2\pi} x\right], \tag{3.21}$$

where $\text{sinc}(x) = \frac{\sin(\pi x)}{\pi x}$.

Therefore, applying (3.21) in (3.20), we obtain a simple approximation for the AFT weighting function, which is furnished by:

$$\hat{w}_F[n; r] = \exp\left[\frac{N-1}{N} \pi j(r-n)\right] \text{sinc}(r-n).$$

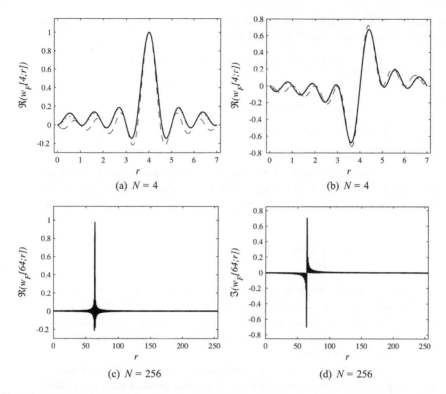

Figure 3.4 Real and imaginary parts of the AFT weighting functions for (a)–(b) $n = 4$ and $N = 8$ and (c)–(d) $n = 64$ and $N = 256$. Solid, dotted, and dashed curves denote $w_F[n; r]$, $\hat{w}_F[n; r]$, and $\tilde{w}_F[n; r]$, respectively.

For large values of N, an even simpler approximation can be derived, since $\frac{N-1}{N} \to 1$, as $N \to \infty$:

$$\tilde{w}_F[n; r] = \exp[\pi j(r - n)] \operatorname{sinc}(r - n),$$

Notice that above approximate expressions connect the ACT interpolation to the sinc function interpolation. Signal interpolation according to the sinc function can be efficiently implemented in the time domain [51, 19]. Additionally, fractional delay FIR filtering methods offer another computational approach [57, 30]. Figure 3.4 displays the real and imaginary parts of the AFT weighting function and its approximations for $n = 4$ and $N = 8$; and for $n = 64$ and $N = 256$. For large N, approximations become indistinguishable from the exact result.

Moreover, as $N \to \infty$, the following limits hold true:

$$\lim_{x \to 0} \operatorname{Diric}_N(x) = 1 \quad \text{and} \quad \lim_{N \to \infty} \operatorname{Diric}_N(x) = 0, \quad x \neq 0.$$

Therefore, for large values of N, the periodic sinc function is approximately equal to one at the vicinity of $n = \text{round}(r)$; and zero elsewhere, where round$(\cdot)$ is the nearest integer function. Thus, inspecting (3.20) reveals that

$$w_F[n; r] \approx \begin{cases} \exp\{\pi j[r - \text{round}(r)]\}, & \text{if } n = \text{round}(r), \\ 0, & \text{otherwise.} \end{cases}$$

Thus, it follows that a crude interpolation for the AFT is furnished by:

$$x_F[r] \approx \begin{cases} \exp\{\pi j[r - \text{round}(r)]\} \cdot x[\text{round}(r)], & \text{if } n = \text{round}(r), \\ 0, & \text{otherwise.} \end{cases}$$

In other words, this interpolation is simply the zero-order interpolation weighted by a complex number over the unitary circle. Usual AFT theory often neglects the complex nature of the weighting function. In [48] the following approximation is suggested:

$$x_F[r] \approx \begin{cases} x[\text{round}(r)], & \text{if } n = \text{round}(r), \\ 0, & \text{otherwise.} \end{cases} \tag{3.22}$$

Above approximation inappropriately ignores the imaginary part of the weighting function. As a result, the AFT can only compute the spectrum of even-symmetric signals. In fact, without the imaginary part, the AFT weighting function collapses to the weighting function associated with the real part of the discrete Fourier transform (discrete Fourier cosine transform).

For a qualitative description of the AFT computation, we generated random real input signal from a stationary first-order Markov stochastic process with correlation $\rho = 0.95$. This type of signal is often employed in image processing analysis [9]. Figure 3.5 shows the AFT magnitude spectrum for $N = 8, 16, 32, 64$. Solid, dotted, and dashed curves indicate the DFT computation according to the AHT by means of $w_F[n; r]$, $\hat{w}_F[n; r]$, and $\tilde{w}_F[n; r]$. Only the first half of the spectra were computed, since the second half is mirrored to first.

3.7.2 AHT INTERPOLATION

The interpolation required for the AHT employs a weighting function that is similar to the AFT weighting function. Considering the DHT kernel and invoking the trigonometric identity $\text{cas}(x) \cdot \text{cas}(y) = \cos(x - y) + \sin(x + y)$, we obtain:

$$\begin{aligned} w_H[n; r] &= \frac{1}{N} \sum_{k=0}^{N-1} \text{cas}\left(\frac{2\pi}{N}nk\right) \text{cas}\left(\frac{2\pi}{N}kr\right) \\ &= \frac{1}{N} \sum_{k=0}^{N-1} \cos\left[\frac{2\pi}{N}k(n - r)\right] + \frac{1}{N} \sum_{k=0}^{N-1} \sin\left[\frac{2\pi}{N}k(n + r)\right] \\ &= \cos\left[\frac{N-1}{N}\pi(n - r)\right] \cdot \text{Diric}_N\left[\frac{2\pi}{N}(n - r)\right] \\ &\quad + \sin\left[\frac{N-1}{N}\pi(n + r)\right] \cdot \text{Diric}_N\left[\frac{2\pi}{N}(n + r)\right]. \end{aligned} \tag{3.23}$$

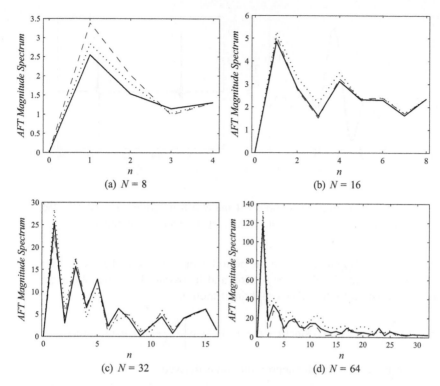

Figure 3.5 Magnitude of AFT computation of random real input signals for $N = 16, 32, 64, 128$. Solid, dotted, and dashed curves denote spectra computed according to $w_F[n; r]$ (exact spectrum), $\hat{w}_F[n; r]$, and $\tilde{w}_F[n; r]$, respectively.

Comparable to the AFT weighting function, the AHT weighting function is also normalized:

$$\sum_{n=0}^{N-1} w_H[n; r] = \sum_{n=0}^{N-1} \frac{1}{N} \sum_{k=0}^{N-1} \text{cas}\left(\frac{2\pi}{N}nk\right) \text{cas}\left(\frac{2\pi}{N}kr\right)$$

$$= \frac{1}{N} \sum_{k=0}^{N-1} \text{cas}\left(\frac{2\pi}{N}kr\right) \sum_{n=0}^{N-1} \text{cas}\left(\frac{2\pi}{N}nk\right)$$

$$= \frac{1}{N} \sum_{k=0}^{N-1} \text{cas}\left(\frac{2\pi}{N}kr\right) \delta_{k,0}$$

$$= 1.$$

The argument of the second instantiation of the $\text{Diric}_N(\cdot)$ function in (3.23) ranges from 0 to 4π. Thus, the $\text{Diric}_N(\cdot)$ can be approximated by $(-1)^N \text{sinc}(\cdot)$ to take into

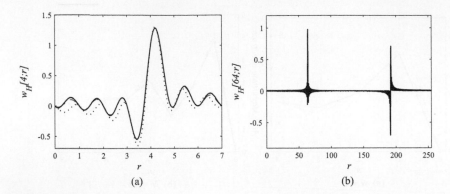

Figure 3.6 AFT weighting functions for (a) $n = 4$ and $N = 8$ and (b) $n = 64$ and $N = 256$. Solid, dotted, and dashed curves denote $w_H[n; r]$, $\hat{w}_H[n; r]$, and $\tilde{w}_H[n; r]$, respectively.

account the main lobe flipping nature of the periodic sinc function when N is even. Thus, taking into account the above discussion and invoking (3.21), we obtain the following approximation for the AHT weighting function:

$$\hat{w}_H[n; r] = \cos\left[\frac{N-1}{N}\pi(n - r)\right] \cdot \text{sinc}(n - r) - (-1)^N \sin\left[\frac{N-1}{N}\pi(n + r)\right] \cdot \text{sinc}(n + r)$$

Additionally, for large N, a simpler approximation is derived:

$$\tilde{w}_H[n; r] = \cos\left[\pi(n - r)\right] \cdot \text{sinc}(n - r) + (-1)^N \sin\left[\pi(n + r)\right] \cdot \text{sinc}(n + r)$$

Figure 3.6 depicts the AHT weighting function and its approximations for $n = 4$ and $N = 8$; and $n = 64$ and $N = 256$. Figure 3.7 shows the AHT spectrum for random input signals for several blocklenghts. The displayed spectra were computed according to $w_H[n; r]$, $\hat{w}_H[n; r]$, and $\tilde{w}_H[n; r]$. Notice that the computation derived from $w_H[n; r]$ results in the *exact* DHT spectrum.

3.7.3 ACT INTERPOLATION

For the DCT, we have that (3.17) yields:

$$
\begin{aligned}
w_C[n; r] &= \frac{2}{N} \sum_{k=0}^{N-1} \alpha_k^2 \cos\left(\frac{\pi k(n + 1/2)}{N}\right) \cos\left(\frac{\pi k(r + 1/2)}{N}\right) \\
&= -\frac{1}{N} + \frac{2}{N} \sum_{k=0}^{N-1} \cos\left(\frac{\pi k(n + 1/2)}{N}\right) \cos\left(\frac{\pi k(r + 1/2)}{N}\right), \quad n = 0, 1, \ldots, N-1.
\end{aligned}
$$

(3.24)

Similar to the AFT and the AHT interpolation weighting functions, the ACT interpolation also furnishes normalized weights. In fact, summing over n in (3.24) leads

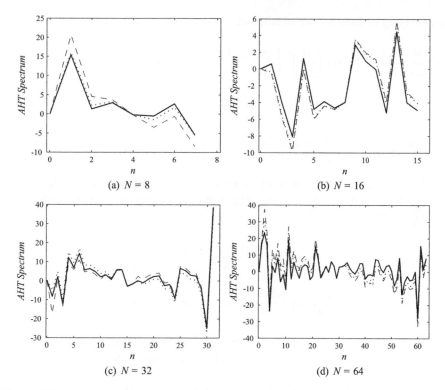

Figure 3.7 AHT computation of random real input signals for $N = 8, 16, 32, 64$. Solid, dotted, and dashed curves denote spectra computed according to $w_H[n; r]$ (exact spectrum), $\hat{w}_H[n; r]$, and $\tilde{w}_H[n; r]$, respectively.

to the following expression:

$$\sum_{n=0}^{N-1} w_C[n; r] = -1 + \frac{2}{N} \sum_{k=0}^{N-1} \cos\left(\frac{\pi k(r + 1/2)}{N}\right) \sum_{n=0}^{N-1} \cos\left(\frac{\pi k(n + 1/2)}{N}\right).$$

For each k, the inner summation of the above expression can be expanded as:

$$\sum_{n=0}^{N-1} \cos\left(\frac{\pi k(n + 1/2)}{N}\right) = \cos\left(\frac{\pi k}{2N}\right) \sum_{n=0}^{N-1} \cos\left(\frac{2\pi(k/2)n}{N}\right)$$
$$- \sin\left(\frac{\pi k}{2N}\right) \sum_{n=0}^{N-1} \sin\left(\frac{2\pi(k/2)n}{N}\right).$$

Since the summation index k runs from 0 to $N - 1$, we have that N never divides $k/2$.

Therefore, applying Lemma 3.1, we obtain that

$$\sum_{n=0}^{N-1} \cos\left(\frac{\pi k(n+1/2)}{N}\right) = \cos\left(\frac{\pi k}{2N}\right) \cdot \left\{ \begin{array}{ll} N, & \text{if } k = 0, \\ 0, & \text{otherwise} \end{array} \right\} = \left\{ \begin{array}{ll} N, & \text{if } k = 0, \\ 0, & \text{otherwise.} \end{array} \right.$$

Finally, returning to the previous double summation, we establish that

$$\sum_{n=0}^{N-1} w_C[n; r] = -1 + \frac{2}{N} \sum_{k=0}^{N-1} \cos\left(\frac{\pi k(r+1/2)}{N}\right) \cdot \left\{ \begin{array}{ll} N, & \text{if } k = 0, \\ 0, & \text{otherwise,} \end{array} \right\} = 1.$$

The ACT weighting function can be given a closed-form expression as detailed below. In fact, invoking the product-to-sum trigonometric identity, we can establish the following relations:

$$w_C[n; r] = -\frac{1}{N} + \frac{1}{N} \sum_{k=0}^{N-1} \cos\left(\frac{\pi k(n+r+1)}{N}\right)$$

$$+ \frac{1}{N} \sum_{k=0}^{N-1} \cos\left(\frac{\pi k(n-r)}{N}\right), \quad n = 0, 1, \ldots, N-1.$$

The above trigonometric summations can be given in terms of the Dirichlet function, since $\text{Diric}_{2N+1}(x) = \frac{1}{2N+1} \cdot \left[1 + 2 \sum_{k=1}^{N} \cos(kx)\right]$. Therefore, it holds that

$$w_C[n; r] = -\frac{1}{N} + \frac{1}{N} \left\{ \frac{1}{2} + \frac{2N-1}{2} \text{Diric}_{2N-1}\left[\frac{\pi}{N}(n+r+1)\right] \right\}$$

$$+ \frac{1}{N} \left\{ \frac{1}{2} + \frac{2N-1}{2} \text{Diric}_{2N-1}\left[\frac{\pi}{N}(n-r)\right] \right\}$$

$$= \frac{2N-1}{2N} \left\{ \text{Diric}_{2N-1}\left[\frac{\pi}{N}(n+r+1)\right] + \text{Diric}_{2N-1}\left[\frac{\pi}{N}(n-r)\right] \right\},$$

Regardless of the considered blocklength, the use of the ACT interpolation results in an exact calculation of the DCT spectrum. Indeed, no approximation was considered in any of the above derivations. Nevertheless, considering the approximation for the periodic sinc function in (3.21), the discussed weighting function is given by the sum of two sinc functions:

$$\hat{w}_C[n; r] = \text{sinc}(n+r+1) + \text{sinc}(n-r).$$

The approximate form $\hat{w}_C[n; r]$ can be further simplified. Notice that the argument of the first instantiation of the sinc function above assumes a minimum value of $1/2$. This is because $0 \leq n \leq N-1$ and $-1/2 \leq r \leq 2N-1/2$. Thus, we have $|\text{sinc}(n+r+1)| \leq 0.6366$, being the maximum absolute value occurring at $n+r = -1/2$. This can only happen if $n = 0$ and $r = -1/2$. As n increases, the magnitude of $\text{sinc}(n+r+1)$ decreases quickly. For $n = 1, 2, 3$, it is approximately equal to $-0.2122, 0.1273$, and -0.0909, respectively. These are the cases in which $\text{sinc}(n+r+1)$ assumes its larger

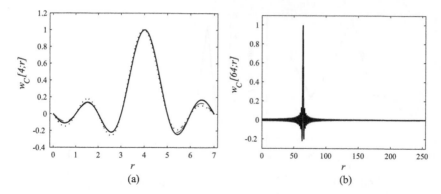

Figure 3.8 ACT weighting functions for (a) $n = 4$ and $N = 8$ and (b) $n = 64$ and $N = 256$. Solid, dotted, and dashed curves denote $w_C[n; r]$, $\hat{w}_C[n; r]$, and $\tilde{w}_C[n; r]$, respectively.

values over the considered range of n and r. Thus, for the vast majority of cases the contribution of $\text{sinc}(n + r + 1)$ to the weighting function is very small, and it could be neglected. Thus, under such conditions, we obtain the following approximation:

$$\tilde{w}_C[n; r] = \text{sinc}(n - r).$$

Figure 3.8 shows the exact and approximate weighting functions for the ACT for $n = 4$ and $N = 8$; and $n = 64$ and $N = 256$. Figure 3.9 displays the DCT computation of random input signals of different blocklengths according to the ACT considering $w_C[n; r]$, $\hat{w}_C[n; r]$, and $\tilde{w}_C[n; r]$. For large values of N, the ACT computation resulting from the approximate weighting functions are very close to the exact DCT spectrum.

3.8 DISCUSSION AND CONCLUSION

To obtain further insight on the nature of the proposed interpolation schemes, we generated an 8-point random input signal and submitted it to (3.17) for $r \in [0, 7]$. We considered the Fourier, Hartley, and cosine kernels. This operation can effectively construct a continuously interpolated signal associated to each transformation. Figure 3.10 shows the resulting interpolated signals. Clearly, the obtained signals are different. This means that the required values at non-integer sampling points are not necessarily available in the original physical/continuous input signal itself. Indeed, the continuous input signal from which the integer samples were originally obtained is not necessarily equal to any of the interpolated signals. On the contrary, it is expected that the input signal assumes totally different values, expect for the integer sampling points. However, methods based on over-sampling assume that the required non-integer samples are available at the original continuous input signals. Equation (3.17) shows that this is not the case.

Additionally, (3.17) shows that the arithmetic transform computation is deeply linked to the interpolation process. In fact, the interpolation dictates which transfor-

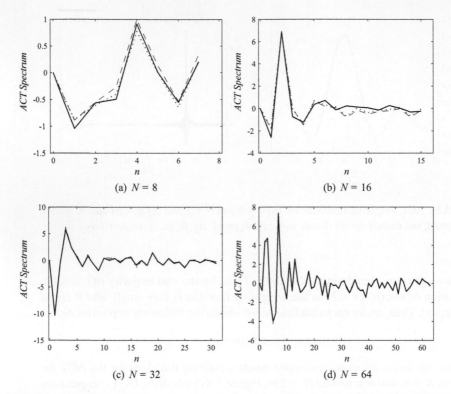

Figure 3.9 ACT computation of random real input signals for $N = 8, 16, 32, 64$. Solid, dotted, and dashed curves denote spectra computed according to $w_C[n; r]$ (exact spectrum), $\hat{w}_C[n; r]$, and $\tilde{w}_C[n; r]$, respectively.

mation is to be computed. Moreover, simple interpolation schemes aimed at approximating data at the non-integer sampling points may not be adequate. For example, early AFT schemes employed zero-order (round-off) interpolation to obtain the required samples [56]. By analyzing (3.22), we demonstrated that zero-order interpolation coincides with the real part of an approximation to the actual interpolation required by the AFT. Therefore, the zero-order interpolation can only compute an approximation to the real part of the DFT.

Figure 3.11 shows the DFT of a 32-point odd-symmetric random signal computed by means of the AFT according to: (i) the exact interpolation as described in (3.20) (solid curve) and (ii) the zero-order interpolation (dashed curve) (cf. (3.22)). Zero-order interpolation effected a null spectrum; totally missing the DFT computation. This particular issue puzzled the researchers in this area. Our contribution aimed at clarifying this particular point.

The arithmetic transform algorithm can be summarized in four major steps:

1. Non-integer index generation;

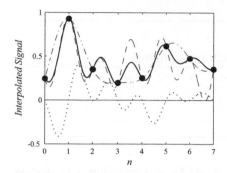

Figure 3.10 Solid, dotted, dashed, and dot-dashed curves denote the real part of the AFT interpolation, the imaginary part of the AFT interpolation, the AHT interpolation, and the ACT interpolation.

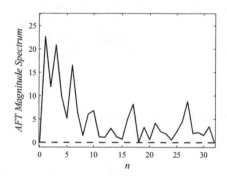

Figure 3.11 AFT of a 32-point odd-symmetric signal computed via exact interpolation (solid curve) and via zero-order interpolation (dashed curve).

2. Computation of the input signal at non-integer sampling points by means of interpolation based on the available integer index samples;
3. Computation of arithmetic averages;
4. Computation of spectrum by means of the generalized Möbius inversion formula.

Current work addressed all above steps. Nevertheless, arithmetic transform theory still leaves open issues to be examined. A still significant disadvantage of the arithmetic transforms is the large number of nonuniform samples. For instance, our computations indicate that the AFT and AHT require approximately $\frac{3}{10}N^2$ nonuniform samples, which is in agreement with the estimate of $\frac{1}{3}N^2$ nonuniform samples shown in [7]. The ACT sampling scheme demands approximately $\frac{3}{20}N^2$ nonuniform samples. Thus, the ACT has roughly half the computational complexity of the AFT/AHT. Table 3.3 displays the number of nonuniform samples required by the AFT/AHT and ACT for selected blocklengths.

Another issue is related to the complexity of the interpolation process. On one

Table 3.3
Number of required nonuniform samples

Method	8	16	24	32	64	128	256	512
AFT/AHT	18	72	172	308	1228	4958	19820	79596
ACT	10	37	87	155	615	2480	9911	39799

hand, the average computation and Möbius inversion steps possess very low computational complexity and fairly identical structures across the discussed transforms. On the other hand, although precisely defined, the interpolation schemes exhibit a noticeable computational complexity (cf. (3.14)). Two approaches could be employed to minimize the complexity of the interpolation step: (i) deriving low-complexity approximations tailored for a given tolerance of computation error and/or (ii) designing fast algorithms for the exact interpolation [44]. We understand that the research on arithmetic transforms may follow the above lines.

ACKNOWLEDGMENTS

This work was partially supported by CNPq and FACEPE.

REFERENCES

1. A. Abo Zaid, A. El-Mahdy, A. O. Attia, and M. M. Selim. A high speed classifier using the arithmetic Fourier transform. In *Proceedings of the 35th Midwest Symposium on Circuits and Systems*, pages 36–39, 1992.
2. G. Andria, V. Di Lecce, and A. Guerriero. An AFT-based virtual instrument for low-cost spectrum analysis. In *Proceedings of IMCT '96*, Brussels, 1996.
3. G. Andria, V. Di Lecce, and M. Savino. Application of the AFT technique for low-cost and accurate measurements. In *8th Mediterranean Electrotechnical Conference*, pages 1347–1350, Bari, Italy, May 1996.
4. T. M. Apostol. *Introduction to Analytic Number Theory*. Springer-Verlag, New York, 1984.
5. V. G. Atlas, D. G. Atlas, and E. I. Bovbel. 2-D arithmetic Fourier transform using the Bruns method. *IEEE Transactions on Circuits and Systems I: Fundamental Theory and Applications*, 44(6):546–551, June 1997.
6. G. Bachman, L. Narici, and E. Beckenstein. *Fourier and wavelet analysis*. Springer-Verlag, New York, 2000.
7. G. Faye Bartels-Boudreaux, Donald W. Tufts, P. Dhir, Angaraih G. Sadasiv, and G. Fischer. Analysis of errors in the computation of Fourier coefficients using the arithmetic Fourier transform (AFT) and summation by parts (SBP). In *Proceedings of the International Conference on Acoustics, Speech, and Signal Processing – ICASSP '89*, pages 1011–1014, Glasgow, U.K., May 1989.
8. P. T. Bateman and H. G. Diamond. *Analytic Number Theory*. World Scientific, New Jersey, 2004.

9. V. Britanak, P. Yip, and K. R. Rao. *Discrete cosine and sine transforms*. Academic Press, Amsterdam, 2007.

10. E. H. Bruns. *Grundlinien des wissenschaftlichnen Rechnens*. Druck und Verlag von B. G. Teubner, Leipzig, 1903.

11. Y. Y. Choi. *Algorithms for Computing the 2-D Arithmetic Fourier Transform*. PhD thesis, Department of Electrical Engineering – Systems, University of Southern California, Los Angeles, CA, 1989.

12. Y. Y. Choi, Irving S. Reed, and Ming-Tang Shih. The new arithmetical approach to Fourier series analysis for a 2-D signal. In *Proceedings of the International Conference on Acoustic, Speech, and Signal Processing – ICASSP '90*, pages 1997–2000, New Mexico, 1990.

13. R. J. Cintra and H. M. de Oliveira. How to interpolate in arithmetic transform algorithms. In *Proceedings of the International Conference on Acoustics, Speech and Signal Processing – ICASSP '02*, Orlando, Florida, May 2002.

14. R. J. Cintra and H. M. de Oliveira. Interpolating in arithmetic transform algorithms. In *6th WSEAS CSCC Multiconference, 2nd WSEAS International Conference on Signal Processing and Computational Geometry And Vision*, Crete Island, Greece, July 2002.

15. R. J. Cintra and V. S. Dimitrov. The arithmetic cosine transform: Exact and approximate algorithms. *Signal Processing, IEEE Transactions on*, 58(6):3076–3085, June 2010.

16. R. J. Cintra and H. M. Oliveira. A short survey on arithmetic transforms and the arithmetic Hartley transform. *Revista da Sociedade Brasileira de Telecomunicações*, 19:68–79, 2004.

17. V. Di Lecce and A. Guerriero. A FT processor based in short AFT module. In *Proceedings of International Symposium on Applied Informatics, IASTED*, Innsbruck, February 1995.

18. V. Di Lecce and A. Guerriero. Spectral estimation by AFT computation. *Digital Signal Processing*, 6(24):213–223, 1996.

19. S. R. Dooley and A. K. Nandi. Notes on the interpolation of discrete periodic signals using sinc function related approaches. *IEEE Transactions on Signal Processing*, 48(4):1201–1203, April 2000.

20. G. Fischer, D. W. Tufts, and A. G. Sadasiv. *VLSI Implementation of the Arithmetic Fourier Transform (AFT): A New Approach to High Speed Communication for Signal Processing*, chapter VLSI Signal Processing III, pages 264–275. IEEE Press, New York, 1989. R. Broderson and H. Moscovitz, Eds.

21. G. Fisher, D. W. Tufts, and R. Unnikrishnan. VLSI implementation of the arithmetic Fourier transform. In *Proceedings of the 32nd Midwest Symposium on Circuits and Systems*, volume 2, pages 800–803, Champaign, IL, August 1989.

22. X.-J. Ge, N.-X. Chen, and Z.-D. Chen. Efficient algorithm for 2-D arithmetic Fourier transform. *IEEE Transactions on Signal Processing*, 45(8):2136–2140, August 1997.

23. R. R. Goldberg and R. S. Varga. Moebius inversion of Fourier transforms. *Duke Mathematical Journal*, 24(4):553–560, December 1956.

24. G. H. Hardy, E. M. Wright, R. Heath-Brown, J. Silverman, and A. Wiles. *An Introduction to the Theory of Numbers*. Oxford University Press, New York, 2008.

25. C.-C. Hsu. *Use of Number Theory and Modern Algebra in the Reed-Solomon Code and the Arithmetic Fourier Transform*. PhD thesis, Department of Electrical Engineering – Systems, University of Southern California, Los Angeles, CA, August 1994.

26. C.-C. Hsu, I. S. Reed, and T. K. Truong. Inverse Z-transform by Möbius inversion and error bounds of aliasing in sampling. *IEEE Transactions on Signal Processing*,

42(10):2823–2831, October 1994.

27. B. T. Kelley and V. K. Madisetti. Efficient VLSI architectures for the arithmetic Fourier transform (AFT). *IEEE Transactions on Signal Processing*, 41(1):365–384, January 1993.

28. L. Knockaert. A generalized Möbius transform and arithmetic Fourier transforms. *IEEE Transactions on Signal Processing*, 42(11):2967–2971, November 1994.

29. L. Knockaert. A generalized Möbius transform, arithmetic Fourier transforms, and primitive roots. *IEEE Transactions on Signal Processing*, 44(5):1307–1310, May 1996.

30. T. I. Laakso, V. Valimaki, M. Karjalainen, and U. K. Laine. Splitting the unit delay [FIR/all pass filters design]. *IEEE Signal Processing Magazine*, 13(1):30–60, January 1996.

31. V. Laohakosol, P. Ruengsinsub, and N. Pabhapote. Ramanujan sums via generalized Möbius functions and applications. *International Journal of Mathematics and Mathematical Sciences*, 2006.

32. W. Li. Fourier analysis using adaptative AFT. In *Proceedings of International Conference on Acoustics, Speech, and Signal Processing – ICASSP '90*, pages 1523–1526, Albuquerque, NM, April 1990.

33. J. B. Lima, R. M. Campello de Souza, H. M. Oliveira, and M. M. Campello de Souza. *Faster DTMF Decoding*, volume 3124/2004 of *Lecture Notes in Computer Science*, chapter Telecommunications and Networking - ICT 2004, pages 510–515. Springer-Verlag Berlin / Heidelberg, 2004.

34. F. P. Lovine and S. Tantaratana. Some alternate realizations of the arithmetic Fourier transform. In *Conference Record of the Twenty-Seventh Asilomar Conference on Signals, Systems and Computers*, pages 310–314, Pacific Grove, November 1993.

35. Matlab Documentation Center. Dirichlet or periodic sinc function, June 2014.

36. A. V. Oppenheim, R. W. Schafer, and J. R. Buck. *Discrete-Time Signal Processing*. Prentice Hall, New Jersey, 1999.

37. H. Park and V. K. Prassana. VLSI architectures for computing the arithmetic Fourier transform. Technical report, Department of Electrical Engineering – Systems, University of Southern California, December 1990.

38. H. Park and V. K. Prassana. Fixed size array architectures for computing arithmetic Fourier transform. In *Conference Record of the Twenty-Fifth Asilomar Conference on Signals, Systems and Computers*, pages 85–89, November 1991.

39. H. Park and V. K. Prassana. Modular VLSI architectures for computing arithmetic Fourier transform. Technical report, IRIS #273, Department of Electrical Engineering – Systems, University of Southern California, August 1991.

40. H. Park and V. K. Prassana. VLSI architectures for computing the arithmetic Fourier transform. In *Proceedings of the International Conference on Acoustics, Speech, and Signal Processing – ICASSP '91*, pages 1029–1032, May 1991.

41. H. Park and V. K. Prassana. Modular VLSI architectures for computing arithmetic Fourier transform. *IEEE Transactions on Signal Processing*, 41(6):2236–2246, June 1993.

42. S.-C. Pei and K.-W. Chang. Odd Ramanujan sums of complex roots of unity. *IEEE Signal Processing Letters*, 14(1):20–23, January 2007.

43. A. D. Poularikas. *The Handbook of Formulas and Tables for Signal Processing*. CRC Press LLC, 1999.

44. N. Rajapaksha, A. Madanayake, R. J. Cintra, J. Adikari, and V. Dimitrov. VLSI computational architectures for the arithmetic cosine transform. *IEEE Transactions on Com-*

puters, 2014.

45. G. Ray and M. Chen. Analog to feature conversion. In *IEEE International Conference on Acoustics, Speech and Signal Processing*, volume 2, pages 365–368, Honolulu, HI, April 2007.

46. I. S. Reed, Y. Y. Choi, and X. Yu. Practical algorithm for computing the 2-D arithmetic Fourier transform. In *Proceedings of The International Society for Optical Engineering (SPIE)*, pages 54–61, January 1989.

47. I. S. Reed, M.-T. Shih, E. Hendon, T. K. Truong, and D. W. Tufts. A VLSI architecture for simplified arithmetic Fourier transform algorithm. In *International Conference on Application Specific Array Processors*, Special-Purpose Systems, pages 542–553, Princeton, U.S.A., September 1990.

48. I. S. Reed, M. T. Shih, T. K. Truong, E. Hendon, and D. W. Tufts. A VLSI architecture for simplified arithmetic Fourier transform algorithm. *IEEE Transactions on Signal Processing*, 40(5):1122–1133, May 1992.

49. I. S. Reed, D. W. Tufts, X. Yu, T. K. Truong, M. T. Shih, and X. Yin. Fourier analysis and signal processing by use of the Möbius inversion formula. *IEEE Transactions on Acoustics, Speech and Signal Processing*, ASSP-38(3):458–470, March 1990.

50. N. Roma and L. Sousa. Efficient hybrid DCT-domain algorithm for video spatial downscaling. *EURASIP Journal on Advances in Signal Processing*, 2007(2):30–30, 2007.

51. T. Schanze. Sinc interpolation of discrete periodic signals. *IEEE Transactions on Signal Processing*, 43(6):1502–1503, June 1995.

52. M. Schroeder. *Number Theory in Science and Communication: With Applications in Cryptography, Physics, Digital Information, Computing, and Self-similarity*. Springer-Verlag, New York, 2008.

53. E. J. Tan, Z. Ignjatovic, and M. F. Bocko. A CMOS image sensor with focal plane discrete cosine transform computation. In *IEEE International Symposium on Circuits and Systems*, pages 2395–2398, New Orleans, LA, May 2007.

54. D. W. Tufts and H. Chen. Iterative realization of the arithmetic Fourier transform. *IEEE Transactions on Signal Processing*, 41(1):152–161, January 1993.

55. D. W. Tufts, Z. Fan, and Z. Cao. Image processing and the arithmetic Fourier transform. In *SPIE High Speed Computing II*, volume 1058, pages 46–53, Los Angeles, CA, May 1989.

56. D. W. Tufts and G. Sadasiv. The arithmetic Fourier transform. *IEEE ASSP Magazine*, 5(1):13–17, 1988.

57. V. Välimäki and T. I. Laakso. *Nonuniform sampling: theory and practice*, chapter Fractional Delay Filters—Design and Applications. Springer-Verlag, New York, 2001.

58. N. Venkateswaran and K. Bharath. Frequency domain testing of general purpose processors at the instruction execution level. In *Second IEEE International Workshop on Electronic Design, Test and Applications*, pages 15–20, Perth, Australia, January 2004.

59. N. Wigley and G. A. Jullien. Sampling reduction for the arithmetic Fourier transform. In *Proceedings of 32nd Midwest Symposium on Circuits and Systems*, pages 841–844, 1990.

60. N. Wigley and G. A. Jullien. On implementing the arithmetic Fourier transform. *IEEE Transactions on Signal Processing*, 40(9):2233–2242, September 1992.

61. A. F. Wintner. An arithmetical approach to ordinary Fourier series. Monograph, 1945.

3.A APPENDIX: DIRICHLET INVERSE OF $\{(-1)^N\}$

In order to derive the Dirichlet inverse of the sequence $a_n = (-1)^n$, for $n = 1, 2, 3, \ldots$, let us examine its associated Dirichlet series. A result from the theory of functions [24, p. 337] states that

$$\sum_{n=1}^{\infty} \frac{(-1)^n}{n^s} = -(1 - 2^{1-s})\zeta(s), \quad \Re(s) > 0,$$

where $\zeta(s) \triangleq \sum_{n=1}^{\infty} 1/n^s$ is the Riemann zeta function. Therefore, we directly have that the closed form of Dirichlet series of $\{a_n\}$, $A(s)$, is

$$A(s) = -(1 - 2^{1-s})\zeta(s).$$

The Dirichlet inverse of $\{a_n\}$ is a sequence $\{b_n\}$ such that its Dirichlet series, $B(s) = \sum_{n=1}^{\infty} b_n/n^s$, is equal to $1/A(s)$. Thus, we can maintain that

$$B(s) = -\frac{1}{1 - 2^{1-s}} \frac{1}{\zeta(s)} = -\frac{1}{1 - 2^{1-s}} \sum_{n=1}^{\infty} \frac{\mu(n)}{n^s}.$$

Before finding $\{b_n\}$, we must identify the Dirichlet series of $1/(1 - 2^{1-s})$. This is necessary in order to put $B(s)$ as a product of two Dirichlet series and apply the convolution theorem for Dirichlet series. Accordingly, we have that

$$\frac{1}{1 - 2^{1-s}} = \sum_{k=0}^{\infty} (2^{1-s})^k = \sum_{k=0}^{\infty} \frac{2^k}{(2^k)^s}.$$

This final expression is already in the Dirichlet series format. Therefore, the sequence associated to $1/(1 - 2^{1-s})$ is simply

$$c_n = \begin{cases} n, & \text{if } n \text{ is a power of two}, \\ 0, & \text{otherwise}. \end{cases}$$

Returning to the expression for $B(s)$, we can write as

$$B(s) = -\sum_{n=1}^{\infty} \frac{c_n}{n^s} \sum_{n=1}^{\infty} \frac{\mu(n)}{n^s} = -\sum_{n=1}^{\infty} \frac{(c \circledast \mu)(n)}{n^s},$$

where $\circledast$ denotes the Dirichlet convolution. By the equivalence property of Dirichlet series, we conclude that $b_n = -(c \circledast \mu)(n)$.

Now let us evaluate $(c \circledast \mu)(n)$. Observe that if n is odd, then the only divisor of n, which is a power of two, is the unit. Therefore, we obtain

$$(c \circledast \mu)(n) = \sum_{d \mid n} c_d \mu(n/d) = c_1 \mu(n) = \mu(n).$$

On the other hand, let us assume that n is even in the form $n = 2^m s$, where s is an odd integer and m is a positive integer. Considering that (i) the Dirichlet convolution of two multiplicative functions results in a multiplicative function [4, p. 35] and (ii) the sequence $\{c_n\}$ is multiplicative with respect to n (a fairly direct result), we maintain that

$$(c \circledast \mu)(n) = (c \circledast \mu)(2^m) \cdot (c \circledast \mu)(s) = (c \circledast \mu)(2^m) \cdot \mu(s).$$

According to the definition of the Dirichlet convolution, the expansion of $(c \circledast \mu)(2^m)$ yields

$$(c \circledast \mu)(2^m) = c_1 \mu(2^m) + c_2 \mu(2^{m-1}) + \cdots + c_{2^{m-1}} \mu(2) + c_{2^m} \mu(1).$$

Due to the Möbius function, only the last two terms are possibly nonnull. Thus, we have

$$(c \circledast \mu)(2^m) = c_{2^{m-1}} \mu(2) + c_{2^m} \mu(1) = 2^{m-1}(-1) + 2^m(1) = 2^{m-1}.$$

Joining the above manipulations, we have that

$$(c \circledast \mu)(n) = \begin{cases} \mu(n), & \text{if } n \text{ is odd,} \\ 2^{m-1} \mu(2^{-m}n), & \text{if } n = 2^m s, \text{ where } s \text{ is odd.} \end{cases}$$

4 Distributed Particle Filtering in Cooperative Networks

MARCELO GOMES DA SILVA BRUNO
Aeronautics Institute of Technology (ITA)

CLAUDIO JOSÉ BORDIN JÚNIOR
Federal University of ABC (UFABC)

STIVEN SCHWANZ DIAS
Embraer Defense and Security

4.1 INTRODUCTION

In several modern engineering applications, multiple agents [20] that are physically dispersed over remote nodes of a sensor or receiver network cooperate to execute a common global task without the need to rely on a global data fusion center. Examples of such applications include target localization and tracking by sensor networks, cooperative equalization of digital communication channels using multiple receivers, and distributed fault detection in industrial plants and critical infrastructure. In most scenarios, the different network nodes have access to measurements generated by their own local sensors and are capable of processing such local data independently of each other. At the same time, however, network nodes can communicate with each other via some form of message passing in order to build, in a collaborative fashion, a joint estimate of a hidden signal or parameter of interest which needs to be inferred from the available network data. Ideally, that collaborative estimate should be equal to, or at least approximate the optimal global estimate that would be generated by a data fusion center that received the measurements from all network nodes and jointly processed them. Compared to the centralized approach, a distributed architecture with in-network data processing has the advantage of robustness and flexibility since there is no risk of a system collapse due to a catastrophic failure at the data fusion center and the network can be adaptively reconfigured if a particular local node suddenly becomes unavailable.

Most of the existing literature in distributed signal processing is based on linear estimation techniques. Some references, e.g., [33] and [11], use distributed linear adaptive filters to obtain the linear least-squares (LLS) or the linear minimum-mean-square-error (LMMSE) estimates of a time-invariant parameter vector using multiple noisy and linearly distorted observations of that same hidden vector that are dispersed over a sensor or receiver network. Other references suggest tracking a time-variant state vector associated with a known linear dynamic model using a distributed Kalman filter, see, e.g., [39], [41], [12], [30].

In scenarios, however, where the observation and/or the state models are nonlinear, and/or the states and observations are not jointly Gaussian, the LMMSE estimate obtained by distributed linear Kalman filters may differ significantly from the (generally nonlinear) global minimum-mean-square error (MMSE) or the global maximum a posteriori (MAP) estimates of the hidden states given the observations. A possible alternative to the linear approach in such scenarios is to use a distributed particle filter (PF) [27]. Particle filters are basically sequential Monte Carlo methods [22], [9] that seek to represent the posterior distribution of the hidden states conditioned on the observations at each time instant by a properly weighted set of random samples, also called particles, from which desired MMSE or MAP estimates can be computed with asymptotic convergence as the number of samples goes to infinity.

In particular, in a distributed Sequential Monte Carlo framework [27], [13], [45], each node in the network runs a local particle filter. Cooperation via internode message passing then allows the different particle filters at each node to generate a weighted sequence of samples which, at each instant k, is a valid Monte Carlo representation ideally for *global* posterior distribution of the hidden state s_k conditioned on the observations at *all* nodes of the network from instant zero up to the present. In an ideal situation, the sampled particles and their respective weights should be identical at all network nodes such that all local node particle filters yield the same global estimate of the hidden state. That latter condition, however, has been relaxed in several recent algorithms found in the literature, e.g., in [27], [4], [28], [29], [15], [16].

Distributed Filtering Strategies We discuss in this chapter different original distributed particle filtering algorithms that operate in networks that are subdivided into two classes, namely: *fully connected networks*, where any chosen node has a point-to-point (single-hop) connection to any other node in the network, and *partially connected networks*, where each node only has direct, single-hop connections to a subset of network nodes in its immediate neighborhood, but there is a multi-hop path between any two nodes.

Given power restrictions, most networks of practical interest that operate with wireless communication channels are only partially connected. The research challenge in distributed filtering and estimation is precisely to develop decentralized algorithms that operate on partially connected networks, but are able to reproduce, either exactly or approximately, the global estimate that would be generated by a data fusion center, and, at the same time, have low internode communication cost.

Different distributed filtering algorithms have been recently proposed in the literature, see, e.g., [28], [29], [16], [23], [44], [5], [34], [35], [6], to obtain samples and global weights in a fully distributed manner, dispensing with the need for message broadcast beyond the immediate neighborhood of any given physical node. Most of those algorithms are based on *consensus* methods [31], which have the advantage of generating the same state estimate at all nodes at each time instant and, furthermore, in some cases, see, e.g., [16], are capable of exactly reproducing the optimal centralized global estimate. The drawback, however, of consensus methods is that they require multiple, iterative internode communication in the time interval between the arrival of two consecutive sensor measurements, implying therefore a communication cost that is normally prohibitive in real-time applications. Several additional suboptimal approximations have been proposed in the literature to reduce the internode communication cost associated with consensus-based distributed particle filters, see, e.g., [4], [28], [6], but all proposed solutions still require sensing and processing (filtering) at different time scales.

An alternative approach to reduce communication cost and enable real-time distributed processing is to use diffusion algorithms [40], which, contrary to consensus methods, do not require multiple iterative internode communication between consecutive sensor measurements. Unlike iterative consensus algorithms, however, diffusion methods are often suboptimal in the sense that they do not converge at each time k to the centralized global state (or parameter) estimate, but rather, at best, only asymptotically approximate the centralized solution as k goes to infinity. Moreover, unlike consensus algorithms, diffusion algorithms generate different estimates at each node at any given instant k, although, ideally, it is desirable that those estimates be identically distributed at all nodes when k goes to infinity (a condition that is referred to in the literature as *weak consensus* [30]). In the context of sequential Monte Carlo methods, distributed particle filters incorporating diffusion techniques were introduced in references [17], [18], [19].

Overview of the Chapter The chapter has six sections and three appendices. Section 4.1 is this introduction. In Section 4.2, we discuss in greater detail the problem of centralized particle filtering with multiple observations that are forwarded to a global data fusion center. In Section 4.3, we introduce distributed implementations of the centralized filter in Section 4.2, assuming respectively fully and partially connected networks. In the case of partially connected networks, we propose fully distributed solutions using both consensus and diffusion strategies. Among consensus solutions, we review methods based respectively on iterative average consensus [46], iterative minimum consensus [16], [47], and flooding [16], [43]. In the class of diffusion methods, we follow a random information dissemination strategy [30] and introduce a general formulation of the Random Exchange Diffusion Particle Filter (ReDif-PF), which had been previously introduced in [17], [18], [19] for a specific signal model. Sections 4.4 and 4.5 provide illustrative applications of the algorithms discussed in Section 4.3 to selected distributed filtering problems of practical interest, namely target tracking in networks of passive sensors and cooperative equalization in receiver networks. Finally, Section 4.6 offers some conclusions, and Appendices 4.A and 4.B present proofs of some key results in the chapter.

4.2 COOPERATIVE PARTICLE FILTERING WITH MULTIPLE OB-SERVERS

In this section, we examine in detail the problem where a single sequence of hidden random vectors $\{s_k\}$, $k \geq 0$, is tracked by a network of R observers, where each observer has built-in sensors capable of recording a set of local measurements $\{z_{k,r}\}$, $k \geq 0$, $r = 1, \ldots, R$. The set of all observations in all network nodes from instant 0 up to instant k is denoted $z_{0:k,1:R}$. Similarly, $s_{0:k}$ denotes the set $\{s_n\}$, $n = 0, \ldots, k$. The probability density function (p.d.f.) of a continuous random vector is denoted $p(.)$ and the probability mass function (p.m.f.) of a discrete random vector is denoted $P(.)$. For simplicity of notation, we use lowercase letters to denote both random variables and realizations (samples) of random variables; the appropriate interpretation is implied in context. In the remainder of the chapter, unless otherwise noted, all state vectors, observations and parameters of interest are assumed continuous and real valued. We denote by N_s the dimension of the random states s_k. Similarly, N_z denotes the dimension of the observation vectors $z_{k,r}$, which, without loss of generality, is assumed to be identical for all nodes r.

In a centralized particle filter architecture, a global data fusion center jointly processes all network measurements $\{z_{k,r}\}$, $r \in \{1, \ldots, R\}$, available at each instant $k \geq 0$, and recursively generates a sequence of samples, also known as particles, $\{s_k^{(j)}\}$, $j = 1, \ldots, N_p$, drawn according to a proposal probability distribution specified by a so-called importance function $q(s_k|s_{0:k-1}, z_{0:k,1:R})$. Each particle $s_k^{(j)}$ is in turn assigned at instant k an importance weight

$$w_k^{(j)} \propto \frac{p(s_{0:k}^{(j)}|z_{0:k,1:R})}{\prod_{n=1}^{k} q(s_n^{(j)}|s_{0:n-1}^{(j)}, z_{0:n,1:R})}$$

such that $\sum_{j=1}^{N_p} w_k^{(j)} = 1$ and $\{(w_k^{(j)}, s_{0:k}^{(j)})\}$ is a properly weighted set in the Monte Carlo sense [22], [9] to represent the posterior distribution of the state sequence $s_{0:k}$ given the observations $z_{0:k,1:R}$. In other words, for any measurable function $\Omega(.)$,

$$\sum_{j=1}^{N_p} w_k^{(j)} \Omega(s_{0:k}^{(j)}) \xrightarrow[N_p \to \infty]{} E\{\Omega(s_{0:k})|z_{0:k,1:R}\} \tag{4.1}$$

where $E\{.|.\}$ in (4.1) denotes conditional expectation and the convergence in (4.1) is according to some statistical criterion, e.g., almost sure (a.s.) or mean square (m.s.) convergence, see details in [22], [14].

4.2.1 OPTIMAL GLOBAL IMPORTANCE FUNCTION

The optimal importance function, which minimizes the variance of the importance weights conditioned on the observations and the simulated particle trajectories is given by [22]

$$q(s_k \mid s_{0:k-1}^{(j)}, z_{0:k,1:R}) = p(s_k \mid s_{0:k-1}^{(j)}, z_{0:k,1:R}). \tag{4.2}$$

The corresponding proper importance weights are then recursively updated by the fusion center by making [9]

$$
w_k^{(j)} \propto w_{k-1}^{(j)} \frac{p(\mathbf{z}_{k,1:R}|\mathbf{s}_k, \mathbf{s}_{0:k-1}^{(j)}, \mathbf{z}_{0:k-1,1:R}) p(\mathbf{s}_k|\mathbf{s}_{0:k-1}^{(j)}, \mathbf{z}_{0:k-1,1:R})}{q(\mathbf{s}_k^{(j)}|\mathbf{s}_{0:k-1}^{(j)}, \mathbf{z}_{0:k,1:R})} \tag{4.3}
$$

$$
= w_{k-1}^{(j)} p(\mathbf{z}_{k,1:R}|\mathbf{s}_{0:k-1}^{(j)}, \mathbf{z}_{0:k-1,1:R}) \tag{4.4}
$$

$$
= w_{k-1}^{(j)} \int_{\mathfrak{R}^{N_s}} p(\mathbf{z}_{k,1:R}|\mathbf{s}_k, \mathbf{s}_{0:k-1}^{(j)}, \mathbf{z}_{0:k-1,1:R})
$$

$$
\cdot p(\mathbf{s}_k|\mathbf{s}_{0:k-1}^{(j)}, \mathbf{z}_{0:k-1,1:R}) \, d\mathbf{s}_k . \tag{4.5}
$$

where the proportionality constant in (4.3) is obtained such that $\sum_j w_k^{(j)} = 1$. It can be shown [22], [25] that, when $N_p \to \infty$, the sum $\sum_j w_k^{(j)} \mathbf{s}_k^{(j)}$ converges almost surely to the global MMSE estimate $\hat{\mathbf{s}}_{k|k} = E\{\mathbf{s}_k|\mathbf{z}_{0:k,1:R}\}$.

Filtering with Unknown Observation Model Parameters Let $\theta_{1:R} = \{\theta_1, \theta_2, \ldots, \theta_R\}$, be a set of unknown, random parameter vectors, which are assumed to take values in $\mathfrak{R}^{N_\theta}$ and specify the conditional distribution of the network observations $\mathbf{z}_{0:k,1:R}$ given the state sequence $\mathbf{s}_{0:k}$. In most applications of interest, the local observations $\mathbf{z}_{0:k,r}$, $r \in \{1, \ldots, R\}$, are assumed conditionally independent given $\theta_{1:R}$ and $\mathbf{s}_{0:k}$, i.e.,

$$
p(\mathbf{z}_{0:k,1:R}|\theta_{1:R}, \mathbf{s}_{0:k}) = \prod_{r=1}^{R} p(\mathbf{z}_{0:k,r}|\theta_r, \mathbf{s}_{0:k}) \tag{4.6}
$$

where $p(.|.)$ denotes a conditional p.d.f. On the other hand, assuming the parameter vectors θ_r, $r \in \{1, \ldots, R\}$, to be a priori independent of each other and also independent of the state sequence $\mathbf{s}_{0:k}$, it follows that

$$
p(\theta_{1:R}|\mathbf{s}_{0:k}) = \prod_{r=1}^{R} p(\theta_r|\mathbf{s}_{0:k}) = \prod_{r=1}^{R} p(\theta_r) . \tag{4.7}
$$

Exploiting conditional independence relations induced by (4.6) and (4.7), it can be verified that [9]

$$
p(\mathbf{z}_{k,1:R}|\mathbf{s}_{0:k}^{(j)}, \mathbf{z}_{0:k-1,1:R}) = \prod_{r=1}^{R} p(\mathbf{z}_{k,r}|\mathbf{s}_{0:k}^{(j)}, \mathbf{z}_{0:k-1,r}). \tag{4.8}
$$

In scenarios where the usual Markovian assumptions also apply, it also follows that

$$
p(\mathbf{z}_{k,r}|\mathbf{s}_{0:k}, \theta_r, \mathbf{z}_{0:k-1,r}) = p(\mathbf{z}_{k,r}|\mathbf{s}_k, \theta_r) \tag{4.9}
$$

and

$$
p(\mathbf{s}_k|\mathbf{s}_{0:k-1}, \mathbf{z}_{0:k-1,1:R}) = p(\mathbf{s}_k|\mathbf{s}_{k-1}) . \tag{4.10}
$$

Using now (4.8) and (4.10), the optimal importance function in (4.2) is rewritten as

$$
\begin{aligned}
p(\mathbf{s}_k|\mathbf{s}_{0:k-1}^{(j)}, \mathbf{z}_{0:k,1:R}) = \\
= \frac{p(\mathbf{z}_{k,1:R}|\mathbf{s}_k, \mathbf{s}_{0:k-1}^{(j)}, \mathbf{z}_{0:k-1,1:R}) \, p(\mathbf{s}_k|\mathbf{s}_{0:k-1}^{(j)}, \mathbf{z}_{0:k-1,1:R})}{\int_{\Re^{N_s}} p(\mathbf{z}_{k,1:R}|\mathbf{s}_k', \mathbf{s}_{0:k-1}^{(j)}, \mathbf{z}_{0:k-1,1:R}) \, p(\mathbf{s}_k'|\mathbf{s}_{0:k-1}^{(j)}, \mathbf{z}_{0:k-1,1:R}) d\mathbf{s}_k'} \\
= \frac{\left[\prod_{r=1}^{R} \lambda_{k,r}^{(j)}(\mathbf{s}_k)\right] p(\mathbf{s}_k|\mathbf{s}_{k-1}^{(j)})}{\int_{\Re^{N_s}} \left[\prod_{r'=1}^{R} \lambda_{k,r'}^{(j)}(\mathbf{s}_k')\right] p(\mathbf{s}_k'|\mathbf{s}_{k-1}^{(j)}) d\mathbf{s}_k'}
\end{aligned}
\tag{4.11}
$$

where, also using (4.9),

$$
\begin{aligned}
\lambda_{k,r}^{(j)}(\mathbf{s}_k) &\triangleq p(\mathbf{z}_{k,r}|\mathbf{s}_k, \mathbf{s}_{0:k-1}^{(j)}, \mathbf{z}_{0:k-1,r}) \\
&= \int_{\Re^{N_\theta}} p(\mathbf{z}_{k,r}|\mathbf{s}_k, \boldsymbol{\theta}_r) \, p(\boldsymbol{\theta}_r|\mathbf{s}_k, \mathbf{s}_{0:k-1}^{(j)}, \mathbf{z}_{0:k-1,r}) d\boldsymbol{\theta}_r.
\end{aligned}
\tag{4.12}
$$

Similarly, plugging (4.8) into (4.5), the weight update rule becomes

$$
w_k^{(j)} \propto w_{k-1}^{(j)} \int_{\Re^{N_s}} \prod_{r=1}^{R} \left[\lambda_{k,r}^{(j)}(\mathbf{s}_k)\right] p(\mathbf{s}_k|\mathbf{s}_{k-1}^{(j)}) \, d\mathbf{s}_k.
\tag{4.13}
$$

In the particular case where the hidden states are discrete valued, but the observation vectors are continuous, Equations (4.11) and (4.13) are changed to [5]

$$
P(\mathbf{s}_k|\mathbf{s}_{0:k-1}^{(j)}, \mathbf{z}_{0:k,1:R}) = \frac{\left[\prod_{r=1}^{R} \lambda_{k,r}^{(j)}(\mathbf{s}_k)\right] P(\mathbf{s}_k|\mathbf{s}_{k-1}^{(j)})}{\sum_{\mathbf{s}_k'} \prod_{r'=1}^{R} \left[\lambda_{k,r'}^{(j)}(\mathbf{s}_k')\right] P(\mathbf{s}_k'|\mathbf{s}_{0:k-1}^{(j)})}
\tag{4.14}
$$

$$
w_k^{(j)} \propto w_{k-1}^{(j)} \sum_{\mathbf{s}_k} \prod_{r=1}^{R} \left[\lambda_{k,r}^{(j)}(\mathbf{s}_k)\right] P(\mathbf{s}_k|\mathbf{s}_{k-1}^{(j)}).
\tag{4.15}
$$

4.2.2 BLIND IMPORTANCE FUNCTION

In situations where it is not possible to compute (4.11) and (4.13) in closed form, a possible alternative is to use the simpler, but suboptimal blind importance function [22]

$$
q(\mathbf{s}_k|\mathbf{s}_{0:k-1}, \mathbf{z}_{0:k,1:R}) = p(\mathbf{s}_k|\mathbf{s}_{0:k-1}, \mathbf{z}_{0:k-1,1:R})
\tag{4.16}
$$

which does not depend on the current observations $\mathbf{z}_{k,1:R}$ at instant k. Assuming a Markovian framework, the blind importance function in (4.16) is simplified further by noting that

$$
p(\mathbf{s}_k|\mathbf{s}_{0:k-1}, \mathbf{z}_{0:k-1,1:R}) = p(\mathbf{s}_k|\mathbf{s}_{k-1}).
$$

Specifically, in a centralized architecture, the global data fusion center draws at each instant k a new set of particles $\{\mathbf{s}_k^{(j)}\}$ such that

$$
\mathbf{s}_k^{(j)} \sim p(\mathbf{s}_k|\mathbf{s}_{k-1}^{(j)}) \qquad j = 1, \ldots, N_p
$$

and the corresponding global importance weights are updated as [22], [9]

$$
\begin{aligned}
w_k^{(j)} &\propto w_{k-1}^{(j)} \; p(\mathbf{z}_{k,1:R}|\mathbf{s}_k^{(j)}, \mathbf{s}_{0:k-1}^{(j)}, \mathbf{z}_{0:k-1,1:R}) \\
&= w_{k-1}^{(j)} \prod_{r=1}^{R} \underbrace{p(\mathbf{z}_{k,r}|\mathbf{s}_k^{(j)}, \mathbf{s}_{0:k-1}^{(j)}, \mathbf{z}_{0:k-1,r})}_{\lambda_{k,r}^{(j)}(\mathbf{s}_k^{(j)})}.
\end{aligned}
\tag{4.17}
$$

4.3 DISTRIBUTED PARTICLE FILTERS

In scenarios where a data fusion center is unavailable or to be avoided, an alternative to the implementation described in Section 4.2 is to use a decentralized architecture where processing of the sensor measurements is distributed over a network described by a graph $G = (\mathcal{R}, \mathcal{E})$, where $\mathcal{R} = \{1, \dots, R\}$ is the set of network nodes and the graph has an edge $(u, v) \in \mathcal{E}$, $(u, v) \in \mathcal{R} \times \mathcal{R}$ if and only if nodes u and v can communicate directly with each other. For simplicity, we assume G to be time-invariant, although that is not required in the derivation of our algorithms.

Note that the terms $\lambda_{k,r}^{(j)}(.)$ in Eq. (4.17), also called *likelihoods*, depend only on local observations and can be computed independently at each network node. In a *fully connected* network where any pair of nodes (u, v) can communicate directly with each other, a straightforward decentralized algorithm that is mathematically equivalent to the global centralized particle filter using the blind importance function can be implemented if all nodes $r \in \mathcal{R}$ *synchronously* draw samples [13] from $p(\mathbf{s}_k|\mathbf{s}_{k-1}^{(j)})$ at instant k so that they all have an identical set of new particles $\{\mathbf{s}_k^{(j)}\}$, $j \in \{1, \dots, N_p\}$ at instant k given an identical set of particles $\{\mathbf{s}_{k-1}^{(j)}\}$ at instant $k-1$. That can be accomplished by sampling from identical pseudo-random generators initialized with the same seed in the same order. Each node then independently computes its local weight update factor $\lambda_{k,r}^{(j)}(\mathbf{s}_k^{(j)})$ and broadcasts it to the remaining nodes in the network. Once all nodes have received all local update factors from the remote nodes, they compute the product in (4.17) to obtain the same global weight update factor throughout the network. Synchronized multinomial resampling according to the global weights [22], [9] or regularization [16] may follow at each node to mitigate particle degeneracy [22].

The decentralized algorithm described in the previous paragraph is referred to as the *Decentralized Particle Filter* (DcPF) in references [15], [16], [2]. DcPF's obvious limitations, however, are that, first, it requires a fully connected network and, second, it has a high internode communication cost since each node has to transmit and receive N_p real numbers per iteration of the algorithm, where N_p (the number of particles) is typically in the order of hundreds of samples to guarantee the asymptotic convergence of the weighted average of the particles to the global MMSE state estimate [14].

As mentioned, however, in Section 4.1, most networks of practical interest in engineering applications are only *partially connected* and alternative *fully distributed* algorithms are required assuming direct communication between nodes only on local neighborhood regions specified by the network graph.

4.3.1   CONSENSUS ALGORITHMS

A possible, albeit rather costly, alternative to implement (4.11)-(4.13) or (4.17) in a fully distributed fashion with local neighborhood communication only is to use iterative average consensus [46]. Specifically, defining

$$\Lambda_{k,r}^{(j)}(\mathbf{s}_k) = \log_e\left(\lambda_{k,r}^{(j)}(\mathbf{s}_k)\right),\tag{4.18}$$

it follows that the product over all the nodes r on the right-hand side of Equations (4.11), (4.13), and (4.17) is given by

$$\prod_{r=1}^{R}\lambda_{k,r}^{(j)}(\mathbf{s}_k) = \exp\left[\sum_{r=1}^{R}\Lambda_{k,r}^{(j)}(\mathbf{s}_k)\right].\tag{4.19}$$

The sum on the right-hand side of Eq. (4.19) can be computed then by iterative average consensus by introducing an auxiliary variable $\tilde{\Lambda}_{k,r}^{(l,j)}(\mathbf{s}_k)$ such that

$$\tilde{\Lambda}_{k,r}^{(0,j)}(\mathbf{s}_k) = R\Lambda_{k,r}^{(j)}(\mathbf{s}_k)\tag{4.20}$$

and

$$\tilde{\Lambda}_{k,r}^{(l,j)}(\mathbf{s}_k) = \tilde{\Lambda}_{k,r}^{(l-1,j)}(\mathbf{s}_k)+\\ \sum_{n\in\mathbf{N}(r)} a_{r,n}\left(\tilde{\Lambda}_{k,n}^{(l-1,j)}(\mathbf{s}_k) - \tilde{\Lambda}_{k,r}^{(l-1,j)}(\mathbf{s}_k)\right).\tag{4.21}$$

In Eq. (4.21), $\mathbf{N}(r)$ denotes the neighborhood of node r, the superscript l is the consensus algorithm iteration index, and $a_{r,n}$ are real coefficients such that $a_{r,n} \geq 0$, $\forall\,(r,n)$, $a_{r,n} = a_{n,r}$, and $\sum_{n\in\mathbf{N}(r)} a_{r,n} < 1$. Stacking up the terms $\tilde{\Lambda}_{k,r}^{(l,j)}$, $r \in \{1, 2, \ldots, R\}$, in a long vector $\tilde{\mathbf{\Lambda}}_{k}^{(l,j)}$ of dimension $R \times 1$, we can rewrite Eq. (4.21) in compact matrix notation as

$$\tilde{\mathbf{\Lambda}}_{k}^{(l,j)}(\mathbf{s}_k) = \mathbf{A}\,\tilde{\mathbf{\Lambda}}_{k}^{(l-1,j)}(\mathbf{s}_k)\tag{4.22}$$

where $\mathbf{A}$ is, by construction, a doubly stochastic matrix [36] of dimension $R \times R$ whose rows and columns both add up to 1. If the coefficients $a_{r,n}$ are additionally chosen such that $\mathbf{A}$ is a primitive matrix [36], then $\mathbf{A}$ will have only one single eigenvalue equal to one, with associated eigenvectors in the linear subspace spanned by $[1\ 1\ \ldots 1]^T$, and all other eigenvalues with magnitude less than one. As $l \to \infty$, the eigenvalues of $\mathbf{A}^l$ with magnitude less than one decay exponentially and $\mathbf{A}^l$ converges [36] to a rank-one matrix whose entries are all equal to $1/R$. It follows then that, with the initialization in (4.20),

$$\lim_{l\to\infty}\tilde{\Lambda}_{k,r}^{(l,j)}(\mathbf{s}_k) = \sum_{r=1}^{R}\Lambda_{k,r}^{(j)}(\mathbf{s}_k)\qquad\forall r.\tag{4.23}$$

However, in actual implementations of the algorithm, where the number l of consensus iterations is finite,

$$\tilde{\Lambda}_{k,r}^{(l,j)}(\mathbf{s}_k) - \sum_{r=1}^{R} \Lambda_{k,r}^{(j)}(\mathbf{s}_k) \neq 0.$$

To ensure then that all nodes have the same weight update factor, some form of quantization step has to be added to the algorithm. For a detailed description of one possible quantization scheme, see [5].

Exact Reproduction of the Centralized Solution using Iterative Algorithms To circumvent the limitations of the average consensus approach, reference [16] computes the product on the right-hand side of Eq. (4.17) exactly in a *finite* number of iterations with no need for additional quantization using iterative *minimum consensus* [47]. Given a set of auxiliary variables ι_r^0 available at remote nodes $r \in \{1, \ldots, R\}$ of a network, iterative minimum consensus computes the global minimum of ι_r^0 over all r's using the recursion

$$\iota_r^{(l+1)} = \min_{i \in \{r\} \cup \mathbf{N}(r)} \iota_i^{(l)} \qquad l \geq 0. \tag{4.24}$$

Denoting by D the diameter of the network graph, i.e., the maximum distance in hops between any two nodes, the auxiliary variable $\iota_r^{(l)}$ in (4.24) is guaranteed to converge to $\min_i \iota_i^0$, $\forall r \in \{1, \ldots, R\}$, in at most D iterations [47].

Reference [16] specifically proposes to run first an iterative minimum consensus protocol to find the global minimum of $\{\lambda_{k,r}^{(j)}(.)\}$ over all nodes r. That global minimum is stored in all nodes and a second iterative minimum consensus protocol is subsequently run with the global minimum replaced in its node of origin by the maximum real number that can be represented in the computer. At the end of the second min consensus protocol, all nodes will have stored then the second smallest $\lambda_{k,r}^{(j)}(.)$. By repeating the aforementioned process R consecutive times, one can build an ordered list of all likelihood coefficients $\{\lambda_{k,r}^{(j)}(.)\}$, $r \in \{1, \ldots, R\}$, which is identical in all nodes. Each node can then locally compute the product of the likelihoods as in Eq. (4.17) and obtain identical, optimal global importance weights $\{w_k^{(j)}\}$. Since each minimum consensus protocol ends, as explained before, in at most D iterations, the maximum number of iteration steps between sensor measurements required for exact computation of the product in (4.17) for each particle label j is equal to $R \times D$. In reference [16], the distributed particle filter incorporating the exact minimum consensus computation of the global importance weights is referred to as CbPFa.

Alternatively, a more efficient way to compute the exact optimal global weights is to use a *flooding* technique [43]. Given a partially connected sensor network, one can simultaneously flood the R distinct likelihoods over the network as follows. First of all, each node r maintains an ordered list of distinct likelihoods. A likelihood in turn is flagged in case it has not been sent yet to some neighbor of node r. Initially, node r stores its local flagged likelihood in its list. At a given iteration, node r sends its lowest flagged likelihood to all neighbors and then unflags it. Conversely, it receives remote likelihoods from nodes $t \in \mathbf{N}(r)$. If a received remote likelihood is not included in node r's list yet, it is inserted with a flag in its list. This procedure is guaranteed to converge in a finite number of iterations as soon as each node has R distinct values in its ordered list of likelihoods. The distributed particle filter incorporating the flooding strategy is referred to as CbPFb in reference [16].

Figure 4.1 illustrates how the flooding strategy iteratively creates an identical ordered list of likelihood functions at each node of a simple toy network with 3 nodes, denoted respectively as N_1, N_2, and N_3, where node N_1 is connected to N_2, node N_2 is connected to N_1 and N_3, and node N_3 is connected only to N_2. Stars are used in the figure to indicate the flagged likelihoods at each iteration of the algorithm. The iteration index is denoted by l and we remind the reader that the flooding iterations l take place between two consecutive sensor measurement arrival instants k and $k + 1$.

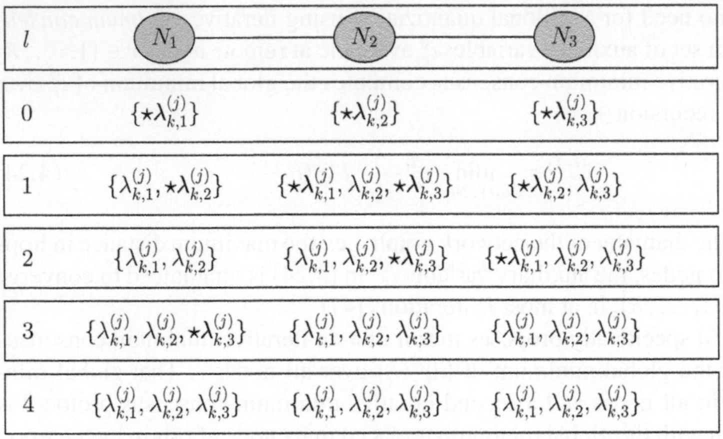

l	N_1	N_2	N_3
0	$\{\star\lambda_{k,1}^{(j)}\}$	$\{\star\lambda_{k,2}^{(j)}\}$	$\{\star\lambda_{k,3}^{(j)}\}$
1	$\{\lambda_{k,1}^{(j)}, \star\lambda_{k,2}^{(j)}\}$	$\{\star\lambda_{k,1}^{(j)}, \lambda_{k,2}^{(j)}, \star\lambda_{k,3}^{(j)}\}$	$\{\star\lambda_{k,2}^{(j)}, \lambda_{k,3}^{(j)}\}$
2	$\{\lambda_{k,1}^{(j)}, \lambda_{k,2}^{(j)}\}$	$\{\lambda_{k,1}^{(j)}, \lambda_{k,2}^{(j)}, \star\lambda_{k,3}^{(j)}\}$	$\{\star\lambda_{k,1}^{(j)}, \lambda_{k,2}^{(j)}, \lambda_{k,3}^{(j)}\}$
3	$\{\lambda_{k,1}^{(j)}, \lambda_{k,2}^{(j)}, \star\lambda_{k,3}^{(j)}\}$	$\{\lambda_{k,1}^{(j)}, \lambda_{k,2}^{(j)}, \lambda_{k,3}^{(j)}\}$	$\{\lambda_{k,1}^{(j)}, \lambda_{k,2}^{(j)}, \lambda_{k,3}^{(j)}\}$
4	$\{\lambda_{k,1}^{(j)}, \lambda_{k,2}^{(j)}, \lambda_{k,3}^{(j)}\}$	$\{\lambda_{k,1}^{(j)}, \lambda_{k,2}^{(j)}, \lambda_{k,3}^{(j)}\}$	$\{\lambda_{k,1}^{(j)}, \lambda_{k,2}^{(j)}, \lambda_{k,3}^{(j)}\}$

Figure 4.1   Illustrative example of likelihood flooding.

The CbPFa and CbPFb algorithms, although optimal in the sense of being capable of reproducing exactly the centralized estimate, have nonetheless the drawback, as explained before, of requiring multiple iterative internode communication in the time interval between sensor measurements, and are therefore inadequate for real-time processing. In particular, when posterior distributions are represented by particle sets as is the case in sequential Monte Carlo frameworks, consensus approaches require in theory that N_p iterative consensus protocols be run in parallel between instants k and $k + 1$. The latter requirement has been however relaxed in the literature, see, e.g., [28], [29], [23], [44], [34], [35], by using additional suboptimal parametric approximations that seek to summarize the messages which are transmitted over the network by dropping their dependence on the particle label j. In the sequel, we pursue, however, a different strategy based on random information dissemination, which is aimed at eliminating altogether multiple iterative internode communication between measurements, thus enabling computation and sensing on the same time scale.

4.3.2 RANDOM INFORMATION DISSEMINATION ALGORITHMS

As mentioned in Section 4.1, a possible way to circumvent the high internode communication cost associated with consensus algorithms is to employ diffusion techniques [40].

Most diffusion algorithms found in the distributed linear filtering literature are based on convex combinations of Kalman filters, see, e.g., [12]. An alternative approach is diffusion by random information dissemination [30], which basically seeks to build over time at each node r different local estimates $\hat{s}_{k|k,r}$ conditioned on different sets of observations $Z_{0:k,r}$ coming from different random network locations.

The approach proposed in [30] for linear distributed filtering may be extended to nonlinear models in a sequential Monte Carlo framework using the *Random Exchange Diffusion Particle Filter* (ReDif-PF) described as follows, see also [17], [18], and [19] for further details.

- Assume that, at instant $k-1$, a certain network node t has a weighted set of particles $(\{w_{k-1,t}^{(j)}, s_{0:k-1,t}^{(j)}\})$ that represent $p(s_{0:k-1}|Z_{0:k-1,t})$ where $Z_{0:k-1,t}$ denotes the set of all measurements assimilated by node t up to instant $k-1$.
- Node t then sends $(\{w_{k-1,t}^{(j)}, s_{k-1,t}^{(j)}\})$ to a neighboring node r and receives in return from node r the alternative weighted set $(\{w_{k-1,r}^{(j)}, s_{k-1,r}^{(j)}\})$.
- At instant k, assuming $\{s_k\}$ to be a first-order Markov process, node r draws new samples $s_{k,r}^{(j)} \sim p(s_k|s_{k-1,t}^{(j)})$ and updates its particle weights as

$$w_{k,r}^{(j)} \propto w_{k-1,t}^{(j)} \, p(\{z_{k,i}\}|s_{k,r}^{(j)}, s_{0:k-1,t}^{(j)}, Z_{0:k-1,t}) \qquad i \in \{r\} \cup \mathbf{N}(r).$$

- It can be shown then, see Appendix 4.A in this chapter, that $(\{w_{k,r}^{(j)}, s_{0:k-1,t}^{(j)}, s_{k,r}^{(j)}\})$ is a proper Monte Carlo representation at node r at instant k for the posterior p.d.f. $p(s_{0:k}|\{z_{k,i}\}, Z_{0:k-1,t})$.

In particular, if the observation model at each node does *not* include unknown parameters θ_r, the importance weight update factor reduces to

$$p(\{z_{k,i}\}|s_{k,r}^{(j)}, s_{0:k-1,t}^{(j)}, Z_{0:k-1,t}) = \prod_{i \in \{r\} \cup \mathbf{N}(r)} p(z_{k,i}|s_{k,r}^{(j)}).$$

In a more challenging scenario, however, the sensor models may be parameterized by a global vector of *unknown* parameters

$$\theta_{1:R} = \begin{bmatrix} \theta_1^T & \theta_2^T & \cdots & \theta_R^T \end{bmatrix}^T \qquad (4.25)$$

where θ_i denotes the vector of unknown parameters associated with the sensor model at node i, $i \in \{1, \ldots, R\}$. Given the observations available at the different sensors, the goal of the cooperative network is then to jointly estimate the sequence of hidden states $\{s_k\}$, $k \geq 0$, and the global vector of unknown parameters $\theta_{1:R}$, which is assumed to be time-invariant.

In order to derive a Rao-Blackwellized version [10] of the diffusion particle filter that is suitable for a scenario with unknown parameters, we assume that, at instant $k - 1$,

$$p(\boldsymbol{\theta}_{1:R}|\mathbf{s}_{0:k-1,t}^{(j)}, \mathbf{Z}_{0:k-1,t}) = \prod_{i=1}^{R} \phi(\boldsymbol{\theta}_i|\boldsymbol{\psi}_{i,k-1,t}^{(j)}) \qquad (4.26)$$

where $\phi(\boldsymbol{\theta}_i|\boldsymbol{\psi}_{i,k-1,t}^{(j)})$ denotes a marginal p.d.f. of the unknown parameter vector $\boldsymbol{\theta}_i$ at node t and instant $k - 1$, specified by the vector of hyperparameters $\boldsymbol{\psi}_{i,k-1,t}^{(j)}$. We can show then, see Appendix 4.B, that

$$p(\{\mathbf{z}_{k,i}\} | \mathbf{s}_{k,r}^{(j)}, \mathbf{s}_{0:k-1,t}^{(j)}, \mathbf{Z}_{0:k-1,t}) = \prod_i \underbrace{p(\mathbf{z}_{k,i}|\mathbf{s}_{k,r}^{(j)}, \mathbf{s}_{0:k-1,t}^{(j)}, \mathbf{Z}_{0:k-1,t})}_{\bar{\lambda}_{k,i}^{(j)}(\mathbf{s}_{k,r}^{(j)})} \qquad (4.27)$$

where $i \in \{r\} \cup \mathbf{N}(r)$ and each factor $\bar{\lambda}_{k,i}^{(j)}(\mathbf{s}_{k,r}^{(j)})$ in the product on the right-hand side of Eq. (4.27) is computed by solving the integral

$$\int_{\mathfrak{R}^{N_\theta}} p(\mathbf{z}_{k,i}|\mathbf{s}_{k,r}^{(j)}, \boldsymbol{\theta}_i) p(\boldsymbol{\theta}_i|\mathbf{s}_{0:k-1,t}^{(j)}, \mathbf{Z}_{0:k-1,t}) \mathrm{d}\boldsymbol{\theta}_i$$

$$= \int_{\mathfrak{R}^{N_\theta}} p(\mathbf{z}_{k,i}|\mathbf{s}_{k,r}^{(j)}, \boldsymbol{\theta}_i) \, \phi(\boldsymbol{\theta}_i|\boldsymbol{\psi}_{i,k-1,t}^{(j)}) \, \mathrm{d}\boldsymbol{\theta}_i, \qquad (4.28)$$

with N_θ in (4.28) denoting the dimension of the samples of the random vectors $\boldsymbol{\theta}_i$, assumed for simplicity of notation, but without loss of generality, identical for all i. In particular, if the marginal p.d.f.'s $\phi(\boldsymbol{\theta}_i|\boldsymbol{\psi}_{i,k-1,t}^{(j)})$ are *conjugate priors* [24] for the likelihood functions $p(\mathbf{z}_{k,i}|\mathbf{s}_{k,r}^{(j)}, \boldsymbol{\theta}_i)$, it follows that the updated posterior p.d.f. of the global vector of unknown network parameters $\boldsymbol{\theta}$ at node r and instant k also has the form

$$p(\boldsymbol{\theta}_{1:R}|\mathbf{s}_{0:k,r}^{(j)}, \mathbf{Z}_{0:k,r}) \quad \propto \quad p(\mathbf{Z}_{k,r}|\mathbf{s}_{k,r}^{(j)}, \boldsymbol{\theta}_{1:R}) \prod_{i=1}^{R} \phi(\boldsymbol{\theta}_i|\boldsymbol{\psi}_{i,k-1,t}^{(j)}) \qquad (4.29)$$

$$= \quad \prod_{i=1}^{R} \phi(\boldsymbol{\theta}_i|\boldsymbol{\psi}_{i,k,r}^{(j)}) \qquad (4.30)$$

where $\mathbf{Z}_{k,r}$ is an alternative notation for $\{\mathbf{z}_{k,i}\}$, $i \in \{r\} \cup \mathbf{N}(r)$, $\mathbf{Z}_{0:k,r} = \{\mathbf{Z}_{0:k-1,t}, \mathbf{Z}_{k,r}\}$, $\mathbf{s}_{0:k,r}^{(j)} = \{\mathbf{s}_{0:k-1,t}^{(j)}, \mathbf{s}_{k,r}^{(j)}\}$, and the new hyperparameters $\boldsymbol{\psi}_{i,k,r}^{(j)}$ are equal to $\boldsymbol{\psi}_{i,k-1,t}^{(j)}$ if $i \notin \{r\} \cup \mathbf{N}(r)$ or, otherwise, if $i \in \{r\} \cup \mathbf{N}(r)$, are recursively updated using $\boldsymbol{\psi}_{i,k-1,t}^{(j)}$, the new particles $\mathbf{s}_{k,r}^{(j)}$ sampled at node r at instant k, and the new observations $\{\mathbf{z}_{k,i}\}$ assimilated by node r at instant k. The exact recursive update equations for the hyperparameters depend on the particular signal model under consideration. An illustrative example in the case of emitter tracking in passive sensor networks is shown in Section 4.4 of this chapter and derived in detail in reference [19].

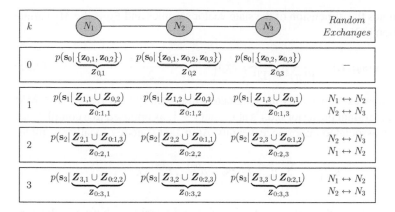

Figure 4.2 Random exchange protocol.

Random Exchange Protocol In order to build, at each instant k and at each node r different Monte Carlo representations of the posterior distribution of the states conditioned on different sets of observations originating from random locations in the entire network, it suffices to implement a protocol where each node r, starting from instant zero, exchanges its particles, weights, and, when applicable, hyperparameters with a randomly chosen neighbor t, propagates the received particles using, e.g., the blind importance function and updates the respective weights using the likelihood function $p(\mathbf{Z}_{k,r}|\mathbf{s}_{0:k,r}^{(j)}, \mathbf{Z}_{0:k-1,t})$.

Unlike random gossip algorithms found in the literature [8], the aforementioned procedure diffuses information by randomly propagating posterior probability distributions over the network. Specifically, as a posterior p.d.f. that is available at a given node r_0 at instant 0 follows a path $\mathcal{P} \triangleq \{r_0, r_1, \ldots, r_n\}$ across the network, it assimilates the measurements $\mathbf{Z}_{k,r}$ that are available at each visited node $r \in \mathcal{P}$ and at their respective neighboring nodes.

Figure 4.2 illustrates the evolution of the marginal posterior at each node in a toy example featuring a linear network with three nodes running the Random Exchange protocol over four time instants. Initially, each node $r \in \{1, 2, 3\}$ has a posterior at instant zero conditioned on the measurements $\mathbf{Z}_{0,r} = \{\mathbf{z}_{0,i}\}$, $i \in \{r\} \cup \mathbf{N}(r)$, in its vicinity only. At each time instant $k \in \{1, 2, 3\}$, network nodes perform the sequence of random exchanges as indicated in the rightmost column of Figure 4.2 and, then, update the received posterior by assimilating measurements in their respective neighborhoods.

Approximate Implementations with Low Internode Communication Cost To reduce the internode communication cost, we can use a Gaussian mixture model (GMM) approximation [42]

$$p(\mathbf{s}_{k-1}|\mathbf{Z}_{0:k-1,t}) \approx \sum_{m=1}^{L} \eta_{k-1,t}^{(m)} \xi_{N_s}(\mathbf{s}_{k-1}|\boldsymbol{\mu}_{k-1,t}^{(m)}, \boldsymbol{\Sigma}_{k-1,t}^{(m)})$$

where N_s denotes the dimension of the state vector samples and ξ_M is the $M-$variate normal function [32]

$$\xi_M(\mathbf{s}|\boldsymbol{\mu}, \boldsymbol{\Sigma}) = \frac{1}{(2\pi)^{M/2}|\boldsymbol{\Sigma}|^{1/2}} \exp\left(-\frac{(\mathbf{s}-\boldsymbol{\mu})^T\boldsymbol{\Sigma}^{-1}(\mathbf{s}-\boldsymbol{\mu})}{2}\right). \quad (4.31)$$

Nodes r and t now exchange only the parameters that specify their respective GMM approximated marginal posterior p.d.f.'s at instant $k-1$, as opposed to a complete set of particles and associated importance weights. At instant k, node r locally resamples

$$\mathbf{s}_{k-1,t}^{(j)} \sim \sum_{m=1}^{L} \eta_{k-1,t}^{(m)} \xi_{N_s}(\mathbf{s}_{k-1}|\boldsymbol{\mu}_{k-1,t}^{(m)}, \boldsymbol{\Sigma}_{k-1,t}^{(m)}),$$

resets $w_{k-1,t}^{(j)}$ to $1/N_p$, and propagates the particles and weights as before in the conventional ReDif-PF algorithm without parametric approximation.

In the particular case, however, where the sensor models include unknown parameters, the approach described in the previous paragraph would in principle require that node r also locally resample all previous particles $\mathbf{s}_{0:k-2,t}^{(j)}$ jointly with $\mathbf{s}_{k-1,t}^{(j)}$ from some parametric approximation to $p(\mathbf{s}_{0:k-1}|\mathbf{Z}_{0:k-1,t})$ and then retroactively recalculate the posterior p.d.f.'s $p(\theta_i| \mathbf{s}_{0:k-1,t}^{(j)}, \mathbf{Z}_{0:k-1,t}), i = 1,\ldots,R$ with the resampled particles. To avoid this curse of dimensionality, it is desirable to introduce an additional parametric approximation to $p(\theta_i| \mathbf{s}_{0:k-1,t}^{(j)}, \mathbf{Z}_{0:k-1,t})$ that eliminates that function's dependence on the particle label j and on the simulated sequence $\mathbf{s}_{0:k-1,t}^{(j)}$.

Specifically, we follow the lead in [4], [15], [18] and, for each $i \in \{1,\ldots,R\}$, we approximate the marginal posterior p.d.f.'s $p(\theta_i|\mathbf{s}_{0:k-1}^{(j)}, \mathbf{Z}_{0:k-1,t})$ for all indices j and all possible sequences $\mathbf{s}_{0:k-1}^{(j)}$ by a new probability density

$$\tilde{p}(\theta_i|\mathbf{Z}_{0:k-1,t}) = \phi(\theta_i|\tilde{\boldsymbol{\psi}}_{i,k-1,t}) \quad (4.32)$$

where ϕ is the same function that appears on the right-hand side of Equation (4.26) and the hyperparameters $\tilde{\boldsymbol{\psi}}_{i,k-1,t}$, *independent* of j, are chosen such that the approximate p.d.f. $\tilde{p}(\theta_i|\mathbf{Z}_{0:k-1,t})$ matches the first- and second-order moments associated with

$$p(\theta_i| \mathbf{Z}_{0:k-1,t}) = \int [p(\theta_i|\mathbf{s}_{0:k-1}, \mathbf{Z}_{0:k-1,t})$$
$$\times \quad p(\mathbf{s}_{0:k-1}|\mathbf{Z}_{0:k-1,t})] \, d\mathbf{s}_{0:k-1} . \quad (4.33)$$

Note that the term on the left-hand side of Eq. (4.33) is precisely the average (or *expected value*) of $p(\theta_i|\mathbf{s}_{0:k-1}, \mathbf{Z}_{0:k-1,t})$ over all possible realizations of $\mathbf{s}_{0:k-1}$ conditioned on the observations $\mathbf{Z}_{0:k-1,t}$. Assuming next that $\{(w_{k-1,t}^{(j)}, \mathbf{s}_{0:k-1,t}^{(j)})\}, j \in \{1,\ldots,N_p\}$, is a properly weighted set of samples available at node t at instant $k-1$ to represent $p(\mathbf{s}_{0:k-1}|\mathbf{Z}_{0:k-1,t})$, we make the Monte Carlo approximation

$$p(\theta_i| \mathbf{Z}_{0:k-1,t}) \approx \sum_{j=1}^{N_p} w_{k-1,t}^{(j)} p(\theta_i|\mathbf{s}_{0:k-1,t}^{(j)}, \mathbf{Z}_{0:k-1,t}) . \quad (4.34)$$

On the other hand, from the assumption that $p(\theta_{1:R}|\,\mathbf{s}^{(j)}_{0:k-1,t},\mathbf{Z}_{0:k-1,t})$ is a separable function which factors as in Eq. (4.26), it follows that

$$p(\theta_i|\mathbf{s}^{(j)}_{0:k-1,t},\mathbf{Z}_{0:k-1,t}) = \phi(\theta_i|\psi^{(j)}_{i,k-1,t})$$

and, therefore,

$$p(\theta_i|\,\mathbf{Z}_{0:k-1,t}) \approx \sum_{j=1}^{N_p} w^{(j)}_{k-1}\phi(\theta_i|\psi^{(j)}_{i,k-1,t}) . \tag{4.35}$$

Illustrative examples of the computation of the modified vector of hyperparameters $\tilde{\psi}_{i,k-1,t}$ such that $\tilde{p}(.)$ in (4.32) matches the first- and second-order moments obtained from the mixture of p.d.f.'s on the right-hand side of (4.35) are shown in Section 4.4 of this chapter in a problem of cooperative emitter tracking in passive sensor networks.

Replacing in the sequel $p(\theta_i|\mathbf{s}^{(j)}_{0:k-1,t},\mathbf{Z}_{0:k-1,t})$ in (4.28) with

$$\tilde{p}(\theta_i|\mathbf{Z}_{0:k-1,t}) = \phi(\theta_i|\tilde{\psi}_{i,k-1,t})$$

in (4.32) for all $j \in \{1,\dots,N_p\}$ and *all possible sequences* $\mathbf{s}^{(j)}_{0:k-1,t}$, we compute new factors $\tilde{\lambda}_{k,i}(.)$ at node r at instant k such that

$$\begin{aligned}
\tilde{\lambda}_{k,i}(\mathbf{s}^{(j)}_{k,r}) &= \int_{\Re^{N_\theta}} p(\mathbf{z}_{k,i}|\mathbf{s}^{(j)}_{k,r},\theta_i)\tilde{p}(\theta_i|\mathbf{Z}_{0:k-1,t})d\theta_i \\
&= \int_{\Re^{N_\theta}} p(\mathbf{z}_{k,i}|\mathbf{s}^{(j)}_{k,r},\theta_i)\phi(\theta_i|\tilde{\psi}_{i,k-1,t})d\theta_i.
\end{aligned} \tag{4.36}$$

Once again, if the functions $\phi(\theta_i|\tilde{\psi}_{i,k-1,t})$ are conjugate priors to the likelihood functions $p(\mathbf{z}_{k,i}|\mathbf{s}^{(j)}_{k,r},\theta_i)$, the integral in (4.36) has a closed-form solution. Illustrative examples are shown in Section 4.4.

The modified importance weight update rule at node r at instant k thus becomes

$$w^{(j)}_{k,r} \propto w^{(j)}_{k-1,t} \prod_{i\in\{r\}\cup\mathbf{N}(r)} \tilde{\lambda}_{k,i}(\mathbf{s}^{(j)}_{k,r}) \qquad j \in \{1,\dots,N_p\}. \tag{4.37}$$

On the other hand, replacing $\phi(\theta_i|\psi^{(j)}_{i,k-1,t})$ in (4.29) with $\phi(\theta_i|\tilde{\psi}_{i,k-1,t})$ independent of j, and assuming further the conjugate prior hypothesis, the updated (approximate) posterior p.d.f. of $\theta_{1:R}$ given $\mathbf{Z}_{0:k,r}$ and $\mathbf{s}^{(j)}_{0:k,r}$ remains of the form

$$\prod_{i=1}^{R} \phi(\theta_i|\psi^{(j)}_{i,k,r}),$$

but the hyperparameters $\psi^{(j)}_{i,k,r}$ arc now updated, for all $j \in \{1,\dots,N_p\}$, using solely $\mathbf{s}^{(j)}_{k,r}$, $\mathbf{Z}_{k,r}$ and the set $\{\tilde{\psi}_{i,k-1,t}\}$, $i \in \{1,\dots,R\}$. Detailed expressions for updating the approximate hyperparameters in specific select applications are presented in

Section 4.4.

Communication Cost in the Approximate Implementation Let N_s be the dimension of the state vector. Using the GMM approximation, node t transmits at each instant k to a random neighbor r only the corresponding GMM parameters $\eta_{k-1,t}^{(m)}$, $\mu_{k-1,t}^{(m)}$, and $\Sigma_{k-1,t}^{(m)}$, $m \in \{1, \ldots, L\}$, i.e., $L \times [N_s + 1 + (N_s^2 + N_s)/2]$ real numbers taking into account the symmetry of the covariance matrices $\Sigma_{k-1,t}^{(m)}$. In addition, using the moment-matching parametric approximation described before, node t also transmits to its neighbor r a set of R hyperparameter vectors $\tilde{\psi}_{i,k-1,t}$, $i \in \{1, \ldots, R\}$, i.e., $R \times N_\psi$ real numbers, where N_ψ is the dimension of the vector $\tilde{\psi}_{k-1,i,t}$ that specifies $\tilde{p}(\theta_i | Z_{0:k-1,t})$ for $i = 1, \ldots, R$. For simplicity of notation, but without loss of generality, we assume N_ψ to be constant for all $i \in \{1, \ldots, R\}$.

In comparison, if the aforementioned GMM and moment-matching parametric approximations were not used, each node would have to transmit to its chosen neighbor at instant k, N_p real vectors of length N_s, N_p (scalar) real weights, and $R \times N_p$ vectors real hyperparameters $\psi_{i,k-1,t}^{(j)}$ of length N_ψ, respectively, for each $j \in \{1, \ldots, N_p\}$ and each $i \in \{1, \ldots, R\}$, that is, a final total of $N_p \times (N_s + 1 + R \times N_\psi)$ real numbers.

4.4 COOPERATIVE EMITTER TRACKING USING PASSIVE SENSORS

In this section, we describe an application where multiple processors dispersed over the nodes of a network of passive *received-signal-strength* (RSS) sensors cooperate to track a single emitter that moves randomly in a two-dimensional (2D) surveillance space. Figure 4.3 shows a possible trajectory of a target of interest in a network with $R = 25$ sensors.

Specifically, we assume a prior dynamic model for the emitter's motion, which is in general nonlinear and described by equation [1]

$$\mathbf{s}_{k+1} = \mathbf{f}(\mathbf{s}_k) + \mathbf{G}\mathbf{u}_k \tag{4.38}$$

where $\mathbf{u}_k$ is the dynamic model noise and $\mathbf{s}_k$ is the hidden state vector whose components are the positions and velocities of the target centroid in each of the two dimensions of the surveillance space. The measurements collected at sensor r at instant k, $r = 1, \ldots, R$, are in turn given by [37], [26], [38]

$$z_{k,r} = \underbrace{P_0 - A_r \, 10 \log(\frac{\|\mathbf{H}\mathbf{s}_k - \mathbf{s}_r\|}{d_0})}_{g_r(\mathbf{s}_k)} + \sqrt{\sigma_r} \, v_{k,r} \tag{4.39}$$

where $\mathbf{s}_r$ denotes the 2D position of sensor r in the surveillance space; $\|.\|$ is the Euclidian norm; $\mathbf{H}$ is a matrix of dimension 2×4 such that $H(1,1) = H(2,3) = 1$ and $H(i,j) = 0$ otherwise; (P_0, d_0, A_r) are known model parameters, see details in [38]; $\{v_{k,r}\}$, $k \geq 0$, $r \in \{1, \ldots, R\}$ is a sequence of independent, identically distributed (i.i.d) Gaussian random variables such that

$$v_{k,r} \sim \mathcal{N}(0, 1) \qquad \forall k \geq 0, \; r \in \{1, \ldots, R\}$$

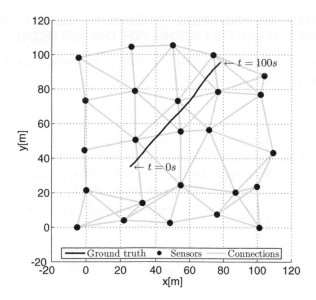

Figure 4.3 Cooperative tracking in sensor networks.

where $\mathcal{N}(0, 1)$ denotes the univariate normal distribution with mean zero and variance one, and $\{\sigma_r^2\}$, $r \in \{1, \ldots, R\}$, are scale factors for the sensor noise variances, which may be deterministic and known, or assumed random and unknown, depending on the underlying sensor model.

The goal of the distributed filter is to recursively compute (or approximate) over the network the global MMSE estimate

$$\hat{\mathbf{s}}_{k|k}^g = E\left\{\mathbf{s}_k|\mathbf{z}_{0:k,1:R}\right\}. \tag{4.40}$$

That problem was investigated in detail, see [15], [16], [17], [18], [19], when $\mathbf{s}_0$, the sequence $\{\mathbf{u}_k\}$, $k \geq 0$, the sequences $\{\mathbf{v}_{k,r}\}$, $k \geq 0$, $r = 1, \ldots, R$, and, when modeled as random objects, the parameters $\sigma_{1:R}^2 = \left[\sigma_1^2 \sigma_2^2 \ldots \sigma_R^2\right]^T$ are assumed *a priori* mutually independent. Due to the nonlinearity of the observation model in (4.39) and possibly of the dynamic model in (4.38), particle filters are used to obtain suitable approximations to MMSE estimate in (4.40).

Different solutions to the problem described in this section were proposed using broadcast algorithms in fully connected networks [15], [16]; average consensus, minimum consensus, and flooding algorithms [16]; and diffusion algorithms [17], [18], [19]. In cases where the sensor noise variance scale factors $\sigma_{1:R}^2$ were assumed random and a priori independent from node to node, e.g., in [15], [16], and [18], we assumed an *inverse Gamma* [24] prior probability distribution for σ_r^2, $\forall r$, such that the posterior p.d.f.'s $p(\sigma_r^2|\mathbf{s}_{0:k}^{(j)}, \mathbf{z}_{0:k,r})$ have a closed-form expression for all $k \geq 1$, all $r = 1, \ldots, R$, and all $j = 1, \ldots, N_p$, see details in [15], [16], [18], and [19].

4.4.1   HYPERPARAMETER POSTERIOR PROBABILITY DENSITY FUNC-TION AND WEIGHT UPDATE FACTORS FOR THE RSS MODEL

Assuming the variance scales $\{\sigma_r^2\}$ to be random variables, we make $\theta_{1:R}$ in (4.25) equal to $\sigma_{1:R}^2$. Specifically, we assume that, a priori,

$$\sigma_r^2 \sim IG(\alpha, \beta) \qquad \forall r \in \{1, \ldots, R\}$$

where $IG(\alpha, \beta)$ denotes the Gamma inverse distribution specified by the hyperparameters α and β, and σ_r^2 is independent of σ_t^2 for all $t \neq r$. It follows then that, using the notation in Eq. (4.26) in Section 4.3.2, the vector of hyperparameters $\psi_{i,k,r}^{(j)}$, $i \in \{1, \ldots, R\}$ at node r at instant k associated with the sampled particle trajectory $s_{0:k}^{(j)}$ has the form $\psi_{i,k,r}^{(j)} = \left[\alpha_{i,k,r}^{(j)} \;\; \beta_{i,k,r}^{(j)} \right]^T$ and the function $\phi(\theta_i | \psi_{i,k,r}^{(j)})$ in (4.30) reduces in this particular case, see a detailed derivation in [19], to

$$\phi(\sigma_i^2 | \alpha_{i,k,r}^{(j)}, \beta_{i,k,r}^{(j)}) = \frac{(\beta_{i,k,r}^{(j)})^{\alpha_{i,k,r}^{(j)}}}{\Gamma(\alpha_{i,k,r}^{(j)})} \sigma_i^{-2(\alpha_{i,k,r}^{(j)}+1)} \exp(-\frac{\beta_{i,k,r}^{(j)}}{\sigma_i^2}) \qquad \sigma_i^2 > 0 \qquad (4.41)$$

where, in the random exchange diffusion algorithm described in Section 4.3.2, when node r communicates at instant k with a neighbor t, $\alpha_{i,k,r}^{(j)}$ and $\beta_{i,k,r}^{(j)}$ are updated from $\alpha_{i,k-1,t}^{(j)}$ and $\beta_{i,k-1,t}^{(j)}$ by the recursions, see also [19],

$$\alpha_{i,k,r}^{(j)} = \alpha_{i,k-1,t}^{(j)} + \frac{1}{2} \qquad\qquad\qquad (4.42)$$

$$\beta_{i,k,r}^{(j)} = \beta_{i,k-1,t}^{(j)} + \frac{1}{2} \left[z_{k,i} - g_i(s_{k,r}^{(j)}) \right]^2, \qquad (4.43)$$

for all $i \in \{r\} \cup \mathbf{N}(r)$; otherwise, if $i \notin \{r\} \cup \mathbf{N}(r)$, $\alpha_{i,k,r}^{(j)} = \alpha_{i,k-1,t}^{(j)}$, and $\beta_{i,k,r}^{(j)} = \beta_{i,k-1,t}^{(j)}$. In (4.43), $g_i(.)$ is the nonlinear observation function shown in Eq. (4.39) and corresponding to node i; vector $s_{k,r}^{(j)} \sim p(s_k | s_{k-1,t}^{(j)})$ is in turn the jth new particle sampled by node r at instant k given the previous sample $s_{k-1,t}^{(j)}$ received from node t, and the symbol $\Gamma(.)$ in (4.41) denotes the Gamma function defined as

$$\Gamma(x) = \int_0^\infty y^{x-1} \exp(-y)\, dy \qquad x > 0.$$

Substituting now

$$p(z_{k,i} | s_{k,r}^{(j)}, \theta_i) = \frac{1}{\sqrt{2\pi \sigma_i^2}} \exp(-\frac{\left[z_i - g_i(s_{k,r}^{(j)}) \right]^2}{2\sigma_i^2}) \qquad (4.44)$$

$$\phi(\theta_i | \psi_{i,k-1,t}^{(j)}) = \frac{(\beta_{i,k-1,t}^{(j)})^{\alpha_{i,k-1,t}^{(j)}}}{\Gamma(\alpha_{i,k-1,t}^{(j)})} \sigma_i^{-2(\alpha_{i,k-1,t}^{(j)}+1)} \exp(-\frac{\beta_{i,k-1,t}^{(j)}}{\sigma_i^2}) \qquad (4.45)$$

into Eq. (4.28), it follows that the weight update rule at node r and instant k is given by

$$w_{k,r}^{(j)} \propto w_{k-1,t}^{(j)} \prod_{i \in \{r\} \cup \mathbf{N}(r)} \bar{\lambda}_{k,i}^{(j)}(\mathbf{s}_{k,r}^{(j)}) \qquad (4.46)$$

where the factors $\bar{\lambda}_{k,i}^{(j)}(\mathbf{s}_{k,r}^{(j)})$ are computed by the expression, see also the derivation in [19],

$$\bar{\lambda}_{k,i}^{(j)}(\mathbf{s}_{k,r}^{(j)}) \propto \frac{\left[\beta_{i,k-1,t}^{(j)}\right]^{\alpha_{i,k-1,t}^{(j)}}}{\Gamma(\alpha_{i,k-1,t}^{(j)})} \frac{\Gamma(\alpha_{i,k,r}^{(j)})}{\left[\beta_{i,k,r}^{(j)}\right]^{\alpha_{i,k,r}}} \qquad (4.47)$$

for all $i \in \{r\} \cup \mathbf{N}(r)$. The proportionality constant in (4.46), independent of j, is computed as before such that $\sum_j w_{k,r}^{(j)} = 1$.

Finally, applying (4.30), we can also write

$$p(\sigma_{1:R}^2 | \mathbf{s}_{0:k,r}^{(j)}, \mathbf{Z}_{0:k,r}) = \prod_{i=1}^{R} \phi(\sigma_i^2 | \alpha_{i,k,r}^{(j)}, \beta_{i,k,r}^{(j)})$$

$$= \prod_{i=1}^{R} \frac{(\beta_{i,k,r}^{(j)})^{\alpha_{i,k,r}^{(j)}}}{\Gamma(\alpha_{i,k,r}^{(j)})} \sigma_i^{-2(\alpha_{i,k,r}^{(j)}+1)} \exp(-\frac{\beta_{i,k,r}^{(j)}}{\sigma_i^2}), \ \sigma_i^2 > 0.$$

4.4.2 MODIFIED HYPERPARAMETERS AND WEIGHT UPDATE FACTORS USING MOMENT-MATCHING APPROXIMATIONS

Following on the other hand the moment-matching parametric approximation methodology in (4.32), the modified hyperparameters at node t at instant $k - 1$,

$$\tilde{\boldsymbol{\psi}}_{i,k-1,t} = \left[\tilde{\alpha}_{i,k-1,t} \ \tilde{\beta}_{i,k-1,t}\right]^T,$$

which, as discussed before, are independent of the particle label j, may be obtained from the weighted sample set $\{(w_{k-1,t}^{(j)}, \mathbf{s}_{k-1,t}^{(j)})\}$ and from the hyperparameter set $\{(\alpha_{i,k-1,t}^{(j)}, \beta_{i,k-1,t}^{(j)})\}, j \in \{1, \dots, N_p\}$ by making [19]

$$\tilde{\alpha}_{i,k-1,t} = 2 + \widehat{E}_{k-1,t}^2\left[\sigma_i^2\right] / \widehat{VAR}_{k-1,t}\left[\sigma_i^2\right] \qquad (4.48)$$

$$\tilde{\beta}_{i,k-1,t} = (\tilde{\alpha}_{i,k-1,t} - 1)\widehat{E}_{k-1,t}\left[\sigma_i^2\right] \qquad (4.49)$$

where

$$\widehat{E}_{k-1,t}\left[\sigma_i^2\right] = \frac{\sum_{j=1}^{N_p} w_{k-1,t}^{(j)} \beta_{i,k-1,t}^{(j)}}{\alpha_{i,k-1,t}^{(j)} - 1} \qquad (4.50)$$

$$\widehat{VAR}_{k-1,t}\left[\sigma_i^2\right] = \frac{\sum_{j=1}^{N_p} w_{k-1,t}^{(j)} (\beta_{i,k-1,t}^{(j)})^2}{(\alpha_{i,k-1,t}^{(j)} - 1)(\alpha_{i,k-1,t}^{(j)} - 2)}$$

$$- \widehat{E}_{k-1,t}^2\left[\sigma_i^2\right],$$

if $\alpha_{i,k-1,t}^{(j)} > 2$, $\forall i \in \{1, \ldots R\}$. Note that Eqs. (4.48) and (4.49) are independently run in parallel at node t and instant $k - 1$ for each index $i \in \{1, \ldots, R\}$.

The approximate weight update factors are computed now, as explained in Section 4.3.2, by replacing the hyperparameter vectors $\boldsymbol{\psi}_{i,k-1,t}^{(j)}$ for all j with the j-invariant vectors, $\tilde{\boldsymbol{\psi}}_{i,k-1,t}$. In other words, we make

$$w_{k,r}^{(j)} \propto w_{k-1,t}^{(j)} \prod_{i \in \{r\} \cup \mathbf{N}(r)} \tilde{\lambda}_{k,i}(\mathbf{s}_{k,r}^{(j)}) \tag{4.51}$$

where the terms $\tilde{\lambda}_{k,i}^{(j)}(\mathbf{s}_{k,r}^{(j)})$ are given by

$$\tilde{\lambda}_{k,i}(\mathbf{s}_{k,r}^{(j)}) \propto \frac{\left[\tilde{\beta}_{i,k-1,t}\right]^{\tilde{\alpha}_{i,k-1,t}}}{\Gamma(\tilde{\alpha}_{i,k-1,t})} \frac{\Gamma(\alpha_{i,k,r}^{(j)})}{\left[\beta_{i,k,r}^{(j)}\right]^{\alpha_{i,k,r}}} \tag{4.52}$$

and, in turn,

$$\alpha_{i,k,r}^{(j)} = \tilde{\alpha}_{i,k-1,t} + \frac{1}{2} \qquad \forall j \tag{4.53}$$

$$\beta_{i,k,r}^{(j)} = \tilde{\beta}_{i,k-1,t} + \frac{1}{2}\left[z_{k,i} - g_i(\mathbf{s}_{k,r}^{(j)})\right]^2, \tag{4.54}$$

for all $i \in \{r\} \cup \mathbf{N}(r)$; otherwise, if $i \notin \{r\} \cup \mathbf{N}(r)$, we make as before, $\alpha_{i,k,r}^{(j)} = \tilde{\alpha}_{i,k-1,t}$ and $\beta_{i,k,r}^{(j)} = \tilde{\beta}_{i,k-1,t}$, $\forall j$.

4.4.3 NUMERICAL RESULTS

The performance of the iterative consensus and random exchange diffusion filters in the specific problem of tracking with RSS sensors were investigated in detail with simulated data in references [16] and [19]. For illustrative purposes, we reproduce in this chapter some simulations results in the test scenario corresponding to Figure 4.3. A detailed description of the simulated scenario is found in [19] and is omitted here for lack of space.

Figure 4.4 shows the evolution over time of the root-mean-square (RMS) error of the norm of the estimated emitter position averaged over all network nodes, respectively for the Rao-Blackwellized random exchange diffusion algorithm with parametric approximations (RB ReDif-PF) as described in Section 4.3.2, for the iterative minimum consensus (CbPFa) and flooding (CbPFb) algorithms described in Section 4.3.1, and for the DcPF algorithm from reference [2]. Additionally, Figure 4.4 also shows the average RMS error curve first for a non-cooperative scheme where each node runs an isolated local particle filter and assimilates only its own local node measurements, and, second, for a limited, local cooperation scheme where, in addition to its own local measurements, a node also assimilates observations from its neighbors, but, unlike in the ReDif-PF algorithm, does not exchange representations of its posterior state or parameter p.d.f.'s with other nodes. The bars shown in Figure 4.4 represent the standard deviation of the RMS error norm across the different

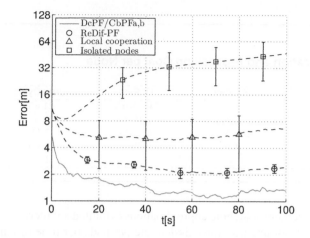

Figure 4.4 Average RMS error over time for different filtering algorithms.

network nodes. There are no bars superimposed to the DcPF, CbPFa, and CbPFb error curves because, as explained in previous sections, those algorithms generate identical state estimates in all nodes. The RB ReDif-PF with GMM approximation was implemented with only one Gaussian mode in the parametric representation of $p(\mathbf{s}_{k-1}|\mathbf{Z}_{0:k-1,t})$ at node t at instant $k-1$.

As theoretically expected, the iterative CbPFa and CbPFb algorithms match the performance of the DcPF tracker since all three aforementioned algorithms reproduce the optimal centralized PF tracker exactly, albeit with different communication and computational costs. On the other hand, as shown in Figure 4.4, the suboptimal RB ReDif-PF tracker has a performance degradation compared to DcPF. However, ReDif-PF offers an improvement in error performance compared to the local cooperation scheme by better diffusing the information across the network. We also note from Figure 4.4 that the standard deviation of the state estimate across the different network nodes is much lower in the ReDif-PF algorithm than in the local cooperation scheme. Finally, as shown in Figure 4.4, we note that, when the local particle filters at each node were isolated and did not cooperate with each other, the network on average was unable to properly track the emitter in the simulated scenario.

Using a four-byte and a one-byte network representation respectively for real and Boolean values, we recorded the total amount of bytes transmitted and received by all nodes over the network while running each tracker in Fig 4.4. Table 4.1 shows the communication cost for each algorithm in terms of average transmission (TX) and average reception (RX) rates per node and also quantifies the processing cost for each algorithm in terms of average duty cycle per node, measured in an Intel Core i5 machine with 4GB RAM. The duty cycle of a given node is defined as the ratio between the total node processing time and the duration of a simulated trajectory of the emitter, i.e., $100s$ in our experiments.

As shown in Table 4.1, the RB ReDif-PF tracker incorporating the previously

Table 4.1

Average Communication and Processing Cost per Node

	RX Rate	TX Rate	Duty Cycle
CbPFa	1.2MB/s	244.1KB/s	21.4%
CbPFb	232.0KB/s	48.7KB/s	22.6%
RB ReDif-PF	531.5B/s	515.7B/s	7.7%
Local Cooperation	4B/s	19.8B/s	9.3%
Isolated Nodes	–	–	2.0%

discussed GMM and moment-matching approximations exhibited an internode communication cost in terms of average transmission rate per node per time instant that is roughly two orders of magnitude lower than the TX rate recorded for iterative minimum consensus (CbPFa) solution.

4.5 COOPERATIVE EQUALIZATION OF DIGITAL COMMUNICATION CHANNELS

In this section, we present another illustrative example where a digital source outputs a random sequence of i.i.d. binary digits $\{b_k\}$, $k \geq 0$, which are differentially encoded [5], [6], into a sequence of symbols $\{s_k\}$, $k \geq 0$. The encoded symbols are in turn transmitted over multiple frequency-selective channels to a physical network of R receivers generating, after appropriate pre-processing, the sequences of observations $\{z_{k,r}\}$, $k \geq 0$, at each receiver $r \in \{1, \ldots, R\}$, such that

$$z_{k,r} = \mathbf{h}_r^T \mathbf{s}_k + v_{k,r}, \qquad (4.55)$$

where $\mathbf{h}_r \in \mathbb{R}^M$ is a time-invariant vector of channel impulse response coefficients, $\mathbf{s}_k \triangleq [s_k \ldots s_{k-M+1}]^T$, and $v_{k,r}$ is a sequence of i.i.d. Gaussian random variables with zero mean and known variance σ_r^2.

Using a Bayesian approach, the vectors of unknown channel coefficients $\mathbf{h}_r$, $r \in \{1 \ldots R\}$, are modeled as random vectors that are independent of each other for two different nodes $r \neq t$ and distributed a priori as

$$\mathbf{h}_r \sim \mathcal{N}_M(\mathbf{0}; I/\varepsilon^2),$$

where $\mathcal{N}_M$ denotes an M-variate Normal distribution and ε is a regularization coefficient assumed to be known in the prior model of the channel.

Since the transmitted symbols are defined on a discrete sample space, it results that the joint posterior distribution of the transmitted symbols and the observations is non-Gaussian and, therefore, the LMMSE estimate generated, e.g., by a Wiener filter [32], even with perfect knowledge of the parameters $\{\mathbf{h}_r\}$ and $\{\sigma_r^2\}$, may be substantially different from the maximum a posteriori (MAP) estimate that minimizes the bit error rate (BER) [32]. Using a particle filter, it is possible, however, to recursively approximate at each instant k the optimal MAP estimate

$$\hat{b}_k = \arg \max_{d \in \{0,1\}} Pr(\{b_k = d\}|\mathbf{z}_{0:k,1:R})$$

where

$$\mathbf{z}_{0:k,1:R} \triangleq \left[\mathbf{z}_{0:k,1}^T \ldots \mathbf{z}_{0:k,R}^T \right]^T,$$

and $Pr(A)$ denotes the probability of an event A.

Considering discrete-valued hidden states, the centralized particle filter sequentially builds a weighted set of samples $\left\{ w_k^{(j)}, b_k^{(j)} \right\}$, $j \in \{1, \ldots, N_p\}$, such that the posterior probability of the source bits given the observations recorded at all network receivers is approximated as [9]

$$Pr(\{b_k = d\}|\mathbf{z}_{0:k,1:R}) \approx \sum_{j=1}^{N_p} w_k^{(j)} \mathcal{I}\{b_k^{(j)} = d\} \qquad d \in \{0, 1\} \qquad (4.56)$$

where $\mathcal{I}\{ \cdot \}$ is the indicator function. Using differential encoding, the particles $\{b_k^{(j)}\}$ corresponding to the original source sequence are obtained univocally by a deterministic mapping from Monte Carlo samples $\{\mathbf{s}_k^{(j)}\}$ of the transmitted symbols that are randomly drawn according to the importance function assumed by the filter. Under the assumptions of Gaussian priors for the channels and Gaussian likelihood functions conditioned on the channel coefficients, it can be shown, see also [3], that the posterior p.d.f.'s

$$p(\mathbf{h}_i|\mathbf{s}_{0:k-1}^{(j)}, \mathbf{z}_{0:k-1,1:R}) = \xi_M(\mathbf{h}_i| \hat{\mathbf{h}}_{k-1,i}^{(j)}, \Sigma_{k-1,i}^{(j)}) \qquad i \in \{1, \ldots, R\} \qquad (4.57)$$

where we recall that ξ_M is the M-variate normal function[1] defined in Eq. (4.31) in Section 4.3.2.

In particular, if the assumed importance function is the probability mass function (p.m.f) of symbol $\mathbf{s}_k$ given $\mathbf{s}_{0:k-1}$, i.e., if

$$s_k^{(j)} \sim P(\mathbf{s}_k|\mathbf{s}_{k-1}^{(j)}),$$

it follows that the optimal global weights, as in the continuous-state case discussed in Section 4.2.2, are updated as

$$w_k^{(j)} \propto w_{k-1}^{(j)} \, p(\mathbf{z}_{k,1:R}|\mathbf{s}_k^{(j)}, \mathbf{s}_{0:k-1}^{(j)}, \mathbf{z}_{0:k-1,1:R})$$

$$= w_{k-1}^{(j)} \prod_{i=1}^{R} \underbrace{p(\mathbf{z}_{k,i}|\mathbf{s}_k^{(j)}, \mathbf{s}_{0:k-1}^{(j)}, \mathbf{z}_{0:k-1,i})}_{\lambda_{k,i}^{(j)}(\mathbf{s}_k^{(j)})} \qquad (4.58)$$

---

[1] For simplicity of notation, if $M = 1$, we denote the univariate normal function simply as ξ, instead of ξ_1.

where the factorization in line (4.58) arises again from the conditional independence assumptions in the signal model. The weight update factors $\lambda_{k,i}^{(j)}(\mathbf{s}_k^{(j)})$ in (4.58) have in turn a closed-form analytic expression given by

$$
\begin{aligned}
\lambda_{k,i}^{(j)}(\mathbf{s}_k^{(j)}) &= \int_{\mathcal{R}^M} p(\mathbf{z}_{k,i}|\mathbf{s}_k^{(j)}, \mathbf{h}_i)\, \xi_M(\mathbf{h}_i|\, \hat{\mathbf{h}}_{k-1,i}^{(j)}, \mathbf{\Sigma}_{k-1,i}^{(j)})\, d\mathbf{h}_i \\
&= \xi\left(z_{k,i}|\, (\hat{\mathbf{h}}_{k-1,i}^{(j)})^T \mathbf{s}_k^{(j)}\, ; \gamma_{k,i}^{(j)}\right)
\end{aligned}
\tag{4.59}
$$

where the hyperparameters $\hat{\mathbf{h}}_{k-1,r}^{(j)}$ and $\gamma_{k,r}^{(j)}$ are sequentially updated by the recursions [3], [7]

$$
\gamma_{k,i}^{(j)} \triangleq \sigma_i^2 + (\mathbf{s}_k^{(j)})^T \mathbf{\Sigma}_{k-1,i}^{(j)} \mathbf{s}_k^{(j)}, \tag{4.60}
$$

$$
e_{k,i}^{(j)} \triangleq z_{k,i} - (\hat{\mathbf{h}}_{k-1,i}^{(j)})^T \mathbf{s}_k^{(j)}, \tag{4.61}
$$

$$
\hat{\mathbf{h}}_{k,i}^{(j)} = \hat{\mathbf{h}}_{k-1,i}^{(j)} + \mathbf{\Sigma}_{k-1,i}^{(j)} \mathbf{s}_k^{(j)} e_{k,i}^{(j)} / \gamma_{k,i}^{(j)}, \tag{4.62}
$$

$$
\mathbf{\Sigma}_{k,i}^{(j)} = \mathbf{\Sigma}_{k-1,i}^{(j)} - \mathbf{\Sigma}_{k-1,i}^{(j)} \mathbf{s}_k^{(j)} (\mathbf{s}_k^{(j)})^T \mathbf{\Sigma}_{k-1,i}^{(j)} / \gamma_{k,i}^{(j)}, \tag{4.63}
$$

with initialization $\hat{\mathbf{h}}_{-1,i}^{(j)} = \mathbf{0}$ and $\mathbf{\Sigma}_{-1,i}^{(j)} = \mathbf{I}\varepsilon^{-2}$, $\forall j$. The updated posterior p.d.f. of the channel coefficients at instant k is, on the other hand, written as [3], [7]

$$
p(\mathbf{h}_i|\mathbf{s}_{0:k}^{(j)}, \mathbf{z}_{0:k,1:R}) = \xi_M(\mathbf{h}_i|\, \hat{\mathbf{h}}_{k,i}^{(j)}, \mathbf{\Sigma}_{k,i}^{(j)}) \qquad i \in \{1, \ldots, R\} \tag{4.64}
$$

4.5.1 CONSENSUS IMPLEMENTATIONS WITH MOMENT-MATCHING PARAMETRIC APPROXIMATIONS

Reference [5] implements (4.58) in a fully distributed fashion using an iterative average consensus algorithm as described in Section 4.3.1. As explained in Section 4.3.1, such implementation requires that N_p consensus protocols (one for each particle j) be run in parallel.

In order to reduce the internode communication cost, reference [7] follows a methodology that is similar to the moment-matching approximation described in Section 4.3.2, see also Eq. (4.35), and replaces $p(\mathbf{h}_i|\mathbf{s}_{0:k-1}^{(j)}, \mathbf{z}_{0:k-1,1:R})$, $i \in \{1 \ldots R\}$, for any sequence $\mathbf{s}_{0:k-1}^{(j)}$, with the alternative probability density function, independent of j, $\tilde{p}(\mathbf{h}_i|\mathbf{z}_{0:k-1,1:R}) = \xi_M(\mathbf{h}_i|\, \tilde{\mathbf{h}}_{k-1,i}, \tilde{\mathbf{\Sigma}}_{k-1,i})$ such that the first- and second-order moments associated with $\tilde{p}$ coincide with those associated with

$$
E\{p(\mathbf{h}_i|\mathbf{s}_{0:k-1}, \mathbf{z}_{0:k-1,1:R}) \mid \mathbf{z}_{0:k-1,1:R}\} \approx \sum_{j=1}^{N_p} w_{k-1,i}^{(j)} \xi_M(\mathbf{h}_i|\, \hat{\mathbf{h}}_{k-1,i}^{(j)}, \mathbf{\Sigma}_{k-1,i}^{(j)}). \tag{4.65}
$$

Specifically, given a weighted set of particles $\{(w_{k-1,i}^{(j)}, \mathbf{s}_{0:k-1}^{(j)}\}, j \in \{1, \ldots, N_p\}$ that represent at node i and instant $k-1$ the posterior p.m.f. $P(\mathbf{s}_{0:k-1}|\mathbf{z}_{0:k-1,1:R})$, the modified hyperparameters that specify $\tilde{p}$ are obtained in a straightforward fashion by making

[7]

$$\tilde{\mathbf{h}}_{k-1,i} = \sum_{j=1}^{N_p} w_{k-1,i}^{(j)} \rho_{k-1,i}^{(j)} \hat{\mathbf{h}}_{k-1,i}^{(j)}, \qquad (4.66)$$

$$\tilde{\boldsymbol{\Sigma}}_{k-1,i} = \sum_{j=1}^{N_p} w_{k-1,i}^{(j)} \left\{ [\hat{\mathbf{h}}_{k-1,i}^{(j)} (\hat{\mathbf{h}}_{k-1,i}^{(j)})^T + \boldsymbol{\Sigma}_{k-1,i}^{(j)}] \right\}$$
$$- \tilde{\mathbf{h}}_{k-1,i} \tilde{\mathbf{h}}_{k-1,i}^T,$$

where $\rho_{k-1,i}^{(j)} \in \{\pm 1\}$ are correction constants designed to circumvent phase ambiguity among particles and defined so that $\rho_{k-1,i}^{(j)} = -\rho_{k-1,i}^{(j)}$ if $\mathbf{s}_{k-1,i}^{(j)} = -\mathbf{s}_{k-1,i}^{(j')}$.

Next, reference [7] proposes to change the weight update rule (4.58) at each node r to

$$w_{k,r}^{(j)} = w_{k-1,r}^{(j)} \tilde{\lambda}_{k,r}^{(j)}(\mathbf{s}_{k,r}^{(j)}) \prod_{i \neq r, i \in \{1,...,R\}} \tilde{\lambda}_{k,i}(\mathbf{s}_{k,r}^{(j)}) \qquad (4.67)$$

where, mimicking (4.59),

$$\tilde{\lambda}_{k,i}(\mathbf{s}_{k,r}^{(j)}) = \int_{\mathcal{R}^M} p(\mathbf{z}_{k,i}|\mathbf{s}_{k,r}^{(j)}, \mathbf{h}_i) \, \xi_M(\mathbf{h}_i | \tilde{\mathbf{h}}_{k-1,i}, \tilde{\boldsymbol{\Sigma}}_{k-1,i}) \, d\mathbf{h}_i$$
$$= \xi \left(z_{k,i} | (\tilde{\mathbf{h}}_{k-1,i})^T \mathbf{s}_{k,r}^{(j)} ; \sigma_i^2 + (\mathbf{s}_{k,r}^{(j)})^T \tilde{\boldsymbol{\Sigma}}_{k-1,i} \mathbf{s}_{k,r}^{(j)} \right).$$

Since $\mathbf{s}_{k,r}^{(j)}$, in the case of transmission of binary symbols, can take only 2^M distinct values, it suffices to compute $\prod_{i \in \{1,...,R\}} \tilde{\lambda}_{k,i}(\mathbf{s}_{k,r}^{(j)})$ for all possible values of $\mathbf{s}_{k,r}^{(j)}$ and, from that table, evaluate (4.67) at node r at instant k for the specific value of $\mathbf{s}_{k,r}^{(j)}$ sampled from $P(\mathbf{s}_k|\mathbf{s}_{k-1,r}^{(j)})$. The required table may be built with only 2^M parallel average consensus protocols, instead of N_p average consensus protocols as in [5]. Alternatively, as explained in Section 4.3.1, minimum consensus or flooding may be used instead of average consensus. In any case, if the channel impulse response length M is much less than $\log_2 N_p$, the reduction in internode communication cost is significant when the moment-matching parametric approximation is used.

4.5.2 NUMERICAL RESULTS

The steady-state performance and communication complexity of distributed blind equalization algorithms were assessed via numerical simulations consisting of 2000 independent Monte Carlo runs, in each of which a random sequence of 300 i.i.d. bits was transmitted, being the first 200 bits discarded to allow for convergence. The simulated system comprises one transmitter and $R = 4$ receiving nodes. The receiving nodes communicate through a static flat topology as shown in Figure 4.5.

The transmission channels between the transmitter and the receivers have $M = 3$ coefficients, and were obtained by sampling independently, in each realization and for each receiver $\mathbf{h}_i \sim \mathcal{N}_M(\mathbf{0}; \mathbf{I})$, and normalizing so that $\|\mathbf{h}_i\| = 1$. The noise variances were determined as $\sigma_i^2 = \|\mathbf{h}_i\|^2/\text{SNR}$, where SNR denotes the desired value of

Figure 4.5 Receiver network topology.

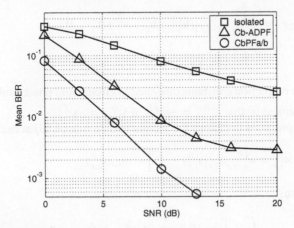

Figure 4.6 Mean bit error rate (BER) recorded for different filtering algorithms as a function of the signal-to-noise ratio (SNR).

signal-to-noise ratio. The model hyperparameter ε was set to 1. For all tested algorithms, the filters employ $N_P = 300$ particles and perform systematic resampling [21] at every time step.

Figure 4.6 displays the mean bit error rates averaged over all network nodes, respectively, for the CbPFa and CbPFb algorithms in Section 4.5, which implement the product in Eq. (4.58) exactly using respectively minimum consensus and flooding, and for the modified Cb-ADPF algorithm, which employs a fully distributed, flooding-based implementation of the approximate weight update factors in Eq. (4.67), see Section 4.5.1. For comparison, Figure 4.6 also displays the performance of isolated receivers that only assimilate local observations and do not cooperate with other network nodes. As expected, CbPFa/b are the top performers since, as discussed in Section 4.4.3, they are equivalent to the optimal centralized PF-based estimator. Cb-ADPF is surpassed by CbPFa/b by a raising margin as the performance of the former converges to a plateau for high SNRs, but still outmatches the isolated receivers and, as discussed in the sequel, has a much lower internode communication cost than CbPFa/b.

Table 4.2 shows the computational and communication effort required by the same set of algorithms averaged over all network nodes. We measured the reception rates (RX) in Kbytes (KB) per processed symbol, which equal the transmission rates (TX) in case no form of point-to-multipoint transmission is employed. Similarly to Section 4.4.3, we used respectively a four-byte and one-byte representation for real and Boolean values. Differently from Section 4.4.3, however, we assumed

Table 4.2

Average Communication and Processing Cost per Symbol per Node

	RX Rate	Flops
CbPFa	21.1 KB/symbol	$O\left(M^2 N_P\right)$
CbPFb	6.23 KB/symbol	$O\left(M^2 N_P\right)$
Cb-ADPF	0.203 KB/symbol	$2\,O\left(M^2 N_P\right)$

that the *flooding*-based algorithms (CbPFb and Cb-ADPF) attach to the transmitted messages a one-byte label with the original node index for use in the ordering operations; this has the effect of turning the complexity of such algorithms independent of the (random) transmitted values.

The computational complexity of CbPFa/b is essentially dictated by the need to run N_P Kalman filter steps per node per symbol. Cb-ADPF, in turn, evaluates Gaussian approximation parameters whose computation demands additional $O(M^2 N_P)$ operations. On the other hand, the communication complexity of Cb-ADPF is about 30 times smaller when compared to CbPFb and about 100 times smaller when compared to CbPFa.

4.6 CONCLUSIONS

We provided in this chapter an overview of different novel distributed particle filtering algorithms for state tracking and parameter estimation in cooperative sensor/receiver networks. The choice of particle filters to solve the distributed tracking/estimation problem is motivated by the poor performance of alternative linear methods such as distributed Kalman filters in applications where the underlying signal model is nonlinear/non-Gaussian, see also an extended discussion about this topic in [9], [19].

The proposed distributed particle filters in this chapter assume partially connected networks, where nodes can only communicate directly with their immediate neighbors, and fall into two main classes, namely consensus and diffusion algorithms. The consensus solutions require iterative internode communication between consecutive sensor measurements and are therefore not very well suited for real-time applications, but, provided that the required bandwidth is available, they are capable of exactly reproducing the optimal centralized PF solution at each time instant. In particular, two exact, fully distributed implementations of the optimal solution, using respectively iterative minimum consensus and flooding, and referred to as the CbPFa and CbPFb algorithms [16], are reviewed in the chapter. Diffusion methods, on the other hand, eliminate the need for multiple internode message passes in the time interval between sensor measurements, thus allowing processing and sensing on the same time scale, but are suboptimal in the sense they do not mimic at each time step the optimal centralized solution. In particular, we follow in this chapter a diffusion strategy based on random information dissemination [30] and review the Random Exchange Diffusion Particle Filter [17], [18], [19], which builds over time, at each node, different Monte Carlo representations of the posterior distributions of the unknown states or parameters conditioned on different subsets of measurements coming from random locations in the entire network and not only from a node's immediate vicinity.

The performance of the proposed algorithms is briefly illustrated in applications of cooperative emitter tracking in passive sensor networks and cooperative blind equalization of digital communication channels in receiver networks. Introducing additional parametric approximations [4], [19], we were able to achieve further internode communication cost savings while retaining acceptable state estimation accuracy.

REFERENCES

1. Y. Bar-Shalom and X. Li. *Multitarget-Multisensor Tracking: Principles and Techniques.* YBS, 1995.
2. C.J. Bordin Jr. and M.G.S. Bruno. Cooperative blind equalization of frequency-selective channels in sensor networks using decentralized particle filtering. In *42nd Asilomar Conf. on Sign., Syst. and Comp.*, pages 1198–1201, October 2008.
3. C.J. Bordin Jr. and M.G.S. Bruno. Particle Filters for Joint Blind Equalization and Decoding in Frequency Selective Channels. *IEEE Transactions on Signal Processing*, 6(56):2395–2405, June 2008.
4. C.J. Bordin Jr. and M.G.S. Bruno. A particle filtering algorithm for cooperative blind equalization using vb parametric approximations. In *IEEE International Conference on Acoustics, Speech and Signal Processing*, pages 3834–3837, March 2010.
5. C.J. Bordin Jr. and M.G.S. Bruno. Consensus-based distributed particle filtering algorithms for cooperative blind equalization in receiver network. In *IEEE International Conference on Acoustics, Speech and Signal Processing*, pages 3968–397, May 2011.
6. C.J. Bordin Jr. and M.G.S. Bruno. A new minimum-consensus distributed particle filter for blind equalization in receiver networks. In *IEEE International Conference on Acoustics, Speech and Signal Processing*, pages 6323–632, May 2013.
7. C.J. Bordin Jr. and M.G.S. Bruno. Distributed particle filtering for blind equalization in receiver networks using marginal non-parametric approximations. In *IEEE International Conference on Acoustics, Speech and Signal Processing*, pages 7984–7987, May 2014.
8. S. Boyd, A. Gosh, B. Prabhakar, and D. Shah. Randomized gossip algorithms. *IEEE Transactions on Information Theory*, 52(6):2508–2530, June 2006.
9. M.G.S. Bruno. *Sequential Monte Carlo methods for nonlinear discrete-time filtering.* Morgan & Claypool Publishers, February 2013.
10. G. Casella and C.P. Robert. Rao-Blackwellisation of sampling schemes. *Biometrika*, 83(1):81–94, 1996.
11. F.S. Cattivelli, C.G. Lopes, and A.H. Sayed. Diffusion Recursive Least-Squares for Distributed Estimation Over Adaptive Networks. *IEEE Transactions on Signal Processing*, 56(5):1865–1877, May 2008.
12. F.S. Cattivelli and A.H. Sayed. Diffusion strategies for distributed Kalman filtering and smoothing. *IEEE Transactions on Automatic Control*, 55(9):2069–2084, September 2010.

13. M.J. Coates. Distributed particle filtering for sensor networks. In *3rd Intl. Symp. on Inf. Proc. in Sensor Networks*, pages 99–107, April 2004.

14. D. Crisan and A. Doucet. A Survey of convergence results on particle filtering methods for practitioners. *IEEE Transactions on Signal Processing*, 50(3):736–746, March 2002.

15. S.S. Dias and M.G.S. Bruno. Cooperative particle filtering for emitter tracking with unknown noise variance. In *IEEE International Conference on Acoustics, Speech and Signal Processing*, pages 2629–2632, March 2012.

16. S.S. Dias and M.G.S. Bruno. Cooperative target tracking using decentralized particle filtering and RSS sensors. *IEEE Transactions on Signal Processing*, 61(14):3632–3646, July 2013.

17. S.S. Dias and M.G.S. Bruno. Distributed emitter tracking using random exchange diffusion particle filters. In *IEEE Int. Con. on Information Fusion*, pages 1251–1257, July 2013.

18. S.S. Dias and M.G.S. Bruno. A rao-blackwellized random exchange diffusion particle filter for distributed emitter tracking. In *IEEE Int. Work. on Comp. Adv. in Multi-sensor Adapt. Process. (CAMSAP)*, pages 348–351, December 2013.

19. S.S. Dias and M.G.S. Bruno. Cooperative emitter tracking using Rao-Blackwellized random exchange diffusion particle filters. *EURASIP Journal on Advances in Signal Processing*, pages 1–18, February 2014.

20. P.M. Djurić, J. Beaudeau, and M.F. Bugallo. Non-centralized target tracking with mobile agents. In *IEEE International Conference on Acoustics, Speech and Signal Processing*, pages 5928–5931, May 2011.

21. R. Douc and O. Cappe. Comparison of resampling schemes for particle filtering. In *4th Int'l. Symp. on Image and Sig. Proc. and Analysis (ISPA)*, pages 64–69, September 2005.

22. A. Doucet, S.J. Godsill, and C. Andrieu. On sequential Monte Carlo sampling methods for Bayesian filtering. *Statistics and Computing*, 10(3):197–208, 2000.

23. S. Farahmand, S.I. Roumeliotis, and G.B. Giannakis. Particle filter adaptation for distributed sensors via set membership. In *IEEE International Conference on Acoustics, Speech and Signal Processing*, pages 3374–3377, March 2010.

24. A. Gelman, J.B. Carlin, H.S. Stern, and D.B. Rubin. *Bayesian Data Analysis*. Texts in Statistical Science. Chapman & Hall/CRC, 2nd edition, 2004.

25. J. Geweke. Bayesian inference in econometric models using Monte Carlo integration. *Econometrica*, 57(6):1317–1339, November 1989.

26. M. Hata. Empirical formula for propagation loss in land mobile radio services. *IEEE Transactions on Vehicular Technology*, 29(3):317–325, 1980.

27. O. Hlinka, F. Hlawatsch, and P.M. Djuric. Distributed Particle Filtering in Agent Networks: A Survey, Classification, and Comparison. *IEEE Signal Processing Magazine*, 30(1):61–81, January 2013.

28. O. Hlinka, O. Sluciak, F. Hlawatsch, P. Djuric, and M. Rupp. Distributed gaussian particle filtering using likelihood consensus. In *IEEE International Conference on Acoustics, Speech and Signal Processing*, pages 3756–3759, May 2011.

29. O. Hlinka, O. Sluciak, F. Hlawatsch, P. Djuric, and M. Rupp. Likelihood consensus and its applications to distributed particle filtering. *IEEE Transactions on Signal Processing*, 60(8):4334–4349, August 2012.

30. S. Kar and J.M.F. Moura. Gossip and distributed Kalman filtering: weak consensus under weak detectability. *IEEE Transactions on Signal Processing*, 59(4):1766–1784, April 2011.

31. S. Kar and J.M.F. Moura. Consensus + innovations distributed inference over networks: cooperation and sensing in networked systems. *IEEE Signal Processing Magazine*, 30(3):99–109, May 2013.

32. S.M. Kay. *Fundamentals of Statistical Signal Processing: Volume I, Estimation Theory*. Prentice Hall Inc, 1993.

33. C.G. Lopes and A.H. Sayed. Incremental Adaptive Strategies Over Distributed Networks. *IEEE Transactions on Signal Processing*, 55(8):4064–4077, August 2007.

34. A. Mohammadi and A. Asif. Consensus-based distributed unscented particle filter. In *2011 IEEE Statistical Signal Processing Workshop*, pages 237–240, June 2011.

35. A. Mohammadi and A. Asif. A consensus/fusion based distributed implementation of the particle filter. In *4th IEEE Int. Work. on Comp. Adv. in Multi-sensor Adapt. Process.*, pages 285–288, December 2011.

36. J.R. Norris. *Markov Chains*. Cambridge Series in Statistical and Probabilistic Mathematics. Cambridge University Press, 1997.

37. Y. Okumura, E. Ohmori, T. Kawano, and K. Fukuda. Field strength and its variability in VHF and UHF land-mobile radio service. *Rev. Elec. Commun. Lab.*, 16(9):825–873, September 1968.

38. N. Patwari, A.O. Hero III, M. Perkins, N.S. Correal, and R.J. O'Dea. Relative location estimation in wireless sensor networks. *IEEE Transactions on Signal Processing*, 51(8):2137–2148, March 2003.

39. A. Ribeiro, G.B. Giannakis, and S.I. Roumeliotis. SOI-KF: distributed Kalman filtering with low-cost communications using the sign of innovations. *IEEE Transactions on Signal Processing*, 54(12):4782–4795, December 2006.

40. A.H. Sayed, S.Y. Tu, J. Chen, and X. Zhao. Diffusion strategies for adaptation and learning over networks: an examination of distributed strategies and network behavior. *IEEE Signal Processing Magazine*, 30(3):155–171, May 2013.

41. I.D. Schizas, G.B. Giannakis, S.I. Roumeliotis, and A. Ribeiro. Consensus in Ad Hoc WSNs with noisy links-Part II: distributed estimation and smoothing of random signals. *IEEE Transactions on Signal Processing*, 56(4):1650–1666, April 2008.

42. X. Sheng, Y.H. Hu, and P. Ramanathan. Distributed particle filter with gmm approximation for multiple targets localization and tracking in wireless sensor network. In *Proc. IPSN*, pages 181–188, April 2005.

43. D. Tsoumakos and N. Roussoupoulos. A comparison of peer-to-peer search methods. In *Proc. of the WebDB*, pages 61–66, June 2003.

44. D. Uztebay, M. Coates, and M. Rabat. Distributed auxiliary particle filter using selective gossip. In *IEEE International Conference on Acoustics, Speech and Signal Processing*, pages 3296–3299, May 2011.

45. M. Vemula, M.F. Bugallo, and P.M. Djurić. Target tracking by fusion of random measures. *Signal, Image and Video Processing*, 1(2):149–161, 2007.

46. L. Xiao and S. Boyd. Fast linear iterations for distributed averaging. *Systems & Control Letters*, 53(1):65–78, September 2004.

47. V. Yadav and M.V. Salapaka. Distributed protocol for determining when averaging consensus is reached. In *45th Annual Allerton Conf.*, pages 715–720, September 2007.

4.A APPENDIX I

In this appendix, we use an importance sampling methodology, see [22], [9], to show that the augmented particle set $\mathbf{s}_{0:k,r}^{(j)} = \{(\mathbf{s}_{0:k-1,t}^{(j)}, \mathbf{s}_{k,r}^{(j)})\}$, $j = 1, \ldots, N_p$ with weights

$\{w_{k,r}^{(j)}\}$ is a properly weighted set to represent the posterior PDF $p(s_{0:k}|Z_{k,r}, Z_{0:k-1,t})$ in the sense that for any measurable function $\Omega(\cdot)$,

$$E\{\Omega(s_{0:k})|Z_{k,r}, Z_{0:k-1,t}\} \approx \sum_{j=1}^{N_p} w_{k,r}^{(j)} \Omega(s_{0:k,r}^{(j)}).$$

Specifically, let $\{s_{0:k-1,t}^{(j)}\}$ with associated weights $\{w_{k-1,t}^{(j)}\}$, $j \in \{1, \ldots, N_p\}$, be a properly weighted set that represents the posterior PDF $p(s_{0:k-1}|Z_{0:k-1,t})$ at node t. Assuming that the particle set $\{s_{0:k-1,t}^{(j)}\}$ was sampled according to some proposal importance function $q(s_{0:k-1}|Z_{0:k-1,t})$, the proper weights $\{w_{k-1,t}^{(j)}\}$ may be written as [22], [9]

$$w_{k-1,t}^{(j)} = \frac{w(s_{0:k-1,t}^{(j)})}{\sum_{m=1}^{N_p} w(s_{0:k-1,t}^{(m)})} \qquad j \in \{1 \ldots N_p\}, \tag{4.68}$$

where

$$w(s_{0:k-1,t}^{(j)}) = \frac{p(s_{0:k-1,t}^{(j)}|Z_{0:k-1,t})}{q(s_{0:k-1,t}^{(j)}|Z_{0:k-1,t})}.$$

Assume next that node t sends its particle set and weights to a neighboring node r that can access at instant k the measurements $Z_{k,r} = \{z_{k,r}\} \cup \{z_{k,i}\}_{i\in N(r)}$. For any measurable function $\Omega(\cdot)$, we note that

$$E\{\Omega(s_{0:k})|Z_{k,r}, Z_{0:k-1,t}\} =$$

$$\frac{\int \Omega(s_{0:k}) \dfrac{p(s_{0:k}|Z_{k,r}, Z_{0:k-1,t})}{p(s_k|s_{k-1})q(s_{0:k-1}|Z_{0:k-1,t})} p(s_k|s_{k-1})q(s_{0:k-1}|Z_{0:k-1,t})\, ds_{0:k}}{\int \dfrac{p(s_{0:k}|Z_{k,r}, Z_{0:k-1,t})}{p(s_k|s_{k-1})q(s_{0:k-1}|Z_{0:k-1,t})} p(s_k|s_{k-1})q(s_{0:k-1}|Z_{0:k-1,t})\, ds_{0:k}}. \tag{4.69}$$

Sampling now at node r new particles $s_{k,r}^{(j)} \sim p(s_k|s_{k-1,t}^{(j)})$ and building the augmented particle trajectories $s_{0:k,r}^{(j)} = (s_{0:k-1,t}^{(j)}, s_{k,r}^{(j)}) \sim p(s_k|s_{k-1}) q(s_{0:k-1}|Z_{0:k-1,t})$ the integral on the right-hand side of (4.69) can be approximated as

$$E\{\Omega(s_{0:k})|Z_{k,r}, Z_{0:k-1,t}\} \approx \frac{\frac{1}{N_p}\sum_{j=1}^{N_p} \Omega(s_{0:k,r}^{(j)}) w(s_{0:k,r}^{(j)})}{\frac{1}{N_p}\sum_{m=1}^{N_p} w(s_{0:k,r}^{(m)})}$$

$$= \sum_{j=1}^{N_p} w_{k,r}^{(j)} \Omega(s_{0:k,r}^{(j)}) \tag{4.70}$$

where

$$w_{k,r}^{(j)} = \frac{w(s_{0:k,r}^{(j)})}{\sum_{m=1}^{N_p} w(s_{0:k,r}^{(m)})} \tag{4.71}$$

and

$$
\begin{aligned}
w(\mathbf{s}_{0:k}) &= \frac{p(\mathbf{s}_{0:k}|\mathbf{Z}_{k,r}, \mathbf{Z}_{0:k-1,t})}{p(\mathbf{s}_k|\mathbf{s}_{k-1})q(\mathbf{s}_{0:k-1}|\mathbf{Z}_{0:k-1,t})} \\
&= \frac{p(\mathbf{Z}_{k,r}|\mathbf{s}_{0:k}, \mathbf{Z}_{0:k-1,t})\, p(\mathbf{s}_k|\mathbf{s}_{0:k-1}, \mathbf{Z}_{0:k-1,t})}{p(\mathbf{s}_k|\mathbf{s}_{k-1})p(\mathbf{Z}_{k,r}|\mathbf{Z}_{0:k-1,t})}\, w(\mathbf{s}_{0:k-1}).
\end{aligned}
\tag{4.72}
$$

Substituting (4.72) into (4.71) and recalling from the model assumptions that $p(\mathbf{s}_k|\mathbf{s}_{0:k-1}, \mathbf{Z}_{0:k-1,t}) = p(\mathbf{s}_k|\mathbf{s}_{k-1})$, we get the recursion

$$
\begin{aligned}
w_{k,r}^{(j)} &= \frac{w(\mathbf{s}_{0:k-1,t}^{(j)})\, p(\mathbf{Z}_{k,r}|\mathbf{s}_{0:k,r}^{(j)}, \mathbf{Z}_{0:k-1,t})}{\sum_{m=1}^{N_p} w(\mathbf{s}_{0:k-1,t}^{(m)})\, p(\mathbf{Z}_{k,r}|\mathbf{s}_{0:k,r}^{(m)}, \mathbf{Z}_{0:k-1,t})} \\
&= \frac{w_{k-1,t}^{(j)}\, p(\mathbf{Z}_{k,r}|\mathbf{s}_{0:k,r}^{(j)}, \mathbf{Z}_{0:k-1,t})}{\sum_{m=1}^{N_p} w_{k-1,t}^{(m)}\, p(\mathbf{Z}_{k,r}|\mathbf{s}_{0:k,r}^{(m)}, \mathbf{Z}_{0:k-1,t})}
\end{aligned}
\tag{4.73}
$$

where, as before,

$$
w_{k-1,t}^{(j)} = \frac{w(\mathbf{s}_{0:k-1,t}^{(j)})}{\sum_{l=1}^{N_p} w(\mathbf{s}_{0:k-1,t}^{(l)})}.
$$

4.B APPENDIX II

Assume, as in Eq. (4.26) in Section 4.3.2, that, at node t at instant $k-1$, the joint posterior p.d.f. $p(\boldsymbol{\theta}_{1:R}|\mathbf{s}_{0:k-1,t}^{(j)}, \mathbf{Z}_{0:k-1,t})$ is factored as

$$
p(\boldsymbol{\theta}_{1:R}|\mathbf{s}_{0:k-1,t}^{(j)}, \mathbf{Z}_{0:k-1,t}) = \prod_{i=1}^{R} \phi(\boldsymbol{\theta}_i|\boldsymbol{\psi}_{i,k-1,t}^{(j)}).
\tag{4.74}
$$

Assume also that node t transmits to a neighboring node r its weighted particle set $\{(w_{k-1,t}^{(j)}, \mathbf{s}_{k-1,t}^{(j)})\}$ and the corresponding hyperparameter vectors $\{\boldsymbol{\psi}_{i,k-1,t}^{(j)}\}$, $j = 1, \ldots, N_p$, $i = 1, \ldots, R$. At instant k, as explained in Section 4.3.2, node r samples a new set of particles $\mathbf{s}_{k,r}^{(j)} \sim p(\mathbf{s}_k|\mathbf{s}_{k-1,t}^{(j)})$ and updates its weights as

$$
\begin{aligned}
w_{k,r}^{(j)} &= w_{k-1,t}^{(j)}\, p(\mathbf{Z}_{k,r}|\mathbf{s}_{k,r}^{(j)}, \mathbf{s}_{0:k-1,t}^{(j)}, \mathbf{Z}_{0:k-1,t}) \\
&= w_{k-1,t}^{(j)} \int_{\mathfrak{R}^{N_\theta}} \cdots \int_{\mathfrak{R}^{N_\theta}} \Big[p(\mathbf{Z}_{k,r}|\mathbf{s}_{k,r}^{(j)}, \boldsymbol{\theta}_{1:R}) \\
&\qquad \times p(\boldsymbol{\theta}_{1:R}|\mathbf{s}_{k,r}^{(j)}, \mathbf{s}_{0:k-1,t}^{(j)}, \mathbf{Z}_{0:k-1,t}) \Big]\, d\boldsymbol{\theta}_{1:R} \\
&= w_{k-1,t}^{(j)} \Bigg\{ \prod_{i \in \widetilde{\mathbf{N}}(r)} \int_{\mathfrak{R}^{N_\theta}} \Big[p(\mathbf{z}_{k,i}|\mathbf{s}_{k,r}^{(j)}, \boldsymbol{\theta}_i) \\
&\qquad \times \phi(\boldsymbol{\theta}_i|\boldsymbol{\psi}_{i,k-1,t}^{(j)}) \Big]\, d\boldsymbol{\theta}_i \Bigg\} \\
&\qquad \times \prod_{i \notin \widetilde{\mathbf{N}}(r)} \underbrace{\int_{\mathfrak{R}^{N_\theta}} \phi(\boldsymbol{\theta}_i|\boldsymbol{\psi}_{i,k-1,t}^{(j)})\, d\boldsymbol{\theta}_i}_{=\,1},
\end{aligned}
\tag{4.75}
$$

where $\widetilde{\mathbf{N}}(r)$ denotes $\{r\} \cup \mathbf{N}(r)$ and, as in Section 4.3.2, $\mathbf{N}(r)$ is a notation for the neighborhood of node r, and $\mathbf{Z}_{k,r}$ is a notation for the set $\{\mathbf{z}_{k,i}\}$ for all $i \in \widetilde{\mathbf{N}}(r)$. In (4.75), we used the fact that

$$
\begin{aligned}
p(\mathbf{Z}_{k,r}|\mathbf{s}_{k,r}^{(j)}, \mathbf{s}_{0:k-1,t}^{(j)}, \boldsymbol{\theta}_{1:R}, \mathbf{Z}_{0:k-1,t}) &= p(\mathbf{Z}_{k,r}|\mathbf{s}_{k,r}^{(j)}, \boldsymbol{\theta}_{1:R}) \\
&= \prod_{i \in \{r\} \cup \mathbf{N}(r)} p(\mathbf{z}_{k,i}|\mathbf{s}_{k,r}^{(j)}, \boldsymbol{\theta}_i)
\end{aligned}
\tag{4.76}
$$

and

$$
\begin{aligned}
&p(\boldsymbol{\theta}_{1:R}|\mathbf{s}_{k,r}^{(j)}, \mathbf{s}_{0:k-1,t}^{(j)}, \mathbf{Z}_{0:k-1,t}) \\
&= \frac{p(\mathbf{s}_{k,r}^{(j)} \mid \mathbf{s}_{0:k-1,t}^{(j)}, \boldsymbol{\theta}_{1:R}, \mathbf{Z}_{0:k-1,t})\, p(\boldsymbol{\theta}_{1:R}|\mathbf{s}_{0:k-1,t}^{(j)}, \mathbf{Z}_{0:k-1,t})}{p(\mathbf{s}_{k,r}^{(j)} \mid \mathbf{s}_{0:k-1,t}^{(j)}, \mathbf{Z}_{0:k-1,t})} \\
&= \frac{p(\mathbf{s}_{k,r}^{(j)} \mid \mathbf{s}_{k-1,t}^{(j)})\, p(\boldsymbol{\theta}_{1:R}|\mathbf{s}_{0:k-1,t}^{(j)}, \mathbf{Z}_{0:k-1,t})}{p(\mathbf{s}_{k,r}^{(j)} \mid \mathbf{s}_{k-1,t}^{(j)})} \\
&= p(\boldsymbol{\theta}_{1:R}|\mathbf{s}_{0:k-1,t}^{(j)}, \mathbf{Z}_{0:k-1,t}),
\end{aligned}
\tag{4.77}
$$

which, in turn, is assumed to be factored as in Equation (4.74).

Similarly, node r at instant k updates the posterior PDF of the unknown variances as

$$
\begin{aligned}
&p(\boldsymbol{\theta}_{1:R}|\mathbf{s}_{k,r}^{(j)}, \mathbf{s}_{0:k-1,t}^{(j)}, \mathbf{Z}_{k,r}, \mathbf{Z}_{0:k-1,t}) \\
&= C_k\, p(\mathbf{Z}_{k,r}|\mathbf{s}_{k,r}^{(j)}, \mathbf{s}_{0:k-1,t}^{(j)}, \boldsymbol{\theta}_{1:R}, \mathbf{Z}_{0:k-1,t})\, p(\boldsymbol{\theta}_{1:R}|\mathbf{s}_{k,r}^{(j)}, \mathbf{s}_{0:k-1,t}^{(j)}, \mathbf{Z}_{0:k-1,t}) \\
&= C_k \left[\prod_{i \in \widetilde{\mathbf{N}}(r)} p(\mathbf{z}_{k,i}|\mathbf{s}_{k,r}^{(j)}, \boldsymbol{\theta}_i)\, \phi(\boldsymbol{\theta}_i|\boldsymbol{\psi}_{i,k-1,t}^{(j)}) \right] \left[\prod_{i \notin \widetilde{\mathbf{N}}(r)} \phi(\boldsymbol{\theta}_i|\boldsymbol{\psi}_{i,k-1,t}^{(j)}) \right]
\end{aligned}
\tag{4.78}
$$

$$
= \prod_{i=1}^{R} \phi(\boldsymbol{\theta}_i|\boldsymbol{\psi}_{i,k,r}^{(j)})
\tag{4.79}
$$

if $\phi(\boldsymbol{\theta}_i|\boldsymbol{\psi}_{i,k-1,t}^{(j)})$ is a conjugate prior p.d.f. [24] for $p(\mathbf{z}_{k,i}|\mathbf{s}_{k,r}^{(j)}, \boldsymbol{\theta}_i)$. Note that, in Eq. (4.79), $\phi(\boldsymbol{\theta}_i|\boldsymbol{\psi}_{i,k,r}^{(j)})$ is equal to $\phi(\boldsymbol{\theta}_i|\boldsymbol{\psi}_{i,k-1,t}^{(j)})$ if $i \notin \widetilde{\mathbf{N}}(r)$. Otherwise, if $i \in \widetilde{\mathbf{N}}(r)$, it is updated according to (4.78) using $\boldsymbol{\psi}_{i,k-1,t}^{(j)}$, the new particles $\mathbf{s}_{k,r}^{(j)}$ sampled at node r at instant k, and the new observations $\{\mathbf{z}_{k,i}\}$ assimilated by node r at instant k. The normalization constant C_k in (4.78) is in turn given by

$$
C_k = \left\{ \prod_{i \in \widetilde{\mathbf{N}}(r)} \left[\int_{\mathfrak{R}^{N_\theta}} p(\mathbf{z}_{k,i}|\mathbf{s}_{k,r}^{(j)}, \boldsymbol{\theta}_i)\phi(\boldsymbol{\theta}_i|\boldsymbol{\psi}_{i,k-1,t}^{(j)})d\boldsymbol{\theta}_i \right] \right\}^{-1}.
\tag{4.80}
$$

Acoustic Signal Processing

Rosângela Fernandes Coelho
Vítor Heloiz Nascimento

5 Empirical Mode Decomposition Theory Applied to Speech Enhancement

Rosângela Fernandes Coelho
Military Institute of Engineering (IME)

Leonardo Augusto Zão
Military Institute of Engineering (IME)

5.1 INTRODUCTION

Empirical Mode Decomposition (EMD) is a theoretical concept and a powerful tool for signal analysis and processing in time domain. The data-driven decomposition is devoted to treat nonlinear and non-stationary sample sequences.

Originally, EMD was defined by Huang *et al.* [27] to be used in conjunction with the Hilbert transform to estimate the instantaneous frequencies of a signal. In [26], the authors proved the effectiveness of the Hilbert spectral analysis by investigating the behavior of water surface waves, i.e., a scenario in which the linearity and stationarity assumptions are inappropriate. Since then, much effort has been made in the literature to enrich the EMD theory. Some of these approaches are noise-assisted EMD methods such as the ensemble empirical mode decomposition (EEMD) [56] and the complete EEMD with adaptive noise (CEEMDAN) [9]. Recently, another class of solutions was proposed to expand the EMD concepts based on convex optimization tools, e.g., sequential variational modal decomposition (Seq-VMD) [43].

By definition, EMD decomposes a signal into a series of oscillatory intrinsic mode functions (IMF) and a residual component. Different from other methods such as wavelets, in the EMD a set of *a priori* basis functions is not required for the decomposition process. The wavelet decomposition is not adaptable to local or temporary variations of non-stationary signals since it uses linear time-invariant filters. Instead, the IMFs are completely based on the local properties of the input signal.

The benefits of the EMD have been demonstrated by the analysis of numerous natural and artificial phenomena of different research fields and applications [25]; for instance, biomedical engineering [29], financial market [18], speech processing [5, 32, 60], and image processing [36, 41]. This chapter presents the EMD theory applied to an important issue of the speech processing field: speech enhancement.

The most recent EMD-based speech enhancement methods are discussed and evaluated under non-stationary acoustic noises. The solutions are compared to techniques based on the traditional Fourier spectral analysis. The results considering different objective measures demonstrate that the EMD-based solutions attained interesting speech quality and intelligibility in several noisy scenarios.

The chapter is organized as follows. Section 5.2 introduces the main concepts and the definition of the EMD. Section 5.2.1 includes the original EMD algorithm [27] and the interpretation of the IMFs in a dyadic filterbank structure. Section 5.2.2 highlights some of the major EMD challenges: the mode mixing problem, the computation of the envelope, and the sampling effects. Alternative EMD algorithms proposed in the literature to overcome these drawbacks are also presented in Section 5.2.2. Section 5.3 aims at discussing the main advantages of adopting the EMD in speech enhancement techniques. The most recent EMD-based approaches and a novel speech enhancement proposal are presented in Section 5.3.1. Some of the traditional spectral speech enhancement techniques are briefly summarized in Section 5.3.2. The spectral and the EMD-based speech enhancement techniques are compared in the experiments described in Section 5.4. Finally, Section 5.5 concludes the chapter.

5.2 EMPIRICAL MODE DECOMPOSITION

One of the major interests of the signal processing area is the analysis of natural signals that are typically nonlinear and non-stationary. The multicomponent AM-FM (amplitude modulation-frequency modulation) model is generally considered to represent such a signal $x(t)$,

$$x(t) = \sum_{k=1}^{K} x_k(t) = \sum_{k=1}^{K} A_k(t) \cos \theta_k(t), \tag{5.1}$$

where $A_k(t)$ is the instantaneous amplitude and $\omega_k(t) = \frac{d}{dt}\theta_k(t)$ is the instantaneous frequency of each component $x_k(t)$.

The decomposition modes obtained with the EMD correspond to the AM-FM components of $x(t)$ in (5.1). The instantaneous amplitudes and frequencies are evaluated directly from the Hilbert transform of the IMFs. The general idea is to analyze the signal between two consecutive extrema (minima or maxima). The fast oscillations are defined as the detail components and the remaining slow fluctuations as the local trend or residual components. For example, the first detail component, $d_1(t)$, is obtained from all the consecutive extrema of $x(t)$, such that

$$x(t) = d_1(t) + a_1(t), \tag{5.2}$$

where $a_1(t)$ denotes the first local trend. This residual is computed as the average of the upper and lower envelopes of $x(t)$, which are achieved by interpolating the local maxima and minima, respectively, using cubic splines. This idea of computing new detail and local trend components is iteratively repeated over the current residual

$a_1(t)$. In general, the k-th iteration is applied over the residual of order $k - 1$ to compute the detail and the local trend of order k, i.e.,

$$a_{k-1}(t) = d_k(t) + a_k(t). \qquad (5.3)$$

This operation is repeated until the current residual is not decomposed into new components.

5.2.1 EMPIRICAL MODE DECOMPOSITION ALGORITHM

The EMD algorithm can be summarized in the following steps:

1. Set $k = 1$ and initialize the temporary variable $a_0(t) = x(t)$;
2. Identify all extrema (local minima and maxima) of $a_{k-1}(t)$;
3. Obtain the upper ($e_{max}(t)$) and lower ($e_{min}(t)$) envelopes by cubic splines interpolation of the local maxima and minima, respectively;
4. Compute the local trend as the average between the upper and lower envelopes, i.e., $a_k(t) = (e_{min}(t) + e_{max}(t))/2$;
5. Calculate the detail component as $d_k(t) = a_{k-1}(t) - a_k(t)$;
6. Set $k = k + 1$ and iterate steps 2-5 on the new residual local trend $a_k(t)$.

According to Huang *et al.* [27], a sample sequence is considered as an IMF if it follows the properties:

1. The average of the envelopes must have zero mean.
2. The number of zero-crossing and the number of extrema must be equal or differ at most by one.

The second condition determines that all the local maxima and the local minima must be positive and negative, respectively. If the detail component $d_k(t)$ obtained in step 5 does not follow these properties, steps 2 to 5 must be repeated with $a_{k-1}(t)$ replaced with $d_k(t)$. This process, named *sifting*, is repeated until the new $d_k(t)$ is considered as a true IMF. If it is the case, then

$$\text{IMF}_k(t) = d_k(t). \qquad (5.4)$$

For the next IMF, the *sifting* process is applied on the new residual $a_k(t) = a_{k-1}(t) - \text{IMF}_k(t)$, and so on.

In [46], the authors argued that the cubic splines are more applicable than linear or polynomial interpolation for the computation of the envelopes. They showed that the polynomial interpolation requires a higher number of *sifting* operations than the cubic splines, which leads to the mode mixing problem.

The EMD algorithm guarantees that the total number of extrema is reduced from a current IMF to the next one. Each EMD iteration can only be applied if there are at least two extrema in the last computed residual $a_k(t)$. Thus, any signal $x(t)$ can be decomposed into a finite number of IMFs. If a total of K IMFs are extracted from

$x(t)$, then the EMD algorithm assures that,

$$x(t) = \sum_{k=1}^{K} \text{IMF}_k(t) + r(t), \qquad (5.5)$$

where $\text{IMF}_k(t)$ denotes the k-th IMF and $r(t)$ is the last residual, i.e., $r(t) = a_K(t)$.

Figure 5.1 illustrates an example of the first five IMFs obtained from the decomposition of a sample speech segment of 500 ms collected from the TIMIT database [16]. The waveform of each IMF is interpreted as a zero-mean AM-FM signal. The first mode (IMF_1) is composed of faster oscillations when compared to the second IMF, which has faster fluctuations than the third one, and so on. This demonstrates that at each time interval, the EMD applies a fast-oscillation versus slow-oscillation separation between IMFs.

Another interesting feature was observed in [15] where it is shown that the EMD, when applied to fractional Gaussian noises (fGn)[1] processes, behaves like a dyadic filterbank with overlapping band-pass filters. In that study, the first IMF is interpreted as the output of a high-pass filter with a non-negligible content in its lower half-band. For the remaining modes, each IMF is roughly composed of the upper half-band part of the residual $a_k(t)$ that resulted from the previous iteration. Figure 5.2 depicts an example of the first five IMFs in a dyadic filterbank structure. In this case, EMD is applied to a fGn with $H = 1/2$, which corresponds to Gaussian white noise. The values plotted in Figure 5.2 are the average magnitude obtained from the Fourier analysis of the IMFs considering 500 fGn sequences each one generated with 1024 samples. It can be seen that the first IMF presents the greatest magnitude values at normalized frequencies above 0.25. Moreover, the magnitude peaks of the other IMFs are located at the upper half-band part of the previous residual.

5.2.2   EMD: CHALLENGES AND RECENT ALTERNATIVES

In the last decade, many studies have identified some drawbacks of the EMD algorithm. The most often discussed questions are: the mode mixing problem [26], the understanding and interpretation of the envelope computation [21], and the sampling frequencies effects [44, 45, 46]. This section examines these important topics and introduces some of the most recent alternatives found in the literature to overcome these issues.

5.2.2.1   Mode Mixing Problem

Mode mixing occurs when a single IMF contains signals with widely disparate oscillations or scales,[2] and also when oscillations of similar degree are decomposed

---

[1]The fGn is defined as the increment of a fractional Brownian motion (fBm) stochastic process, whose autocorrelation function is indexed by its Hurst exponent ($0 < H < 1$) [28]. When $H = 1/2$, fGn corresponds to a Gaussian white noise. Noise generators based on the fBm processes considering different values of H can be found in [47, 59].

[2]A signal scale is related to the time spacing between consecutive zero-crossing locations or, alternatively, between successive extrema locations. Thus, fast and slow oscillations have different scales.

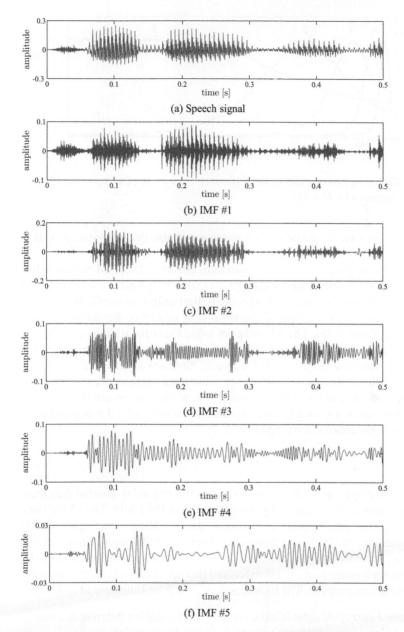

Figure 5.1 Segment of a speech signal spoken by a male speaker and the first five intrinsic mode functions obtained with the EMD: (a) Speech signal; (b)-(f) IMFs.

into different IMFs. The mode mixing problem was first detected in [26] when the authors applied the EMD to a low-frequency sine wave superimposed to some intermittent abrupt oscillations. The first IMF has both sine wave and fast fluctuations

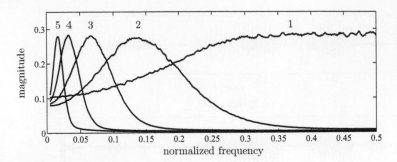

Figure 5.2 The average magnitudes of IMFs obtained from the application of the EMD to fGn sample sequences with $H = 1/2$.

components. Moreover, the single scale sine wave was also decomposed into different modes.

The ensemble empirical mode decomposition (EEMD) [56] and the complete EEMD with adaptive noise (CEEMDAN) [9, 51] were proposed in the literature as noise-assisted methods to overcome this problem. Both solutions use Gaussian white noise sequences to corrupt the input signal in order to guarantee that different scales will appear in different IMFs. The advantage of using Gaussian white noise relies on the fact that it assures that comparable scales of the input signal are concentrated in only one mode. This is achieved due to the dyadic filterbank structure of the ensembled modes. Additionally, since the white Gaussian noise samples are uncorrelated, the IMFs with the same index obtained from different realizations are also uncorrelated. Therefore, the noise samples are expected to cancel while the original input signal persists in the ensemble average IMFs.

a) EEMD

The ensemble empirical mode decomposition is the first noise-assisted data analysis method described to overcome the mode mixing of the EMD. The EEMD first obtains multiple corrupted versions of the input data by adding to the input samples different realizations of Gaussian white noise. Then, the EMD algorithm is applied to decompose each version of the corrupted or noisy data. The corresponding modes obtained from each realization are averaged and this set of IMFs is then considered as the ultimate decomposition. The EEMD algorithm can be summarized as follows:

1. Generate I corrupted signals $x^i(t)$, $i = 1, \ldots, I$, by adding different realizations of Gaussian white noise $w^i(t)$ to the input data samples $x(t)$, i.e.,

$$x^i(t) = x(t) + w^i(t); \tag{5.6}$$

2. Decompose each corrupted signal $x^i(t)$, $i = 1, \ldots, I$, into K modes $\text{IMF}^i_k(t)$, where $k = 1, \ldots, K$ indicates the mode index;

3. Calculate the ensemble of the average of each IMF,

$$\overline{\text{IMF}}_k(t) = \frac{1}{I} \sum_{i=1}^{I} \text{IMF}_k^i(t), k = 1, \ldots, K, \tag{5.7}$$

that is the definitive or final decomposition of $x(t)$.

b) CEEMDAN

The CEEMDAN is an adapted version of the EEMD. By denoting the CEEMDAN modes by $\widetilde{\text{IMF}}_k(t)$, the first IMF is computed as in the EEMD, i.e.,

$$\widetilde{\text{IMF}}_1(t) = \overline{\text{IMF}}_1(t) = \frac{1}{I} \sum_{i=1}^{I} \text{IMF}_1^i(t), \tag{5.8}$$

where $\text{IMF}_1^i(t)$, $i = 1, \ldots, I$, are obtained by applying the EMD to each version of the corrupted signal $x^i(t)$. The main difference between the CEEMDAN and EEMD algorithms relies on the computation of the residuals or local trends during each iteration. In the CEEMDAN, the first residual is computed as

$$a_1(t) = x(t) - \widetilde{\text{IMF}}_1(t). \tag{5.9}$$

The second IMF is computed by averaging the first mode of the decomposition of a new ensemble of $a_1(t)$ with different realizations of Gaussian white noise. The second residual is then given by

$$a_2(t) = a_1(t) - \widetilde{\text{IMF}}_2(t). \tag{5.10}$$

This procedure is iteratively applied to all the remaining modes.

Define an operator $E_k\{\cdot\}$ as the k-th IMF of a given signal, obtained with the original EMD. The CEEMDAN algorithm can be summarized as follows:

1. Apply the EMD to obtain the first IMF of the I corrupted sample sequences generated as in (5.6), i.e., $\text{IMF}_1^i(t)$, $i = 1, \ldots, I$;
2. Compute the first mode $\widetilde{\text{IMF}}_1(t)$ and the first residual $a_1(t)$ using (5.8) and (5.9), respectively;
3. Obtain the first IMF of each realization of $a_1(t) + E_1\{w^i(t)\}$, $i = 1, \ldots, I$. The second IMF is given by

$$\widetilde{\text{IMF}}_2(t) = \frac{1}{I} \sum_{i=1}^{I} E_1 \left\{ a_1(t) + E_1\{w^i(t)\} \right\}, \tag{5.11}$$

while the second residual $a_2(t)$ is computed as in (5.10);
4. For $k = 3, \ldots, K$, calculate the k-th mode as

$$\widetilde{\text{IMF}}_k(t) = \frac{1}{I} \sum_{i=1}^{I} E_1 \left\{ a_{k-1}(t) + E_{k-1}\{w^i(t)\} \right\}, \tag{5.12}$$

and the k-th residual as

$$a_k(t) = a_{k-1}(t) - \widetilde{\text{IMF}}_k(t). \qquad (5.13)$$

Equations (5.12) and (5.13) are iteratively applied until the current residual $a_K(t)$ does not have at least two extrema, i.e., can no longer be decomposed;

5. Compute the last residual as

$$r(t) = x(t) - \sum_{k=1}^{K} \widetilde{\text{IMF}}_k(t). \qquad (5.14)$$

In [9], the authors compared the EEMD and the CEEMDAN in the separation of a pure tone from different types of corrupting noises. The results showed that the CEEMDAN outperforms the EEMD in the sense that it needs less Gaussian white noise realizations to achieve similar reconstruction errors. Moreover, the EEMD reconstruction error showed to be dependent on the input signal-to-noise ratios (SNR). On the other hand, the CEEMDAN led to similar values of the reconstruction error for different SNR levels.

5.2.2.2 Envelope Computation

The cubic splines interpolation is generally adopted to obtain the envelope for the computation of the local trends. However, the use of cubic splines may lead to undesired envelope properties. For example, the upper and lower extremities may appear crossed. Two broad classes of solutions were found in this problem: the application of other interpolation techniques and the calculation of the local trend without the use of any interpolation.

In the last few years, many research studies have been focused on evaluating the first class of solutions. For instance, the cubic spline was replaced by the B-spline interpolation in [6], by the trigonometric interpolation in [19], and by higher-order polynomials in [57]. In [21], a new algorithm for the computation of envelopes was derived based on piecewise Hermite polynomial interpolation.

From the second class of solutions, some works have obtained the envelopes without directly using the minimum and maximum points. Some examples are the methods based on partial differential equations [11] and the genetic algorithm [34]. More recently, convex optimization tools have also been used to define the upper and lower envelopes [24, 43]. The sequential variational modal decomposition (Seq-VMD) [43] is a recent alternative to avoid the envelope computation. This decomposition uses non-smooth convex optimization to sequentially determine the IMFs and residuals without envelopes. The iterative steps of the EMD are replaced by an adaptation of texture-geometry decomposition methods.

a) Seq-VMD

Seq-VMD is based on a variational approach and convex optimization methods. Albeit the convergence is not established in the original EMD, in Seq-VMD the convergence can be guaranteed by choosing appropriate tools to solve the optimization problem.

The Seq-VMD also requires K sequential iterations to decompose an input signal $x(t)$ into K IMFs and a final residual. At each iteration, the current residual is split into new detail $d_k(t)$ and local trend $a_k(t)$ components, such that,

$$a_{k-1}(t) \approx d_k(t) + a_k(t), \quad k = 2, \ldots, K. \tag{5.15}$$

and $a_0(t) = x(t) \approx d_1(t) + a_1(t)$.

Seq-VMD algorithm obtains an EMD-like decomposition by solving the optimization problem

$$(a_k(t), d_k(t)) \in \arg \min_{(a,d)} \|a_{k-1} - d - a\|_2^2, \quad k = 1, \ldots, K. \tag{5.16}$$

The following conditions are applied to guarantee that $a(t)$ and $d(t)$ in (5.16) are the detail and the local trend components of the EMD iterations:

1. $d(t)$ must have a zero-mean average envelope.
2. $a(t)$ is smoother than $d(t)$, i.e., $a(t)$ is composed of slower oscillations than $d(t)$.
3. $d(t)$ is quasi-orthogonal to $d_j(t)$, for $1 \leq j < k$.

To ensure the first and most challenging condition, let the location of all extrema of $a_{k-1}(t)$ be denoted as $(t_k[l])_{1 \leq l \leq L_k}$. The first constraint is then approximated by

$$\left| d(t_k[l]) + \frac{\alpha_l \, d(t_k[l-1]) + \beta_l \, d(t_k[l+1])}{\alpha_l + \beta_l} \right| < \epsilon_{k,l}, \tag{5.17}$$

where $\alpha_l = t_k[l+1] - t_k[l], \beta_l = t_k[l] - t_k[l-1]$ and $\epsilon_{k,l} > 0$.

The second condition is achieved by

$$\|Aa(t)\|_p^p \leq \nu_k, \quad k = 1, \ldots, K, \tag{5.18}$$

where A is the derivative operator, $p \geq 1$ and $\nu_k > 0$.

Finally, the third constraint is established by considering, for each $k = 1, \ldots, K$,

$$\left\| \langle d(t), d_j(t) \rangle \right\|_p^p \leq \zeta_{k,j}, \quad j < k, \tag{5.19}$$

with $\zeta > 0$. This guarantees that $d(t)$ is quasi-orthogonal to all the previous detail sequences.

The optimization problem (5.16) constrained to (5.17)-(5.19) was solved with the primal-dual proximal algorithm derived in [4]. The Seq-VMD showed interesting results on decomposing artificial and real data samples, when compared to the original EMD and to the optimization-based mode decomposition (OMD) proposed in [42]. In general, the Seq-VMD presented lower computational cost than the OMD, but higher than the original EMD. However, in some experiments Seq-VMD and EMD showed similar computational cost with the proper choice of the parameters.

5.2.2.3 Sampling Effects

In the description of the EMD algorithm, the input signal $x(t)$ is defined in continuous time. However, only discrete-time signals can be decomposed in practice. Since the EMD is a nonlinear method, the IMFs obtained from a continuous-time signal may differ from its discrete-time version. The study of these sampling effects has been the focus of some investigations [44, 45, 46]. The idea was to define a new upper bound for the difference between the IMFs of continuous-time and discrete-time signals as a function of the sampling frequency.

In [46], it was shown that when the EMD is applied to a single tone signal, the decomposition error is minimized if the tone period is a multiple of $2T_s$, where T_s is the sampling period. For such tones, the sampling error was found to be upper bounded by a function proportional to f_s^{-2}, where f_s is the sampling frequency. In [44], the authors studied the sampling effects on the EMD for more general signals. The results demonstrated that for low sampling frequencies the error is generally upper bounded by a function proportional to f_s^{-2}, which is similar to the single tone case. For higher sampling frequencies, the upper bound appears to be proportional to f_s^{-1}.

The occurrence of all the extrema is the minimum requirement to guarantee that the EMD leads to similar results when applied to a continuous-time signal and its discrete-time version. For this purpose, the sampling period must be less than one half of the minimum distance between consecutive extrema [44]. To avoid the sampling effects, it is also recommended to decompose the signals with a large amount of oversampling. However, it is important to highlight that the studies of the sampling effects have not yet considered the EMD applied to real data samples, e.g., speech signals.

5.3 SPEECH ENHANCEMENT

The presence of background acoustic noise may severely degrade the perceived quality and intelligibility of speech signals. Consequently, the need for intelligent acoustic noise suppression algorithms is a fundamental research issue. The main challenge concerns the fact that speech and acoustic noise are, usually, non-stationary signals. This section discusses the main issues of using the EMD theory for speech enhancement in non-stationary noise environment. The most recent EMD-based speech enhancement solutions are presented in Section 5.3.1: EMD-based filtering (EMDF) [5], EMD detrending (EMD-DT) [14], EMD and Hurst-based solution (EMDH) [60], and a novel alternative based on the Seq-VMD and the Hurst exponent (SeqVMD-H). Section 5.3.2 briefly describes some spectral speech enhancement techniques considered as baseline in the study. Finally, Section 5.3.3 introduces some objective measures used to evaluate the speech enhancement solutions in terms of speech quality and intelligibility.

5.3.1 SPEECH ENHANCEMENT WITH EMD

Different from the traditional spectral speech enhancement, EMD-based techniques avoid the explicit estimation of the noise statistics. Due to this advantage, EMD has been preferred over the traditional Fourier analysis in many speech enhancement procedures [5, 32, 60]. The main issue of the EMD-based solutions is to identify the number of IMFs to recompose the speech signal and exclude the noise components. In [14], EMD-based detrending (EMD-DT) was adopted to separate a signal from an additive interference with slow oscillations. EMD-Shrinkage [32] applies "hard" and "soft" shrinkage to the IMFs in order to enhance speech signals corrupted with white noise. In [5], a post-enhancement EMD-based filtering (EMDF) approach was introduced to remove low-frequency noise. Recently, EMDH [60] used the Hurst exponent to select on a frame-by-frame basis the IMFs that are most affected by non-stationary acoustic noise. This section also introduces the SeqVMD-H, a speech enhancement technique that employs the Seq-VMD to decompose the noisy signals.

5.3.1.1 EMDF

EMD-based filtering [5] was proposed as a post-processing solution to improve the quality of speech signals previously enhanced with the optimally modified log-spectral amplitude (OMLSA) [8] technique. The authors showed that the energy of a clean speech signal is mostly concentrated in the first four IMFs. Hence, at least four IMFs must be used to proceed the speech signal reconstruction. To remove the low-frequency noise components, a selection criteria was defined based on the IMF variances.

The EMDF algorithm is described as follows:

1. Decompose the speech signal $x(t)$ using EMD;
2. Compute the variance from the amplitude samples of each IMF, i.e.,

$$V(k) = \sigma_k^2 = (1/T) \sum_{t=1}^{T} [\mathrm{IMF}_k(t)]^2 , \qquad (5.20)$$

   where T is the total number of speech samples;
3. Identify the first peak[3] value of $V(k)$ and denote by k_p the IMF index that corresponds to this value, such that $k_p > 4$;
4. Find the minimum value of $V(k)$ that occurs prior to k_p. By denoting the corresponding IMF index by k_t, for $k_t < k_p$, then $V(k_t) < V(k_t + 1)$, and $V(k_t) < V(k_t - 1)$;
5. Reconstruct the speech signal using the first N IMFs,

$$\tilde{x}(t) = \sum_{k=1}^{N} \mathrm{IMF}_k(t) , \qquad (5.21)$$

---

[3] A peak $V(k)$ is defined as the maximum variance value when compared to the variance of the adjacent IMFs, i.e., $V(k-1) < V(k)$ and $V(k+1) < V(k)$.

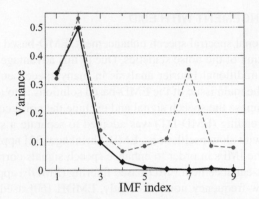

Figure 5.3 The continuous line indicates the variance of IMFs obtained from a clean speech utterance collected from TIMIT database. The dashed line represents the values from the same speech segment corrupted by Factory noise with SNR of 0 dB.

where $N = k_t$. The index N corresponds to the IMF with minimum variance value that occurs prior to the variance peak.

Figure 5.3 presents an example of the IMFs variance values estimated from two segments of speech signals. The continuous and dashed lines indicate the variance estimated from clean and noisy speech utterances,[4] respectively. Figure 5.3 shows that the modes with the highest indexes ($k > 4$) of the clean speech signal have lower energy values than the first ones. The dashed line demonstrates an abrupt variance increase from IMFs 5 to 9. This fact is explained by the low-frequency characteristics of the acoustic noise [31]. In the example of Figure 5.3, the identified indexes are $k_p = 7$ and $k_t = 4$.

5.3.1.2   EMD-DT

The EMD-based detrending and denoising [14] method was defined as a simple criteria to identify the slow oscillations of a certain signal. In this chapter, the EMD-DT is applied as a speech enhancement technique to remove low-frequency components from noisy speech signals.

After the decomposition of the noisy signal, the normalized mean of each IMF is computed as

$$\text{Mean}_{\text{norm}}(k) = \frac{\frac{1}{T} \sum_{t=1}^{T} \text{IMF}_k(t)}{\sqrt{\frac{1}{T} \sum_{t=1}^{T} [\text{IMF}_k(t)]^2}}, \quad k = 1, \ldots, K. \quad (5.22)$$

---

[4]The clean speech utterance was collected from the TIMIT database. The noisy signal corresponds to the same speech segment corrupted with a real Factory noise, extracted from NOISEX-92 database [52], with SNR of 0 dB.

Assuming that IMFs are expected to have zero mean, the idea is to identify the first IMF with index $N + 1$ ($N \geq 4$) for which the value $\text{Mean}_{\text{norm}}(N + 1)$ significantly differs from zero. For that purpose, the $\text{Mean}_{\text{norm}}(k)$ values are compared to the root mean square of $\text{Mean}_{\text{norm}}(k)$ calculated from the first four modes. The last IMF that is considered in the signal reconstruction is

$$
|\text{Mean}_{\text{norm}}(N)| \leq \zeta \sqrt{\frac{1}{4} \sum_{k=1}^{4} \text{Mean}_{\text{norm}}^2(k)} , \tag{5.23}
$$

where the threshold[5] is set to $\zeta \approx 2$. The signal is finally reconstructed as in (5.21) using only the first N IMFs, $N \geq 4$.

5.3.1.3 EMDH

EMDH [60] is an EMD-based technique in which the Hurst exponent (H) [28] is adopted as a criteria to identify the most corrupted modes. The IMFs selection and the speech signal reconstruction are performed on a frame-by-frame basis. In each frame, the Hurst exponent identifies the low-frequency noise components of each IMF. The selection criteria defined in [60] is applied to remove IMF frames whose value of H is above a given threshold H_{th}. The remaining modes are then used to reconstruct the enhanced version of the speech signal.

Let the speech signal be represented by a stochastic process $x(t)$. The normalized autocorrelation coefficient function ($\rho(\tau)$) is computed as

$$
\rho(\tau) = \frac{E\left[(x(t) - \mu_x)(x(t + \tau) - \mu_x)\right]}{E\left[(x(t) - \mu_x)^2\right]} , \tag{5.24}
$$

where μ_x is the mean of $x(t)$ and τ is the time lag. For a fractional Gaussian noise, $\rho(\tau)$ is given by [38]

$$
\rho(\tau) = \frac{1}{2}\left(|\tau - 1|^{2H} - 2|\tau|^{2H} + |\tau + 1|^{2H}\right) , \tag{5.25}
$$

where $0 \leq H \leq 1$ is the Hurst exponent of $x(t)$. The H value is defined by the decaying rate of $\rho(\tau)$, whose asymptotic behavior is

$$
\rho(\tau) \sim H(2H - 1)\tau^{2(H-1)}, \quad \tau \to \infty . \tag{5.26}
$$

The Hurst exponent expresses the time-dependence or scaling degree of $x(t)$ and is related to its spectral characteristics. Within the whole range $]0, 1[$, the power spectral density $S_x(f)$ can be shown to be proportional to f^{1-2H} when $f \to 0$ [38]. For $H = 1/2$, $S_x(f)$ is constant over the whole frequency spectrum (e.g., white noise), whereas low frequencies are prominent in the case where $H > 1/2$, and in particular when $H \to 1$ ($1/f$ or pink noise). In [48], a vector of Hurst exponent and

---

[5]The value of threshold ζ was empirically defined in [60].

the extraction method were introduced to serve as speech feature. This feature was successfully evaluated for speaker recognition [48], acoustic emotion identification [58], robust speaker verification in non-stationary noisy conditions [54], and also for robust acoustic source localization [12].

The EMDH algorithm can be summarized as follows:

1. Decompose the speech signal using EMD;
2. Window each IMF into a number Q of non-overlapping short-time frames, denoted as w-IMF, i.e.,

$$\text{w-IMF}_{k,q}(t) = \begin{cases} \text{IMF}_k(t + qT_d) & , t \in [0, T_d], \\ 0 & , \text{elsewhere}, \end{cases} \tag{5.27}$$

where $q \in \{0, \ldots, Q - 1\}$ is the frame index and T_d is a fixed time-duration of the frames;

3. For each frame $q = 1, \ldots, Q$, estimate the Hurst exponent of each windowed IMF (w-IMF$_{k,q}(t)$, $k = 1, \ldots, K$) and compose a K-dimensional vector $\mathbf{H}_q(k)$ of Hurst values;
4. Determine, for each frame q, the index N_q of the last windowed IMF whose value is less than a given threshold, i.e., $\mathbf{H}_q(N_q) < H_{\text{th}}$;
5. Reconstruct each frame $\hat{x}_q(t)$ of the enhanced speech signal as

$$\hat{x}_q(t) = \sum_{k=1}^{N_q} \text{w-IMF}_{k,q}(t), \quad q = 0, \ldots, Q - 1. \tag{5.28}$$

Finally, the complete enhanced speech signal $\hat{x}(t)$ is

$$\hat{x}(t) = \sum_{q=0}^{Q-1} x_q(t - qT_d). \tag{5.29}$$

In step 3, the Hurst exponent is estimated with the wavelet-based method [53], using the Daubechies filters [10] with 12 coefficients and the 3-12 scales.

Figure 5.4 depicts the average of the H values estimated from the IMFs of a clean (continuous line) and a noisy (dashed line) speech signal. These are the same speech utterances used to compute the variance in Figure 5.3. The H exponent is obtained from non-overlapping frames of 512 samples, which corresponds to 32 ms with a sampling rate of 16 kHz. It can be seen that the first IMFs have $H < 1/2$, which represents the fast oscillations, i.e., high-frequency components. For IMFs 6-9, the H values are closer to the unity and correspond to the low-frequency content. The dashed line demonstrates that the presence of the low-frequency noise significantly increases the H values of the last IMFs. Thus, the estimation of the Hurst exponent proved to adequately select the low-frequency noise components.

The frame-by-frame analysis of the EMDH is motivated by the presence of abrupt changes in the power spectrum of non-stationary noises. The definition of the threshold H_{th} is crucial to determine the portion of the low-frequency noise that is removed

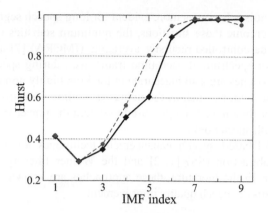

Figure 5.4   The continuous line indicates the Hurst exponent of IMFs obtained from a clean speech utterance collected from TIMIT database. The dashed line represents the H values from the same speech segment corrupted by Factory noise with SNR of 0 dB.

from each frame. Since the energy of real acoustic noises are generally concentrated at the lower frequencies [31, 55], these noise components are suppressed from the speech signal if $H_{th} \approx 1$. Hence, the threshold was set to $H_{th} = 0.9$.

5.3.1.4   SeqVMD-H

SeqVMD-H solution is introduced here as an alternative speech enhancement technique. The main difference to the EMDH relies on the Seq-VMD application, instead of the original EMD, to obtain the IMFs of the noisy speech signals. The main issue is to evaluate the contribution of the Seq-VMD in the analysis of speech signals corrupted by non-stationary noises. The selection of the IMFs that compose the enhanced speech signal follows the same criteria as in the EMDH. It means that the IMFs are also selected based on the Hurst exponent in a frame-by-frame basis, and the threshold adopted for the IMFs selection is also $H_{th} = 0.9$.

5.3.2   SPECTRAL SPEECH ENHANCEMENT: AN OVERVIEW

Spectral speech enhancement techniques are commonly composed of three main steps: i) application of the short-time Fourier transform (STFT) to the noisy speech signal to estimate the acoustic noise power spectrum, ii) subtraction of the noise components from the noisy spectrum, and iii) reconstruction of the enhanced speech signal in the time-domain using the inverse Fourier transform.

The accurate estimation of the noise power spectrum is crucial for the speech enhancement techniques. The classical noise estimators [2] are based on voice activity detectors (VAD). The noise power spectrum is then computed as the average of the noisy signal power spectrum values obtained during the speech pauses. This procedure shows reasonable accuracy for stationary background noise. However, time-varying spectra are not precisely evaluated with such estimator. The difficulty in tracking non-stationary noises becomes more evident for long speech segments and low SNR values. To overcome these situations, the minimum statistics (MS) [40] and the improved minima controlled recursive averaging (IMCRA) [7] algorithms estimate the noise power spectrum at each time frame even during speech activity. However, these approaches are also inaccurate in tracking highly non-stationary noises [39]. Recent contributions, such as the unbiased minimum mean-square error (UMMSE) [17] algorithm, have been proposed to obtain the power spectrum of non-stationary noises with shorter delays.

Following, two STFT-based speech enhancement techniques are briefly described: the spectral subtraction (SS) [1, 2] and the Wiener filtering [49] with UMMSE [17] estimator. In the literature, these approaches are known to achieve interesting results in terms of speech quality improvement.

5.3.2.1 Spectral Subtraction

Let $y(t)$ be a speech utterance corrupted by an additive noise $\eta(t)$, i.e., $y(t) = x(t) + \eta(t)$, where $x(t)$ is the clean speech signal. Applying the STFT,

$$Y(\kappa, \tau) = X(\kappa, \tau) + \mathcal{N}(\kappa, \tau), \tag{5.30}$$

where κ and τ are, respectively, the frequency bin and the time frame index.

The first step of the SS [1, 2] is to compute the noise power spectrum $|\hat{\mathcal{N}}(\kappa, \tau)|^2$ using the classical VAD-based estimator. Then, the clean speech power spectrum is given by

$$|\hat{X}|^2 = \max \left\{ |Y|^2 - \alpha |\hat{\mathcal{N}}|^2, \beta |\hat{\mathcal{N}}|^2 \right\}, \tag{5.31}$$

where α represents the time-varying "oversubtraction" factor and β is the spectral floor. The spectrum of the enhanced signal is then estimated using the phase of the noisy speech signal. The enhanced speech signal $\hat{x}(t)$ is finally reconstructed by overlapping and adding the inverse Fourier transform of each time frame.

5.3.2.2 Wiener/UMMSE

The Wiener/UMMSE speech enhancement technique obtains the noise components using the UMMSE [17] estimator. The enhanced version of the speech signal is then achieved using the Wiener filtering approach presented in [49]. The UMMSE noise estimator combines speech presence uncertainty to the algorithm originally presented in [20]. The authors found that the estimation of the noise power spectrum $|\hat{\mathcal{N}}(\kappa, \tau)|^2$ can be updated every time frame via the recursive smoothing

$$|\hat{\mathcal{N}}(\kappa, \tau)|^2 = \alpha_p |\hat{\mathcal{N}}(\kappa, \tau - 1)|^2 + (1 - \alpha_p) E\left(|\mathcal{N}|^2|Y\right), \tag{5.32}$$

where α_p is a "smoothing factor." The noise periodogram estimate $E\left(|\mathcal{N}|^2|Y\right)$ depends on the speech presence and absence probabilities, and on the noise power spectrum estimated from the last frame. The UMMSE algorithm avoids the bias compensation factor and the minimum search within a given number of past frames. It leads to shorter delays in the noise estimation when compared to MS or IMCRA.

The UMMSE noise estimator is followed by the speech enhancement technique proposed in [49]. The Wiener filtering gain $G_W(\kappa, \tau)$ is computed as

$$G_W(\kappa, \tau) = \frac{\xi(\kappa, \tau)}{1 + \xi(\kappa, \tau)},$$

(5.33)

where $\xi(\kappa, \tau)$ is the *a priori* SNR. The estimation of $\xi(\kappa, \tau)$ is obtained with the decision-directed approach applied in [13], i.e.,

$$\hat{\xi}(\kappa, \tau) = \alpha_W G_W^2(\kappa, \tau - 1) \gamma(\kappa, \tau - 1) + (1 - \alpha_W) \max\{\gamma(\kappa, \tau) - 1, 0\}.$$

(5.34)

The parameter α_W is generally set to 0.98.

5.3.3 OBJECTIVE MEASURES: QUALITY AND INTELLIGIBILITY

This section introduces five objective measures to evaluate the speech enhancement techniques: the segmental SNR (SegSNR) and the overall quality composite measure (OQCM) [22] are used to assess the speech quality; the frequency-weighted SegSNR (fwSegSNR) [23], the short-time objective intelligibility (STOI) [50], and the coherence speech intelligibility index (CSII) [30] are considered for intelligibility.

5.3.3.1 SegSNR

The segmental SNR is defined in time-domain as

$$\text{SegSNR} = \frac{10}{Q} \sum_{\tau=0}^{Q-1} \log \frac{\sum_{t=\tau T_{\text{sh}}}^{\tau T_{\text{sh}} + T_d - 1} x^2(t)}{\sum_{t=\tau T_{\text{sh}}}^{\tau T_{\text{sh}} + T_d - 1} [x(t) - \hat{x}(t)]^2},$$

(5.35)

where T_d is the frame length (in samples), T_{sh} is the frame shift, Q is the total number of frames, and $x(t)$ and $\hat{x}(t)$ represent the clean and enhanced speech signals, respectively. The SegSNR values are computed with frame size of 32 ms with 75% overlapping corresponding to the values $T_d = 512$ and $T_{\text{sh}} = 128$ samples with 16 kHz sampling rate. The SNR of each frame is limited between -10 dB and 35 dB to avoid the need to discard frames where speech is absent. The results are shown in terms of the SegSNR improvement, which is here defined as the SegSNR measured from the enhanced speech subtracted from the SegSNR values obtained from the noisy signals. The same definition is adopted for the OQCM (Section 5.3.3.2) and fwSegSNR (Section 5.3.3.3) improvement.

5.3.3.2 OQCM

In [22], the authors evaluated the correlation between five objective measures and three subjective rating scores: signal distortion, background noise distortion, and overall quality. In order to achieve higher correlation with the scores obtained from the subjective listening tests, three composite measures were introduced as the linear combination of the existing objective measures. The overall quality composite measure was defined as

$$\text{OQCM} = 1.594 + 0.805 \, \text{PESQ} - 0.512 \, \text{LLR} - 0.007 \, \text{WSS},$$

(5.36)

where PESQ is the perceptual evaluation of speech quality, LLR is the log-likelihood ratio and WSS is the weighted spectral slope distance [33]. For the results presented here, the OQCM (5.36) is computed considering the wideband version of PESQ, as defined by the ITU-T recommendation P.862.2.

5.3.3.3 fwSegSNR

For the frequency-weighted SegSNR, the spectra of the clean ($|X(j, \tau)|$) and enhanced ($|\hat{X}(j, \tau)|$) speech signals are calculated by dividing their entire bandwidth into $J = 25$ frequency bands using Gaussian-shaped filters. Then, the fwSegSNR is computed as

$$\text{fwSegSNR} = \frac{10}{Q} \sum_{\tau=0}^{Q-1} \frac{\sum_{j=1}^{J} W(j, \tau) \log \frac{|X(j,\tau)|^2}{(|X(j,\tau)|-|\hat{X}(j,\tau)|)^2}}{\sum_{j=1}^{J} W(j, \tau)}, \tag{5.37}$$

where τ and j are the frame and frequency band index, respectively. The weighting function [23] is signal-dependent and defined by $W(j, \tau) = |X(j, \tau)|^{(0.2)}$. Once again, the SNR values computed at each frame and each frequency band are limited to the range of $[-10, 35]$. In [37], the authors demonstrated that the fwSegSNR is highly correlated to the speech intelligibility scores obtained in subjective listening tests.

5.3.3.4 STOI

The short-time objective intelligibility measure [50] was proposed as a correlation-based method to evaluate the speech intelligibility degradation caused by the speech enhancement procedures. The STOI showed high and very close correlation to the subjective intelligibility rates obtained with speech signals enhanced by noise-reduction algorithms. For the STOI computation, the clean and noisy versions of the speech signal are divided into short-time frames and grouped into 15 one-third octave bands. For each frame τ and each band j, the intermediate intelligibility measure is defined as the correlation coefficient of the temporal envelope vectors computed from the clean and the noisy speech signals. Finally, the STOI measure is the average of the intermediate values evaluated from the 15 one-third octave bands and all Q speech frames.

5.3.3.5 CSII

The CSII [30] is an extension to the speech intelligibility index (SII).[6] The SII evaluates the SNR of the enhanced speech signal in different frequency bands. A critical band formulation that models the auditory periphery is used to define the SII frequency bands. The intelligibility measure is estimated as a weighted sum of the SNR values of all frequency bands. Since in the SII the power spectrum of speech and noise are calculated using long-term averages, the intelligibility prediction may be inaccurate for non-stationary noises.

---

[6]The SII is defined by standard ANSI S3.5-1997.

In the CSII, the standard speech SNR estimate is replaced by the signal-to-distortion ratio (SDR) obtained from the magnitude-squared coherence (MSC). In order to achieve a higher correlation to the intelligibility scores from subjective listening tests, the CSII is evaluated separately for low-, medium-, and high-level segments of each sentence. The highest correlation was computed with a weighted sum of the corresponding measures (CSII$_{Low}$, CSII$_{Med}$, and CSII$_{High}$) considering the following weights

$$CSII = 0.155CSII_{Low} + 0.845CSII_{Med} + 0.0CSII_{High} . \qquad (5.38)$$

This same set of weights is here adopted for the CSII results.

5.4   EMD-BASED SPEECH ENHANCEMENT RESULTS

This section discusses the main results of the EMD-DT, EMDF, EMDH, and SeqVMD-H techniques. The speech enhancement solutions are evaluated in terms of quality (SegSNR and OQCM) and intelligibility (fwSegSNR, STOI, and CSII). The SS and Wiener/UMMSE approaches are considered as baseline. The Wiener/UMMSE is hereinafter denoted as UMMSE. In the experiments, a subset of 24 speakers (16 male and 8 female) is selected from the TIMIT speech database [16]. It leads to a total of 240 speech segments, 10 per speaker, with a sampling rate of 16 kHz and an average time duration of 3 seconds.

5.4.1   NOISE DATABASE

Four acoustic noises (Babble, Chainsaw, Factory, and Train) are selected to corrupt the speech signals. The noises are collected from the NOISEX-92 [52] (Babble and Factory), Freesound.org[7] (Train), and Freesfx.co.uk[8] (Chainsaw) databases.

These four noises are chosen since they are non-stationary and present quite different indexes of non-stationarity (INS) [3]. The INS is adopted to objectively examine the non-stationarity of the acoustic noise. Figure 5.5 depicts the INS values evaluated from 3-second segments of the four noises. The time scale T_h/T is the ratio of the length of the short-time spectral analysis (T_h), and the total time duration ($T = 3$ seconds) of the noise sample sequences. For the stationarity test, the INS value is compared to the threshold γ, which is also shown in the dashed lines of Figure 5.5. Thus,

$$INS \begin{cases} \leq \gamma & \text{, noise is stationary;} \\ > \gamma & \text{, noise is non-stationary.} \end{cases} \qquad (5.39)$$

The selected noises are here classified according to their INS values: highly non-stationary (INS $\geq 100\gamma$), non-stationary ($10\gamma \leq$ INS $< 100\gamma$), and moderately non-stationary (INS $< 10\gamma$). In this case, the Chainsaw noise (Figure 5.5(a)) is described

---

[7] http://www.freesound.org.

[8] http://www.freesfx.co.uk.

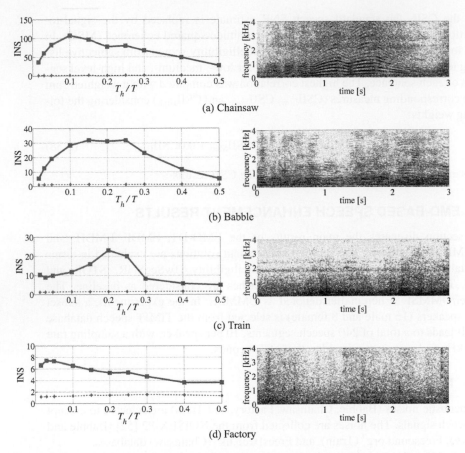

Figure 5.5 INS and spectrogram segments of acoustic noise: 3 seconds duration. The dashed lines at the left indicate the values of the threshold γ for the stationarity tests.

as highly non-stationary. Babble and Train (Figure 5.5(b-c)) are designated as non-stationary since their INS values are greater than 30 and 20, respectively. Since the INS values are lower than 8 for all time scales, Factory noise (Figure 5.5(d)) is considered as moderately non-stationary.

Non-stationarity of the four noises may also be perceived by the spectrogram plots in Figure 5.5. It can be seen that the selected noises present spectral components fluctuating over the entire voice frequency band (0-4 kHz).

5.4.2 SPEECH QUALITY RESULTS

This section shows the SegSNR and OQCM results to examine the speech enhancement techniques in terms of speech quality. The selected acoustic noises are used to corrupt the utterances considering five SNR values: -10 dB, -5 dB, 0 dB, 5 dB, and

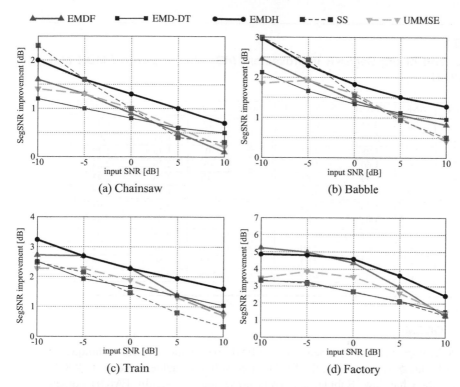

Figure 5.6 The SegSNR improvement (dB) obtained with the EMD-based and the baseline speech enhancement techniques.

10 dB.

5.4.2.1 SegSNR

Figure 5.6 depicts the SegSNR improvement (in dB) obtained with EMD-DT, EMDF, EMDH, and the baseline techniques: SS and UMMSE. In general, EMDH and EMDF outperform the baseline spectral solutions for almost all the noise conditions. EMDH achieves the highest SegSNR improvement of the EMD-based approaches for three sources: the highly non-stationary Chainsaw, and the non-stationary Babble and Train noises. It is worth to mention that the Babble noise is particularly challenging since this noise is composed of several speech signals. For the moderately non-stationary Factory noise, EMDH and EMDF reaches the highest SegSNR gain: EMDH for SNR ≥ 0 dB and EMDF for SNR < 0 dB.

Table 5.1 shows the SegSNR gain of the SeqVMD-H approach. The results obtained with EMDH are also included since it achieves, in general, the best SegSNR improvement (refer to Figure 5.6). SeqVMD-H and EMDH present similar speech quality improvement for the highly non-stationary Chainsaw and the non-stationary Babble noises.

Table 5.1

Comparison of the SegSNR improvement (dB) obtained with EMDH and SeqVMD-H

Noise	SNR	EMDH	SeqVMD-H	Noise	SNR	EMDH	SeqVMD-H
Chainsaw	10	0.74	0.82	Babble	10	1.28	1.39
	5	0.99	0.92		5	1.51	1.53
	0	1.28	1.12		0	1.83	1.79
	-5	1.60	1.40		-5	2.29	2.24
	-10	1.99	1.82		-10	2.96	2.93
	Aver.	1.32	1.22		Aver.	1.97	1.98
Train	10	1.60	1.44	Factory	10	2.44	1.81
	5	1.94	1.65		5	3.62	2.45
	0	2.27	1.91		0	4.58	2.97
	-5	2.69	2.30		-5	4.81	3.21
	-10	3.24	2.93		-10	4.86	3.48
	Aver.	2.35	2.05		Aver.	4.06	2.78

5.4.2.2 OQCM

The OQCM improvement obtained with the EMD-based and spectral techniques is shown in Figure 5.7. In general, the EMD-based solutions achieve the best OQCM values for the Chainsaw, Babble, and Train noise sources. Again, EMDH outperforms EMDF and EMD-DT in terms of speech quality for all the noise sources. EMDH also provides the best OQCM results in comparison to the spectral approaches for the highly non-stationary Chainsaw noise. For Babble and Train, the EMDH leads to the highest OQCM gain for almost noise scenarios, i.e., Babble with SNR ≤ 5 dB and Train with all SNR values. For the moderately non-stationary Factory noise, the best quality is attained by EMDH for SNR ≤ 0 dB.

Figure 5.7 also demonstrates that SS and UMMSE provide negative OQCM gain for the three most non-stationary noises with severe degradation (SNR < 0 dB). These poor results can be explained by the inaccurate estimation of the time-varying power spectra of the acoustic noise. Even the UMMSE estimator, that was designed to capture the rapid changes in the noise power spectrum, failed to accurately suppress the noise components from the noisy speech signal. The EMD-based techniques can avoid such phenomenon since they do not require the estimation of the noise components.

5.4.3 SPEECH INTELLIGIBILITY RESULTS

The speech enhancement techniques are evaluated in terms of speech intelligibility with fwSegSNR, STOI, and CSII. The STOI and CSII measures are applied to predict the percentage of correctly recognized words in subjective intelligibility experiments.

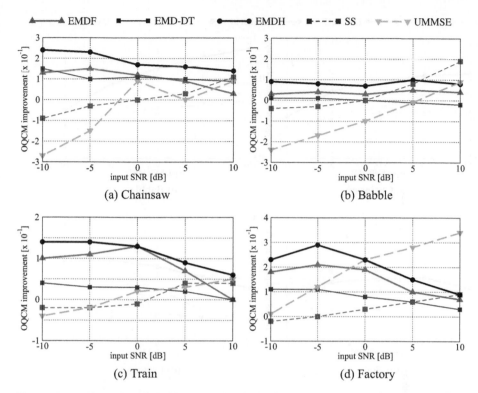

Figure 5.7   The OQCM improvement obtained with the EMD-based and the spectral speech enhancement techniques.

5.4.3.1 fwSegSNR

The fwSegSNR gain obtained with the EMD-based and the spectral speech enhancement techniques is illustrated in Figure 5.8. EMDH achieves the best fwSegSNR results for the highly non-stationary Chainsaw and the non-stationary Babble noises. For the Train noise, EMDH outperforms the EMDF and EMD-DT techniques for SNR of -5 dB and 10 dB. Regarding the moderately non-stationary Factory noise, the EMDH once again leads to the highest fwSegSNR improvement of the EMD-based solutions. When the spectral approaches are considered, EMDH also outperforms the UMMSE for SNR of -10 dB and 10 dB for the Train and Factory noises.

Table 5.2 shows the comparison of the fwSegSNR improvement from the EMDH and the SeqVMD-H techniques. Similarly to the SegSNR results, the SeqVMD-H achieves promising fwSegSNR values for the highly non-stationary Chainsaw and also for the non-stationary Babble. For instance, the SeqVMD-H improves the EMDH results for most of the input SNR considering the Babble noise.

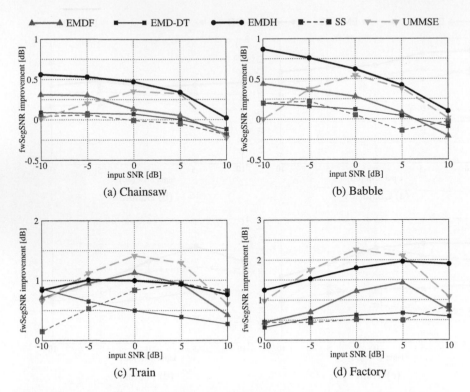

Figure 5.8 The fwSegSNR improvement (dB) obtained with the EMD-based and the spectral speech enhancement techniques.

5.4.3.2 STOI

The STOI is used here to examine the speech enhancement techniques in terms of intelligibility. A monotonic nonlinear mapping is applied to the STOI results to obtain a prediction of the percentage of correct words achieved in subjective listening tests. The predicted intelligibility scores are here obtained by applying the same mapping function adopted in [50], i.e.,

$$f(d) = \frac{100}{1 + \exp(a\,d + b)}, \tag{5.40}$$

where d is the value of the objective measure. The predicted percentage of correct words provides a more practical way to evaluate the intelligibility of the speech enhancement procedures.

Table 5.3 shows the intelligibility prediction rates calculated with the STOI followed by the mapping function (5.40) with $a = -13.45$ and $b = 9.36$. These values are chosen to fit the STOI to the subjective listening scores presented in [35]. Table 5.3 also includes the STOI results for the alternative SeqVMD-H. The highest and lowest average STOI values for all the speech enhancement procedures are computed

Table 5.2

Comparison of the fwSegSNR improvement (dB) obtained with EMDH and SeqVMD-H

Noise	SNR	EMDH	SeqVMD-H	Noise	SNR	EMDH	SeqVMD-H
Chainsaw	10	0.02	0.07	Babble	10	0.10	0.37
	5	0.34	0.18		5	0.42	0.49
	0	0.47	0.26		0	0.62	0.62
	-5	0.53	0.32		-5	0.76	0.77
	-10	0.56	0.37		-10	0.87	0.88
	Aver.	0.38	0.24		Aver.	0.55	0.63
Train	10	0.76	0.59	Factory	10	1.90	0.85
	5	0.94	0.72		5	1.96	1.09
	0	1.00	0.81		0	1.80	1.04
	-5	1.01	0.87		-5	1.53	1.05
	-10	0.84	0.76		-10	1.25	0.96
	Aver.	0.91	0.75		Aver.	1.69	1.00

for the moderately non-stationary Factory and the highly non-stationary Chainsaw noises, respectively. The EMD-based solutions achieve the best intelligibility values for three noise sources: Chainsaw, Babble and Factory. The best average score for the highly non-stationary Chainsaw noise is attained by SeqVMD-H, which is 3.1 percentage points (p.p.) greater than UMMSE. For Babble, SeqVMD-H and EMDF lead to similar average results, 44.1%, and outperform the spectral techniques in more than 2 p.p. in average. Even for the Train noise, where the UMMSE reaches the best average intelligibility rate, EMDH and SeqVMD-H provide the highest STOI results for SNR = 0 dB and SNR > 0 dB, respectively. The highest average score for the Factory noise is obtained with SeqVMD-H, which is 0.6 p.p. higher than the UMMSE.

5.4.3.3   CSII

Table 5.4 presents the intelligibility prediction scores with the CSII. Here, the mapping function (5.40) is applied with $a = 10.088$ and $b = 4.654$ to fit the intelligibility rates in [35]. Table 5.3 also includes the results for the proposed SeqVMD-H solution. Once again, the EMD-based techniques obtain the best average CSII values for the Chainsaw and Babble noises. For the highly non-stationary Chainsaw noise, both EMDH and SeqVMD-H achieve the highest intelligibility prediction results: 31.6% in average. It corresponds to 1.1 p.p. and 2.1 p.p. higher than the EMDF and UMMSE, respectively. The EMDH also leads to the highest intelligibility scores for the Babble noise, 1.1 p.p. greater than the UMMSE. Even for the Train and Factory noises the EMDH outperforms the other solutions for SNR of 10 dB. Finally, it may

Table 5.3
Intelligibility prediction rate (%) obtained with the STOI

Noise	SNR	SS	UMMSE	EMDF	EMDH	SeqVMD-H
Chainsaw	10	86.6	85.7	84.4	88.2	88.4
	5	55.1	57.0	59.0	61.8	62.4
	0	16.7	19.3	24.4	25.1	24.8
	-5	2.8	3.5	5.2	5.0	5.1
	-10	0.7	1.1	1.5	1.5	1.5
	Aver.	32.4	33.3	34.9	36.3	36.4
Babble	10	89.6	88.8	88.6	89.0	89.1
	5	69.8	72.0	73.9	73.4	73.9
	0	28.8	37.6	42.7	42.0	42.5
	-5	5.9	9.2	12.9	12.5	12.5
	-10	1.1	1.6	2.5	2.6	2.6
	Aver.	39.0	41.8	44.1	43.9	44.1
Train	10	90.9	90.2	89.4	90.0	90.9
	5	81.5	80.8	80.1	81.0	81.5
	0	62.3	63.5	63.5	63.6	62.4
	-5	23.4	35.9	34.7	35.3	32.8
	-10	4.8	11.2	10.4	10.8	9.9
	Aver.	52.6	56.3	55.6	56.1	55.5
Factory	10	92.9	93.5	89.8	94.6	94.7
	5	81.0	87.5	85.7	89.0	89.0
	0	48.3	73.5	74.7	75.4	75.3
	-5	18.2	46.5	46.5	46.7	46.5
	-10	6.0	17.3	14.7	15.7	15.6
	Aver.	49.3	63.7	62.3	64.3	64.2

be noticed that SeqVMD-H and EMDH reach similar CSII values for most of the noise conditions.

5.4.4   SUMMARY

The objective quality and intelligibility results presented in this section showed that the EMD-based techniques increase the time and frequency-domain SegSNR and also achieve promising intelligibility prediction scores. Particularly for the highly non-stationary Chainsaw and non-stationary Babble noises, the EMDH solution shows the best results for most of the experimental scenarios. The results also demonstrated that the proposed SeqVMD-H technique is a valuable alternative to the EMDH, mainly for the intelligibility scores. Thus, the EMD-based approaches are suitable for improving the quality and the intelligibility of speech signals corrupted by non-stationary acoustic noises.

Table 5.4

Intelligibility prediction rate (%) obtained with the CSII

Noise	SNR	SS	UMMSE	EMDF	EMDH	SeqVMD-H
Chainsaw	10	82.5	80.6	80.5	82.3	82.3
	5	44.0	45.6	46.5	49.1	49.0
	0	14.7	15.3	17.9	18.8	18.7
	-5	4.3	4.2	5.4	5.8	5.8
	-10	1.7	1.6	2.0	2.1	2.1
	Aver.	29.4	29.5	30.5	31.6	31.6
Babble	10	93.0	92.2	92.1	92.2	92.2
	5	68.1	71.4	71.2	71.5	71.5
	0	29.1	34.6	36.3	36.4	36.4
	-5	9.7	10.0	12.6	12.6	12.6
	-10	3.0	2.8	4.0	4.0	4.0
	Aver.	40.6	42.2	43.2	43.4	43.3
Train	10	97.9	97.2	97.5	97.6	97.5
	5	92.3	90.7	90.6	90.8	90.8
	0	69.0	70.4	67.9	67.8	67.8
	-5	28.8	35.7	33.1	32.8	32.8
	-10	8.6	11.3	11.4	11.4	11.3
	Aver.	59.3	61.1	60.1	60.1	60.1
Factory	10	97.5	97.5	97.4	97.9	97.9
	5	86.4	91.6	91.0	92.0	91.8
	0	55.6	74.3	69.0	70.3	69.6
	-5	26.7	40.3	34.1	34.5	33.9
	-10	9.7	14.5	11.6	11.6	11.5
	Aver.	55.2	63.7	60.6	61.3	60.9

5.5 CONCLUSION

This chapter presented the empirical mode decomposition as a nonlinear time-domain adaptive method for analyzing nonlinear and non-stationary signals. Different from other time-frequency methods, such as the wavelet decomposition, the EMD is completely data driven and does not require a set of basis functions. Besides the introduction of the EMD algorithm, some of the main challenges of the EMD theory are also discussed: the mode mixing problem, the computation of the envelopes, and the sampling effects. The analysis of alternative decomposition solutions such as the EEMD, the CEEMDAN, and the Seq-VMD was also included in this work.

Due to its characteristics, the EMD has been widely adopted in the literature for the speech signals analysis, particularly for speech enhancement. Some of the most important EMD-based speech enhancement techniques were discussed in this chapter. These approaches showed to be efficient alternatives to the traditional spectral solutions to improve the quality and the intelligibility of speech signals corrupted by non-stationary acoustic noises. However, it is important to mention that the application of the EMD to speech signals is not restricted to the quality and the intelligibility improvement shown in this chapter. For instance, in [60] the EMD was adopted in a pre-processing step to improve the robustness of speaker identification systems. Thus, the EMD is expected to continue in the focus of many research works in the speech processing area.

ACKNOWLEDGMENTS

R. Coelho was partially supported by the National Council for Scientific and Technological Development (CNPq) under the 304254/2012-6 research grant.

REFERENCES

1. M. Berouti, R. Schwartz, and J. Makhoul. Enhancement of speech corrupted by acoustic noise. *Proceedings of the IEEE International Conference on Acoustics, Speech and Signal Processing (ICASSP 79)*, 4:208–211, April 1979.

2. S. Boll. Suppression of acoustic noise in speech using spectral subtraction. *IEEE Transactions on Acoustics, Speech and Signal Processing*, 27(2):113–120, April 1979.

3. P. Borgnat, P. Flandrin, P. Honeine, C. Richard, and Jun Xiao. Testing stationarity with surrogates: A time-frequency approach. *IEEE Transactions on Signal Processing*, 58(7):3459–3470, July 2010.

4. L. Briceño Arias and P. Combettes. A monotone + skew splitting model for composite monotone inclusions in duality. *SIAM Journal on Optimization*, 21(4):1230–1250, October 2011.

5. N. Chatlani and J. Soraghan. EMD-based filtering (EMDF) of low-frequency noise for speech enhancement. *IEEE Transactions on Audio, Speech, and Language Processing*, 20(4):1158–1166, May 2012.

6. Q. Chen, N. Huang, S. Riemenschneider, and Y. Xu. A b-spline approach for empirical mode decompositions. *Advances in Computational Mathematics*, 24(1-4):171–195, 2006.

7. I. Cohen. Noise spectrum estimation in adverse environments: improved minima controlled recursive averaging. *IEEE Transactions on Speech and Audio Processing*, 11(5):466–475, September 2003.

8. I. Cohen and B. Berdugo. Speech enhancement for non-stationary noise environments. *Signal Processing*, 81(11):2403–2418, November 2001.

9. M. Colominas, G. Schlotthauer, M. Torres, and P. Flandrin. Noise-assisted EMD methods in action. *Advances in Adaptive Data Analysis*, 04(04):1250025.1–1250025.11, 2012.

10. I. Daubechies. *Ten lectures on wavelets*. Society for Industrial and Applied Mathematics, Philadelphia, USA, 1992.

11. E. Delechelle, J. Lemoine, and O. Niang. Empirical mode decomposition: an analytical approach for sifting process. *IEEE Signal Processing Letters*, 12(11):764–767, November 2005.

12. E. Dranka and R. Coelho. Robust maximum likelihood acoustic energy based source localization in correlated noisy sensing environments. *IEEE Journal of Selected Topics in Signal Processing*, 9(2):259–267, March 2015.

13. Y. Ephraim and D. Malah. Speech enhancement using a minimum mean square error short-time spectral amplitude estimator. *IEEE Transactions on Acoustics, Speech, and Signal Processing*, 32(6):1109–1121, December 1984.

14. P. Flandrin, P. Gonçalves, and G. Rilling. Detrending and denoising with empirical mode decompositions. *Proceedings of the European Signal Processing Conference (EUSIPCO 2004)*, pages 1581–1584, September 2004.

15. P. Flandrin, G. Rilling, and P. Gonçalvès. Empirical mode decomposition as a filter bank. *IEEE Signal Processing Letters*, 11(2):112–114, February 2004.

16. J. Garofolo, L. Lamel, W. Fisher, J. Fiscus, D. Pallett, N. Dahlgren, and V. Zue. TIMIT acoustic-phonetic continuous speech corpus. *Linguistic Data Consortium, Philadelphia*, 1993.

17. T. Gerkmann and R. Hendriks. Unbiased MMSE-based noise power estimation with low complexity and low tracking delay. *IEEE Transactions on Audio, Speech, and Language Processing*, 20(4):1383–1393, May 2012.

18. K. Guhathakurta, I. Mukherjee, and A. Chowdhury. Empirical mode decomposition analysis of two different financial time series and their comparison. *Chaos, Solitons & Fractals*, 37(4):1214–1227, 2008.

19. S. Hawley, L. Atlas, and H. Chizeck. Some properties of an empirical mode type signal decomposition algorithm. *IEEE Signal Processing Letters*, 17(1):24–27, January 2010.

20. R. Hendriks, R. Heusdens, and J. Jensen. MMSE based noise PSD tracking with low complexity. pages 4266–4269, March 2010.

21. X. Hu, S. Peng, and W-L. Hwang. EMD revisited: A new understanding of the envelope and resolving the mode-mixing problem in am-fm signals. *IEEE Transactions on Signal Processing*, 60(3):1075–1086, March 2012.

22. Y. Hu and P Loizou. Evaluation of objective measures for speech enhancement. *Proceedings of INTERSPEECH 2006*, pages 1–4, September 2006.

23. Y. Hu and P Loizou. Evaluation of objective quality measures for speech enhancement. *IEEE Transactions on Audio, Speech and Language Processing*, 16(1):229–238, January 2008.

24. B. Huang and A. Kunoth. An optimization based empirical mode decomposition scheme. *Journal of Computational and Applied Mathematics*, 240:174–183, March 2013.

25. N. Huang and S. Shen, eds. *Hilbert-Huang Transform and Its Applications*. World Scientific Publishing Co. Pte. Ltd., 2005.

26. N. Huang, Z. Shen, and S. Long. A new view of nonlinear water waves: The Hilbert spectrum. *Annual Review of Fluid Mechanics*, 31:417–457, January 1999.

27. N. Huang, Z. Shen, S. Long, M. Wu, H. Shih, Q. Zheng, N. Yen, C. Tung, and H. Liu. The empirical mode decomposition and the hilbert spectrum for nonlinear and non-stationary time series analysis. *Proceedings of the Royal Society of London. Series A: Mathematical, Physical and Engineering Sciences*, 454(1971):903–995, March 1998.

28. E. Hurst. Long-term storage capacity of reservoirs. *Transactions of the American Society of Civil Engineers*, (11):770–799, April 1951.

29. A. Karagiannis and P. Constantinou. Noise-assisted data processing with empirical mode decomposition in biomedical signals. *IEEE Transactions on Information Technology in Biomedicine*, 15(1):11–18, January 2011.

30. J. Kates and K. Arehart. Coherence and the speech intelligibility index. *The Journal of the Acoustical Society of America*, 117(4):2224–2237, April 2005.

31. M.S. Keshner. $1/f$ noise. *Proceedings of the IEEE*, 70(3):212–218, March 1982.

32. K. Khaldi, A. Boudraa, A. Bouchikhi, and M. Alouane. Speech enhancement via EMD. *EURASIP Journal on Advances in Signal Processing*, 2008(1):873204.1–873204.8, May 2008.

33. D. Klatt. Prediction of perceived phonetic distance from critical-band spectra: A first step. *Proceedings of the IEEE International Conference on Acoustics, Speech, and Signal Processing (ICASSP '82)*, 7:1278–1281, May 1982.

34. Y. Kopsinis and S. Mclaughlin. Investigation and performance enhancement of the empirical mode decomposition method based on a heuristic search optimization approach. *IEEE Transactions on Signal Processing*, 56(1):1–13, January 2008.

35. P. Loizou and Y. Hu. A comparative intelligibility study of single-microphone noise reduction algorithms. *The Journal of the Acoustical Society of America*, 22(3):1777–1786, September 2007.

36. D. Looney and D. Mandic. Multiscale image fusion using complex extensions of emd. *IEEE Transactions on Signal Processing*, 57(4):1626–1630, April 2009.

37. J. Ma, Y. Hu, and P Loizou. Objective measures for predicting speech intelligibility in noisy conditions based on new band-importance functions. *The Journal of the Acoustical Society of America*, 125(5):3387–3405, May 2009.

38. B. Mandelbrot and J. Van Ness. Fractional brownian motions, fractional noises and applications. *SIAM Review*, 10(4):422–437, October 1968.

39. K. Manohar and P. Rao. Speech enhancement in nonstationary noise environments using noise properties. *Speech Communication*, 48:96–109, January 2006.

40. R. Martin. Noise power spectral density estimation based on optimal smoothing and minimum statistics. *IEEE Transactions on Speech and Audio Processing*, 9(5):504–512, July 2001.

41. J. Nunes, Y. Bouaoune, É. Deléchelle, O. Niang, and P. Bunel. Image analysis by bidimensional empirical mode decomposition. *Image and Vision Computing*, 21(12):1019–1026, 2003.

42. T. Oberlin, S. Meignen, and V. Perrier. An alternative formulation for the empirical mode decomposition. *IEEE Transactions on Signal Processing*, 60(5):2236–2246, May 2012.

43. N. Pustelnik, P. Borgnat, and P. Flandrin. Empirical mode decomposition revisited by multicomponent non-smooth convex optimization. *Signal Processing*, 102:313–331, 2014.

44. G. Rilling and P. Flandrin. On the influence of sampling on the empirical mode decomposition. 3:III–444–III–447, May 2006.

45. G. Rilling and P. Flandrin. Sampling effects on the empirical mode decomposition. *Advances in Adaptive Data Analysis*, 1(1):43–59, 2009.

46. G. Rilling, P. Flandrin, and P. Goncalvès. On empirical mode decomposition and its algorithms. *Proceedings of the IEEE-EURASIP Workshop on Nonlinear Signal and Image Processing (NSIP '03)*, pages 444–447, 2003.

47. R. Santana and R. Coelho. Low-frequency ambient noise generator with application to automatic speaker classification. *EURASIP Journal on Advances in Signal Processing*, 2012(175):1–7, August 2012.

48. R. Sant'Ana, R. Coelho, and A. Alcaim. Text-independent speaker recognition based on the hurst parameter and the multidimensional fractional brownian motion model. *IEEE Transactions on Audio, Speech, and Language Processing*, 14(3):931–940, May 2006.

49. P. Scalart and J. Filho. Speech enhancement based on a priori signal to noise estimation. *Proceedings of the IEEE International Conference on Acoustics, Speech and Signal*

Processing (ICASSP 1996), 32(6):629–632, December 1996.

50. C. Taal, R. Hendriks, R. Heusdens, and J. Jensen. An algorithm for intelligibility prediction of time-frequency weighted noisy speech. *IEEE Transactions on Audio, Speech and Language Processing*, 19(7):2125–2136, September 2011.

51. M. Torres, M. Colominas, G. Schlotthauer, and P. Flandrin. A complete ensemble empirical mode decomposition with adaptive noise. In *Proceedings of the IEEE International Conference on Acoustics, Speech and Signal Processing (ICASSP 2011)*, pages 4144–4147, May 2011.

52. A. Varga and H. Steeneken. Assessment for automatic speech recognition: II. NOISEX-92: a database and an experiment to study the effect of additive noise on speech recognition systems. *Speech Communications*, 12(3):247–251, July 1993.

53. D. Veitch and P. Abry. A wavelet-based joint estimator of the parameters of long-range dependence. *IEEE Transactions on Information Theory*, 45(3):878–897, April 1999.

54. A. Venturini, L. Zão, and R. Coelho. On speech features fusion, α-integration gaussian modeling and multi-style training for noise robust speaker classification. *IEEE/ACM Transactions on Audio, Speech, and Language Processing*, 22(12):1951–1964, December 2014.

55. R. Voss and J. Clarke. $1/f$ noise in music: Music from $1/f$ noise. *The Journal of the Acoustical Society of America*, 63(1):258–263, January 1978.

56. Z. Wu and N. Huang. Ensemble empirical mode decomposition: a noise-assisted data analysis method. *Advances in Adaptive Data Analysis*, 1(1):1–41, 2009.

57. Z. Xu, B. Huang, and K. Li. An alternative envelope approach for empirical mode decomposition. *Digital Signal Processing*, 20(1):77–84, January 2010.

58. L. Zão, D. Cavalcante, and R. Coelho. Time-frequency feature and AMS-GMM mask for acoustic emotion classification. *IEEE Signal Processing Letters*, 21(5):620–624, May 2014.

59. L. Zão and R. Coelho. Low-frequency optical noise generator using fractional statistics. *Electronics Letters*, 46(15):1072–1074, July 2010.

60. L. Zão, R. Coelho, and P. Flandrin. Speech enhancement with EMD and Hurst-based mode selection. *IEEE/ACM Transactions on Audio, Speech, and Language Processing*, 22(5):899–911, May 2014.

6 Acoustic Imaging Using the Kronecker Array Transform

Vítor Heloiz Nascimento
University of São Paulo (USP)

Bruno Sanches Masiero
University of Campinas (UNICAMP)

Flávio Protasio Ribeiro
Applied Sciences Group, Microsoft

The acoustic imaging problem consists of mapping the directions and intensities of sound sources using a microphone array. These maps are used, e.g., to design airplanes, cars, and trains that are quieter and more aerodynamically efficient, and also to analyze structures such as concert halls and turbines. In this chapter we describe ways to accelerate the computation of acoustic images, in particular the Kronecker array transform (KAT). We start by giving a short description of the problem of acoustic imaging, and the main state-of-the-art methods for solving it, from the standard beamforming method, through more accurate solutions such as DAMAS and covariance-fitting. We proceed by describing the KAT and how it can be applied to accelerate these methods, or to make possible the application of even more powerful methods, such as recent sparse estimation techniques, which without the KAT would be too computer intensive to be used in acoustic imaging.

6.1 INTRODUCTION

An *acoustic image* is generated when acoustic levels are coded into a colormap, generating an image of sound level as a function of direction of arrival. Acoustic images are commonly associated with the problem of detecting and characterizing acoustic sources.

Acoustic images can be superimposed over photographs, for instance, to identify unknown sound sources or to compare the relative sound power emitted by a set of sound sources. They may be used for noise reduction and analysis, typically present in the prototyping stages of machine and vehicle development [15], and in the analysis of wind-tunnel measurements [16], turbine noise [22], and in vortex-borne noise detection [4].

Acoustic levels are associated with a sound field, which is an acoustic wave field resulting from the interaction between an acoustic or a vibration source and an elastic medium, such as air or water. Acoustic levels at a point in the sound field can be either directly measured or, when that is not possible, estimated from measurements elsewhere in the same sound field. Such measurements are usually taken using an acoustic sensor array, such as, for example, a microphone or a hydrophone array.

A straightforward method to detect the presence of a sound source is to scan a measurement grid over a closed surface with an intensity probe [10]. If the total average intensity leaving this surface is greater than zero, then there is at least one sound source present inside this surface [17].

If one wishes, however, to more precisely characterize the sound source or even estimate the source's surface velocity out of acoustic measurements, then near field acoustic holography (NAH) should be used [38]. By adequately sampling a surface containing all of the sources that generate the desired sound field, NAH allows the extrapolation of the field's behavior in other regions of the source-free space or even allows one to identify, separate, and characterize the sources that generated the wave field.

As with any wave field, an acoustic wave field can be decomposed into its *active* and *reactive* components [30] and only the active component can transport the radiated acoustic energy far away from the sound source—the far field. The reactive component is composed of evanescent waves whose energy strongly decay while still in the vicinity of the sound source—the near field. That is the reason why if complete reconstruction of the sound field near the sound source is desired, near field measurements have to be conducted. A recently proposed method, however, can image the sound pressure field using laser tomography measurements conducted in the source's far field [24].

Acoustic imaging, on the other hand, focuses on a set of methods that can estimate the sound levels arriving at a point in space from different directions, that is, no attempt is made to estimate the whole sound field. Using these methods, one can verify the presence of sound sources and their directions in relation to a microphone array. The sound level arriving from each direction can be estimated through the use of spatial filters matched to both the array geometry and the source directions, or through the solution of a global optimization problem. Although these methods can be designed to suit several design criteria, they are all based on specific models for the signals received at the microphones, and their performance will depend on how well the models used correspond to reality. The plane wave model is most commonly used and is often adequate for sources in the far field. Under this assumption it is possible to estimate the sound intensity arriving at the microphone array from each source in the sound field.

The geometry of the microphone array will directly influence the quality of the acoustic image obtained. A measurement taken with a microphone array can be understood as a spatial sampling of the sound field. Traditional imaging techniques will pass these sampled signals through a spatial filter that acts as a window function convolved with the impinging sound field [18]. Microphone arrays usually have a reduced number of sensors, which results in window functions with a wide beamwidth and, consequently, in a smeared acoustic image.

Several methods have been proposed to increase image resolution without increasing the number of sensors in the array, either by changing the geometry of the array to reduce sidelobes [13, 40]; applying deconvolution techniques to eliminate the effect of the convolution with the response function [8, 9, 36]; or using regularized optimization [41]. These methods improve results, but have the drawback of increased computational costs as they involve the iterative solution of an optimization problem containing products of vectors with rather large matrices.

Furthermore, any imperfections in microphone positioning and gain will result in a different response function, mismatched to the designed spatial filters, and consequently resulting in errors in the estimated acoustic image. The influence of these imperfections can be countered by the calibration of the array, which will be briefly discussed at the end of the chapter.

This chapter describes three methods for accelerating the calculation of vector-matrix products required for all of the above mentioned methods: the *non-equispaced in time and frequency fast Fourier transform* (NNFFT), the *non-equispaced fast Fourier transform* (NFFT), and the new *Kronecker array transform* (KAT). The NNFFT is the most general, but slowest, acceleration method. It can be employed for any array geometry or space parametrization (that is, any choice of directions toward which the array will "look"). The NFFT is faster, but is restricted to a uniformly sampled choice of look directions. Finally, the KAT is the fastest method, but both the array and the look directions must be organized in a separable geometry (i.e., a possibly non-uniform rectangular grid). The KAT however, unlike the other methods, can be extended to the calculation of acoustic images with sources closer to the microphone array, when some of the far field approximations are no longer valid [26, 27].

The most important advantage of the acceleration methods, and of the KAT in particular, is that they allow one to use more advanced reconstruction algorithms, such as sparse or regularized methods, for larger problems (i.e., with more microphones and look directions).

6.2 SIGNAL MODEL

The acoustic imaging techniques discussed in this chapter are all model-based techniques, i.e., they use different strategies to solve an inverse problem based on a wave propagation model. The signal model that will be used throughout the chapter is based on the following assumptions. First, we assume that all sources lay in the far field and thus each wave front that arrives at the microphone array is a plane wave. Second, we assume that the sound intensity at the array is low enough that superposition applies (this is not a restrictive assumption in general). Finally, we assume that all sources are statistically uncorrelated (this assumption is not true in general, but is necessary to keep the problem computationally feasible).

6.2.1 WAVE PROPAGATION

The linearized acoustic wave equation in Cartesian coordinates is [38]

$$\frac{\partial^2 p}{\partial x^2} + \frac{\partial^2 p}{\partial y^2} + \frac{\partial^2 p}{\partial z^2} = \frac{1}{c^2}\frac{\partial^2 p}{\partial t^2}, \tag{6.1}$$

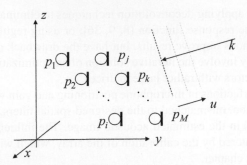

Figure 6.1 Plane wave propagating in direction k and M-element microphone array.

where c is the speed of sound in air. The solution to (6.1) can be decomposed at a point $\boldsymbol{x} = [x \ \ y \ \ z]^T$ sufficiently far from all sources as a superposition of plane waves propagating in different directions [35]. A single plane wave traveling along the direction $\boldsymbol{k}$ assumes the form

$$x(t, \boldsymbol{p}) = f(t - \boldsymbol{k}^T \boldsymbol{p}, \boldsymbol{u}) = f(t + \omega \boldsymbol{u}^T \boldsymbol{p}/c, \boldsymbol{u}),$$

where $\boldsymbol{k} = [k_x \ \ k_y \ \ k_z]^T$ is the *wavenumber* vector, a vector that points in the direction of propagation of the wave, with magnitude $\|\boldsymbol{k}\|_2 = \omega/c$ ($\| \cdot \|_2$ is the Euclidean norm), and $\boldsymbol{u} = -c\boldsymbol{k}/\omega$ is a unit-length vector that points towards the direction *from which* the wave is arriving at the array (see Figure 6.1). If the waveform $f(t, \boldsymbol{u})$ has a single frequency, $f(t, \boldsymbol{u}) = F(\omega, \boldsymbol{u})e^{j\omega t}$, then the signal at a point $\boldsymbol{p}$ in space has the form $x(t, \boldsymbol{p}) = X(\omega, \boldsymbol{p})e^{j\omega t}$, with[1]

$$X(\omega, \boldsymbol{p}) = F(\omega, \boldsymbol{u})e^{j\frac{\omega}{c}\boldsymbol{u}^T \boldsymbol{p}} = F(\omega, \boldsymbol{u})e^{-j\boldsymbol{k}^T \boldsymbol{p}}. \tag{6.2}$$

We now use an array with M microphones to sample this sound field. The position of each microphone is given by $\boldsymbol{p}_m = [x_m \ \ y_m \ \ z_m]^T$. We write the signal sensed by each microphone when the plane wave arrives at the array from direction $-\boldsymbol{u}$ (as shown in Figure 6.1) as

$$\boldsymbol{X}(\omega) = \begin{bmatrix} X(\omega, \boldsymbol{p}_1) \\ X(\omega, \boldsymbol{p}_2) \\ \vdots \\ X(\omega, \boldsymbol{p}_M) \end{bmatrix} = \begin{bmatrix} A_1(\omega, \boldsymbol{u})e^{j\omega \boldsymbol{u}^T \boldsymbol{p}_1/c} \\ A_2(\omega, \boldsymbol{u})e^{j\omega \boldsymbol{u}^T \boldsymbol{p}_2/c} \\ \vdots \\ A_N(\omega, \boldsymbol{u})e^{j\omega \boldsymbol{u}^T \boldsymbol{p}_M/c} \end{bmatrix}. \tag{6.3}$$

$A_m(\omega, \boldsymbol{u}) = G_m(\omega, \boldsymbol{u}) \cdot F(\omega, \boldsymbol{u})$ has a component $G_m(\omega, \boldsymbol{u})$ proportional to the microphone's sensitivity and directivity in addition to the component $F(\omega, \boldsymbol{u})$ relative to

---

[1]We remind the reader that model (6.2) in general is an approximation, valid for a single plane wave in the vicinity of a point $\boldsymbol{p}$. Since it does not consider attenuation or other distortions suffered by the signal, $f(t)$ is *not* the signal actually emitted by the source, but rather the signal *arriving* at point $\boldsymbol{p}$ from a certain direction $\boldsymbol{u}$, using the phase at the origin $\boldsymbol{p} = \boldsymbol{0}$ as a time reference. If the source is sufficiently far, the waveform $f(\cdot)$ will not change on a neighborhood of $\boldsymbol{p}$, only the phase at each point.

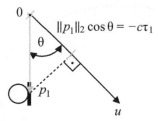

Figure 6.2 Relative delay for a plane wave arriving from direction $k = -u$. Note that in this example, since the wavefront arrives at p_1 before it reaches the origin 0, the delay τ_1 is negative.

the arriving signal. For the time being we assume that all microphones are omnidirectional and that they are all adequately calibrated, so that $G_m(\omega, u) = 1$, and (6.3) simplifies to

$$X(\omega) = F(\omega, u) \begin{bmatrix} e^{j\omega u^T p_1/c} \\ e^{j\omega u^T p_2/c} \\ \vdots \\ e^{j\omega u^T p_M/c} \end{bmatrix} \triangleq F(\omega, u)v(u), \qquad (6.4)$$

where $v(u)$ is known as the *array manifold vector*. It contains the relative delays with which the plane wave propagating from direction $-u$ reaches each of the array's sensors. A microphone array will be able to differentiate between plane waves arriving from different directions by the relative delays between the signals at each microphone. The array manifold vector has a fundamental role in acoustic imaging because it is a concise and convenient way of representing these delays, as we show next. Consider the microphone at position p_1. Then

$$k^T p_1 = -\frac{\omega}{c} u^T p_1 = -\frac{\omega}{c} \|p_1\|_2 \cos \theta \triangleq \omega \tau_1, \qquad (6.5)$$

where θ is the angle between u and p_1, as shown in Figure 6.2, and τ_1 is the delay at p_1, using the phase at the origin as reference, for a wave arriving from direction $-u$. Note that the product of $F(\omega, u)$ and each entry of $v(u)$ is of the form $e^{-j\omega \tau_i} F(\omega, u)$, and thus corresponds to the application of a delay to $f(t)$.

6.2.2 SUPERPOSITION OF SOUND SOURCES

If the sound field next to the array is linear and all sources are far enough according to the maximum wavelength and maximum array dimension, the signal received at the microphones will be the superposition of an infinite number of plane waves (for a detailed discussion, see [35]). Of course, in general, a discrete approximation is computed, which corresponds to approximating the signals received at the array as a superposition of a finite number of plane waves coming from certain *previously chosen* directions, as shown in Figure 6.3. This sampling in u-space corresponds to

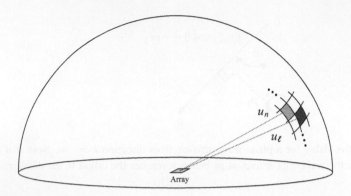

Figure 6.3   Sampling in $\boldsymbol{u}$-space.

choosing a finite number of "look" directions $\boldsymbol{u}_1 \ldots \boldsymbol{u}_N$ towards which the array will be steered, resulting in a model

$$X(\omega) = \sum_{n=1}^{N} F(\omega, \boldsymbol{u}_n) \cdot \boldsymbol{v}(\boldsymbol{u}_n) = \begin{bmatrix} \boldsymbol{v}(\boldsymbol{u}_1) & \cdots & \boldsymbol{v}(\boldsymbol{u}_N) \end{bmatrix} \begin{bmatrix} F(\omega, \boldsymbol{u}_1) \\ \vdots \\ F(\omega, \boldsymbol{u}_N) \end{bmatrix}. \tag{6.6}$$

We are assuming for the time being that the signals coming from all directions are deterministic. In this case, the (discretized) acoustic image we want to estimate is defined to be the square power for each frequency and direction of the incoming signal, i.e., $Y(\omega, \boldsymbol{u}_n) \stackrel{\Delta}{=} |F(\omega, \boldsymbol{u}_n)|^2$.

We now need to expand our model to include more general kinds of signals, letting the signals $f(t)$ arriving from each direction be stationary random processes. In this case, a direct model like (6.6) would not be available (since stationary processes do not have finite energy, only the power spectrum is defined). A detailed and precise explanation would take too long, so we refer the reader to [35]. A way around the technical difficulties is to use the discrete Fourier transform (DFT). Define the DFTs of the microphone and source signals over a window of length K, using a proper sampling rate Δt as

$$\hat{F}(\omega_k, \boldsymbol{u}) = \sum_{p=0}^{K-1} f(p\Delta t, \boldsymbol{u}) e^{-j2\pi kp/K}, \tag{6.7}$$

$$\hat{X}(\omega_k) = \sum_{p=0}^{K-1} x(p\Delta t) e^{-j2\pi kp/K}, \ 0 \le k \le K - 1, \tag{6.8}$$

where $\omega_k \stackrel{\Delta}{=} 2\pi p/K$. With these definitions, (6.6) still holds approximately with the approximation becoming better as the window length K grows [35].

Define $V = [\boldsymbol{v}(\boldsymbol{u}_1) \ \cdots \ \boldsymbol{v}(\boldsymbol{u}_N)]^T$. In this case, the autocorrelation matrix of $\hat{X}(\omega_k)$

can be written as

$$R_x(\omega_k) = \mathrm{E}\left\{\hat{X}(\omega_k)\hat{X}^H(\omega_k)\right\} = VR_F(\omega_k)V^H, \tag{6.9}$$

where $\mathrm{E}\{\cdot\}$ is the expected value, and the source autocorrelation matrix is

$$R_F(\omega_k) = \mathrm{E}\left\{\begin{bmatrix}\hat{F}(\omega_k,u_1)\\ \vdots \\ \hat{F}(\omega_k,u_N)\end{bmatrix}\begin{bmatrix}\hat{F}^*(\omega_k,u_1) & \cdots & \hat{F}^*(\omega_k,u_N)\end{bmatrix}\right\}. \tag{6.10}$$

Note that the acoustic image corresponds to the diagonal entries of $R_F(\omega_k)$:

$$Y(\omega_k,u_n) = [R_F(\omega_k)]_{n,n} = \mathrm{E}\left\{\left|\hat{F}(\omega_k,u_n)\right|^2\right\}. \tag{6.11}$$

In general, $R_F(\omega_k)$ will be a full matrix (meaning that signals arriving from different directions may be correlated). However, if N is large (as we would like it to be, in order to compute an acoustic image with good resolution), taking account of all the $N(N-1)/2$ different correlations would not be feasible, so it is usual to assume that there is no correlation. This is, of course, an approximation, which may create artifacts in the estimated acoustic image.

From now on we omit the frequency ω_k in order to simplify the notation. Under the assumption of uncorrelated signals, the expression for R_x simplifies to

$$R_x = \mathrm{E}\left\{\hat{X}\hat{X}^H\right\} = \sum_{n=1}^N Y(u_n) \cdot v(u_n)v^H(u_n), \tag{6.12}$$

so there is a linear relationship between the autocorrelation matrix of the microphone signals (which can be estimated directly from observations) and the desired acoustic image.

This linear relationship becomes more apparent if we rewrite (6.12) as follows. For matrices $A = [a_{ij}]$ and B of any dimensions, define

$$\mathrm{vec}(A) \triangleq \begin{bmatrix}a_{11}\\a_{21}\\\vdots\\a_{12}\\a_{22}\\\vdots\end{bmatrix}, \qquad A \otimes B \triangleq \begin{bmatrix}a_{11}B & a_{12}B & \cdots\\a_{21}B & a_{22}B & \cdots\\\vdots & \vdots & \ddots\end{bmatrix}.$$

The vec($\cdot$) operator therefore corresponds to stacking the columns of a matrix one on top of the other, and the *Kronecker product* $A \otimes B$ of two matrices corresponds to a block matrix in which each block entry is an element of A multiplied by B. A property of Kronecker products is that [14]

$$\mathrm{vec}\,(ACB) = \left(B^T \otimes A\right)\mathrm{vec}\,(C), \tag{6.13}$$

for any matrices A, B, and C for which the product ACB is defined.

Applying (6.13) to each term $v(u_n)Y(u_n)v^H(u_n)$ in (6.12), we obtain

$$
r_x \overset{\Delta}{=} \text{vec}(R_x) = \underbrace{\begin{bmatrix} v^*(u_1) \otimes v(u_1) & \cdots & v^*(u_N) \otimes v(u_N) \end{bmatrix}}_{\overset{\Delta}{=}A} \underbrace{\begin{bmatrix} Y(u_1) \\ \vdots \\ Y(u_N) \end{bmatrix}}_{\overset{\Delta}{=}y}, \qquad (6.14)
$$

where we defined the vector y whose elements are the acoustic image pixels, the vector r_x representing the microphone correlations, and the matrix A that relates them.

6.2.3 ADDITIVE NOISE

Real measurements are always corrupted by noise, such as thermal noise at the sensors or quantization noise after analog-to-digital conversion. We model the presence of noise as an additive term, replacing (6.6) by

$$
\hat{X} = \sum_{n=1}^{N} \hat{F}(u_n) \cdot v(u_n) + \hat{z}, \qquad (6.15)
$$

where $\hat{z}$ is a vector with the additive noise at each microphone at frequency ω_k. We assume that the noise is uncorrelated with our signals of interest and, consequently, the autocorrelation matrix of $\hat{X}$ is updated to

$$
R_x = \sum_{n=1}^{N} \left[Y(u_n) \cdot v(u_n)v^H(u_n) \right] + R_z, \qquad (6.16)
$$

where $R_z = \text{E}\left\{\hat{z}\hat{z}^H\right\}$ is the autocorrelation matrix of the noise component. If the noise is uncorrelated between the sensors, then $R_z = \sigma_z^2 I$, where I is the identity matrix.

6.3 METHODS FOR ACOUSTIC IMAGING

There are many methods to estimate an acoustic image, with increasing levels of sophistication. All methods mentioned below estimate the sound levels arriving from different directions, assuming the array to be in the far field of all sources.

Common array applications such as antenna arrays, radar, and sonar work with narrowband signals. In this case, a single manifold vector matched to the central frequency of the signal can be used to model the plane wave. Acoustic signals, however, are commonly broadband in nature. All of the methods that will be presented in this section calculate the acoustic image using the manifold vector, which depends on ω. Therefore, the acoustic image must be calculated independently for several narrowband-filtered versions of the signal and, if desired, later added together to form a broadband image.

6.3.1 SPATIAL FILTERING

A spatial filter is implemented as a weighted sum of the signals captured by the sensors, such that

$$Z = w^H \hat{X}, \tag{6.17}$$

where $w = [w_1 \ w_2 \ \cdots \ w_M]^T$ is a complex weight vector.

6.3.1.1 Deterministic Beamformer

There are several ways to calculate w, the most straightforward manner being deterministic beamformers. In the *Bartlett* beamformer the spatial filter w is chosen so that the filter output power is maximized when the array is excited by a plane wave arriving from $-u$ [21]. Thus, we aim to solve

$$\arg \max_{w} \mathrm{E}\{|Z(\omega)|^2\}. \tag{6.18}$$

Substituting (6.15) and (6.17) into (6.18) and assuming that the sound field is composed of a single plane wave propagating in direction $-u$, we obtain the cost function

$$J = \left|w^H v(u)\right|^2 R_S + \|w\|^2 \sigma_n^2. \tag{6.19}$$

To avoid the trivial solution $\|w\| \to \infty$, the Barlett beamformer adds the restriction $\|w\| = 1$ and uses the Cauchy-Schwarz inequality to maximize J, which results in

$$w_{\mathrm{BF}}(u) = \frac{v(u)}{\|v(u)\|}. \tag{6.20}$$

Thus, the Bartlett beamformer acts by applying a delay to the signals captured by each sensor, so that the signals arriving from $-u$ are aligned in time and thus constructively added. Note that the Bartlett beamformer is a deterministic method since its weights do not depend on the statistics of the incoming signal, but only on the "listening" direction and the geometry of the array.

Another very common deterministic beamformer is the *delay-and-sum* (DAS) beamformer [35]. Similarly to the Bartlett beamformer, the DAS beamformer seeks to compensate for the relative delay at each sensor and then averages the resulting signals, thus

$$w_{\mathrm{DAS}}(u) = \frac{1}{M} v(u) = \frac{v(u)}{v^H(u)v(u)}. \tag{6.21}$$

The DAS beamformer is equivalent to the Bartlett beamformer except for a scalar gain.

To obtain the acoustic image, we need to estimate the sound intensity coming from each direction u in a pre-defined grid, to obtain a vector of estimates $\hat{y}$ to y. Using a fixed beamformer for each direction in the grid and assuming a perfect estimate of the signal, we have

$$\hat{Y}(u_n) = \mathrm{E}\{|w^H(u_n)\hat{X}|^2\} = w^H(u_n)R_x w(u_n). \tag{6.22}$$

The expected value is approximated usually by estimating $\hat{X}$ for a number L of (possibly overlapping) length-K windows, and taking the average of $|w^H \hat{X}|^2$ over the L windows.

Remark that using (6.13) and (6.14) with the DAS beamformer gives us,

$$
\hat{Y}(u_n) = \frac{1}{M^2} v^H(u_n) R_x v(u_n)
$$
$$
= \frac{1}{M^2} \left(v^T(u_n) \otimes v^H(u_n) \right) \text{vec}(R_x) = \frac{1}{M^2} [A]_n^H r_x, \tag{6.23}
$$

where $[A]_n$ denotes the n-th column of A, and thus the DAS beamformer is equivalent to the operation

$$
\hat{y} = \frac{1}{M^2} A^H r_x, \tag{6.24}
$$

while the Bartlett beamformer corresponds to

$$
\hat{y} = \frac{1}{M} A^H r_x. \tag{6.25}
$$

The artifacts in the image resulting from using these beamformers are better understood thinking in terms of array point-spread functions (PSFs), and their 2D convolution with the true acoustic image [35], as we mention further ahead in Section 6.3.2. However, it is also useful to compare (6.24) and (6.14). We see that $\hat{y}$ is equal to y only if $A^{-1} = 1/M^2 A^H$, which would only be true if the columns of A were orthogonal (and $N \leq M^2$ for the inverse to exist). Since in general these conditions are not met, we can expect the image estimated using beamforming to have large artifacts.

6.3.1.2 Optimal Beamformers

The conventional beamformer features a very simple implementation but has the drawback of low image resolution, i.e., if two sound sources are placed too close to each other acoustic images obtained with the conventional beamformer will not be able to resolve both sources. Numerous methods have been proposed as an attempt to improve the image resolution and an important group of such methods are the statistically optimal beamformers. Using the statistics of the sound field, represented by the signals' autocorrelation matrix, these methods can reduce the influence of the neighboring sources resulting in an image with better resolution. As a trade-off these methods are more computationally expensive as they all include the inversion of an autocorrelation matrix as a step to calculate the optimal weights.

In [35, ch. 6] it is demonstrated that for the wide class of optimal beamformers—a class which includes the *minimum variance distortionless response* (MVDR) beamformer, the *minimum power distortionless response* (MPDR) beamformer, the *minimum mean square error* (MMSE) beamformer, and the *maximum signal to noise* (SNR) beamformer—the resulting spatial filter is an MVDR beamformer followed by a scalar filter dependent on the optimization criterion.

The MVDR beamformer minimizes the variance of the received signal $Z(\omega)$ while keeping a unitary gain at the listening direction, i.e., $w^H v(u) = 1$. This optimization problem can be solved using Lagrange multipliers [35] and results in

$$w_{\text{MVDR}}^H(u) = \frac{v^H(u)R_z^{-1}}{v^H(u)R_z^{-1}v(u)}. \tag{6.26}$$

In practical implementations it may be difficult to estimate R_z if a signal is always present in the direction of interest u. In this case, the MPDR beamformer can be used with weights given by

$$w_{\text{MPDR}}^H(u) = \frac{v^H(u)R_x^{-1}}{v^H(u)R_x^{-1}v(u)}. \tag{6.27}$$

Note that R_x tends to be ill conditioned and Thikonov regularization is usually necessary:

$$w_{\text{MPDR}}^H(u) \approx \frac{v^H(u)\,[R_x + \lambda I]^{-1}}{v^H(u)\,[R_x + \lambda I]^{-1}\,v(u)}, \tag{6.28}$$

where $\lambda \geq 0$ is the regularization parameter, which must be carefully chosen for good performance.

The acoustic image is generated in the same manner as with the conventional beamforming using (6.22), that is, $\hat{Y}(u_n) = \mathrm{E}\left\{|w_{\text{MPDR}}^H(u_n)X|^2\right\}$. It is not difficult to verify that this is equivalent to

$$\hat{y} = \frac{1}{(v^H(u)[R_x + \lambda I]^{-1}v(u))^2}A^H\,\mathrm{vec}\left([R_x + \lambda I]^{-1}R_x[R_x + \lambda I]^{-1}\right). \tag{6.29}$$

6.3.2 DECONVOLUTION METHODS

Consider a single plane wave traveling along direction $-u$. The estimated pixel corresponding to a generic look direction u_n is given by (6.22). Considering the discrete model (6.11), we have[2]

$$\hat{Y}(u_n) = Y(u) \cdot w^H(u_n)v(u)v^H(u)w(u_n) \overset{\Delta}{=} Y(u) \cdot P(u_n, u). \tag{6.30}$$

For a fixed look direction u_n, the term $P(u_n, u)$, considered as a function of u, is the array's *point spread function* (PSF), which describes the gain applied by the array to an input plane wave arriving from direction $-u$ [35]. $P(u_n, u)$ is defined over the entire space and can be interpreted as a spatial sampling function that should ideally be maximally sharp, that is (again in our discrete model), equal to

$$P(u_n, u) = \delta_{u_n, u}, \tag{6.31}$$

where $\delta_{u_n, u} = 1$ if $u = u_n$ and zero otherwise. However, as microphone arrays have a limited number of sensors, their typical PSF will present a larger beamwidth and consequently a smeared acoustic image.

---

[2]We restrict ourselves again to a discrete spatial distribution of sources, to avoid the long detour necessary to explain adequately the continuous model.

Now, calculating the acoustic image using (6.22) and considering a superposition of N sources as in (6.12) results in

$$\hat{Y}(u_n) = \sum_{\ell=1}^{N} Y(u_\ell) \cdot P(u_n, u_\ell). \qquad (6.32)$$

Equation (6.32) can be interpreted as a spatial convolution [38], i.e., when calculating an acoustic image with conventional or optimal beamformers the result is, in fact, the convolution of the actual acoustic image with the array PSF. This is a second way of explaining the smeared images produced by standard beamformers (compare with (6.24)).

To reduce the smearing observed in beamforming, several deconvolution techniques have been proposed [8, 9, 29, 36]. They use as inputs the PSF and the image obtained with the DAS beamformer, and generally produce a better approximation of the original source distribution.

6.3.2.1 DAMAS2

One of the most popular deconvolution methods is the *deconvolution approach for the mapping of acoustic sources* (DAMAS) [3], later improved in [8] and named DAMAS2. Denote by Y the 2-D acoustic image (i.e., y rearranged as a two-dimensional image), and similarly by $\hat{Y}$ the estimated image. DAMAS2 calculates a better approximation $\hat{Y}$ for Y given the DAS estimate (denoted below by $\check{Y}$) by iterating

$$\hat{Y}^{(k+1)} = \max\left\{\hat{Y}^{(k)} + \frac{1}{a}\left[\check{Y} - \left(P * \hat{Y}^{(k)}\right)\right], 0\right\}, \qquad (6.33)$$

where $*$ denotes 2-D convolution, $\hat{Y}^{(k)}$ is the reconstructed image at iteration k with $\hat{Y}^{(0)} = 0$, P is the discretized PSF also arranged as a 2-D array, $a = \sum_{i,j}|P|_{i,j}$, and $\max\{\cdot, \cdot\}$ returns the pointwise maximum. This function is used to guarantee strictly positive power estimates. Convolution can be implemented as a multiplication in the wavelength domain [38] and, therefore, the DAMAS2 algorithm is able to efficiently produce a deconvolved or "clean" acoustic image.

6.3.3 COVARIANCE FITTING

Note that even though DAMAS2 is a state-of-the-art method for computationally efficient acoustic imaging, it does not use any regularization other than forcing pointwise non-negativity, i.e., it does not incorporate a prior model of the source distribution.

Regularized signal reconstruction has been a topic of interest for many decades, and gained significant momentum with the popularity of compressive sensing [5, 7]. Indeed, many image reconstruction problems can be recast as convex optimization problems, which can be solved with computationally efficient iterative methods. While many of these techniques were designed for imaging applications, they have remained limited to fields such as medical image reconstruction. Therefore, most of these developments have not yet been applied to acoustic imaging.

Considering the presence of noise we rewrite (6.15) in the matrix form as

$$r_x = Ay + \text{vec}\{R_z\} = Ay + \sigma^2 \text{vec}\{I\}, \tag{6.34}$$

assuming spatially uncorrelated noise. Note that the transfer matrix A has usually more columns than rows, so (6.34) is underdetermined. Prior models of the source distribution can be incorporated as constraints that allow the underdetermined system of equations to be solved.

6.3.3.1 ℓ_1-Regularized Least Squares

Assume that the acoustic field arriving at the microphone array was generated by only a few compact sources, that is, that the source distribution is *sparse*. In this case we can apply a sparsity constraint to regularize the inversion problem, as suggested in [41], where the following convex optimization problem is proposed

$$\begin{aligned} \underset{\hat{y},\sigma^2}{\text{minimize}} \quad & \left\| r_x - A\hat{y} - \sigma^2 \text{vec}\{I\} \right\|_2^2 \\ \text{subject to} \quad & \hat{y}_{i,j} \geq 0, \ \sigma^2 \geq 0, \ \text{and} \ \|\hat{y}\|_1 \leq \lambda. \end{aligned} \tag{6.35}$$

The ℓ_1 constraint $\|\hat{y}\|_1 \leq \lambda$ serves to regularize the problem while forcing sparsity. λ is a regularization parameter.

Thanks to the ℓ_1 regularization, the authors of [41] show, using numerical examples, that by solving (6.35) one can indeed reconstruct sparse images with very high accuracy. Their proposal outperforms DAMAS2 regarding reconstruction accuracy due to the use of regularization and because no deconvolution was involved.

Another option is to recast (6.35) as a basis pursuit with denoising problem (BPDN), which has the form

$$\begin{aligned} \underset{\hat{y}}{\text{minimize}} \quad & \|\hat{y}\|_1 \\ \text{subject to} \quad & \|r_x - A\hat{y}\|_2 \leq \sigma, \end{aligned} \tag{6.36}$$

a kind of optimization problem that has been studied in detail in the compressive sensing literature [34].

6.3.3.2 Total Variation Regularized Least Squares

To address scenarios where the acoustic images are not sparse in their canonical representations, another possibility is to reconstruct acoustic images with total variation (TV) regularization.

The isotropic total variation norm is defined as

$$\|Y\|_{\text{TV}} = \sum_{i,j} \sqrt{[\nabla_x Y]_{i,j}^2 + [\nabla_y Y]_{i,j}^2} \tag{6.37}$$

where ∇_x and ∇_y are the first difference operators along the x and y dimensions with periodic boundaries, and i and j are the indices in the x and y dimensions, respectively.

The following optimization problem can then be solved

$$
\begin{aligned}
\underset{\hat{Y}}{\text{minimize}} \quad & \left\|\hat{Y}\right\|_{TV} + \mu \left\|\mathbf{r}_x - A\hat{y}\right\|_2^2 \\
\text{subject to} \quad & \left[\hat{Y}\right]_{i,j} \geq 0.
\end{aligned}
\tag{6.38}
$$

The first term measures how much an image oscillates. Therefore, it is smallest for images with plateaus and monotonic transitions, and tends to privilege simple solutions with small amounts of noise. The second term ensures a good fit between the reconstructed image and the measured data. This formulation was first proposed for image denoising [28], and was later generalized and applied successfully to many image reconstruction problems. This method provides accurate and stable image reconstructions with guaranteed convergence.

6.4 EFFICIENT CALCULATION METHODS

Using covariance-fitting or deconvolution methods, it is possible to obtain acoustic images with good resolution using moderate-sized arrays. These methods are iterative, requiring repeated computation of matrix-vector products of the form $A\hat{y}$ and/or $A^H\hat{s}$. Matrix A is, however, rather large in practice: for a 64-element array and a 128×128-pixel image, matrix A in (6.14) would be $64^2 \times 128^2 = 4\,096 \times 16\,384$, making the more advanced methods very time consuming. We now show how the structure of A can be used to compute matrix-vector products more efficiently.

There are three strategies for accelerating the computation of acoustic images. When each method can be employed depends on the array geometry (i.e., how the microphones are distributed in space) and on the sampling scheme (i.e., the choice of look directions $\mathbf{u}_n$). The NNFFT (non-equispaced in time and frequency fast Fourier transform), applicable to any array geometry or sampling scheme; the NFFT (non-equispaced fast Fourier transform), valid for any *planar* array geometry, and *uniform* sampling of the look directions; and the Kronecker array transform, valid for *planar and separable* array geometries and *separable* sampling of the look directions. By separable we mean that microphones and look directions must be arranged in a rectangular grid, not necessarily uniform, as the example shown in Figure 6.4.

The KAT provides the largest gain in computational cost, under the constraint of separable geometry and sampling. It can also be combined with the other transforms to further decrease the cost. We describe the three transforms next, additional details can be found in [26].

6.4.1 NNFFT

Given a sequence of points h_n, the NNFFT is an approximate algorithm for computing expressions of the form [19]

$$
\hat{h}_m = \sum_{n=1}^{N} h_n e^{-j2\pi b_n^T D c_m}, \quad 1 \leq m \leq M^2,
\tag{6.39}
$$

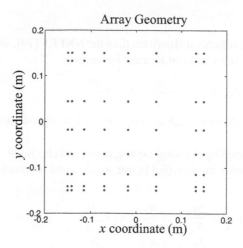

Figure 6.4   Non-uniform separable geometry. The dots represent positions of microphones.

where D is a diagonal matrix and b_n, $1 \leq n \leq N$, c_m, $1 \leq m \leq M^2$ are vectors whose entries satisfy

$$-\frac{1}{2} \leq b_{\ell,n} < \frac{1}{2}, \qquad\qquad -C_\ell \leq c_{\ell,m} < C_\ell, \qquad (6.40)$$

for positive constants C_ℓ.

To see that the NNFFT can be used for acoustic imaging, let the entries of u_n be $u_{x,n}$, $u_{y,n}$, and similarly for vector p_m. Note that the direction vector is of the form $u_n = [u_{x,n} \; u_{y,n} \; u_{z,n}]^T$, with $u_{x,n}^2 + u_{y,n}^2 + u_{z,n}^2 \leq 1$. Therefore, the entries are in the range $-1 \leq u_{x,n}, u_{y,n}, u_{z,n} \leq 1$.

Take for example the second entry of a product of the form $\hat{r} = A\hat{y}$. The second row of A has the form

$$\left[v_1^*(u_1)v_2(u_1) \quad \ldots \quad v_1^*(u_N)v_2(u_N) \right]$$
$$= \left[e^{-j\omega u_1^T(p_1 - p_2)/c} \quad \ldots \quad e^{-j\omega u_N^T(p_1 - p_2)/c} \right].$$

The product of this row by a vector $\hat{y}$ has therefore the form (6.39), with $b_n = u_n/2$, $c_2 = 2\omega(p_1 - p_2)/(2\pi c)$, and $D = I_3$ (the 3×3 identity matrix). The other elements of $\hat{r}$ have similar form, but now with $c_m = \omega(p_i - p_\ell)/(\pi c)$, for a particular pair (i, ℓ) satisfying $1 \leq i, \ell \leq M$. Choosing an ordering of the differences $p_i - p_\ell$, we can use the NNFFT algorithm to compute the product $\hat{r} = A\hat{y}$. The inputs h_n to the NNFFT are the entries of $\hat{y}$ and the outputs $\hat{h}_i$ are the entries of the product $\hat{r}$.

Note that if the array is planar, we can define the coordinate system so that the array lies in the x, y plane, such that $p_{z,m} = 0$ for all microphones. In this case (6.39) reduces to a two-dimensional transform (similarly, for a linear array only a one-dimensional transform is necessary). Note that in these cases there will be ambiguities between look directions: for example, a planar array cannot distinguish between signals coming from its front or its back.

6.4.2 NFFT

The NFFT is a faster, but less general algorithm than the NNFFT [19], with a restriction on vector b_n in (6.39): the entries of b_n must be integers

$$b_{\ell,n} \in \mathbb{Z}: \quad -\frac{N_\ell}{2} \le b_{\ell,n} < \frac{N_\ell}{2}, \tag{6.41}$$

for even $N_\ell \in \mathbb{N}$. To satisfy these restrictions, we need to choose adequate look directions u_n.

If one is interested in sampling the whole space, one could choose $N = N_x N_y$ with even N_x and N_y, and, in order to obey (6.41), we would need to choose a uniform sampling:

$$u_{x,n} = \frac{2n_x}{N_x}, \quad -\frac{N_x}{2} \le n_x < \frac{N_x}{2}, \tag{6.42a}$$

$$u_{y,n} = \frac{2n_y}{N_y}, \quad -\frac{N_y}{2} \le n_y < \frac{N_y}{2}. \tag{6.42b}$$

Defining $b_{x,n} = N_x u_{x,n}/2$, $b_{y,n} = N_y u_{y,n}/2$, we would obey (6.41) for directions x and y. However, u_n must have unit length, so for direction z we would need $u_{z,n} = \pm \sqrt{1 - u_{x,n}^2 - u_{y,n}^2}$, and the sampling in the z-direction would not be uniform.

A solution is to restrict ourselves to planar arrays, and choose a coordinate system so that the x, y plane corresponds to the array plane. With this choice, the microphone coordinates satisfy $p_{z,m} = 0$ and the $u_{z,n}$ entries vanish in the dot products $u_n^T p_m$. The NFFT algorithm can then be used, with

$$b_n = \begin{bmatrix} \frac{N_x}{2} & 0 & 0 \\ 0 & \frac{N_y}{2} & 0 \end{bmatrix} u_n, \quad c_m = \frac{\omega}{2\pi c} \begin{bmatrix} \frac{2}{N_x} & 0 & 0 \\ 0 & \frac{2}{N_y} & 0 \end{bmatrix} p_m. \tag{6.43}$$

The fact that (6.42) allows $u_{x,n}^2 + u_{y,n}^2 > 1$ means that the transform would compute images for directions that do not in fact exist. This results in a performance loss (since we compute values that we do not need), but for large N_x, N_y there is a net gain. Of course, one could choose $u_{x,n}, u_{y,n}$ restricted to the interval $\left[-\sqrt{2}/2, \sqrt{2}/2 \right]$, thereby guaranteeing that only values for true look directions are evaluated, but losing information that might come from the edges of the array. See Figure 6.5.

Acceleration with the FFT

One can verify that, if the microphones are placed in a uniform rectangular grid, the look directions u_n are chosen through uniform sampling of u_x and u_y, and if the frequency of interest ω is such that the distance between consecutive microphones is half the wavelength, then the NFFT just described reduces to the (much faster) FFT. Unfortunately for acoustic imaging this observation is of little use, given that acoustic signals are broadband. In addition, when using covariance fitting methods with regularization, non-uniform microphone arrays lead to better results (see [25] and Section 6.6).

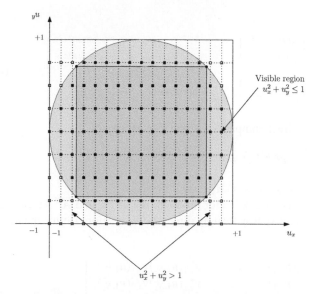

Figure 6.5 Choice of look directions u_n according to (6.42), with $N_x = 16$, $N_y = 8$. The blue circle is the "visible" region, for which $u_x^2 + u_y^2 \leq 1$ (black squares). If one wishes to avoid computing unnecessary points with $u_x^2 + u_y^2 > 1$ (white squares), it suffices to choose (u_x, u_y) only inside the gray square, changing the definition of b_n and c_m accordingly. This diagram corresponds to Figure 6.3 viewed from the top, with the array at the center.

6.4.3 KRONECKER ARRAY TRANSFORM

Consider a planar array, with microphones placed in a rectangular grid such as that shown in Figure 6.4. Assume that the coordinate system is chosen so that the array lies in the x, y plane, and that the look directions u_n are chosen such that

$$u_{\ell+(i-1)N_y} = \begin{bmatrix} u_{1,i} \\ u_{2,\ell} \\ \sqrt{1 - u_{1,i}^2 - u_{2,\ell}^2} \end{bmatrix}, \quad 1 \leq i \leq N_x, \ 1 \leq \ell \leq N_y,$$

where $-1 \leq u_{1,1} < u_{1,2} < \cdots < u_{1,N_x} \leq 1$, $-1 \leq u_{2,1} < u_{2,2} < \cdots < u_{2,N_y} \leq 1$. In the notation of the previous sections, we are choosing $u_{x,n} \in \{u_{1,1}, \ldots, u_{1,N_x}\}$, $u_{y,n} \in \{u_{2,1}, \ldots, u_{2,N_y}\}$, so that $u_{x,n} = u_{1,i}$, $u_{y,n} = u_{2,\ell}$ if $n = \ell + (i-1)N_y$. Similarly, the microphone positions are such that

$$p_{s+(r-1)M_y} = \begin{bmatrix} p_{1,r} \\ p_{2,s} \\ 0 \end{bmatrix}, \quad 1 \leq r \leq M_x, \ 1 \leq s \leq M_y.$$

In this case, the array manifold vector can be decomposed as follows:

$$v(u_{\ell+(i-1)N_y}) = v_x(u_{1,i}) \otimes v_y(u_{2,\ell}), \tag{6.44}$$

where

$$
v_x(u_{1,i}) \triangleq \begin{bmatrix} e^{j\omega u_{1,i} p_{1,1}/c} \\ e^{j\omega u_{1,i} p_{1,2}/c} \\ \vdots \\ e^{j\omega u_{1,i} p_{1,M_x}/c} \end{bmatrix}, \qquad v_y(u_{2,\ell}) \triangleq \begin{bmatrix} e^{j\omega u_{2,\ell} p_{2,1}/c} \\ e^{j\omega u_{2,\ell} p_{2,2}/c} \\ \vdots \\ e^{j\omega u_{2,\ell} p_{2,M_y}/c} \end{bmatrix}. \tag{6.45}
$$

This can be verified by direct comparison with (6.4), since

$$
e^{j\omega u_n^T p_m/c} = e^{j\omega u_{1,i} p_{1,r}/c} e^{j\omega u_{2,\ell_u} p_{2,s}/c},
$$

for $n = \ell + (i - 1)N_y$, $m = s + (r - 1)M_y$.

Under these conditions, matrix-vector products $A\hat{y}$, $A^H\hat{r}$, and $A^H A\hat{y}$ can be obtained in a more cost-effective way than by direct multiplication. Define the matrices V_x and V_y as follows:

$$
V_x \triangleq \begin{bmatrix} v_{x,1}^*(u_{1,1})v_{x,1}(u_{1,1}) & \cdots & v_{x,1}^*(u_{1,N_x})v_{x,1}(u_{1,N_x}) \\ v_{x,1}^*(u_{1,1})v_{x,2}(u_{1,1}) & \cdots & v_{x,1}^*(u_{1,N_x})v_{x,2}(u_{1,N_x}) \\ \vdots & & \vdots \\ v_{x,M_x}^*(u_{1,1})v_{x,M_x}(u_{1,1}) & \cdots & v_{x,M_x}^*(u_{1,N_x})v_{x,M_x}(u_{1,N_x}) \end{bmatrix} \in \mathbb{C}^{M_x^2 \times N_x}, \tag{6.46}
$$

$$
V_y \triangleq \begin{bmatrix} v_{y,1}^*(u_{2,1})v_{y,1}(u_{2,1}) & \cdots & v_{y,1}^*(u_{2,N_y})v_{y,1}(u_{2,N_y}) \\ v_{y,1}^*(u_{2,1})v_{y,2}(u_{2,1}) & \cdots & v_{y,1}^*(u_{2,N_y})v_{y,2}(u_{2,N_y}) \\ \vdots & & \vdots \\ v_{y,M_y}^*(u_{2,1})v_{y,M_y}(u_{2,1}) & \cdots & v_{y,M_y}^*(u_{2,N_y})v_{y,M_y}(u_{2,N_y}) \end{bmatrix} \in \mathbb{C}^{M_y^2 \times N_y}. \tag{6.47}
$$

Given a vector $\hat{y}$, define $\hat{Y}$ such that $\hat{y} = \text{vec}(\hat{Y})$, and $Z = V_y\hat{Y}V_x^T$. It can be verified by direct computation that there exists a permutation matrix H such that [26]

$$
A\hat{y} = \hat{r} = H\,\text{vec}(Z) = H\,\text{vec}(V_y\hat{Y}V_x^T). \tag{6.48}
$$

This is the Kronecker array transform. Taking advantage of the fact that $\hat{y}$ and $\hat{Y}$ are real, computing the product $A\hat{y}$ directly requires $0.5M_x^2 M_y^2 N_x N_y$ complex multiply-and-accumulate (MAC) operations, while using (6.48) the required number of operations reduces to $0.5M_y^2 N_x N_y + M_x^2 N_x N_y$ complex MACs if we compute $V_y\hat{Y}$ first, or to $0.5M_x^2 N_x N_y + M_x^2 M_y^2 N_y$ if we compute $\hat{Y}V_x^T$ first.

The products $\hat{y} = A^H\hat{r}$ and $\bar{y} = A^H A\hat{y}$ can be similarly obtained. Define $\bar{Z}$ such that $\text{vec}(\bar{Z}) = H^T\hat{r}$, and $\bar{Y}$ such that $\bar{y} = \text{vec}(\bar{Y})$, then

$$
\text{vec}(A^H\hat{r}) = \text{vec}(V_y^H \bar{Z} V_x^*), \qquad \bar{Y} = (V_y^H V_y)\hat{Y}(V_x^T V_x^*). \tag{6.49}
$$

Note that, since $V_y^H V_y$ and $V_x^T V_x^*$ can be pre-calculated, the use of the second form of (6.49) is more efficient than computing $V_y^H(V_y\hat{Y}V_x^T)V_x^*$.

Table 6.1

Asymptotic complexity for different implementations of $A\hat{y}$, assuming $M_x = M_y$, $N_x = N_y$, and $M^2 < N$.

Transform	Computational cost
KAT with matrix multiplication	$O(MN + M^2 N^{1/2})$
KAT with 1-D NFFTs	$O(N \log N + MN^{1/2})$
2-D NFFT	$O(N \log N + M^2)$
2-D NNFFT	$O(N \log N + M^2)$
Matrix multiplication	$O(M^2 N)$

Simultaneous application of the KAT and NFFT or NNFFT

Since the entries of V_x and V_y are complex numbers with modulus equal to one, the NFFT or the NNFFT can be used to compute the products in (6.48) and (6.49), providing further acceleration to the KAT when the number of microphones and look directions are large enough.[3]

6.5 COMPUTIONAL COST

In Table 6.1 we list the asymptotic cost of different methods for the case $M_x = M_y$ (i.e., $M = M_x^2$), $N_x = N_y$ ($N = N_x^2$), and assuming $M^2 < N$. It is important however to remember that the asymptotic costs in Table 6.1 do not show the constants multiplying each entry, so from the table one cannot see for example that the NFFT is much faster than the NNFFT.

For a practical application it is also important to consider memory requirements—for example, to directly store A, we would need $M^2 N$ complex variables; while storing V_x and V_y requires $2MN^{1/2}$ complex variables (again in the case of $M_x = M_y$ and $N_x = N_y$), so using the KAT also reduces memory storage considerably.

The expressions for computational cost give an idea of the advantages of using the KAT, but a full comparison should take into account not only the number of arithmetic operations, but also issues such as memory access and the particular hardware in which the methods are implemented. Figure 6.6 compares the time required to compute a product $A\hat{y}$, for different dimensions of A. The computations were performed on a 64-bit Intel Core 2 Duo T9400 processor using a single core. The permutation H was implemented in ANSI C, the NFFT also used a C implementation with default optimization, as used in [19]. The remaining routines were implemented as m-files in Matlab 2008b.

---

[3]Note however that $V_y^H V_y$ and $V_x^T V_x^*$ do not have only entries in the unit circle, so to use the NFFT or the NNFFT to compute $A^H A\hat{y}$, one would need to compute $V_y^H (V_y \hat{Y} V_x^T) V_x^*$.

$$N_x = N_y = 256$$

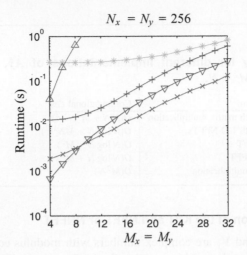

Figure 6.6 Comparison of runtimes for computation of $A\hat{y}$ in Matlab. $\triangledown$: KAT implemented with matrix multiplication; $\times$: KAT implemented with 1-D NFFTs replacing matrix multiplication; $+$: direct NFFT implementation; $*$: direct NNFFT implementation; $\triangle$: matrix multiplication. From [26].

Table 6.2

x and y coordinates of a separable microphone array with 64 microphones (shown in Figure 6.4). For all microphones, $z = 0$.

x and y coordinates (m)							
−0.1500	−0.1412	−0.1147	−0.0706	−0.0176	0.0441	0.1324	0.1500

6.6 EXAMPLES

We present now a few examples of simulated acoustic image reconstructions. In all cases the microphone's coordinates p_m are as given in Figure 6.4. The exact values of the coordinates are given in Table 6.2.

The examples (following [27]) are simulations of reconstructions of the test images shown in Figure 6.7. The first test image simulates a sparse source distribution. We compare the results obtained using DAS, DAMAS2, and covariance fitting with ℓ_1 and TV regularization. The regularized optimization problems were solved using package routines SPGL1 [34] for solving (6.36), using 200 iterations, with $\sigma = 0.01\|R_x\|_F$.[4] For the solution of (6.38) we used TVAL3 [23] using 100 iterations with $\mu = 10^3$. DAMAS2 used 1 000 iterations. The simulations were performed assuming an 8×8 microphone array, with microphone positions as in Figure 6.4 and

---

[4]$\| \cdot \|_F$ is the Frobenius norm.

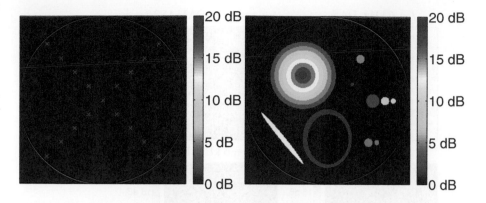

Figure 6.7 Test images. On the left, a sparse pattern, and on the right, a smoother test image.

Table 6.2 ($M_x = M_y = 8$). The look directions were sampled uniformly in u_x, u_y, using the entire range $[-1, 1)$, with 256 points in each direction ($N_x = N_y = 256$). Note that this means that the algorithms compute values also outside the visible region (i.e., for points with $u_x^2 + u_y^2 > 1$). A good test of the quality of reconstruction is to see only small (blue) values outside the visible region. In the next figures, the visible region boundary ($u_x^2 + u_y^2 = 1$) is marked with a black circle. The signals at the microphones were simulated using the ideal image and model (6.6), assuming a frequency of 6 kHz, and a noise variance set for an SNR of 20 dB.

The results for the sparse pattern can be seen in Figure 6.8, and for the smoother pattern, in Figure 6.9. As expected, ℓ_1 regularization has the best performance for the sparse pattern, while TV regularization gives the best results for the smoother pattern.

Finally, we present a comparison of results obtained using separable array geometries with those obtained using a multi-arm logarithmic spiral geometry, specially designed to reduce the sidelobes in the PSF [32], see Figure 6.10. As can be seen in Figure 6.11, the result obtained with delay-and-sum algorithm using a 50 cm, 63-element logarithmic spiral array is indeed better, compared with the delay-and-sum reconstruction in Figure 6.9. However, the results obtained with the other methods are comparable.

6.7 PRACTICAL CONSIDERATIONS

All methods presented in Section 6.3 assumed perfect knowledge of the relative position of the sensors in the array, and that these sensors present the same sensitivity response in terms of gain and phase. However, depending on how the array is constructed, exact positioning of the microphones cannot be guaranteed. Furthermore, microphones present a variation in their sensitivity response, even when using microphones of the same model.

It has been shown that both variations in gain and phase, as well as microphone positioning errors, result in distortions at the observed source's direction and sound

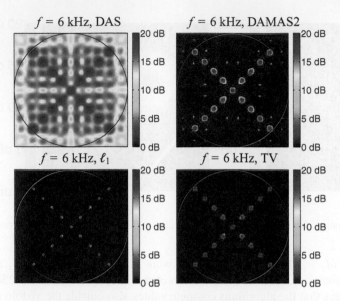

Figure 6.8 Results for (clockwise from top left) Delay-and-sum, DAMAS2, ℓ_1 regularization, and TV regularization, for the sparse pattern. From [27].

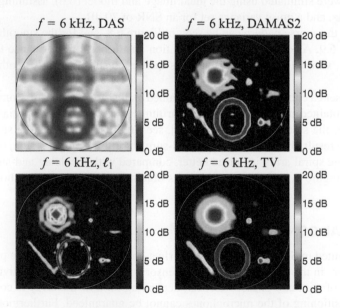

Figure 6.9 Results for (clockwise from top left) Delay-and-sum, DAMAS2, ℓ_1 regularization, and TV regularization, for the smooth pattern. From [27].

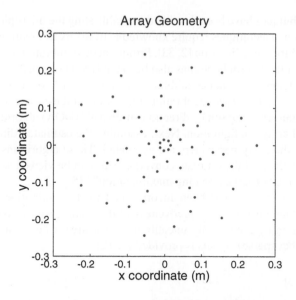

Figure 6.10   Multi-arm logarithmic spiral array, 50 cm in diameter, with 63 elements.

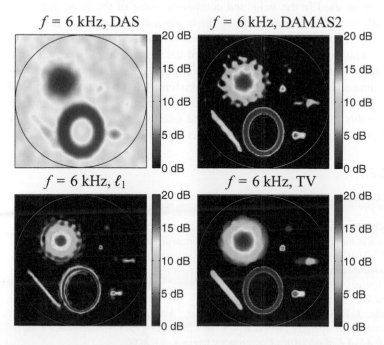

Figure 6.11   Results for (clockwise from top left) Delay-and-sum, DAMAS2, ℓ_1 regularization, and TV regularization, for the smooth pattern using the logarithmic spiral array of Figure 6.10. One can see, comparing with Figure 6.9, that the results here are better for DAS, but equivalent for the other methods. From [27].

level [20]. Several techniques have been described for calibrating the microphone location. More conventional techniques require knowledge of the exact position of the reference sources used for the calibration [2, 33]. Simultaneous calibration of the position and sensitivity of the microphones has also been analyzed previously [31, 39].

There have been attempts to conduct calibration without prior knowledge of the reference sources [11, 37]. This *blind* calibration algorithms first estimate the position of the reference sources using some "direction-of-arrival" (DOA) algorithm and assuming the nominal array configuration. Next, assuming the estimated directions of arrival are correct, the array parameters are estimated. These algorithms repeat these two steps iteratively until the estimate converge. This can be viewed as a joint estimation problem using "group alternating maximization" [35].

Recently, the use of sparsity has been incorporated to the calibration procedure [6]. Simulations demonstrate the effectiveness of the compressive sensing approach to the calibration of even highly uncalibrated measures, when a sufficient number of (unknown but sparse) signals is provided [1, 12].

6.8 CONCLUSION

This chapter provides an introduction to acoustic imaging, describing the main models and assumptions used in the field, and comparing some of the most important algorithms available in the literature. Since the computation of acoustic images is a computationally demanding task, we emphasized recent methods for speeding up computations, taking advantage of the structure of the array manifold vectors in the far field. The three methods, in order of least to highest acceleration, are the non-equispaced in time and frequency fast Fourier transform (NNFFT), non-equispaced fast Fourier transform (NFFT), and the Kronecker array transform (KAT). Although not described in this chapter, the KAT can also be extended to work when some of the far field approximations are no longer valid, as described in [27].

REFERENCES

1. C. Bilen, G. Puy, R. Gribonval, and L. Daudet. Blind phase calibration in sparse recovery. In *21st European Signal Processing Conference*, Marrakech, Morocco, 2013.
2. S. T. Birchfield and A. Subramanya. Microphone array position calibration by basis-point classical multidimensional scaling. *IEEE Transactions on Speech and Audio Processing*, 13(5):1025–1034, 2005.
3. T. F. Brooks and W. M. Humphreys. A deconvolution approach for the mapping of acoustic sources (DAMAS) determined from phased microphone arrays. *Journal of Sound and Vibration*, 294(4):856–879, 2006.
4. L. Brusniak, J. R. Underbrink, and R. W. Stoker. Acoustic imaging of aircraft noise sources using large aperture phased arrays. In *12th AIAA/CEAS Aeroacoustics Conference (27th AIAA Aeroacoustics Conference)*, pages 1–20, 2006.
5. E. J. Candès and M. B. Wakin. An Introduction To Compressive Sampling. *IEEE Signal Processing Magazine*, 25(2):21–30, 2008.

6. V. Cevher and R. Baraniuk. Compressive sensing for sensor calibration. In *2008 5th IEEE Sensor Array and Multichannel Signal Processing Workshop*, pages 175–178, 2008.

7. D. Donoho. Compressed sensing. *IEEE Transactions on Information Theory*, 52(4):1289–1306, Apr. 2006.

8. R. P. Dougherty. Extensions of DAMAS and Benefits and Limitations of Deconvolution in Beamforming. In *11th AIAA/CEAS Aeroacoustics Conference (26th AIAA Aeroacoustics Conference)*, pages 1–13, 2005.

9. K. Ehrenfried and L. Koop. Comparison of Iterative Deconvolution Algorithms for the Mapping of Acoustic Sources. *AIAA Journal*, 45(7):1584–1595, 2007.

10. F. J. Fahy. *Sound Intensity*. Elsevier, London, 1987.

11. B. P. Flanagan and K. L. Bell. Array self-calibration with large sensor position errors. *Signal Processing*, 81(10):2201–2214, Oct. 2001.

12. R. Gribonval, G. Chardon, and L. Daudet. Blind calibration for compressed sensing by convex optimization. In *Acoustics, Speech and Signal Processing. ICASSP. IEEE International Conference on*, pages 2713–2716, 2012.

13. J. A. Högbom. Aperture synthesis with a non-regular distribution of interferometer baselines. *Astronomy and Astrophysics Supplement*, 15:417–426, 1974.

14. R. A. Horn and C. R. Johnson. *Matrix Analysis*. Cambridge University Press, 1987.

15. W. C. Horne, K. D. James, T. K. Arledge, P. T. Soderman, N. Burnside, and S. M. Jaeger. measurement of 26%-scale 777 airframe noise in the NASA Ames 40- by 80 foot wind tunnel. In *11th AIAA/CEAS Aeroacoustics Conference (26th AIAA Aeroacoustics Conference)*, pages 23 – 25, 2005.

16. W. M. Humphreys and T. F. Brooks. Noise spectra and directivity for a scale-model landing gear. *International Journal of Aeroacoustics*, 8(5):409–443, 2009.

17. F. Jacobsen. Intensity Probe. In D. Havelock, S. Kuwano, and M. Vorländer, editors, *Handbook of Signal Processing in Acoustics*, chapter 58. Springer New York, New York, USA, 2008.

18. D. H. Johnson and D. E. Dudgeon. *Array Signal Processing: Concepts and Techniques*. Prentice Hall, Englewood-Cliffs N.J., 1993.

19. J. Keiner, S. Kunis, and D. Potts. Using NFFT3–a software library for various nonequispaced fast fourier transforms. *ACM Transactions on Mathematical Software (TOMS)*, 36(4):19, 2009.

20. A. W. H. Khong and M. Brookes. The effect of calibration errors on source localization with microphone arrays. In *Acoustics, Speech and Signal Processing. ICASSP. IEEE International Conference on*, volume 1, 2007.

21. H. Krim and M. Viberg. Two decades of array signal processing research: the parametric approach. *IEEE Signal Processing Magazine*, 13(4):67–94, 1996.

22. S. S. Lee. Phased-Array Measurement of Modern Regional Aircraft Turbofan Engine Noise. In *12th AIAA/CEAS Aeroacoustics Conference (27th AIAA Aeroacoustics Conference)*, pages 8–10, 2006.

23. C. Li. *An Efficient Algorithm For Total Variation Regularization with Applications to the Single Pixel Camera*. Master thesis, Rice University, 2009.

24. Y. Oikawa. Spatial Information of Sound Fields. In D. Havelock, S. Kuwano, and M. Vorländer, editors, *Handbook of Signal Processing in Acoustics*, chapter 76, pages 1403–1421. Springer New York, New York, USA, 2nd edition, 2008.

25. F. P. Ribeiro and V. H. Nascimento. Computationally efficient regularized acoustic imaging. In *Acoustics, Speech and Signal Processing. ICASSP. IEEE International Confer-

ence on, pages 2688–2691, 2011.

26. F. P. Ribeiro and V. H. Nascimento. Fast transforms for acoustic imaging—part I: Theory. *IEEE Transactions on Image Processing*, 20(8):2229–2240, 2011.

27. F. P. Ribeiro and V. H. Nascimento. Fast transforms for acoustic imaging—part II: Applications. *IEEE Transactions on Image Processing*, 20(8):2241–2247, 2011.

28. L. I. Rudin, S. Osher, and E. Fatemi. Nonlinear total variation based noise removal algorithms. *Physica D: Nonlinear Phenomena*, 60:259–268, 1992.

29. P. Sijtsma. CLEAN based on spatial source coherence. *International Journal of Aeroacoustics*, 6(4):357–374, 2009.

30. D. Stanzial. Reactive acoustic intensity for general fields and energy polarization. *Journal of the Acoustic Society of America*, 99(4):1868, 1996.

31. I. Tashev. Gain self-calibration procedure for microphone arrays. In *Multimedia and Expo. (ICME '04). IEEE International Conference on*, volume 2, pages 983–986, 2004.

32. J. R. Underbrink and R. P. Dougherty. Array Design for Non-Intrusive Measurement of Noise Sources. In *National Conference on Noise Control Engineering*, pages 757–762, 1996.

33. S. D. Valente, M. Tagliasacchi, F. Antonacci, P. Bestagini, A. Sarti, and S. Tubaro. Geometric calibration of distributed microphone arrays from acoustic source correspondences. In *Multimedia Signal Processing (MMSP), 2010 IEEE International Workshop on*, pages 13–18, 2010.

34. E. van den Berg and M. P. Friedlander. Probing the Pareto Frontier for Basis Pursuit Solutions. *SIAM Journal on Scientific Computing*, 31(2):890–912, 2008.

35. H. L. van Trees. *Optimum Array Processing: Part IV of Detection, Estimation and Modulation Theory*. John Wiley & Sons, 2002.

36. Y. Wang, J. Li, P. Stoica, M. Sheplak, and T. Nishida. Wideband RELAX and wideband CLEAN for aeroacoustic imaging. *The Journal of the Acoustical Society of America*, 115(2):757–767, 2004.

37. A. J. Weiss and B. Friedlander. Eigenstructure methods for direction finding with sensor gain and phase uncertainties. *Circuits, Systems, and Signal Processing*, 9(3):271–300, 1990.

38. E. G. Williams. *Fourier Acoustics: sound radiation and nearfield acoustical holography*. Academic Press, 1999.

39. H. Xiao, H.-z. Shao, and Q.-c. Peng. A New Calibration Method for Microphone Array with Gain, Phase, and Position Errors. *Journal Of Electronic Science And Technology Of China*, 5(3):248–251, 2007.

40. S. Yan, Y. Ma, and C. Hou. Optimal array pattern synthesis for broadband arrays. *The Journal of the Acoustical Society of America*, 122(5):2686–2696, 2007.

41. T. Yardibi, J. Li, P. Stoica, and L. N. Cattafesta. Sparsity constrained deconvolution approaches for acoustic source mapping. *The Journal of the Acoustical Society of America*, 123(5):2631–2642, 2008.

7 Automatic Evaluation of Acoustically Degraded Full-Band Speech

LUIZ WAGNER PEREIRA BISCAINHO
Federal University of Rio de Janeiro (UFRJ)

LEONARDO DE OLIVEIRA NUNES
Microsoft

Evaluation of speech quality in modern telecommunication systems has become a primary concern. While novel coding and networking strategies provide increasingly fast, high-quality, and reliable services, acoustically induced effects like echo, reverberation, and some types of noise remain difficult issues to control. This chapter gives a brief overview of automatic assessment of acoustically degraded high-quality speech signals. After a basic overview of different quality assessment strategies, the chapter highlights relevant literature and standards on the topic, with focus on ITU documents. The text proceeds with recent contributions by the authors research group in full- and no-reference tools for automatic quality assessment of acoustically degraded full-band speech.

7.1 INTRODUCTION

7.1.1 QUALITY – AUDIO AND SPEECH

The abstract concept of **quality** may be linked to criteria as diverse as:

- utility—e.g., how well a tool fits to some purpose;
- financial value—e.g., how much income an investment generates;
- aesthetic pleasure—e.g., how beautiful a work of art is;
- accuracy—e.g., how exactly a model describes a real-world object.

In any case, one often refers to high or low quality, thus implying some kind of **quantitative measure**. When it comes to sound, quality results directly from human (sensory/psychological) perception, which places the subject at an entirely subjective dimension. This twofold nature of **sound quality assessment** (QA) is quite challenging.

What are the topics surrounding this subject? The following classification, although loose, can be useful for the purposes of this discussion:

1. At a general entertainment level, i.e., regarding general **audio**, the main target is to preserve **fidelity** along the stages of recording, storage, transmission, and reproduction. Historically, means of capture (acoustical, electrical), forms of representation (analog, digital), media types (record, tape, CD, memory), broadcast services (radio, TV, Internet), and reproduction equipment (amplifier, loudspeakers, headphones) have been objects of concern as determinant of the overall audio quality. Indirect measures, either mechanical (e.g., component accuracy and alignment) or electrical (e.g., circuit stability and noise power) have been useful tools before the only fair judge (the listener) can assess audio fidelity against a reference. If along the path from the analog to the digital era the quality focus changed from hi-fi systems to digital codecs, at the same time digital processing boosted the design of (increasingly necessary) tools for automatic QA.

2. At an interpersonal communication level, i.e., regarding information exchange by **speech**, the goal is to guarantee **intelligibility** after the stages of capture, storage, transmission, and reproduction. Along the development of telecommunication resources, means of capture (electrical), forms of representation (analog, digital), transmission networks (telephone, internet), and reproduction (amplifier, loudspeakers, headphones) have been carefully addressed as instrumental factors in providing acceptable speech quality levels. Again, electrical indirect measures (e.g., circuit stability and noise power) supplement auditory tests in the pursuit of comfortable mutual comprehension during conversation. From analog to digital communications, the users traveled "at light speed" from PSTN (public switched telephone network) to VOIP (voice over Internet protocol); and once more digital processing was crucial in the development of automatic QA tools to tackle specific (sometimes quite tough) telecommunication scenarios.

Today, the large processing and memory resources available weakens the boundaries between these two classes: very high-quality speech brings elements of fidelity to the evaluation of speech quality, while the fidelity of highly compressed audio retrogresses from the extraordinary degree of audio quality already achieved. Of course, economical interests will always be found behind the evolutionary coupling between technological achievements and human needs: low price, fast production, and wide distribution are clearly in opposition to quality. Finding out the best compromise between the two tendencies is a hard but rewarding task. Standardization efforts try to keep an equilibrium among developers', market's, and users' interests by regulating aspects as compatibility and quality. In the context of audio and speech, three international organizations are especially influential for their encompassing areas of activity:

- ISO – International Organization for Standardization;
- IEC – International Electrotechnical Commission;
- ITU – International Telecommunication Union.

One should also mention the AES – Audio Engineering Society, in audio, as well as the (regional) EBU – European Broadcasting Union and ETSI – European Telecommunication Standards Institute, in telecommunications. All of them have produced relevant documents related to audio and/or speech quality.

7.1.2 AUDIO/SPEECH QUALITY ASSESSMENT

The fairest method to assess the quality of audio or speech is critically listening to it. But what are the factors involved in this task, and how difficult can it be?

Suppose, for instance, one wants to evaluate a new lossy audio coder A. The test goal is to quantify how transparent the codec chain is. To this end, the most direct strategy is to ask a group of individuals to listen to and compare the original audio signal against its coded/decoded version. The stated verdict can be either binary (perceptually equal/different), or graded (e.g., from imperceptibly to annoyingly different). When designing and analyzing these **subjective** tests [3], the following issues must be taken into account:

- Since individual opinions exhibit an intrinsically large variance, a large group of volunteers is necessary to produce a useful average opinion. Inviting trained specialists may help to reduce this requirement.
- Auditory perception can be influenced by factors that range from environmental to psychological conditions, some of them quite difficult to control.
- The technical setup (headphones or loudspeakers, amplifiers, etc.) must be kept invariant in order to assure repeatability.
- The way the questions are formulated as well as textual labels to grade the quality itself (both language-dependant) can produce different results. Also the use of a (continuous/discrete) scale must be carefully considered.
- Which and how many different signals should be employed in the tests? They must be statistically representative of the universe of target signals. The working database is very hard to define and bound.
- The duration of each test session cannot induce fatigue in the volunteers, which directly biases their opinions.

Obviously, the better such issues are addressed, the more expensive/longer the overall process becomes. And if another coder B is to be tested and compared with coder A afterwards, how can one guarantee they are tested under the same circumstances? This example gives an idea of the problems involved in subjective assessment of audio quality.

Different applications bring different concerns. Broadcast systems should provide continuous transmission and quality; multimedia calls for combined sound/image tests; and interpersonal communication introduces a new component: interaction. Natural and comfortable conversation can be hindered by effects like brief interruptions, delays, or echo. Furthermore, for maintenance reasons, telecommunication services may need to be tested while in operation, which almost always means "without a reference signal." This wide range of needs / problems justify permanent efforts towards standardization of signal databases, equipments, acoustic conditions, test procedures, analysis methods, etc., for each target application.

The obvious advantages of alternative means to automatically assess audio/speech quality without resorting to people have led to the study and development of **objective** test strategies to fulfill this necessity. The idea is to employ mathematical relations to emulate the grades that would be attributed by auditory tests. How much information a simple measure as the SNR (signal-to-noise ratio) conveys on sound quality? If the SNR is exceptionally high, one can ascertain that no disturbance will be noticed; but on the other hand, a low-SNR audio signal may sound perfectly clean, as long as the large distortion is masked by the signal itself. Thus, modeling auditory perception (the object of psychoacoustics) can be of great use in the design of automatic methods for QA. If a lossy coder builds a perceptual model for the audio signal in order to discard imperceptible information, an objective method to assess its transparency can simply compare the original and coded/decoded versions of the signal in the perceptual domain. Such strategy is expected to recognize a seemingly clean low-SNR signal as perceptually identical to its non-distorted version.

Fortunately, digital processing enables the design of powerful solutions tailored to specific scenarios/problems, which do not necessarily need to employ explicit modeling of human hearing. One can devise implicit high-level models based on direct machine learning that, once trained, are capable of mapping the signal under test to the desired grade according to an abstract sequence of mathematical operations.

As said before, QA in communications can be especially challenging. Speech degradation can be roughly classified as acoustically induced (e.g., by locally produced reverberation, noise, or echo) or network-induced (e.g., by packet loss or electromagnetic noise). Parametric QA tools (capable of predicting speech quality from network performance measures) [22] are better suited to the latter category, rather than signal-based ones. One of the most difficult tasks to be tackled by automatic tools is signal-based QA when no reference signal is available: a possible approach to that is to first enhance the signal under test to artificially produce a reference it can be compared with [18]. Even more than subjective methods, objective QA methods have been the object of a large collection of application-specific standards, and also a hot research topic in search for representativeness, robustness, and reliability. Grades yielded by a good objective QA method are expected to be highly correlated with their corresponding subjective grades.

Some examples of the jargon found in the literature on speech QA for telecommunications:

- **Full-reference** (or **double-ended**) QA tools compare the signal under test with an available reference. **No-reference** (or **single-ended**) tools are based only on the signal under test.
- **Intrusive** QA tools require the system operation being interrupted, while **non-intrusive** ones can be used during normal operation. (Although not strictly correct, their respective association with double-ended and single-ended tools is usual.)
- **SDG** (subjective difference grade)/**ODG** (objective difference grade) [12]

= output of a full-reference subjective/objective QA tool.

- **DCR** (degradation category rating)/**ACR** (absolute category rating) [15]= standardized full-reference/no-reference subjective method for speech QA.
- **MOS** (mean opinion score) [15] = average of grades attributed to a given signal by all subjects in a particular subjective QA test.
- Speech quality can be assessed in three main contexts [11]: **listening** (how an individual listens to other person's voice), **talking** (how an individual listens to her/his own voice), **conversational** (how an individual acts/feels in a conversation)

Still in recent years, some topics needed to be further addressed, namely:

- High-quality speech – modern VOIP systems can extend the speech bandwidth much beyond the telephone-channel 4 kHz, thus bringing to scene elements of fidelity rather than just intelligibility.
- Acoustic degradations – Noise, reverberation, and echo locally originated at each voice terminal can modify speech characteristics to the point of hindering conversation, and yet (except for echo) are virtually out of the network's control.
- Multi-degradations – Different acoustic degradations require dedicated tools, but can concurrently affect speech.
- No-reference QA – Single-ended diagnosis may help to keep the system in continuous operation as long as fast actions succeed in mitigating the detected issues.

This chapter focuses on digital processing methods for objective speech quality assessment in high-quality telecommunications networks subjected to acoustically induced degradations. In particular, some attempts from the authors' research group to fill some of these gaps are reviewed in the following sections.

7.2   HIGHLIGHTED STANDARDS AND REFERENCES

Although specific literature on audio/speech quality assessment is relatively scarce, along the last 10 years some interesting references have appeared, among which one can cite the following:

- [24] the book by Jekosch discusses the elements involved in voice/speech quality perception and how they can yield usable assessment strategies, with some emphasis on synthetic speech;
- [3] the book by Bech and Zacharov covers in detail almost every aspect of subjective evaluation of speech and audio quality, including theoretical and practical aspects as well as related standards;
- [11] the chapter by Grancharov and Kleijn gives a very good overview of subjective/objective speech QA;
- [26] the chapter by Loizou is another extensive overview of subjective/objective speech QA.

A general overview on quality assessment of signals can also be found in [6]. And the present chapter is heavily based on the tutorial presented in [31] (in Portuguese).

From ITU standards, some recommendations especially relevant to the subject in question can be singled out.

1. Related to audio QA:
   - [13] ITU-R Rec. BS.1284-1: Defines general guidelines for subjective assessment of audio QA.
   - [14] ITU-R Rec. BS.1116-2: Describes subjective methods for full-reference evaluation of audio degraded by small impairments.
   - [12] ITU-R Rec. BS.1387-1: Describes the PEAQ (Perceptual Evaluation of Audio Quality), a psycho-acoustically based objective method for full-reference evaluation of audio quality.

2. Related to speech QA:
   - [15] ITU-T Rec. P.800: Defines the methods for systematic subjective assessment of transmitted speech QA.
   - [16] ITU-T Rec. P.862: Describes the PESQ (Perceptual Evaluation of Speech Quality), a psycho-acoustically based objective method for full-reference evaluation of narrowband (300-3400 Hz) speech quality.
   - [19] ITU-T Rec. P.862.2: Describes the W-PESQ (wideband PESQ), an extension of PESQ for wideband (50-7000 Hz) speech.
   - [23] ITU-T Rec. P. 863: Describes the POLQA (Perceptual objective listening quality assessment), a completely revised psycho-acoustically based objective method for full-reference evaluation of narrowband, wideband, and super-wideband (50-14,000 Hz) speech quality.
   - [18] ITU-T Rec. P.563: Describes an objective method for no-reference evaluation of speech quality in narrowband channels.
   - [22] ITU-T Rec. G.107: Describes the E-model, which maps network parameters into the predicted quality of conversational speech transmitted through narrowband channels.
   - [20] ITU-T Rec. G.107.1: Describes the wideband E-model, an extension of the E-model for wideband channels.

It should be noted that QA methods like PEAQ, PESQ, and POLQA are under license control.

7.3 MODELS FOR A TELEPRESENCE SYSTEM

In this section, models for how a speech signal is acoustically degraded in a telepresence system are described. Such models, originally proposed in [1] and [29], can be used to identify measurement points for the signals to be submitted to QA, elicit which degradations may be present at each of these measurement points, and simulate degradations on speech signals that are meaningful in practice.

After a general telepresence model, two specific models that define possible measurement points for intrusive and non-intrusive QA, respectively, are presented.

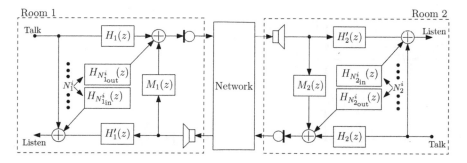

Figure 7.1 General model for telepresence systems. Figure adapted from [1].

7.3.1 GENERAL MODEL

The starting point for the scenarios described in the next sections is the general telepresence model shown in Figure 7.1, which considers only the case of a conversation between two participants [1]. In the figure, it is assumed that Person 1 in Room 1 is in conversation with Person 2 in Room 2. Speech is being emitted by Person 1, and received by

1. his/her ears, where only the direct transmission is considered for simplicity sake;
2. the microphone located in Room 1, modified by the room transfer function $H_1(z)$.

Each noise signal N_1^i from noise source i, such as an air conditioner or computer fan, inside Room 1 arrives at the ears of Person 1 and is also picked up by the microphone inside the room after being modified by the room responses $H_{N_{1_{in}}^i}(z)$ and $H_{N_{1_{out}}^i}(z)$ associated with their respective paths. The signal that arrives at the loudspeakers located in Room 1 impinges on the ears of Person 1 modified by the response $H_1'(z)$ and is acoustically fed to the microphone located in Room 1 after being modified by the response $M_1(z)$. An analogous description can be made for the model in Room 2. The "Network" box in the diagram summarizes all degradations caused by the signal transmission, such as delays, attenuations, packet losses, and coding artifacts.

The model described above can be simplified, leading to the one seen in Figure 7.2, by considering the following modifications.

- Only the transmission and reception cycle related to Person 1 is considered;
- A single additive noise source, modeled as a combination of the aforementioned sources, is used.

In this simplified model, it is assumed that only one participant is talking. Therefore, any QA method derived using this model must be used in association with a double-talk detector [21]. The new noise signal can be described as a function of the previous

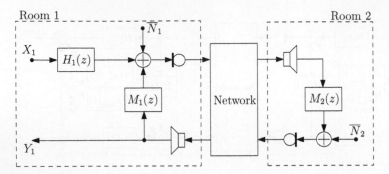

Figure 7.2 Simplified model for telepresence systems. Only transmission and reception paths for the speech Person 1 emits in Room 1 are considered. Figure adapted from [29].

noise sources as

$$\overline{N}_1(z) = \sum_{i=0}^{M_1-1} H_{N_{1_{in}}^i}(z)N_1^i(z) \tag{7.1}$$

and

$$\overline{N}_2(z) = \sum_{i=0}^{M_2-1} H_{N_{2_{in}}^i}(z)N_2^i(z), \tag{7.2}$$

where M_1 and M_2 are the number of noise sources in Rooms 1 and 2, respectively.

This model will be used in the development of the intrusive and non-intrusive monitoring scenarios described in the next sections.

7.3.2 INTRUSIVE MODEL

The intrusive scenarios described in this section have a twofold objective [1]:

1. Model the degradations so as to allow for a generation of a database with signals with a controlled level of impairment that will aid in the development of QA tools;
2. Elicit the measurement points for both the reference and degraded speech signal that are meaningful for both objective tools and subjective listening tests.

The model shown in Figure 7.2 can be further simplified to independently characterize degradations caused by echo, noise, and reverberation in each room. This can be achieved by splitting the model into two parts: a local and a remote scenario.

Figure 7.3 exhibits the local scenario, which models acoustic degradations originated in the room where the talker is located: background noise generated by local noise sources ($\overline{N}_1$) and reverberation induced by the local room impulse response.

Figure 7.4 displays the remote scenario, which models acoustic degradations originated in the room remote to the talker: mainly acoustic echo induced by the microphone-loudspeaker coupling combined with transmission delays. Degradations

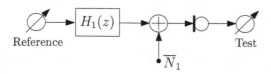

Figure 7.3 Local scenario [1].

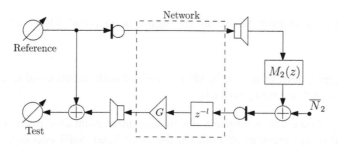

Figure 7.4 Remote scenario [1].

generated in the room where the talker is located are discarded (hence $H_1(z) = 1$ and $\overline{N}_1 = 0$ is assumed). Losses induced by the signal transmission are also ignored: the overall effect of transmitting the signal through the network is reduced to just an attenuation G and a delay l. The noise source $\overline{N}_2$ located in the room remote to the talker and the response $M_2(z)$ of this room allow for a realistic modeling of the acoustic echo effect.

7.3.3 NON-INTRUSIVE MODEL

In [29], a model with similar objectives as the model described in the previous section, but suited for non-intrusive evaluation of telepresence systems, is described. Analogously to the steps taken for intrusive QA tools, two scenarios are proposed based on the simplified model of Figure 7.2: one for the direct path, and the other for the feedback path.

The direct path scenario, displayed in Figure 7.5, isolates degradations that can happen irrespective of any signal feedback between the two rooms, that is, the direct path signal between the talker in Room 1 and the signal being played back by the loudspeaker located in Room 2.

Two possible measurement points, MP1 and MP2, can be identified in this scenario. The first one allows the evaluation of the signal to be transmitted to Room 2, while the second one allows the evaluation of the signal being received in Room 2. In each measurement point, the following degradations can be evaluated:

-   MP1: acoustic degradations originated in Room 1 (reverberation and background noise) and possible non-linear distortions caused by the audio capture equipment (microphone and analog-to-digital converter), such as signal clipping;

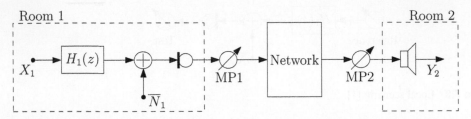

Figure 7.5 Direct path scenario. MP stands for "measurement point." Figure adapted from [29].

- MP2: all degradations measured in MP1 combined with network-induced degradations, such as coding artifacts.

MP1 of the direct path scenario is similar to the measurement point used in the local scenario described in the previous section. On the other hand, MP2 encompasses both acoustic- and network-induced degradations.

In the feedback path scenario, shown in Figure 7.6, only degradations associated with the feedback signal caused by the microphone-loudspeaker coupling are considered. In this case, the acoustic echo canceller (AEC) located in Room 2 is explicitly shown. The acoustic degradations happening in Room 2 and the microphone-loudspeaker coupling of Room 1 have been disconsidered.

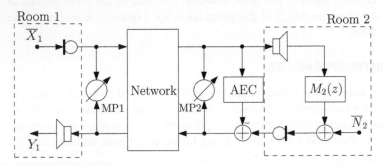

Figure 7.6 Feedback path model. MP stands for "measurement point." Figure adapted from [29].

For this scenario, measurement points MP1 and MP2 enable the evaluation of the following elements:

- MP1: acoustic echo and background noise generated in Room 2.
- MP2: the AEC of Room 2.

It should be noted that both measurement points are double-ended, allowing the access to both reference and degraded signals. MP1 is very similar to the reference-test path in remote scenario of the intrusive model, if the network is simplified to delay and gain. On the other hand, MP2 allows the evaluation of AEC operating conditions.

7.3.4 USING THE MODELS IN SUBJECTIVE LISTENING TESTS

One of the main applications of the models described in the previous sections is the development of speech signals databases for subjective listening tests. The results of such tests associated with the databases can then be employed in the development of novel automatic QA tools. In this section, a brief discussion of the procedures involved in the generation of such databases using the telepresence models is made.

The following steps should be taken into account when developing a speech database for subjective listening tests [1]:

1. Specification of the desired characteristics of the degraded signals;
2. Specification of the degradations to be employed by selecting the appropriate model and type of subjective listening test;
3. Specification of the severity of the degradations to be imposed on the signals;
4. Generation of the set of degraded signals based on the chosen model and level(s) of degradation.

A description of appropriate reference speech signals for subjective listening tests can be found in ITU Recommendation P.800 [15]. According to this standard, each signal under test should be composed of short sentences, each one with a duration between 2 and 3 s. The sentences should be taken out of non-technical publications and newspapers, should not be semantically related, and should be chosen as to guarantee phonetic variability within the signals. Moreover, they should be emitted by persons of both genders. Depending on the type of subjective listening tests, between 2 and 5 sentences intercalated by 0.5-s silence intervals should be used per signal. Care should also be taken to equalize the loudness of each sentence before combining them into the reference signals. In [1], a procedure for loudness adjustment is described.

Once the reference speech signals are available, the degradations to be studied as well as the type of subjective listening test to be performed must be defined. Generally speaking, Absolute Category Rating (ACR) [15]) should be employed when the non-intrusive model is used, whereas Degradation Category Rating (DCR) [15] is preferable for the intrusive model. The scenario variation chosen, on the other hand, is determined by the degradation to be evaluated:

- Background noise and reverberation: local scenario (intrusive), or MP1 in the direct path scenario (non-intrusive);
- Acoustic echo and background noise: remote scenario (intrusive), or MP1 in the feedback path scenario (non-intrusive).

In both scenarios of the non-intrusive model, MP2 can be used if one wants to measure network-induced impairments in the signal.

For the intrusive models, the degraded signals can be measured at the points indicated as "Test" in Figures 7.3 and 7.4. For the direct path scenario of the non-intrusive model, measurement points MP1 and MP2 can be used to obtain the degraded signal. On the other hand, for the feedback path scenario of the non-intrusive model, measurement points MP1 and MP2 cannot be directly used in subjective listening tests since the echo effect is only perceived when the return signal is acoustically added to the talker's speech; in this case, the signal $Y_1 + \overline{X}_1$ should be employed in the subjective listening test.

7.4   QUALITY EVALUATION TOOLS

In this section, quality evaluation tools intended for three types of acoustic degradation are described: (1) **background noise** [5], (2) **acoustic echo** [2], and (3) **reverberation** [8].

7.4.1   QA TOOL FOR SIGNALS CONTAMINATED WITH BACKGROUND NOISE

In telepresence systems, speech may be easily impaired by acoustic background noise due to the presence of sources within (or even outside) the rooms where the talkers are located. An adequate QA tool to tackle such kind of degradation should be signal based, since the perceived quality is directly dependent on spectral characteristics of the noise and speech signals. For this same reason, a method that is solely based on signal-to-noise measures is not able to capture the actual perceived quality.

In [5], a method for QA of high-quality speech signals contaminated by background noise that extends PESQ and W-PESQ (already mentioned in Section 7.2) is presented. In the next section these two standards will be briefly described, followed by a review of the method proposed in [5].

7.4.1.1   PESQ and W-PESQ

The well-known quality evaluation method called PESQ was designed to deal with narrow-band speech signals (from 300 Hz until 3.4 kHz) in telephony systems. In Figure 7.7 one can see a block diagram of PESQ in which the algorithm is split into three parts: pre-processing, and perceptual and cognitive models.

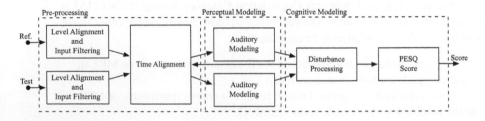

Figure 7.7   PESQ Block Diagram. Adapted from [5].

The pre-processing step conditions the two input signals so that they can be submitted to the perceptual model. They are first adjusted to the same power level, then filtered according to the frequency response of a typical telephone handset, and at last aligned in time. More information regarding these procedures can be found in [35].

The perceptual model aims to mimic the transformations suffered by the signals along the human auditory system. To this effect, the signals are first divided into time blocks, which are then Fourier transformed. Afterwards, the frequency components are mapped to the Bark [27] scale, which follows the human perceptual frequency resolution, and then transformed into a loudness [27] measure.

The cognitive model computes two metrics of perceived distortion based on the outputs of the perceptual model. The first metric computes the difference between what is heard in the signal under test and in the reference signal. This difference takes into account cognitive aspects such as the fact that the same amounts of missing or additional auditory information impact differently the perceived quality. The second metric takes into account the alignment between the signals by looking for misaligned frames, which can be due, for example, to loss of synchrony. At the end, both metrics are combined to generate a single PESQ score.

In order to be directly compared to the MOS scale defined for subjective tests [15], the PESQ scores needs to be mapped to an adequate scale; this is the object of the ITU standard [17].

W-PESQ [19] is similar to PESQ, but was designed for wideband (from 50 Hz to 7 kHz) communications. For this reason, the initial PESQ filtering operation is replaced by a high-pass filter with cutoff frequency at 100 Hz.

7.4.1.2 EW-PESQ

In this section, the method proposed in [5] for full band (20 kHz and above) speech contaminated by broadband noise is briefly described. Designed to extend the W-PESQ operation range to the complete audio frequency band, it is called Extra-Wideband PESQ (EW-PESQ).

The main action required is to modify the perceptual model of W-PESQ in order to include frequencies (above 7 kHz). Two possible strategies were devised to this end: (1) in EW-PESQ(E), the direct extension of the original model is performed; (2) EW-PESQ(R) borrows the perceptual model for audio signals up to 18 kHz employed in PEAQ [12, 36] (already mentioned in Section 7.2), an automatic quality evaluation tool for full-band audio signals.

Lastly, the mapping function of W-PESQ's cognitive model was also modified, as will be described in the following.

7.4.1.3 Signal Database and Subjective Listening Tests

In order to assist in the development and assess the performance of EW-PESQ, a signal database with speech signals degraded by background noise was created. For this, a reference database was generated by concatenating 2 sentences (selected from a set of 200 [28]) intercalated by 0.5 s of silence.

The degraded signals were obtained by adding three types of background noise to the reference signals: air-conditioner, computer-fan, and white Gaussian noise. The signal-to-noise ratio (SNR) was varied from 55 down to 25 dB in steps of 5 dB. Five degraded signals were generated for each combination of noise type and SNR for four different speakers (2 male and 2 female). Overall, 112 signals were generated, each sampled at 48 kHz with a precision of 24 bits.

The subjective evaluation of these signals was carried out through a DCR test following Recommendation [15]. The subjects were instructed to mark down the score in a scale that went from 1 up to 5, with a resolution of 0.1. After 21 subjects had judged each prepared signal, their corresponding MOS was computed.

7.4.1.4 Mapping of the PESQ Score to MOS Scale

The difference computed by the cognitive model of PESQ (and by extension, of W-PESQ and EW-PESQ) is not in the usual MOS scale. In order to harmonize the two scales, ITU recommends the use of the following function

$$y = 0.999 + \frac{4}{1 + e^{\alpha_1 x + \alpha_2}}, \tag{7.3}$$

where x is the score in the original scale and y is its corresponding value in the MOS scale. Parameters α_1 and α_2 were optimized for PESQ [17] and W-PESQ [19] to yield the best fit for a large number of signals.

Analogously, these two parameters were optimized for EW-PESQ. First, with the knowledge of the subjective MOS and the EW-PESQ scores given to half of the signals described in the previous section, an optimization algorithm searched for the values of α_1 and α_2 that minimized the quadratic error between subjective MOS and mapped (via Equation (7.3)) EW-PESQ scores scores. At the end of the optimization, the following values were found: $\alpha_1 = -1.9902$ and $\alpha_2 = 6,6007$ for EW-PESQ(E); and $\alpha_1 = -1.9027$ and $\alpha_2 = 6,1833$ for EW-PESQ(R).

7.4.1.5 Performance Evaluation

The signals not used in the optimization of the training function were employed in the performance evaluation of EW-PESQ. For this, the Pearson correlation coefficient between MOS values predicted by EW-PESQ (both E and R version) and actual MOS values was computed (a correlation of 100% indicates that one set can be predicted by an affine transformation of the other set). Correlation coefficients of around 97% were obtained for both (E and R) versions of EW-PESQ. Figures 7.8 and 7.9 show for each signal under test the comparisons between actual MOS values and their counterparts predicted by EW-PESQ(E) and EW-PESQ(R), respectively.

As can be gathered, both versions of EW-PESQ achieved a good performance, being capable of predicting MOS with high accuracy for signals degraded by background noise. Taking the fact that EW-PESQ(E) maintains compatibility with PESQ and W-PESQ without any degradation in performance w.r.t. EW-PESQ(R), the former version is preferred over the latter one.

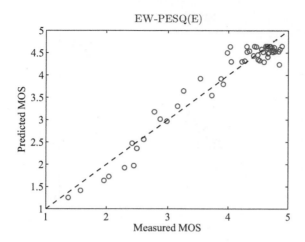

Figure 7.8 EW-PESQ(E): actual x predicted values. Each signal is denoted by "o". Adapted from [5].

7.4.2 QA TOOLS FOR SIGNALS DEGRADED BY ECHO

In [4], an objective QA tool was proposed for full-band speech signals degraded by echo in conditions similar to that found in telepresence systems. Following that work, the proposed tool was shown to also be able to handle signals degraded by acoustic echo and background noise in [2]. In the following sections, both works will be described. It should be noted that an alternative approach which estimates quality from the basic echo parameters (gain and delay) is described in [30].

7.4.2.1 Signals and Subjective Listening Tests

To begin the development of the QA tool, a database of speech signals degraded by echo was created, based on the remote scenario described in Section 7.3.2 with $M_2(z) = 1$ (i.e., ignoring the loudspeaker-microphone path effect inside the remote room). The reference signals were created from a set of 200 sentences [28] recorded by four speakers (two female and two male), according to [15, 9]. In order to simulate a conversation, the reference signals were created using 4 sentences alternately uttered by 2 different speakers: sentence 1 by speaker 1, sentence 2 by speaker 2, sentence 3 by speaker 1, and sentence 4 by speaker 2. A silence interval was inserted between sentences, in such a way that the echo from one sentence would not overlap with the beginning of the next sentence. The degraded signals were created by echoing sentences 1 and 3 according to the following conditions: delays of {100, 200, 300, or 400} ms and attenuations of {40, 35, 30, 25, or 20} dB w.r.t the emitted signal. Overall, 108 degraded signals were generated, thus ensuring that each combination of delay and attenuation value was present in at least 5 degraded signals. All signals were sampled at 48 kHz with 24 bits.

Using the prepared signal database, DCR listening tests were conducted following

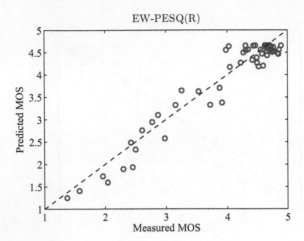

Figure 7.9 EW-PESQ(R): actual x predicted values. Each signal is denoted by "o". Adapted from [5].

Recommendation [15]. Each of the 20 participants was asked to rate the 20 signals focusing solely on the perceived degradation due to echo. Their grades were annotated in a scale ranging from 0 (not noticeable) to 5 (very annoying). At the end of the test, the MOS value was computed for each test signal.

7.4.2.2 QAfE

The starting point to the development of the echo QA tool, called QAfE (Quality Assessment for Echo), was the standardized audio QA tool called PEAQ [12] (see Section 7.2), which is divided into two parts: the computation of intermediate scores called MOVs (model output variables), and their combination into a single score based on a cognitive model that uses a neural network. Considering that this cognitive model was trained for evaluating audio signals degraded by coding artifacts, it was not employed by QAfE. Instead, the authors opted for choosing the more appropriate among the individual MOVs to be combined according to a cognitive model tailored for the target degradation.

In the first development step, the relation between PEAQ's MOVs and MOS values obtained in subjective listening tests was studied. Figure 7.10 shows this relation for MOV 5 from PEAQ, called "average distorted blocks." As can be gathered, there is a very strong tendency to associate high MOS values with low MOV values, and vice versa. This fact led to the selection of this single MOV to be used in a QA tool for speech signals degraded by acoustic echo.

By observing Figure 7.10, one can see that ADB encompasses the range from 0 (high quality) to 2.5 (low quality) in a non-linear relation with respect to MOS values. Hence, if the ADB values are to be used to predict MOS, a mapping function is necessary: the same equation (7.3) employed in the PESQ family was selected for this task. A least-square optimization routine was used to find the best parameters

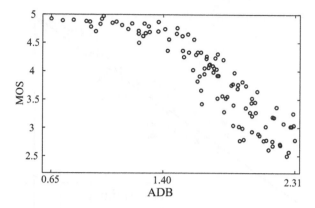

Figure 7.10 ADB (MOV 5) values versus MOS for speech signals degraded by echo. Adapted from [4].

α_1 and α_2, considering the paired values of measured and predicted (by ADB) MOS values. The optimization routine only had access to half of the speech signals used in the development of the algorithm. After optimization, the following values were found: $\alpha_1 = 2.7$ and $\alpha_2 = -5.741$.

The remaining half of the speech signals was used to evaluate the performance of the QA tool, which attained a correlation of 94% between actual and predicted MOS. Figure 7.11 illustrates their comparison for the complete set of signals.

7.4.2.3   Acoustic Echo and Background Noise

In [2], three new subjective listening tests were performed to assess different effects over the perception of echo. As a result, it was found that QAfE could deal with a more realistic model for the remote room and with background noise, besides the echo.

All new experiments used degradation category ratings (DCR) and a continuous MOS scale of 1 to 5 to average individual subjective grades. The databases were generated using the remote scenario described in Section 7.3.2. Below, a brief description of each experiment is given.

-   In experiment 1, 300 signals degraded by acoustic echo and background noise were generated under a wide range of conditions: the delay was varied between 50 and 300 ms in steps of 50 ms; the gain and the SNR were chosen from the sets {-20, -30, -40} dB and {20, 30, 40} dB, respectively; and the impulse response of a room with a reverberation time of 400 ms was chosen. For each combination of gain and delay, 18 different signals were generated (all other parameters were randomly selected) and tested.
-   Experiment 2 focused on the effect of the remote room impulse response and a larger set of delays. Hence, the SNR was fixed at 40 dB and the echo

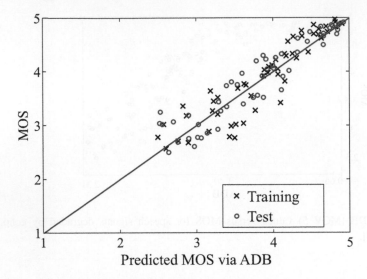

Figure 7.11 Predicted and actual MOS values for the partitions of the database of signals degraded by acoustic echo. Adapted from [4].

gain at -20 dB, with delays ranging from 100 to 500 ms in steps of 100 ms. Regarding reverberation, responses from 3 different rooms with reverberation times of respectively 300, 600, and 1200 ms were used. In total, 60 signals were evaluated in this experiment.

- Experiment 3 focused on changing the coloration of noise and evaluating the effect of very low-SNR signals. Two types of noise sources were considered: natural (recorded from either an air conditioner, a computer, or a projector device) and pink noise; they were evaluated under SNRs of either 0 or 10 dB. The remote room reverberation time was fixed at 400 ms, echo gain was either -20 or -30 dB, and echo delay was chosen among the set {50, 150, 250} ms. In total, 48 signals were generated and assessed.

Once the listening experiments had been conducted, half of the signals of experiment 1 were used to re-train the mapping curve between ADB and MOS, again according to the functional form in Equation (7.3): the new optimum parameter values resulted $\alpha_1 = 2.5$ and $\alpha_2 = -5.1$. For the remaining signals of experiment 1 not used in the optimization step, the MOS values predicted by this curve can be seen in Figure 7.12.

Using the same mapping curve, the evaluation of the signals used in experiments 2 and 3 by PEAQ attained correlations of 0.86 and 0.96, respectively, with their subjective counterparts. Hence, one can conclude that QAfE can tackle signals degraded by acoustic echo and background noise with a simple re-adjustment of its mapping function. Moreover, since some of the conditions used in experiments 2 and 3 were more severe than those used in experiment 1, it can also be inferred that the proposed method is able to correctly predict the MOS even under conditions more stringent

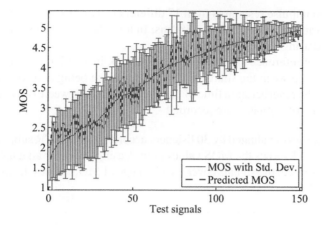

Figure 7.12   Actual and predicted MOS for experiment 1. Figure from [2].

than those for which its mapping function was optimized.

7.4.3   QA TOOL FOR SIGNALS CONTAMINATED WITH REVERBERATION

In this section, recent developments in evaluation of speech signals degraded by reverberation[1] are described, focusing on the algorithm described in [8], which received continuations in [33] and [34] (non-intrusive). Following the same outline as in the previous sections, initially the database used to develop the algorithm is described, then an overview of the QA tool is made, and finally the performance of the tool is assessed.

7.4.3.1   Database

The employed database was created based on the local scenario of Section 7.3.2 where only the effect of the local room was considered. The starting point of the database is 4 anechoic signals recorded by 2 speakers (one female, one male) sampled at 48 kHz. The reference signals were then created as the concatenation of 2 among these signals intercalated by a silent interval of approximately 1.7 s, totaling 8.4 s on average.

These reference signals were then contaminated with reverberation by three different methods:

- Simulated room impulse responses with reverberation time between 200 and 700 ms in steps of 100 ms, respectively. In total, 24 degraded signals were generated.

---

[1] An overview of reverberation assessment methods can be found in [7].

- Recorded room impulse responses from 17 different rooms [25] with reverberation time varying from 120 ms to 780 ms. In total, 68 degraded signals
  were generated using these responses.
- Recording of the reference signals being played back in 7 different rooms
  with 4 distances between the source and the microphone being considered
  in each room. The reverberation time of each room varied between 140 and
  920 ms. In total, 108 signals were recorded.

Each of the 204 signals was evaluated by 30 listeners according to an ACR subjective
listening test. Figure 7.13 shows the MOS values in increasing order for all database
signals. As can be observed, the signals cover a wide range of MOS values, from low
to high perceived quality.

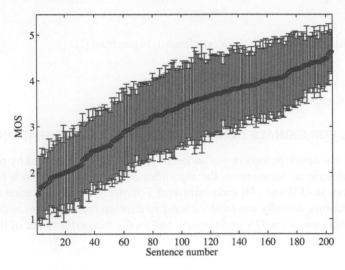

Figure 7.13 MOS for the database of signals corrupted with reverberation (standard deviations denoted as vertical bars).

7.4.3.2 Q_{MOS} **Measure**

In order to assess the quality of signals degraded by reverberation, a new quality measure called Q_{MOS} was proposed in [8]. This measure is based on both reference and
degraded speech signals, i.e., is double-ended, and is based on a feature computed
from an estimate of the impulse response affecting the degraded signal.

Figure 7.14 shows the main processing blocks required to calculate the measure.
In its first step of the proposed method, a pre-processing stage removes any DC
bias present in both the reference and the degraded signal. Subsequently, the impulse response of the room is estimated via de-convolution of the degraded signal by
the reference one: first, the room transfer function is obtained as the division of the

spectrum of the degraded signal by the spectrum of the reference signal; the room impulse response (RIR) can then be obtained as the inverse Fourier transform of the room transfer function.

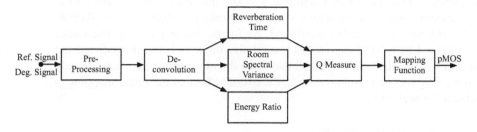

Figure 7.14 Block diagram of a reverberation QA tool. Adapted from [8].

Once the room impulse response is estimated, three features are computed:

- Reverberation time – defined as the interval of time from the cessation of an acoustic stimulus until the measured sound pressure decays by 60 dB from its initial level;
- Room spectral variance – defined as the variance of the magnitude response in dB of the room transfer function;
- Energy ratio – ratio between the energies computed at the beginning and final portions of the RIR.

These are well-known features commonly used to characterize RIRs, which are able to capture important perceptual attributes of a room.

The next step in the tool consists of combining these three features into a single measure of the perceived quality of the room, the Q measure, defined as [8]

$$Q = -\frac{T_{60}\sigma_l^2}{R\gamma},$$

(7.4)

where T_{60}, σ_l, and R are the room reverberation time, spectral variance, and energy ratio, respectively; and γ is a parameter that needs to be adjusted.

The final stage of the algorithm consists in mapping the Q measure to a MOS value, yielding Q_{MOS}. This mapping is performed in two steps. Initially, a third-order polynomial, also employed in [18], produces an intermediate measure $\overline{Q}$:

$$\overline{Q} = x_1 Q^3 + x_2 Q^2 + x_3 Q + x_4.$$

(7.5)

Finally, test-specific linear mapping is employed to fit different grading scales and reverberation ranges related to different evaluations. Q_{MOS} is then obtained as

$$Q_{MOS} = \alpha \overline{Q} + \beta.$$

(7.6)

In [8], a methodology is described to adjust each of these open parameters. Because of their different natures, rather than sharing a single optimization routine, each parameter was chosen according to the method's performance given the signals in the database. Overall, it was found that $\gamma = 0.3$ maximized the correlation between subjective and objective scores when coefficients $x_1 = 0.0017$, $x_2 = 0.0598$, $x_3 = 0.7014$, and $x_4 = 4.5387$ were used. As for the linear mapping, the values $\alpha = 1$ and $\beta = 1.85 \times 10^{-10}$ were found to minimize the error between predicted and actual MOS values. It should be noted that since the other parameters have already been optimized for the current database, the overall effect of the linear mapping is expected to be negligible.

7.4.3.3 Performance Evaluation

Figure 7.15 shows predicted and actual MOS values for all signals in the database, between which a correlation of 91% was found. When considering separately the different ways the signals were generated, correlations of 91%, 96%, and 88% were found for simulated, recorded, and actual reverberation, respectively. Hence, the proposed tool is able to correctly predict the quality of signals degraded by reverberation.

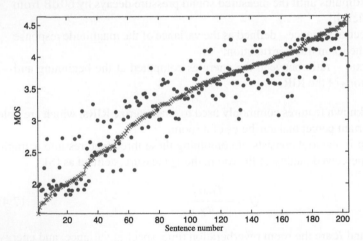

Figure 7.15   Actual ($\times$) and predicted ($\bullet$) MOS values for signals degraded by reverberation.

7.5   DEGRADATION TYPE CLASSIFICATION

Usually, automatic tools for speech QA are developed for one specific kind of degradation, as was the case of the tools presented in the previous section. In practice, however, degradations of different natures may happen concurrently, or the specific degradation affecting a given speech signal might be unknown. In [10], a possible solution for this problem is presented in the form of the scheme shown in Figure 7.16, where a previous stage of classification is responsible for identifying which degradations are present in a given speech signal and then activating the appropriate QA tool. For this, a degradation type classifier (DTC) must be developed.

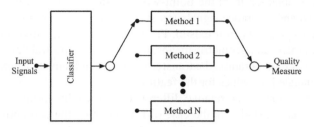

Figure 7.16 Scheme for a multi-degradation quality assessment tool. Figure adapted from [10].

Similar solutions were also proposed in [37, 38]: in these works, the classifier is responsible for selecting the appropriate cognitive model; and the resulting overall system is shown to outperform solutions where no degradation identification is employed. A classification step is also included in Recommendation P.563 [18] for non-intrusive speech quality assessment.

In [32], a double-ended classifier for acoustic degradations usually found in telepresence systems was described. This classifier is capable of identifying five different kinds of degradations: background noise (N), acoustic echo (E), reverberation (R), and combinations of background noise with acoustic echo (E+N) and reverberation (R+N). The system operates in two steps detailed in the next sections: in the first, a feature vector is computed from both degraded and reference systems; then, this feature vector is submitted to a classifier responsible for identifying the degradations present in the signal under test.

7.5.1 FEATURE EXTRACTION

At the core of the proposed classification strategy, a set of features was jointly computed from both the reference signal and the signal under test. Such features, described in [32], were intended to explore the way different degradations influence the signals: either as an addition in the case of noise or as a convolution, as is the case of reverberation and echo. In total, ten features were proposed, among which the four most relevant ones are briefly described in this section.

The first feature is defined as the minimum value of the residue envelope. To arrive at this feature, initially the reference signal is subtracted from the signal under test. Afterwards, the energy envelope of this residue signal is computed and, lastly, its minimum value is taken. This feature is appropriate for echo detection, whose isolated occurrence produces values close to zero.

The second feature is the kurtosis [32] of the residue signal. The kurtosis measures how "peaky" a given distribution of values is. In the case of the residue, for signals degraded by background noise, values spread out around zero are expected (hence with a low kurtosis), whereas for the other degradations a very prominent peak is expected around zero (hence showing high kurtosis values). Taking this into account, this feature helps to identify the presence of background noise.

The third feature is obtained from the point-wise division of the Fourier transforms of the degraded and reference signals (as an estimate of the "transfer function" relating the two signals, when it exists). From the result of this division, the ratio between the arithmetic and geometric means, usually called spectral flatness, is computed. In the case of reverberation, the transfer function is effectively a model of the room and will generate low values for the feature.

The fourth and last feature is the relative difference between the norms of reference and degraded signals. Usually, in the case of reverberation the two norms are similar, whereas for signals degraded by background noise and echo the norm of the degraded signal tends to be larger.

7.5.2 SIGNAL DATABASE

In order to train and assess the performance of the classifiers described in the next section, a signal database was created using the intrusive scenario described in Section 7.3.2, considering specifically: the local scenario for the signals degraded by background noise, reverberation, and reverberation plus background noise; and the remote scenario for echo and echo plus background noise. The reference signals were chosen as isolated sentences (taken from a set with 200 sentences [28]) uttered by 4 different speakers (two female and two male), recorded in a professional studio with sampling rates of 48 kHz and 24 bits of precision.

In total, 1000 impaired signals were generated for each degradation type, with different parameters. In the case of reverberation, different room impulse responses were employed, covering a large gamut of possible rooms. For background noise, both the noise source and the SNR were varied. For echo, both the remote room impulse response and the transmission delay and gain were varied.

7.5.3 CLASSIFICATION

Having computed the feature set for each signal in the database, there is still the task of deciding which degradations are present in the signal under test. For this, two supervised classification schemes were proposed in [32]: one employing a random forest (RF) classifier and another using a support vector machine. In this section, only the RF solution will be described.

In order to train the classifiers, the signals were split into two equal-sized sets: one for training and the other for validation. Moreover, 10 such partitions were generated, with signals being randomly split between training and validation sets. In Table 7.1 the confusion matrix[2] for a classifier with 5 trees can be seen, including the average results (and standard deviation) for the 10 partitions. As can be observed, the classifier attained excellent results when using the 4 features described in the previous

---

[2]The confusion matrix displays how many signals degraded by x were identified as degraded by y.

Table 7.1

Confusion matrix obtained for the RF classifier with 5 trees. The values shown are in % with standard deviations displayed within parentheses.

Actual \ Identified	E	R	N	E+N	R+N
E	100.0 (0.1)	0.0 (0.0)	0.0 (0.0)	0.0 (0.1)	0.0 (0.0)
R	0.0 (0.0)	100.0 (0.0)	0.0 (0.0)	0.0 (0.0)	0.0 (0.0)
N	0.0 (0.0)	0.0 (0.0)	99.9 (0.0)	0.0 (0.0)	0.1 (0.0)
E+N	0.0 (0.0)	0.0 (0.0)	0.0 (0.0)	100.0 (0.0)	0.0 (0.0)
R+N	0.0 (0.0)	0.2 (0.2)	0.0 (0.0)	0.0 (0.0)	99.8 (0.2)

section, with only 0.2 % of the signals wrongly classified on average, in the worst case. Similar results were found for other classification schemes.

7.6 CONCLUDING REMARKS

This chapter's main subject was quality assessment of acoustically degraded full-band speech signals in the context of telepresence systems. After a brief introduction to the topic of quality assessment and related relevant literature, the authors described which degradations are usually found in a telepresence system and how they can be modeled. The chapter proceeded by reviewing different QA tools designed to deal with specific degradations, namely background noise, acoustic echo, and reverberation. Finally, a tool capable of identifying which degradations are present in a speech signal was described.

REFERENCES

1. F. R. Ávila, L. W. P. Biscainho, L. O. Nunes, A. F. Tygel, B. Lee, A. Said, T. Kalker, and R. W. Schafer. A teleconference model with acoustic impairments suitable for speech quality assessment. In *Anais do XXVII Simpósio Brasileiro de Telecomunicações*, Blumenau, Brazil, October 2009. SBrT. Paper 58318.

2. F. R. Ávila, L. O. Nunes, L. W. P. Biscainho, A. F. Tygel, and B. Lee. Objective quality assessment of echo-impaired full-band speech signals. In *Proceedings of the International Telecommunications Symposium*, pages 1–5, August 2014.

3. S. Bech and Z. Zacharov. *Perceptual Audio Evaluation - Theory, Method and Application*. Wiley, Chichester, UK, 2006.

4. L. W. P. Biscainho, P. A. A. Esquef, F. P. Freeland, L. O. Nunes, A. F. Tygel, B. Lee, A. Said, T. Kalker, and R. W. Schafer. An objective method for quality assessment of ultra-wideband speech corrupted by echo. In *Proceedings of the 2009 IEEE International Workshop on Multimedia Signal Processing*, Rio de Janeiro, Brazil, October 2009. Paper 149.

5. B. C. Bispo, P. A. A. Esquef, L. W. P. Biscainho, A. A. de Lima, F. P. Freeland, R. A. de Jesus, A. Said, B. Lee, R. W. Schafer, and T. Kalker. EW-PESQ: A quality assessment

method for speech signals sampled at 48 kHz. *Journal of the Audio Engineering Society*, 58(4):251–268, April 2010.

6. A. A. de Lima, F. P. Freeland, B. C. Bispo, L. W. P. Biscainho, S. L. Netto, A. Said, A. Kalker, R. W. Schafer, B. Lee, and M. Jam. On the quality assessment of sound signals. In *Proceedings of the IEEE International Symposium on Circuits and Systems*, pages 416–419, Seattle, May 2008.

7. A. A. de Lima, F. P. Freeland, P. A. A. Esquef, L. W. P. Biscainho, B. C. Bispo, R. A. de Jesus, S. L. Netto, R. W. Schafer, A. Said, B. Lee, and A. Kalker. Reverberation assessment in audioband speech signals for telepresence systems. In *Proceedings of the International Conference on Signal Processing and Multimedia Applications*, pages 257–262, Porto, Portugal, May 2008.

8. A. A. de Lima, T. de M. Prego, S. L. Netto, B. Lee, A. Said, R. W. Schafer, T. Kalker, and M. Fozunbal. On the quality-assessment of reverberated speech. *Speech Communication*, 54(3):393–401, March 2012.

9. EBU Technical. *EBU Tech.3276: Listening conditions for the assessment of sound programme material*. European Broadcast Union, Geneva, Switzerland, 2nd edition, May 1998.

10. P. A. A. Esquef, L. W. P. Biscainho, L. O. Nunes, B. Lee, A. Said, T. Kalker, and R. W. Schafer. Quality assessment of audio: Increasing applicability scope of objective methods via prior identification of impairment types. In *Proceedings of the 2009 IEEE International Workshop on Multimedia Signal Processing*, Rio de Janeiro, Brazil, October 2009. Paper 148.

11. V. Grancharov and W. B. Kleijn. Speech quality assessment. In J. Benesty, M. M. Sondhi, and Y. Huang, eds., *Springer Handbook of Speech Processing*, chapter 5, pages 83–100. Springer, Berlin, Germany, 2008.

12. ITU-R. *Rec. BS.1387-1: Method for Objective Measurements of Perceived Audio Quality*. International Telecommunication Union, Geneva, Switzerland, November 2001.

13. ITU-R. *Rec. BS.1294-1: General Methods for the Subjective Assessment of Sound Quality*. International Telecommunication Union, Geneva, Switzerland, December 2003.

14. ITU-R. *Rec. BS.1116-2: Methods for the Subjective Assessment of Small Impairments in Audio Systems*. International Telecommunication Union, Geneva, Switzerland, June 2014.

15. ITU-T. *Rec. P.800: Methods of Subjective Determination of Transmission Quality*. International Telecommunication Union, Geneva, Switzerland, August 1996.

16. ITU-T. *Rec. P.862: Perceptual Evaluation of Speech Quality (PESQ): Objective Method for End-to-end Speech Quality Assessment of Narrow Band Telephone Networks and Speech Codecs*. International Telecommunication Union, Geneva, Switzerland, February 2001.

17. ITU-T. *Rec. P.862.1: Mapping Function for Transforming P.862 Raw Result Scores to MOS-LQO*. International Telecommunication Union, Geneva, Switzerland, November 2003.

18. ITU-T. *Rec. P.563: Single-ended Method for Objective Speech Quality Assessment in Narrow-band Telephony Applications*. International Telecommunication Union, Geneva, Switzerland, May 2004.

19. ITU-T. *Rec. P.862.2: Wideband Extention to Recommendation P.862 for the Assessment of Wideband Telephone Networks and Speech Codecs*. International Telecommunication Union, Geneva, Switzerland, November 2007.

20. ITU-T. *Rec. G.107.1: Wideband E-model*. International Telecommunication Union,

Geneva, Switzerland, December 2011.

21. ITU-T. *Rec. P.56: Objective Measurement of Active Speech Level.* International Telecommunication Union, Geneva, Switzerland, December 2011.

22. ITU-T. *Rec. G.107: The E-model: a computational model for use in transmission planning.* International Telecommunication Union, Geneva, Switzerland, February 2014.

23. ITU-T. *Rec. P.863: Perceptual Objective Listening Quality Assessment.* International Telecommunication Union, Geneva, Switzerland, September 2014.

24. U. Jekosch. *Voice and Speech Quality Perception: Assessment and Evaluation.* Signals and Communication Technology. Springer, Berlin, Germany, 2005.

25. M. Jeub, M. Schäfer, and P. Vary. A binaural room impulse response database for the evaluation of dereverberation algorithms. In *Proceedings of the 16th International Conference on Digital Signal Processing*, pages 1–5, Santorini, Greece, 2009.

26. P. C. Loizou. Speech quality assessment. In W. Lin, D. Tao, J. Kacprzyk, Z. Li, E. Izquierdo, and H. Wang, eds., *Multimedia Analysis, Processing and Communications*, volume 346 of *Studies in Computational Intelligence*, part III: Communications Related Processing, pages 623–654. Springer, Berlin, Germany, 2011.

27. B. C. J. Moore. *An Introduction to the Psychology of Hearing.* Emerald, Bingley, UK, 5th edition, 2003.

28. J. A. Moraes, A. Alcaim, and J. A. Solewics. Freqüência de ocorrência dos fones e lista de frases foneticamente balenceadas no português do Rio de Janeiro. *Revista da Sociedade Brasileira de Telecomunicações*, 7(1):23–41, December 1992. In Portuguese.

29. L. O. Nunes, F. R. Ávila, L. W. P. Biscainho, B. Lee, A. Said, and R. W. Schafer. Monitoring scenario for non-intrusive quality assessment of teleconference systems. In *Anais do XXIX Simpósio Brasileiro de Telecomunicações*, Curitiba, Brazil, October 2011. SBrT. Paper 85448.

30. L. O. Nunes, F. R. Ávila, A. F. Tygel, L. W. P. Biscainho, B. Lee, A. Said, and R. W. Schafer. A parametric objective quality assessment tool for speech signals degraded by acoustic echo. *IEEE Transactions on Audio, Speech, and Language Processing*, 20(8):2181–2190, May 2012.

31. L. O. Nunes and L. W. P. Biscainho. Revisão de métodos de avaliação de qualidade de voz para sistemas de telepresença. In *Anais do 12o. Congresso de Engenharia de Áudio da AES-Brasil*, pages 28–39, São Paulo, Brazil, May 2012. In Portuguese.

32. L. O. Nunes, L. W. P. Biscainho, B. Lee, A. Said, T. Kalker, and R. W. Schafer. Degradation type classifier for full band speech contaminated with echo, broadband noise, and reverberation. *IEEE Transactions on Audio, Speech, and Language Processing*, 19(8):2516–2526, November 2011.

33. T. de M. Prego, A. A. de Lima, and S. L. Netto. Perceptual analysis of higher-order statistics in estimating reverberation. In *Proceedings of the 5th International Symposium on Communications Control and Signal Processing*, pages 1–4, Rome, Italy, May 2012.

34. T. de M. Prego, A. A. de Lima, S. L. Netto, B. Lee, A. Said, R. W. Schafer, and T. Kalker. A blind algorithm for reverberation-time estimation using subband decomposition of speech signals. *Journal of the Acoustical Society of America*, 131(4):2811–2816, April 2012.

35. A. Rix, M. Hollier, A. Hekstra, and J. G. Beerends. Perceptual evaluation of speech quality (PESQ), the new ITU standard for end-to-end speech quality assessment, Part I – Time-Delay Compensation. *Journal of Audio Engineering Society*, 50(10):755–764, October 2002.

36. T. Thiede, W. C. Treurniet, R. Bitto, C. Schmidmer, T. Sporer, J. G. Beerends,

C. Colomes, M. Keyhl, G. Stoll, K. Brandenburg, and B. Feiten. PEAQ—the ITU standard for objective measurement of perceived audio quality. *Journal of the Audio Engineering Society*, 48(1/2):3–29, January/February 2000.

37. H. Yuan, T. H. Falk, and W.-Y. Chan. Classification of speech degradations at network endpoints using psychoacoustic features. In *Proceedings of the Canadian Conference on Electrical and Computer Engineering*, pages 1602–1605, Vancouver, Canada, April 2007.

38. H. Yuan, T. H. Falk, and W.-Y. Chan. Degradation-classification assisted single-ended quality measurement of speech. In *Proceedings of the INTERSPEECH*, pages 1689–1692, Antwerp, Belgium, August 2007.

8 Models for Speech Processing

MIGUEL ARJONA RAMÍREZ
University of São Paulo (USP)

MÁRIO MINAMI
Federal University of ABC (UFABC)

THIPPUR VENKATANARASAIAH SREENIVAS
Indian Institute of Science (IISc)

ABSTRACT

Spectral and time-frequency representations and production and synthesis models for speech coding, speech synthesis, speech recognition, and speech enhancement are presented, enabling an understanding of the basic pillars of current speech technology and providing hints towards future improvements. Classical quasi-stationary representations such as the short-time Fourier transform (STFT) and the associated speech spectrogram are presented, and linear prediction (LP) or autoregressive (AR) models are introduced. In this process, some important parameters arise in the development of LP analysis, providing links with speech production models and with source-filter models. Next some successful classifiers for application in speech recognition and speaker identification are presented such as Markovian (HMM) models and Gaussian (GMM) models. The important issues of model duration and adaptation are also discussed in connection with speech recognition. Even though the speech signal is intrinsically time-varying, quasi-time-invariant models have been proven useful if associated with proper segmentation of the signal suitable to the application at hand. However, more stringent application requirements demand a time-varying outlook in handling the models which depends on their adaptation. This perspective is introduced by means of time-varying amplitude modulation-frequency modulation (AM-FM) models.

8.1 INTRODUCTION

Speech is a very complex nonstationary signal. Considering its bandwidth and accepted human auditory perception, it may be represented by rates ranging from 64 kbit/s for narrowband telephone speech to 256 kbit/s for wideband speech. However, if its lexical information content is targeted, speech could be represented by text at rates around 50 bit/s. It should be granted that speech conveys much more information than text such as prosody, speaker identity, and emotional cues. That is why different models may fit the speech signal, selectively capturing relevant features. From the viewpoint of waveform production, articulatory models [30] were the first to be used and are the basic ones for phonetic description. However, their computation is an ill-posed inverse problem.

It is interesting to consider that articulatory models are mostly quasi-stationary because the muscles react much more slowly than signal variations so that the speech may be considered stationary for analyses of a few tens of milliseconds. But this hypothesis may fail, for instance in the case of plosive sounds, and time-varying models should be considered for better approximation. Anyway, quasi-stationarity is a very important simplifying hypothesis which spectral and time-frequency representations such as the spectrogram and models such as linear prediction and sinusoidal models have also inherited.

While linear prediction models are parametric, time-frequency representations (TFRs) are mostly non-parametric representations, such as the most popular STFT (short-time Fourier transform), i.e., the spectrogram. Other TFRs such as the Wigner-Ville distribution, Wavelet transform, are also non-parametric in nature and they have not provided as much insight into speech properties as the STFT. The wideband spectrogram, i.e., magnitude of STFT with a short window and the narrowband spectrogram which is magnitude STFT with a long window have revealed the two popular complementary parametric models of speech representation: (i) speech consists of time-varying resonances, showing as prominent fluctuating bands of energy in the wideband spectrogram, (ii) narrowband spectrogram showing time-varying harmonic frequency components with varying strength (amplitude) and varying frequency. The former observation led to the formulation of the linear time-invariant (LTI) model of speech, excited by a periodic excitation or noise excitation, over a short duration of the signal of around 30-50 ms. The latter observation led to the sinusoidal model of speech wherein, again over the short duration, the sinusoids are assumed to be constant in frequency and amplitude. The most popular LPC (linear prediction coefficients) is the parametric representation of the LTI formulation, which is treated in Section 8.4 and Quatieri's sinusoidal synthesis model [28] is representative of the second approach. The parametric nature of both of these models is well utilized in speech coding (i.e., for compact digital communication, analysis of parameters at the transmitter, and synthesis using the quantized parameters at the receiver). For speech synthesis, the parametric models have shown limited success, which is indicative of the limitation of the models. Of course various auditory perceptual criteria have been used to alter the parameters, but with only limited success.

The non-parametric representations, such as STFT or Wigner-Ville distribution can at best be alternate representations of the signal for further processing; they do not provide a parsimonious representation of speech that can lead us to the all-important information-carrying parameters, such as the phonetic content or speaker identity, mood, or the language spoken. Because of the limitation of the quasi-stationary (or quasi-time-invariant) analysis, there is significant variability of the parameters estimated and hence we are forced to use stochastic models for representing the basic units of sound. The stochastic models also accommodate speaker variability or acoustic environment variability, but primarily we are yet unable to solve for a smooth parametric representation that corresponds to the smooth perceived information in the signal.

8.2 TIME-FREQUENCY MODELS

The most successful representation of the speech signal is the spectrogram, which is widely used by phoneticians and phonoaudiologists [23]. It is based on the short-time Fourier transform (STFT), which may represent the signal in high definition and allow the discrimination of important features if suitable frame length and frequency resolution or filter bandwidth are configured; so much so that people may be trained to read spectrograms [34].

Spectrograms are taken at two equivalent filter bandwidths, namely, narrowband and wideband spectrograms as shown in Figure 8.1. Wideband spectrograms mark formant bandwidths as dark and wide horizontal bands and pitch periods as vertical stripes whereas narrowband spectrograms display fundamental frequency harmonics as horizontal stripes. Using analog electronics, spectrograms used to be implemented as a filterbank, which could even be a single tuned filter used for printing a horizontal stripe at each center frequency setting. Nowadays, the STFT is implemented one frame at a time, generating a vertical stripe for each frame.

The filterbank implementation was used before the invention of the spectrograph for speech synthesis with the vocoder by Homer Dudley [13]. However, the vocoder had only ten filters in the filterbank and could be compared to a wideband spectrographic representation. In order to reconstruct the speech, power was provided to the filterbank by either a buzz or a hiss for voiced and unvoiced sounds, respectively. This construction embodies some features of a source-filter model to be presented in Section 8.3.

In the limit of very short frames, the spectrogram becomes the amplitude-modulation-frequency-modulation (AM-FM) model with instantaneous frequency tracks.

8.3 PRODUCTION MODELS

The speech signal is generated in the vocal tract by an actuator, represented by the lungs or the vocal folds, and is spectrally shaped by the resonating cavities above, which are contained in the mouth and nose and whose shapes and couplings are determined by the positions and attitudes of articulators such as the tongue, the uvula, and the lips.

Speech is a very complex nonstationary signal. From a systems point of view, input signals, output signals, and dynamic rules must be identified that transform the input signals into the output signals. We may simplify the picture by considering the speech signal $s(n)$ as the single output as shown in Figure 8.2. As for the input signal $x(n)$, alternative sources may be envisaged and considered to be active alternately. Making up the bulk of the system, the dynamic rules may be enclosed into the

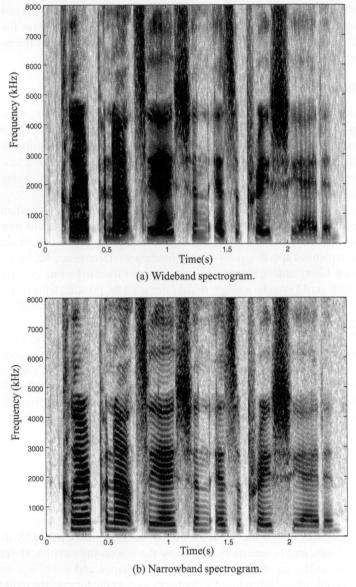

(a) Wideband spectrogram.

(b) Narrowband spectrogram.

Figure 8.1 Wideband and narrowband spectrograms of the sentence "Clear pronunciation is appreciated." uttered by a female speaker.

operation of the filter $H(z)$, representing the effects brought about by the cavities of the vocal tract and their couplings.

As discussed in Section 8.1, quasistationarity is a very important hypothesis for rather simple speech models. In conjunction with it, another hypothesis that has proven very useful is the independence between source of excitation and vocal tract

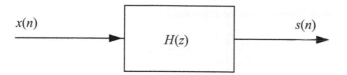

$x(n)$ $H(z)$ $s(n)$

Figure 8.2 Speech synthesis model.

within the systems approach. The first complete proposal of a source-filter model was put forward by Gunnar Fant in the 1960s. It used a synthesis filter made of second-order sections, each one standing for a formant frequency. However, while cascade associations of second-order sections provide a simple control of relative formant amplitudes, parallel associations are needed for fricative and plosive bursts [1].

The resonant frequencies or peaks in the frequency response of the filter are responsible for the timbre of the corresponding speech sound. They are the result of the combined action of the various poles or natural frequencies of the vocal tract. Each formant has a central frequency and a bandwidth while each pole has a real part and an imaginary part in continuous time representation, or an absolute value and a phase in discrete time representation. Assuming the poles to be rather far apart, each pair of poles will represent a single formant and may be placed in a different second-order section. However, formant frequency and bandwidth estimation is a difficult task when based on articulatory models. A simpler approach is possible by using linear prediction techniques as will be seen in Section 8.4.

The accuracy of *source-filter* models rests upon the independence between source and filter, which holds approximately for a wide range of sounds and purposes, even more so because the split into source and filter need not be a physical one for the model to be useful.

From a conceptual point of view, the source-filter model provides a very clear-cut localization of the parameters describing speech for synthesis. For instance, pitch may be ascribed to the source, which provides the fundamental frequency f_0 and its harmonics $k f_0$ for $k = 2$ and so on up to the maximum below the Nyquist frequency, which is half the sampling frequency of the system. Also, the formants f_k are associated to the resonance frequencies in the frequency response of the filter for $k = 1$ and so on, usually up to $k = 4$ for narrowband speech.

The simplest model for the voiced source is the train of impulses

$$x(n) = G \sum_{k=0}^{K_p-1} \delta(n - kP_0), \qquad (8.1)$$

where G is an amplitude scale factor related to loudness, P_0 is the pitch period, assumed to be an integer number of samples, and K_p is the number of pulses that fit within the speech frame. Alternatively, Eq. (8.1), by means of a discrete Fourier transform (DFT) and its inverse, may be expressed as

$$x(n) = \frac{G}{P_0} \sum_{l=0}^{P_0-1} \exp(j2\pi l f_0 n). \tag{8.2}$$

The simplest model for the unvoiced source is a properly scaled version

$$x(n) = G u_N(n) \tag{8.3}$$

of the wide-sense stationary random process $u_N(n)$ normally distributed as $N(0, 1)$ and with unit constant power spectral density.

The voiced and unvoiced sources may be combined to approximate mixed excitation segments of speech. Also, the scaling factor and the pitch period may be varied along the synthesis for a more natural reconstruction.

Basically, the source-filter model allows for perfect reconstruction if we simply filter the speech signal with the inverse filter, generating a residual signal which may feed the synthesis filter to precisely reconstruct the original signal in the absence of numerical errors only. Simply put, this is only of conceptual value, but, by using proper quantization methods, it allows the transmission of the residual signal in a number of speech coders used for communications.

Alternatively, for sinusoidal coders, the excitation signal is broadly defined as

$$x(n) = \sum_{l=1}^{K} A_l \Re\{\exp(j(2\pi f_l n + \theta_l))\}, \tag{8.4}$$

where the sinusoidal frequencies f_l may be in a harmonic relation or not and their phases θ_l may have a random uniformly distributed component for unvoiced or mixed excitation segments. Besides, the sinusoidal model for the excitation may produce a perfectly periodic train of impulses, just as expressed by either Eq. (8.2) or Eq. (8.1), when the amplitudes A_l for $l = 1, 2, \ldots, K$ are all the same.

More flexible excitation models are employed in quality low-bit-rate coders [2] and codecs for packet voice communications [3].

8.4 LINEAR PREDICTION

From the speech signal, acoustic models may be estimated by resting on some of its features. Considering that the vocal tract has a natural inertia, it should act upon the acoustic signal as a filter with an overall nonflat characteristic. This aspect suggests that the speech signal is redundant so that the current sample $s(n)$ depends on previous samples. Assuming linear dependency, we can get the predicted value

$$\hat{s}(n) = -\sum_{i=1}^{p} a_i s(n - i) \tag{8.5}$$

for a prediction order p, which should be defined beforehand to suit our modeling purposes, and with predictor coefficients $a_1, a_2, \ldots, a_p$, which must be computed for each quasistationary speech segment or frame under prescribed conditions.

The prediction order p is defined by considering some physical conditions prevailing in speech production. The mean vocal tract length for a male adult speaker is about 17 cm and the speed of sound in the vocal tract is about the same as in the air or 340 m/s so that this relaxed vocal tract is able to resonate at any integer multiple of 1 kHz [27]. This requires two units of order per kHz in the baseband. Also two units of order should be used to account for glottal and radiation effects. Combining these requirements, prediction orders of 10 or 12 are used for narrowband speech and prediction order of 16 is more usual for wideband speech [36]. The prediction coefficients are determined for minimizing the mean square error

$$\varepsilon = E\left[(s(n) - \hat{s}(n))^2\right]. \tag{8.6}$$

Under wide-sense stationarity, by setting to zero the gradient of ε with respect to $a = \begin{bmatrix} a_1 & a_2 & \cdots & a_p \end{bmatrix}^T$, we obtain the set of normal equations

$$\sum_{j=1}^{p} a_i R(i - j) = -R(i) \tag{8.7}$$

for $i = 1, 2, \ldots, p$, where $R(m) = E\left[s(n)s(n + m)\right]$ is the autocorrelation function of signal $s(n)$. These are the normal equations for the autocorrelation method [19] and an alternative set of normal equations may be obtained for the nonstationary case that leads to the covariance method [4, 27]. After zero-padding the segment of speech at both ends, the autocorrelation function may be estimated as

$$R(m) = \sum_{n=-\infty}^{\infty} s(n)s(n + m). \tag{8.8}$$

The normal equations for the autocorrelation method may be set in matrix notation as

$$Ra = -r, \tag{8.9}$$

where R is the $p \times p$ autocorrelation matrix with entries $[R]_{ij} = R(i - j)$ and r is the $p \times 1$ autocorrelation vector r with entries $r(i) = R(i)$. The autocorrelation matrix is a symmetric Toeplitz matrix, that is, its elements in the main diagonal are constant and also the elements in each superdiagonal and in each subdiagonal are the same.

Further, the positive definite property of the autocorrelation matrix guarantees the proper definition of the inner product

$$\left\langle z^{-i}, z^{-j} \right\rangle = R(i - j) \tag{8.10}$$

for i and j sample delay operators represented by the elementary transfer functions z^{-i} and z^{-j}. Then, the distributive property of the inner product over vector addition extends the definition of this inner product to any pair $A(z)$ and $B(z)$ of polynomials of degree up to and including p.

Now, by using the inner product above, the set of normal equations (8.7) may be expressed as

$$\left\langle A(z), z^{-i} \right\rangle = 0 \tag{8.11}$$

for $i = 1, 2, \ldots, p$, whose solution determines the inverse filter by the transfer function

$$A(z) = 1 + \sum_{j=1}^{p} a_j z^{-j}. \tag{8.12}$$

The normal equations (8.11), under the conditions for the autocorrelation method, may be solved by the Durbin algorithm [27], which starts by a trivial zero-order solution $A_0(z) = 1$ with an auxiliary backward prediction inverse filter $B_0(z) = z^{-1}$, whose squared norms are

$$\begin{aligned} \varepsilon_0 &= \langle A_0(z), A_0(z) \rangle \\ &= R(0) \\ \beta_0 &= \langle B_0(z), B_0(z) \rangle \\ &= R(0). \end{aligned} \tag{8.13}$$

The prediction order m is increased by steps $m = 1, 2, \ldots, p$ and for each mth normal equation the solution

$$A_m(z) = A_{m-1}(z) + k_m B_{m-1}(z) \tag{8.14}$$

is postulated, preserving the solutions to the previous equations and requiring just the determination of the partial correlation coefficient, which may be found as

$$k_m = -\frac{\langle A_{m-1}(z), z^{-m} \rangle}{\varepsilon_{m-1}} \tag{8.15}$$

thus determining $A_m(z)$ by Eq. (8.14). Also, by the Toeplitz structure of the autocorrelation matrix, the backward prediction inverse filter may be found by reversing the order of the prediction coefficients as

$$B_m(z) = z^{-(m+1)} A_m(1/z) \tag{8.16}$$

and the mean square prediction error for prediction order m is found as

$$\begin{aligned} \varepsilon_m &= \langle A_m(z), A_m(z) \rangle \\ &= (1 - k_m^2) \varepsilon_{m-1}. \end{aligned} \tag{8.17}$$

After solving for order $m = p$, the solution to the set of equations (8.11) is found as $A(z) = A_p(z)$ with minimum square prediction error $\varepsilon = \varepsilon_p$, which is useful for adjusting the level of the excitation signal. Assuming segment of length L, for periodic excitation, as a first approximation, the excitation gain in Eq. (8.1) may be computed as

$$G = \sqrt{\frac{\varepsilon}{K_p}} \tag{8.18}$$

with the number of pulses in the segment given by

$$K_p = \left\lceil \frac{L}{P_0} \right\rceil \tag{8.19}$$

whereas, for the random excitation in Eq. (8.3), the excitation gain may be estimated by

$$G = \sqrt{\frac{\varepsilon}{L}} \qquad (8.20)$$

Along the resolution above, the PARCOR coefficients stand out, linking minimum square errors as the prediction order increases by Eq. (8.17) and shedding light on the stability conditions

$$|k_m| < 1 \qquad (8.21)$$

for $m = 1, 2, \ldots, p$. If any PARCOR coefficient happens to be ± 1, it is a case of perfect prediction. Also, these coefficients give rise to a new filter structure, the lattice filter, initially proposed by Itakura and Saito [19] and projected in an alternative way by Burg [10].

Even though the PARCOR coefficients are very convenient due to their bounded range, their values do not distribute uniformly making them hard to quantize. But a more readily manageable set of coefficients for quantization and interpolation were proposed by Itakura, called the line spectral frequencies (LSFs) or the line spectral pair (LSP) representation [18].

The main properties of the LSF coefficients may be understood by considering the allpass rational function [36]

$$F(z) = \frac{B(z)}{A(z)} \qquad (8.22)$$

and by virtually extending the analysis one step further to produce a forced condition of perfect prediction. There are two possible paths to perfect prediction by taking either $k_{p+1} = 1$ or $k_{p+1} = -1$ and obtaining, respectively, by applying Eq. (8.14), the inverse filters for forward prediction

$$\begin{aligned} P(z) &= A(z) + B(z) \\ Q(z) &= A(z) - B(z). \end{aligned} \qquad (8.23)$$

Further, the roots of $P(z)$ occur when $F(z) = -1$ and the roots of $Q(z)$ occur when $F(z) = 1$. Since for an allpass function the condition $|F(z)| = 1$ occurs only on the unit circle, we conclude that the roots of $P(z)$ and $Q(z)$ lie on the unit circle. So they may be written as $z_l = e^{j\omega_l}$ for $l = 1, 2, \ldots, 2p + 2$. By discarding two trivial roots at 1 and -1 and by observing that the remaining roots are complex conjugate pairs, the essential p roots are such that their complex arguments satisfy the relations

$$0 < \omega_1 < \omega_2 < \ldots < \omega_p < \pi \qquad (8.24)$$

and ω_l for $l = 1, 2, \ldots, p$ are called the line spectral frequencies.

Finally, as a sign of the good fit of the LP models to the representations of the speech signal, a great number of speech coders are based on linear prediction techniques [3, 2].

8.5 SPECTRAL REPRESENTATIONS AND MODELS

The most intuitive and widespread spectral representation for signals is based on the Fourier transform, defined for a stationary discrete-time signal $s(n)$ as

$$S\left(e^{j\omega}\right) = \sum_{n=-\infty}^{\infty} s(n)e^{-j\omega n}. \tag{8.25}$$

As mentioned before, a speech signal is not stationary so that the Fourier transform may not be directly applied in a meaningful way. However, considering the speech taken within a quasistationary segment by means of a window function $w(n)$, a windowed signal located around time n is isolated as

$$s_w(m, n) = s(m)w(n - m) \tag{8.26}$$

whose discrete-time Fourier transform (DTFT), for a fixed time n is now obtained as

$$S\left(e^{j\omega}, n\right) = \sum_{m=-\infty}^{\infty} s_w(n - m)e^{-j\omega m}. \tag{8.27}$$

The window function essentially vanishes outside the given segment so that the short-time transform will now be able to represent the local spectral features of the signal. When the time location n is a variable, Eq. (8.27) describes the short-time Fourier transform (STFT), whose magnitude is the spectrogram. Two such spectrograms are depicted in Figure 8.1 for different length windows.

Starting at 0.9 s, a segment is taken from the speech signal previously represented in Figure 8.1 and shown before windowing in Figure 8.3. After having a Hamming window applied, the magnitude spectrum of this segment is depicted in Figure 8.4 in a solid line that displays the pitch harmonics. Also, the dashed line superimposed is the spectral envelope obtained with the frequency response of the synthesis filter resulting from a 16th-order LP autocorrelation analysis, which highlights the formant frequencies by its smooth peaks.

The frequency response for the LP synthesis filter is further shown in Figure 8.5 together with its line spectral frequencies, which may be observed to cluster around the formant frequencies and to be closer together when the formant bandwidth is narrower and further apart when it is wider.

8.6 STOCHASTIC MODELS

In previous sections we presented some acoustic speech models and now we are focused on how to conjugate those short-time models to phonetic-linguistic and language models, using stochastic probabilistic modeling with Markov Models.

8.6.1 HIDDEN MARKOV MODELS

In order to bridge the gap of information between the acoustic representations of the speech signal and the meaning conveyed by this sound, speech researchers began to

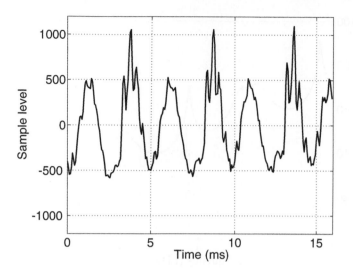

Figure 8.3   Speech segment starting at 0.9 s from speech utterance represented in Figure 8.1.

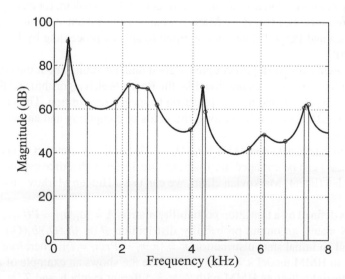

Figure 8.4   Fourier magnitude spectrum (solid line) and 16th-order LP envelope (dashed line) for speech segment starting at 0.9 s from speech utterance represented in Figure 8.1.

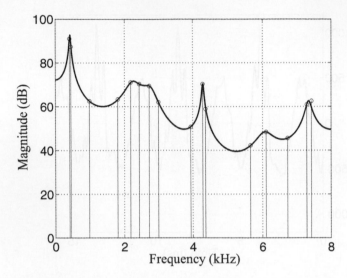

Figure 8.5 Magnitude frequency response for 16th-order LP synthesis filter (solid line) with line spectral frequencies (vertical lines) superimposed for speech segment starting at 0.9 s from speech utterance represented in Figure 8.1.

study and to apply hidden (or latent) variables to modeling this problem, for example illustrated in Figure 8.6. The theory of Markov chains with hidden variables was developed by Baum and Petrie [7] and first applied to speech processing by Baker [5] and Jelinek [20].

Furthermore, Levinson's work [25] coupled a good understanding about the convergence of algorithms for model estimation, like the Baum-Welch algorithm and the forward-backward algorithm, and their application to speech recognition. The great success of CMU SPHINX system [24] linked the fields of computing and engineering with linguistics.

If observable stochastic data items $\mathbf{O} = \{O_1, O_2, ..., O_T\}$ of the discrete set of symbols $v = \{v_1, v_2, ..., v_L\}$ can be arranged in a discrete-time series with a time evolution modeled by a hidden (latent) Markovian chain, we can use a Hidden Markov Model (HMM).

This HMM λ is defined by a transition probability matrix, $A = \{a_{ij}|a_{ij} = Pr(s_{t+1} = j|s_t = i)\}$ with N states, an output probability distribution $B = \{b_j(O_t)|b_j(O_t) = Pr(O_t|s_t = j)\}$, and an initial state distribution $\pi = \{\pi_i|\pi_i = Pr(s_i = i)\}$. Therefore, a compact notation to HMM model $\lambda = \{\pi, A, B\}$. Figure 8.6 shows an example of an $N = 4$ state left-to-right discrete HMM with $L = 3$ different symbols and $T = 10$

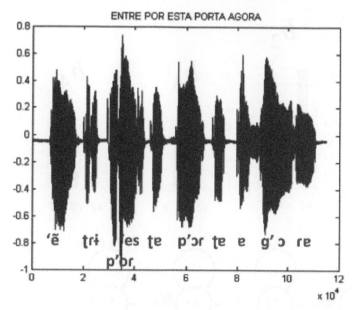

Figure 8.6 Waveform of spoken phrase "Entre por esta porta agora" and the respective phonetic transcription (bottom).

observations, with the initial, transition and observation matrices:

$$
\pi = \begin{bmatrix} 1 & 0 & 0 & 0 \end{bmatrix} \quad
A = \begin{bmatrix} 0.3 & 0.6 & 0.1 & 0 \\ 0 & 0.7 & 0.1 & 0.2 \\ 0 & 0 & 0.5 & 0.5 \\ 0 & 0 & 0 & 1 \end{bmatrix} \quad
B = \begin{bmatrix} 0.25 & 0.5 & 0.25 \\ 0.7 & 0.2 & 0.1 \\ 0.2 & 0.3 & 0.5 \\ 0.45 & 0.1 & 0.45 \end{bmatrix} \tag{8.28}
$$

In this example, as usual for a left-to-right HMM, the process starts only in the
first state s_1 as we choose the π initial probability distribution and we have a discrete
output observation probability for each state s_i, $1 \leq i \leq 4$ so that the output observation probability B is a matrix with four rows(one for each state) and three columns,
where the observed symbol v_k is one of three different coloured balls. The experiment consists of ten trials. The observation probabilities can be continuous densities
too (continuous HMM) and we will present this modeling soon.

After this formal definition we are able to present the three main problems in
HMM modeling [35]:

(i) How does one evaluate the probability of an observed data sequence $\mathbf{O}$ ($\mathbf{O} =$
$[O_1, O_2, ..., O_{10}]$ in the Figure 8.7 of the known model $\lambda = (\pi, A, B)$, or evaluate
$Pr(\mathbf{O}|\lambda)$?

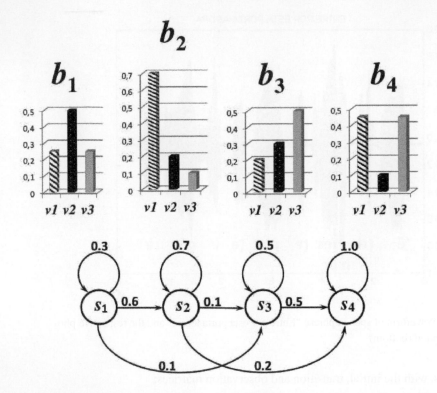

$$\mathbf{O} = [O_1=v_2,\ O_2=v_1,\ O_3=v_1,\ O_4=v_1,\ O_5=v_3,\ O_6=v_3,\ O_7=v_3,\ O_8=v_2,\ O_9=v_1,\ O_{10}=v_1]$$

HIDDEN State Sequence:

$$S_1,\quad S_2,\quad S_2,\quad S_2,\quad S_3,\quad S_4,\quad S_4,\quad S_4,\quad S_4,\quad S_4$$

Figure 8.7 Left-to-right HMM, $N_{st} = 4$, $L_{sym} = 3$, $1 \leq t \leq T = 10$.

(ii) Given the observed sequence $\mathbf{O} = [O_1, O_2, ..., O_{10}]$ and the model $\lambda = (\pi, A, B)$, how does one discover the hidden state sequence?

(iii) How does one estimate the model parameters $\lambda = (\pi, A, B)$ to maximize $Pr(\mathbf{O}|\lambda)$?

The direct form to answer the (i) problem or to evaluate the probability of an observation is the enumeration of all possible hidden state sequences $H = h_1, h_2, ..., h_T$ of length T:

$Pr(\mathbf{O}|\mathbf{H}, \lambda) = b_{h_1}(O_1)b_{h_2}(O_2) \cdots b_{h_T}(O_T)$

and the probability of the sequence of states $\mathbf{H}$, given the model λ is

$Pr(\mathbf{H}|\lambda) = \pi_{h_1}a_{h_1 h_2}a_{h_2 h_3} \cdots a_{h_{t-1} h_T}$

thus, the joint probability for this sequence:

$Pr(\mathbf{O}, \mathbf{H}|\lambda) = Pr(\mathbf{O}|\mathbf{H}, \lambda).Pr(\mathbf{H}|\lambda)$

finally, for all possible sequences the summation over all hidden cases:

$$Pr(\mathbf{O}, \mathbf{H}|\lambda) = \sum_{all\ \mathbf{H}} Pr(\mathbf{O}|\mathbf{H}, \lambda)Pr(\mathbf{H}|\lambda) \qquad (8.29)$$

and starting with $a_{h_0 h_1} = \pi_{h_1}$ we can write

$$Pr(\mathbf{O}, \mathbf{H}|\lambda) = \sum_{all\ \mathbf{H}} \prod_{t=1}^{T} a_{h_{t-1} h_t} b_{h_t}(O_t) \qquad (8.30)$$

for example, in Figure 8.7 the hidden state sequence is

$S = [s_1, s_2, s_2, s_2, s_3, s_4, s_4, s_4, s_4, s_4]$

with output probability

$Pr(\mathbf{O}|\mathbf{S}, \lambda) = b_{s_1}(v_2)b_{s_2}(v_1)b_{s_2}(v_1)b_{s_2}(v_1)b_{s_3}(v_3)b_{s_4}(v_3)b_{s_4}(v_3)b_{s_4}(v_2)b_{s_4}(v_1)b_{s_4}(v_1)$

$Pr(\mathbf{O}|\mathbf{S}, \lambda) = (0.5)(0.7)(0.7)(0.7)(0.5)(0.45)(0.45)(0.1)(0.45)(0.45) = 0.000352$

the state sequence $\mathbf{S}$ probability

$Pr(\mathbf{S}|\lambda) = \pi_{s_1}a_{s_1 s_2}a_{s_2 s_2}a_{s_2 s_2}a_{s_2 s_3}a_{s_3 s_4}a_{s_4 s_4}a_{s_4 s_4}a_{s_4 s_4}a_{s_4 s_4}$

$Pr(\mathbf{S}|\lambda) = (1)(0.6)(0.7)(0.7)(0.1)(0.5)(1)(1)(1)(1) = 0.0147$

the joint probability of $\mathbf{O}$ and $\mathbf{S}$ is the simple product between this two terms:

$Pr(\mathbf{O}, \mathbf{S}|\lambda) = Pr(\mathbf{O}|\mathbf{S}, \lambda).Pr(\mathbf{S}|\lambda)$

$Pr(\mathbf{O}, \mathbf{S}|\lambda) = 0.000352 * 0.0147 = 5.17 * 10^{-6}$

It is easy to note that this direct form of calculation (Eq. 8.30) has a large computational complexity (N states, T observations that results in order of $O(N^T)$ calculations). The well-known algorithm to reduce this complexity is the forward-backward procedure [25].

Let us define the forward $\alpha_t(i)$ (from initial to state s_i in time t) and backward $\beta_t(i)$ (from state s_t to final state s_T) probabilities as:

$$\alpha_t(i) = Pr(O_1 O_2 \cdots O_t, s_t = i|\lambda) \qquad (8.31)$$

$$\beta_t(i) = Pr(O_{t+1} O_{t+2} \cdots O_T, s_t = i|\lambda) \qquad (8.32)$$

Forward Algorithm

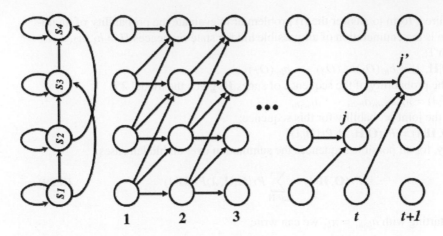

Figure 8.8   Forward procedure example.

Initialization: $\alpha_1(i) = \pi_i b_i(O_1)$, $1 \le i \le N$
Recursion:

$$\alpha_{t+1}(j) = \left[\sum_{i=1}^{N} \alpha_t(i) a_{ij} \right] b_j(O_{t+1}), \quad 1 \le t \le T-1, \ 1 \le j \le N \qquad (8.33)$$

Final:

$$Pr(\mathbf{O}|\lambda) = \sum_{i=1}^{N} \alpha_T(i). \qquad (8.34)$$

The initialization procedure sets up the forward probabilities with the initial probability of each state. Next, in the recursion procedure, Eq. (8.33) shows that state j at time $t + 1$ can be reached by all the possible states i at time t. Provided that $\alpha_t(i)$ is the probability of the joint event that the sequence $O_1 O_2 \cdots O_t$ is observed at time t in state i, the product $\alpha_t(i) a_{ij}$ is the probability that this joint event reaches the state j through state i. Then, summing this factor over all possible states will result in the probability that state j being reached at time $t+1$ through all partial observations. The multiplication by $b_j(O_{t+1})$, the probability that observation O_{t+1} results from state j, updates the probability of the next observation sequence $O_1 O_2 \cdots O_t O_{t+1}$ at time t and state j.

Finally, Eq. (8.34) gives the computation of $Pr(\mathbf{O}|\lambda)$ as the sum of the final forward variables $\alpha_T(i)$ at final states. The complexity order of calculations is $O(N^2 T)$. Figure 8.8 shows all computation times measured in ticks 1, 2, and 3, when possible states are scanned starting in $j = 1$ through $j = 4$. In time t only two possibilities are allowed when state $j = 2$ and on time $t + 1$ three computations are allowed when $j' = 3$.

In fact, for the solution of the first problem, only the forward algorithm is necessary. On the other hand, the third problem, which is the estimation of the parameters for the model $\lambda = (\pi, A, B)$ parameters, requires the backward variables to be solved. We have the following similar procedure to compute the backward probabilities:

Backward Algorithm

Initialization: $\beta_T(j) = 1, \quad 1 \leq j \leq N$
Recursion:

$$\beta_t(i) = \sum_{j=1}^{N} a_{ij} b_j(O_{t+1}) \beta_{t+1}(j), \quad T - 1 \geq t \geq 1 \tag{8.35}$$

Final:

$$Pr(\mathbf{O}|\lambda) = \sum_{i=1}^{N} \pi_i b_i(O_1) \beta_1(i). \tag{8.36}$$

In the first step, Initialization, the last backward variables (β_T) are set up to 1 (arbitrary) for each state $1 \leq i \leq N$. In the Recursion step, since the variables have been evaluated in the state s_j at time t, we need to consider all possibilities of transitions at time $t + 1$. This is why the computations are performed in reverse order, $T - 1, T - 2, ..., t + 1, t, t - 1, ..., 2, 1$ for each possible state $1 \leq j \leq N$.

We must recall the second problem, which seeks the hidden state sequence, and be advised that its solution is not unique due to the optimization criteria chosen [35]. We can join both variables and use the forward-backward procedure to calculate the sum of all possible state sequences probability $(Pr(\mathbf{O}|\lambda))$ but in many practical cases, including speech recognition, the best approach is the search for the state sequence that maximizes the observed sequence, i.e. $\max_{H}[Pr(\mathbf{O}, H|\lambda)]$, known as Viterbi algorithm [40][15].

In HMM the Viterbi algorithm searches the best hidden state sequence $\tilde{H} = \{\tilde{h}_1, \tilde{h}_2, ..., \tilde{h}_T\}$ for the given observation sequence $O = O_1, O_2, ..., O_T$, using an auxiliary quantity

$$\delta_t(i) = \max_{h_1 h_2 \cdots h_T} Pr[h_1 h_2 \cdots h_T = i, O_1 O_2 \cdots O_T|\lambda] \tag{8.37}$$

that save the highest probability of a single path H respect to observations $O_1 O_2 \cdots O_t$, at time t, until the state h_i. To recover the path, the other auxiliary array $\psi_t(j)$ saves the argument of each maximum probability.

Viterbi algorithm

Initialization:

$$\delta_1(i) = \pi_i b_i(O_1), \quad 1 \leq i \leq N$$
$$\psi_1(i) = 0 \tag{8.38}$$

Recursion:

$$\delta_t(j) = \max_{1 \le i \le N}[\delta_{t-1}(i)a_{ij}]b_j(O_t), \quad 2 \le t \le T$$

$$1 \le j \le N \tag{8.39}$$

$$\psi_t(j) = \arg\max_{1 \le i \le N}[\delta_{t-1}(i)a_{ij}], \quad 2 \le t \le T$$

$$1 \le j \le N \tag{8.40}$$

Final:

$$\tilde{Pr} = \max_{1 \le j \le N}[\delta_T(i)]$$

$$\tilde{H}_T = \arg\max_{1 \le i \le N}[\delta_T(i)]. \tag{8.41}$$

Hidden state revealed (backtracking):

$$\tilde{h}_t = \psi_{t+1}(\tilde{h}_{t+1}), \quad t = T-1, T-2, \cdots, 1. \tag{8.42}$$

At this point, before the presentation of the final algorithm that solves the HMM third problem (how to estimate the model $\lambda = (\pi, A, B)$), we will resume the discussion about discrete HMM modeling, as in the example shown in Figure 8.7) and (8.28 and the continuous HMM modeling approach, mainly when the observed processes are continuous variables and require a more accurate statistical modeling. Even though it is possible to quantize the parameters, in some applications like continuous speech language processing (streaming video, broadcasting TV and radio), there might be great degradation in performance if quantization is applied. That is why probability density function (pdf) modeling for the observation processes at the output is usual and the common solution is a mixture of Gaussian pdfs, because conveniently selected Gaussian mixture densities can approximate any probability function in the sense of a measure of error between pdfs.

Formally, a continuous HMM with mixture Gaussian pdfs of observation $\mathbf{O}$ in the j-th state is given by

$$b_j(\mathbf{O}) = \sum_{k=1}^{M} c_{jk}b_{jk}(\mathbf{O})$$

$$= \sum_{k=1}^{M} c_{jk}\mathcal{N}(\mathbf{O}|\mu_{jk}, \Sigma_{jk}) \tag{8.43}$$

where $\mathcal{N}(O|\mu, \Sigma)$ is a D-dimensional Gaussian pdf of observation O with mean vector μ and covariance matrix Σ:

$$\mathcal{N}(\mathbf{O}|\mu, \Sigma) = \frac{1}{(2\pi)^{D/2}|\Sigma|^{1/2}} \exp\{-\frac{1}{2}(O-\mu)'\Sigma^{-1}(O-\mu)\} \tag{8.44}$$

where D is the dimension of observation vector O, M is the number of mixture components and c_{jk} are the weight for the k-th mixture with the constraints

$$\sum_{k=1}^{M} c_{jk} = 1. \tag{8.45}$$

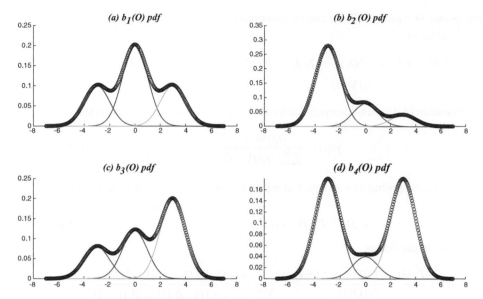

Figure 8.9 Univariate weighted pdfs as a sum of Gaussian k-Mixtures, $k = 1, 2, 3$: (a) state s_1 weights $c_{1k} = [0.25, 0.5, 0.25]$; (b) state s_2, $c_{2k} = [0.7, 0.2, 0.1]$; (c) state s_3, $c_{3k} = [0.3, 0.2, 0.5]$ and (d) state s_4, $c_{4k} = [0.45, 0.1, 0.45]$.

Thus,

$$\int_{-\infty}^{\infty} b_j(\mathbf{O})d\mathbf{O} = 1. \tag{8.46}$$

Figure 8.9 shows the continuous versions of the pdfs (circled resulting pdf and continuous HMM) related at the states $s_1, ..., s_4$ shown in the Figure 8.7 (discrete HMM), where we used three Gaussian mixtures (with means $\mu_1 = -3, \mu_2 = 0, \mu_3 = 3$ and variances $\sigma_1 = \sigma_2 = \sigma_3 = 1$) and a univariate case:

$$\begin{aligned}
b_1(\mathbf{O}) &= 0.25 * \mathcal{N}(\mathbf{O}| -3, 1) + 0.5 * \mathcal{N}(\mathbf{O}|0, 1) + 0.25 * \mathcal{N}(\mathbf{O}|3, 1) \\
b_2(\mathbf{O}) &= 0.7 * \mathcal{N}(\mathbf{O}| -3, 1) + 0.2 * \mathcal{N}(\mathbf{O}|0, 1) + 0.1 * \mathcal{N}(\mathbf{O}|3, 1) \\
b_3(\mathbf{O}) &= 0.3 * \mathcal{N}(\mathbf{O}| -3, 1) + 0.2 * \mathcal{N}(\mathbf{O}|0, 1) + 0.5 * \mathcal{N}(\mathbf{O}|3, 1) \\
b_4(\mathbf{O}) &= 0.45 * \mathcal{N}(\mathbf{O}| -3, 1) + 0.1 * \mathcal{N}(\mathbf{O}|0, 1) + 0.45 * \mathcal{N}(\mathbf{O}|3, 1).
\end{aligned} \tag{8.47}$$

Now, we can present the Baum-Welch algorithm [35] that is a version of Expectation-Maximization (EM) algorithm for HMM [8][17].

Denoting the probability of being in hidden state i at time t for the state sequence $\mathbf{O}$ as:

$$\gamma_i(t) \triangleq Pr(H_t = i|\mathbf{O}, \lambda) \tag{8.48}$$

and manipulating conditional probabilities as

$$Pr(H_t = i|\mathbf{O}, \lambda) = \frac{Pr(\mathbf{O}, H_t = i|\lambda)}{Pr(\mathbf{O}|\lambda)} = \frac{Pr(\mathbf{O}, H_t = i|\lambda)}{\sum_{j=1}^{N} Pr(\mathbf{O}, H_t = j|\lambda)}, \tag{8.49}$$

the numerator in equation (8.49) can be computed by forward (8.31) and backward (8.32) variables as

$$
\begin{aligned}
Pr(\mathbf{O}, H_t = i|\lambda) &= Pr(O_1 \cdots O_t, H_t = i|\lambda)Pr(O_{t+1} \cdots O_T, H_t = i|\lambda) \\
&= \alpha_i(t)\beta_i(t)
\end{aligned}
\tag{8.50}
$$

so that Equation (8.48) can be expressed by forward and backward probabilities as

$$
\gamma_i(t) = \frac{\alpha_i(t)\beta_i(t)}{\sum_{j=1}^{N} \alpha_j(t)\beta_j(t)}.
\tag{8.51}
$$

We also define the probability of being in state i at time t and being in state j at time $t + 1$ by

$$
\xi_{ij}(t) \overset{\Delta}{=} Pr(H_t = i, H_{t+1} = j|\mathbf{O}, \lambda)
\tag{8.52}
$$

which can be expanded as

$$
\xi_{ij}(t) = \frac{Pr(H_t = i, H_{t+1} = j, \mathbf{O}|\lambda)}{Pr(\mathbf{O}|\lambda)} = \frac{\alpha_i(t)a_{ij}b_j(O_{t+1})\beta_j(t + 1)}{\sum_{i=1}^{N}\sum_{j=1}^{N} \alpha_i(t)a_{ij}b_j(O_{t+1})\beta_j(t + 1)}
$$

or as

$$
\begin{aligned}
\xi_{ij}(t) &= \frac{Pr(H_t = i|\mathbf{O})Pr(O_{t+1} \cdots O_T, H_{t+1} = j|H_t = i, \lambda)}{Pr(O_{t+1} \cdots O_T|H_t = i, \lambda)} \\
&= \frac{\gamma_i(t)a_{ij}b_j(O_{t+1})\beta_j(t + 1)}{\beta_i(t)}
\end{aligned}
$$

Using the definitions in (8.48) and (8.52) we can relate $\gamma_i(t)$ to ξ_{ij} by summation over j, given

$$
\gamma_i(t) = \sum_{j=1}^{N} \xi_{ij}(t)
\tag{8.53}
$$

and if we sum these quantities over across time, results in useful values:

$$
\sum_{t=1}^{T-1} \gamma_i(t) = \text{expected number of times in state } i
\tag{8.54}
$$

$$
= \text{expected number of transitions away from state } i \text{ for } \mathbf{O}, \text{ and:}
$$

$$
\sum_{t=1}^{T-1} \xi_{ij}(t) = \text{expected number of transitions from state } i \text{ to state } j \text{ for } \mathbf{O}.
\tag{8.55}
$$

Actually, a good form to see these results is the use of indicators random variables:

$$
I_t(i) = \begin{cases} 1 & \text{if we are in state } i \text{ at time } t \\ 0 & \text{otherwise} \end{cases}
\tag{8.56}
$$

and

$$I_t(i, j) = \begin{cases} 1 & \text{when we move from state } i \text{ to state } j \text{ after time } t \\ 0 & \text{otherwise} \end{cases} \qquad (8.57)$$

that can represent the latent (hidden) random variables of the stochastic process with the transitions probabilities given by the transition matrix A.[1]. Straightaway, the sums over time are the expectations of $I_t(i)$ and $I_t(i, j)$:

$$\sum_{t=1}^{T-1} \gamma_i(t) = \sum_{t=1}^{T-1} E[I_t(i)] = E\left[\sum_{t=1}^{T-1} I_t(i)\right]$$
$$\sum_{t=1}^{T-1} \xi_{ij}(t) = \sum_{t=1}^{T-1} E[I_t(i, j)] = E\left[\sum_{t=1}^{T-1} I_t(i, j)\right].$$

Now, we can use the definitions (8.48) and (8.52) with these expectations (8.54) and (8.55) to obtain the estimations of HMM parameters [8] $\lambda = (\pi, A, B)$.

First, the expected initial frequency (or at time $t = 1$) in state i:

$$\hat{\pi}_i = \gamma_i(1). \qquad (8.58)$$

Second, the expected number of transitions from state i to state j, normalized by expected total number of transitions away from state i:

$$\hat{a}_{ij} = \frac{\sum_{t=1}^{T-1} \xi_{ij}(t)}{\sum_{t=1}^{T} \gamma_i(t)}. \qquad (8.59)$$

Finally, for discrete HMMs, the expected number of times that the output observation at time t have been equal to v_k through state j, normalized by expected total number of times visiting state j (the $\delta_{x,y}$ is the Kronecker delta):

$$\hat{b}_j(k) = \frac{\sum_{t=1}^{T} \delta_{O_t, v_k} \gamma_j(t)}{\sum_{t=1}^{T} \gamma_j(t)} \qquad (8.60)$$

The proof for these equations in the solution of estimation problem (the third HMM problem) can be found in [35] or [17].

For continuous HMMs (with Gaussian mixtures), we can define the probability that the k-th component of the i-th mixture and observed observation O_t:

$$\gamma_{ik}(t) \equiv \gamma_i(t) \frac{c_{ik} b_{ik}(O_t)}{b_i(O_t)} = Pr(H_t = i, Z_{it} = k | \mathbf{O}, \lambda) \qquad (8.61)$$

with the random variable Z_{it} indicating the mixture component at time t for state i. Using the probability (8.61), and the notation of (8.43) we have the estimation formulas for each Observation sequence of size T modeled by continuous HMM:

$$c_{ik} = \frac{\sum_{t=1}^{T} \gamma_{ik}(t)}{\sum_{t=1}^{T} \gamma_i(t)} \qquad 1 \le i \le N, \text{ } k\text{-th mixture weights} \qquad (8.62)$$

---

[1]$A_{ij} \equiv Pr[I_t(i) = 1 | I_{t+1}(j) = 1]$

$$\mu_{ik} = \frac{\sum_{t=1}^{T} \gamma_{ik}(t) O_t}{\sum_{t=1}^{T} \gamma_{ik}(t)} \qquad 1 \leq i \leq N, \text{ } k\text{-th mixture means} \qquad (8.63)$$

$$\Sigma_{ik} = \frac{\sum_{t=1}^{T} \gamma_{ik}(t)(O_t - \mu_{ik})(O_t - \mu_{ik})'}{\sum_{t=1}^{T} \gamma_{ik}(t)} \qquad 1 \leq i \leq N, \text{ } k\text{-th mixture covariance} \quad (8.64)$$

At this point, using the indicator random variables (8.56), we can represent the HMM as a Markov chain of latent (hidden) variables, working with sequential observable data. The latent $I_t(i)$ variables form a Markov chain, giving rise to the graphical probabilistic representation known as *state space model* [9], shown on top of Figure 8.10a, where the uncolored states represents latent (hidden) variables and colored states represents observable variables, the output sequences **O** $= O_1, O_2, ..., O_{t-1}, O_t, O_{t+1}$. As a matter of fact, this figure represents the random variable in circled states with double line format and this change in notation from previous figures, where we used solid lines to represent states of our model, is meant to highlight the fact that they are not really random variables, but only realizations of random variables. Using the latent indicator random variable I_t connected with observable O_t the probabilistic stochastic model for an HMM is complete.

The advantage of this graphical representation of HMM is illustrated in Figure 8.10b, where an extension of an HMM with a correlation of observable random variables is shown, the autoregressive hidden Markov model (ARHMM) [9], that has applications in speech enhancement and recognition [14]. In order to emphasize that we need additional processes, in this bottom figure the correlations between observable variables are shown in dotted lines along with Baum-Welch re-estimation equations (8.59), (8.60), (8.58), (8.62), (8.63), and (8.64) to achieve the estimation of the ARHMM (see, for example, [35] and [14]).

In an HMM only the observable process can be continuous or discrete, and the latent process must be discrete. Put another way, this graphical model representation to sequential data has a clear connection to *Linear Dynamical Systems* [9] where both stochastic processes (latent and observable) could be continuous so that it could be used in modeling other more complicated problems.

8.6.2 GAUSSIAN MIXTURES MODELS

In the previous section we discussed how to model temporal evolution of the speech using hidden Markov models (Markov chains) and both discrete and continuous probabilistic distributions to the observed data are possible. However, there are some applications where the temporal structure of the speech signal doesn't matter, for example in speaker recognition, both identification and verification. For these applications, Gaussian Mixtures Models (GMM), stochastic models without temporal evolution, have achieved good performance.

The study of finite mixtures of probabilities has a long history, since the middle of the 18th century . Mclachlan [29] presents a historical development of these studies

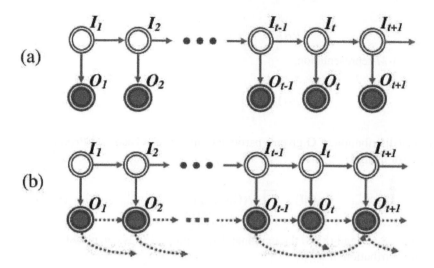

Figure 8.10 State space model as a representation of an HMM. (a) Top - Sequential data of a Markov chain of latent (hidden) variables I_t (uncolored) with observations variables O_t (colored) connected one-to-one to a latent variable. (b) Bottom - Autoregressive correlations between observable random variables (ARHMM) shown in dotted lines, using this graphical probabilistic model.

and the paper by Dempster [12] was a milestone in the evolution of the computational solution for the parameters of this representation using the EM algorithm.

The formal definition of a GMM was done by equations (8.43) through (8.46) because in a continuous HMM the observed output probability of each hidden state can be modelled by a GMM. Then, if we fix an HMM model with a single state, it ends up in a GMM and, in order to reach this result in a formal way, we follow Dempster's [12] scheme.

First, remember the probability of the Eq. (8.43), cutting the j subscript index, we have the probability of the M Gaussian mixtures:

$$Pr(\mathbf{O}) = \sum_{k=1}^{M} c_k \mathcal{N}(\mathbf{O}|\mu_k, \Sigma_k) \tag{8.65}$$

and

$$\sum_{k=1}^{M} c_k = 1, \qquad 0 \le c_k \le 1 \tag{8.66}$$

Even in the absence of the Markovian state sequence we will continue to use the latent variables to support the observable process, introducing the M-dimensional latent (hidden) binary random variable $\mathbf{I}$, an indicator variable, i.e., only one value equal to 1 and all other 0, therefore, $i_k \in \{0, 1\}$ and $\sum_k i_k = 1$. The marginal distribution over $\mathbf{I}$ is given by the mixing coefficients:

$$Pr(i_k = 1) = c_k$$

or, using the 1-of-M representation:

$$Pr(\mathbf{I}) = \prod_{k=1}^{M} c_k^{i_k}. \tag{8.67}$$

The conditional distribution of $\mathbf{O}$ given a particular value i_k is a Gaussian $Pr(\mathbf{O}|i_k = 1) = \mathcal{N}(\mu_k, \Sigma_k)$ or using $\mathbf{I}$:

$$Pr(\mathbf{O}|\mathbf{I}) = \prod_{k=1}^{M} \mathcal{N}(\mathbf{O}|\mu_k, \Sigma_k)^{i_k}. \tag{8.68}$$

Using equations (8.67) and (8.68) the marginal distribution of $\mathbf{O}$ is calculated summing the joint distribution of $\mathbf{I}$:

$$Pr(\mathbf{O}) = \sum_{k} Pr(\mathbf{I})Pr(\mathbf{O}|\mathbf{I}) = \sum_{k=1}^{M} \left[\prod_{k=1}^{M} c_k^{i_k} \right] \left[\prod_{k=1}^{M} \mathcal{N}(\mathbf{O}|\mu_k, \Sigma_k)^{i_k} \right]$$

$$= \prod_{k=1}^{M} c_k \mathcal{N}(\mathbf{O}|\mu_k, \Sigma_k)$$

as our definition in Eq. (8.65). The graphical representation of a mixture model is illustrated in Figure 8.11a with a single mixture and in Figure 8.11b with M-Gaussian mixtures, weights c_k, means μ_k and covariances matrices Σ_k, $1 \leq k \leq M$. Comparing this Figure 8.11a with the graphical representation of HMM in Figure 8.10a we notice that an HMM can be viewed as a Markovian time sequence of GMM (the more accurate representation would be a sequence of Figure 8.11b but the reader can easily understand this simplification to represent the HMM state sequence model).

The basic idea of a GMM was illustrated in the previous section in Figure 8.9 and one strategy to view the *play the role* of the each Gaussian in the mixture is to compute the *responsibility* quantity [9]:

$$\gamma(I_k) \equiv Pr(I_k = 1|\mathbf{O}) = \frac{Pr(I_k = 1)Pr(\mathbf{O}|I_k = 1)}{\sum_{j=1}^{M} Pr(I_j = 1)Pr(\mathbf{O}|I_j = 1)}$$

$$= \frac{c_k \mathcal{N}(\mathbf{O}|\mu_k, \Sigma_k)}{\sum_{j=1}^{M} \mathcal{N}(\mathbf{O}\mu_j, \Sigma_j)}. \tag{8.69}$$

In Figure 8.12a we have an override plot of three different Gaussian data with 200 samples in each one. After mixing the data we have a plot in Figure 8.12b, with 600 samples. Using the responsibilities of Eq. (8.69) to set the color of each plotted data point (proportion of R,G,B according to the responsibility of each $k = 1, 2, 3$ mixture) we have the result of Figure 8.12c.

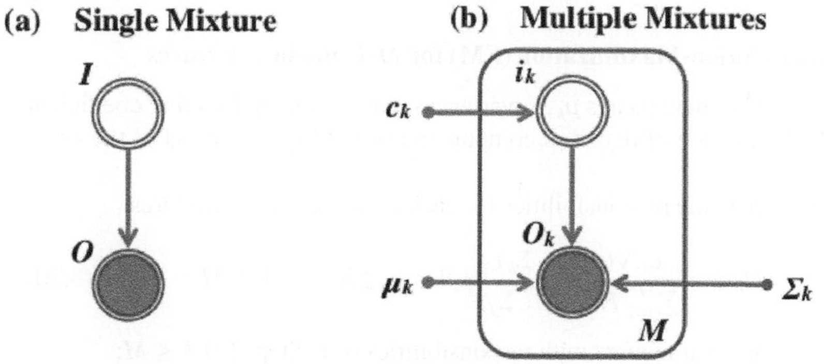

(a) Single Mixture   **(b) Multiple Mixtures**

Figure 8.11 Graphical model representation of Gaussian Mixtures: (a) A simple mixture representation of joint probability $Pr(\mathbf{O}, \mathbf{I}) = Pr(\mathbf{I})Pr(\mathbf{O}|\mathbf{I})$ (white circle latent variable, colored circle observable variable); (b) Graphical Gaussian Mixture Model (GMM) representation of Eq. (8.65) for a set of M mixtures, observations $\{O_k\}$, and latent variables $\{i_k\}$, $k = 1, 2, ..., M$.

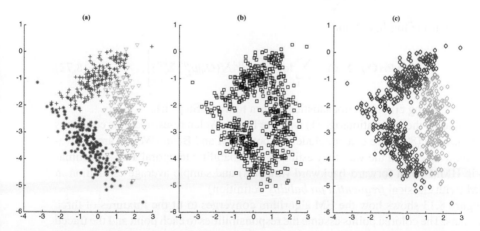

Figure 8.12 Bi-dimensional Gaussian mixtures. (a) Samples of three bi-dimensional Gaussians, 3 times 200 samples with +, ∗ or ∇ marks;(b) 600 mixed samples of a bi-dimensional 3-Gaussian mixture; (c) Identification of the 600 mixed samples, where each one was colored according to the computed responsibilities.

Expectation-Maximization (EM) for M-Gaussian mixtures

(i) Initialization: Compute means μ_k, covariances matrices Σ_k and mixing coefficient c_k at each M-th partition of data and compute the initial log likelihood of the complete N data.

(ii) **E-Step.** Compute the responsibilities for each data and for all mixtures:

$$\gamma(I_{nk}) = \frac{c_k \mathcal{N}(O_n|\mu_k, \Sigma_k)}{\sum_{j=1}^{M} \mathcal{N}(O_n|\mu_j, \Sigma_j)} \quad 1 \le n \le N, \ 1 \le k \le M \tag{8.70}$$

(iii) **M-Step.** Update parameters with responsibilities of E-Step, $1 \le k \le M$:

$$N_{tot}^{k^{th}} = \sum_{n=1}^{N} \gamma(I_{nk})$$

$$\mu_k^{up} = \frac{\sum_{n=1}^{N} \gamma(I_{nk})O_n}{N_{tot}^{k^{th}}}$$

$$\Sigma_k^{up} = \frac{\sum_{n=1}^{N} \gamma(I_{nk})(O_n - \mu_k^{up})(O_n - \mu_k^{up})'}{N_{tot}^{k^{th}}} \tag{8.71}$$

$$c_k^{up} = \frac{N_{tot}^{k^{th}}}{N}$$

(iv) Compute the log-likelihood

$$\log Pr(\mathbf{O}|\mu, \Sigma, \mathbf{c}) = \sum_{n=1}^{N} \log \left[\sum_{k=1}^{M} c_k^{up} \mathcal{N}(O_n|\mu_k^{up}, \Sigma_k^{up}) \right]. \tag{8.72}$$

If the stopping criteria is not satisfied, return to the E-Step (ii).

Comparing these equations (8.71) with the Baum-Welch re-estimations equations (8.62) through (8.64) we can conclude that both (EM and Baum-Welch) are equivalent, differing only in the way they estimate γ probability: temporal average estimation in HMM (with forward-backward variables) and sample average estimation in GMM (with classical *frequentist probability* definition).

Figure 8.13 shows how the EM algorithm converges to fit the mixtures of three Gaussians. The colored points assume the responsibilities to each point and the traced lines represent the adjusted Gaussian contour after 10, 20, 30, and 40 iterations; after this the log-likelihood stabilizes at a maximum value. The initialization procedure was started with random local means and diagonal unitary covariance matrices.

For some speaker recognition applications we can use pre-trained models called Universal Background Models (UBM) [37] estimated by EM algorithm over a known database and after using a maximum a posteriori (MAP) procedure to adapt the model to the new collected data set.

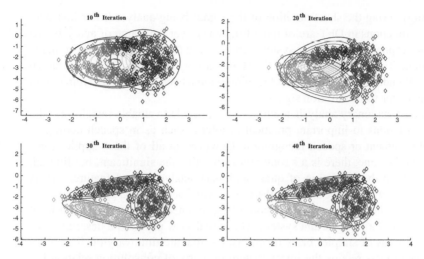

Figure 8.13 Example of EM convergence for 3-mixture bi-variate Gaussians adjusted in a 600 samples data set. The solid lines represent contour of each Gaussian adjusted after the number of iterations shown on top.

8.7   TIME-VARYING MODELS

It is but natural that speech signals being meant for effective information transfer are highly time-varying in their characteristics; i.e., signal parameters change with time, to enable representation of a variety of symbols for information communication. This is in contrast to the oscillations of the signal itself, which are necessary for the transfer of energy required for the expression of each symbol. Several mathematical tools in engineering, both deterministic and stochastic, such as the Fourier transform or the probability density function, have been developed to characterize the asymptotic properties of a signal or a process. However, to analyze information rich signals, such as speech or even music, we recognize they are time-varying or non-stationary in their properties (signal properties), but we do not know where the symbols begin or end and in some cases even what the symbols are! That is, the signals are not only continuous, but they are a continuum at the information representation level also (symbols); for instance, one phoneme in speech blends into the adjoining phonemes, providing a smooth perception of different sounds, referred to as coarticulation. Mostly, the units of information also vary in duration causing major challenges for detection, estimation, or any processing of speech signals, without causing artifacts.

The quasi-stationary, or short-time, analysis that is the "bread and butter" of speech processing and such other non-stationary signals, viz, in SONAR, RADAR, etc. Even spatial signals such as Image/Video are analyzed in a "short-time", i.e., blocked manner. A lot of modern digital signal processing techniques (epitomized by the DFT) have grown and matured to enable quasi-stationary analysis of signals. In quasi-stationary analysis, the signal parameters are assumed to be constant (not time-varying) during the short duration of the signal being analyzed. For instance, (i) the basis functions in DFT are of fixed frequency over the frame of analysis, (ii) the impulse response of the linear system representing the quasi-stationary signal is time-invariant. This is mainly because the stationary or time-invariant mathematical tools are well understood and efficient techniques have been developed to implement the techniques for short-duration signals.

These quasi-stationary analysis methods have provided valuable insight into the signal and solutions to important practical problems such as in speech coding or speech enhancement or speech recognition. However, in all of these applications, especially the last one, there is a strong realization that the significant, but limited, success is due to the limitation of quasi-stationary analysis-based feature vectors since they are later post-processed for the gradient and acceleration parameters.

The time-varying model of speech could be examined either at the signal level or as the output of a time-variant system. The signal view provides a description of the signal in terms of simpler basis functions, but that are time-varying. The time-varying system view retains the linear system property of convolution relation between a certain input and the system function, but the system function itself is considered time-varying. These issues have been addressed in the literature sparingly, but we can consider them to be important for any further breakthrough in speech parameterization and more effective speech recognition, coding, or enhancement. Parametric speech synthesis is also an important application that would benefit from better speech parameterization models, although much of its present success is realized through waveform concatenation techniques. A couple of examples of time-varying signal models in the literature are: (i) AM-FM decomposition by the Hilbert transform approach [31] or the discrete energy separation algorithm (DESA) [26], (ii) modulation signal decomposition [11], and (iii) harmonic AM-FM decomposition [21]. Similarly, examples of time-varying system models are: (i) time-varying linear prediction [16] and (ii) time-varying lattice filter [38, 22].

The harmonic AM-FM model [21] synthesizes the speech signal as the harmonic combination

$$s(t) = \sum_{k=1}^{K} \Re \{a_k(t) \exp(jk\phi_0(t))\},$$ (8.73)

where $a_k(t)$ is the instantaneous complex amplitude function for the kth harmonic component, K is the number of harmonic components, and $\phi_0(t)$ is the instantaneous phase function, given by

$$\phi_0(t) = 2\pi \int_0^t f_0(\tau)d\tau$$ (8.74)

where $f_0(t)$ is the instantaneous fundamental frequency function.

Alternatively, a bandpass AM-FM model [32, 33, 39] reconstructs the speech signal as the combination of resonant signals

$$s(t) = \sum_{k=1}^{M} \Re \{a_k(t) \exp(j\phi_k(t))\},$$ (8.75)

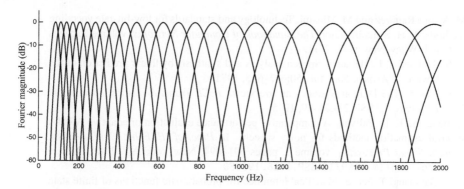

Figure 8.14 Magnitude frequency responses of the mel-spaced Gabor filterbank used for Multibank Demodulation Analysis (MDA).

where $a_k(t)$ is the instantaneous complex amplitude function for the kth bandpass component, M is the number of formants, and $\phi_k(t)$ is the instantaneous phase function for the kth bandpass component, given by

$$\phi_k(t) = 2\pi \left(f_{ck}t + \int_0^t q_k(\tau)d\tau \right) \tag{8.76}$$

where f_{ck} is the central formant frequency and $q_k(t)$ is the frequency modulating signal for the kth bandpass component. The frequency responses of the mel-spaced Gabor filterbank used for Multibank Demodulation Analysis (MDA) in the AM-FM representation described in [32] are shown in Figure 8.14.

The previous models can be considered deterministic models of speech representation. Among the stochastic models of speech are the commonly used GMM (Gaussian mixture model) and HMM (hidden Markov model), extensively applied in speech recognition as presented in Section 8.6, and also Kalman filtering which is a time-varying linear stochastic model, which has been formulated for speech enhancement [6].

8.8 ACKNOWLEDGMENT

The authors acknowledge the support received from Fundação de Amparo à Pesquisa do Estado de São Paulo (FAPESP) under Grant no. 2012/24789-0.

REFERENCES

1. J. Allen, M. S. Hunnicutt, and D. Klatt. *From text to speech: MITalk system*. Cambridge University Press, Cambridge, 1987.
2. M. Arjona Ramírez and M. Minami. Low bit rate speech coding. In J. G. Proakis, editor, *Wiley Encyclopedia of Telecommunications*, volume 3, pages 1299–1308. Wiley, New York, 2003.

3. M. Arjona Ramírez and M. Minami. Technology and standards for low-bit-rate vocoding methods. In H. Bidgoli, editor, *The Handbook of Computer Networks*, volume 2, pages 447–467. Wiley, New York, 2011.

4. B. S. Atal and S. L. Hanauer. Speech analysis and synthesis by linear prediction of the speech wave. *J. Acoust. Soc. Am.*, 50:637–655, 1971.

5. J. K. Baker. The Dragon System - an overview. *IEEE Trans. Acoustics, Speech and Signal Process.*, (ASSP-23):24–29, 1975.

6. A. Basu and K. K. Paliwal. A comparative performance evaluation of adaptive ARMA spectral estimation methods for noisy speech. In *Proc. of IEEE Int. Conf. Acoust., Speech, Signal Processing*, volume 1, pages 691–694, New York, NY, 1988. http://dx.doi.org/10.1109/ICASSP.1988.196680.

7. L. E. Baum and T. Petrie. Statistical inference for probabilistic functions of finite state Markov chains. *Ann. Math. Stat.*, (37):1554–1563, 1966.

8. J. A. Bilmes. A gentle tutorial of the EM algorithm and its application to parameter estimation for Gaussian Mixture and Hidden Markov Models. Technical report, International Computer Science Institute, U. C. Berkeley, April 1998.

9. C. M. Bishop. *Pattern Recognition and Machine Learning*. Springer, 2006.

10. J. P. Burg. A new analysis technique for time series data. In *Proc. NATO Advanced Institute on Signal Proc.*, Enschede, Netherlands, 1968. NATO.

11. P. Clark and L. Atlas. A sum-of-products model for effective coherent modulation filtering. In *Proc. of IEEE Int. Conf. Acoust., Speech, Signal Processing*, pages 4485–4488, Taipei, Taiwan, 2009. http://dx.doi.org/10.1109/ICASSP.2009.4960626.

12. A. P. Dempster, N. M. Laird, and D. B. Rubin. Maximum likelihood from incomplete data via the EM algorithm. *Journal of the Royal Statistical Society*, 39(1):1–38, 1977.

13. H. W. Dudley. Remaking speech. *J. Acoust. Soc. Am.*, 11(2):169–177, Oct. 1939.

14. Y. Ephraim and W. Roberts. Revisiting autoregressive Hidden Markov modeling of speech signals. *IEEE Signal Processing Letters*, 12(2):166–169, february 2005.

15. G. D. Forney. The Viterbi algorithm. *Proceedings of the IEEE*, 61:268–278, March 1973.

16. M. G. Hall, A. V. Oppenheim, and A. S. Willsky. Time-varying parametric modeling of speech. *Signal Process.*, 5:267–285, 1983.

17. X. D. Huang, Y. Akiki, and M. A. Jack. *Hidden Markov Models for Speech Recognition*. Edinburgh University Press, Edinburgh, Great Britain, 1990.

18. F. Itakura. Line spectrum representation of linear prediction coefficients of speech signals. *J. Acoust. Soc. Am.*, 57:535, 1975.

19. F. Itakura and S. Saito. Analysis-synthesis telephony based on the maximum likelihood method. In Y. Kohasi, editor, *Proc. 6th International Congress Acoustics*, pages C–17–C–20, Tokyo, Aug 1968. International Union of Pure and Applied Physics of UNESCO, Maruzen, Elsevier. Paper C-5-5.

20. F. Jelinek. Continuous speech recognition by statistical methods. *IEEE Proc.*, (64):532–556, 1976.

21. G. P. Kafentzis1, G. Degottex, O. Rosec, and Y. Stylianou. Time-scale modifications based on a full-band adaptive harmonic model. In *Proc. of IEEE Int. Conf. Acoust., Speech, Signal Processing*, pages 8193–8197, Vancouver, BC, Canada, 2013. http://dx.doi.org/10.1109/ICASSP.2013.6639262.

22. E. Karlsson and M. H. Hayes. Least squares ARMA modeling of linear time-varying systems: Lattice filter structures and fast RLS algorithms. ASSP-35(7):994–1014, July 1987. http://dx.doi.org/10.1109/TASSP.1987.1165246.

23. W. Koenig, H. K. Dunn, and L. Y. Lacey. The sound spectrograph. *J. Acoust. Soc. Am.*, 18:19–49, 1946.

24. K.-F. Lee. *Automatic Speech Recognition: The Development of the SPHINX System*. The Springer International Series in Engineering and Computer Science. Kluwer Academic Publishers, 1989.

25. S. E. Levinson, L. R. Rabiner, and M. M. Sondhi. An introduction to the application of the theory of probabilistic functions of a Markov process to automatic speech recognition. *Bell Sys. Tech. Journal*, (62):1035–1074, 1983.

26. P. Maragos, J. F. Kaiser, and T. F. Quatieri. Energy separation in signal modulations with application to speech analysis. 41(10):3024–3051, Oct. 1993.

27. J. D. Markel and A. H. Gray,Jr. *Linear Prediction of Speech*. Springer, Berlin, 1976.

28. R. J. McAulay and T. F. Quatieri. Sinusoidal coding. In W. Bastiaan Kleijn and K. K. Paliwal, editors, *Speech Coding and Synthesis*, pages 121–173. Elsevier Science, Amsterdam, 1995.

29. G. McLachlan and D. Peel. *Finite Mixture Models*. John Wiley and Sons, 2000.

30. P. Mermelstein. Articulatory model for the study of speech production. *J. Acoust. Soc. Am.*, 53(4):1070–1082, 1973.

31. A. Potamianos and P. Maragos. A comparison of the energy operator and the hilbert transform approach to signal and speech demodulation. *Signal Process.*, 37(1):95–120, 1994.

32. A. Potamianos and P. Maragos. Speech formant frequency and bandwidth tracking using multiband energy demodulation. *Journal of the Acoust. Soc. of America*, (6):3795–3806, June 1996.

33. A. Potamianos and P. Maragos. Speech analysis and synthesis using an AM-FM modulation model. *Speech Communication*, 28:195–209, 1999.

34. R. K. Potter, G. A. Kopp, and H. C. Green. *Visible speech*. D, Van Nostrand Co., New York, 1947.

35. L. R. Rabiner. A tutorial on Hidden Markov Models and selected applications in speech recognition. *Proc. of the IEEE*, 77(2):257–286, 1989.

36. L. R. Rabiner and R. W. Schafer. *Theory and Applications of Digital Speech Processing*. Pearson Higher Education, Upper Saddle River, NJ 07458, 2011.

37. D. A. Reynolds, T. F. Quatieri, and R. B. Dunn. Speaker verification using adapted Gaussian mixture models. *Digital Signal Processing*, 10(1):19–41, 2000.

38. K. Schnell and A. Lacroix. Model-based analysis of speech and audio signals for real-time processing based on time-varying lattice filters. In *Proc. of IEEE Int. Conf. Acoust., Speech, Signal Processing*, pages 3973–3976, Taipei, Taiwan, 2009. http://dx.doi.org/10.1109/ICASSP.2009.4960498.

39. S. C. Sekhar and T. V. Sreenivas. Novel approach to AM-FM decomposition with applications to speech and music analysis. In *Proc. of IEEE Int. Conf. Acoust., Speech, Signal Processing*, volume 2, pages II–753–II–756, Montreal, Quebec, Canada, 2004. http://dx.doi.org/10.1109/ICASSP.2004.1326367.

40. A. J. Viterbi. Error bounds for convolutional codes and an asymptotically optimal decoding algorithm. *IEEE Trans. Information Theory*, 13:260–269, April 1967.

Image Processing

Ricardo Lopes de Queiroz

9 Energy-Aware Video Compression

Ricardo Lopes de Queiroz
University of Brasília (UnB)

Tiago Alves da Fonseca
University of Brasília (UnB)

Over the past years, multimedia communication technologies have demanded higher computing power availability and, therefore, higher energy consumption. In order to meet the challenge to provide software-based video encoding solutions with reduced consumption, we adopted a software implementation of a state-of-the-art video encoding standard and optimized its implementation in the energy (E) sense. Thus, besides looking for the coding options which lead to the best fidelity in a rate-distortion (RD) sense, we constrain the video encoding process to fit within a certain energy budget, i.e., an RDE optimization. We considered energy by integrating power measurements from the system power supply unit.

We present an RDE-optimized framework which allows for software-based real-time video compression, meeting the desired targets of electrical consumption, hence, controlling carbon emissions. The system can be made adaptive, dynamically tracking changes in image contents and in energy demands. We show results of real-time energy-constrained compression wherein one can save as much as 31% of the power consumption with a small impact on RD performance.

9.1 INTRODUCTION

Historically, processor manufacturers have responded to the demand for more processing power primarily with faster processor speeds. Higher clock speeds imply higher power consumption and heat [34]. Image and video processing are driving forces behind this computational power pursuit. The *state-of-the-art* video compression standard, H.264/AVC [20],[31],[10], is a computation-hungry application used throughout the industry. Nevertheless, energy usage and carbon emissions are a major concern today. Data centers are substantially strained by electricity costs and power dissipation is a major concern in portable, battery-operated devices [30],[37],[5]. Governments are providing incentives to save energy and to promote the use of renewable energy resources. Individuals, companies, and organizations moving towards energy-efficient products as energy costs have grown to be a major factor. Saving energy has become a leading design constraint for computing devices through new energy-efficient architectures and algorithms [1].

As a result of this new design trend, we observe the emergence of new energy efficiency technologies [3] which provide subsystems that are able to scale the processor frequency and voltage in order to reduce the power demands.[1,2] Apart from the scalability of voltage and clock (dynamic voltage and frequency scaling, or DVFS), CPU manufacturers can turn off parts of the CPU which are not being used (power gating), resulting in further savings in energy consumption and lower heat dissipation. All these technologies allow for modern processors to correlate computation throughput with energy consumption.

Traditionally, complexity can be considered as a measure of the effort to accomplish certain computation tasks and can be accounted either as the amount of memory, the time, or the number of operations it takes to perform some computation [13]. We propose to evaluate energy demand instead of complexity [14], since energy is a fundamental resource that can be directly mapped to operational costs, and we will show that complexity estimation is not always a reliable indicator of energy consumption.

The present work suggests new strategies in the direction of saving energy in real-time computation. We present a fidelity-energy (ΦE) optimization strategy to constrain the energy demanded by an application in a real-time scenario. In a video encoder, fidelity Φ can be evaluated in terms of the rate-distortion (RD) performance [33, 41]. Then, the optimized parameters are used to implement an RDE-optimized real-time encoding framework. We chose an open-source high-performance encoder, x264 [26], as the H.264/AVC software implementation due to its excellent encoding speed and good rate distortion (RD) performance. The proposed approach suits, for example, mobile communication systems where energy efficiency is still a major bottleneck [36]. The system can be made adaptive, dynamically tracking changes in image contents and in energy demands.

The present work is similar to another [35] in the aspect of optimizing a video encoder constrained to energy expenditure. However there are significant variations. There is also work [15] proposing a power-rate-distortion model for wireless video communications under energy constraints, and the dissimilarities to both works will be discussed in the next section.

Our framework allows for real-time software-based energy-constrained video coding. We provide a management module capable of delivering the user-demanded encoding speed while spending less energy and smoothly affecting the RD-performance. Part of the novelty of our approach is that we take a standard video encoder to achieve significant encoding energy savings (up to 31% less energy) on SD and HD video (rather than CIF and QCIF), without resorting to frame-skipping or resolution changes. Further novelty is that we analyze the encoder within a global RDE trade-off, wherein encoding is performed in groups of frames and the energy is actually measured. Also, it can all be done within a closed-loop-adaptive framework. We have not found these features elsewhere.

---

[1] AMD® Cool'n Quiet[tm]: http://www.amd.com/us/products/technologies/cool-n-quiet/Pages/cool-n-quiet.aspx

[2] Intel EIST®: http://www.intel.com/technology/product/demos/eist/demo.htm

The proposed encoding framework can be considered a true example of green computing where the same task is accomplished in the same hardware system with much less energy consumption, reducing the carbon footprint of video compression systems.

9.2 BACKGROUND ON H.264/AVC IMPLEMENTATION

The H.264/AVC is a hybrid video codec, i.e., along with a transform module, it has a prediction module, a differential stage, and a feedback loop [33]. The H.264/AVC prediction module has techniques which can be categorized in two classes: temporal ("Inter-prediction") and spatial ("Intra-prediction") techniques. AVC brought significant advances in inter-prediction in comparison to earlier video standards, which include the support for a wide range of block sizes (16×16-pixels and smaller), multiple reference frames, and refined motion vectors (quarter-sample resolution for the luminance component). In intra-prediction, the predicted block can have different sizes (besides 16×16-pixel size macroblock, blocks of 8×8 and 4×4-pixel size are also allowed) and is formed based on planar extrapolation of previously encoded blocks in the same frame. The prediction residue is transformed and quantized through the use of integer transforms [28].

The data set composed by block size and intra (spatial extrapolation) choice or inter parameters, like motion vectors and reference frames, forms the "prediction mode" of a block. The encoder typically selects the prediction mode that minimizes the difference between the predicted block and the block to be encoded, constrained to a given bitrate.

In order to scale the encoder complexity, one may modify the prediction stage, which is one of the most computationally intensive steps in digital video encoding, as the numbers in Table 9.1 suggest. These results are for Platform 1 and x264 implementation (see Section 9.3) set to High Profile [40]. [3] Similar tables can be verified in [9] and [17] for the reference software implementation.

There are many studies into managing H.264/AVC complexity. Some explore prediction techniques for reducing computations with small *RD* penalties [6, 23, 21]. Assuming a correlation between computations and demanded energy, reducing the computations can help in reducing the energy demands. A recent work provides substantial H.264/AVC complexity reduction [27] using the reference software as baseline. Nevertheless, much of the complexity scaling would not be perceived if the framework was implemented using faster algorithms, high-performance libraries, and platform dependent resources [29, 18]. Other works [22, 39] developed complexity models. Their results are evaluated using the reference H.264/AVC software (which is not optimized in terms of encoding complexity) and are tested on low-resolution material. There are recent investigations on providing complexity scalability to a high-performance encoder [7] within a somewhat short range. Energy-awareness in video compression was first presented by Sachs et al. [32], who propose

---

[3]We analyzed encoder executions using *gprof*, an open source profiler. Available at http://www.gnu.org/software/binutils.

Table 9.1

X264 relative computational complexity for encoding "Mobile" (CIF) and "Mobcal" (720p) sequences.

Coding Stage	Resolution	
	CIF	720p
Predictions	91.24%	90.42%
Encoding	6.07%	6.13%
Other Stages	2.69%	3.45%
Total	100.0%	

a proprietary video encoder for general purpose processors that trade computational complexity for compression efficiency in order to minimize total system energy. As we mentioned, the present work is similar to the one by Shafique et al. [35] in many aspects. Nevertheless, while the focus is on the motion estimation (ME) stage of the video coder (varying the search patterns and the motion vector precision), we cover the whole prediction stage and its different parameters. Any change in pattern can be easily re-trained and we incorporate many other parameters such as number and types (I, P, or B) of reference frames and multi-threading. Furthermore, that work [35] uses lower-resolution content (the largest frame-size tested was CIF), focuses on a hardware implementation, and relies on energy consumption estimation. We, however, focus on real-time software-based standard-definition (SD) and high-definition (HD) video coding on general purpose computers and we use actual energy measurements. Additionally, our framework is adaptive to changes in video contents and power targets. He et al. [15] proposed a power-rate-distortion model for wireless video communications under energy constraints. They analyzed the encoding mechanism of typical video coding systems and developed a parametric video encoding architecture which is fully scalable in a computational sense, focusing only on DVFS and stock processors. The baseline video encoder was H.263 [19] applied to low-resolution (QCIF, i.e., 176×144-pixel) frames of head and shoulder sequences and allowing for frame dropping.

9.3 OUR H.264/AVC TEST SYSTEMS

A software-based video solution implies platform-dependent results. Nevertheless, the collected data suggests that, even for different processors and underlying hardware for different PCs, the power profile can be well characterized to reduce consumption in the mean power sense for a group of frames. Analyzing hardware implementations is beyond the scope of this paper and we use two systems as our test platforms (PCs): Platform 1 has an Intel® Core i7 CPU 950 processor in an Asus® P6X58D-E motherboard, while Platform 2 has an AMD® Phenom II X6 1055T processor in an Asus® M4A78LT-M motherboard. Both systems have 8GB RAM

DDR3, a solid-state disk Corsair® CSSD-F115GB2-A, and no monitors are attached.

Both platforms run LINUX Operating System (Debian 2.6.32) in multi-user mode and the coding processes run at maximum priority, set to real-time scheduling. All unnecessary processes are made inactive and we assume that only one user requests the coding of video frames.

Nowadays, there are different H.264/AVC encoder implementations. The reference H.264/AVC standard implementation, also known as JM,[4] tries to provide the most complete encoder/decoder implementation. The Intel® Performance Primitives (IPP) library [18] has a proprietary implementation of the H.264/AVC video codec built upon its high performance primitives. Even though we can control complexity within an IPP implementation [8], we feel that x264, an open-source H.264/AVC standard-compliant implementation [26] is better suited for the present work, yielding better performance. Hence, we opted for only using x264 in our tests. x264 uses assembly-optimized routines for the most complexity-intensive operations [26] and explores "early stop" tests during rate-distortion optimization, yielding a 50-times speed-up over JM without significantly sacrificing RD performance. We ran x264 in H.264/AVC High profile: 64×64-pixel motion-estimation window, 5 reference-frames, refined RD-optimization in all macroblock predictions, quarter-pixel-precision motion vectors, uneven multi-hexagon search, 8×8 integer transform, and CABAC entropy coder.

9.4 POWER AND ENERGY IN COMPUTING SYSTEMS

In the scope of computing, work is related to activities associated with running programs (the microprocessor instructions involved in certain computation), power (P) is the rate at which the computer consumes electrical energy while performing these activities, and E is the accumulated electrical energy demanded by the computer during a certain time interval. Complexity [14] can be expressed as the number of iterations of an algorithm, or as the amount of memory or even the time necessary to execute it.

The distinction among energy, power, and complexity is important because optimizing for one does not always ensure the others will be optimized. For example, an application can be implemented using specific instructions provided by the execution platform. This can raise the instantaneous power demand, but should reduce the execution time, perhaps bringing energy savings. So, in this example, compared to not using the specific instructions, the second implementation would have greater complexity (in the spent time sense), lower power profile, but reduced energy. This could be an issue for a mobile battery-operated platform. For a high-performance server, the temperature profile is an issue, so that power surges should be avoided [44, 16]. Power consumption can be addressed at different levels.

---

[4]"JM," available at http://iphome.hhi.de/suehring/tml

9.4.1 ADDRESSING POWER CONSUMPTION AT THE DEVICE LEVEL

CMOS technology prevails in modern electronic devices [34] and is usually profiled according to two power models: static and dynamic. The static (leakage) power profile is composed by the leakage currents that occur while keeping circuits polarized, regardless of clock rates and usage. This static power is mainly determined by the type of transistors and the fabrication process technology. Reduction of the static power requires changes at the low-level system design.

The dynamic power profile is created by circuit activity (transistors switching, memory components varying their states, etc.) and depends on the usage. It has two sources: short-circuit current and switched capacitance. The short-circuit current causes only 10-15% of total power consumption and there is no effective way to reduce it without compromising the performance [45]. Switched capacitance is the primary source of dynamic power consumption described as

$$P_{dynamic} \propto aC_{phys.} V^2 f, \tag{9.1}$$

where $C_{phys.}$ is the physical capacitance, V is voltage, f is the clock frequency, and a is an activity factor. In order to change physical capacitance, changes in low-level system design and fabrication methodologies are required. The combined reduction of f and V is achieved with widely adopted DVFS, which intentionally down-scales the CPU performance, when it is not fully demanded. DVFS should ideally change dynamic power dissipation in a cubic factor because dynamic power is quadratically affected by voltage and is linearly affected by clock frequency [3].

9.4.2 ADDRESSING POWER CONSUMPTION AT THE INFRA-STRUCTURE LEVEL

Studies show that the main part of the energy consumed by a server is drawn by the CPU, followed by the memory and by losses due to the power supply unit (PSU) inefficiency [38, 25]. Nowadays, the systems can dynamically enable low-power CPU modes, saving resources. Current desktop and server CPUs can consume less than 30% of their peak power at low-activity modes, leading to a dynamic power range of up to 70% of peak power [2]. In contrast, dynamic power ranges of all other server's components are much narrower: less than 50% for DRAM, and 25% for disk drives [11]. The reason is that many components cannot be partially switched off and may have current surges while transitioning from inactivity.

9.4.3 ADDRESSING POWER CONSUMPTION AT THE APPLICATION LEVEL

The application software can also allow for power reduction using compiler tools such as statistical optimizations and dynamic compilation [45]. Holistic approaches give applications a large role in power management decisions. Some works adopted an "architecture centric" view of applications that allows for some high-level transformations capable of reducing the system power demand [43]. Sachs et al. [32]

explored a different adaptation method which involves trading the accuracy of computations for reduced energy consumption in video encoding.

The energy consumption of a computing device is not only determined by the efficiency of its physical devices, but it is also dependent on resource management and on applications usage patterns [38, 44, 12, 30].

9.5 ENERGY VS. COMPLEXITY

9.5.1 SAVING ENERGY IN A PC-BASED PLATFORM

We first define idle and full-power states. In idle state only the basic operations are executed and the scheduler keeps the processor "sleeping" almost at all times. In full-power state the processor carries intensive operations and the scheduler never allows the processor to "sleep."

Because of energy management techniques like DVFS, when in idle state, our Platforms 1 and 2 demand 105W and 80W, respectively. When the computation workload increases, the power demand also increases. When in full-power state, our Platforms 1 and 2 drive 240W and 180W, respectively, of active power to provide the currents to feed the increased gate switching, and to keep up with higher access rates to memory, hard-disks, buses, and other components.

We consider a real-time *clocked* video coding scenario, where frames are periodically made available to be encoded at a given rate f_a, e.g., at 30 Hz, or 30 frames/second (fps). Hence, we have $T_a = 1/f_a$ seconds to encode each frame, and that is the period that governs the compression system. If we use only T_p seconds to encode each frame, in the remaining time ($T_i = T_a - T_p$) the processor may go idle. If we let P_i and P_{fp} be the power demanded in idle and full-power states, respectively, such a power profile can be illustrated as in Figure 9.1(a). It is also useful to define the processing (or encoding) speed as $f_p = 1/T_p$, which indicates the speed (in fps) the encoder would be capable of encoding frames if they are available at once, say off-line. What should be clear from Figure 9.1(a) is that we can save energy consumption if we reduce T_p, i.e., if we increase the encoder speed f_p. In this way, the sooner the encoder is done encoding a frame, the longer the processor goes idle (higher T_i).

In this binary utilization model, in which the processor is either fully idle or fully busy, one can save energy by increasing the encoding speed, i.e., reducing T_p, as in Figure 9.1(b). An increase in encoding speed is typically obtained at the expense of RD performance. While the profile in Figure 9.1(b) would demand less energy than the one in Figure 9.1(a), one could also use dynamic frequency/voltage scaling to slow down the processor and do the same task as in Figure 9.1(b) but at a lower pace [24]. In the case depicted in Figure 9.1(c) the processor would run longer using less power. Here, we are not examining this case, but rather considering the energy savings provided in Figure 9.1(b) by increasing the encoding speed.

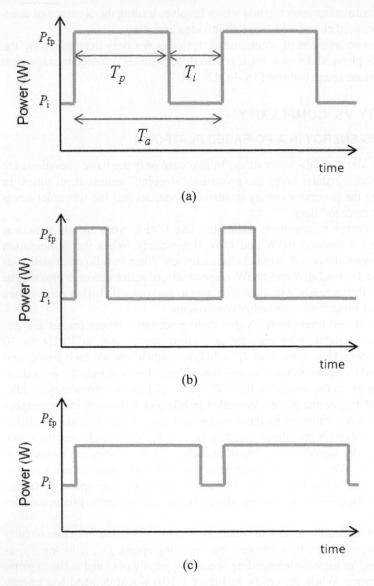

Figure 9.1 Power profile for video coding. (a) Frames are available in T_a intervals. The frame is encoded in T_p seconds and the processor returns to idle state for $T_i = T_a - T_p$ seconds until a new frame arrives. (b) Profile for reduced consumption by making the encoder faster. (c) Profile for reduced consumption by making the processor consume less and slower.

9.5.2   MEASURING ENERGY

Energy consumption is here measured in two ways. The computer is connected by itself (no monitor or other peripherals) to a wattmeter and from there to the local power supply. We can read the energy consumption from the wattmeter on another computer at every second, through a USB port, as shown in Figure 9.2. This is sufficient for steady-state tests.

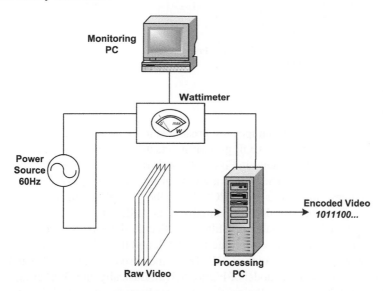

Figure 9.2   On-line measurement setup.

However, in order to investigate the energy consumption behavior at very fast cycles (e.g., 30 Hz or 60 Hz video), which are comparable to the voltage cycles of the energy provided by our local power company (60 Hz), we resorted to oscillography. For these tests we used an Elspec G4500 BlackBox and a California Instruments 5001ix sinusoidal power supply, as illustrated in Figure 9.3.

Time measurements can be disturbed by the OS scheduler in real-time systems. We used a 250 Hz scheduler frequency and we can expect experimental errors of ±2 ms. Considering the encoding speeds provided by our platforms, which can allow the encoding of SD and 720p video sequences of up to 250 fps, the measurement of short time intervals used to encode a video frame can be compromised. One way to overcome the scheduler-induced variances is the grouping of frames in GOPs (Group of Pictures).

The GOP grouping of frames can also affect the demanded power waveforms. To illustrate this, we monitored our test platform, while compressing 900 high-definition (720p 30-Hz) frames in real time, at different GOP sizes. As the processor is faster than necessary to guarantee real-time coding, the processor can "sleep" from the time it is done compressing a GOP until the next GOP is available for compression. The power waveform is registered by measuring the demanded power according to

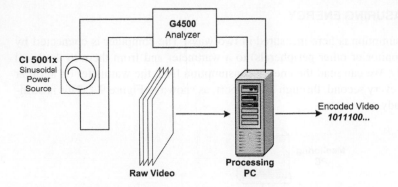

Figure 9.3 Off-line measurement setup.

Figure 9.3. The results from oscillography are presented in Figure 9.4.

The waveforms show distinctive GOP grouping signatures. The rapid processor switching between idle and busy states in Figure 9.4(a) is represented by an irregular sequence of peaks and valleys. The plot is a zoom of the process of compressing 24 frames. When measuring power at the PC's PSU, the "sleep" moments are not well determined, as the processor does not remain in the "idle" state for a long period. This is the result of various factors: a filtering effect at the PSU related to the AC-DC conversion; the processor scaling due to DVFS; and the ACPI activity over other PC components [34, 3, 25]. As the GOP size is increased (Figures 9.4(b) and (c)), the waveforms approach the model of Figure 9.1(a). Basically, the GOP grouping reduces the processor state oscillation and the waveform frequency, giving the energy efficiency embedded to the PC enough time to put the processor (and other subsystems) in the "idle" state. We chose to use a 50-frame GOP to conform the waveform to the model from Figure 9.1 and also to avoid OS scheduling jitter in the time measurements required by our framework. The oscillography of such a setup is presented in Figure 9.5. Furthermore, the overall energy consumption is 3% lower in a GOP of 50 than it is for a single-frame GOP.

9.5.3 COMPLEXITY ISSUES

Computers are very complex systems, where many simultaneous events are treated by the CPU while it interacts with the user and with all the peripherals. Most applications are multi-threaded to guarantee the proper handling of all events. Complexity evaluation in a single task situation is not very precise. Nevertheless, complexity is still useful to perform comparisons of memory and time requirements [38, 3].

The precision of complexity estimation in terms of operations and time measurements is disturbed by the variances induced by computing speedup techniques, caching, compiling optimizations, and the availability of multi-core CPUs. All these enhancement techniques, albeit improving performance, are sources of high unpredictability in time measurements, which, in turn, are also affected by the operating

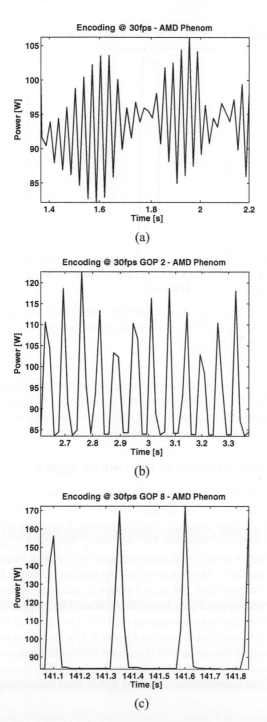

Figure 9.4 Power waveforms for the encoding of 24 720p-video frames made available (and compressed) at 30 Hz and grouped in different GOPs configurations: (a) 1-frame GOP; (b) 2-frame GOP and (c) 8-frame GOP. As the GOP size is increased, the waveform tends to the Figure 9.1(a) model.

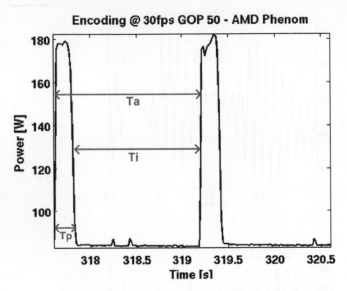

Figure 9.5 Power waveforms for the encoding of 100 720p-frames recorded and compressed at 30 Hz and grouped in a 50-frame GOP. In red, we highlight the time intervals of interest for the Figure 9.1(a) power model: T_a, T_p, and T_i.

system activities and the concurrency of other executing applications. More variance is induced by DVFS and ACPI [45]. Therefore, the accounting of computing effort only in terms of the number of computations is imprecise and can be considered unsuitable in evaluating critical real-time applications.

9.5.4 ENERGY AS A COMPUTATIONAL EFFORT MEASURE

We just argued that complexity measurements based on estimates of operations can be very imprecise. Furthermore, we also argued that energy consumption is completely defined by f_p for a binary utilization model. However, while there are some applications where the correlation between the processing frequency f_p and power can be linear, for more complex tasks that relationship is not so well behaved, as illustrated in Figure 9.6. This figure shows typical results relating f_p and P for a video coding task, where we compute both the power demand and speed. Note that the curve is not very linear (as expected in a logarithmic scale plot) and there are dispersed points in the lower encoding speed interval ($< 10^2$). The reason is because the real power profile is never as well behaved as in Figure 9.1, which does not account for imperfections and oscillations caused by the many hardware nuances involved. f_p cannot be easily measured with small GOP sizes as in Figure 9.4. Because of that, we decided to measure real energy/power demand rather than estimating it in any way.

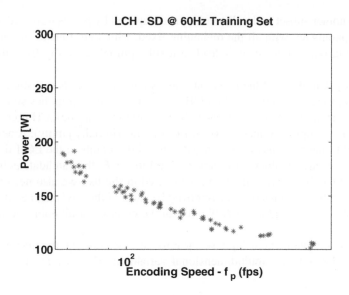

Figure 9.6   Correlation of demanded power and compression speed (f_p).

9.6   ENERGY-AWARE OPTIMIZATION

9.6.1   RDE OPTIMIZATION

Typical optimization tasks deal with cost functions or success measures. Let a software encoder execute its job for which we can somehow measure its cost. For signal compression, the cost measure can be a measure of quality, like distortion (D) or the bit-rate (R) or a combination of both. The compression is assumed parameterized, i.e., one has the freedom to chose the values of N parameters $\{P_i\}_{i=1,...,N}$. Let $\mathbf{P}$ be the vector with all P_i. The encoder runs on a given set of data Z that may be different at every instantiation. For every choice of $\mathbf{P}$ and Z, we can have a measure C of the encoder cost. In essence, we can have a mapping

$$C = f(\mathbf{P}, Z).$$

Another attribute we can derive from each instantiation is the effort taken to execute the encoding task, which can be measured as demanded energy $E = g(\mathbf{P}, Z)$. It is expected that some parameters like number of iterations, data sizes, etc., would influence the demanded energy while some others would not. The central idea in this chapter derives from the fact that the correlation of E and C is different for different parameters. We will use this to find points that minimize the energy consumption. The idea is illustrated in Figure 9.7, which depicts a cloud of points in the cost-energy space of all achievable $\mathbf{P}$ at a given system and input data. Along with the cloud, the figure highlights a subset after optimization, the lower convex hull (LCH) of all points, represented by green square points. Points that lie on the LCH represent instantiations that yield the lowest energy for a given cost, and is where we would like to operate. Another subset in the illustration is composed by points traversed as we increase one parameter, with all the remaining fixed, which are illustrated with red stars. Changing one parameter may lead to a sub-optimal set, away from the LCH.

Out of the many definitions of the LCH, one easy solution that leads to a slightly non-convex set is to include a point in the LCH such that no other point has simultaneously lower C and lower E than it. Hence, the algorithm to find the LCH points, in this case, is rather simple. We make a list of LCH points (initially empty). A new candidate point P to the LCH has to be compared to all the points in the LCH list. If no point in the list has simultaneously lower C and lower E, the candidate point is inserted in the LCH list. Before the point is inserted in the list, we also need to check if any point in the LCH needs to be removed because of the new one, i.e., if it has simultaneously lower C and lower E. We repeat the process for all points in the cloud.

Despite the easier explanation using a scalar cost, in video coding, the mapping is conveniently addressed by a multidimensional variable as $\mathbf{C} = [R, D]$. Hence, $\mathbf{C} = f(\mathbf{P}, Z)$.

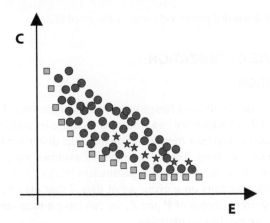

Figure 9.7 Cloud of points in the energy vs. cost space. The LCH points are indicated by the green squares. A suboptimal path is achieved, for example, by varying just one parameter, as illustrated by the red stars.

$\mathbf{P}$ and Z are mapped to R, D, and E, adding the energy dimension to the usual RD optimization problem. We measure active power from which we can derive accumulated energy consumption. We want to find the parameters that allow us to operate on the LCH in RDE space. In this manner, we can be assured that no configuration would yield lower energy consumption for a given cost value. Conversely, we can assure that, for a given energy consumption level, no other configuration would achieve better RD performance. Figure 9.8 illustrates the LCH in RDE space.

Our approach is to use training data sets. Let $\{\mathbf{P}_k\}$ be the set of all parameter

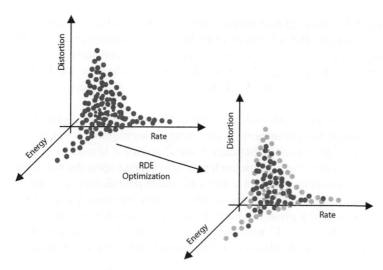

Figure 9.8   Illustration on the set of *RDE* points that compose the Pareto front. The visible green points belong to the lower convex hull; some points are hidden due to the viewpoint.

choices, ordered in some fashion. Let also $\mathbf{P}_k$ have elements P_{kn}. If we use a representative data set $\hat{Z}$, we can span $\{\mathbf{P}_k\}$, computing E, R, and D for each choice and identifying the points that belong to the LCH of $E \times R \times D$. If the n-th point belongs to the LCH, we record $\mathbf{Q}_n = [E_n, R_n, D_n, \mathbf{P}_n]$, which contains the optimal points for the set $\hat{Z}$, but which are also assumed good enough for other data. The off-line training algorithm is:

1. Input a representative data set $\hat{Z}$ and create an empty list Q.
2. For all k, compute $E_k = g(\mathbf{P}_k, \hat{Z})$ and $[R_k, D_k] = f(\mathbf{P}_k, \hat{Z})$. If the point belongs to LCH, record $\mathbf{Q}_k = [E_k, R_k, D_k, \mathbf{P}_k]$ into Q.
3. Output a list Q of points in the LCH.

After finding the N_q points which belong to LCH, we sort Q in ascending order of energy, i.e., $\{E_i\}$ in Q in non-decreasing. When running on-line, the parameter finding algorithm is as follows. Initially, consider a target bit-rate R^r (channel constraint) and a desired energy target E^r. Then:

1. Input a list Q of points in the LCH, the energy target E^r and the rate target R^r. Create an empty list L.
2. Span Q, for $k = 1, \ldots, N_q$. If $|R_k - R^r| < \epsilon$ insert $\mathbf{Q}_k$ into L.
3. Count N_l, the number of items in L. Note that the items in L are still in ascending order of energy and all parameters are supposed to achieve similar bit-rate.
4. Span L, for $k = 1, \ldots, N_l$, until $E_k \leq E^r \leq E_{k+1}$, then stop.
5. Find $\mathbf{P}'$ as a proportional interpolation of $\mathbf{P}_k$ and $\mathbf{P}_{k+1}$ in L.
6. Output parameter vector $\mathbf{P}'$.

Parameter set $\mathbf{P}'$ is then used to compress data set Z. We used energy targets E^r constrained to a bitrate R^r, but it is trivial to replace it with a distortion target D^r. Of course, many parameters do not assume continuous values and some action has to be taken to properly assign them. For example, that being the case for the m-th parameter, one can use the value from P_{km} if $E^r - E_k < E_{k+1} - E^r$, otherwise use the value from $P_{k+1,m}$.

If feedback control is turned on, one can monitor the system energy consumption and continuously adjust the parameters. If the energy consumption is not as predicted, it is because of discrepancies between Z and $\hat{Z}$, so that $\hat{Z}$ is not as representative as one would assume. Such a mismatch may also depend upon the non-linear mapping g. One solution is to start with a target E^r and to periodically measure the energy $E(n)$. We then adapt the parameters in order to control the energy expenditure (or cost). Assume that at any given instant n, $\mathbf{P}'$ is taken somewhere as an interpolation of $\mathbf{P}_j$ and $\mathbf{P}_{j+1}$. If $E(n) < E^r$ one should move $\mathbf{P}'$ towards $\mathbf{P}_{j+1}$ or even $\mathbf{P}_{j+2}$. Conversely, if $E(n) > E^r$ one should move in the opposite direction, i.e., towards $\mathbf{P}_j$ or even $\mathbf{P}_{j-1}$.

The control loop enjoys all the properties of trivial adaptive systems and there are many techniques to choose adaptation steps and to deal with convergence issues.

9.6.2 PRACTICAL APPROACH

We use x264 as our H.264/AVC encoder, with $\mathbf{P}$ being the aggregation of the following parameters: the number of B-frames (#B), the number of references frames (#Refs), the motion vectors precision (MVP) used in motion compensation, the mode decision technique (MD), the quantization parameter (QP), and the number of encoding threads (#Thrds). Hence,

$$\mathbf{P} = \{\#B, \#Refs, MVP, MD, QP, \#Thrds\}.$$

The first step to optimize the H.264/AVC in the E-sense is to determine a representative training set $\hat{Z}$ from where we will derive the encoder Pareto front. To build $\hat{Z}$, we opted to use standard definition (SD, 704×576-pixels) video sequences recorded at 60Hz and high definition (720p, 1280×720-pixels) video sequences recorded at 30Hz and at 50Hz. The SD training sequence was obtained by concatenating the sequences "Harbour," "Crew," and "Soccer." The 50Hz HD training set is composed by sequences "Parkrun," "Stockholm," and "Tractor." The 30Hz HD training set is composed by videoconference sequences 5, 6, and 17.

For each resolution, we encode the training set and, for each encoder instantiation, we record the bitrate, the resulting distortion, and the demanded electric energy. We also measure encoding speed in order to allow for real-time compression. Those values of $\mathbf{P}_k$ which are not capable of delivering $f_p \geq f_a$ are disregarded, in such a way that the optimized codec will only accept setups which allow for real-time encoding.

The maximum number of references frames (#Refs) and of threads (#Thrds) were set to 5 and 8, respectively. The maximum number of B-frames (#B) is restricted

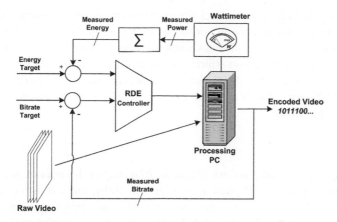

Figure 9.9   Energy controller scheme. The closed-loop framework guarantees to follow the energy target required by the user. Deviations from the requested value are minimized by the framework which adjust the encoder settings in order to vary its energy demand.

by x264 which bounds the maximum number of B-frames between P-frames to the number of reference frames. The other **P** components (QP, MD, and MVP) are freely varied in their ranges. In summary, in the training stage,[5] we focused on finding the fastest settings leading to lower energy demand, assuring that $f_p > f_a$, by varying the motion vectors precision, the mode decision technique (the level of optimization effort in RDO), the QP, the number of reference frames, and the number of B-frames.

The simulations results delivered an *RDE*-point cloud from which we derive the LCH. Once the LCH for the representative sequences is found, we derived look-up tables from where we can adaptively control the encoder energy demands. These tables are inserted in the energy controller framework, whose diagram is depicted in Figure 9.9. The closed-loop controller in Figure 9.9 manages the desktop computer power profile as discussed in Section 9.5.1. It measures the actual encoding energy and adjusts settings. The central idea is to scale the ratio T_p/T_a in order to adjust the demanded energy to the desired target. The closed-loop adjusts the codec to different P_{fp} and P_i levels and guarantees the target energy. If the encoder is spending more energy than it should be, the control module adjusts the encoding parameters to a less energy/power demanding setup which, in turn, yields inferior *RD*-performance. If there is any surplus, the encoder is allowed to use parameters which are more energy/power demanding, but also yield better performance in terms of *RD*.

The resulting parameters are platform dependent but the method is not, just requiring retraining once for each platform, which is not excessively complex in light of a continuous real-time operation.

---

[5]This stage is done once for each processor. Its derived parameters are used in a closed-control framework, which tries to cope with any deviation from expected reference levels. As the system is trained once, we opted to not account for the total energy spent in this stage.

An important issue is the human sensitivity to variations in quality over time. Such variations can be made smooth enough not to cause impairing. We expect higher variations, perhaps visible, at lower bit-rates when tracking large energy savings. Of course, there may be curious situations which would cause rapid oscillating behavior in quality control and cause noticeable flickering. However, our one-measurement-per-second setup in Figure 9.2 only provides for very slow transitions and we have not observed any impairment.

9.7 RESULTS

At every sequence that is compressed we obtain an *RDE* triplet. In order to display results in 2D, we can use the *RD* plots as in Figure 9.10(b), one curve for each energy (power) level. It is important to note that not all points in a curve indicate the same power consumption. We simply labeled the curve by its average as shown Figure 9.10(a), which indicates the actual power consumption as the controller tracks the demanded energy target for various bit-rates.

RD curves for encoding an SD sequence at different power levels are shown in Figure 9.10(b) and Figure 9.11. Similar plots are shown in Figures 9.12 and 9.13 for 720p sequences at 50 Hz and 30Hz, respectively. The controller acts by forcing the energy demand to comply to the available budget. The higher baseline speed, required to handle 50Hz and 60Hz sequences, demands increased power compared to the compression of videoconferencing sequences, recorded at 30Hz.

The curves in Figures 9.10 to 9.13 are close to each other. In order to compare them, it is convenient to analyze averaged PSNR differences between two *RD* curves as described in [4]. For each sequence, each *RD*-curve is compared to the best *RD*-performance setup, which, in turn, has the highest averaged power expenditure. Power expenditure is also presented in relative numbers. The averaged results are illustrated in Figure 9.14(a) for SD video sequences. The general behavior suggests that, as we reduce the available power (and energy) used to encode a video sequence, the performance penalties increase. In Figure 9.14(b) the results are shown for 720p sequences.

The main result is an energy-controlled framework which allows the user to choose the desired energy budget while real-time encoding HD and SD[6] video sequences. As expected, the RD-perfomance tends to be penalized as the encoding speed is raised. However, the curves are close to each other and the worst case is represented by high-motion high-frequency (50Hz) detailed sequences ("Shields" and "Mobcal"). For less demanding video sequences, like those in 30 Hz videoconferencing ("Seq15" and "Seq21"),[7] PSNR reduction is less than 1.3dB on average while providing up to 31% of mean power and energy savings. The SD results, besides the

---

[6]"Soccer" is present in the training and in the evaluation steps; however, the frames used to evaluate the encoder are from a different set from those used to build the training sequence.

[7]"Seq15" and "Seq21" are scenes where there are a couple of speakers on a table: the background is plain on "Seq15" and is detailed on "Seq21."

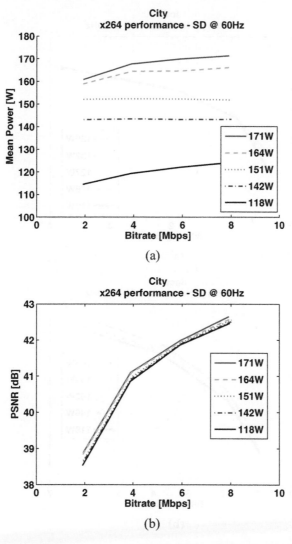

Figure 9.10 Energy scaling for compressing SD sequence "City." A range of 10% of deviation is allowed for both bitrate and power. (a) Actual demanded power for various bit-rates and several target power demands. (b) RD curves for real-time compression.

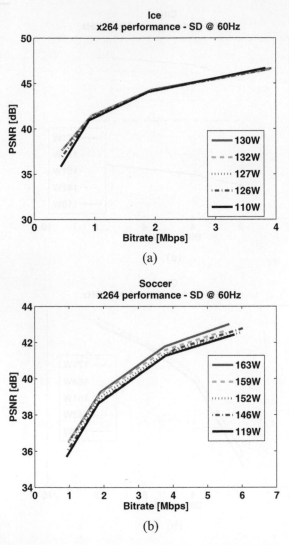

Figure 9.11 RD curves for sequences (a) "Ice" and (b) "Soccer" encoded at different averaged power levels.

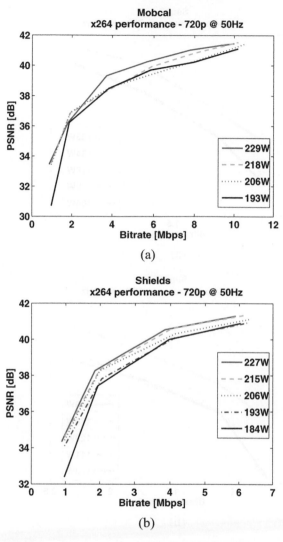

Figure 9.12 RD curves for sequences (a) "Mobcal" and (b) "Shields" encoded at different averaged power levels. These sequences were trained and evaluated in the Intel® Core™ i7-powered PC.

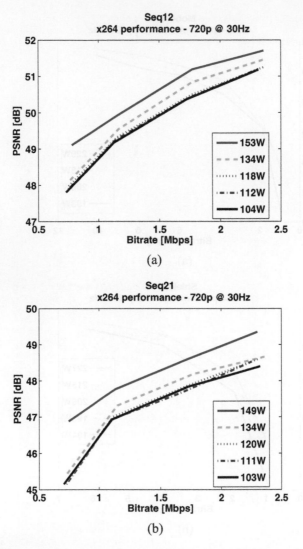

Figure 9.13 RD curves for videoconference sequences (a) "Seq12" and (b) "Seq21" encoded at different averaged power levels.

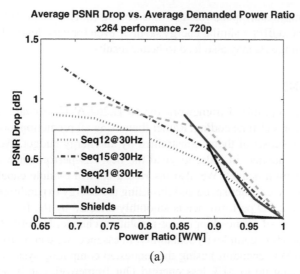

(a)

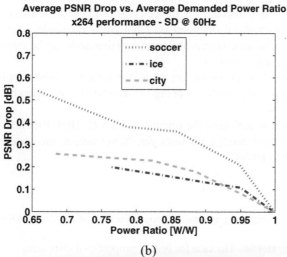

(b)

Figure 9.14 PSNR drop vs. mean power ratio for (a) SD and (b) 720p video sequences. Video quality increases as we increase the power budget. A energy ratio of 1.0W/W represents the case of best *RD*-performance for real-time coding.

increased baseline compression speed for real-time coding (60Hz), delivered lower PSNR drops (less than 0.6dB) for similar energy savings, even for very detailed video sequences. Better training sets may also lead to better results.

9.8 CONCLUSIONS

We proposed an energy-optimized framework for an H.264/AVC software implementation that allows for real-time coding. Rather than using all prediction tools, we can optimally choose a subset of them, constrained by an energy budget. We have trained and adjusted parameters in order to yield the best *RD*-performance within a given power consumption budget. We also inserted a control module capable of continuously adjusting the encoder speed and throttling the energy expenditure. Our tests have shown that the *RD* performance is smoothly affected by the framework, which does not make use of frame-skipping or resolution change. Nevertheless, it provides significant encoding complexity scalability. In essence, we can perform the requested task (H.264/AVC encoding) using the requested computing system (software and hardware) using up to 31% less energy! Our framework can be readily used to build PC-based video encoder appliances that can adjust themselves to the available *RDE* conditions without the need for changing the decoder implementation. Eventual changes in image contents and in energy demands can be dynamically tracked by the adaptive control system.

This is a true example of green computing where the same task is accomplished in the same hardware system with much less energy consumption, incurring only small *RD* performance penalties.

Algorithms and implementation of the upcoming HEVC (High Efficiency Video Coding) [42] are not mature enough for tests yet. Nevertheless, the concepts here discussed apply as well to HEVC.

REFERENCES

1. S. Albers. Energy-efficient algorithms. *Communications of the ACM*, 53(5):86–96, 2010.
2. L.A. Barroso and U. Holzle. The case for energy-proportional computing. *Computer*, 40(12):33–37, 2007.
3. A. Beloglazov, R. Buyya, Y.C. Lee, A. Zomaya, et al. A taxonomy and survey of energy-efficient data centers and cloud computing systems. *Advances in Computers*, 82(2):47–111, 2011.
4. G. Bjontegaard. Calculation of average PSNR differences between RD-curves. *VCEG-M33*, April 2001.
5. A. Carroll and G. Heiser. An analysis of power consumption in a smartphone. In *Proceedings of the 2010 USENIX conference on USENIX annual technical conference*, pages 21–21. USENIX Association, 2010.
6. Z. Chen, P. Zhou, and Y. He. Hybrid unsymmetrical-cross multi-hexagon-grid search strategy for integer pel motion estimation in H.264. *Proc. Picture Coding Symposium*, April 2003.

7. M.C. Chien, J.Y. Huang, and P.C. Chang. Complexity Control for H.264 Video Encoding over Power-Scalable Embedded Systems. In *Proc. 13th IEEE Internatinal Symposium on Consumer Electronics, ISCE2009*, pages 221–224, 2009.

8. T. A. da Fonseca, R. L. de Queiroz, and D. Mukherjee. Complexity-scalable H. 264/AVC in an IPP-based video encoder. In *Image Processing (ICIP), 2010 17th IEEE International Conference on*, pages 2885–2888. IEEE, 2010.

9. T.A. da Fonseca and R.L. de Queiroz. Complexity Reduction Techniques for the Compression of High-Definition Video. *Journal of Communications and Information Systems*, 24(1), 2009.

10. R. L. de Queiroz, R. S. Ortis, A. Zaghetto, and T. A. Fonseca. Fringe benefits of the H.264/AVC. In *Proc. International Telecommunication Symposium*, pages 208–212, 2006.

11. X. Fan, W.D. Weber, and L.A. Barroso. Power provisioning for a warehouse-sized computer. *ACM SIGARCH Computer Architecture News*, 35(2):13–23, 2007.

12. J. Flinn and M. Satyanarayanan. Managing battery lifetime with energy-aware adaptation. *ACM Transactions on Computer Systems (TOCS)*, 22(2):137–179, 2004.

13. L. Fortnow and S. Homer. A short history of computational complexity. *Bulletin of the EATCS*, 80:95–133, 2003.

14. J. Hartmanis and R.E. Stearns. On the computational complexity of algorithms. *Transactions of the American Mathematical Society*, 117(5):285–306, 1965.

15. Z. He, Y. Liang, L. Chen, I. Ahmad, and D. Wu. Power-rate-distortion analysis for wireless video communication under energy constraints. *IEEE Transactions on Circuits and Systems for Video Technology*, 15(5):645–658, May 2005.

16. W. Huang, M. Allen-Ware, J.B. Carter, M.R. Stan, K. Skadron, and E. Cheng. Temperature-Aware Architecture: Lessons and Opportunities. *IEEE Micro*, 31(3):82–86, 2011.

17. Y.-Y. Huang, B.-Y. Hsieh, S.-Y. Chien, S.-Y. Ma, and L.-G. Chen. Analysis and complexity reduction of multiple reference frames motion estimation in H.264/AVC. *IEEE Transactions on Circuits and Systems for Video Technology*, 16(4):507–522, Apr. 2006.

18. Intel. Intel Integrated Performance Primitives. *http://software.intel.com/en-us/intel-ipp/*.

19. ITU-T. ITU-T Recommendation H.263, Video coding for low bit rate communication. Technical report, November 2000.

20. JVT of ISO/IEC MPEG and ITU-T VCEG. Advanced Video Coding for Generic Audiovisual Services. Technical Report 14496-10:2005, March 2005.

21. C. S. Kannangara, Y. Zhao, I. E. Richardson, and M. Bystrom. Complexity control of H.264 based on a Bayesian framework. *Proc. Picture Coding Symposium*, November 2007.

22. C.S. Kannangara, I.E. Richardson, M. Bystrom, and Y. Zhao. Complexity control of H.264/AVC based on mode-conditional cost probability distributions. *IEEE Transactions on Multimedia*, 11(3):433–442, 2009.

23. B.G. Kim, S.-K. Song, and C.-S. Cho. Efficient inter-mode decision based on contextual prediction for the P-slice in H.264/AVC video coding. *Proc. IEEE International Conference on Image Processing*, pages 1333–1336, September 2006.

24. E.G. Larsson and O. Gustafsson. The Impact of Dynamic Voltage and Frequency Scaling on Multicore DSP Algorithm Design [Exploratory DSP]. *Signal Processing Magazine, IEEE*, 28(3):127–144, 2011.

25. A. Mahesri and V. Vardhan. Power consumption breakdown on a modern laptop. *Power-aware computer systems*, pages 165–180, 2005.

26. L. Merritt and R. Vanam. Improved rate control and motion estimation for H.264 encoder. In *Proc. IEEE International Conference on Image Processing*, volume 5, 2007.

27. M. Moecke and R. Seara. Sorting Rates in Video Encoding Process for Complexity Reduction. *IEEE Transactions on Circuits and Systems for Video Technology*, 20(1):88–101, 2010.

28. J. Ostermann, J. Bormans, P. List, D. Marpe, M. Narroschke, F. Pereira, T. Stockhammer, and T. Wedi. Video coding with H.264/AVC: tools, performance, and complexity. *IEEE Circuits and Systems Magazine*, pages 7–28, January-March 2004.

29. K. Park, N. Singhal, M. H. Lee, and S. Cho. Efficient Design and Implementation of Visual Computing Algorithms on the GPU. *Proc. IEEE Intl. Conf. on Image Processing*, pages 2321–2324, November 2009.

30. G. Procaccianti, A. Vetro', L. Ardito, and M. Morisio. Profiling power consumption on desktop computer systems. *Information and Communication on Technology for the Fight against Global Warming*, pages 110–123, 2011.

31. I. E. G. Richardson. *H.264 and MPEG-4 Video Compression*. John Wiley & Sons Ltd., 2003.

32. D.G. Sachs, S.V. Adve, and D.L. Jones. Cross-layer adaptive video coding to reduce energy on general-purpose processors. In *Proceedings of 2003 International Conference on Image Processing*, volume 3, pages III–109. IEEE, 2003.

33. K. Sayood. *Introduction to Data Compression*. Morgan Kauffmann Publishers, 2000.

34. A.S. Sedra and K.C. Smith. *Microelectronic circuits*, volume 1. Oxford University Press, USA, 1998.

35. M. Shafique, L. Bauer, and J. Henkel. enBudget: A Run-Time Adaptive Predictive Energy-Budgeting scheme for energy-aware Motion Estimation in H.264/MPEG-4 AVC video encoder. In *Design, Automation & Test in Europe Conference & Exhibition (DATE), 2010*, pages 1725–1730. IEEE, 2010.

36. O. Silven and K. Jyrkkä. Observations on Power-Efficiency Trends in Mobile Communication Devices. *EURASIP Journal on Embedded Systems*, 2007(1):17–27, 2007.

37. P. Somavat, S. Jadhav, and V. Namboodiri. Accounting for the energy consumption of personal computing including portable devices. In *Proceedings of the 1st International Conference on Energy-Efficient Computing and Networking*, pages 141–149. ACM, 2010.

38. S. Song, R. Ge, X. Feng, and K.W. Cameron. Energy profiling and analysis of the HPC challenge benchmarks. *International Journal of High Performance Computing Applications*, 23(3):265–276, 2009.

39. L. Su, Y. Lu, F. Wu, S. Li, and W. Gao. Complexity-constrained H.264 video encoding. *IEEE Transactions on Circuits and Systems for Video Technology*, 19(4):477–490, 2009.

40. G. J. Sullivan, P. Topiwala, and A. Luthra. The H.264/AVC Advanced Video Coding Standard: Overview and Introduction to the Fidelity Range Extensions. *Proc. SPIE Conference on Applications of Digital Image Processing XXVII*, August 2004.

41. Gary J. Sullivan and Thomas Wiegand. Rate-Distortion Optimization for Video Compression. *IEEE Signal Processing Magazine*, 15(6):74–90, November 1998.

42. G.J. Sullivan and J.R. Ohm. Recent developments in standardization of high efficiency video coding (HEVC). *SPIE Applications of Digital Image Processing XXXIII, Proc. SPIE*, 7798, 2010.

43. T.K. Tan, A. Raghunathan, and N.K. Jha. Software architectural transformations: A new approach to low energy embedded software. In *Proceedings of the conference on*

Design, Automation and Test in Europe-Volume 1, page 11046. IEEE Computer Society, 2003.

44. Q. Tang, S.K.S. Gupta, D. Stanzione, and P. Cayton. Thermal-aware task scheduling to minimize energy usage of blade server based datacenters. In *Dependable, Autonomic and Secure Computing, 2nd IEEE International Symposium on*, pages 195–202. IEEE, 2006.

45. V. Venkatachalam and M. Franz. Power reduction techniques for microprocessor systems. *ACM computing surveys*, 37(3):195–237, 2005.

10 Rotation and Scale Invariant Template Matchings

Hae Yong Kim

University of São Paulo (USP)

ABSTRACT

Template matching is a technique widely used for finding patterns in digital images. It is very desirable that template matching should be able to detect template instances that have undergone geometric transformations like rotation and scaling. The obvious "brute force" solution executes a series of conventional template matchings between the search image and the query template rotated by every angle, translated to every position, and scaled by every factor. Clearly, this takes too long and thus is not practical. In this chapter, we describe two rotation and scale invariant template matchings that substantially accelerate this process. The first is called Ciratefi and consists of three cascaded filters that successively exclude pixels that have no chance of matching the template. This algorithm is especially suited for parallel implementations, because similar sequence of operations is applied to all pixels. The second is called Forapro and uses the complex coefficients of the discrete Fourier transform of circular and radial projections to efficiently compute rotation-invariant local features. Once the local features are computed, many different templates can be quickly localized in the search image. Thus, this algorithm is suitable for: multi-scale searching; searching many different query templates in a search image; and partial occlusion-robust template matching.

10.1 INTRODUCTION

Template matching is a classical problem in computer vision. It consists in detecting a given query template Q in a search image to analyze A. This task becomes more complex with the invariance to rotation (R), scale (S) and translation (T), and robustness to brightness (B) and contrast (C) changes.

In the literature, there are many techniques to solve this problem. The most obvious solution is the "brute-force" algorithm. It makes a series of conventional BC-invariant template matchings between A and Q rotated by many angles and scaled by many factors. Conventional BC-invariant template matching usually uses the normalized cross-correlation (NCC). Computation of NCC can be accelerated using fast Fourier transform (FFT) and integral images [14] or bounded partial correlation [23]. However, even using fast NCC, the brute-force rotation-invariant algorithm is slow because fast NCC must be applied repeatedly for many angles and scales.

Some techniques solve rotation-invariant template matching using previous segmentation/binarization, for example [9, 18]. Given a grayscale image to analyze A, they first convert it into a binary image using some segmentation/thresholding algorithm. Then, they separate each connected component from the background. Once the shape is segmented, they obtain scale-invariance by normalizing the shape's size. Next, they compute some rotation-invariant features for each component. These features are compared with the template's features. The most commonly used rotation-invariant features include Hu's seven moments [8] and Zernike moments [24]. Unfortunately, in many practical cases, images Q and A cannot be converted into binary images and thus this method cannot be applied.

Some approaches achieve RST-invariance using detection of interest points and edges, including: generalized Hough transform [2]; geometric hashing [12]; graph matching [13]; and curvature scale space [20].

Ullah and Kaneko use local gradient orientation histogram to obtain rotation-discriminating template matching [26]. Marimon and Ebrahimi present a much faster technique that also uses gradient orientation histogram [17]. The speedup is mainly due to the use of integral histograms. The gradient orientation histograms are not intrinsically rotation invariant and a "circular shifting" is necessary to find the best matchings.

Many other techniques use circular projections (also called ring projections) for the rotation-invariant template matching, for example [25, 15]. The underlying idea of these techniques is that the features computed over circular or annular regions are intrinsically rotation invariant.

The algorithms based on scale and rotation-invariant keypoints and local features, like SIFT (Scale Invariant Feature Transform) [16], GLOH (Gradient Location and Orientation Histogram) [19], and SURF (Speeded Up Robust Features) [3] have been widely used for image matching tasks. SIFT is very popular and has proven to be one of the most efficient methods to extract local features from images. It extracts some scale-invariant keypoints and computes their associated features based on local gradient orientations. Then, it finds the correspondences between the keypoints of Q and A based on the distances of the features. SIFT (followed by a Hough transform to identify clusters belonging to a single object) can be used for object recognition or template matching. These algorithms are particularly appropriate for finding query images with rich textures. However, they can fail to find some simple shapes with little grayscale variations or if the search image A has many repeating textures.

Fourier-Mellin transform is a popular technique for RST-invariant image registration, where image Q may appear rotated, scaled, and translated in A [22, 4]. However, this technique supposes implicitly that the non-intersecting areas of Q and A are small. So, it cannot be directly applied for template matching, where template Q and image A usually have large non-intersecting areas. If Fourier-Mellin is applied in this case, the correlation peaks become weak, making it difficult to detect the geometric transformation parameters.

There are also some commercial RSTBC-invariant template matchings, such as

Open eVision EasyMatch.[1] However, as they are commercial software we do not know exactly which algorithm is implemented inside.

In this chapter, we first present the conventional brute-force RSTBC-invariant template matching. Then, we present two algorithms that can accelerate this process, named Ciratefi (Circular, Radial, and Template-Matching Filter) and Forapro (Fourier coefficients of Radial Projections).

10.2 CONVENTIONAL BC-INVARIANT TEMPLATE MATCHING

10.2.1 CLASSIC TEMPLATE MATCHING

Conventional template matching (not RS invariant) uses some difference (or similarity) measuring function to evaluate how well the template Q matches the region around a given position (x, y) of image A. This function is evaluated for every pixel (x, y) of A, yielding a processed image P.

A post-processing algorithm uses the processed image P to find the occurrences of Q in A. Usually, the pixels of P with values below (above) some user-defined threshold are considered the occurrences of Q in A. The thresholded pixels usually form clusters, and only one representative pixel of each cluster should be considered an occurrence of Q. Many times, it is not easy to find a threshold that is appropriate for all situations. So, whenever possible, the use of threshold should be avoided. If the user somehow knows that Q occurs exactly once in A, the post-processing algorithm may simply search for the lowest (highest) value in P, instead of using a threshold. Similarly, if the user knows that Q occurs exactly n times in A and that any pair of instances are separated by distance at least d, the post-processing algorithm may use this information to find the n instances.

Let $\mathbf{q}$ be the columnwise vector obtained by copying the grayscales of Q's pixels and let $\mathbf{a}$ be the vector obtained by copying the grayscales of A's pixels around the pixel (x, y). Then, sum of absolute differences (SAD) or sum of squared differences (SSD) can be used to measure the difference between $\mathbf{q}$ and $\mathbf{a}$:

$$SAD(\mathbf{q}, \mathbf{a}) = \sum |\mathbf{q}_i - \mathbf{a}_i| \qquad (10.1)$$

$$SSD(\mathbf{q}, \mathbf{a}) = \sum [\mathbf{q}_i - \mathbf{a}_i]^2. \qquad (10.2)$$

Desirably, a template matching should be invariant to changes of brightness (B) and contrast (C), because the brightness and contrast of an observed object may vary according to the illumination condition. The brightness (or bias) of an image is related to its mean value, while the contrast (or gain or amplitude) is related to its standard deviation. We define that two image pieces $\mathbf{q}$ and $\mathbf{a}$ are equivalent under BC variation if there are a brightness correction factor α and a contrast correction factor $\beta \neq 0$ such that $\mathbf{a} = \beta\mathbf{q} + \alpha\mathbf{1}$, where $\mathbf{1}$ is the vector of 1s. Clearly, SAD and SSD are neither brightness nor contrast invariant. Thus, using these difference measuring

---

[1] http://www.euresys.com/Products/MachineVisionSoftware/MachineVision.asp

functions, the template matching will not find the template Q if it appears under different illumination conditions in the search image A.

In order to obtain BC-invariant template matching, let us consider using the dot product of the mean-corrected vectors $\tilde{\mathbf{q}} \cdot \tilde{\mathbf{a}}$ as the similarity measuring function, where $\tilde{\mathbf{q}} = \mathbf{q} - \bar{\mathbf{q}}$ is the mean-corrected vector, and $\bar{\mathbf{q}}$ is the mean of $\mathbf{q}$. Similar definitions are applicable to $\mathbf{a}$. Dot product is directly proportional to the cosine of the two argument vectors and can be used to measure their "similarity." When the window slides throughout A, we obtain "sliding dot product" or "cross correlation" (CC). Note that $\tilde{\mathbf{q}} \cdot \tilde{\mathbf{a}} = \tilde{\mathbf{q}} \cdot \mathbf{a}$, so it is necessary to mean-correct only the query image Q.

CC is invariant to brightness (because the argument vectors are mean corrected) but it is not invariant to contrast (because it is directly proportional to the amplitudes of the two vectors). Dot product becomes invariant to contrast (besides being invariant to brightness) if we divide it by the norm of the two vectors, obtaining correlation coefficient $\mathbf{X}(\mathbf{q}, \mathbf{a}) = \tilde{\mathbf{q}} \cdot \tilde{\mathbf{a}}/(\| \tilde{\mathbf{q}} \| \| \tilde{\mathbf{a}} \|)$. When the window slides throughout A, we obtain normalized cross correlation (NCC), a BC-invariant template matching. NCC always ranges from -1 to +1, unless in case of a division by zero. If all pixels of $\mathbf{q}$ or $\mathbf{a}$ have the same value, the denominator will be zero, and NCC will receive an undefined value. In practice, we do not have to worry about single-valued $\mathbf{q}$, because it does not make sense to try finding a single-valued image in another. However, we have to worry about regions of A with constant grayscale.

It is widely known that the computation of CC can be accelerated using FFT (Fast Fourier Transform). The computation of NCC can be accelerated using FFT and integral image [14].

10.2.2   CONTRAST INVARIANCE EXAMPLE

Figure 10.1 depicts an example that highlights the difference between using CC and NCC in template matching. In search image A (Figure 10.1(b)), there are four instances of the bear image Q (Figure 10.1(a)), each one with different brightness and contrast. Figure 10.1(c) depicts the output of the template matching using CC. Actually, we computed the absolute value of CC (to match both positive and negative instances) and mapped the highest value to white and the lowest value to black. Thresholding Figure 10.1(c), we obtain Figure 10.1(d), where we missed an instance of bear (false negative) and detected a non-occurring bear (false positive). Figure 10.1(e) depicts the output of the template matching using NCC. Again, we computed the absolute value of NCC output and mapped the lowest (highest) value to black (white). Figure 10.1(f) was obtained thresholding Figure 10.1(e). Note that all four instances of bear were correctly detected, regardless of their brightness and contrast.

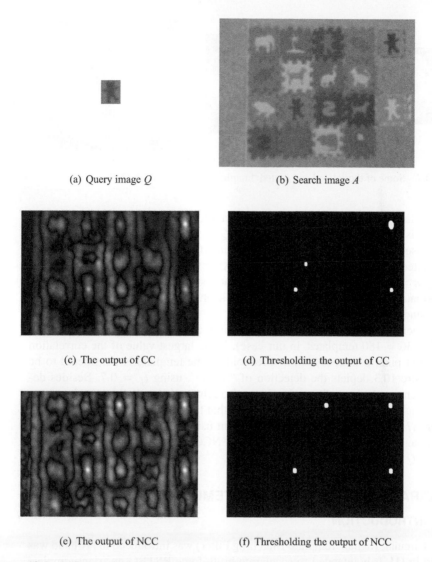

(a) Query image Q (b) Search image A

(c) The output of CC   (d) Thresholding the output of CC

(e) The output of NCC   (f) Thresholding the output of NCC

Figure 10.1 Difference between using CC and NCC in template matching. CC did not succeed detecting the four instances of the bear, while NCC succeeded.

10.3 BRUTE FORCE RSTBC-INVARIANT TEMPLATE MATCHING

In this section, we describe the "brute force" RSTBC-invariant template matching obtained by applying repeatedly the conventional BC-invariant template matching, using templates rotated by many angles and scaled by many factors.

To obtain the brute force RSTBC-invariant template matching, we said above that

Figure 10.2   Some of the rotated and scaled "maple" template images Q.

the query shape Q must be rotated by every angle and scaled by every factor. In practice, it is not possible to rotate and scale Q by every angle and scale, but only by some discrete set of angles and scales. Figure 10.2 depicts some of the "maple" image rotated in $m = 36$ different angles ($\alpha_0 = 0, \alpha_1 = 10, ..., \alpha_{35} = 350$) and scaled by $n = 5$ different factors. To avoid that a small misalignment may cause a large mismatching, a low-pass filter (for example, the Gaussian filter) smoothes both images A and Q.

Then, each pixel p of A is tested for matching against all the transformed templates ($5 \times 36 = 180$ templates, in our case). If the largest value of the correlation coefficient at pixel p is above some threshold t_f, the template is considered to be found. Figure 10.3 depicts the detection of "maple", using $t_f = 0.7$. Besides detecting the shape, the brute force algorithm also returns the scale factor and rotation angle for each matching. The only problem is that this process takes 950s, using an Intel Core i7-4500U 1.8GHz computer (without using multi-threading, FFT and/or integral image to accelerate the computation of NCC; image A has 347×471 pixels and image Q has 71×71 pixels).

10.4   CIRATEFI: RSTBC-INVARIANT TEMPLATE MATCHING

10.4.1   INTRODUCTION

Ciratefi (Circular, Radial, Template-Matching Filter) was introduced in [10] and was improved in [1]. It is intended to accelerate brute force RSTBC-invariant template matching, that is, to find a query template image Q in a larger search image to analyze A, invariant to rotation (R), scaling (S), and translation (T) and with controlled robustness to brightness (B) and contrast (C) changes.

Ciratefi consists of three cascaded filters. Each filter successively excludes pixels that have no chance of matching the template from further processing, while keeping the "candidate pixels" that can match the template to further refined classifications. The first filter, called Cifi (circular sampling filter), uses the projections of images A and Q on circles to divide the pixels of A in two categories: those that have no chance of matching the template Q (to be discarded) and those that have some chance (called first-grade candidate pixels). This filter is responsible for determining the

Figure 10.3 "Maple" detected by the brute force RSTBC-invariant template matching.

scale without isolating the shapes. It determines a "probable scale factor" for each first-grade candidate pixel. The second filter, called Rafi (radial sampling filter), uses the projections of images A and Q on radial lines and the "probable scale factors" computed by Cifi to upgrade some of the first-grade candidate pixels to second grade. It also assigns a "probable rotation angle" to each second-grade candidate pixel. The pixels that are not upgraded are discarded. The third filter, called Tefi (template matching filter), is a conventional template matching using correlation coefficient. The second-grade candidate pixels are usually few in number and Cifi and Rafi have already computed their probable scales and rotation angles. Thus, the conventional template matching can quickly categorize all the second-grade candidate pixels in true and false matchings.

10.4.2 BASIC CIRATEFI

10.4.2.1 Circular Sampling Filter

The first filter, called Cifi (Circular sampling Filter) uses the projections of the images A and Q on a set of circular rings (Figure 10.4) to detect the "first-grade candidate pixels" and their "best matching scales." Given an image B, let us define the circular sampling as the average grayscale of the pixels of B situated at distance r from pixel (x, y):

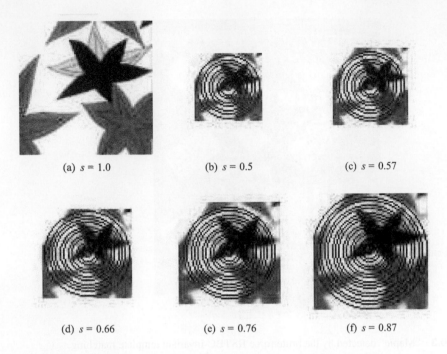

(a) $s = 1.0$ (b) $s = 0.5$ (c) $s = 0.57$

(d) $s = 0.66$ (e) $s = 0.76$ (f) $s = 0.87$

Figure 10.4 The query image and the circular projections at different scales.

$$S_B^\Omega(x, y, r) = \frac{1}{2\pi r} \int_0^{2\pi} B(x + r\cos\theta, y + r\sin\theta)d\theta. \qquad (10.3)$$

We use the superscript Ω to indicate circular sampling.

Given the template Q and the set of n scales $s_0, s_1, ..., s_{n-1}$, the image Q is resized to each scale s_i, obtaining the resized templates $Q_0, Q_1, ..., Q_{n-1}$. Then, each resized template Q_i is circularly sampled at a set of circle rings with l predefined radii $r_0, r_1, ..., r_{l-1}$, yielding a 2D table of multi-scale rotation-invariant features with n rows (scales) and l columns (radii):

$$C_Q[i, k] = S_{Q_i}^\Omega(x_0, y_0, r_k), 0 \le i < n \text{ and } 0 \le k < l \qquad (10.4)$$

where (x_0, y_0) is the central pixel of Q. In small scales, some of the outer circles may not fit inside the resized templates. These circles are represented by a special value in table C_Q (say, -1) and are not used to compute the correlations.

Given the image to analyze A, we build the 3D image that contains l circular projections for each pixel (x, y):

$$C_A[x, y, k] = S_A^\Omega(x, y, r_k), 0 \le k < l, (x, y) \in \text{domain}(A) \qquad (10.5)$$

Cifi uses matrices C_Q and C_A to detect the circular sampling correlation at the best matching scale for each pixel (x, y):

Figure 10.5 Radial projections at the selected scale.

$$X_{A,Q}^{\Omega}(x,y) = \max_{i=0}^{n-1} [\mathbf{X}(C_Q[i], C_A[x,y])], \qquad (10.6)$$

where $\mathbf{X}(\mathbf{a}, \mathbf{b})$ is the correlation coefficient between vectors $\mathbf{a}$ and $\mathbf{b}$. A pixel (x, y) is classified as a "first-grade candidate pixel" if $X_{A,Q}^{\Omega}(x,y) \geq t_1$ for some threshold t_1. The appropriate value for t_1 depends on the application. Assigning small value for t_1 makes the algorithm slower, and assigning large value for t_1 decreases the algorithm's accuracy. In Section 10.4.3, we explain how to choose this parameter automatically. The "best matching scale" of a first-grade candidate pixel (x, y) is the argument that maximizes the correlation:

$$G_{A,Q}^{\Omega}(x,y) = \operatorname*{argmax}_{i=0}^{n-1} [\mathbf{X}(C_Q[i], C_A[x,y])] \qquad (10.7)$$

10.4.2.2 Radial Sampling Filter

The second filter, called Rafi (Radial sampling Filter), uses the projections of images A and Q on a set of radial lines (Figure 10.5) to upgrade some of the first-grade candidate pixels to the second grade. The pixels that are not upgraded are discarded. It also assigns the "best matching rotation angle" to each second-grade candidate pixel. Given an image B, let us define the radial sampling as the average grayscale of the pixels of B located on the radial line with one vertex at (x, y), length λ and inclination α:

$$S_B^{\Phi}(x, y, \lambda, \alpha) = \frac{1}{\lambda} \int_0^{\lambda} B(x + t\cos\alpha, y + t\sin\alpha)dt. \qquad (10.8)$$

We use the superscript Φ to indicate radial sampling.

Given the template Q and the set of m angles $\alpha_0, \alpha_1, ..., \alpha_{m-1}$, Q is radially sampled using r_{l-1}, the radius of the largest sampling circle that fits inside Q, yielding a vector with m features:

$$R_Q[j] = S_Q^{\Phi}(x_0, y_0, r_{l-1}, \alpha_j), 0 \leq j < m \qquad (10.9)$$

where (x_0, y_0) is the central pixel of Q.

For each first-grade candidate pixel (x, y), A is radially sampled at its probable scale $i = G_{A,Q}^{\Omega}(x, y)$:

$$R_A[x, y, j] = S_A^{\Phi}(x, y, s_i r_{l-1}, \alpha_j), 0 \leq j < m \tag{10.10}$$

where $s_i r_{l-1}$ is the radius of the scaled template Q_i.

At each first-grade candidate pixel (x, y), Rafi uses vectors $R_A[x, y]$ and R_Q to detect the radial sampling correlation at the best matching angle:

$$X_{A,Q}^{\Phi}(x, y) = \max_{j=0}^{m-1} [\mathbf{X}(R_A[x, y], \mathrm{cshift}_j(R_Q))] \tag{10.11}$$

where $\mathbf{X}(\mathbf{a}, \mathbf{b})$ is the correlation coefficient between vectors $\mathbf{a}$ and $\mathbf{b}$ and "cshift_j" means circular shifting (or element-wise rotation) j positions of the argument vector. A first-grade pixel (x, y) is upgraded to the second grade if $X_{A,Q}^{\Phi}(x, y) \geq t_2$ for some threshold t_2. In Section 10.4.3, we explain how to select this parameter automatically. The probable rotation angle at a second-grade candidate pixel (x, y) is the angle that maximizes the correlation:

$$G_{A,Q}^{\Phi}(x, y) = \underset{j=0}{\overset{m-1}{\mathrm{argmax}}} [\mathbf{X}(R_A[x, y], \mathrm{cshift}_j(R_Q))] \tag{10.12}$$

10.4.2.3   Template Matching Filter

The third filter, called Tefi (Template matching Filter), computes the correlation coefficient between the neighborhood of each second-grade candidate pixel and the template scaled and rotated using the scale and angle determined respectively by Cifi and Rafi.

Tefi first resizes and rotates template Q to all m angles and n scales and stores them in a table named T_Q. Let (x, y) be a second-grade candidate pixel, with its probable scale $i = G_{A,Q}^{\Omega}(x, y)$ and probable angle $j = G_{A,Q}^{\Phi}(x, y)$. Tefi computes the correlation coefficient between the image A at pixel (x, y) and $T_Q[i, j]$ (the template image Q at scale s_i and angle j). Actually, to make the algorithm more robust, Tefi tests scales $i - 1$, i, $i + 1$ and angles $j - 1$, j, $j + 1$ (in this case, subtraction and addition are computed modulus m) and takes the greatest correlation coefficient. If the greatest coefficient is above some threshold t_3, the template is considered to be found at pixel (x, y), at the scale and the angle that yielded the greatest correlation. If the user knows that Q appears only once in A, the threshold t_3 is not used. Instead, a single pixel with the highest correlation is chosen.

10.4.2.4   Considerations

Cifi, Rafi, and Tefi cannot be executed in a different order, because Cifi detects the probable scale, to be used by Rafi, and Rafi detects the probable angle, to be used by Tefi. Rafi cannot be executed without the probable scale, and Tefi cannot be executed without the probable scale and angle.

Figure 10.6   A query image with ambiguous scale and rotation angle.

Some query images may have "ambiguous" scale and/or rotation angle. Consider the query image depicted in Figure 10.6. All the circular and radial projections yield the same average grayscale. So, Ciratefi cannot search for this image. However, even in this case, Ciratefi can search for an off-centered subimage of this query image.

10.4.3   IMPROVED CIRATEFI

In basic Ciratefi, there are many parameters left to be adjusted by hand. In this section, we introduce some improvements to automate the choice of the parameters.

10.4.3.1   Thresholds

The appropriate values for t_1, t_2, and t_3 are difficult to set by hand and can vary from applcation to application. Choosing small thresholds (great number of candidate pixels) increases the accuracy but makes the process slower. Choosing large thresholds may discard true matchings. So, we replaced these parameters by p_1, p_2, and p_3, where p_1 and p_2 are the percentage of the first- and second-grade candidate pixels in relation to the original number of the pixels of A; and p_3 is the total number of occurrences of Q in A. We also defined d_1, d_2, and d_3, where d_i is the minimal distance that separates the candidate pixels of the degree i. Experimentally, we chose $p_1 = 2\%$, $p_2 = 1\%$, and $p_3 = 1$ pixel, for applications where Q appears once in A. This means that 2% of pixels of A are first-grade candidate, 1% of pixels of A are second-grade candidate, and one single pixel is chosen as the final matching. If Q appears more than once, it is possible to define a minimal distance between matchings. For example, specifying $p_3 = 4$ pixels and $d_3 = 50$ pixels, the program will choose 4 matchings separated by at least 50 pixels. In practice, we can find quickly the candidate pixels that satisfy the conditions above by using the data structure known as "priority queue" or "heap."

10.4.3.2   Scale Range

It is also difficult to choose the appropriate scale range. So, we decided to build a pyramidal structure for A. This structure is widely used to obtain scale-invariance, for example, in [16]. Figure 10.7 depicts an example of this structure, obtained by

Figure 10.7   The scale range expands to infinite by building a pyramidal structure for A.

concatenating the original A with its reduced versions at scales $0.5, 0.25$, etc. For the query image Q, we chose to use a set of 5 scales in geometric progression $0.5, 0.57, 0.66, 0.76, 0.87$. With this choice, Ciratefi becomes scale invariant in the range $[0.5, \infty]$. For example, suppose that Q at scale 0.66 was detected in A at scale 0.25. This means that Q scale $(1/0.25) \times 0.66 = 2.64$ appears in original A. In our implementation, we took care to not detect the template in the boundaries of different scales of pyramidal A.

In Cifi, we chose to use 16 circles whose radii increase in arithmetic progression from zero to the radius of the query image at the greatest scale. With this choice, the query image at the smallest scale (0.5) has 9 circles inside, enough to compute the correlation coefficient with some precision. Query image Q must be large enough in order to have 16 distinct circles inside the query image at scale 0.87. This happens when the size of Q is larger than 39×39. In the experiments, we use query images with typical size 61×61. Too large query Q makes the algorithm slower without increasing the accuracy. In this case, we suggest extracting and finding a sub-image of Q or resizing down both Q and A. In Rafi, we chose to use 36 angles.

With these alterations and choices, we obtained an implementation where the standard parameters can be used in all experiments.

10.4.3.3   Structural Similarity

Structural similarity (SSIM) index is an image distortion metric for grayscale images designed to emulate the human visual perceptual system [27]. It separates the image distortion in three independent components: luminance (or brightness), contrast, and structure (or correlation coefficient). Then, each component receives a weight that depends on the application. Using structural similarity, we can assign different weights to changes in brightness, contrast, and correlation coefficient, generalizing the NCC-based template matching (that takes into account only the correlation coefficient).

To assess the perceptual similarity between the query image Q with content $\mathbf{q}$ and a region $\mathbf{a}$ around pixel (x, y) of image to analyze A, the local statistics $\mu_{\mathbf{q}}$, $\mu_{\mathbf{a}}$, $\sigma_{\mathbf{q}}$, $\sigma_{\mathbf{a}}$, $\sigma_{\mathbf{qa}}$ are computed (where $\mu_{\mathbf{q}}$ is the mean of $\mathbf{q}$, $\sigma_{\mathbf{q}}$ is the standard deviation of $\mathbf{q}$, and $\sigma_{\mathbf{qa}}$ is the covariance of $\mathbf{q}$ and $\mathbf{a}$). Then, three similarity functions are computed:

$$l(\mathbf{q}, \mathbf{a}) = \frac{2\mu_{\mathbf{q}}\mu_{\mathbf{a}} + C_1}{\mu_{\mathbf{q}}^2 + \mu_{\mathbf{a}}^2 + C_1} \tag{10.13}$$

$$c(\mathbf{q}, \mathbf{a}) = \frac{2\sigma_{\mathbf{q}}\sigma_{\mathbf{a}} + C_2}{\sigma_{\mathbf{q}}^2 + \sigma_{\mathbf{a}}^2 + C_2} \tag{10.14}$$

$$s(\mathbf{q}, \mathbf{a}) = \frac{\sigma_{\mathbf{qa}} + C_3}{\sigma_{\mathbf{q}}\sigma_{\mathbf{a}} + C_3} \tag{10.15}$$

where l measures the lightness (brightness) similarity, c measures the contrast similarity, and s measures the structural similarity. The constants C_i are small numbers introduced to avoid numerical instability when the denominators are close to zero. Note that, disregarding C_3, $s(\mathbf{q}, \mathbf{a})$ is actually the correlation coefficient. The structural similarity is defined:

$$\mathrm{SSIM}(\mathbf{q}, \mathbf{a}) = [l(\mathbf{q}, \mathbf{a})]^\alpha [c(\mathbf{q}, \mathbf{a})]^\beta [s(\mathbf{q}, \mathbf{a})]^\gamma \tag{10.16}$$

where α, β and γ are parameters to adjust the relative importance of the three components. Using SSIM as the similarity measure, the user can assign different weights to brightness, contrast, and structure depending on the application. To obtain a complete invariance to brightness/contrast, one can set $\alpha = \beta = 0$ and $\gamma = 1$. However, this is not a good choice because regions of A with constant grayscale will match any template (SSIM will be one). In (almost) BC-invariant applications, we use $\alpha = \beta = 0.01$ and $\gamma = 0.98$. If brightness/contrast changes only slightly in an application, higher weights may be assigned to α and β to increase the accuracy.

In the improved grayscale Ciratefi, we replace the correlation coefficient by SSIM as the similarity measure in all the three filters.

10.4.4 EXPERIMENTAL RESULTS

We compare improved Ciratefi[2] with SIFT (Scale Invariant Feature Transform) proposed and implemented by Lowe[3] and EasyMatch 1.1,[4] a template matching tool of Euresys Open eVision.

Lowe's SIFT implementation finds the keypoint correspondences, but does not find the template matching locations. To find Q in A using the keypoint correspondences, we implemented the generalized Hough transform [2] to identify clusters, as suggested by Lowe. Although both SIFT and Ciratefi can be used for image matching, they are quite different. We enumerate two differences below:

---

[2]http://www.lps.usp.br/~hae/software/cirateg/

[3]http://www.cs.ubc.ca/~lowe/keypoints/

[4]http://www.euresys.com/Products/MachineVisionSoftware/MachineVision.asp

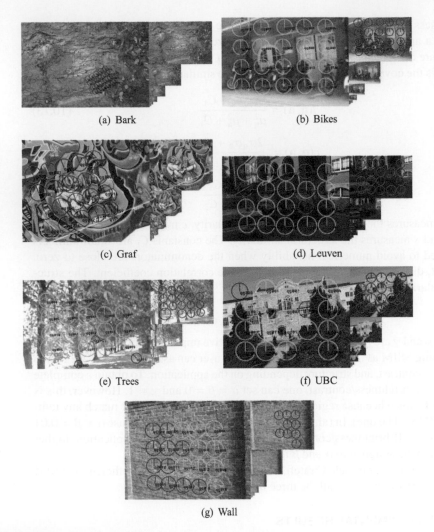

(a) Bark

(b) Bikes

(c) Graf

(d) Leuven

(e) Trees

(f) UBC

(g) Wall

Figure 10.8 We use Mikolajczyk database of natural images to compare Ciratefi with SIFT and EasyMatch.

1. SIFT is based on keypoints and local gradient orientations while Ciratefi directly compares grayscales of regions of Q and A. This makes Ciratefi more reliable than SIFT when the images are blurred, have large areas with constant grayscales, suffer JPEG compression, have few textures, or suffer large brightness/contrast changes. SIFT is more reliable than Ciratefi when the images have many small textures. A small misalignment of the tiny textures may decrease the correlation used by Ciratefi to find the template, decreasing its accuracy.

2. Template matching using SIFT followed by Hough transform is robust to

Table 10.1

Error rates of each algorithm searching for 120 patches of Mikolajczyk database. * means that the experiment was not done.

	Ciratefi	SIFT	EasyMatch
Bark (zoom/rotation)	0%	0%	*
Bikes (focus blur)	0%	31%	53%
Graf (viewpoint)	38%	55%	*
Leuven (cam. apert.)	3%	23%	55%
Trees (focus blur)	13%	38%	86%
UBC (JPEG)	3%	13%	49%
Wall (viewpoint)	30%	33%	*
Average	12%	27%	60%

partial occlusions, while Ciratefi alone is not. However, Ciratefi (as well as any template matching algorithm) can become robust to partial occlusions by taking some sub-templates of Q, finding them all in A, and combining the results by a Hough transform.

10.4.4.1 Mikolajczyk's Image Database

We compared the algorithms using Mikolajczyk's image database.[5] In all sets except Bark, we reduced the images to 50% of the original sizes, extracted twenty 61×61 templates uniformly distributed in the first image and searched for them in the six reduced images. In set Bark, we reduced the first image to 50% of the original size, extracted twenty 61×61 templates uniformly distributed in the first image and searched for them in the six original non-reduced images. All algorithms knew that there was only one instance of Q inside each A. The Hough transform that follows SIFT was programmed to detect the template even if there is only one keypoint correspondence between Q and A. In EasyMatch, we set the range of scales from 50% to 200%. We tested EasyMatch only in some cases, due to the impossibility to run automated tests. The results are in Table 10.1 and some of the processed images are in Figure 10.8. Overall, Ciratefi made less errors than SIFT and EasyMatch. Ciratefi clearly outperforms SIFT and EasyMatch in situations with focus blur (Bikes), large brightness/contrast change (Leuven), and JPEG compression (UBC).

10.4.4.2 ALOI Image Database

ALOI (Amsterdam Library of Object Images) is an image collection of small objects.[6][7] In order to capture the sensory variation in object recordings, the authors

---

[5]http://www.robots.ox.ac.uk/~vgg/research/affine/

[6]http://staff.science.uva.nl/~aloi/

Table 10.2

Errors rates obtained searching for 80 objects in ALOI database. * means that the experiment was not done.

	Ciratefi	SIFT	EasyMatch
Color-A	1%	5%	60%
Color-B	0%	8%	*
Illum-A	13%	38%	53%
Illum-B	26%	48%	*
View-A	29%	53%	65%
View-B	24%	56%	*
Blur-A	0%	31%	*
Blur-B	0%	28%	*
Jpeg-A	3%	19%	*
Jpeg-B	0%	14%	*
Average	9%	30%	59%

systematically varied viewing angle, illumination angle, and illumination color for each object.

We took the images with 4 different illumination colors (with the illuminating lamp temperatures 3075K, 2750K, 2475K, and 2175K) of the first 20 objects, reduced them by 2, and glued the images with the same illumination temperatures together, obtaining four 288×480 images. We searched the objects in the first image (3075K), cropped to 61×61 pixels, in the four images. The errors are depicted in row Color-A of Table 10.2. We repeated the experiment using the next 20 objects (row Color-B).

We took the images with 4 different illumination directions (identified as l8c1, l7c1, l6c1, l4c1 in the database) and searched the objects in the first image (l8c1) in the four images, obtaining rows Illum-A and Illum-B.

We searched the unrotated objects in images with the objects rotated in 4 different angles (0, 20, 40, and 60 degrees), obtaining rows View-A and View-B. Figure 10.9 depicts two of the images obtained in this experiment.

We searched the unblurred objects in images distorted with Gaussian blur with kernels 1×1, 3×3, 5×5, and 7×7 ($\sigma = 0$, 0.95, 1.25, and 1.55), obtaining rows Blur-A and Blur-B.

We searched the uncompressed objects in JPEG-compressed images with qualities 100%, 75%, 50%, and 25%, obtaining rows JPEG-A and JPEG-B.

Let us analyze the results of Table 10.2. Ciratefi has lower error rates than SIFT and EasyMatch in all tests. This superiority is especially evident in: illumination color variation, blurring, and JPEG compression. EasyMatch was the worst algorithm in all tests.

To find an object, typically Ciratefi takes 9s, SIFT takes 2s, and EasyMatch less than 1s. We remark that our Ciratefi implementation is not optimized.

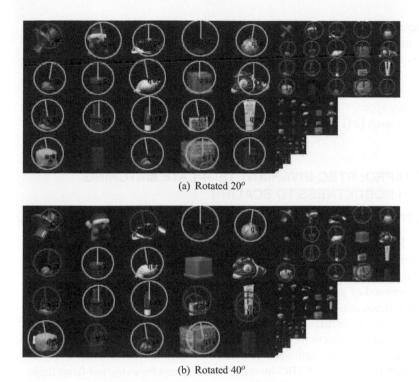

(a) Rotated 20°

(b) Rotated 40°

Figure 10.9 We took the images of 20 unrotated objects and searched for them in images of the objects rotated by 20° and by 40°.

10.4.5 CONSIDERATIONS

Our experiments indicate that Ciratefi is more accurate than SIFT and EasyMatch. Nevertheless, SIFT has many practical advantages over Ciratefi:

1. SIFT is faster than Ciratefi. SIFT is especially fast when searching for many different templates in an unchanging image A, because most of the processing time is spent in computing the keypoints and the features of A (that can be done only once).
2. The query image in Ciratefi must contain only the searching pattern. The query image in SIFT may contain the searching pattern among many other "junk" background patterns, because it searches the occurrences of the keypoints of Q in A instead of the whole query image.
3. Template matching using SIFT followed by Hough transform is robust to partial occlusions, while Ciratefi by itself is not.
4. SIFT is wholly scale invariant, while even the improved Ciratefi is scale invariant only from 0.5 to ∞. As a truly scale-invariant method, SIFT can find a small or large template. Meanwhile, Ciratefi is better suited for finding relatively small templates, because a large template may be time consuming.

In our opinion, even if Ciratefi is not practical right now for some applications, it deserves to be more thoroughly studied because of its superior accuracy. There remains the challenge of designing an algorithm as practical as SIFT and as accurate as Ciratefi.

Ciratefi repeats exactly the same series of simple operations for each pixel, making it especially appropriate for highly parallel implementation. The author has participated in research [21] to implement Ciratefi in FPGA (Field Programmable Gate Array).

10.5 FORAPRO: RTBC-INVARIANT TEMPLATE MATCHING WITH ROBUSTNESS TO SCALING

10.5.1 INTRODUCTION

The second RSTBC-invariant template matching we consider is named Forapro (Fourier coefficients of Radial Projections) [11]. Forapro (like Ciratefi) uses circular and radial projections. Circular and radial projections reduce the 2D information of the neighborhood of a pixel into unidimensional vectors. It is possible to reduce even more the dimension of features by computing the first low-frequency complex Fourier coefficients of circular and radial projections. Note that the original vector of circular and radial projections can be roughly reconstructed using the first low-frequency complex Fourier coefficients and so the Fourier coefficients aptly represent the original vectors. The RTBC-invariant features can be extracted from those coefficients. These features are used to construct RSTBC-invariant template matching.

10.5.2 FORAPRO

Choi and Kim [5] have proposed an interesting approach to accelerate circular projection-based rotation-invariant template matching. Their method computes, for each pixel (x, y) in A, circular projections of its neighborhood, forming a unidimensional vector $C(A(x, y))$. The circular projection reduces the 2D information of the neighborhood of $A(x, y)$ into a unidimensional rotation-invariant vector, accelerating the processing. To reduce the data even more, the method computes the first low-frequency complex Fourier coefficients $c_1, c_2, ...$ of $C(A(x, y))$, and uses them as the rotation-invariant features. Actually, this technique computes the Fourier coefficients directly, without explicitly computing the circular projections, by convolving A with appropriate kernels via FFT. The features of Q are compared with the features of each pixel $A(x, y)$ to test for possible matchings.

Forapro improves Choi and Kim's algorithm by using rotation-invariant and rotation-discriminating features derived from radial projections (together with the features derived from circular projections). We compute, for each pixel (x, y) in A, the radial projections of its neighborhood, forming a unidimensional vector of radial projections $R(A(x, y))$. Then, we compute the first low-frequency complex inverse

Fourier coefficients $r_1, r_2, \dots$ of $R(A(x, y))$. Actually, we do not compute the radial projections, but the Fourier coefficients directly. Convolutions in the frequency domain are used to compute quickly the Fourier coefficients, employing appropriate kernels and FFT. In order to obtain B-invariance, the DC coefficient r_0 is not used. In order to obtain C-invariance, the magnitudes of the radial coefficients are normalized. It is possible to derive many rotation-invariant features and a rotation-discriminating feature from the magnitudes and angles of radial coefficients.

Template matchings based on pre-computed RTBC-invariant features are advantageous principally when the algorithm must search an image A for a large number of templates T_i. In this case, the vector of RTBC-invariant features $v_f(A(x, y))$ is computed only once for each pixel $A(x, y)$. Then, each template T_i can be found quickly in A by computing the vector of features $v_f(T_i(x_0, y_0))$ at the central pixel $T_i(x_0, y_0)$ and comparing it with $v_f(A(x, y))$. If the distance is below some threshold, then the neighborhood of $A(x, y)$ is "similar" (in RTBC-invariant sense) to the template image T_i and (x, y) is considered a candidate for the matching. This property makes our algorithm suitable for: finding many different query images in A, multi-scale searching, and partial occlusion-robust template matching. If the number N of template images is very large, it is possible to use some special data structure to accelerate the searching even more, for example, some variation of the kd-tree [6].

Our compiled programs and some test images are available at www.lps.usp.br/ ~hae/software/forapro.

10.5.3   NEW RTBC-INVARIANT FEATURES

10.5.3.1   Radial IDFT Coefficients

Given a grayscale image A, let us define the radial projection $S_A^\Phi(x, y, \lambda, \alpha)$ as the average grayscale of the pixels of A located on the radial line with one vertex at (x, y), length λ and inclination α:

$$S_A^\Phi(x, y, \lambda, \alpha) = \frac{1}{\lambda} \int_0^\lambda A(x + t\cos\alpha, y + t\sin\alpha)dt. \tag{10.17}$$

We use the superscript Φ to indicate circular sampling. In practice, a sum must replace the integral, because digital images are spatially discrete. The vector of m discrete radial projections at pixel $A(x, y)$ with radius λ can be obtained by varying the angle α:

$$R_{A(x,y)}^\lambda[m] = S_A^\Phi(x, y, \lambda, 2\pi m/M), \ 0 \le m < M. \tag{10.18}$$

The vector of radial projections $R_{A(x,y)}^\lambda[m]$ characterizes the neighborhood of $A(x, y)$ of radius λ. If A rotates, then this vector shifts circularly. This property is illustrated in Figure 10.10.

The k-th Fourier coefficient of a vector of radial projections R is (we omit indices $A(x, y)$ and λ):

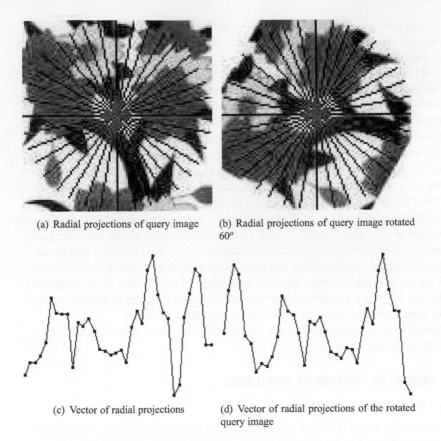

(a) Radial projections of query image

(b) Radial projections of query image rotated 60°

(c) Vector of radial projections

(d) Vector of radial projections of the rotated query image

Figure 10.10   Rotation of the image causes circular shifting in the vector of radial projections.

$$r[k] = \sum_{m=0}^{M-1} R[m] \exp(-j2\pi km/M),\ 0 \le k < M. \tag{10.19}$$

The Fourier coefficients of a vector of radial projections can be computed directly convolving A with an appropriate kernel K, without explicitly calculating the radial projections. Figure 10.11(a) depicts the "sparse DFT kernel" K (with $M = 8$ angles) such that the convolution $A * \check{K}$ yields the first Fourier coefficient of the radial projections (where $\check{K}(x, y) = K(-x, -y)$ is the double reflection of K or 180 degree rotation):

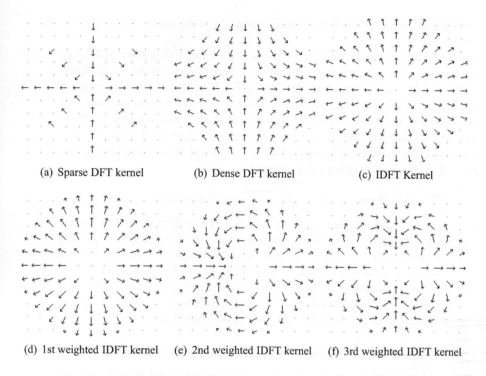

(a) Sparse DFT kernel   (b) Dense DFT kernel   (c) IDFT Kernel

(d) 1st weighted IDFT kernel   (e) 2nd weighted IDFT kernel   (f) 3rd weighted IDFT kernel

Figure 10.11   Radial kernels.

$$(A * \check{K})(x, y)$$
$$= \sum_p \sum_q A(p, q)\check{K}(x - p, y - q)$$
$$= \sum_p \sum_q A(p, q)K(p - x, q - y). \qquad (10.20)$$

It is well known that the convolution $A * \check{K}$ can be computed by multiplications in the frequency domain:

$$A * \check{K} \Leftrightarrow \mathbf{A}\check{\mathbf{K}} \qquad (10.21)$$

where $\mathbf{A}$ and $\check{\mathbf{K}}$ are respectively the discrete Fourier transforms of A and $\check{K}$.

The sparse kernel in Figure 10.11(a) does not take into account most of the outer pixels and consequently it does not yield accurate features. To overcome this problem, the "dense DFT kernel" depicted in Figure 10.11(b) can be used instead. It fills all empty kernel elements, except the very central pixel. The non-zero elements of this kernel are defined as:

$$\text{denseRadialDftKernel}_k[x, y] = \exp\left(-jk\angle(x + yj)\right) \qquad (10.22)$$

where k is the order of the Fourier coefficient and $\angle(.)$ is the angle of the complex number.

The linear filter using this kernel has no intuitive meaning. Using inverse DFT, the result of the convolution acquires a meaning: it becomes analogous to the gradient. Figure 10.11(c) depicts the kernel obtained using IDFT and $k = 1$. In order to make the kernel more "stable," that is, to make the result of the convolution less sensitive to small perturbations like sub-pixel translation or rotation, we assign less weight to the pixels in the outer and central regions, resulting the weighted kernel depicted in Figure 10.11(d). We tested empirically a number of weight distributions and chose the most stable one. The resulting radial kernel is:

$$\text{radialKernel}_k[x, y] = \sqrt{r(\lambda - r)} \exp\left(jk\angle(x + yj)\right) \qquad (10.23)$$

where $r = \sqrt{x^2 + y^2}$ and λ is the radius of the kernel. The weights of the five central pixels are zeroes. In order to make the weights sum to one, all weights may be divided by the sum of the weights. The kernels used to obtain the inverse Fourier coefficients with $k = 2$ and $k = 3$ are depicted respectively in Figures 10.11(e) and 10.11(f).

We will call the convolution of $A(x, y)$ with the double reflection of the k-th radial kernel "k-th radial coefficient" and denote it $r_k(A(x, y))$ or simply r_k. We will call $\angle r_k$ and $|r_k|$ respectively "k-th radial angle" and "k-th radial magnitude."

10.5.3.2   Rotation-Discriminating "Canonical Orientation"

The rotation-discriminating feature, or the "canonical orientation," is the first radial angle:

$$\text{canonicalOrientation} = \angle r_1. \qquad (10.24)$$

Canonical orientation $\angle r_1(A(x, y))$ indicates the local direction of $A(x, y)$ within a neighborhood of radius λ. If A rotates θ radians, then the vector of radial projections $R^\lambda_{A(x,y)}[m]$ shifts circularly θ radians and, by the time shift property of IDFT, $r_1(A(x, y))$ is multiplied by $\exp(j\theta)$. In other words, if A rotates, the first radial angle $\angle r_1(A(x, y))$ rotates by the same angle. Moreover, an illumination change does not alter the canonical orientation.

10.5.3.3   Rotation-Invariant "Vector of Radial Magnitudes"

Radial magnitudes are invariant to rotation because if A rotates, then the vector of radial projections $R^\lambda_{A(x,y)}[m]$ shifts circularly, and a circular shifting does not change the magnitudes of the IDFT coefficients. Radial magnitudes $|r_k|, k \geq 1$, are also invariant to brightness because a brightness alteration only affects the DC coefficient r_0. Finally, the ratios between radial magnitudes are invariant to contrast (besides being rotation and brightness invariant), because a contrast alteration multiplies by the same factor all the radial coefficients. For example, let r_1 and r_k be the first and the k-th radial coefficients ($k \geq 2$). Then, the ratio of their magnitudes $|r_k|/|r_1|$ is invariant to rotation and illumination. In our implementation, instead of ratio, we use the following vector of radial magnitudes v_{rm} that takes into account the magnitudes of all radial coefficients up to degree K:

$$v_{rm} = \mathrm{L}^1\text{-versor}[|r_1|, |r_2|, ..., |r_K|] \tag{10.25}$$

where L^1-versor consists of dividing each element of the vector by its L^1-length $|r_1| + |r_2| + ... + |r_K|$. We define the distance function between two v_{rm}'s as:

$$
\begin{aligned}
&\text{distance}\left(v_{rm}(A(x,y)), v_{rm}(T(x_0, y_0))\right) \\
&= \frac{1}{2} \parallel v_{rm}(A(x,y)) - v_{rm}(T(x_0, y_0)) \parallel_1 .
\end{aligned}
\tag{10.26}
$$

This distance is limited to interval $[0, 1]$. We use L^1-distance instead of Euclidean L^2-distance because the computation of L^1-norm is faster than L^2-norm.

10.5.3.4   Rotation-Invariant "Vector of Radial Angles"

If A rotates θ radians, then the vector of radial projections $R^\lambda_{A(x,y)}[m]$ shifts circularly θ radians and, by the time shift property of IDFT, the k-th radial coefficient $r_k(A(x,y))$ is multiplied by $\exp(jk\theta)$. Moreover, a brightness or contrast change does not alter $\angle r_k$. Thus, the difference between $\angle r_k$ and $k\angle r_1$ is RTBC-invariant. We call this feature "difference of radial angles" k and 1:

$$\mathrm{dra}_k = \mathrm{mod}(\angle r_k - k\angle r_1, 2\pi), k \geq 2. \tag{10.27}$$

This feature is computed modulo 2π, because it is an angle. In our implementation, we packed the dra's up to order $K \geq 2$ in a structure that we named vector of radial angles v_{ra}:

$$v_{ra} = [\mathrm{dra}_2, \mathrm{dra}_3, ..., \mathrm{dra}_K]. \tag{10.28}$$

We define the distance between two v_{ra}'s as the weighted average of the angle differences:

$$
\begin{aligned}
&\text{distance}\left(v_{ra}(A(x,y)), v_{ra}(T(x_0, y_0))\right) \\
&= (w_2/(\pi w_t))\,\mathrm{difang}\left(\mathrm{dra}_2(A(x,y)), dra_2(T(x_0, y_0))\right) + \\
&\quad (w_2/(\pi w_t))\,\mathrm{difang}\left(\mathrm{dra}_3(A(x,y)), dra_3(T(x_0, y_0))\right) + \\
&\quad ... \\
&\quad (w_K/(\pi w_t))\,\mathrm{difang}\left(\mathrm{dra}_K(A(x,y)), dra_K(T(x_0, y_0))\right)
\end{aligned}
\tag{10.29}
$$

where $w_k = 1/k, 2 \leq k \leq K$, $w_t = w_2 + w_3 + ... + w_K$ and "difang" is the difference between two angles, defined as:

$$\mathrm{difang}(a, b) = \min(\mathrm{mod}(a - b, 2\pi), 2\pi - \mathrm{mod}(a - b, 2\pi)). \tag{10.30}$$

This distance function is limited to interval $[0, 1]$. We assign lighter weights to high order dra's, because they are more easily affected by image distortions.

(a) Truncated integer circular (b) Floating-point circular ker- (c) Second-order circular kernel.
kernel.   nel.

Figure 10.12   Circular kernels.

10.5.3.5   Rotation-Invariant "Vector of Circular Features"

Choi and Kim have used the DFT coefficients of the circular projections as the rotation-invariant features. We will continue using these features, together with the radial features. However, we introduce two small alterations in the circular kernels. First, to be consistent with the radial features, we use IDFT instead of DFT. Second, instead of using the original truncated integer radius $r = (\text{int})\lfloor \sqrt{x^2 + y^2} \rfloor$, we use the floating-point radius $r = \sqrt{x^2 + y^2}$ because experiments indicate that the latter (Figure 10.12(b)) is more stable than the former (Figure 10.12(a)). The resulting circular kernel is:

$$\text{circularKernel}_l[x, y] = \begin{cases} 0, \text{ if } \lambda \le r \\ \frac{1}{2\pi r} \exp\left(\frac{jlr}{\lambda}\right), \text{ if } 0 < r < \lambda \\ |\sum_{r \ne 0} \text{circularKernel}_l[x, y]|, \text{ if } r = 0 \end{cases} \qquad (10.31)$$

where $r = \sqrt{x^2 + y^2}$ and λ is the radius of the kernel. The weight for $r = 0$ is computed to distribute evenly the angles of the complex image resulting from the convolution. In order to make the weights sum to one, all weights must be divided by the sum of the weights. Figure 10.12(c) depicts the kernel used to obtain the second circular coefficients.

We will call the convolution of $A(x, y)$ with the double reflection of the l-th circular kernel "l-th circular coefficient" and denote it $c_l(A(x, y))$ or simply c_l. In our implementation, we use the following "vector of circular features" that take into account the real and imaginary components of all circular coefficients up to degree L:

$$v_{cf} = L^1\text{-versor}[\text{re}(c_1), \text{im}(c_1), \text{re}(c_2), \text{im}(c_2), ...\text{re}(c_L), \text{im}(c_L)] \qquad (10.32)$$

where "re" and "im" are respectively the real and imaginary parts of the complex number. The distance between two v_{cf}'s is based on L^1-distance:

$$\text{distance}\,(v_{cf}(A(x,y)), v_{cf}(T(x_0, y_0)))$$

$$= \frac{1}{2}\parallel v_{cf}(A(x,y)) - v_{cf}(T(x_0, y_0)) \parallel_1 . \tag{10.33}$$

This distance is limited to interval $[0, 1]$.

10.5.3.6 Combining Rotation-Invariant Features

In previous Sections, we obtained three rotation-invariant classes of features, using up to K radial and L circular coefficients and packed them in three vectors: v_{rm}, v_{ra}, and v_{cf}. We will group these three vectors into another structure named "vector of features":

$$v_f = (v_{rm}, v_{ra}, v_{cf}) \tag{10.34}$$

We define the distance function between two vectors of features as the weighted average of the three constituent vectors of features:

$$\begin{aligned}
\text{distance}\,(&v_f(A(x,y)), v_f(T(x_0, y_0)))\\
=\ & (w_m/w_t)\,\text{distance}\,(v_{rm}(A(x,y)), v_{rm}(T(x_0, y_0)))\\
+\ & (w_a/w_t)\,\text{distance}\,(v_{ra}(A(x,y)), v_{ra}(T(x_0, y_0)))\\
+\ & (w_c/w_t)\,\text{distance}\,(v_{cf}(A(x,y)), v_{cf}(T(x_0, y_0)))
\end{aligned} \tag{10.35}$$

where $w_m = w_a = K - 1$, $w_c = 2L - 1$, and $w_t = w_m + w_a + w_c$. These three weights are proportional to the number of "independent variables" in each constituent vector. This distance is limited to interval $[0, 1]$.

10.5.4 NEW TEMPLATE MATCHING ALGORITHM

Using vector of features v_f, we can detect the matching candidates: if the distance between $v_f(A(x,y))$ and $v_f(T(x_0, y_0))$ is below a threshold t_c, $A(x,y)$ is considered a matching candidate. Yet, a matching candidate is not necessarily a true template matching, because:

1. Many non-equivalent templates can generate the same features. Thus, false positive errors can occur.
2. Some templates may not have clear features. For example, it is possible that the first radial magnitude is (almost) zero. In this case, it is impossible to determine the canonical orientation and it is also impossible to determine the difference of radial angles (for $k \geq 2$), because they depend on the canonical orientation. Similarly, if the k-th radial magnitude is (almost) zero, it is impossible to compute the difference of radial angles k and 1.
3. In some parts of the image, the features may be "unstable," that is, a sub-pixel rotation or translation may completely change the features.

10.5.4.1 Generalized Hough Transform

To overcome the problems above, we use an idea inspired in the generalized Hough transform [2]. We extract a set of circular templates $T_1, ..., T_N, T_i \subset Q$. Then, we search them all in A. It is possible that a single sub-template may generate an error. However, if many sub-templates agree to point a pixel as the matching point, the probability of error is minimized. Our feature-based algorithm is appropriate for this approach, because finding N templates in the same image to analyze A is much faster than finding N templates in different images.

After extracting the sub-templates $T_1, ..., T_N$, we extract the canonical orientations and compute the vectors of features for all pixels in A. We also compute the canonical orientation and the feature vector at the centers of the templates $T_i(x_0, y_0)$. We calculate the "matching distance images" D_i, whose grayscale at pixel $D_i(x, y)$ is the distance between the vectors of features $v_f(T_i(x_0, y_0))$ and $v_f(A(x, y))$. If $D_i(x, y)$ is below some chosen threshold t_c, pixel $A(x, y)$ is a matching candidate. However, the appropriate thresholding level uses to vary from template to template. So, we prefer to use n_c pixels with the smallest matching distances as the matching candidates.

Using the canonical orientation and the position of each T_i in Q, we find the center (x, y) of Q pointed by each matching candidate pixel and increment the Hough transform accumulator array $H(x, y)$ by $1 - D_i(x, y)$. We increment the accumulator based on the matching distance, to provide a tie-breaking criterion. In accumulator increment, the imprecision of the canonical orientation must be taken into account. The true matchings are the pixels of H with the largest values.

We can take advantage of the small number of matching candidate pixels to accelerate the Hough transform, substituting the accumulator array H (of the same size as the image A) by a small matrix V. We will name this algorithm matrix Hough transform. Let us suppose that we use N templates and n_c matching candidates for each template. Then, we construct a matrix V with N rows and n_c columns, where each element $V[i, j]$ has four entries: $[x, y, w, a]$. Let the j-th matching candidate of the i-th template T_i be located at $A(x', y')$. Then, the template T_i must be rotated a radians to match A at pixel (x', y'), where:

$$a = \angle r_1(A(x', y')) - \angle r_1(T(x_0, y_0)). \tag{10.36}$$

This information, together with the position of T_i in Q, allows us to compute the pixel $A(x, y)$ that matches the center of Q. The weight w is defined as $1 - D_i(x', y')$. To find the best matching, we construct a set S of elements extracting one (or none) element of V per row, such that all elements of S are "near" one another and the sum of their weights is maximal. The numerical example in Table 10.3 clarifies these ideas.

In this example, there are three elements that are "near" one another: $V[1, 1]$, $V[1, 2]$, $V[2, 2]$. The best matching is $S = \{V[1, 1], V[2, 2]\}$, because this set takes one element per row and maximizes the sum of the weights. The spatial position of the matching is defined as the weighted average of the positions (x, y) in S, in the example $x = 123.00$ and $y = 136.04$. The rotation angle of the matching is the weighted average of the angles a in S. To compute the weighted average of angles,

Table 10.3

A numerical example of matrix V in matrix Hough transform

123, 137, 0.95, 0.18	125, 137, 0.89, 0.25	165, 4, 0.83, -2.45
228, 36, 0.90, 1.11	123, 135, 0.88, 0.04	1, 22, 0.84, 2.92

Table 10.4

The observed number of errors using the Hough transform, varying the number N and the size $n_T \times n_T$ of the templates

Errors (maximum=96)	$N=2$	$N=4$	$N=6$
$n_T = 21$	2	0	0
$n_T = 31$	0	0	0
$n_T = 41$	0	0	0
$n_T = 51$	0	0	0

we transform each angle a with weight w in a complex number $w\cos(a) + jw\sin(a)$, add them up, and compute the angle of the resulting sum. In the example, the rotation angle is 0.113.

Using the Hough transform, the template matching can become robust to partial occlusions up to a certain degree, because the algorithm can locate the query image even if some of its sub-templates cannot be located.

10.5.4.2 Robustness to Scaling

The proposed algorithm can become robust to scale changes, that is, it can find the query image even if it appears in another scale in A (within a predefined range of scale factors). The matching is tested using the query image rescaled by a set of scale factors, making it a kind of "brute force" approach. However, as the features of A are computed only once, it is not slow. The algorithm robust to rescaling is: Given a query image Q, rescale it by a set of predefined scales, obtaining S rescaled query images $Q_1, ..., Q_S$. Find them all in A. Take the best matchings as the true matchings. The robustness to scaling can be further improved by constructing a pyramid structure for A, as we did for Ciratefi.

10.5.5 EXPERIMENTAL RESULTS

10.5.5.1 Forapro with Generalized Hough Transform

We took 24 game cards with 12 different figures. We scanned these cards 8 times, so that all the 12 figures appear only once in each scanned image (Figure 10.13(a)). We

(a) One of the eight images to (b) The output of Forapro algo- (c) Two of the twelve 51 × 51
analyze   rithm   query images

Figure 10.13   Example of Forapro algorithm.

Table 10.5
Robustness against partial occlusions

Errors (maximum=96)	N=4	N=6	N=8	N=10	N=12
$n_T = 15$	9	0	0	0	0
$n_T = 21$	2	0	0	0	0
$n_T = 25$	5	1	0	0	0

placed magazines behind the cards to obtain non-blank backgrounds. These 8 images
(each one with approximately 530×400 pixels) will be our images to analyze A_1, ...,
A_8. From one of these images, we extracted the 12 query images Q_1, ..., Q_{12}, each
one with 51×51 pixels (Figure 10.13(c)).

Our program successfully localized all 12×8 query images. We used $K = L = 4$
and varied the size $n_T \times n_T$ of the sub-templates and the number N of templates.
We always took 10 best matching pixels of each template to compute the Hough
transform. Table 10.4 shows that, using 4 or more templates or using templates with
at least 31×31 pixels, no error occurs.

Our implementation takes 1.1s in a Intel Core i7-4500U 1.8GHz computer to
find the 12 templates in an image to analyze A, subdivided in: 0.24s to compute the
radial and circular coefficients of A, 0.66s to derive the features from the coefficients,
and 0.18 seconds to compute the 12 distance images and find the 12 pixels with
the smallest distance. Our implementation uses FFT/IFFT functions of the OpenCV
library. However, the rest of our program is written in plain C++ and is not optimized.

Figure 10.14   We copied 15×15 square blocks randomly to test robustness to partial occlusions.

10.5.5.2   Robustness to Partial Occlusion

To test the robustness against partial occlusions, we copied 15×15 square blocks from 80 randomly chosen positions to other 80 random positions in each image to analyze. Figure 10.14 depicts such an image. Then, we tested the matching using the Hough transform, $L = K = 4$, and 10 candidate pixels. We varied the number of templates N and the size of the templates $n_T \times n_T$. The results are depicted in Table 10.5. Using $N = 8$ or more templates, no error was observed.

10.5.5.3   Robustness to Scaling

To test robustness to scaling, we took 9 query images, resized them by scale factors chosen randomly in the range [0.7, 1.4], rotated them randomly, and pasted them in a random non-overlapping locations over a background image, resulting eight 512×512 images to analyze (Figure 10.15). We kept constants the size of templates 17×17, the maximum order $K = L = 4$, and the number of the matching candidates $n_c = 10$. We varied the number of scales n_s and the number of templates N used in each Hough transform. The observed number of errors is depicted in Table 10.6. Using $N = 4$ or more templates, no error was observed, even using the number of scales n_s as small as 3. Our implementation takes 3.86s to find 9 query images in an image to analyze.

Table 10.6

Robustness to scaling using 10 candidate pixels, $n_T = 17, K = L = 4$

Errors (maximum=72)	$N = 3$	$N = 4$	$N = 5$
$n_s = 3$	1	0	0
$n_s = 4$	0	0	0
$n_s = 5$	1	0	0
$n_s = 6$	0	0	0

(a) Search image to analyze A

(b) The output of Forapro

Figure 10.15   Robustness to scaling test.

(a) Forapro found correctly all 48 queries

(b) SIFT made 6 errors (out of 48). The image depicts SIFT's tentative to find "S"

Figure 10.16   Comparison of Forapro and SIFT.

10.5.5.4   Brief Comparison with SIFT

SIFT (Scale Invariant Feature Transform) [16] is a very popular, accurate, and efficient image matching algorithm. It extracts some scale-invariant key-points and computes their associated local features (scale, rotation, and local textures). Then, to find a query image Q in A, it is necessary only to match the key-points. We evaluated SIFT using the three sets of images used so far and SIFT made no error, like our algorithm. The processing times also were of similar magnitudes (typically SIFT takes 1.5 seconds to compute the key-points of an image to analyze).

Figure 10.16 presents an example created especially to explore SIFT's weakness. As SIFT is based on key-points and local textures, it may fail to find simple shapes with constant grayscales. We tested finding 12 query images in 4 different images to analyze. Our algorithm successfully found all the query images (Figure 10.16(a)). Meanwhile, SIFT made 6 errors (out of 48). Figure 10.16(b) depicts one of SIFT's failures, where it could not find the key-points corresponding to the dark shape "S." To find 12 query images in each image to analyze, both our algorithm and SIFT take approximately 3 seconds. However, note that our algorithm is making one-scale searching, while SIFT is making scale-invariant matching.

10.6 CONCLUSIONS

This chapter has presented RSTBC-invariant (Rotation, Scaling, Translation, Brightness, and Contrast invariant) template matching techniques. We first presented the classic BC-invariant template matching based on normalized cross correlation. Then, we obtained brute-force RSTBC-invariant template matching by executing classic template matching many times, for templates rotated by many angles and scaled by many factors. Then, we presented two template matchings that can accelerate the brute-force RSTBC-invariant algorithm: Ciratefi and Forapro. Ciratefi consists of three cascaded filters that successively exclude pixels that have no chance of matching the template. These filters use circular and radial projections. We have compared Ciratefi with SIFT and EasyMatch concluding that our technique is, most of the time, more accurate but slower than the others. As the original Ciratefi has many adjustable parameters, we have presented a methodology to automate the choice of all parameters. Then, we presented Forapro, that also uses circular and radial projections. However, Forapro does not compute these projections explicitly, but computes the low-frequency complex Fourier coefficients of these projections directly, using FFT. Forapro computes rotation-invariant features from these coefficients and uses them in RSTBC-invariant template matching. Once the rotation-invariant features are computed, many different template images can be found quickly. This property has been used to obtain scale invariance and robustness to partial occlusions, using the generalized Hough transform. Experimental results show that Forapro is accurate and fast.

REFERENCES

1. S. A. Araujo and H. Y. Kim, Color-Ciratefi: A color-based RST-invariant template matching algorithm, In: Proceedings of 17th Int. Conf. Systems, Signals and Image Processing, 2010, pp. 101-104.

2. D. H. Ballard, Generalizing the Hough transform to detect arbitrary shapes, Pattern Recognition, vol. 13, n. 2, 1981, pp. 111-122.
3. H. Bay, A. Ess, T. Tuytelaars, and L. Van Gool, SURF: Speeded Up Robust Features, Computer Vision and Image Understanding, vol. 110, n. 3, 2008, pp. 346-359.
4. Q. Chen, M. Defrise, and F. Deconinck, "Symmetric phase-only matched filtering of Fourier-Mellin transforms for image registration and recognition." IEEE Transactions on Pattern Analysis and Machine Intelligence, 16(12):1156-1168, December 1994.
5. M.S. Choi and W.Y. Kim, A novel two stage template matching method for rotation and illumination invariance, Pattern Recognition, 35(1) (2002) 119-129.
6. J.H. Friedman, J.L. Bentley, and R.A. Finkel, An algorithm for finding best matches in logarithmic expected time, ACM Trans. Mathematical Software, 3(3) (1977) 209-226.
7. J. M. Geusebroek, G. J. Burghouts, and A. W. M. Smeulders, The Amsterdam library of object images, Int. Journal of Computer Vision, vol. 61, n. 1, 2005, pp. 103-112.
8. M.K. Hu, Visual pattern recognition by moment invariants. IRE Trans. Inform. Theory, 1(8) (1962) 179-187.
9. W.Y. Kim and P. Yuan, A practical pattern recognition system for translation, scale and rotation invariance, Proc. Int. Conf. Comput. Vis. Pattern Recognit., (1994) 391-396.
10. H. Y. Kim and S. A. de Araújo, Grayscale template-matching invariant to rotation, scale, translation, brightness and contrast, In: Proceedings of the 2nd Pacific Rim conference on Advances in image and video technology, LNCS, vol. 4872, 2007, pp. 100-113.
11. H. Y. Kim, Rotation-discriminating template matching based on Fourier coefficients of radial projections with robustness to scaling and partial occlusion, Pattern Recognition, vol. 43, n. 3, 2010, pp. 859-872.
12. Y. Lamdan and H. J. Wolfson, Geometric hashing: a general and efficient model-based recognition scheme. In: Proceedings of 2nd Int. Conference on Computer Vision, 1988, pp. 238-249.
13. T. K., Leung, M. C. Burl, and P. Perona, Finding faces in cluttered scenes using random labeled graph matching. In: Proceedings of 5th Int. Conference on Computer Vision, 1995, pp. 637-644.
14. J. P. Lewis, Fast normalized cross-correlation, Vision Interface, 1995, pp. 120-123.
15. Y. Lin and C. Chen, "Template matching using the parametric template vector with translation, rotation and scale invariance," Pattern Recognition, vol. 41, no. 7, July 2008, pp. 2413-2421.
16. D. G. Lowe, Distinctive image features from scale-invariant keypoints, Int. Journal of Computer Vision vol. 60, n. 2, 2004, pp. 91-110.
17. D. Marimon, T. Ebrahimi, Efficient rotation-discriminative template matching, 12th Iberoamerican Congress on Pattern Recognition, Lecture Notes in Computer Science 4756 (2007) 221-230.
18. L.A. Torres-Méndez, J.C. Ruiz-Suárez, L.E. Sucar, G. Gómez, Translation, rotation and scale-invariant object recognition, IEEE Trans. Systems, Man, and Cybernetics - part C: App. and Reviews 30(1) (2000) 125-130.
19. K. Mikolajczyk, A performance evaluation of local descriptors, IEEE Trans. on Pattern Analysis and Machine Intelligence, vol. 27, n. 10, 2005, pp. 1615-1630.
20. F. Mokhtarian, A. K. Mackworth, "A theory of multi-scale, curvature based shape representation for planar curves," IEEE Trans. on Pattern Analysis and Machine Intelligence, vol. 14, n. 8, 1992, pp. 789-805.
21. H. P. A. Nobre and H. Y. Kim, Automatic VHDL generation for solving rotation and scale-invariant template matching in FPGA. In: Proceedings of 5th Southern Pro-

grammable Logic Conference, 2009, pp. 229-231.

22. S. Reddy and B. N. Chatterji, "An FFT-based technique for translation, rotation, and scale-invariant image registration." IEEE Trans. on Image Processing, 3(8):1266-1270, August 1996.

23. L.D. Stefano, S. Mattoccia, and F. Tombari, ZNCC-based template matching using bounded partial correlation, Pattern Recognition Letters (26) 2005 2129-2134.

24. C.H. Teh and R.T. Chin, On image analysis by the methods of moments. IEEE Trans. Pattern Analysis Machine Intelligence 10(4) (1988) 496-513.

25. D. M. Tsai and Y. H. Tsai, Rotation-invariant pattern matching with color ring-projection, Pattern Recognition, vol. 35, n. 1, 2002, pp. 131-141.

26. F. Ullah, and S. Kaneko, Using orientation codes for rotation-invariant template matching. Pattern Recognition, vol 37, n. 2, 2004, pp. 201-209.

27. Z. Wang, A. C. Bovik, H. R. Sheikh, and E. P. Simoncelli, Image quality assessment: from error visibility to structural similarity, IEEE Trans. on Image Processing, vol. 13, n. 4, 2004, pp. 600-612.

11 Three-Dimensional Television (3DTV)

CARLA LIBERAL PAGLIARI
Military Institute of Engineering (IME)

EDUARDO ANTÔNIO BARROS DA SILVA
Federal University of Rio de Janeiro (UFRJ)

ABSTRACT

Three-dimensional television (3DTV) aims to bring the "real world" sensation to the viewer. However, there are several problems to tackle before providing services for 3DTV. 3D content can be delivered as stereoscopic or multi-view video in a frame-compatible or depth-based format. As large data volumes are produced to generate stereo or multi-view content, a number of compression methods have been proposed to efficiently store and/or transmit the data. Researchers are also striving to successfully develop 3DTV displays that provide good image quality (and depth sensation) without the need for glasses and without visual discomfort.

As the 3D content has to be delivered to the consumers, standards are important to ensure that there is support from the industry. Hence, these compression schemes and formats need to consider the constraints imposed by different storage and transmission systems, as well as backward compatibility.

This chapter presents an overview of stereo and multi-view coding techniques, including frame-compatible and depth-based video formats. International standard organizations, such as MPEG, ISO/IEC/ and ITU-T that have already standardized video coding schemes encompassing 3D content compression are also addressed.

11.1 3D BASICS

High definition (HD), ultra-high-definition (UHD) 4K, and 8K (also known as Super Hi-Vision) television aim to add realism to 2D displays, while developments of autosterescopic displays are struggling to franchise the advent of 3D television without special glasses.

Television sets displaying 2D images do not confer on the realism that a 3D rendering could provide. This creates the necessity of the depth perception offered by stereoscopic video systems which employ, at least, two views of a scene. It could bring the "real world" to many applications such as 3D movies, medicine surgery, video-conferencing, virtual reality, multimedia, remote operations, surveillance, industrial robotics, stereomicroscopy, and human-computer interactions, amongst others.

Humans have two frontally positioned eyes, each one having a slightly different view of the scene [46]. The resulting difference in the images is called binocular disparity. In a similar way, a stereo system acquires views from two cameras slightly displaced. Therefore, 3D shooting requires the knowledge of stereo camera geometry in order to align the cameras for best results. In addition, the processing chain must include coding, transmission, rendering, and display of the 3D content [41]. Needless to stress that each of these presents several challenges:

- 3D acquisition (stereo and/or multi-view) can be performed using different techniques such as stereo (or multiple) cameras, depth cameras, and 2D-to-3D conversion algorithms. All of them require knowledge of camera geometry and its parameters [40], [37].
- 3D compression schemes exploit not only spatial and temporal similarities, but also the correlation between views [36].
- the transmission of 3D content needs to consider 2D devices. Various representation formats for 3D video have been proposed. These formats include the traditional simulcast, frame compatible formats, service-compatible 3D video formats, and depth-enhancement video formats [35].
- accurate 3D content rendering is another challenge. 3D display technology creates the sensation of depth by presenting a different image to each eye. There are stereoscopic techniques that use passive polarized lenses and active shutter glasses, and autostereoscopic techniques that render 3D content without the need for special glasses [35].

The generation of 3D content using a stereo camera set-up demands the knowledge of the camera geometry and camera calibration. This process limits post-production adjustments. Another way to shoot 3D scenes is using 3D cameras that capture the texture views and generate their associated depth maps [34]. Moreover, a large number of content was shot from a single point of view only. Therefore, 3D content could be generated by 2D-to-3D conversion. Since a good depth sensation is what is needed, the conversion methods do not necessarily preserve depth precision. However, this assumption generates several artefacts [43], [42] .

One of the biggest business challenges is that the specification for 3DTV uses the available bandwidth for the "monocular channels". Besides the rate-distortion (R-D) compromise, the industry has to guarantee compatibility with 2D monitors. Although there are several compression and display techniques for stereo/multi-views videos, it is important to provide standards to the consumers. 3D compression schemes are addressed in the next section.

Various methods have been proposed to provide 3DTV services over a terrestrial broadcasting network. However, the basic problem is how to provide quality 3D content to consumers using the same bandwidth used to deliver 2D content. Research efforts generated proposals to overcome this issue. In [33] a hybrid 3DTV broadcasting system, employing a terrestrial broadcast network and a broadband network is presented. The two elementary streams (ES) of each stereo view are terrestrial broadcasted (left view) and transmitted over a broadband network (right view) maintaing the synchronism. Another hybrid method is proposed in [32] for DVB-T2 fixed and mobile broadcasting services. Both views are encoded using the H.264/MPEG-4 AVC standard [24], where one is transmitted over the main HD channel and the other is coveyed through the mobile/handheld service channel. The MPEG-2 Transport Stream (MPEG-2 TS) [22] is the tool used for multiplexing the ESs of the compressed views.

The recent DVB-3DTV effort specifies frame-compatible and service-compatible services to deliver 3DTV to consumers. The document [12] specifies the delivery system for frame compatible plano-stereoscopic 3DTV services, enabling service providers to utilize their existing HDTV infrastructures to deliver 3DTV services that are compatible with 3DTV-capable displays already in the market. [13] specifies the delivery system for HDTV-service-compatible plano-stereoscopic 3DTV services.

The representation formats, such as simulcast, frame-compatible, multi-view, and texture-plus-depth (2D+depth) are detailed in the next section.

The majority of 3D displays can be classified as stereo pair based, volumetric, and holographic. The display consumer market is flooded by stereo pair based monitors. This type of technology distributes the left and right views of a scene in a way that reaches a viewer's left and right eyes respectively. To perceive a 3D scene on a 3D-stereo-pair-based display requires usually temporal-multiplexing or polarization methods. Special glasses are not very popular among the viewers, and autostereoscopic displays might be a better solution [35]. Autostereoscopic (also stereo-pair-based) displays project a different viewpoint of the scene depending on the position of the viewer. The drawback is that the transmitted/received video has to provide multiple point-of-views of the same scene, considerably increasing the required bitrate. This problem could be minimized through the generation of virtual views, that will require the transmission of their associated depth maps, which need a more modest bitrate.

In general, volumetric 3D displays rely on optical techniques, such as moving mirrors, to project/reflect light at points in space. A survey of such methods can be found in [29].

The holographic 3D displays [28] are based on the holography principle. It uses laser beams aided by interference to record the scene on a medium, such as a photorefractive polymer. Some desired properties of recording media are fast recording, image persistency, and resistance to optical and electrical damage [27]. 3D displays history is detailed in [30] and [31].

Depth rendering is a fundamental step in 3D representation, yet a very difficult one. The human visual system combines depth information from monocular and binocular cues. As monocular cues one can cite the shading of the objects, the linear perspective, the relative size of the object, among others. All of them contribute to the process of depth perception. Although the human visual system can infer depth from a number of visual cues in the retina, the binocular disparity is the most critical one [45].

The success of 3DTV, multimedia, and other applications of 3D imaging depends strongly on the acceptance of the technology. The new displays could be welcome provided there is no harm and no discomfort to the viewers.

Unfortunately, stereo perception involves more than "the good will to accept". As 3D displays present their content on a 2D surface, the viewers tend to perceive distorted 3D scenes. Moreover, undesirable symptoms have also been reported such as headaches, visual fatigue, general vision discomfort, dizziness, and blurred vision [44]. These issues are so important that there are ITU-T studies related to display quality and visual fatigue assessment [47], [50]. In addition, subjective tests were conducted in order to establish 3D assessment factors, such as depth resolution, depth motion, and visual comfort limits [49]. Other factors, including sharpness, color, and resolution, applied to conventional 2D displays, are also addressed.

Visual comfort is a key issue when dealing with 3D acceptance. While 3D (glasses or glasses-free) displays present images/videos on one 2D flat screen, we receive different depth cues in the, real, 3D world. One has to look, turning the eyes (and head), to an object to focus. Such an eye movement is called vergenge, targeting to locate the region of interest on the fovea. The muscles of the eyes adjust to focus the same area of interest. Vergence (convergence) eye movements are interrelated to the ability of the eye to adjust its focal length (accommodation) when viewing a real world 3D scene [44]. Conventional 3D displays may create conflicts between vergence and accommodation yielding visual discomfort.

In general, the viewer experiences the 3D sensation from a fixed distance of a stereoscopic display. The device displays two 2D images on its 2D surface, and the viewer's eyes have to focus (accommodate) to the physical location of the device, and converge his/her eyes to the rendered 3D content. It is important to stress that the scene content may present different depths, with different binocular disparities, leading the eyes to move towards different depths of the rendered 3D content. As the vergence is uncoupled from the accommodation (fixed at the distance between the viewer and the display) the 3D rendering may be disrupted [44].

Hence, one could list the binocular parallax (binocular disparity) and the duo vergence-accommodation as critical depth cues. Another depth cue is the motion parallax caused by the relative position change between the viewer and the objects. This depth cue is rather critical when watching stereo-pair-based displays.

The disparity, or the displacement between the respected projected images of a 3D point of the two views, is inversely proportional to the depth, so that small disparity values correspond to large depth distances (from the viewer to the object). There is a geometry that defines the relationship between two corresponding pixels (homologous points) and the depth of a 3D point [39], [38]. Depth maps could be either estimated from disparity maps, that have to match homologous points, or obtained from depth cameras. Figure 11.1 shows the view (texture) and its correspondent depth map. Each gray intensity value is associated to a depth value. Lighter gray values represent objects closer to the camera (viewer), while darker ones represent objects that are farther away from the camera.

Figure 11.1 Poznan Street, Camera 03, Frame 0 (from [2]).

11.2 3D COMPRESSION

Techniques for digital compression of the increased amount of data generated by (at least) two images while maintaining compatibility with normal video decoding and display systems are being developed.

The most straightforward solution is called the simulcast approach, where both views are independently encoded ignoring the inherent similarity between the left and right views. Other schemes have been proposed exploiting the stereo redundancy.

Interframe prediction using elements from other frames that are displaced in space, has proven to be effective in compressing moving video sequences. Such a prediction is called motion compensation. Block-matching is performed and motion vectors are generated. The motion vectors applied to the previous frame generate a predicted image. The arithmetic difference between the original and the predicted image is called the differential, residual, or motion compensated difference image. Several coding schemes including H.264/MPEG-4 AVC [24] and HEVC [23] use motion compensated prediction. Similarly, some stereo image coding schemes use the inherent redundancy between the left and right views to compress both views. For each left view frame there is one, correspondent in time, right view frame. Although they have been acquired at the same time, the differences between them are given by the disparity/depth map. Therefore, the redundancy between the left and right views can be exploited by a disparity compensated prediction. Figure 11.2 shows an example of inter-view and temporal predictions.

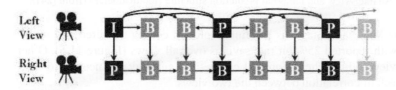

Figure 11.2 The inter-view and temporal prediction structure (stereo) (from [35]).

The H.264/MPEG-4 AVC received an extension called Multiview Video Coding (MVC) [21]. Similarly, HEVC has its multi-view extension, 3D-HEVC [20]. Nevertheless, researches have been investigating stereo/multi-view coding well before that. Unfortunately, the coding efforts had to wait for the maturity of the 3D display technology. Stereo coding schemes were proposed in [15], [16], and [17]. Even the international video coding standard H.262/MPEG-2 [19] had support of stereo/multi-view video coding. More details are given in [18].

A significant impact was provided by the stereo and multi-view video coding extensions of the (monoscopic) H.264/MPEG-4 AVC standard, the Multiview Video Coding (MVC). Besides the known spatial and temporal predictions, it introduces an inter-view prediction structure that allows frames of multiple cameras to be referenced. Backward compatibility with 2D schemes is guaranteed by the coding/decoding of the base view as a monocular independent video [21]. The disparity compensation prediction scheme proved to be efficient with multiple camera views. Profiles confine the MVC supported coding tools. There are two MVC profiles with support for two or more views: the stereo high profile and the multi-view high profile. The former is restricted to two views, not supporting interlaced coding tools, while the latter supports multiple views and does not support interlaced coding tools.

Figure 11.3 illustrates the MVC temporal and inter-view prediction structure. Figure 11.2 is a stereo version of the actual MVC coding scheme. Figure 11.4 displays six views of the same scene.

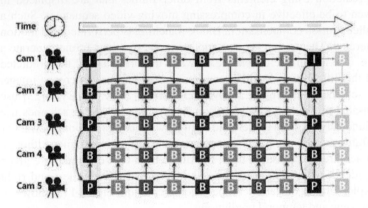

Figure 11.3   An inter-view and temporal prediction structure (multicameras) (from [24]).

One of the MVC advantages is the full-resolution coding of stereo and multi-view video, with reported 25% bit rate savings over all views (Figure 11.5). Other stereo/multi-view coding schemes are less competitive and may produce coding artefacts due to lack of correlation between the two views.

Frame-compatible formats pack two stereo views into a single coded frame or sequence of frames. Both views are spatially (Figure 11.6) or temporally multiplexed (views are interleaved as alternating frames or fields). Each coded view has half the resolution of the full coded frame. Figure 11.6 depicts the most common frame-compatible formats.

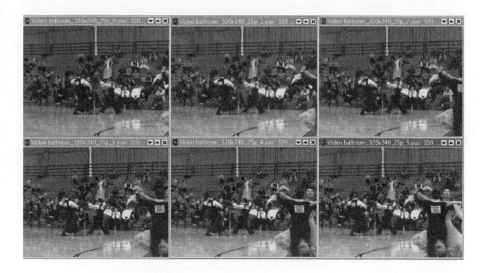

Figure 11.4 Multiview Ballroom sequence (from [14]).

Figure 11.7 displays a time multiplexed format. This format alternates even and odd frames of each view (views interleaved as alternating frames or fields).

Figure 11.8 shows an example of a side-by-side (SbS) format for the texture information and their associated depth maps.

The most obvious benefit of frame-compatible formats is the straightforward introduction of stereoscopic services through existing 2D infrastructure. Moreover, the frame-packed video can be coded by monoscopic encoders, such as H.264/MPEG-4 AVC and transmitted through existing channels. One problem is how to demultiplex the frame-packed videos in order to display them in 2D monitors [35]. H.264/MPEG-4 AVC has standadized the signaling for a complete set of frame-compatible formats as supplemental enhancement information (SEI) messages. Hence, the decoder needs to understand the SEI message to decode and display the stereo content. Another drawback is that some spatial and/or temporal resolution is lost, leading to a lower correlation between views that may generate coding artefacts. In [11] a series of experiments into frame-compatible systems is reported, listing the advantages and disadvantages of the available formats. Another issue to consider is whether the source material is interlaced. As the interlaced field is already half the resolution of a frame, the top-bottom format (Figure 11.6) imposes a resolution loss in the vertical dimension. Therefore, the side-by-side format is generally preferred over the top-bottom format for interlaced content [10].

Until now this chapter showed several ways in which 3D content can be coded and transmitted. Although the MVC scheme is efficient in terms of rate-distortion, it still requires a large amount of views. So, a coding scheme that uses less views is highly welcome. The 2D-plus-Depth also known as 2D+Z format combines the

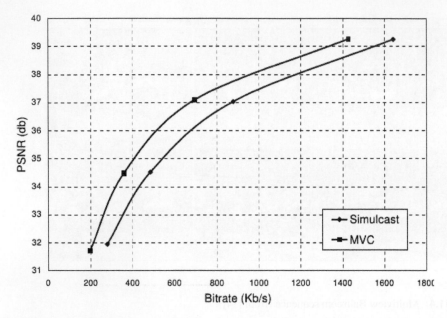

Figure 11.5 MVC coding performance - Multiview Ballroom sequence (8 views) (from [14]).

texture information (the view itself) with its grayscale representation of the depth map. The depth information is used to render the 3D representation on a 3D display. The depth map values indicate the position of each 2D image pixel on the Z (depth) axis in or out of the screen plane.

2D+Depth was standardized as ISO/IEC 23002-3, MPEG-C Part 3: *Representation of auxiliary video and supplemental information.* It defines auxiliary video streams as data coded as video sequences and supplementing a primary video sequence. Depth maps and parallax maps are types of auxiliary video streams, relating to stereoscopic-view video content. Hence, the standard specifies syntax and semantics for conveying information describing the interpretation of these auxiliary video streams. 2D+Depth provides good compatibility with existing 2D systems, with limited rendering capability as it uses only one texture view.

It is possible to create an inexistent (virtual) view using the texture views and depth maps. The depth data provides information that combined with its associatedated texture view can create its correspondent view. From the left view and its depth map, it is possible to generate the right view of the same scene. Of course, occluded points cannot be mapped, and some kind of solution must fill the unmapped "holes" at the synthesized view. If the system provides more than one view, the precision/quality of the virtual view increases. For example, instead of coding and transmitting 3 (three) views, a compression method can encode only 2 (two) views and their associated depth maps (that demand much lower bitrates) and synthesize a third view, known as virtual or intermediate view. Using projective geometry [38], [39],

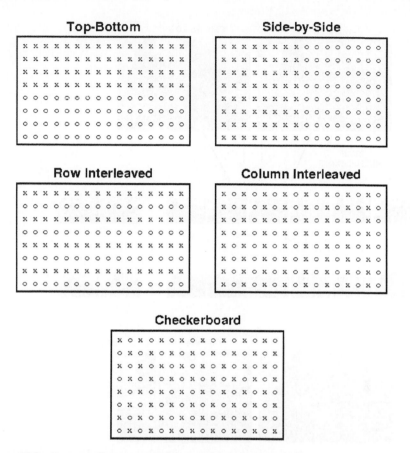

Figure 11.6 Common frame-compatible formats, 'x' represents the samples from one view and 'o' represents the samples from the other view (from [35]).

arbitrary intermediate views can be generated via 3D projection or 2D warping from original (or decoded) camera texture views by means of so-called depth-image-based rendering (DIBR) techniques [9]. These view generation techniques map the 2D data to the 3D space utilizing the respective depth data. Thereafter, these 3D points are projected into the image plane of a virtual camera, located at the required viewing position (Figure 11.9).

High-quality view synthesis is continously being pursued [6], [7], [8].

Stereo and multi-view videos are not supported by the 2D+depth format, as decoders/receivers would be required to generate the second view from the 2D video plus depth data for a stereo monitor. Moreover, autostereoscopic displays often employ DIBR techniques, where depth maps play an important role in 3D quality rendering. Another benefit of this kind of coding scheme is to allow the adjustment of the depth perception level according to the display size and the viewing distance [25].

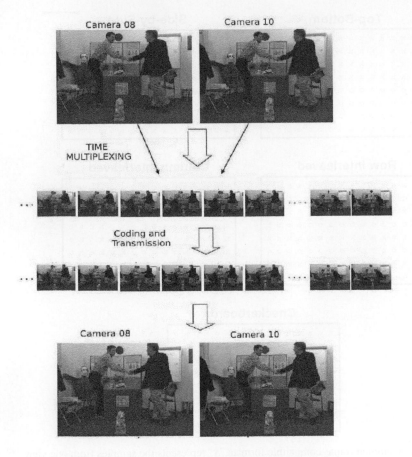

Figure 11.7   BookArrival, Cameras 08 and 10, Time Multiplexed Format.

A multi-view video plus depth (MVD) format with original input views and associated depth maps is under development. HEVC extensions for the efficient compression of stereo and multi-view video are being developed by JCT-3V [5], namely the 3D-HEVC. In the 3D-HEVC extension 3D video is represented using the Multiview Video plus Depth (MVD) format. Figure 11.10 depicts its encoding scheme [5]. The texture and depth components are encoded by HEVC-based encoders, with the resulting bitstreams multiplexed to form the 3D video bitstream. This structure is very interesting as the base/independent view is encoded by the monoscopic HEVC encoder. Hence, it can be independently decoded by any HEVC decoder. The same principle applies when decoding the dependent views and associated depth maps. However, for the dependent views, the use of modified HEVC codecs is necessary. These codecs include additional coding tools and inter-component prediction techniques that can be configured to independently decode the texture information (with-

Figure 11.8   Poznan Street, Camera 03, Frame 0, Side-by-Side (SbS) Format.

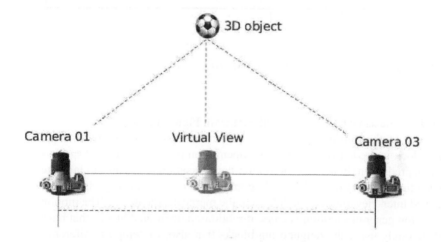

Figure 11.9   Virtual camera view position and original cameras 01 and 03.

out the depth data). The new syntax signals the prediction dependencies between different views, identifying which pictures belong to each view, as well as how to extract the base/independent view.

The encoder scheme allows the dependent views/texture to be encoded using different types of predictions: intra-prediction. inter-view (motion-compensation prediction), inter-view residual prediction, disparity-compensated prediction, and virtual-view synthesis prediction. The Advanced Residual Prediction (ARP) was designed to further improve the coding efficiency of inter-view residual prediction. It takes into account the correlation between the motion-compensated residual signal of two views [3]. In addition, an adaptive weighting factor is applied to the residue signal so that the prediction error is further reduced.

The disparity-compensated prediction (DCP), also used in MVC, has been added to 3D-HEVC. The DCP is an inter-picture prediction that uses already coded pictures

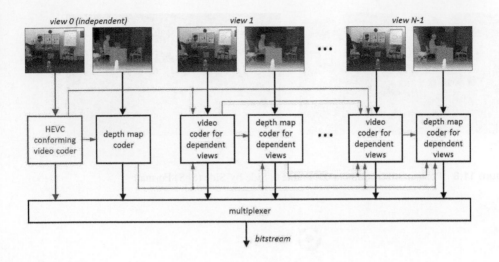

Figure 11.10   High-level encoder structure of the 3D-HEVC (from [5]).

of other views in the same access unit as illustrated in Figure 11.11.

In Figure 11.11 the transmitted reference picture index (R) signals whether an inter-coded block is predicted by motion-compensation prediction or by disparity-compensation prediction. In the case of motion-compensation, it has also to indicate the reference list (L0, L1) [23]. This coding architecture provides disparity vector candidates and inter-view motion candidates to the merge candidate list. The motion vector prediction generates motion vectors of motion-compensated blocks that have been predicted only using the neighboring blocks that also use temporal reference pictures. In a similar manner, the disparity vectors of disparity-compensated blocks are predicted by only using the neighboring blocks that also use inter-view reference pictures [5]. Therefore, the 3D video encoder can select the best coding method for each block from a set of conventional and additional texture coding tools.

As, in general, MVD systems consist of a large amount of data, a view synthesis prediction (VSP) technique can be used to further reduce inter-view redundancy employing virtual (synthetic) views as predictors. The idea of VSP methods is to generate a block predictor for the target block by warping pixel-by-pixel values using the reference view texture and depth. The VSP candidate is then inserted into the merge candidate list with the derived neighboring block disparity vector and the associated reference view index [5].

The 3D-HEVC codec includes a number of coding tools for depth map compression. This is so because texture-based codings tools are not adequate to compress depth maps. As previously described, depth maps represent the distance between the viewer and the 3D objects. Eventual coding artefacts on the graylevel representation of the depth data may lead to erroneously render the 3D object on a 3D display. Moreover, the artefacts may cause the mispositioning of the affected pixels when cre-

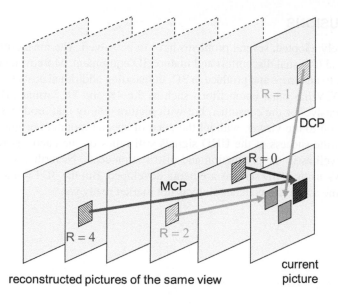

Figure 11.11   Disparity-compensated prediction (from [5]).

ating a virtual view. In general, depth maps are characterized by large homogeneous areas and sharp edges (that may indicate the boundaries of an object). Hence, edge preservation is rather important since inaccurate edge reconstruction may lead to distortions and perceptual artefacts on synthesized views. To address this issue, four new intra prediction modes for depth coding were added to the 3D-HEVC model:

- Mode 1: Explicit wedgelet signaling;
- Mode 2: Intra-predicted wedgelet partitioning;
- Mode 3: Restricted signalling and inter-component prediction of wedgelet partitions;
- Mode 4: Inter-component-predicted contour partitioning.

In all four modes, a depth block is approximated by a model that partitions the block area into two non-rectangular regions, where each region is represented by a constant value [5].

At the decoder side it is possible to decode the texture and depth data. Moreover, one may use DIBR techniques to create intermediate views to display the 3D content on autostereoscopic displays. Therefore, camera parameters, necessary to DIBR methods, are also conveyed by the bitstream. The syntax signals view number, texture, and depth identifiers [5], [4].

The 3D-HEVC is a scalable stereo/multi-view encoding scheme that combines texture and depth data. Its integration with the view synthesis algorithms tends to reduce the bit rate as it provides optimal rate allocation between texture and depth.

11.3 CONCLUSIONS

For 3D to be widely adopted, several problems have to be solved. The most critical are the ones related to visual discomfort and mature 3D equipment. Moreover, content creators have to advertise and produce in 3D, demanding additional costs to TV networks. But 3DTV has other competitors, such as the 4K and 8K formats. Both are still very expensive for the consumer, but with picture quality that produces an immersive sensation to the viewer, without the need of special monitors or glasses [1]. Nevertheless, the success of the UHD signals still relies on the encoders R-D performance, as well as in the production and distribution costs. Nevertheless, autostereoscopic TVs using UHD screens are being developed. But the 3DTV market will need more time as autostereo technology is not market ready yet.

REFERENCES

1. Komine K., Tsushima Y., Hiruma N. (2013). Higher-resolution image enhances subjective depth sensation in natural scenes. *European Conference on Visual Perception.*
2. Common test conditions of 3DV core experiments. *Joint Collaborative Team on 3D Video Coding Extension Development of ITU-T SG 16 WP3 and ISO/IEC JTC 1/SC 29/WG*, Jan 2014.
3. 3D-HEVC Draft Text 2, ISO/IEC JTC1/SC29/WG11, MPEG13/m27784, JCT3V-C0049, Geneva, Switzerland, October/November 2013.
4. 3D-HEVC Draft Text 2, ISO/IEC JTC1/SC29/WG11, MPEG13/m31738, JCT3V-F1001, Geneva, Switzerland, October/November 2013.
5. 3D-HEVC Test Model 4, ISO/IEC JTC1/SC29/WG11, MPEG13/m29533, JCT3V-D1005, Incheon, South Korea, April 2013.
6. M. Koppel et alli (2012). Depth Image-based Rendering with Spatio-Temporally Consistent Texture Synthesis for 3-D Video with Global Motion. *Proceedings IEEE International Conference on Image Processing (ICIP)*, Orlando, USA, September 2012.
7. M. Koppel et alli (2012). Consistent Spatio-Temporal Filling of Disocclusions in the Multiview-Video-Plus-Depth Format. *Proceedings IEEE International Workshop on Multimedia Signal Processing (MMSP)*, Banff, Canada, September 2012.
8. E. Bosc et alli (2011). Towards a New Quality Metric for 3D Synthesized Views Assessment. *IEEE Journal of Selected Topics in Signal Processing Special Issue on Emerging Technologies for Video Compression*, 5(7), 1332-1343.
9. L. McMillan (1997). An Image-Based Approach to Three-Dimensional Computer Graphics. Ph.D. thesis, University of North Carolina at Chapel Hill, Chapel Hill, NC, USA, 1997.
10. Cable Television Laboratories (2010). Content Encoding Profiles 3.0 Specification OC-SP-CEP3.0-I01-100827, version 101.
11. Dolby Laboratories (2010). Dolby Open Specification for Frame-Compatible 3D Systems Issue 1.
12. ETSI TS 101 547-2 (2012). Plano-stereoscopic 3DTV, Part 2: Frame Compatible Plano-stereoscopic 3DTV. Digital Video Broadcasting (DVB).
13. ETSI TS 101 547-3 (2012). Plano-stereoscopic 3DTV, Part 3: HDTV Service Compatible Plano-stereoscopic 3DTV. Digital Video Broadcasting (DVB).
14. ISO/IEC JTC1/SC29/WG11 MPEG05/M12077 (2005). Multiview Video Test Sequences from MERL.
15. M. E. Lukacs (1996). Predictive coding of multi-viewpoint image sets. *Proc. IEEE Int.Conf. Acoust. Speech Signal Process.*, Tokyo, Japan, 521-524.
16. I. Dinstein, G. Guy, J. Rabany, J. Tzelgov, and A. Henik (1989). On the compression of stereo images: Preliminary results. Signal Processing Image Communication, 17(4), 373-382.
17. M. G. Perkins (1992). Data compression of stereo pairs. *IEEE Transactions on Communications*, 40(4), 684-696.
18. Tseng, B. e Anastassiou, D. (1996). A multi-viewpoint video coding with MPEG-2 compatibility. *IEEE Transactions on Circuits and Systems for Video Technology*, 6, 414-418.
19. ITU-T and ISO/IEC JTC 1 (1994). Generic coding of moving pictures and associated audio information-Part 2: Video. *ITU-T Recommendation H.262 and ISO/IEC 13818-2 (MPEG-2 Video)*.
20. G. Tech, K. Wegner, Y. Chen, and S. Yea (2013). 3D-HEVC Draft Text 2, JCT-3V document JCT3V-F1001. JCT-3V.
21. A. Vetro, T. Wiegand, and G. J. Sullivan (2011). Overview of the Stereo and Multiview

Video Coding Extensions of the H.264/MPEG-4 AVC Standard. *Proceedings of the IEEE*, 99(4), 626-642.

22. ISO/IEC 13818-1:2000 (2000). Information Technology-Generic Coding of Moving pictures and Associated Audio Information- Part1: Systems. *ISO/IEC*.

23. High Efficiency Video Coding, Rec. ITU-T H.265 and ISO/IEC 23008-2 (2013). *ITU-T & ISO/IEC*.

24. Advanced Video Coding for Generic Audiovisual Services, Rec. ITU-T H.264 and ISO/IEC 14496-10 (MPEG-4 AVC) (2012). *ITU-T & ISO/IEC*.

25. G. J. Sullivan, J. M. Boyce, Y. Chen, J.-R. Ohm, C. A. Segall, and A. Vetro (2013). Standardized Extensions of High Efficiency Video Coding (HEVC). *IEEE JOURNAL OF SELECTED TOPICS IN SIGNAL PROCESSING*, 7(6), 1001-1016.

26. A. Buchowicz (2013). Video coding and transmission standards for 3D television - a survey. *Opto-Electronics Review*, 21(1), 39-51.

27. Savas T. et alli.. An updatable holographic three-dimensional display. *Nature*, 451, 694-698 .

28. L. Xinan et alli (2009). 3D holographic display with optically addressed spatial light modulator. *3DTV Conference: The True Vision - Capture, Transmission and Display of 3D Video*.

29. B. Blundell and A. Schwarz (2000). Volumetric Three Dimensional Display Systems. JohnWiley and Sons, New York, NY.

30. T. Okoshi (1976). Three-dimensional imaging techniques. *Academic Press, NY*.

31. D. F. McAllister (Editor) (1993). Stereo Computer Grahics and Other True 3D Technologies. Princeton U. Press, Princeton, NJ.

32. S.-H. Kim, Jooyoung Lee, S. Jeong, J. Choi, and J. Kim (2013). Development Of Fixed And Mobile Hybrid 3DTV For Next Generation Terrestrial DTV. *3DTV-Conference: The True Vision-Capture, Transmission and Dispaly of 3D Video (3DTV-CON)*.

33. I. J. Lee, K. Yun, and K. Kim (2013). A 3DTV Broadcasting Scheme for High-Quality Stereoscopic Content Over a Hybrid Network. *IEEE TRANSACTIONS ON BROADCASTING*, 59 (2), 281-289.

34. L. M. J. Meesters, W. A. IJsselsteijn, and P. J. H. SeuntiÃ«ns (2004). A Survey of Perceptual Evaluations and Requirements of Three-Dimensional TV. *IEEE Transactions on Circuits and Systems for Video Technology*, 14(3), 381-391.

35. ISO/IEC JTC1/SC29/WG11 N13364 (2013). White Paper on State of the Art in compression and transmission of 3D Video.

36. M. Strintzis and S. Malassiotis (1998). Review of methods for object-based coding of stereoscopic and 3D image sequences. *Proc. IEEE Int. Symp. Circuits and Systems*, 5, 510-513.

37. G. Iddan and G. Yahav (2001). 3D imaging in the studio (and elsewhere...). *Proceedings of SPIE*, 4298, 48-55.

38. O. Faugeras (1993). Three-dimensional Computer Vision: A Geometric Viewpoint. MIT Press.

39. R. Hartley and A. Zisserman (2004). Multiple View Geometry in Computer Vision. Cambridge University Press, 2nd Edition.

40. P. Harman, J. Flack, S. Fox, and M. Dowley (2002). Rapid 2D to 3D conversion. *Proceedings of SPIE*, 4660, 78-86.

41. A.Kubota, A. Smolic, M. Magnor, M. Tanimoto, Ts. Chen and Ch. Zhang (2007). Multiview Imaging and 3DTV. *Signal Processing Magazine IEEE*, 24(6).

42. M. Barkowsky, R. Cousseau and P. Le Callet (2009). Influence of depth rendering on the

quality of experience for an autostereoscopic display. *iProc. Int. Workshop QoMEX*, July, 192-197.

43. M. T. M. Lambooija, W. A. IJsselsteijn and I. Heynderickx (2007). Visual Discomfort in Stereoscopic Displays: A Review. *Stereoscopic Displays and Virtual Reality Systems XIV*, (6490), Proc. of SPIE-IS&T Electronic Imaging.

44. D. M. Hoffman, A. R. Grishick, K. Akeley, and M. S. Banks (2008). Vergence-accommodation conflicts hinder visual performance and cause visual fatigue. *Journal of Vision*, 8(3), 1-30 .

45. N. Qian (1997). Binocular Disparity and the Perception of Depth. *Neuron*, 18, 359-368.

46. I. P. Howard and B. J. Rogers (1995). Binocular Vision and Stereopsis, Oxford University Press, New York.

47. ITU-T Draft New Recommendation J.3D-fatigue (2012). Assessment methods of visual fatigue and safety guideline for 3D video. *ITU-T*.

48. Recommendation ITU-R BT.2021 (2012). Subjective methods for the assessment of stereoscopic 3DTV systems. *ITU-R*.

49. ITU-T Draft New Recommendation P.3D-sam (2012). Subjective assessment methods for 3D Video Quality. *ITU-T*.

50. ITU-T Draft New Recommendation J.3D-disp-req (2012). Display requirements for 3D video quality assessment. *ITU-T*.

51. Video Quality Experts Group (2012). Test plan for evaluation of video quality models for us with stereoscopic three-dimensional television content. *VQEG*.

Signal Processing in Communications

João Marcos Travassos Romano
Charles Casimiro Cavalcante

12 Overview of Tensor Decompositions with Applications to Communications

ANDRÉ LIMA FERRER DE ALMEIDA
Federal University of Ceará (UFC)

GÉRARD FAVIER
University of Nice Sophia Antipolis (UNS)

JOÃO CESAR MOURA MOTA
Federal University of Ceará (UFC)

JOÃO PAULO CARVALHO LUSTOSA DA COSTA
University of Brasília (UnB)

12.1 INTRODUCTION

The evolution of communication systems and networks implies establishing new paradigms to meet satisfactorily and comfortably the technological and human requirements to enable access to information anytime, anywhere, and in any form. The systemic panorama becomes more complex due to the growing expansion of quality and quantity of services, users, and environmental conditions. In this context, the design, deployment, and operation of a communication system, under either regular or critical operation, require a good strategic planning to ensure maximum integrity and reliability.

Currently, in 4G systems, also known as Long Term Evolution (LTE), multiple antenna based communication systems are employed [1]. Moreover, airborne radar systems are also equipped with antenna arrays. Recently antenna array systems have been incorporated in GNSS systems in order to mitigate multipath, jammers, and spoofers effects [2, 3].

Due to the rapid improvement of the hardware along the last few decades, a large amount of digital information can be processed nowadays. According to Moore's law, which was first published in 1965 [4], every two years the number of transistors to be placed in an integrated circuit is approximately doubled. Therefore, with the rising development of faster processors and memories with higher capacities, it has become common to process considerable amounts of data even in real-time applications [5], which years ago would have been prohibitive. This huge amount of data in many fields turns out to have a multidimensional structure, which can be exploited by applying tensor tools. Note that a variety of other applications including radar, mobile communications, sonar, and seismology may have multidimensional data, which can also be exploited by multidimensional solutions, also known as tensor-based solutions [6]. When considering wireless communication systems, signal processing strategies to improve system reliability by inserting a controlled degree of redundancy in multiple signaling dimensions (space, time, frequency, and/or code) are of paramount importance and tensor-based signal processing is a powerful mathematical tool to deal with this problem [7].

This chapter is interested in wireless communication systems that exploit the redundancies contained in the temporal, spatial, and frequency processing domains in order to guarantee availability, integrity, and intelligibility of useful information. Most signal processing techniques in commercial wireless communications systems nowadays use signal models based on linear algebra, in order to capture the changes and/or transformations made by the transmitter (encoding, beamforming, pre-filtering), by the communication channel (temporal, frequency, and spatial spreading), as well as by the receiving architectures. Current approaches have several limitations in the treatment of the inherent complexities for new wireless communication systems, since they do not exploit the natural structure of the data. Therefore, the redundancies contained in the received signals in communications systems have not been exhaustively utilized in practice.

In this perspective, other domains can be used also to bear the information to add resiliency to transmission in harmful environments such as the domain of coding, which can, in turn, be associated with other domains: space, time, frequency, and more domains in the systems: modulation, transmission rates, sampling rates, pulse shapes, among others. More specifically, to promote compliance with the specifications of multi-user digital communications systems, multicarrier and multiantenna, other processing domains gained importance and are being introduced in the chain of transmission and reception, in order to add more degrees of freedom and diversity to the system, and thus more effectively combat the degrading effects of multiple access interference in frequency selective fading and temporal spread of information.

From a signal processing point of view, the most common approach currently found in commercial communication systems is to combine two or more domains so that they can use algebraic models for matrix manipulation of signals at the level of the transmitter and receiver. Such an approach has been growing lately, since it preserves the multidimensional nature of the data and of the data statistics by making use of multilinear algebra. In this case, the tensor decompositions may be used to model the received signal, establishing a direct association between the physical parameters of the communication link (transmitted symbols, channel attenuation, spatial factor matrices, and precoding factor matrix). [7]. Each field of information is mapped to one dimension of the tensor, and its dimensionality depends on the physical system parameters (number of transmit and receive antennas, data block size, code length, number of parallel transmission frequency, etc.). The received data in such systems is a tensor, whose order is, therefore, the number of domains considered in the model. These aspects are crucial to ensure uniqueness in recovery of information and system parameters, and will be explored throughout this chapter.

The use of tensor-based signal processing in wireless communication systems has received great attention over the past decade. In most communication systems, the received signal generally has a multidimensional nature and can be algebraically modeled as a third- or fourth-order tensor, where each dimension is associated with a particular type of systematic variation of the received signal. In such a higher-dimensional space, each dimension of the received signal tensor can be interpreted as a particular form of signal "diversity." In most of cases, two of these three axes account for space and time dimensions. The space dimension generally corresponds to the number of transmit/receive antennas while the time dimension corresponds to the length of the data block to be processed at the receiver or to the number of symbol periods spanning each data block. The third (and possibly fourth) dimension of the received signal tensor depends on the particular wireless communication system and is generally linked to the type of coding that is applied at the transmitter. For instance, in MIMO communication systems the third (or fourth) dimension can be associated with signal coding across frequency (subcarriers) [7, 8, 9].

The practical motivation for tensor modeling of wireless communication systems comes from the fact that one can simultaneously benefit from multiple (more than two) forms of diversity to perform multiuser signal separation/equalization and channel estimation under model identifiability requirements more relaxed than with conventional matrix-based approaches. These benefits come from the fundamental uniqueness property of tensor decompositions [10, 11, 12, 13].

New communication systems based on tensor decompositions have been studied over the past 15 years [14, 9]. The multilinear or tensor algebra favors the appropriate integration of tensor signal processing to those domains whose processing strategies appropriately exploit the presence of redundancy of information in unsupervised or blind estimation systems. In this chapter, tensor algebra will be explored in strategies that include: space-time and space-time-frequency coding, cooperative (relay-based) systems, and multidimensional array processing. Our aim is to exploit the redundancies of latent useful information on the received signals for allowing blind/semi-blind joint channel and/or symbol estimation.

The chapter is structured as follows: Section 12.3 presents the tensor decomposition and its most important results. Section 12.4 is reserved for tensor application to MIMO systems with space-time and space-time-frequency codings. Section 12.5 is dedicated to the tensor applications in cooperative systems. Section 12.6 applies tensor in multidimensional array processing. The chapter is concluded in Section 12.7 with a discussion on some trends on the topic.

12.2 NOTATIONS

Scalars, column vectors, matrices, and tensors are denoted by lowercase, boldface lowercase, boldface uppercase, and calligraphic uppercase letters, e.g., a, $\mathbf{a}$, $\mathbf{A}$, and

$\mathcal{A}$, respectively. The vector $\mathbf{a}_{i\cdot}$ (resp. $\mathbf{a}_{\cdot j}$) represents the ith row (resp. j^{th} column) of $\mathbf{A}$. $a_{i,j}^{(n)}$ and $x_{i_1,\cdots,i_N}$ denote the entries (i,j) of the matrix $\mathbf{A}^{(n)} \in \mathbb{C}^{I\times J}$, and $(i_1,\cdots,i_N)$ of the tensor $\mathcal{X} \in \mathbb{C}^{I_1\times\cdots\times I_N}$, respectively. Each index i_n ($i_n = 1,\cdots,I_N$, for $n = 1,\cdots,N$) is associated with the mode-n of $\mathcal{X}$, and I_n represents its mode-n dimension.

$\mathbf{I}_R$ stands for the identity matrix of dimensions $R \times R$. The identity tensor of order N and dimensions $R\times\cdots\times R$, denoted by $\mathcal{I}_{N,R}$ or simply $\mathcal{I}$, is a diagonal hypercubic tensor whose elements $\delta_{r_1,\cdots,r_N}$ are defined by means of the generalized Kronecker delta, i.e., $\delta_{r_1,\cdots,r_N} = \begin{cases} 1 & \text{if } r_1 = \cdots = r_N \\ 0 & \text{otherwise} \end{cases}$, and $R_n = R, \forall n = 1,\cdots,N$.

$\mathbf{A}^T$, $\mathbf{A}^*$, $\mathbf{A}^H$, and $r(\mathbf{A})$ denote the transpose, the conjugate, the conjugate transpose, and the rank of $\mathbf{A}$, respectively. The operator $diag(.)$ forms a diagonal matrix from its vector argument. Similarly, $D_i(\mathbf{A}) \in \mathbb{C}^{J\times J}$ denotes the diagonal matrix formed with the ith row of $\mathbf{A} \in \mathbb{C}^{I\times J}$. The operator $vec(.)$ transforms a matrix into a column vector by stacking the columns of its matrix argument. The Kronecker, Khatri-Rao, and Hadamard products are denoted by $\otimes$, $\diamond$, and $\odot$, respectively. The Khatri-Rao product of $\mathbf{A} \in \mathbb{C}^{I\times J}$ with $\mathbf{B} \in \mathbb{C}^{K\times J}$ can be written as follows

$$\mathbf{A} \diamond \mathbf{B} = \begin{bmatrix} \mathbf{B}D_1(\mathbf{A}) \\ \vdots \\ \mathbf{B}D_K(\mathbf{A}) \end{bmatrix} \in \mathbb{C}^{IK\times J} \tag{12.1}$$

$$= [\mathbf{A}_{\cdot 1} \otimes \mathbf{B}_{\cdot 1}, \cdots, \mathbf{A}_{\cdot J} \otimes \mathbf{B}_{\cdot J}]. \tag{12.2}$$

12.3 TENSOR MODELS

In this section, we present an overview of the main tensor models used in this chapter.

A Tucker-N model of an Nth-order tensor $\mathcal{X} \in \mathbb{C}^{I_1\times\cdots\times I_N}$ is defined by means of the following scalar equation

$$x_{i_1,\cdots,i_N} = \sum_{r_1=1}^{R_1} \cdots \sum_{r_N=1}^{R_N} g_{r_1,\cdots,r_N} \prod_{n=1}^{N} a_{i_n,r_n}^{(n)} \tag{12.3}$$

with $i_n = 1,\cdots,I_n$ for $n = 1,\cdots,N$, where $g_{r_1,\cdots,r_N}$ is an element of the core tensor $\mathcal{G} \in \mathbb{C}^{R_1\times\cdots\times R_N}$ and $a_{i_n,r_n}^{(n)}$ is an element of the matrix factor $\mathbf{A}^{(n)} \in \mathbb{C}^{I_n\times R_n}$. This model can be written in terms of mode-n products as

$$\begin{aligned} \mathcal{X} &= \mathcal{G}\times_1\mathbf{A}^{(1)}\times_2\mathbf{A}^{(2)}\times_3\cdots\times_N\mathbf{A}^{(N)} \\ &= \mathcal{G}\times_{n=1}^{N}\mathbf{A}^{(n)}, \end{aligned} \tag{12.4}$$

where the mode-n product of $\mathcal{G}$ with $\mathbf{A}^{(n)}$, denoted by $\mathcal{G}\times_n\mathbf{A}^{(n)}$, gives the tensor $\mathcal{Y}$ of order N and dimensions $R_1 \times \cdots \times R_{n-1} \times I_n \times R_{n+1} \times \cdots \times R_N$, such as [15]

$$y_{r_1,\cdots,r_{n-1},i_n,r_{n+1},\cdots,r_N} = \sum_{r_n=1}^{R_n} a_{i_n,r_n}g_{r_1,\cdots,r_{n-1},r_n,r_{n+1},\cdots,r_N}. \tag{12.5}$$

Different models can be derived from the Tucker model (12.3)-(12.4). In [16], we recently introduced Tucker-(N_1, N) models which correspond to the case where $N - N_1$ factor matrices are equal to identity matrices. For instance, assuming that $\mathbf{A}^{(n)} = \mathbf{I}_{I_n}$, which implies $R_n = I_n$, for $n = N_1 + 1, \cdots, N$, Eq. (12.3) and (12.4) become

$$x_{i_1, \cdots, i_N} = \sum_{r_1=1}^{R_1} \cdots \sum_{r_{N_1}=1}^{R_{N_1}} g_{r_1, \cdots, r_{N_1}, i_{N_1+1}, \cdots, i_N} \prod_{n=1}^{N_1} a_{i_n, r_n}^{(n)} \tag{12.6}$$

$$\mathcal{X} = \mathcal{G} \times_{n=1}^{N_1} \mathbf{A}^{(n)} \tag{12.7}$$

with $\mathcal{G} \in \mathbb{C}^{R_1 \times \cdots \times R_{N_1} \times I_{N_1+1} \times \cdots \times I_N}$.

Examples of Tucker-N and Tucker-(N_1, N) models are given in Table 12.1 for third-order ($N = 3$) and fourth-order ($N = 4$) tensors, with $N_1 = 2$.

PARAFAC-N models can be deduced from Tucker-N models in choosing the identity tensor $\mathcal{I}_{N,R}$ for the core tensor, which implies $R_1 = \cdots = R_N = R$. Eq. (12.3)-(12.4) then become, respectively

$$x_{i_1, \cdots, i_N} = \sum_{r=1}^{R} \prod_{n=1}^{N} a_{i_n, r}^{(n)} \tag{12.8}$$

$$\mathcal{X} = \mathcal{I}_{N,R} \times_{n=1}^{N} \mathbf{A}^{(n)} \tag{12.9}$$

with the factor matrices $\mathbf{A}^{(n)} \in \mathbb{C}^{I_n \times R}, n = 1, \cdots, N$. When R is minimal in (12.8), it is called the rank of $\mathcal{X}$. PARAFAC-3 models are detailed in Table 12.1. In section 12.6, some methods are reviewed for tensor rank estimation.

An important property of PARAFAC models is their essential uniqueness, i.e., uniqueness of their factor matrices up to column permutation and scaling. In other words, any N-uplet $(\bar{\mathbf{A}}^{(1)}, \cdots, \bar{\mathbf{A}}^{(N)})$ is related to another N-uplet $(\mathbf{A}^{(1)}, \cdots, \mathbf{A}^{(N)})$ via the relation $\bar{\mathbf{A}}^{(n)} = \mathbf{A}^{(n)} \mathbf{\Pi} \mathbf{\Lambda}^{(n)}$, where $\mathbf{\Pi}$ is a permutation matrix and $\mathbf{\Lambda}^{(n)}$, $n = 1, \cdots, N$, are diagonal matrices such as $\prod_{n=1}^{N} \mathbf{\Lambda}^{(n)} = \mathbf{I}_R$. A sufficient condition for essential uniqueness of the PARAFAC-N model (12.8)-(12.8) is given by [13], [17]

$$\sum_{n=1}^{N} k_{\mathbf{A}^{(n)}} \geq 2R + N - 1 \tag{12.10}$$

where $k_{\mathbf{A}^{(n)}}$ denotes the k-rank of $\mathbf{A}^{(n)}$, i.e., the largest integer such that any set of $k_{\mathbf{A}^{(n)}}$ columns of $\mathbf{A}^{(n)}$ is linearly independent.

Generalized Tucker-(N_1, N) models were recently introduced in [9]. They correspond to the case where the matrix factors $\mathbf{A}^{(n)}$ are replaced by tensor factors $\mathcal{A}^{(n)}$. Two other families of tensor models are now briefly described: PARATUCK and nested PARAFAC models.

PARATUCK-(N_1, N) models which were for the first time proposed in [18], can be viewed as constrained Tucker-(N_1, N) models [16]. They correspond to particular choices of the core tensor. For instance, PARATUCK-(2,3) models, commonly known as PARATUCK-2 models [19], correspond to Tucker-(2,3) models

$$x_{i_1,i_2,i_3} = \sum_{r_1=1}^{R_1} \sum_{r_2=1}^{R_2} g_{r_1,r_2,i_3} a_{i_1,r_1}^{(1)} a_{i_2,r_2}^{(2)} \qquad (12.11)$$

with the core tensor defined as $g_{r_1,r_2,i_3} = w_{r_1,r_2} \phi_{r_1,i_3} \psi_{r_2,i_3}$, where ϕ_{r_1,i_3} and ψ_{r_2,i_3} are entries of two constraint matrices $\Phi \in \mathbb{C}^{R_1 \times I_3}$ and $\Psi \in \mathbb{C}^{R_2 \times I_3}$ which define interactions between columns of the factor matrices $(\mathbf{A}^{(1)}, \mathbf{A}^{(2)})$, along the mode-3 of $\mathcal{X}$, while the matrix $\mathbf{W} \in \mathbb{C}^{R_1 \times R_2}$ contains the weights of these interactions. In the context of wireless communications, $\mathbf{W}$ corresponds to the code matrix, while Φ and Ψ define the allocation tensor $\mathcal{A} \in \mathbb{C}^{R_1 \times R_2 \times I_3}$ such as

$$a_{r_1,r_2,i_3} = \phi_{r_1,i_3} \psi_{r_2,i_3}. \qquad (12.12)$$

The core tensor $\mathcal{G} \in \mathbb{C}^{R_1 \times R_2 \times I_3}$ of the associated Tucker-(2,3) model can then be rewritten as $g_{r_1,r_2,i_3} = w_{r_1,r_2} a_{r_1,r_2,i_3}$, or equivalently as the Hadamard product of $\mathbf{W}$ and $\mathcal{A}$ along their common modes $\{r_1, r_2\}$

$$\mathcal{G} = \mathbf{W} \underset{\{r_1,r_2\}}{\odot} \mathcal{A}. \qquad (12.13)$$

Similarly, in the case of a fourth-order tensor $\mathcal{X} \in \mathbb{C}^{I_1 \times I_2 \times I_3 \times I_4}$, PARATUCK-(2,4) models, used in [18] - [20] for the tensor space-time (TST) coding system that will be presented in the next section, are particular Tucker-(2,4) models with the core tensor defined as $g_{r_1,r_2,i_3,i_4} = w_{r_1,r_2,i_4} \phi_{r_1,i_3} \psi_{r_2,i_3}$. In this case, the core tensor $\mathcal{G} \in \mathbb{C}^{R_1 \times R_2 \times I_3 \times I_4}$ can be expressed as the Hadamard product of the code tensor $\mathcal{W} \in \mathbb{C}^{R_1 \times R_2 \times I_4}$ and the allocation tensor $\mathcal{A} \in \mathbb{C}^{R_1 \times R_2 \times I_3}$ along their common modes $\{r_1, r_2\}$

$$\mathcal{G} = \mathcal{W} \underset{\{r_1,r_2\}}{\odot} \mathcal{A}$$

where $\mathcal{A}$ is defined as in (12.12).

These PARATUCK models are presented in Table 12.1, with two extensions corresponding to generalized PARATUCK-(2,4) and PARATUCK-(2,5) models of fourth- and fifth-order tensors. Note that for these generalized PARATUCK models, some factors are tensors. See [9] for a more detailed presentation of such tensor models.

To complete this overview of tensor models, we present nested PARAFAC models for a fourth-order tensor, defined by means of the following scalar equation [21]

$$x_{i_1,i_2,i_3,i_4} = \sum_{r_1=1}^{R_1} \sum_{r_2=1}^{R_2} a_{i_1,r_1}^{(1)} a_{i_2,r_1}^{(2)} u_{r_1,r_2} b_{i_3,r_2}^{(1)} b_{i_4,r_2}^{(2)}. \qquad (12.14)$$

Let us define the third-order tensors $\mathcal{W} \in \mathbb{C}^{I_3 \times I_4 \times R_1}$ and $\mathcal{Z} \in \mathbb{C}^{I_1 \times I_2 \times R_2}$ such as

$$w_{i_3,i_4,r_1} = \sum_{r_2=1}^{R_2} b_{i_3,r_2}^{(1)} b_{i_4,r_2}^{(2)} u_{r_1,r_2} \tag{12.15}$$

$$z_{i_1,i_2,r_2} = \sum_{r_1=1}^{R_1} a_{i_1,r_1}^{(1)} a_{i_2,r_1}^{(2)} u_{r_1,r_2} \tag{12.16}$$

or equivalently in terms of mode-n products

$$\mathcal{W} = \mathcal{I}_{3,R_2} \times_1 \mathbf{B}^{(1)} \times_2 \mathbf{B}^{(2)} \times_3 \mathbf{U} \tag{12.17}$$

$$\mathcal{Z} = \mathcal{I}_{3,R_1} \times_1 \mathbf{A}^{(1)} \times_2 \mathbf{A}^{(2)} \times_3 \mathbf{U}^T. \tag{12.18}$$

Combining the last two modes of $\mathcal{X}$ and the first two ones, by means of the variable changes $k_1 = (i_4-1)I_3 + i_3$ and $k_2 = (i_2-1)I_1 + i_1$, the fourth-order nested PARAFAC model (12.14) can be rewritten as two third-order PARAFAC models of the tensors $\mathcal{X}^{(1)} \in \mathbb{C}^{I_1 \times I_2 \times K_1}$ and $\mathcal{X}^{(2)} \in \mathbb{C}^{K_2 \times I_3 \times I_4}$, where $K_1 = I_3 I_4$ and $K_2 = I_1 I_2$

$$x_{i_1,i_2,k_1}^{(1)} = \sum_{r_1=1}^{R_1} a_{i_1,r_1}^{(1)} a_{i_2,r_1}^{(2)} w_{k_1,r_1} \tag{12.19}$$

$$x_{k_2,i_3,i_4}^{(2)} = \sum_{r_2=1}^{R_2} z_{k_2,r_2} b_{i_3,r_2}^{(1)} b_{i_4,r_2}^{(2)} \tag{12.20}$$

These tensors $\mathcal{X}^{(1)}$ and $\mathcal{X}^{(2)}$ which are two contracted forms of the original tensor $\mathcal{X}$ highlight the nesting of the two PARAFAC models (12.19) and (12.20) via the tensors $\mathcal{W}$ and $\mathcal{Z}$ whose proper PARAFAC models (12.15) and (12.16) share the matrix factor $\mathbf{U}$. The five factor matrices $(\mathbf{A}^{(1)}, \mathbf{A}^{(2)}, \mathbf{U}, \mathbf{B}^{(1)}, \mathbf{B}^{(2)})$ of the nested PARAFAC model (12.14) can be estimated using a five-step ALS algorithm as shown in [22].

12.4   APPLICATION TO MIMO COMMUNICATION SYSTEMS

In this section, we present, in a unified way, seven tensor-based systems. Four systems (DS-CDMA, KRST, DKSTF, TSTF) are presented in more detail, and three other ones (STF, TST, ST) are briefly introduced as particular cases of the TSTF system. The tensor models of all the considered systems are given in Table 12.2. Due to a lack of space, for a description of semi-blind deterministic receivers developed for these systems, we postpone the reader to the references quoted in Table 12.2. Moreover, to simplify the presentation, we consider the noiseless case.

12.4.1 DS-CDMA SYSTEMS

Direct-sequence code-division multiple access (DS-CDMA) systems were the first ones to be modeled by means of a PARAFAC model, in 2000 [14]. Such systems are composed of K receive antennas receiving the information bearing signals transmitted by M users. The wireless channel between user m and receive antenna k is

Table 12.1

Some examples of tensor models

Models	Scalar writings	Ref.
Tucker-3	$x_{i_1,i_2,i_3} = \sum\limits_{r_1=1}^{R_1} \sum\limits_{r_2=1}^{R_2} \sum\limits_{r_3=1}^{R_3} g_{r_1,r_2,r_3} a_{i_1,r_1}^{(1)} a_{i_2,r_2}^{(2)} a_{i_3,r_3}^{(3)}$	[23]
PARAFAC-3	$x_{i_1,i_2,i_3} = \sum\limits_{r=1}^{R} a_{i_1,r}^{(1)} a_{i_2,r}^{(2)} a_{i_3,r}^{(3)}$	[11], [12]
Tucker-(2,3)	$x_{i_1,i_2,i_3} = \sum\limits_{r_1=1}^{R_1} \sum\limits_{r_2=1}^{R_2} g_{r_1,r_2,i_3} a_{i_1,r_1}^{(1)} a_{i_2,r_2}^{(2)}$	[16]
Paratuck-2	$x_{i_1,i_2,i_3} = \sum\limits_{r_1=1}^{R_1} \sum\limits_{r_2=1}^{R_2} g_{r_1,r_2,i_3} a_{i_1,r_1}^{(1)} a_{i_2,r_2}^{(2)}$ $g_{r_1,r_2,i_3} = w_{r_1,r_2} \phi_{r_1,i_3} \psi_{r_2,i_3}$	[19]
Tucker-(2,4)	$x_{i_1,i_2,i_3,i_4} = \sum\limits_{r_1=1}^{R_1} \sum\limits_{r_2=1}^{R_2} g_{r_1,r_2,i_3,i_4} a_{i_1,r_1}^{(1)} a_{i_2,r_2}^{(2)}$	[16]
Paratuck-(2,4)	$x_{i_1,i_2,i_3,i_4} = \sum\limits_{r_1=1}^{R_1} \sum\limits_{r_2=1}^{R_2} g_{r_1,r_2,i_3,i_4} a_{i_1,r_1}^{(1)} a_{i_2,r_2}^{(2)}$ $g_{r_1,r_2,i_3,i_4} = w_{r_1,r_2,i_4} \phi_{r_1,i_3} \psi_{r_2,i_3}$	[18],[20]
Nested PARAFAC	$x_{i_1,i_2,i_3,i_4} = \sum\limits_{r_1=1}^{R_1} \sum\limits_{r_2=1}^{R_2} a_{i_1,r_1}^{(1)} a_{i_2,r_1}^{(2)} u_{r_1,r_2} b_{i_3,r_2}^{(1)} b_{i_4,r_2}^{(2)}$	[21]
Generalized Paratuck-(2,4)	$x_{i_1,i_2,i_3,i_4} = \sum\limits_{r_1=1}^{R_1} \sum\limits_{r_2=1}^{R_2} g_{r_1,r_2,i_3,i_4} a_{i_1,r_1,i_3}^{(1)} a_{i_2,r_2}^{(2)}$ $g_{r_1,r_2,i_3,i_4} = w_{r_1,r_2} \phi_{r_1,i_3,i_4} \psi_{r_2,i_3,i_4}$	[24]
Generalized Paratuck-(2,5)	$x_{i_1,i_2,i_3,i_4,i_5} = \sum\limits_{r_1=1}^{R_1} \sum\limits_{r_2=1}^{R_2} g_{r_1,r_2,i_3,i_4,i_5} a_{i_1,r_1,i_3}^{(1)} a_{i_2,r_2,i_4}^{(2)}$ $g_{r_1,r_2,i_3,i_4,i_5} = w_{r_1,r_2,i_3,i_4,i_5} a_{r_1,r_2,i_3,i_4}$	[9]

assumed to be quasi-static and flat fading, with fading coefficient $h_{k,m}$, assumed to be time-invariant over N symbol periods. At the symbol period n, each user m transmits a symbol $s_{n,m}$ encoded using a spreading code $w_{j,m}$ of length J. The M encoded signals $u_{m,n,j} = s_{n,m} w_{j,m}$ are transmitted through the channel $\mathbf{H} \in \mathbb{C}^{K \times M}$, and then summed at the receive antenna k, to give the following discrete-time baseband-equivalent model for the received signals

$$x_{k,n,j} = \sum_{m=1}^{M} h_{k,m} u_{m,n,j} = \sum_{m=1}^{M} h_{k,m} s_{n,m} w_{j,m} \tag{12.21}$$

with $k = 1, \cdots, K; n = 1, \cdots, N; j = 1, \cdots, J$. The tensor $\mathcal{X} \in \mathbb{C}^{K \times N \times J}$ of received signals satisfies a PARAFAC model of rank M, with factor matrices $(\mathbf{H}, \mathbf{S}, \mathbf{W})$. Such a system exploits three diversities (space, time, code) associated with the indices (k, n, j). The matrix factors can be jointly and semi-blindly estimated using a standard alternating least squares (ALS) algorithm [14].

12.4.2   KRST SYSTEMS

Khatri-Rao space-time coding systems, proposed in [25], consist in transmitting during each time block n, one data stream $\mathbf{s}_{n.} \in \mathbb{C}^{1 \times M}$ composed of M symbols, using M transmit antennas, two coding matrices $\mathbf{\Theta} \in \mathbb{C}^{M \times M}$ and $\mathbf{W} \in \mathbb{C}^{J \times M}$ are used for encoding the information symbols. The first one allows to linearly combine the M symbols of $\mathbf{s}_{n.}$ onto each transmit antenna m, which gives the pre-coded signal $v_{n,m} = \sum_{l=1}^{M} \theta_{m,l} s_{n,l}$, whereas the aim of the second one is to spread the pre-coded signal $v_{n,m}$ over J slots. The resulting encoded signals $u_{m,n,j} = v_{n,m} w_{j,m}$ define a third-order tensor $\mathcal{U} \in \mathbb{C}^{M \times N \times J}$ whose matrix slice $\mathbf{U}_{.,n,.} \in \mathbb{C}^{M \times J}$ is given by $\mathbf{U}_{.,n,.} = diag(\mathbf{\Theta} \mathbf{s}_{n,.}^{T}) \mathbf{W}^{T}$. This equation illustrates the space spreading of each data stream $\mathbf{s}_{n.}$ owing to pre-coding $(\mathbf{\Theta})$, and its time spreading across J time slots by means of post-coding $(\mathbf{W})$. These coding operations can be represented by means of the following equations

$$\mathbf{V} = \mathbf{S}\mathbf{\Theta}^{T} \in \mathbb{C}^{N \times M} \;, \quad \mathcal{U} = \mathbf{V} \underset{m}{\odot} \mathbf{W}. \tag{12.22}$$

The encoded signals are transmitted through the channel $\mathbf{H} \in \mathbb{C}^{K \times M}$, assumed to be flat fading and time-invariant during N time blocks. The signal received by the receive antenna k, during the slot j of the time block n, satisfies the following third-order PARAFAC model

$$x_{k,n,j} = \sum_{m=1}^{M} h_{k,m} u_{m,n,j} = \sum_{m=1}^{M} \sum_{l=1}^{M} h_{k,m} s_{n,l} \theta_{m,l} w_{j,m} \tag{12.23}$$

$$= \sum_{m=1}^{M} h_{k,m} v_{n,m} w_{j,m}. \tag{12.24}$$

Comparing this equation with the PARAFAC model (12.21) of a DS-CDMA system, we can conclude that the KRST system which uses a space-time (ST) coding instead of a CDMA coding technique, relies also on a PARAFAC model whose factor matrices $(\mathbf{H}, \mathbf{V}, \mathbf{W})$ can be estimated by means of an ALS algorithm. However, we have to note that, once the pre-coded signal matrix $(\mathbf{V} = \mathbf{S\Theta}^T)$ is estimated, a sphere de-coding is needed at the receiver for performing detection of the transmitted symbols, i.e., for estimating the symbol matrix $(\mathbf{S})$. Consequently, the introduction of space spreading by means of pre-coding induces a supplementary decoding step.

REMARK

Noting that $\mathbf{v}_{n,.} = \mathbf{s}_{n,.}\mathbf{\Theta}^T$, the transpose of the matrix slice $\mathbf{U}_{.,n,.}$ of the encoded signals tensor $\mathcal{U}$ can be rewritten as $\mathbf{U}_{.,n,.}^T = \mathbf{W}D_n(\mathbf{V})$. Applying the property (12.1) of the Khatri-Rao product, the matrix unfolding $\mathbf{U}_{NJ \times M}$ of $\mathcal{U}$ is given by

$$
\mathbf{U}_{NJ \times M} = \begin{bmatrix} \mathbf{U}_{.,1,.}^T \\ \vdots \\ \mathbf{U}_{.,N,.}^T \end{bmatrix} = \begin{bmatrix} \mathbf{W}D_1(\mathbf{V}) \\ \vdots \\ \mathbf{W}D_N(\mathbf{V}) \end{bmatrix} \tag{12.25}
$$

$$
= \mathbf{V} \diamond \mathbf{W} = \mathbf{S\Theta}^T \diamond \mathbf{W} \in \mathbb{C}^{NJ \times M}. \tag{12.26}
$$

This matrix unfolding of $\mathcal{U}$ highlights the space-time coding under the form of the Khatri-Rao product of the pre-coded signal matrix $\mathbf{V}$ with the time spreading code matrix $\mathbf{W}$.

12.4.3 DKSTF SYSTEMS

An extension of the KRST system including a space-time-frequency coding was pro-posed in [21]. A multiple input multiple output (MIMO) orthogonal frequency divi-sion multiplexing (OFDM) system is considered with K receive antennas and M transmit antennas. The transmitter uses F neighboring sub-carriers for transmitting encoded signals which define a fourth-order tensor $\mathcal{U} \in \mathbb{C}^{M \times N \times J \times F}$ with the follow-ing entry

$$
u_{m,n,j,f} = \sum_{l=1}^{M} s_{n,l}\theta_{m,l}w_{j,m}b_{f,l}. \tag{12.27}
$$

This coding operation is composed of two parts: a space-frequency pre-coding and a time post-coding. The matrix slice $\mathbf{U}_{.,n,.,f} \in \mathbb{C}^{M \times J}$ associated with the time block n and the sub-carrier frequency f, is given by $\mathbf{U}_{.,n,.,f} = diag(\mathbf{\Theta}diag(\mathbf{b}_{f,.})\mathbf{s}_{n,.}^T)\mathbf{W}^T$. In this expression, the term $diag(\mathbf{b}_{f,.})\mathbf{s}_{n,.}^T$ represents the frequency pre-coding which pro-vides the frequency pre-coded signal $d_{n,f,l} = s_{n,l}b_{f,l}$, whereas the term $diag(\mathbf{\Theta}\mathbf{d}_{n,f,.})$ corresponds to the space pre-coding applied to the frequency pre-coded signal vector $\mathbf{d}_{n,f,.}$.

The channel being assumed to be constant and independent of the frequency, dur-ing N time blocks, the frequency-domain version of Eq. (12.23) for the received signals becomes

$$x_{k,n,j,f} = \sum_{m=1}^{M} h_{k,m} u_{m,n,j,f} \tag{12.28}$$

$$= \sum_{m=1}^{M} \sum_{l=1}^{M} h_{k,m} s_{n,l} \theta_{m,l} w_{j,m} b_{f,l}. \tag{12.29}$$

Comparing this equation with (12.14), we conclude that the fourth-order tensor of received signals $X \in \mathbb{C}^{K \times N \times J \times F}$ satisfies a nested PARAFAC model with the following correspondences

$$(I_1, I_2, I_3, I_4, R_1, R_2) \quad \leftrightarrow \quad (K, N, J, F, M, M) \tag{12.30}$$
$$(\mathbf{A}^{(1)}, \mathbf{A}^{(2)}, \mathbf{U}, \mathbf{B}^{(1)}, \mathbf{B}^{(2)}) \quad \leftrightarrow \quad (\mathbf{H}, \mathbf{S}, \mathbf{\Theta}, \mathbf{W}, \mathbf{B}). \tag{12.31}$$

Assuming that the code matrices $(\mathbf{\Theta}, \mathbf{W}, \mathbf{B})$ are known at the receiver, the channel and symbol matrices $(\mathbf{H}, \mathbf{S})$ can be estimated by means of a two-step ALS algorithm [21].

REMARKS

- Define the space-frequency pre-coded signal tensor as $q_{m,n,f} = \sum_{l=1}^{M} \theta_{m,l} s_{n,l} b_{f,l}$. This tensor $Q \in \mathbb{C}^{M \times N \times F}$ of rank M satisfies a PARAFAC model with factor matrices $(\mathbf{\Theta}, \mathbf{S}, \mathbf{B})$. Its matrix unfolding $\mathbf{Q}_{NF \times M}$ is given by $(\mathbf{S} \diamond \mathbf{B})\mathbf{\Theta}^T$, and the unfolding $\mathbf{U}_{NFJ \times M}$ of the encoded signals tensor $\mathcal{U}$ can be deduced as

$$\mathbf{U}_{NFJ \times M} = (\mathbf{S} \diamond \mathbf{B})\mathbf{\Theta}^T \diamond \mathbf{W}. \tag{12.32}$$

This expression, to be compared with (12.26), highlights the DKSTF coding as a double Khatri-Rao coding, the first one corresponding to a space-frequency pre-coding, whereas the second one corresponds to a time post-coding.

- Unlike DS-CDMA and KRST systems which exploit three diversities represented by the indices (k, n, j), the DKSTF system is based on four diversities corresponding to the indices (k, n, j, f). The use of an extra diversity associated with the sub-carrier frequency allows us not only to improve the receiver performance in terms of symbol recovery, but also to simplify the symbol detection by avoiding the decoding step at the receiver. Indeed, capitalizing on the nested PARAFAC model for the received signals allows us to directly estimate the symbol matrix, jointly with the channel estimation.

12.4.4 TSTF SYSTEMS

A MIMO OFDM-CDMA system equipped with M transmit antennas and K receive antennas was recently proposed in [9]. This system uses a tensor space-time-frequency (TSTF) coding for transmitting R data streams composed of N information symbols each, which constitute the symbol matrix $\mathbf{S} \in \mathbb{C}^{N \times R}$. The transmission is decomposed into P time blocks of N symbol periods, each one being composed of J chips. At each symbol period n of the pth block, the transceiver transmits a linear combination of the nth symbols of certain data streams, using a set of transmit antennas and sub-carriers. The transmit antennas and the sub-carriers that are used, and the data streams that are transmitted during each time block p are determined by the allocation tensor $\mathcal{A} \in \mathbb{R}^{M \times R \times F \times P}$ such as $a_{m,r,f,p} = 1$ means that the data stream r is transmitted using the transmit antenna m, with the sub-carrier f, during the time block p. The coding is carried out by means of a fifth-order code tensor $\mathcal{W} \in \mathbb{C}^{M \times R \times F \times P \times J}$ such as the transmitted encoded symbols are defined as

$$u_{m,n,f,j} = \sum_{r=1}^{R} w_{m,r,f,p,j} s_{n,r} a_{m,r,f,p}.$$

Assuming flat Rayleigh fading propagation channels for each frequency f, with fading coefficient $h_{k,m,f}$ between the transmit (m) and receive (k) antennas, the discrete-time baseband-equivalent model for the signal received at the kth receive antenna during the jth chip period of the nth symbol period of the pth block, and associated with the fth sub-carrier, is given by

$$\begin{aligned} x_{k,n,f,p,j} &= \sum_{m=1}^{M} h_{k,m,f} u_{m,n,f,p,j} \\ &= \sum_{m=1}^{M} \sum_{r=1}^{R} w_{m,r,f,p,j} h_{k,m,f} s_{n,r} a_{m,r,f,p}. \end{aligned} \tag{12.33}$$

The received signals define a fifth-order tensor $\mathcal{X} \in \mathbb{R}^{K \times N \times F \times P \times J}$ which satisfies a generalized PARATUCK-(2,5) model as defined in Table 12.1, with the following correspondences

$$(I_1, I_2, I_3, I_4, I_5, R_1, R_2) \quad \Leftrightarrow \quad (K, N, F, P, J, M, R) \tag{12.34}$$

$$(\mathcal{A}^{(1)}, \mathcal{A}^{(2)}, \mathcal{W}, \mathcal{A}) \quad \Leftrightarrow \quad (\mathcal{H}, \mathbf{S}, \mathcal{W}, \mathcal{A}). \tag{12.35}$$

REMARKS

- Assuming the code and allocation tensors $(\mathcal{W}, \mathcal{A})$ known at the receiver, and exploiting the PARATUCK-(2,5) model (12.33), it is possible to jointly and semi-blindly estimate the channel tensor $(\mathcal{H} \in \mathbb{C}^{K \times M \times F})$ and the symbol matrix $(\mathbf{S})$ by means of a two-step ALS algorithm. A Kronecker product least squares (KPLS) based receiver is also proposed in [9]. This closed-form and low-complexity solution which exploits the Kronecker product of the symbol matrix with an unfolded matrix of the channel tensor, can be used in a supervised context, i.e., with the use of a short sequence of pilot symbols, or in a semi-blind configuration.

- Two main characteristics distinguish the TSTF system from the systems previously presented. The first one concerns the use of a fifth-order tensor ($\mathcal{W}$) for space-time-frequency coding, instead of one, two, or three code matrices, as is the case of DS-CDMA, KRST, and DKSTF systems, respectively. The second one is linked to the use of a fourth-order tensor ($\mathcal{A}$) for resource allocation. The five dimensional coding tensor allows us to increase the number of diversities, represented by the indices (k, n, f, p, j), that are exploited. That facilitates performance/complexity tradeoffs in all the signaling dimensions. Moreover, the use of an allocation tensor instead of several allocation matrices, as is the case of other tensor-based systems, provides higher allocation flexibility. This allocation structure can be useful when data streams have different levels of coding redundancy over space, time, and/or frequency, as illustrated by means of computer simulations in [9]. It is also to be noted that, like the STF system [24], the TSTF one takes into account the channel dependency with respect to the sub-carrier frequency, which is not the case of the DKSTF system.

Three other tensor-based systems, which can de derived as particular cases of the TSTF system, are presented in Table 12.2. For instance, the TSTF system is a CDMA extension of the STF system from the coding point of view, the STF system using only a bi-dimensional coding [24]. The TSTF system can also be viewed as an OFDM extension of the TST system [20] with a multicarrier transmission and a tensor space-time-frequency (TSTF) coding instead of a third-order tensor space-time (TST) coding. This last coding leads to a PARATUCK-(2,4) model for the received signals, with the exploitation of four diversities represented by the indices (k, n, p, j). Finally, the ST system of [26] corresponds to a particular TSTF system employing a space-time coding matrix, with two allocation matrices, which leads to a PARATUCK-2 model with only three diversities associated with indices (k, n, p). This system is close to a DS-CDMA system in terms of diversity, with the incorporation of resource allocation.

12.5   APPLICATION TO COOPERATIVE COMMUNICATIONS

12.5.1   BACKGROUND AND MOTIVATION

The concept of cooperative communications has recently emerged as a promising solution to future wireless communication standards due to the promised gains (such as, spatial diversity, enhanced coverage, and increased capacity) that can be attained without necessarily using multiple antennas at the terminals [27, 28]. Relaying strategies are classified according to the forwarding protocol (i.e., amplify-and-forward

Table 12.2

Comparison of tensor-based systems:core tensors-received signals

Systems	Received signals	Models Core tensors	Ref
DS-CDMA	$x_{k,n,j} = \sum_{m=1}^{M} h_{k,m} s_{n,m} w_{j,m}$	PARAFAC-3	[14]
KRST	$x_{k,n,j} = \sum_{m=1}^{M} h_{k,m} v_{n,m} w_{j,m}$ $v_{n,m} = \sum_{l=1}^{M} s_{n,l} \theta_{m,l}$	PARAFAC-3	[25]
DKSTF	$x_{k,n,j,f} = \sum_{m=1}^{M} \sum_{l=1}^{M} h_{k,m} s_{n,l} \theta_{m,l} w_{j,m} b_{f,l}$	Nested PARAFAC	[21]
TSTF	$x_{k,n,f,p,j} = \sum_{m=1}^{M} \sum_{r=1}^{R} g_{m,r,f,p,j} h_{k,m,f} s_{n,r}$	Generalized PARATUCK-(2,5) $g_{m,r,f,p,j} = w_{m,r,f,p,j} c_{m,r,f,p}$	[9]
STF	$x_{k,n,f,p} = \sum_{m=1}^{M} \sum_{r=1}^{R} g_{m,r,f,p} h_{k,m,f} s_{n,r}$	Generalized PARATUCK-(2,4) $g_{m,r,f,p} = w_{m,r} \phi_{f,p,m} \psi_{f,p,r}$	[24]
TST	$x_{k,n,p,j} = \sum_{m=1}^{M} \sum_{r=1}^{R} g_{m,r,p,j} h_{k,m} s_{n,r}$	PARATUCK-(2,4) $g_{m,r,p,j} = w_{m,r,j} \phi_{p,m} \psi_{p,r}$	[18], [20]
ST	$x_{k,n,p} = \sum_{m=1}^{M} \sum_{r=1}^{R} g_{m,r,p} h_{k,m} s_{n,r}$	PARATUCK-2 $g_{m,r,p} = w_{m,r} \phi_{p,m} \psi_{p,r}$	[26]

(AF), decode-and-forward (DF), and others), and also to the network topology (i.e. one-way and two-way) [29, 30, 31]. The AF protocol avoids decoding at the relays and, therefore, is often preferable when complexity and/or latency issues are of importance. Among the existing hoping schemes, two-hop relaying is well known as an efficient way to extend the coverage area and to overcome impairments such as fading, shadowing, and path loss, in wireless channels [28, 32]. When the direct links between the sources (co-channel users) and the destination (base station) are deeply faded, relay stations are used to improve the communication.

With the aim of simplifying the computational burden at the relay stations, a receiver algorithm can be used at the destination only. In the context of two-hop relaying systems, the reliability of receivers strongly depends on the accuracy of channel state information (CSI) for the different communication links. Various recent works have tackled the problem of channel estimation in non-regenerative relaying systems [33, 34, 35, 36, 37, 38, 39, 40, 41]. In [33], traditional single input single output (SISO) channel estimation methods are generalized to the MIMO case. The work [36] proposes a two-stage training sequence based channel estimation method. Other methods deal with the channel estimation problem in cooperative communications by using singular value decomposition (SVD) based algorithms [37, 40, 41]. It is worth mentioning that the use of precoding techniques at the source and/or the relay generally requires the instantaneous CSI knowledge of both hops to carry out transmit optimization [42, 43, 44]. However, in real-life communication systems, the CSI is unknown, and therefore, has to be estimated. Conventional point-to-point pilot-based strategies do not provide channel estimation of both hops separately at the destination. The approach considered in this work aims to jointly estimate the transmitted symbols and the aforementioned channels, with a reduced computational complexity at the relay station.

12.5.2 BRIEF OVERVIEW OF TENSOR-BASED SOLUTIONS

Recently, tensor decompositions have also proven useful in cooperative communications, where tensor-based semi-blind receivers have been developed for channel estimation and symbol detection in cooperative relay-assisted communication systems [45, 37, 46, 47, 48, 49, 50, 51]. The common feature of these tensor-based approaches is the possibility of jointly estimating the partial channels (source-relay and relay-destination channels) at destination, under more relaxed identifiability conditions than conventional matrix-based estimation methods.

The work [45] was the first to use tensor-based signal processing in the context of cooperative communications, followed by its extended version presented in [37]. Therein, the focus was on two-way relaying systems and a training sequence based channel estimation algorithm was proposed for the estimation of the partial channels. During the first phase, two sources send pilot sequences to the MIMO relay. The training phase is divided into P phases, each one composed of N symbol periods. Let M_i, M_R, and M_D denote, respectively, the number of antennas at the i-th source, $i = 1, 2$, and relay and destination. In the second phase, the relay combines the received signals and transmits the result back to the two sources. The noiseless part of the signal received at the i-th source, can be written as the following Tucker2 model [37]:

$$\mathcal{X}_i = \mathcal{G} \times_1 \mathbf{H}_i \times_2 (\mathbf{SH}) \in \mathbb{C}^{M_i \times N \times P}, \quad i = 1, 2, \tag{12.36}$$

where $\mathbf{H}_i \in \mathbb{C}^{M_i \times M_R}$ is the channel matrix linking the relay to the i-th source, $\mathbf{H} \doteq [\mathbf{H}_1^T \ \mathbf{H}_2^T]^T \in \mathbb{C}^{\bar{M} \times M_R}$ is the compound channel, $\mathbf{S}_i \in \mathbb{C}^{N \times M_i}$ is the pilot symbol matrix transmitted by the i-th user, and $\mathbf{S} \doteq [\mathbf{S}_1 \ \mathbf{S}_2] \in \mathbb{C}^{N \times \bar{M}}$ is the compound pilot symbol matrix, with $\bar{M} = M_1 + M_2$, and $\mathcal{G} \in \mathbb{C}^{M_R \times M_R \times P}$ is the AF tensor. The so-called TENCE algorithm [37] relies on a PARAFAC modeling of $\mathcal{G}$ as a function of three AF matrices, i.e., $\mathcal{G} \doteq \mathcal{I}_Q \times_1 \mathbf{G}_1 \times_2 \mathbf{G}_2 \times_3 \mathbf{G}_3$, where Q is the rank of the AF tensor, which allows to rewrite (12.36) as

$$\mathcal{X}_i = \mathcal{I}_Q \times_1 (\mathbf{H}_i \mathbf{G}_1) \times_2 (\mathbf{SHG}_2) \times_3 \mathbf{G}_3, \quad i = 1, 2, \tag{12.37}$$

which corresponds to a third-order PARAFAC model with factor matrices $\mathbf{A} \doteq \mathbf{H}_i \mathbf{G}_1 \in \mathbb{C}^{M_i \times Q}$, $\mathbf{B} \doteq \mathbf{SHG}_2 \in \mathbb{C}^{N \times Q}$, and $\mathbf{C} \doteq \mathbf{G}_3 \in \mathbb{C}^{P \times Q}$. By imposing a PARAFAC decomposition structure for the received signal tensor, the authors perform channel estimation by means of a least squares Khatri-Rao factorization algorithm combined with a refinement stage.

By focusing on one-way two-hop MIMO relaying, a PARAFAC-based method was proposed later in [46] to estimate the partial channel matrices. The receiver algorithm is based on a bilinear alternating least squares procedure. As in [37], the approach of [46] resorts to P training phases. The source node is equipped with M_S antennas and the multiple-antenna relay makes use of a diagonal AF matrix. The received signal tensor in absence of noise is given by:

$$\mathcal{X} = \mathcal{I}_{M_R} \times_1 \mathbf{H}_{RD} \times_2 \left(\mathbf{SH}_{SR}^T\right) \times_3 \mathbf{F} \in \mathbb{C}^{M_D \times N \times P}, \tag{12.38}$$

where $\mathbf{H}_{RD} \in \mathbb{C}^{M_D \times M_R}$ and $\mathbf{H}_{SR} \in \mathbb{C}^{M_R \times M_S}$ are the relay-destination and source-relay channel matrices, respectively, $\mathbf{S} \in \mathbb{C}^{N \times M_S}$ is the pilot symbol matrix transmitted by the source, and $\mathbf{F} \in \mathbb{C}^{P \times M_R}$ is the AF relay matrix, which is known at the destination. Under the assumption of orthogonal pilot sequences, i.e., $\mathbf{S}^H \mathbf{S} = \mathbf{I}_{M_S}$, it follows that

$$\mathcal{Y} \doteq \mathcal{X} \times_2 \mathbf{S}^H = \mathcal{I}_{M_R} \times_1 \mathbf{H}_{RD} \times_2 \mathbf{H}_{SR}^T \times_3 \mathbf{F} \in \mathbb{C}^{M_D \times M_S \times P}, \tag{12.39}$$

where $\mathcal{Y}$ denotes the filtered received signal tensor after mode-2 multiplication by the pilot symbol matrix. The PARAFAC-based receiver of [46] exploits the mode-1 and mode-2 unfoldings of $\mathcal{Y}$ to estimate the partial channels $\mathbf{H}_{RD}$ and $\mathbf{H}_{SR}$ using a bilinear alternating least squares (BALS) algorithm. The channel matrix that models the direct (source-destination) link is not included into the tensor model. The authors of [46] propose to estimate this matrix separately by means of a standard LS algorithm.

The works [47] and [48] tackle the multiuser cooperative communications and are based on the PARAFAC model. Contrarily to the previously discussed approaches, the focus on these works is on a semi-blind joint channel and symbol estimation, by avoiding the use of pilot sequences. In [47], a system with M_U co-channel users is

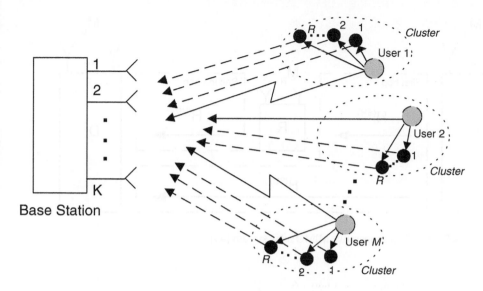

Figure 12.1 Illustration of an uplink cooperative communication system [47]

considered, where each user is assisted by R single-antenna relays. The R relays of a given user multiplexed in time and are located in a cluster, which is remote to the destination. This assumption implies that the signal received at a relay located within the cluster of a given user contains no significant contribution from the other users. Figure 12.1 provides an illustration of the uplink cooperative communication system considered in [47].

In contrast to the other existing works on the topic, [47] further assumes a physical (directional) model for the channels and the key feature of that work is on the use cooperative diversity to build up the third mode of the received signal tensor.

Since the R relays transmit sequentially, i.e., in orthogonal (disjoint) time slots, the overall data received at the destination span $R+1$ time slots, one for the direct link (if available) and R for the relay-assisted links. Thus, the noiseless signal received at the destination can be written as the following PARAFAC model:

$$\mathcal{X} = \mathcal{I}_{M_U} \times_1 \mathbf{A} \times_2 \mathbf{S} \times_3 \mathbf{H} \in \mathbb{C}^{M_D \times N \times (R+1)}, \tag{12.40}$$

where $\mathbf{A} \in \mathbb{C}^{M_D \times M_U}$ is a matrix collecting the steering vectors associated with each cluster/user, $\mathbf{S} \in \mathbb{C}^{N \times M_U}$ is the symbol matrix, and $\mathbf{H} \in \mathbb{C}^{R+1 \times M}$ is the cooperative channel matrix, whose first row $\mathbf{H}_{1.} \in \mathbb{C}^{1 \times M}$ contains the fading channel coefficients linking the M users to the destination in the first time slot (source-destination link), while its R remaining rows contain the channel coefficients linking the users to the destination in the subsequent time slots (relays-destination link). In [47], the authors use a Levenberg-Marquardt (LM) algorithm to jointly estimate the spatial signatures, channel gains, and symbol matrices. It is also worth mentioning that the authors show

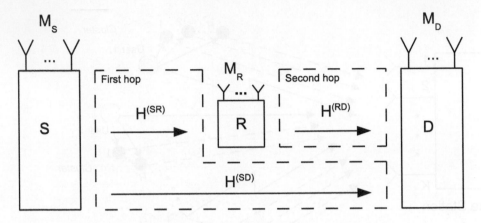

Figure 12.2 Two-hop cooperative MIMO relay system [49]

that model (12.40) unifies the formulation of the received signal in amplify-and-forward (AF), fixed decode-and-forward (FDF), and selective decode-and-forward (SDF) relaying protocols.

The approach of [48] also considers the uplink of a multiuser cooperative communication system, but assumes that DS-CDMA transmission is used at the relays. In contrast to [47] where the relays of a cluster transmit over orthogonal channels, the proposed receiver operates in a more attractive scenario where all the relays simultaneously transmit towards the base station without requiring orthogonal codes. The received signal follows the PARAFAC (12.40), the difference being on the definition and structure of the mode-3 factor matrix $\mathbf{H}$, which concatenates the direct link channel with the effective channel (source-relay and relay-to-destination) multiplied the the spreading code matrix (see [48] for details).

Still considering a one-way two-hop relaying scenario, but resorting to space-time block coding at the source, the work [49] considers a two-hop AF cooperative scheme, where the source, relay, and destination nodes are equipped with multiple antennas. The general system model is illustrated in Figure 12.2. The authors adopt a simplified Khatri-Rao space-time (KRST) coding at the source and show that the third-order tensor of signals received by the destination node satisfies a PARAFAC decomposition for the direct link, and a PARATUCK2 decomposition for the relay-assisted link. The signal received at the destination via the direct and relay-assisted links can be respectively written in PARAFAC and Tucker3 formats as

$$\mathcal{X}_{\text{SD}} \;=\; \mathcal{I}_{M_S} \times_1 \mathbf{H}_{\text{SD}} \times_2 \mathbf{S} \times_3 \mathbf{C} \;\in\; \mathbb{C}^{M_D \times N \times P}, \tag{12.41}$$

$$\mathcal{X}_{\text{SRD}} \;=\; \mathcal{T}_{\text{SR}} \times_1 \mathbf{H}_{\text{RD}} \times_2 \left(\mathbf{G}^T \diamond \mathbf{S}^T\right)^T \times_3 \mathbf{C} \;\in\; \mathbb{C}^{M_D \times N \times P}, \tag{12.42}$$

where $\mathcal{T}_{\text{SR}} \in \mathbb{C}^{M_R \times M_S \times M_R M_S}$ is a channel tensor whose mode-2 unfolding is equal to $[\mathcal{T}_{\text{SR}}]_{(2)} = \text{diag}(\text{vec}(\mathbf{H}_{\text{SR}})) \in \mathbb{C}^{M_R M_S \times M_R M_S}$, $\mathbf{C} \in \mathbb{C}^{P \times M_S}$ is the space-time coding matrix, $\mathbf{G} \in \mathbb{C}^{N \times M_R}$ is the relay amplification matrix, and $\mathbf{S} \in \mathbb{C}^{N \times M_S}$ is the symbol matrix transmitted by the source node. It is worth mentioning that (12.42) can also be written as a PARATUCK2 model. By combining the tensors $\mathcal{X}_{\text{SD}}$ and $\mathcal{X}_{\text{SRD}}$, the work [49] develops algorithms for joint semi-blind estimation of transmitted symbols and channels of both hops. Semi-blind receivers that combine these two tensors in sequential and parallel forms are formulated and discussed therein. Particularly, it is shown that the "tensor combination" of the direct link with the relay-assisted link yields to improved symbol estimation performance.

All the aforementioned works consider two-hop relaying, and the channel estimation problem is concerned with the joint estimation of the channel matrices of both hops. To further extend coverage and combat channel impairments such as path-loss and shadowing, it may be advantageous to introduce additional hops while exploiting multiple relay links, whenever they are available. Toward this goal, tensor-based channel estimators have been presented in [51] and [50] and for one-way three-hop cooperative multi-relay systems. Both approaches rely on the use of training sequences, following the idea of [37]. The approach proposed in [51] is based on a PARATUCK2 tensor model of the data collected at the destination, which results from the use of diagonal channel training matrices at the first and second relay tiers.

Two receiver algorithms are formulated to solve the channel estimation problem. The first one is an iterative channel estimation method based on alternating least squares while the second one is a closed-form solution based on a Kronecker least squares factorization. The work [51] also discusses the generalization of the tensor signal model to the two-way relaying case.

In the method proposed in [50], the channel matrices involved in the communication process are jointly estimated by coupling the PARAFAC and Tucker3 tensor models associated with the multiple hops using a combined alternating least squares algorithm.

12.6 APPLICATION TO MULTIDIMENSIONAL ARRAY PROCESSING

The term sensor array refers to a set of sensors that captures signals, and processes them jointly using array signal processing techniques. For instance, in MIMO communication systems, the transmitted signals are received by different antennas at the receiver. By processing these received signals jointly, four gains can be achieved, namely, the array gain, the diversity gain, the interference reduction gain, and the spatial multiplexing gain.

The array gain is equal to the amount of antennas. However, in order to obtain the array gain, the signals received at each antenna should be electronically shifted so that they are all in phase with respect to each other as a constructive interference. Such phase variation and further combination of all signals is known as beamforming. Therefore, note that the irradiation diagram of an antenna array system can vary electronically.

The diversity gain assumes the presence of different paths, i.e., multipath components, between the transmitter and receiver. These non-line-of-sight (NLOS) paths can be used to transmit copies of the signals transmitted through the line-of-sight

(LOS) path. Note that if one path is blocked and attenuated, the other remaining paths allow reliable communication.

The interference reduction gain can be obtained also with beamforming. Assuming that from a certain direction comes an interfering signal, beamforming can be created such that the signals from the interference direction have no influence. This is referred to as creating a null on the direction of the interference signal.

Besides the traditional four gains of the standard array signal processing, multidimensional array signal processing schemes have additional advantages, since they are more appropriate to handle the natural multidimensional structure of the data. The first advantage is the increased identifiability, which means in a telecommunication context that more subscribers can be served. Consequently, a tensor based system is more efficient in terms of resource usage than a matrix based system. A second advantage of tensors is that multilinear mixtures can be unmixed uniquely without imposing additional constraints. For matrices, i.e., bilinear problems, this requires posing additional artificial constraints, such as orthogonality leading to Principal Component Analysis (PCA) or statistical independence leading to Independent Component Analysis (ICA), which may not be physically reasonable. Finally, the third advantage is the "tensor gain" in terms of an improved accuracy, since we can reject the noise more efficiently, filtering out everything that does not fit to the structure [52].

Due to the mentioned advantages of multidimensional array signal processing schemes, we dedicate this section to them. In Figure 12.3, we summarize the fundamental steps for the parameter estimation. Note that the first step is the estimation of the model order, which means the estimation of the amount of parameters. Further if the noise is colored, a subspace prewhitening step can be included in order to reduce the degradation of the colored noise. Finally, the parameter can be properly estimated.

This section is divided into four subsections. In Section 12.6.1, we present the PARAFAC-3 data model that is going to be considered along this whole section. Note that the extension of the mentioned solutions to the PARAFAC-R, in case of a data model with R dimensions, can be done without loss of generality. In Section 12.6.2, the objective is to estimate the factor matrices as well as their parameters, while in Section 12.6.3, we refer to the model order selection problem, whose goal is to estimate d, i.e., the amount of rank one tensor that originated the tensor $\mathcal{X}_0$ given that only the noisy measurements $\mathcal{X}$ are available. Finally, in Section 12.6.4, we refer to multidimensional prewhitening schemes, which reduce the degradation caused by the colored noise.

12.6.1 GENERAL DATA MODEL FOR MULTIDIMENSIONAL ARRAY PROCESSING

In the literature, there are several types of multidimensional decompositions and each decomposition assumes a certain multidimensional structure for the data. For instance, we refer to Table 12.1 in Section 12.3 about tensor models. In this section, we only consider the PARAFAC model. As shown in earlier sections of this chapter,

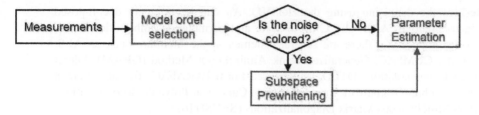

Figure 12.3 Multidimensional model order selection and parameter estimation.

the PARAFAC model is applied today in several and different fields such as chemometrics [53, 54], array signal processing [55], communications [14, 8], biomedical signal processing [56], image compression [57], data mining [58], and numerical mathematics [59].

The PARAFAC-3 model introduced in (12.8) and from Table 12.1 in Section 12.3, can be rewritten in the following fashion by using the n-mode product representation.

$$\mathcal{X}_0 = \mathcal{I}_d \times_1 \mathbf{A} \times_2 \mathbf{B} \times_3 \mathbf{C}, \tag{12.43}$$

where $\mathbf{A} \in \mathbb{C}^{M_1 \times d}$, $\mathbf{B} \in \mathbb{C}^{M_2 \times d}$ and $\mathbf{C} \in \mathbb{C}^{M_3 \times d}$ are the first-, second- and third-mode factor matrices. Given $\mathcal{X}_0$, we desire to find the factor matrices $\mathbf{A}$, $\mathbf{B}$ and $\mathbf{C}$. The identifiability of (12.43) is only fulfilled if the Kruskal's condition given by (12.10) is satisfied.

In practice, the tensor $\mathcal{X}_0$ is contaminated by noise samples. We assume that only additive noise is present. Therefore, the data to be processed has the following structure

$$\mathcal{X} = \mathcal{X}_0 + \mathcal{N}^{(c)}, \tag{12.44}$$

where $\mathcal{N}^{(c)}$ is the colored noise tensor, whose each element is zero mean circularly symmetric (ZMCS) Gaussian random variable. In Subsection 12.6.2, we overview schemes to estimate the factor matrices as well as their parameters, while in Subsection 12.6.3, we refer to the model order selection schemes to find d. In Subsection 12.6.4, we show prewhitening schemes that can be integrated in case that the samples of the tensor $\mathcal{N}^{(c)}$ are colored.

12.6.2 PARAMETER ESTIMATION STRATEGIES FOR MULTIDIMENSIONAL ARRAY PROCESSING

Once Kruskal's condition is fulfilled and once the model order d is known, an estimate of the tensor $\mathcal{X}_0$ given by $\hat{\mathcal{X}}_0$ can be obtained by applying a Higher Order Singular Value Decomposition (HOSVD) based low rank approximation [60]. This approximation process is also known as denoising. Note that in practice d is unknown and it should be estimated by using multidimensional model order selection schemes as summarized in Section 12.6.3.

Besides the denoising using the HOSVD, $\hat{\mathcal{X}}_0$ can be uniquely decomposed into factor matrices $\hat{\mathbf{A}}$, $\hat{\mathbf{B}}$, and $\hat{\mathbf{C}}$ up to scale and permutation ambiguities. In order to decompose $\hat{\mathcal{X}}_0$, there are several schemes in the literature such as ALS, PARAFAC, COMFAC, Generalized Rank Annihilation Method (GRAM), Direct Trilinear Decomposition (DTLD) and Closed-Form PARAFAC [61] also known as SEmi-algebraic framework for approximate Canonical Polyadic decompositions based on SImultaneous Matrix Diagonalizations (SECSI) [62].

Depending on the application, the dimensions of the tensor $\hat{\mathcal{X}}_0$ can be translated into space, frequency, snapshots, frames, and code. For instance, in MIMO communication systems, space is related to the antenna arrays, frequency to the frequency bins and time to the time snapshots. In case of electroencephalography (EEG) [63] and microphone array applications [64], there are only space and time dimensions due to the sensors and due to the snapshots, respectively. However, the frequency dimension can be created by applying a certain Time-Frequency Analysis (TFA) technique.

From now on we consider a classical array signal processing problem which is the extraction of MIMO channel parameters from multipath components. In a MIMO communication channel, each multipath component is composed of physical parameters such as direction of departure associated with the transmit array, direction of arrival associated with the receive array, time delay of arrival associated with the frequency bins, and Doppler frequency associated with time. These parameters are generically mapped into spatial frequencies.

$$\mathcal{X} = \mathcal{A} \times_3 \mathbf{S}^{\mathrm{T}} + \mathcal{N}^{(\mathrm{c})}, \tag{12.45}$$

where $\mathcal{A} \in \mathbb{C}^{M_1 \times M_2 \times d}$ is known as the steering tensor and contains the information about all spatial frequencies, while $\mathbf{S} \in \mathbb{C}^{d \times N}$ are the transmitted symbols. Note that in comparison with (12.43), $\mathcal{A} = \mathcal{I}_d \times_1 \mathbf{A}_1 \times_2 \mathbf{A}_2$, where $\mathbf{A} = \mathbf{A}_1$ and $\mathbf{B} = \mathbf{A}_2$, and $\mathbf{C} = \mathbf{S}^{\mathrm{T}}$. Each vector $\mathbf{a}_i^{(r)}$ is mapped into a spatial frequency $\mu_i^{(r)}$, where r varying from 1 to 2 indicates the dimension and i varying from 1 to d indicates the source or the multipath component.

The blind estimation of these parameters may be performed by different families of schemes. The Maximum Likelihood (ML) family includes the ML scheme with its prohibitive computational complexity as well as the Expectation-Maximization (EM) [65] and Space Alternating Generalized EM (SAGE) [66]. Although EM and SAGE have a low computational complexity for one-dimensional search problems, if the amount of dimensions increases the computational complexity also becomes prohibitive. The schemes of ML family are known by achieving the best accuracy which is the closest to the Cramer-Rao Lower Bound (CRLB) [67] and also these schemes allow to achieve the maximum identifiability of the tensor.

Another important family of schemes is based on the subspaces such as the Multiple Signal Classification (MUSIC) which uses the noise subspace to estimate the spatial frequencies [68]. We define the signal subspace as the d eigenvectors corresponding to the d greatest eigenvalues, while the noise subspace is given by the remaining eigenvectors. Also in case of multidimensional search problems, multidimensional MUSIC has a very high computational complexity. In order to avoid such multidimensional searches and if and only if the tensor $\mathcal{A}$ perfectly fulfills the shift invariance property, then schemes such as R-D Estimation of Signal Parameters via Rotational Invariance Techniques (ESPRIT) [69]. The great advantage of the R-D ESPRIT is that the multidimensional problem is transformed into a closed-form simultaneous diagonalization problem. Therefore, by using R-D ESPRIT all the estimated parameters are already paired. The R-D Tensor ESPRIT is an improvement of the R-D ESPRIT through the incorporation of the HOSVD based low rank approximation [60]. In addition, other additional steps such as Spatial Smoothing (SPS) [60] and Forward Backward Averaging (FBA) [70, 60] can be included as a preprocessing step before the application of parameter estimation schemes. Note that FBA can only be applied in case of centro-symmetric arrays.

There is also a family of parameter based estimation schemes based on the PARAFAC decomposition [71, 72]. By applying the PARAFAC decomposition, the tensor X is decomposed directly into the matrices $\mathbf{A}_1$, $\mathbf{A}_2$, and $\mathbf{S}$. The spatial frequencies can be directly estimated by considering each individual vector $\mathbf{a}_i^{(r)}$ of the matrices $\mathbf{A}_1$ and $\mathbf{A}_2$. Therefore, after applying the PARAFAC decomposition, d multidimensional search problems can be represented as $R \cdot d$ one-dimensional search problems. Since R one-dimensional search problems have a negligible computation complexity in comparison with a multimensional search problem, it is worth it to apply the PARAFAC decomposition.

12.6.3 MULTIDIMENSIONAL MODEL ORDER SELECTION SCHEME FOR THE TENSOR RANK ESTIMATION

In Section 12.6.2, we assumed that the correct model order was correctly estimated. In order to perform this task, we refer to the following multidimensional model order selection schemes: Core Consistency Analysis (CORCONDIA) [73, 74], R-D Exponential Fitting Test (R-D EFT) [75, 77, 76], R-D Akaike Information Criterion (R-D AIC) [75, 77, 76], R-D Minimum Description Length (R-D MDL) [75, 77, 76], and Closed-Form PARAFAC based Model Order Selection (CFP-MOS) [75, 72]. Since the multidimensional model order selection schemes take into account the multidimensional structure of the data, they can estimate the model order d even in scenarios with low signal-to-noise (SNR) ratio.

The CORCONDIA is based on the distance between the tensor $\mathcal{G}_k$ given by

$$\mathcal{G}_k = X \times_1 \hat{\mathbf{A}}^+ \times_2 \hat{\mathbf{B}}^+ \times_3 \hat{\mathbf{C}}^+, \tag{12.46}$$

where $^+$ is the pseudo-inverse operator of a matrix, and the identity tensor $\mathcal{I}_k$ for different candidate values of the model order k. Note that in the noiseless case and if the model order and the factor matrices are correctly estimated, then $\mathcal{G}_d = \mathcal{I}_d$. We have shown in [77] that the eigenvalue based model order schemes achieve a much higher probability of correct detection of the model order in comparison with CORCONDIA. However, by including a two-step threshold selection based also on the reconstruction error, the performance of CORCONDIA has been considerably improved as shown in [74].

The R-D EFT, the R-D AIC, and the R-D MDL are schemes based on the concept of global eigenvalues. Since a tensor can be represented with different unfoldings, for each unfolding, the sample covariance matrix can be computed. Therefore, for each unfolding, a vector of eigenvalues can be obtained. The global eigenvalues are the combination of these vectors of eigenvalues obtained from each dimension. Once the global eigenvalues are obtained they can be plugged into the model order selection schemes. In this case, the parameters of the model order selection schemes should be adjusted to the global eigenvalues as shown in [75, 77, 76].

To the best of our knowledge, the R-D EFT [75, 77, 76] is the most robust multidimensional model order selection scheme in the literature for data contaminated with white Gaussian noise. As shown in [75, 77, 76], the R-D EFT can achieve a very high probability of correct detection of the model order even for extremely low SNR regimes. One drawback of the EFT based schemes is the computation of the threshold coefficients which may require a very high computational complexity. However, the threshold coefficients can be computed offline and then included into a look-up table taking into account different scenarios. Note that the threshold coefficients depend only on the array size, which in general is constant since it is related to the system configuration.

As shown in [75, 77, 76], R-D AIC and R-D MDL are multidimensional extensions of the traditional AIC and MDL by incorporating the concept of global eigenvalues. These extensions result in a considerable improvement in terms of probability of correct detection of the model order for low SNR regimes.

In contrast to the EFT, AIC, and MDL, that were proposed for scenarios where the amount of time snapshots are greater than the amount of sensors, Modified EFT (M-EFT), 1-D AIC, and 1-D MDL are applicable for any array size [75, 77, 76]. Therefore, we highlight that, for massive MIMO applications, where the number of antennas is huge and in general much greater than the amount of time snapshots, M-EFT, 1-D AIC, and 1-D MDL should be used.

The CFP-MOS scheme is based on the Closed-Form PARAFAC which generates several redundant solutions of the factor matrices. By comparing the redundant solutions, it is possible to estimate the model order. The CFP-MOS is also based on threshold coefficients, which require a very high computational complexity. As shown in [75, 77, 76], the CFP-MOS is suggested for scenarios where the noise is colored and the data has multidimensional structure.

12.6.4 MULTIDIMENSIONAL PREWHITENING

In several applications, the noise is colored and if the colored noise has a tensor as shown in (12.44), multidimensional prewhitening schemes can be applied.

Independent on the structure of the colored noise, the degradation of the estimated parameters caused by colored noise is much worse than the degradation caused by white noise assuming the same SNR. As shown in Figure 12.4, in terms of subspace, the colored noise subspace is more concentrated on the signal subspace, i.e. on the d first eigenvectors corresponding to the d greatest eigenvalues, which are the eigenvectors that correspond to the signal subspace [78]. Consequently, the signal

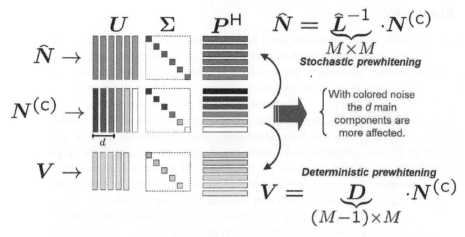

Figure 12.4 SVD analysis of the noise before and after the stochastic, and the deterministic prewhitening

subspace is degraded due to the colored noise. Note that in the literature as well as in Figure 12.4 there are two common matrix prewhitening schemes: the stochastic prewhitening [79] and the deterministic prewhitening [78]. In this section, we refer only to the tensor extension of the stochastic prewhitening.

Commonly in the literature, stochastic prewhitening schemes are applied in order to compensate the effect of the colored noise [79]. Such schemes do not take into account the multidimensional structure that may be also present in the noise tensor

$$\mathcal{N}^{(c)} = \mathcal{N} \times_1 \mathbf{L}_1 \times_2 \mathbf{L}_2 \times_3 \mathbf{L}_3, \tag{12.47}$$

where the matrices $L_r \in \mathcal{R}e^{M_r \times M_r}$ for $r = 1, \ldots, 3$ are known as the correlation matrices. Note that if L_r for $r = 1, \ldots, 3$ is equal to the identity matrix, then the tensor $\mathcal{N}^{(c)}$ is white. In this particular case, $\mathcal{N}$ is the white noise tensor. The multidimensional structure of the noise tensor in (12.47) is common in applications such as EEG [80] and MIMO [81].

In order to handle multidimensional colored noise, the Sequential Generalized Singular Value Decomposition (S-GSVD) has been proposed in [82, 84]. However, the S-GSVD as well as the stochastic prewhitening schemes require noise only samples, which are not available in several applications. Therefore, for such scenarios, the Iterative S-GSVD (I-S-GSVD) has been proposed in [83, 84]. The I-S-GSVD jointly estimates the data and the colored noise in an iterative fashion. Therefore, no noise only samples are required. In addition, the I-S-GSVD requires just a few iterations in order to converge and once the signals and the noise present some multidimensional structure as shown in (12.43) and in (12.47), respectively, no additional information is required to apply it.

12.7 SUMMARY

Tensors decompositions are powerful tools for modeling wireless communication systems and for solving problems such as blind/semi-blind equalization, source/user separation, and channel estimation. Tensors have received great attention over the past decade. In the context of communication systems, the basic motivation for resorting to tensor-based signal processing comes from the multidimensional nature of transmitted and received signals (typical dimensions are symbol periods, time blocks, space, frequency, coding, etc) which translates into more powerful uniqueness and identifiability properties compared with matrix-based signal processing. This chapter has overviewed tensor signal models for three types of systems. First, we have discussed a tensor-based space-time-frequency (TSTF) coding model for MIMO systems, which unifies the model of different space-time and space-time-frequency schemes derived previously. Second, we have presented some tensor models for cooperative communication systems with AF-based relaying. Third, the application of tensors to multidimensional array processing has been discussed, with a focus on the problems of multidimensional model order selection, prewhitening, and parameter estimation.

As already said, tensor models/decompositions are very useful for representing multidimensional data in various fields of application. Among the hot research topics on tensors for future works, we can mention the development of methods for sparse low-rank tensor approximation using compressive sensing techniques, with possible application to sparse massive MIMO channel estimation in the context of cooperative communications.

REFERENCES

1. C. Lim, T. Yoo, B. Clerckx, B. Lee, and B. Shim,"Recent trend of multiuser MIMO in LTE-advanced," *IEEE Communications Magazine*, Vol. 51, Iss. 3, Mar. 2013.
2. M. A. M. Marinho, F. Antreich, and J. P. C. L. da Costa, "Improved array interpolation for reduced bias in DOA estimation for GNSS," in Proc. of Institute of Navigation (ION) GNSS+, link, Jan 2014, San Diego, USA.
3. A. Konovaltsev, M. Cuntz, C. Haettich, and M. Meurer, "Performance analysis of joint multi-antenna spoofing detection and attitude estimation," in Proc. of the 2013 International Technical Meeting of The Institute of Navigation, ION ITM 2013, San Diego, Jan. 2013, pp. 27-29.
4. G. E. Moore, Cramming more components onto integrated circuits, *Electronics Magazine*, Iss. 4, 1965.
5. D. Kuhling, A. Ibing, and V. Jungnickel, "MIMO-OFDM realtime implementation for 3GPP LTE+ on a cell processor," in Proc. of Wireless Conference, June 2008, pp. 15.
6. P. Comon, "Tensors: A brief introduction," *IEEE Signal Processing Magazine*, Vol. 31, Iss. 3, 2014.
7. A. L. F. de Almeida, "Tensor modeling and signal processing for wireless communication systems," Ph.D. dissertation, University of Nice Sophia-Antipolis, France, 2007.
8. K. Liu, J. P. C. L. da Costa, A. L. F. de Almeida and H. C. So, "Semi-blind receivers for joint symbol and channel estimation in space-time-frequency diversity based MIMO-

OFDM systems," *IEEE Transactions on Signal Processing*, Vol. 61, Iss. 21, Nov. 2013, pp. 5444-5457.

9. G. Favier and A. L. F. de Almeida, "Tensor space-time-frequency coding with semi-blind receivers for MIMO wireless communication systems," *IEEE Trans. Signal Process.*, vol. 62, Iss. 22, pp. 5987-6002, Nov. 2014.

10. F. L. Hitchcock, "The expression of a tensor or a polyadic as a sum of products," *Journal of Mathematics and Physics*, Vol. 6, pp. 164-189, 1927.

11. R. A. Harshman, "Foundations of the PARAFAC procedure: Model and conditions for an "explanatory" multi-mode factor analysis," *UCLA Working Papers in Phonetics*, Vol. 16, pp. 1-84, Dec. 1970.

12. J. D. Carroll and J. J. Chang, "Analysis of individual differences in multidimensional scaling via an N-way generalization of "Eckart-Young" decomposition," *Psychometrika*, Vol. 35, Iss. 3, pp. 283-319, 1970.

13. J.B. Kruskal, Three-way arrays: Rank and uniqueness of trilinear decompositions with application to arithmetic complexity and statistics, *Linear Algebra Appl.*, Vol. 18, Iss. 2, pp. 95-138, 1977.

14. N. D. Sidiropoulos, G. B. Giannakis, and R. Bro, "Blind PARAFAC receivers for DS-CDMA systems," *IEEE Trans. Signal Process.*, Vol. 48, Iss. 3, pp. 810-823, March 2000.

15. J. D Carroll, S. Pruzansky, and J. B Kruskal, "Candelinc: a general approach to multidimensional analysis of many-way arrays with linear constraints on parameters", *Psychometrika*, Vol. 45, Iss. 1, pp. 3-24, 1980.

16. G. Favier and A. L. F. de Almeida, "Overview of constrained PARAFAC models," *EURASIP J. Advances in Signal Process.*, Vol. 62, Iss. 14, 2014.

17. N. D Sidiropoulos, and R. Bro, "On the uniqueness of multilinear decomposition of N-way arrays", *J. Chemometrics*, Vol. 14, pp. 229-239, 2000.

18. G. Favier, M. N. da Costa, A. L. F. de Almeida, and J. M. T. Romano, "Tensor coding for CDMA-MIMO wireless communication systems," in Proc. of European Sign. Proc. Conf. (EUSIPCO'2011), Barcelona, Spain, Aug. 2011.

19. R. A. Harshman and M. E. Lundy, "Uniqueness proof for a family of models sharing features of Tucker's three-mode factor analysis and PARAFAC/CANDECOMP," *Psychometrika*, Vol. 61, pp. 133-154, 1996.

20. G. Favier, M. N. da Costa, A. L. F. de Almeida, and J. M. T. Romano, "Tensor space-time (TST) coding for MIMO wireless communication systems," *Signal Processing*, Vol. 92, Iss. 4, pp. 1079-1092, 2012.

21. A. L. F. de Almeida and G. Favier, "Double Khatri-Rao space-time-frequency coding using semi-blind PARAFAC based receiver," *IEEE Signal Processing Letters*, Vol. 20, Iss. 5, pp. 471-474, May 2013.

22. L. R. Ximenes, G. Favier, and A. L. F. de Almeida, "Semi-blind receivers for non-regenerative cooperative MIMO communications based on nested PARAFAC modeling," Submitted to *IEEE Trans. Signal Process.*, Sept. 2014.

23. L. R. Tucker, "Some mathematical notes on three-mode factor analysis," *Psychometrika*, Vol. 31, Iss. 3, pp. 279-311, 1966.

24. A. L. F. de Almeida, G. Favier, and L. R. Ximenes, "Space-time-frequency (STF) MIMO communication systems with blind receiver based on a generalized PARATUCK2 model," *IEEE Trans. Signal Process.*, Vol. 61, Iss. 8, pp. 1895-1909, 2013.

25. N. D. Sidiropoulos and R. Budampati, "Khatri-Rao space-time codes," *IEEE Trans. Signal Process.*, Vol. 50, Iss. 10, pp. 2396-2407, 2002.

26. A. L. F. de Almeida, G. Favier, and J. C. M. Mota, "Space-time spreading-multiplexing for

MIMO wireless communication systems using the PARATUCK-2 tensor model," *Signal Processing*, Vol. 89, Iss. 11, pp. 2103-2116, Nov. 2009.

27. J. N. Laneman, D. N. C. Tse, and G. W. Wornell, "Cooperative diversity in wireless networks: efficient protocols and outage behavior," *IEEE Trans. Inf. Theory*, Vol. 50, Iss. 12, pp. 3062–3080, Dec. 2004.

28. K. J. Ray Liu, A. K. Sadek, W. Su and A. Kwasinski, "Cooperative communications and networking," Cambridge University Press, 2009.

29. A. Nosratinia, T. E. Hunter, and A. Hedayat, "Cooperative communication in wireless networks," *IEEE Communications Magazine*, Vol. 42, Iss. 10, pp. 74–80, Oct. 2004.

30. P. Kumar and S. Prakriy, *Bidirectional Cooperative Relaying*. InTech, Apr. 2010.

31. T. Unger and A. Klein, "Duplex schemes in multiple antenna two-hop relaying," *EURASIP Journal on Advances in Signal Processing*, Vol. 2008, Iss. 92, pp. 1–18, Jan. 2008.

32. M. Dohler and Y. Li, *Cooperative communications: hardware, channel and PHY*, John Wiley & Sons, 2010.

33. F. Gao, T. Cui, and A. Nallanathan, "On channel estimation and optimal training design for amplify and forward relay networks," *IEEE Transactions on Wireless Communications*, Vol. 7, Iss. 5, pp. 1907–1916, 2008.

34. A. Lalos, A. Rontogiannis, and K. Berberidis, "Channel estimation techniques in amplify-and-forward relay networks," in IEEE 9th Workshop on Signal Processing Advances in Wireless Communications (SPAWC), Recife, Brazil, 2008, pp. 446–450.

35. B. Gedik and M. Uysal, "Two channel estimation methods for amplify-and-forward relay networks," in *CCECE Canadian Conference on Electrical and Computer Engineering*, 2008, pp. 615–618.

36. T. Kong and Y. Hua, "Optimal channel estimation and training design for MIMO relays," in *IEEE Conference on Signals, Systems and Computers (ASILOMAR)*, Pacific Grove, CA, USA, Nov. 2010, pp. 663–667.

37. F. Roemer and M. Haardt, "Tensor-based channel estimation and iterative refinements for two-way relaying with multiple antennas and spatial reuse," *IEEE Transactions on Signal Processing*, Vol. 58, Iss. 11, pp. 5720–5735, Nov. 2010.

38. Z. Fang, J. Shi, and H. Shan, "Comparison of channel estimation schemes for MIMO two-way relaying systems," in in Proc. of Cross Strait Quad-Regional Radio Science and Wireless Technology Conference (CSQRWC), Vol. 1, Harbin, China, 2011.

39. H. Ren, J. Zhang, and W. Xu, "An LMMSE receiver scheme for amplify-and-forward relay systems with imperfect channel state information," in Proc. of IEEE 13th International Conference on Communication Technology (ICCT), 2011, pp. 61–65.

40. X. Yu and Y. Jing, "SVD-based channel estimation for MIMO relay networks," in Proc. of IEEE Vehicular Technology Conference (VTC Fall), Quebec City, Canada, 2012, pp. 1–5.

41. P. Lioliou, M. Viberg, and M. Coldrey, "Efficient channel estimation techniques for amplify and forward relaying systems," *IEEE Transactions on Wireless Communications*, Vol. 60, Iss. 11, pp. 3150–3155, Oct. 2012.

42. M. Shariat, M. Biguesh, and S. Gazor, "Relay design for SNR maximization in MIMO communication systems," in Proc. of 5th International Symposium on Telecommunications (IST), Tehran, Iran, 2010, pp. 313–317.

43. Y. Rong, X. Tang, and Y. Hua, "A unified framework for optimizing linear nonregenerative multicarrier MIMO relay communication systems," *IEEE Transactions on Signal Processing*, Vol. 57, Iss. 12, pp. 4837–4851, Dec. 2009.

44. Y. Rong, "Optimal joint source and relay beamforming for MIMO relays with direct link," *IEEE Communications Letters*, Vol. 14, Iss. 5, pp. 390–392, May 2010.

45. F. Roemer and M. Haardt, "Tensor-based channel estimation (TENCE) for two-way relaying with multiple antennas and spatial reuse," in Proc. of IEEE International Conference on Acoustics, Speech and Signal Processing (ICASSP), Taipei, Taiwan, Apr. 2009, pp. 3641–3644.

46. Y. Rong, M. Khandaker, and Y. Xiang, "Channel estimation of dual-hop MIMO relay system via parallel factor analysis," *IEEE Transactions on Wireless Communications*, Vol. 11, Iss. 6, pp. 2224–2233, Jun. 2012.

47. C. A. R. Fernandes, A. L. F. de Almeida, and D. B. Costa, "Unified tensor modeling for blind receivers in multiuser uplink cooperative systems," *IEEE Signal Processing Letters*, Vol. 19, Iss. 5, pp. 247 –250, May 2012.

48. A. L. F. de Almeida, C. A. R. Fernandes, and D. B. Costa, "Multiuser detection for uplink DS-CDMA amplify-and-forward relaying systems," *IEEE Sig. Proc. Letters*, Vol. 20, Iss. 7, pp. 697–700, 2013.

49. L. R. Ximenes, G. Favier, A. L. F. Almeida, and Y. C. B. Silva, "PARAFAC-PARATUCK semi-blind receivers for two-hop cooperative MIMO relay systems," *IEEE Transactions on Signal Processing*, Vol. 62, Iss. 14, pp. 3604–3615, July 2014.

50. I. V. Cavalcante, A. L. F. de Almeida, and M. Haardt, "Tensor-based approach to channel estimation in amplify-and-forward MIMO relaying systems," in Proc. of IEEE Sensor Array and Multichannel Signal Processing Woskshop (SAM 2014), La Coruña , Spain, 2014, pp. 1–4.

51. Xi, Han, A. L. F. de Almeida, Zhen, Yang, "Channel estimation for MIMO multi-relay systems using a tensor approach," *EURASIP Journal on Advances in Signal Processing* Vol. 2014, Iss. 163, 2014.

52. F. Roemer, "Advanced Algebraic Concepts for Efficient Multi-Channel Signal Processing," Ph.D. dissertation, Ilmenau University of Technology, 2013.

53. C. M. Andersen and R. Bro, "Practical aspects of PARAFAC modeling of fluorescence excitation-emission data," *Journal of Chemometrics*, Vol. 17, pp. 200–215, 2003.

54. R. Bro, "Multi-way analysis in the food industry: models, algorithms and applications," Ph.D. Dissertation, University of Amsterdam, Denmark, 1998.

55. N. D. Sidiropoulos, R. Bro, and G. B. Giannakis, "Parallel factor analysis in sensor array processing," *IEEE Transactions on Signal Processing*, Vol. 48, Iss. 8, pp. 2377–2388, Aug. 2000.

56. J. MÃ¶cks, "Topographic components model for event-related potentials and some biophysical considerations," *IEEE Transactions on Biomedical Engineering*, Vol. 35, pp. 482–484, 1988.

57. A. Shashua and A. Levin, "Linear image coding for regression and classification using the tensor-rank principle," in Proc. of the 2001 IEEE Computer Society Conference on Computer Vision and Pattern Recognition (CVPR 2001), Kauai, HI, USA, Dec. 2001, pp. 42–49.

58. T. Kolda and B. Bader, "The TOPHITS model for higher-order web link analysis," in Proc. of the SIAM Data Mining Conference and the Workshop on Link Analysis, Counterterrorism and Security, Bethesda, MD, Apr. 2006.

59. W. Hackbusch, B. N. Khoromskij, and E. E. Tyrtyshnikov, "Hierarchical Kronecker tensor-product approximations," *Journal of Numerical Mathematics*, Vol. 13, pp. 119–156, 2005.

60. M. Haardt, F. Roemer, and G. Del Galdo, "Higher-order SVD based subspace estimation to improve the parameter estimation accuracy in multi-dimensional harmonic retrieval problems," *IEEE Transactions on Signal Processing*, Vol. 56, pp. 3198-3213, July 2008.

61. F. Roemer and M. Haardt, "A closed-form solution for multilinear PARAFAC decompo-

sitions," in Proc. 5-th IEEE Sensor Array and Multichannel Signal Processing Workshop (SAM 2008), (Darmstadt, Germany), pp. 487-491, July 2008.

62. F. Roemer and M. Haardt, "A semi-algebraic framework for approximate CP decompositions via Simultaneous Matrix Diagonalizations (SECSI)," *Signal Processing*, Vol. 93, pp. 2722-2738, Sept. 2013.

63. M. Weis, D. Jannek, T. Guenther, P. Husar, F. Roemer, and M. Haardt, "Space-time-frequency component analysis of visual evoked potentials based on the PARAFAC2 model," IWK Internationales Wissenschaftliches Kolloquium (IWK'10), Sept. 2010.

64. M. A. Silveira, C. P. Schroeder, J. P. C. L. da Costa, C. G. de Oliveira, J. A. Apolinário Junior, A. M. Rubio Serrano, P. Quintiliano, and R. T. de Sousa Júnior, "Convolutive ICA-Based Forensic Speaker Identification Using Mel Frequency Cepstral Coefficients and Gaussian Mixture Models," *International Journal of Forensic Computer Science* (IJoFCS), 2013.

65. M. I. Miller e D. R. Fuhrmann, "Maximun Likelihood Narrow-Band Direction Finding and the EM Algorithm," *IEEE Trans. on Acoustics, Speech and Signal Processing*, Vol. 38, Sept., 1990, pp. 1560-1577.

66. B. H. Fleury, M. Tschudin, R. Heddergott, D. Dahlhaus, and K. I. Pedersen, "Channel parameter estimation in mobile radio environments using the SAGE algorithm," *IEEE J. Select. Areas Commun.*, Vol. 17, Iss. 3, pp. 434-450, Mar. 1999.

67. F. Roemer. Advances in subspace-based parameter estimation: Tensor-ESPRIT type methods and non-circular sources. Diploma thesis, Ilmenau University of Technology, Communications Research Lab, October 2006.

68. R. O. Schmidt, "Multiple Emitter Location and Signal Parameter Estimation," *IEEE Trans. Antennas Propagation*, Vol. AP-34 (March 1986), pp.276-280

69. R. Roy and T. Kailath, "ESPRIT-estimation of signal parameters via rotational invariance techniques," *IEEE Transactions on Acoustics, Speech and Signal Processing*, Vol. 37, Iss. 7, 1989.

70. A. Lee, "Centrohermitian and skew-centrohermitian matrices," Linear Algebra and Its Appl., Vol. 29, pp. 205–210, 1980.

71. X. Liu and N. Sidiropoulos, "PARAFAC techniques for high-resolution array processing," in High-Resolution and Robust Signal Processing, Y. Hua, A. Gershman, and Q. Chen, Eds. 2004, pp. 111–150, Marcel Dekker, New York, NY, Chapter 3.

72. J. P. C. L. da Costa, F. Roemer, M. Weis, and M. Haardt, "Robust R-D parameter estimation via closed-form PARAFAC," in Proc. of ITG Workshop on Smart Antennas (WSA'10), (Bremen, Germany), Feb. 2010.

73. R. Bro and H. A. L. Kiers, A new efficient method for determining the number of components in PARAFAC models, *Journal of Chemometrics*, Vol. 17, pp. 274–286, 2003.

74. K. Liu, H. C. So, J. P. C. L. da Costa, and H. Lei, "Core consistency diagnostic aided by reconstruction error for accurate enumeration of the number of components in PARAFAC models," in Proc. 38th International Conference on Acoustics, Speech, and Signal Processing (ICASSP), May 2013, Vancouver, Canada.

75. J. P. C. L. da Costa, F. Roemer, M. Haardt, and R. T. de Sousa Jr., "Multi-Dimensional Model Order Selection," *EURASIP Journal on Advances in Signal Processing*, Vol. 26, Iss. 2011, July 2011.

76. J. P. C. L. da Costa, M. Haardt, F. Roemer, and G. Del Galdo, "Enhanced model order estimation using higher-order arrays," in Proc. of 41st Asilomar Conf. on Signals, Systems, and Computers, (Pacific Grove, CA, USA), pp. 412-416, Nov. 2007.

77. J. P. C. L. da Costa, M. Haardt, and F. Roemer, "Robust methods based on the HOSVD for

estimating the model order in PARAFAC models," in Proc. of 5th IEEE Sensor Array and Multichannel Signal Processing Workshop (SAM 2008), (Darmstadt, Germany), pp. 510 - 514, July 2008.

78. J. P. C. L. da Costa, F. Roemer, and M. Haardt, "Deterministic prewhitening to improve subspace parameter estimation techniques in severely colored noise environments"., in Proc. of 54th International Scientific Colloquium (IWK), (Ilmenau, Germany), Sept. 2009.

79. M. Haardt, R. S. Thomae, and A. Richter, "Multidimensional high-resolution parameter estimation with applications to channel sounding," in High-Resolution and Robust Signal Processing, Y. Hua, A. Gershman, and Q. Chen, Eds., Chapter 5, pp. 255–338, Marcel Dekker, New York, NY, 2004.

80. H. M. Huizenga, J. C. de Munck, L. J. Waldorp, and R. P. P. P. Grasman, "Spatiotemporal EEG/MEG source analysis based on a parametric noise covariance model," *IEEE Transactions on Biomedical Engineering*, Vol. 49, Iss. 6, pp. 533–539, June 2002.

81. B. Park and T. F. Wong, "Training sequence optimization in MIMO systems with colored noise," in Proc. of Military Communications Conference (MILCOM 2003), Gainesville, USA, Oct. 2003.

82. J. P. C. L. da Costa, M. Haardt, and F. Roemer, "Sequential GSVD based prewhitening for multidimensional HOSVD based subspace estimation," in Proc. of International ITG Workshop on Smart Antennas (WSA 2009), (Berlin, Germany), Feb. 2009.

83. J. P. C. L. da Costa, F. Roemer, and M. Haardt, "Iterative sequential GSVD (I-S-GSVD) based prewhitening for multidimensional HOSVD based subspace estimation without knowledge of the noise covariance information," in Proc. of ITG Workshop on Smart Antennas (WSA 2010), (Bremen, Germany), pp. 151–155, Feb. 2010.

84. J. P. C. L. da Costa, K. Liu, H. C. So, F. Roemer, M. Haardt, and S. Schwarz, "Multidimensional prewhitening for enhanced signal reconstruction and parameter estimation in colored noise with Kronecker correlation structure," *Signal Processing*, Vol. 93, Iss. 11, pp. 3209–3226, Nov. 2013.

13 Signal Detection and Parameter Estimation in Massive MIMO Systems

RODRIGO CAIADO DE LAMARE
Pontifical Catholic University of Rio de Janeiro (PUC-Rio)
University of York

RAIMUNDO SAMPAIO-NETO
Pontifical Catholic University of Rio de Janeiro (PUC-Rio)

13.1 INTRODUCTION

Future wireless networks will have to deal with a substantial increase of data transmission due to a number of emerging applications that include machine-to-machine communications and video streaming [1]-[4]. This very large amount of data exchange is expected to continue and rise in the next decade or so, presenting a very significant challenge to designers of fifth-generation (5G) wireless communications systems [4]. Amongst the main problems are how to make the best use of the available spectrum and how to increase the energy efficiency in the transmission and reception of each information unit. 5G communications will have to rely on technologies that can offer a major increase in transmission capacity as measured in bits/Hz/area but do not require increased spectrum bandwidth or energy consumption.

Multiple-antenna or multi-input multi-output (MIMO) wireless communication devices that employ antenna arrays with a very large number of antenna elements, which are known as massive MIMO systems, have the potential to overcome those challenges and deliver the required data rates, representing a key enabling technology for 5G [5]-[8]. Among the devices of massive MIMO networks are user terminals, tablets, machines, and base stations which could be equipped with a number of antenna elements with orders of magnitude higher than current devices. Massive MIMO networks will be structured by the following key elements: antennas, electronic components, network architectures, protocols, and signal processing. The network architecture, in particular, will evolve from homogeneous cellular layouts to heterogeneous architectures that include small cells and the use of coordination between cells [9]. Since massive MIMO will be incorporated into mobile cellular networks in the future, the network architecture will necessitate special attention on how to manage the interference created [10] and measurements campaigns will be of fundamental importance [11]-[13]. The coordination of adjacent cells will be necessary due to the current trend towards aggressive reuse factors for capacity reasons, which inevitably leads to increased levels of inter-cell interference and signaling. The need to accommodate multiple users while keeping the interference at an acceptable level will also require significant work in scheduling and medium-access protocols.

Another important aspect of massive MIMO networks lies in signal processing, which must be significantly advanced for 5G. In particular, MIMO signal processing will play a crucial role in dealing with the impairments of the physical medium and in providing cost-effective tools for processing information. Current state-of-the-art in MIMO signal processing requires a computational cost for transmit and receive processing that grows as a cubic or super-cubic function of the number of antennas, which is clearly not scalable with a large number of antenna elements. We advocate the need for simpler solutions for both transmit and receive processing tasks, which will require significant research effort in the next years. Novel signal processing strategies will have to be developed to deal with the problems associated with massive MIMO networks like computational complexity and its scalability, pilot contamination effects, RF impairments, coupling effects, delay, and calibration issues.

In this chapter, we focus on signal detection and parameter estimation aspects of massive MIMO systems. We consider both centralized antenna systems (CAS) and distributed antenna systems (DAS) architectures in which a large number of antenna elements are employed and focus on the uplink of a mobile cellular system. In particular, we focus on the uplink and receive processing techniques that include signal detection and parameter estimation problems, and discuss specific needs of massive MIMO systems. We review the optimal maximum likelihood detector, nonlinear and linear suboptimal detectors and discuss potential contributions to the area. We also describe iterative detection and decoding algorithms, which exchange soft information in the form of log likelihood ratios (LLRs) between detectors and channel decoders. Another important area of investigation includes parameter estimation techniques, which deal with methods to obtain the channel state information, compute the parameters of the receive filters, and the hardware mismatch. Simulation results illustrate the performance of detection and estimation algorithms under scenarios of interest. Key problems are discussed and future trends in massive MIMO systems are pointed out.

This chapter is structured as follows: Section 13.2 reviews the signal models with CAS and DAS architectures and discusses the application scenarios. Section 13.3 is dedicated to detection techniques, whereas Section 13.4 is devoted to parameter estimation methods. Section 13.5 discusses the results of some simulations and Section 13.6 presents some open problems and suggestions for further work. The conclusions of this chapter are given in Section 13.7.

13.2   SIGNAL MODELS AND APPLICATION SCENARIOS

In this section, we describe signal models for the uplink of multiuser massive MIMO systems in mobile cellular networks. In particular, we employ a linear algebra approach to describe the transmission and how the signals are collected at the base station or access point. We consider both CAS and DAS [16, 15] configurations. In the CAS configuration a very large array is employed at the rooftop or at the façade of a building or even at the top of a tower. In the DAS scheme, distributed radio heads are deployed over a given geographic area associated with a cell and these radio devices are linked to a base station equipped with an array through either fiber optics or dedicated radio links. These models are based on the assumption of a narrow-band signal transmission over flat fading channels which can be easily generalized to broadband signal transmission with the use of multi-carrier systems.

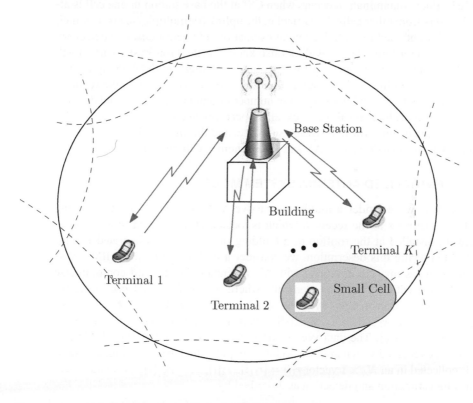

Figure 13.1 Mobile cellular network with a CAS configuration.

The scenario we are interested in this work is that of mobile cellular networks beyond LTE-A [2] and 5G communications [4], which is illustrated in Figure 13.1. In such networks, massive MIMO will play a key role in the deployment of hundreds of antenna elements at the base station using CAS or using DAS over the cell of interest, coordination between cells, and a more modest number of antenna elements at the user terminals. At the base station, very large antenna arrays could be deployed on the roof or on the façade of buildings. With further development in the area of compact antennas and techniques to mitigate mutual coupling effects, it is likely that the number of antenna elements at the user terminals (mobile phones, tables, and other gadgets) might also be significantly increased from 1-4 elements in current terminals to 10-20 in future devices. In these networks, it is preferable to employ time-division-duplexing (TDD) mode to perform uplink channel estimation and obtain downlink CSI by reciprocity for signal processing at the transmit side. This operation mode will require cost-effective calibration algorithms. Another critical requirement is the uplink channel estimation, which employs non-orthogonal pilots and, due to the existence of adjacent cells and the coherence time of the channel, needs to reuse the pilots [63]. Pilot contamination occurs when CSI at the base station in one cell is affected by users from other cells. In particular, the uplink (or multiple-access channel) will need CSI obtained by uplink channel estimation, efficient multiuser detection, and decoding algorithms. The downlink (also known as the broadcast channel) will require CSI obtained by reciprocity for transmit processing and the development of cost-effective scheduling and precoding algorithms. A key challenge in the scenario of interest is how to deal with a very large number of antenna elements and develop cost-effective algorithms, resulting in excellent performance in terms of the metrics of interest, namely, bit error rate (BER), sum-rate, and throughput. In what follows, signal models that can describe CAS and DAS schemes will be detailed.

13.2.1 CENTRALIZED ANTENNA SYSTEM MODEL

In this section, we consider a multiuser massive MIMO system with CAS using N_A antenna elements at the receiver, which is located at a base station of a cellular network installed at the rooftop of a building or a tower, as illustrated in Figure 13.1. Following this description, we consider a multiuser massive MIMO system with K users that are equipped with N_U antenna elements and communicate with a receiver with N_A antenna elements, where $N_A \geq KN_U$. At each time instant, the K users transmit N_U symbols which are organized into a $N_U \times 1$ vector $s_k[i] = [s_{k,1}[i], s_{k,2}[i], \ldots, s_{k,N_U}[i]]^T$ taken from a modulation constellation $A = \{a_1, a_2, \ldots, a_N\}$. The data vectors $s_k[i]$ are then transmitted over flat fading channels. The received signal after demodulation, pulse-matched filtering, and sampling is collected in an $N_A \times 1$ vector $r[i] = [r_1[i], r_2[i], \ldots, r_{N_A}[i]]^T$ with sufficient statistics for estimation and detection as described by

$$r[i] = \sum_{k=1}^{K} \gamma_k H_k s_k[i] + n[i]$$

$$= \sum_{k=1}^{K} G_k s_k[i] + n[i],$$

(13.1)

where the $N_A \times 1$ vector $n[i]$ is a zero mean complex circular symmetric Gaussian noise with covariance matrix $E[n[i]n^H[i]] = \sigma_n^2 I$. The data vectors $s_k[i]$ have zero mean and covariance matrices $E[s_k[i]s_k^H[i]] = \sigma_{s_k}^2 I$, where $\sigma_{s_k}^2$ is the user k transmit signal power. The elements $h_{i,j}^k$ of the $N_A \times N_U$ channel matrices H_k are the complex channel gains from the jth transmit antenna of user k to the ith receive antenna. For a CAS architecture, the channel matrices $\boldsymbol{H}_k$ can be modeled using the Kronecker channel model [14] as detailed by

$$\boldsymbol{H}_k = \boldsymbol{\Theta}_R^{1/2} \boldsymbol{H}_k^o \boldsymbol{\Theta}_T^{1/2}, \tag{13.2}$$

where $\boldsymbol{H}_k^o$ has complex channel gains obtained from complex Gaussian random variables with zero mean and unit variance, $\boldsymbol{\Theta}_R$ and $\boldsymbol{\Theta}_T$ denote the receive and transmit correlation matrices, respectively. The components of correlation matrices $\boldsymbol{\Theta}_R$ and $\boldsymbol{\Theta}_T$ are of the form

$$\boldsymbol{\Theta}_{R/T} = \begin{pmatrix} 1 & \rho & \rho^4 & \cdots & \rho^{(N_a-1)^2} \\ \rho & 1 & \rho & \cdots & \vdots \\ \rho^4 & \rho & 1 & \vdots & \rho^4 \\ \vdots & \vdots & \vdots & \vdots & \vdots \\ \rho^{(N_a-1)^2} & \cdots & \rho^4 & \rho & 1 \end{pmatrix} \tag{13.3}$$

where ρ is the correlation index of neighboring antennas and N_a is the number of antennas of the transmit or receive array. When $\rho = 0$ we have an uncorrelated scenario and when $\rho = 1$ we have a fully correlated scenario. The channels between the different users are assumed uncorrelated due to their geographical location.

The parameters γ_k represent the large-scale propagation effects for user k such as path loss and shadowing which are represented by

$$\gamma_k = \alpha_k \beta_k, \tag{13.4}$$

where the path loss α_k for each user is computed by

$$\alpha_k = \sqrt{\frac{L_k}{d_k^\tau}}, \tag{13.5}$$

where L_k is the power path loss of the link associated with user k with respect to the base station, d_k is the relative distance between the user and the base station, and τ is the path loss exponent chosen between 2 and 4 depending on the environment.

The log-normal shadowing parameter β_k is given by

$$\beta_k = 10^{\frac{\sigma_k v_k}{10}}, \tag{13.6}$$

where σ_k is the shadowing spread in dB and v_k corresponds to a real-valued Gaussian random variable with zero mean and unit variance. The $N_A \times N_U$ composite channel matrix that includes both large-scale and small-scale fading effects is denoted as $\boldsymbol{G}_k$.

13.2.2 DISTRIBUTED ANTENNA SYSTEM MODEL

In this section, we consider a multiuser massive MIMO system with a DAS configuration using N_B antenna elements at the base station and L remote radio heads each

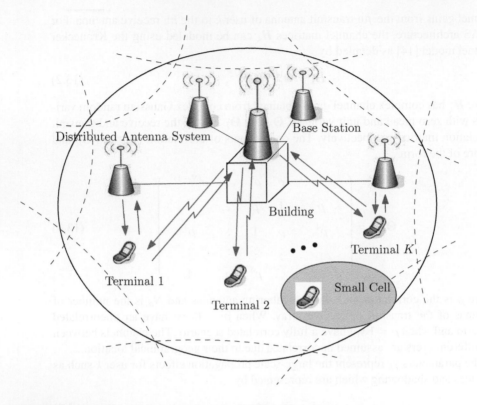

Figure 13.2 Mobile cellular network with a DAS configuration.

with Q antenna elements, which are distributed over the cell and linked to the base station via wired links, as illustrated in Figure 13.2. Following this description, we consider a multiuser massive MIMO system with K users that are equipped with N_U antenna elements and communicate with a receiver with a DAS architecture with a total of $N_A = N_B + LQ$ antenna elements, where $N_A \geq KN_U$. In our exposition, when the number of remote radio heads is set to zero, i.e., $L = 0$, the DAS architecture reduces to the CAS scheme with $N_A = N_B$.

At each time instant, the K users transmit N_U symbols which are organized into a $N_U \times 1$ vector $s_k[i] = [s_{k,1}[i], \; s_{k,2}[i], \; \ldots, \; s_{k,N_U}[i]]^T$ taken from a modulation constellation $A = \{a_1, \; a_2, \; \ldots, \; a_N\}$. The data vectors $s_k[i]$ are then transmitted over flat fading channels. The received signal after demodulation, pulse-matched filtering, and sampling is collected in an $N_A \times 1$ vector $r[i] = [r_1[i], \; r_2[i], \; \ldots, \; r_{N_A}[i]]^T$ with sufficient statistics for estimation and detection as described by

$$
\begin{aligned}
r[i] &= \sum_{k=1}^{K} \Gamma_k H_k s_k[i] + n[i] \\
&= \sum_{k=1}^{K} G_k s_k[i] + n[i],
\end{aligned}
\tag{13.7}
$$

where the $N_A \times 1$ vector $n[i]$ is a zero mean complex circular symmetric Gaussian noise with covariance matrix $E[n[i]n^H[i]] = \sigma_n^2 I$. The data vectors $s_k[i]$ have zero mean and covariance matrices $E[s_k[i]s_k^H[i]] = \sigma_{s_k}^2 I$, where $\sigma_{s_k}^2$ is the user k signal power. The elements $h_{k,j}$ of the $N_A \times N_U$ channel matrices H_k are the complex channel gains from the jth transmit antenna to the ith receive antenna. Unlike the CAS architecture, in a DAS setting the channels between remote radio heads are less likely to suffer from correlation due to the fact that they are geographically separated. However, for the antenna elements located at the base station and at each remote radio head, the $L+1$ submatrices of H_k can be modeled using the Kronecker channel model [14] as detailed in the previous section. Another major difference between CAS and DAS schemes lies in the large-scale propagation effects. Specifically, with DAS the links between the users and the distributed antennas experience on average lower path loss effects because of the reduced distance between their antennas. This helps to create better wireless links and coverage of the cell. Therefore, the large-scale propagation effects are modeled by an $N_A \times N_A$ diagonal matrix given by

$$
\Gamma_k = \text{diag}\left(\underbrace{\gamma_{k,1} \cdots \gamma_{k,1}}_{N_B} \; \underbrace{\gamma_{k,2} \cdots \gamma_{k,2}}_{Q} \; \cdots \; \underbrace{\gamma_{k,L+1} \cdots \gamma_{k,L+1}}_{Q} \right),
\tag{13.8}
$$

where the parameters $\gamma_{k,j}$ for $j = 1, \ldots, L + 1$ denote the large-scale propagation effects like shadowing and pathloss from the kth user to the jth radio head. The parameters $\gamma_{k,j}$ for user k and distributed antenna j are described by

$$
\gamma_{k,j} = \alpha_{k,j} \beta_{k,j}, \quad j = 1, \ldots, L+1
\tag{13.9}
$$

where the path loss $\alpha_{k,j}$ for each user and antenna is computed by

$$
\alpha_{k,j} = \sqrt{\frac{L_{k,j}}{d_{k,j}^\tau}},
\tag{13.10}
$$

where $L_{k,j}$ is the power path loss of the link associated with user k and the jth radio head, $d_{k,j}$ is the relative distance between the user and the radio head, and τ is the path loss exponent chosen between 2 and 4 depending on the environment. The log-normal shadowing $\beta_{k,j}$ is given by

$$
\beta_{k,j} = 10^{\frac{\sigma_k v_{k,j}}{10}},
\tag{13.11}
$$

where σ_k is the shadowing spread in dB and $v_{k,j}$ corresponds to a real-valued Gaussian random variable with zero mean and unit variance. The $N_A \times N_U$ composite channel matrix that includes both large-scale and small-scale fading effects is denoted as G_k.

13.3 DETECTION TECHNIQUES

In this section, we examine signal detection algorithms for massive MIMO systems. In particular, we review various detection techniques and also describe iterative detection and decoding schemes that bring together detection algorithms and error control coding.

13.3.1 DETECTION ALGORITHMS

In the uplink of the multiuser massive MIMO systems under consideration, the signals or data streams transmitted by the users to the receiver overlap and typically result in multiuser interference at the receiver. This means that the interfering signals cannot be easily demodulated at the receiver unless there is a method to separate them. In order to separate the data streams transmitted by the different users, a designer must resort to detection techniques, which are similar to multiuser detection methods [17].

The optimal maximum likelihood (ML) detector is described by

$$\hat{s}_{ML}[i] = \arg \min_{s[i] \in A} \|r[i] - Gs[i]\|^2 \tag{13.12}$$

where the $KN_U \times 1$ data vector $s[i]$ has the symbols of all users stacked and the $N_A \times KN_U$ channel matrix $G = [G_1 \ldots G_K]$ contains the channels of all users concatenated. The ML detector has a cost that is exponential in the number of data streams and the modulation order which is too costly for systems with a large number of antennas. Even though the ML solution can be alternatively computed using sphere decoder (SD) algorithms [19]-[22] that are very efficient for MIMO systems with a small number of antennas, the cost of SD algorithms depends on the noise variance, the number of data streams to be detected, and the signal constellation, resulting in high computational costs for low SNR values, high-order constellations, and a large number of data streams.

The high computational cost of the ML detector and the SD algorithms in scenarios with large arrays have motivated the development of numerous alternative strategies for MIMO detection, which are based on the computation of receive filters and interference cancellation strategies. The key advantage of these approaches with receive filters is that the cost is typically not dependent on the modulation, the receive filter is computed only once per data packet and performs detection with the aid of decision thresholds. Algorithms that can compute the parameters of receive filters with low cost are of central importance to massive MIMO systems. In what follows, we will briefly review some relevant suboptimal detectors, which include linear and decision-driven strategies.

Linear detectors [23] include approaches based on the receive matched filter (RMF), zero forcing (ZF), and minimum mean-square error (MMSE) designs that are described by

$$\hat{s}[i] = Q(W^H r[i]), \tag{13.13}$$

where the receive filters are

$$W_{\text{RMF}} = G, \text{ for the RMF}, \tag{13.14}$$

$$W_{\text{MMSE}} = G(G^H G + \sigma_n^2/\sigma_s^2 I)^{-1}, \text{ for the MMSE design}, \tag{13.15}$$

and

$$W_{\text{ZF}} = G(G^H G)^{-1}, \text{ for the ZF design}, \tag{13.16}$$

and $Q(\cdot)$ represents the slicer used for detection.

Decision-driven detection algorithms such as successive interference cancellation (SIC) approaches used in the Vertical-Bell Laboratories Layered Space-Time (VBLAST) systems [24]-[28] and decision feedback (DF) [29]-[46] detectors are techniques that can offer attractive trade-offs between performance and complexity. Prior work on SIC and DF schemes has been reported using DF detectors with SIC (S-DF) [24]-[28], and DF receivers with parallel interference cancellation (PIC) (P-DF) [39, 40, 44], and combinations of these schemes and mechanisms to mitigate error propagation [43, 46, 47].

SIC detectors [24]-[28] apply linear receive filters to the received data followed by subtraction of the interference and subsequent processing of the remaining users. Ordering algorithms play an important role as they significantly affect the performance of SIC receivers. Amongst the existing criteria for ordering are those based on the channel norm, the SINR, the SNR, and on exhaustive search strategies. The performance of exhaustive search strategies is the best followed by SINR-based ordering, SNR-based ordering, and channel norm-based ordering, whereas the computational complexity of an exhaustive search is by far the highest, followed by SINR-based ordering, SNR-based ordering, and channel norm-based ordering. The data symbol of each user is detected according to:

$$\hat{s}_k[i] = Q(w_k^H r_k[i]), \tag{13.17}$$

where the successively cancelled received data vector that follows a chosen ordering in the k-th stage is given by

$$r_k[i] = r[i] - \sum_{j=1}^{k-1} g_j \hat{s}_j[i], \tag{13.18}$$

where g_j corresponds to the columns of the $N_A \times K N_U$ composite channel matrix given by

$$G = [g_1 \ g_2 \ \cdots \ g_{KN_U}]. \tag{13.19}$$

After subtracting the detected symbols from the received signal vector, the remaining signal vector is processed either by an MMSE or a ZF receive filter for the data estimation of the remaining users. The computational complexity of the SIC detector based on either the MMSE or the ZF criteria is similar and requires a cubic cost in KN_U ($O((KN_U)^3)$) although the performance of MMSE-based receive filters is superior to that of ZF-based detectors.

A generalization of SIC techniques, the multi-branch successive interference cancellation (MB-SIC) algorithm, employs multiple SIC algorithms in parallel branches. The MB-SIC algorithm relies on different ordering patterns and produces multiple candidates for detection, approaching the performance of the ML detector. The ordering of the first branch is identical to a standard SIC algorithm and could be based on the channel norm or the SINR, whereas the remaining branches are ordered by shifted orderings relative to the first branch. In the ℓ-th branch, the MB-SIC detector successively detects the symbols given by the vector $\hat{s}_\ell[i] = [\hat{s}_{\ell,1}[i], \hat{s}_{\ell,2}[i], \ldots, \hat{s}_{\ell,K}[i]]^T$. The term $\hat{s}_\ell[i]$ represents the $K \times 1$ ordered estimated symbol vector, which is detected according to the ordering pattern $\mathbf{T}_\ell, \ell = 1, \ldots, S$ for the ℓ-th branch. The interference cancellation performed on the received vector $r[i]$ is described by:

$$r_{\ell,k}[i] = r[i] - \sum_{j=1}^{k-1} g_{\ell,j}\hat{s}_{\ell,j}[i] \qquad (13.20)$$

where the transformed channel column $g_{\ell,j}$ is obtained by $g_{\ell,j} = \mathbf{T}_\ell g_j$ and $\hat{s}_{\ell,k}$ denotes the estimated symbol for each data stream obtained by the MB-SIC algorithm.

At the end of each branch we can transform $\hat{s}_\ell[i]$ back to the original order $\tilde{s}_\ell[i]$ by using $\mathbf{T}_\ell$ as $\tilde{s}_\ell[i] = \mathbf{T}_\ell^T \hat{s}_\ell[i]$. The MB-SIC algorithm selects the candidate branch with the minimum Euclidean distance according to

$$\ell_{opt} = \arg \min_{1 \le \ell \le S} C(\ell) \qquad (13.21)$$

where $C(\ell) = \|r[i] - \mathbf{T}_\ell \mathbf{G}\tilde{s}_\ell[i]\|$ is the Euclidean distance for the ℓ-th branch. The final detected symbol vector can be obtained from $\hat{s}_{\ell_{opt}}$ which corresponds to

$$\hat{s}_j[i] = Q(w_{\ell_{opt},j}^H r_{\ell_{opt},j}[i]), \quad j = 1, \ldots, KN_U. \qquad (13.22)$$

The MB-SIC algorithm can bring a close-to-optimal performance; however, the exhaustive search of $S = KN_U!$ branches is not practical. Therefore, a reduced number of branches must be employed. In terms of computational complexity, the MB-SIC algorithm requires S times the complexity of a standard SIC algorithm. However, it is possible to implement it using a multi-branch decision feedback structure [42, 27] that is equivalent in performance but which only requires a single matrix inversion as opposed to K matrix inversions required by the standard SIC algorithm and SK matrix inversions required by the MB-SIC algorithm.

DF detectors employ feedforward and feedback matrices that perform interference cancellation as described by

$$\hat{s} = Q(\mathbf{W}^H r[i] - \mathbf{F}^H \hat{s}_o[i]), \qquad (13.23)$$

where $\hat{s}_o$ corresponds to the initial decision vector that is usually performed by the linear section represented by $\mathbf{W}$ of the DF receiver (e.g., $\hat{s}_o = Q(\mathbf{W}^H r)$) prior to the application of the feedback section F, which may have a strictly lower triangular structure for performing successive cancellation or zeros on the main diagonal when performing parallel cancellation. The receive filters W and F can be computed using various parameter estimation algorithms which will be discussed in the next section. Specifically, the receive filters can be based on the RMF, ZF, and MMSE design criteria.

An often criticized aspect of these sub-optimal schemes is that they typically do not achieve the full receive-diversity order of the ML algorithm. This led to the investigation of detection strategies such as lattice-reduction (LR) schemes [30, 31], QR decomposition, M-algorithm (QRD-M) detectors [33], probabilistic data association (PDA) [34, 35], multi-branch [42, 45] detectors, and likelihood ascent search techniques [48, 49], which can approach the ML performance at an acceptable cost for moderate to large systems. The development of cost-effective detection algorithms for massive MIMO systems is a challenging topic that calls for new approaches and ideas in this important research area.

13.3.2   ITERATIVE DETECTION AND DECODING TECHNIQUES

Iterative detection and decoding (IDD) techniques have received considerable attention in the last few years following the discovery of Turbo codes [50] and the application of the Turbo principle to interference mitigation [50, 51, 52, 53, 54, 55, 56, 57, 58, 29]. More recently, work on IDD schemes has been extended to low-density parity-check codes (LDPC) [54] and [57] and their extensions which compete with Turbo codes. The goal of an IDD system is to combine an efficient soft-input soft-output (SISO) detection algorithm and a SISO decoding technique as illustrated in Figure 13.3. Specifically, the detector produces log-likelihood ratios (LLRs) associated with the encoded bits and these LLRs serve as input to the decoder. Then, in the second phase of the detection/decoding iteration, the decoder generates a posteriori probabilities (APPs) after a number of (inner) decoding iterations for encoded bits of each data stream. These APPs are fed to the detector to help in the next iterations between the detector and the decoder, which are called outer iterations. The joint process of detection/decoding is then repeated in an iterative manner until the maximum number of (inner and outer) iterations is reached. In mobile cellular networks, a designer can employ convolutional, Turbo, or LDPC codes in IDD schemes for interference mitigation. LDPC codes exhibit some advantages over Turbo codes that include simpler decoding and implementation issues. However, LDPC codes often require a higher number of decoding iterations which translate into delays or increased complexity. The development of IDD schemes and decoding algorithms that perform message passing with reduced delays are of fundamental importance in massive MIMO systems because they will be able to cope with audio and 3D video which are delay sensitive.

The massive MIMO systems described at the beginning of this chapter are considered here with convolutional codes and an iterative receiver structure consists of the following stages: A soft-input-soft-output (SISO) detector and a maximum *a posteriori* (MAP) decoder. Extensions to other channel codes are straightforward. These

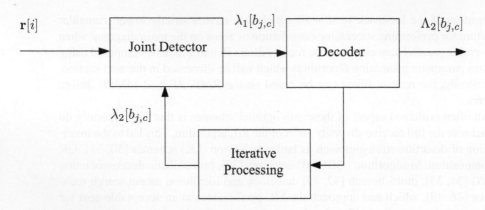

Figure 13.3 Block diagram of an IDD scheme.

stages are separated by interleavers and deinterleavers. The soft outputs from the detector are used to estimate LLRs which are interleaved and serve as input to the MAP decoder for the convolutional code. The MAP decoder [52] computes *a posteriori* probabilities (APPs) for each stream's encoded symbols, which are used to generate soft estimates. These soft estimates are subsequently used to update the receive filters of the detector, de-interleaved and fed back through the feedback filter. The detector computes the *a posteriori* log-likelihood ratio (LLR) of a symbol (+1 or −1) for every code bit of each data stream in a packet with P symbols as given by

$$\Lambda_1[b_{j,c}[i]] = \log \frac{P[b_{j,c}[i] = +1|r[i]]}{P[b_{j,c}[i] = -1|r[i]]},$$

$$j = 1, \ldots, KN_U, \ c = 1, \ldots, C, \tag{13.24}$$

where C is the number of bits used to map the constellation. Using Bayes' rule, the above equation can be written as

$$\Lambda_1[b_{j,c}[i]] = \log \frac{P[r|b_{j,c}[i] = +1]}{P[r[i]|b_{j,c}[i] = -1]} + \log \frac{P[b_{j,c}[i] = +1]}{P[b_{j,c}[i] = -1]} \tag{13.25}$$

$$= \lambda_1[b_{j,c}[i]] + \lambda_2^p[b_{j,c}[i]],$$

where $\lambda_2^p[b_{j,c}[i]] = \log \frac{P[b_{j,c}[i]=+1]}{P[b_{j,c}[i]=-1]}$ is the *a priori* LLR of the code bit $b_{j,c}[i]$, which is computed by the MAP decoder processing the jth data/user stream in the previous iteration, interleaved and then fed back to the detector. The superscript p denotes the quantity obtained in the previous iteration. Assuming equally likely bits, we have $\lambda_2^p[b_{j,c}[i]] = 0$ in the first iteration for all streams/users. The quantity $\lambda_1[b_{j,c}[i]] = \log \frac{P[r[i]|b_{j,c}[i]=+1]}{P[r[i]|b_{j,c}[i]=-1]}$ represents the *extrinsic* information computed by the SISO detector based on the received data $r[i]$, and the prior information about the code bits $\lambda_2^p[b_{j,c}[i]]$, $j = 1, \ldots, KN_U$, $c = 1, \ldots, C$ and the ith data symbol. The extrinsic information $\lambda_1[b_{j,c}[i]]$ is obtained from the detector and the prior information provided by the MAP decoder, which is de-interleaved and fed back into the MAP decoder of the jth data/user stream as the *a priori* information in the next iteration.

For the MAP decoding, we assume that the interference plus noise at the output $z_j[i]$ of the receive filters is Gaussian. This assumption has been reported in previous works and provides an efficient and accurate way of computing the extrinsic information. Thus, for the jth stream/user and the qth iteration the soft output of the detector is

$$z_j^{(q)}[i] = V_j^{(q)} s_j[i] + \xi_j^{(q)}[i], \qquad (13.26)$$

where $V_j^{(q)}[i]$ is a scalar variable equivalent to the magnitude of the channel corresponding to the jth data stream and $\xi_j^{(q)}[i]$ is a Gaussian random variable with variance $\sigma^2_{\xi_j^{(q)}}[i]$. Since we have

$$V_j^{(q)} = E[s_j^*[i] z_j^{(q)[i]}] \qquad (13.27)$$

and

$$\sigma^2_{\xi_j^{(q)}} = E[|z_j^{(q)}[i] - V_j^{(q)}[i] s_j[i]|^2], \qquad (13.28)$$

the receiver can obtain the estimates $\hat{V}_j^{(q)}$ and $\hat{\sigma}^2_{\xi_j^{(q)}}$ via corresponding sample averages over the received symbols. These estimates are used to compute the *a posteriori* probabilities $P[b_{j,c}[i] = \pm 1 | z_{j,l}^{(q)}[i]]$ which are de-interleaved and used as input to the MAP decoder. In what follows, it is assumed that the MAP decoder generates APPs $P[b_{j,c}[i] = \pm 1]$, which are used to compute the input to the receiver. From (13.26) the extrinsic information generated by the iterative receiver is given by

$$\lambda_1[b_{j,c}][i] = \log \frac{P[z_j^{(q)} | b_{j,c}[i] = +1]}{P[z_j^{(q)} | b_{j,c}[i] = -1]}, \quad i = 1, \ldots, P,$$

$$= \log \frac{\sum\limits_{S \in S_c^{+1}} \exp\left(-\frac{|z_j^{(q)} - V_j^{(q)} S|^2}{2\sigma^2_{\xi_j^{(q)}}}\right)}{\sum\limits_{S \in S_c^{-1}} \exp\left(-\frac{|z_j^{(q)} - V_j^{(q)} S|^2}{2\sigma^2_{\xi_j^{(q)}}}\right)}, \qquad (13.29)$$

where S_c^{+1} and S_c^{-1} are the sets of all possible constellations that a symbol can take on such that the cth bit is 1 and -1, respectively. Based on the trellis structure of the code, the MAP decoder processing the jth data stream computes the *a posteriori* LLR of each coded bit as described by

$$\Lambda_2[b_{j,c}[i]] = \log \frac{P[b_{j,c}[i] = +1 | \lambda_1^P[b_{j,c}[i]; \text{decoding}]}{P[b_{j,c}[i] = -1 | \lambda_1^P[b_{j,c}[i]; \text{decoding}]}$$

$$= \lambda_2[b_{j,c}[i]] + \lambda_1^P[b_{j,c}[i]], \qquad (13.30)$$

$$\text{for } j = 1, \ldots, KN_U, \ c = 1, \ldots, C.$$

The computational burden can be significantly reduced using the max-log approximation. From the above, it can be seen that the output of the MAP decoder is the sum of the prior information $\lambda_1^p[b_{j,c}[i]]$ and the extrinsic information $\lambda_2[b_{j,c}[i]]$ produced by the MAP decoder. This extrinsic information is the information about the coded bit $b_{j,c}[i]$ obtained from the selected prior information about the other coded bits $\lambda_1^p[b_{j,c}[i]]$, $j \neq k$. The MAP decoder also computes the *a posteriori* LLR of every information bit, which is used to make a decision on the decoded bit at the last iteration. After interleaving, the extrinsic information obtained by the MAP decoder $\lambda_2[b_{j,c}[i]]$ for $j = 1, \ldots KN_U$, $c = 1, \ldots, C$ is fed back to the detector, as prior information about the coded bits of all streams in the subsequent iteration. For the first iteration, $\lambda_1[b_{j,c}[i]]$ and $\lambda_2[b_{j,c}[i]]$ are statistically independent and as the iterations are computed they become more correlated and the improvement due to each iteration is gradually reduced. It is well known in the field of IDD schemes that there is no performance gain when using more than 5-8 iterations.

The choice of channel coding scheme is fundamental for the performance of iterative joint detection schemes. More sophisticated schemes than convolutional codes such as Turbo or LDPC codes can be considered in IDD schemes for the mitigation of multi-beam and other sources of interference. LDPC codes exhibit some advantages over Turbo codes that include simpler decoding and implementation issues. However, LDPC codes often require a higher number of decoding iterations which translate into delays or increased complexity. The development of IDD schemes and decoding algorithms that perform message passing with reduced delays [60, 61, 62] are of great importance in massive MIMO systems.

13.4 PARAMETER ESTIMATION TECHNIQUES

Amongst the key problems in the uplink of multiuser massive MIMO systems are the estimation of parameters such as channels gains and receive filter coefficients of each user as described by the signal models in Section 13.2. The parameter estimation task usually relies on pilot (or training) sequences, the structure of the data for blind estimation, and signal processing algorithms. In multiuser massive MIMO networks, non-orthogonal training sequences are likely to be used in most application scenarios and the estimation algorithms must be able to provide the most accurate estimates and to track the variations due to mobility within a reduced training period. Standard MIMO linear MMSE and least-squares (LS) channel estimation algorithms [67] can be used for obtaining CSI. However, the cost associated with these algorithms is often cubic in the number of antenna elements at the receiver, i.e., N_A in the uplink. Moreover, in scenarios with mobility the receiver will need to employ adaptive algorithms [66] which can track the channel variations. Interestingly, massive MIMO systems may have an excess of degrees of freedom that translates into a reduced-rank structure to perform parameter estimation. This is an excellent opportunity that massive MIMO offers to apply reduced-rank algorithms [72]-[82] and further develop these techniques. In this section, we review several parameter estimation algorithms and discuss several aspects that are specific for massive MIMO systems such as TDD operation, pilot contamination, and the need for scalable estimation algorithms.

13.4.1 TDD OPERATION

One of the key problems in modern wireless systems is the acquisition of CSI in a timely way. In time-varying channels, TDD offers the most suitable alternative to obtain CSI because the training requirements in a TDD system are independent of the number of antennas at the base station (or access point) [6, 63, 64] and there is no need for CSI feedback. In particular, TDD systems rely on reciprocity by which the uplink channel estimate is used as an estimate of the downlink channel. An issue in this operation mode is the difference in the transfer characteristics of the amplifiers and the filters in the two directions. This can be addressed through measurements and appropriate calibration [65]. In contrast, in a frequency division duplexing (FDD) system the training requirements are proportional to the number of antennas and CSI feedback is essential. For this reason, massive MIMO systems will most likely operate in TDD mode and will require further investigation in calibration methods.

13.4.2 PILOT CONTAMINATION

The adoption of TDD mode and uplink training in massive MIMO systems with multiple cells results in a phenomenon called pilot contamination. In multi-cell scenarios, it is difficult to employ orthogonal pilot sequences because the duration of the pilot sequences depends on the number of cells and this duration is severely limited by the channel coherence time due to mobility. Therefore, non-orthogonal pilot sequences must be employed and this affects the CSI employed at the transmitter. Specifically, the channel estimate is contaminated by a linear combination of channels of other users that share the same pilot [63, 64]. Consequently, the detectors, precoders, and resource allocation algorithms will be highly affected by the contaminated CSI. Strategies to control or mitigate pilot contamination and its effects are very important for massive MIMO networks.

13.4.3 ESTIMATION OF CHANNEL PARAMETERS

Let us now consider channel estimation techniques for multiuser massive MIMO systems and employ the signal models of Section 13.2. The channel estimation problem corresponds to solving the following least-squares (LS) optimization problem:

$$\hat{G}[i] = \arg\min_{G[i]} \sum_{l=1}^{i} \lambda^{i-l} \|r[l] - G[i]s[l]\|^2, \tag{13.31}$$

where the $N_A \times KN_U$ matrix $G = [G_1 \ldots G_K]$ contains the channel parameters of the K users, the $KN_U \times 1$ vector contains the symbols of the K users stacked, and λ is a forgetting factor chosen between 0 and 1. In particular, it is common to use known pilot symbols $s[i]$ in the beginning of the transmission for estimation of the channels and the other receive parameters. This problem can be solved by computing the gradient terms of (13.31), equating them to a zero matrix and manipulating the terms which yields the LS estimate

$$\hat{G}[i] = \hat{Q}[i]\hat{R}_s^{-1}[i], \tag{13.32}$$

where $\hat{Q}[i] = \sum_{l=1}^{i} \lambda^{i-l} r[l] s^H[l]$ is a $N_A \times KNU$ matrix with estimates of cross-correlations between the pilots and the received data $r[i]$ and $\hat{R}[i] = \sum_{l=1}^{i} \lambda^{i-l} s[i] s^H[i]$ is an estimate of the auto-correlation matrix of the pilots. When the channel is static over the duration of the transmission, it is common to set the forgetting factor λ to one. In contrast, when the channel is time-varying one needs to set λ to a value that corresponds to the coherence time of the channel in order to track the channel variations.

The LS estimate of the channel can also be computed recursively by using the matrix inversion lemma [66, 67], which yields the recursive LS (RLS) channel estimation algorithm [27] described by

$$P[i] = \lambda^{-1} P[i-1] - \frac{\lambda^{-2} P[i-1] s[i] s^H[i] P[i-1]}{1 + \lambda^{-1} s^H[i] P[i-1]} s[i], \quad (13.33)$$

$$T[i] = \lambda T[i-1] + r[i] s^H[i], \quad (13.34)$$

$$\hat{G}[i] = T[i] P[i], \quad (13.35)$$

where the computational complexity of the RLS channel estimation algorithm is $N_A(KN_U)^2 + 4(KN_U)^2 + 2N_A(KN_U) + 2KN_U + 2$ multiplications and $N_A(KN_U)^2 + 4(KN_U)^2 - KN_U$ additions [27].

An alternative to using LS-based algorithms is to employ least-mean square (LMS) techniques [68], which can reduce the computational cost. Consider the mean-square error (MSE)-based optimization problem:

$$\hat{G}[i] = \arg \min_{G[i]} E\|r[i] - G[i] s[i]\|^2], \quad (13.36)$$

where $E[\cdot]$ stands for expected value. This problem can be solved by computing the instantaneous gradient terms of (13.36), using a gradient descent rule, and manipulating the terms which results in the LMS channel estimation algorithm given by

$$\hat{G}[i+1] = \hat{G}[i] + \mu e[i] s^H[i], \quad (13.37)$$

where the error vector signal is $e[i] = r[i] - \hat{G}[i] s[i]$ and the step size μ should be chosen between 0 and $2/tr[R_s]$, where $R_s = E[s[i] s^H[i]]$ [66]. The cost of the LMS channel estimation algorithm in this scheme is $N_A(KN_U)^2 + N_A(KN_U) + KN_U$ multiplications and $N_A(KN_U)^2 + N_A KN_U + N_A - KN_U$ additions. The LMS approach has a cost that is one order of magnitude lower than the RLS but the performance in terms of training speed is worse. The channel estimates obtained can be used in the ML rule for ML detectors and SD algorithms, and also to design the receive filters of ZF and MMSE type detectors outlined in the previous section.

13.4.4   ESTIMATION OF RECEIVE FILTER PARAMETERS

An alternative to channel estimation techniques is the direct computation of the receive filters using LS techniques or adaptive algorithms. In this section, we consider the estimation of the receive filters for multiuser massive MIMO systems and employ again the signal models of Section 13.2. The receive filter estimation problem corresponds to solving the LS optimization problem described by

$$w_{k,o}[i] = \arg\min_{w_k[i]} \sum_{l=1}^{i} \lambda^{i-l} |s_k[l] - w_k^H[i]r[l]|^2, \tag{13.38}$$

where the $N_A \times 1$ vector w_k contains the parameters of the receive filters for the kth data stream, and the symbol $s_k[i]$ contains the symbols of the kth data stream. Similarly to channel estimation, it is common to use known pilot symbols $s_k[i]$ in the beginning of the transmission for estimation of the receiver filters. This problem can be solved by computing the gradient terms of (13.38), equating them to a null vector and manipulating the terms which yields the LS estimate

$$w_{k,o}[i] = \hat{R}_r^{-1}[i]\hat{p}_k[i], \tag{13.39}$$

where $R_r[i] = \sum_{l=1}^{i} \lambda^{i-l} r[i]r^H[i]$ is an estimate of the autocorrelation matrix of the received data and $p_k[i] = \sum_{l=1}^{i} \lambda^{i-l} r[l]s_k^H[l]$ is an estimate of the $N_A \times 1$ vector with cross-correlations between the pilots and the received data $r[i]$. When the channel is static over the duration of the transmission, it is common to set the forgetting factor λ to one. Conversely, when the channel is time-varying one needs to set λ to a value that corresponds to the coherence time of the channel in order to track the channel variations. In these situations, a designer can also compute the parameters recursively, thereby taking advantage of the previously computed LS estimates and leading to the RLS algorithm [66] given by

$$k[i] = \frac{\lambda^{-1} P[i-1]r[i]}{1 + \lambda^{-1} r^H[i]P[i-1]r[i]}, \tag{13.40}$$

$$P[i] = \lambda^{-1} P[i-1] - \lambda^{-1} k[i]r^H[i]P[i-1], \tag{13.41}$$

$$w_k[i] = w_k[i-1] - k[i]e_{k,a}^*[i], \tag{13.42}$$

where $e_{k,a}[i] = s_k[i] - w_k^H[i-1]r[i]$ is the *a priori* error signal for the kth data stream. Several other variants of the RLS algorithm could be used to compute the parameters of the receive filters [69]. The computational cost of this RLS algorithm for all data streams corresponds to $KN_U(3N_A^2 + 4N_A + 1)$ multiplications and $KN_U(3N_A^2 + 2N_A - 1) + 2N_A KN_U$ additions.

A reduced complexity alternative to the RLS algorithms is to employ the LMS algorithm to estimate the parameters of the receive filters. Consider the mean-square error (MSE)-based optimization problem:

$$w_{k,o}[i] = \arg\min_{w_k[i]} E[|s_k[i] - w_k^H[i]r[i]|^2]. \tag{13.43}$$

Similarly to the case of channel estimation, this problem can be solved by computing the instantaneous gradient terms of (13.43), using a gradient descent rule, and manipulating the terms which results in the LMS estimation algorithm given by

$$\hat{w}_k[i+1] = \hat{w}_k[i] + \mu e_k^*[i]r[i], \tag{13.44}$$

where the error signal for the kth data stream is $e_k[i] = s_k[i] - \mathbf{w}_k^H[i]\mathbf{r}[i]$ and the step size μ should be chosen between 0 and $2/tr[\mathbf{R}_s]$ [66]. The cost of the LMS estimation algorithm in this scheme is $KN_U(N_A + 1)$ multiplications and KN_UN_A additions.

In parameter estimation problems with a large number of parameters such as those found in massive MIMO systems, an effective technique is to employ reduced-rank algorithms which perform dimensionality reduction followed by parameter estimation with a reduced number of parameters. Consider the mean-square error (MSE)-based optimization problem:

$$[\bar{\mathbf{w}}_{k,o}[i], \mathbf{T}_{D,k,o}[i]] = \arg\min_{\bar{\mathbf{w}}_k[i], \mathbf{T}_{D,k}} E[|s_k[i] - \bar{\mathbf{w}}_k^H[i]\mathbf{T}_{D,k}^H[i]\mathbf{r}[i]|^2], \qquad (13.45)$$

where $\mathbf{T}_{D,k}[i]$ is an $N_A \times D$ matrix that performs dimensionality reduction and $\bar{\mathbf{w}}_k[i]$ is a $D \times 1$ parameter vector. Given $\mathbf{T}_{D,k}[i]$, a generic reduced-rank RLS algorithm [82] with D-dimensional quantities can be obtained from (39)-(41) by substituting the $N_A \times 1$ received vector $\mathbf{r}[i]$ by the reduced-dimension $D \times 1$ vector $\bar{\mathbf{r}}[i] = \mathbf{T}_{D,k}^H[i]\mathbf{r}[i]$.

A central design problem is how to compute the dimensionality reduction matrix $\mathbf{T}_{D,k}[i]$ and several techniques have been considered in the literature, namely:

- Principal components (PC): $\mathbf{T}_{D,k}[i] = \boldsymbol{\phi}_D[i]$, where $\boldsymbol{\phi}_D[i]$ corresponds to a unitary matrix whose columns are the D eigenvectors corresponding to the D largest eigenvectors of an estimate of the covariance matrix $\hat{\mathbf{R}}[i]$.
- Krylov subspace techniques: $\mathbf{T}_{D,k}[i] = [\mathbf{t}_k[i]\hat{\mathbf{R}}[i]\mathbf{t}_k[i]\dots\hat{\mathbf{R}}^{D-1}[i]\mathbf{t}_k[i]$, where $\mathbf{t}_k[i] = \frac{\mathbf{t}_k[i]}{\|\mathbf{p}_k[i]\|}$, for $k = 1, 2, \dots, D$ correspond to the bases of the Krylov subspace [70]-[74].
- Joint iterative optimization methods: $\mathbf{T}_{D,k}[i]$ is estimated along with $\bar{\mathbf{w}}_k[i]$ using an alternating optimization strategy and adaptive algorithms [75]-[83].

13.5 SIMULATION RESULTS

In this section, we illustrate some of the techniques outlined in this article using massive MIMO configurations, namely, a very large antenna array, an excess of degrees of freedom provided by the array, and a large number of users with multiple antennas. We consider QPSK modulation, data packets of 1500 symbols, and channels that are fixed during each data packet that are modeled by complex Gaussian random variables with zero mean and variance equal to unity. For coded systems and iterative detection and decoding, a non-recursive convolutional code with rate $R = 1/2$, constraint length 3, generator polynomial $g = [7\ 5]_{oct}$, and 4 decoding iterations is adopted. The numerical results are averaged over 10^6 runs. For the CAS configuration, we employ $L_k = 0.7, \tau = 2$, the distance d_k to the BS is obtained from a uniform discrete random variable between 0.1 and 0.95, the shadowing spread is $\sigma_k = 3$ dB and the transmit and receive correlation coefficients are equal to $\rho = 0.2$. The signal-to-noise ratio (SNR) in dB per receive antenna is given by SNR $= 10\log_{10}\frac{KN_U\sigma_{s_r}^2}{RC\sigma^2}$, where $\sigma_{s_r}^2 = \sigma_s^2 E[|\gamma_k|^2]$ is the variance of the received symbols, σ_n^2 is the noise variance, $R < 1$ is the rate of the channel code and C is the number of bits used to represent the constellation. For the DAS configuration, we use $L_{k,j}$ taken from a uniform random variable between 0.7 and 1, $\tau = 2$, the distance $d_{k,j}$ for each link to an antenna is obtained from a uniform discrete random variable between 0.1 and 0.5, the shadowing spread is $\sigma_{k,j} = 3$ dB and the transmit and receive correlation index for the co-located antennas that are equal to $\rho = 0.2$. The signal-to-noise ratio (SNR) in dB per receive antenna for the DAS configuration is given by SNR $= 10 \log_{10} \frac{KN_U \sigma_{s_r}^2}{RC\,\sigma^2}$, where $\sigma_{s_r}^2 = \sigma_s^2 E[|\gamma_{k,j}|^2]$ is the variance of the received symbols.

In the first example, we compare the BER performance against the SNR of several detection algorithms, namely, the RMF with multiple users and the RMF in the presence of a single user which is denoted as single user bound, the linear MMSE detector [17], the SIC-MMSE detector using a successive interference cancellation [26], and the multi-branch SIC-MMSE (MB-SIC-MMSE) detector [42, 27, 45]. We assume perfect channel state information and synchronization. In particular, a scenario with $N_A = 64$ antenna elements at the receiver, $K = 32$ users and $N_U = 2$ antenna elements at the user devices is considered, which corresponds to a scenario without an excess of degrees of freedom with $N_A \approx KN_U$. The results shown in Figure 13.4 indicate that the RMF with a single user has the best performance, followed by the MB-SIC-MMSE, the SIC-MMSE, the linear MMSE, and the RMF detectors. Unlike previous works [6] that advocate the use of the RMF, it is clear that the BER performance loss experienced by the RMF should be avoided and more advanced receivers should be considered. However, the cost of linear and SIC receivers is dictated by the matrix inversion of $N_A \times N_A$ matrices which must be reduced for large systems. Moreover, it is clear that a DAS configuration is able to offer a superior BER performance due to a reduction of the average distance from the users to the receive antennas and a reduced correlation amongst the set of N_a receive antennas, resulting in improved links.

In the second example, we consider the coded BER performance against the SNR of several detection algorithms with a DAS configuration using perfect channel state information, as illustrated in Figure 13.5. The results show that the BER is much lower than that obtained for an uncoded systems as indicated in Figure 13.4. Specifically, the MB-SIC-MMSE algorithm obtains the best performance followed by the SIC-MMSE, the linear MMSE and the RMF techniques. Techniques like the MB-SIC-MMSE and SIC-MMSE are more promising for systems with a large number of antennas and users as they can operate with lower SNR values and are therefore more energy efficient. Interestingly, the RMF can offer a BER performance that is acceptable when operating with a high SNR that is not energy efficient and has the advantage that it does not require a matrix inversion. If a designer chooses stronger channel codes like Turbo and LDPC techniques, this choice might allow the operation of the system at lower SNR values.

In the third example, we assess the estimation algorithms when applied to the analyzed detectors. In particular, we compare the BER performance against the SNR of several detection algorithms with a DAS configuration using perfect channel state information and estimated channels with the RLS and the LMS algorithms. The channels are estimated with 250 pilot symbols which are sent at the beginning of packets

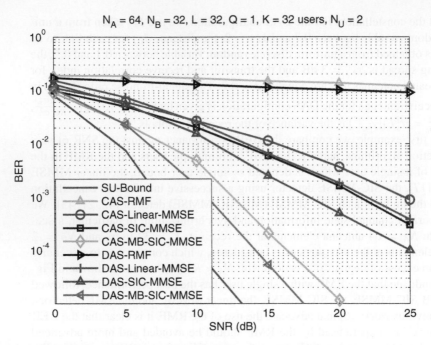

Figure 13.4 BER performance against SNR of detection algorithms in a scenario with $N_A = 64$, $N_B = 32$, $L = 32$, $Q = 1$, $K = 32$ users, and $N_U = 2$ antenna elements.

with 1500 symbols. The results shown in Figure 13.6 indicate that the performance loss caused by the use of the estimated channels is not significant as it remains within 1-2 dB for the same BER performance. The main problems of the use of the standard RLS and LMS is that they require a reasonably large number of pilot symbols to obtain accurate estimates of the channels, resulting in reduced transmission efficiency.

In the fourth example, we evaluate the more sophisticated reduced-rank algorithms to reduce the number of pilot symbols for the training of the receiver filters. In particular, we compare the BER performance against the number of received symbols for a SIC type receiver using a DAS configuration and the standard RLS [66], the Krylov-RLS [71], JIO-RLS [82], and the JIDF-RLS [80] algorithms. We provide the algorithms pilots for the adjustment of the receive filters and assess the BER convergence performance. The results shown in Figure 13.7 illustrate that the performance of the reduced-rank algorithms is significantly better than the standard RLS algorithm, indicating that the use of reduced-rank algorithms can reduce the need for pilot symbols. Specifically, the best performance is obtained by the JIDF-RLS algorithm, followed by the JIO-RLS, the Krylov-RLS, and the standard RLS techniques. In particular, the reduced-rank algorithms can obtain a performance comparable to the standard RLS algorithm with a fraction of the number of pilot symbols required by the RLS algorithm. It should be remarked that for larger training periods the

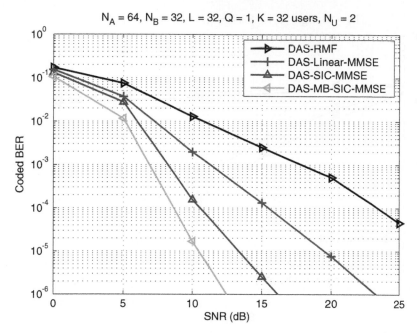

Figure 13.5 Coded BER performance against SNR of detection algorithms with DAS in a scenario with $N_A = 64$, $N_B = 32$, $L = 32$, $Q = 1$, $K = 32$ users, $N_U = 2$ antenna elements and 4 iterations.

standard RLS algorithm will converge to the MMSE bound and the reduced-rank algorithms might converge to the MMSE bound or to higher MSE values depending on the structure of the covariance matrix R and the choice of the rank D.

13.6   FUTURE TRENDS AND EMERGING TOPICS

In this section, we discuss some future signal detection and estimation trends in the area of massive MIMO systems and point out some topics that might attract the interest of researchers. The topics are structured as:

- Signal detection:

   → Cost-effective detection algorithms: Techniques to perform dimensionality reduction [71]-[83] for detection problems will play an important role in massive MIMO devices. By reducing the number of effective processing elements, detection algorithms could be applied. In addition, the development of schemes based on RMF with non-linear interference cancellation capabilities might be a promising option that can close the

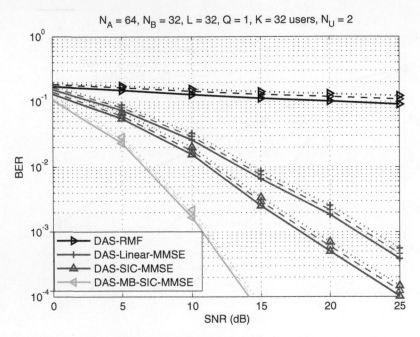

Figure 13.6 BER performance against SNR of detection algorithms in a scenario with channel estimation, $N_A = 64$, $N_B = 32$, $L = 32$, $Q = 1$, $K = 32$ users, and $N_U = 2$ antenna elements. Parameters: $\lambda = 0.999$ and $\mu = 0.05$. The solid lines correspond to perfect channel state information, the dashed lines correspond to channel estimation with the RLS algorithm, and the dotted lines correspond to channel estimation with the LMS algorithm.

complexity gap between RMF and more costly detectors.

→ Decoding strategies with low delay: The development of decoding strategies for DAS configurations with reduced delay will play a key role in applications such as audio and video streaming because of their delay sensitivity. Therefore, novel message passing algorithms with smarter strategies to exchange information should be investigated along with their application to IDD schemes [60, 61, 62].

→ Mitigation of impairments: The identification of impairments originated in the RF chains of massive MIMO systems, delays caused by DAS schemes will need mitigation by smart signal processing algorithms. For example, I/Q imbalance might be dealt with using widely-linear signal processing algorithms [87, 74, 88].

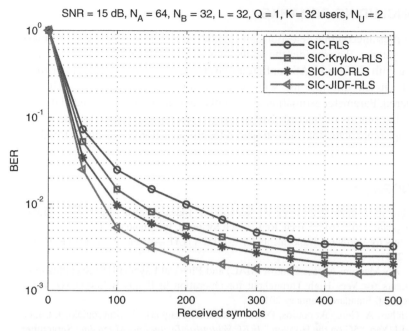

Figure 13.7 BER performance against the number of received symbols for SIC receivers with a DAS architecture operating at SNR = 15 dB in a scenario with the estimation of the receive filters, $N_A = 64$, $N_B = 32$, $L = 32$, $Q = 1$, $K = 32$ users, and $N_U = 2$ antenna elements. Parameters: $\lambda = 0.999$, $D = 5$ for all reduced-rank methods, and interpolators with $I = 3$ parameters, and 12 branches for the JIDF scheme.

→ Detection techniques for multicell scenarios: The development of detection algorithms for scenarios with multiple and small cells requires approaches which minimize the need for channel state information from adjacent cells and the decoding delay [84]-[86].

- Parameter estimation:

  → Blind algorithms: The development of blind estimation algorithms for the channel and receive filter parameters is important for mitigating the problem of pilot contamination [89]-[94].

  → Reduced-rank and sparsity-aware algorithms: The development of reduced-rank and sparsity-aware algorithms that exploit the mathematical structure of massive MIMO channels is an important topic for the future along with features that lend themselves to implementation [70]-[83].

13.7   CONCLUDING REMARKS

This chapter presented signal detection and estimation techniques for multiuser massive MIMO systems. We consider the application to cellular networks with massive MIMO along with CAS and DAS configurations. Recent signal detection algorithms have been discussed and their use with iteration detection and decoding schemes been considered. Parameter estimation algorithms have also been reviewed and studied in several scenarios of interest. Numerical results have illustrated some of the discussions on signal detection and estimation techniques along with future trends in the field.

REFERENCES

1. Cisco and/or its affiliates,"Cisco Visual Networking Index: Global Mobile Data Traffic Forecast Update, 2012-2017," Tech. Rep., Cisco Systems, Inc., January 2013.
2. Requirements for Further Advancements for E-UTRA (LTE-Advanced), 3GPP TR 36.913 Standard, 2011.
3. Wireless LAN Medium Access Control (MAC) and Physical Layer (PHY) Specifications: Enhancements for Very High Throughput for Operation in Bands Below 6GHz, IEEE P802.11ac/D1.0 Standard., January 2011.
4. P. Demestichas, A. Georgakopoulos, D. Karvounas, K. Tsagkaris, V. Stavroulaki, J. Lu, C. Xiong and J. Yao, "5G on the Horizon," *IEEE Vehicular Technology Magazine*, September 2013.
5. T. L. Marzetta, "Noncooperative cellular wireless with unlimited numbers of base station antennas," *IEEE Transactions on Wireless Communications*, vol. 9, no. 11, pp. 3590-3600, November 2010.
6. F. Rusek, D. Persson, B. Lau, E. Larsson, T. Marzetta, O. Edfors, and F. Tufvesson, "Scaling up MIMO: Opportunities, and challenges with very large arrays," *IEEE Signal Processing Mag.*, vol. 30, no. 1, pp. 40-60, January 2013.
7. R. C. de Lamare, "Massive MIMO Systems: Signal Processing Challenges and Future Trends," *URSI Radio Science Bulletin*, December 2013.
8. J. Nam, J.-Y. Ahn, A. Adhikary, and G. Caire, "Joint spatial division and multiplexing: Realizing massive MIMO gains with limited channel state information," *Proc. 46th Annual Conference on Information Sciences and Systems (CISS)*, 2012.
9. R. Combes, Z. Altman and E. Altman, "Interference coordination in wireless networks: A flow-level perspective," *Proc. IEEE INFOCOM 2013*, pp. 2841-2849, 14-19 April 2013.
10. R. Aggarwal, C. E. Koksal, and P. Schniter, "On the design of large scale wireless systems," *IEEE Journal of Selected Areas in Communications*, vol. 31, no. 2, pp. 215-225, February 2013.
11. C. Shepard, H. Yu, N. Anand, L. E. Li, T. L. Marzetta, R. Yang, and L. Zhong, "Argos: Practical many-antenna base stations," *Proc. ACM Int. Conf.Mobile Computing and Networking (MobiCom)*, Istanbul, Turkey, August 2012.
12. J. Hoydis, C. Hoek, T. Wild, and S. ten Brink, "Channel measurements for large antenna arrays," *Proc. IEEE International Symposium on Wireless Communication Systems (ISWCS)*, Paris, France, August 2012.
13. X. Gao, F. Tufvesson, O. Edfors, and F. Rusek, "Measured propagation characteristics for very-large MIMO at 2.6 GHz," *Proc. of the 46th Annual Asilomar Conference on Signals, Systems, and Computers*, Pacific Grove, California, USA, November 2012.

14. J.P. Kermoal, L. Schumacher, and K.I. Perdersen et al., "A stochastic MIMO radio channel model with experimental validation", *IEEE Journal on Selected Areas in Communications*, vol. 20, no. 6, pp. 1211-1226, August 2002.

15. W. Choi, J. G. Andrews, "Downlink performance and capacity of distributed antenna systems in a multicell environment," *IEEE Transactions on Wireless Communications*, vol.6, no.1, pp.69-73, January 2007.

16. L. Dai, "An Uplink Capacity Analysis of the Distributed Antenna System (DAS): From Cellular DAS to DAS with Virtual Cells," *IEEE Transactions on Wireless Communications*, vol.13, no.5, pp. 2717-2731, May 2014.

17. S. Verdu, *Multiuser Detection*, Cambridge, 1998.

18. E. Viterbo and J. Boutros, "A universal lattice code decoder for fading channels," *IEEE Transactions on Information Theory*, vol. 45, no. 5, pp. 1639-1642, July 1999.

19. M. O. Damen, H. E. Gamal and G. Caire, "On maximum likelihood detection and the search for the closest lattice point," *IEEE Transactions on Information Theory*, vol. 49, pp. 2389-2402, October 2003.

20. Z. Guo and P. Nilsson, "Algorithm and Implementation of the K-Best Sphere Decoding for MIMO Detection," *IEEE Journal on Selected Areas in Communications*, vol. 24, no. 3, pp. 491-503, March 2006.

21. C. Studer, A. Burg, and H. Bolcskei, "Soft-output sphere decoding: algorithms and VLSI implementation," *IEEE Journal on Selected Areas in Communications*, vol. 26, pp. 290-300, February 2008.

22. B. Shim and I. Kang, "On further reduction of complexity in tree pruning based sphere search," *IEEE Transactions on Communications*, vol. 58, no. 2, pp. 417-422, February 2010.

23. A. Duel-Hallen, "Equalizers for Multiple Input Multiple Output Channels and PAM Systems with Cyclostationary Input Sequences," emphIEEE Journal on Selected Areas in Communications, vol. 10, pp. 630-639, April, 1992.

24. G. D. Golden, C. J. Foschini, R. A. Valenzuela and P. W. Wolniansky, "Detection algorithm and initial laboratory results using V-BLAST space-time communication architecture," *Electronics Letters*, vol. 35, no.1, January 1999.

25. J. Benesty, Y. Huang, and J. Chen, "A fast recursive algorithm for optimum sequential signal detection in a BLAST system," *IEEE Transactions on Signal Processing*, vol. 51, pp. 1722-1730, July 2003.

26. A. Rontogiannis, V. Kekatos, and K. Berberidis, "A Square-Root Adaptive V-BLAST Algorithm for Fast Time-Varying MIMO Channels," *IEEE Signal Processing Letters*, vol. 13, No. 5, pp. 265-268, May 2006.

27. R. Fa, R. C. de Lamare, "Multi-Branch Successive Interference Cancellation for MIMO Spatial Multiplexing Systems," *IET Communications*, vol. 5, no. 4, pp. 484-494, March 2011.

28. P. Li, R. C. de Lamare and R. Fa, "Multiple Feedback Successive Interference Cancellation Detection for Multiuser MIMO Systems," *IEEE Transactions on Wireless Communications*, vol. 10, no. 8, pp. 2434 - 2439, August 2011.

29. J. H. Choi, H. Y. Yu and Y. H. Lee, "Adaptive MIMO decision feedback equalization for receivers with time-varying channels," *IEEE Transactions on Signal Processing*, vol. 53, no. 11, pp. 4295-4303, November 2005.

30. C. Windpassinger, L. Lampe, R.F.H. Fischer, and T.A Hehn, "A performance study of MIMO detectors," *IEEE Transactions on Wireless Communications*, vol. 5, no. 8, August 2006, pp. 2004-2008.

31. Y. H. Gan, C. Ling, and W. H. Mow, "Complex lattice reduction algorithm for low-complexity full-diversity MIMO detection," *IEEE Transactions on Signal Processing*, vol. 56, no. 7, July 2009.

32. Q. Zhou and X. Ma, "Element-Based Lattice Reduction Algorithms for Large MIMO Detection," *IEEE Journal on Selected Areas in Communications*, vol. 31, no. 2, pp. 274-286, February 2013.

33. K. J. Kim, J. Yue, R. A. Iltis, and J. D. Gibson, "A QRD-M/Kalman filter-based detection and channel estimation algorithm for MIMO-OFDM systems," *IEEE Transactions on Wireless Communications*, vol. 4,pp. 710-721, March 2005.

34. Y. Jia, C. M. Vithanage, C. Andrieu, and R. J. Piechocki, "Probabilistic data association for symbol detection in MIMO systems," *Electronics Letters*, vol. 42, no. 1, pp. 38-40, January 2006.

35. S. Yang, T. Lv, R. Maunder, and L. Hanzo, "Unified Bit-Based Probabilistic Data Association Aided MIMO Detection for High-Order QAM Constellations," *IEEE Transactions on Vehicular Technology*, vol. 60, no. 3, pp. 981-991, March 2011.

36. M. K. Varanasi, "Decision feedback multiuser detection: A systematic approach," *IEEE Transactions on Information Theory*, vol. 45, pp. 219-240, January 1999.

37. J. F. Rößler and J. B. Huber, "Iterative soft decision interference cancellation receivers for DS-CDMA downlink employing 4QAM and 16QAM," *Proc. 36th Asilomar Conf. Signal, Systems and Computers*, Pacific Grove, CA, November 2002.

38. J. Luo, K. R. Pattipati, P. K. Willet, and F. Hasegawa, "Optimal User Ordering and Time Labeling for Ideal Decision Feedback Detection in Asynchronous CDMA," *IEEE Transactions on Communications*, vol. 51, no. 11, November, 2003.

39. G. Woodward, R. Ratasuk, M. L. Honig, and P. Rapajic, "Minimum Mean-Squared Error Multiuser Decision-Feedback Detectors for DS-CDMA," *IEEE Transactions on Communications*, vol. 50, no. 12, December, 2002.

40. R. C. de Lamare and R. Sampaio-Neto, "Adaptive MBER decision feedback multiuser receivers in frequency selective fading channels," *IEEE Communications Letters*, vol. 7, no. 2, February 2003, pp. 73 - 75.

41. F. Cao, J. Li, and J. Yang, "On the relation between PDA and MMSE-ISDIC," *IEEE Signal Processing Letters*, vol. 14, no. 9, September 2007.

42. R.C. de Lamare and R. Sampaio-Neto, "Minimum mean-squared error iterative successive parallel arbitrated decision feedback detectors for DS-CDMA systems, " *IEEE Transactions on Communications*, vol. 56, no. 5, May 2008, pp. 778-789.

43. Y. Cai and R. C. de Lamare, "Adaptive Space-Time Decision Feedback Detectors with Multiple Feedback Cancellation," *IEEE Transactions on Vehicular Technology*, vol. 58, no. 8, October 2009, pp. 4129 - 4140.

44. P. Li and R. C. de Lamare, "Adaptive Decision-Feedback Detection With Constellation Constraints for MIMO Systems," *IEEE Transactions on Vehicular Technology*, vol. 61, no. 2, 853-859, February 2012.

45. R. C. de Lamare, "Adaptive and Iterative Multi-Branch MMSE Decision Feedback Detection Algorithms for Multi-Antenna Systems," *IEEE Transactions on Wireless Communications*, vol. 14, no. 10, October 2013.

46. M. Reuter, J.C. Allen, J. R. Zeidler, and R. C. North, "Mitigating error propagation effects in a decision feedback equalizer," *IEEE Transactions on Communications*, vol. 49, no. 11, November 2001, pp. 2028 - 2041.

47. R.C. de Lamare, R. Sampaio-Neto, and A. Hjorungnes, "Joint iterative interference cancellation and parameter estimation for cdma systems," *IEEE Communications Letters*, vol.

11, no. 12, December 2007, pp. 916-918.

48. K. Vardhan, S. Mohammed, A. Chockalingam, and B. Rajan, "A low-complexity detector for large MIMO systems and multicarrier CDMA systems," *IEEE Journal on Selected Areas in Communications*, vol. 26, no. 3, pp. 473-485, April 2008.

49. P. Li and R. D. Murch, "Multiple Output Selection-LAS Algorithm in Large MIMO Systems," *IEEE Communications Letters*, vol. 14, no. 5, pp. 399-401, May 2010.

50. C. Berrou and A. Glavieux, "Near optimum error-correcting coding and decoding: Turbo codes," *IEEE Transactions on Communications*, vol. 44, October 1996.

51. C. Douillard et al., "Iterative correction of intersymbol interference: Turbo equalization," *European Transactions on Telecommunications*, vol. 6, no. 5, pp. 507-511, September-October 1995.

52. X. Wang and H. V. Poor, "Iterative (turbo) soft interference cancellation and decoding for coded CDMA," *IEEE Transactions on Communications*, vol. 47, pp. 1046-1061, July 1999.

53. M. Tuchler, A. Singer, and R. Koetter, "Minimum mean square error equalization using a priori information," *IEEE Transactions on Signal Processing*, vol. 50, pp. 673-683, March 2002.

54. B. Hochwald and S. ten Brink, "Achieving near-capacity on a mutliple- antenna channel," *IEEE Transactions on Communications*, vol. 51, pp. 389-399, March 2003.

55. J. Hou, P. H. Siegel, and L. B. Milstein, "Design of multi-input multi-output systems based on low-density Parity-check codes," *IEEE Transactions on Communications*, vol. 53, no. 4, pp. 601- 611, April 2005.

56. H. Lee, B. Lee, and I. Lee, "Iterative detection and decoding with an improved V-BLAST for MIMO-OFDM Systems," *IEEE Journal on Selected Areas in Communications*, vol. 24, pp. 504-513, March 2006.

57. J. Wu and H.-N. Lee, "Performance Analysis for LDPC-Coded Modulation in MIMO Multiple-Access Systems," *IEEE Transactions on Communications*, vol. 55, no. 7, pp. 1417-1426, July 2007.

58. X. Yuan, Q. Guo, X. Wang, and Li Ping, "Evolution analysis of low-cost iterative equalization in coded linear systems with cyclic prefixes," *IEEE Journal on Selected Areas in Communications*, vol. 26, no. 2, pp. 301-310, February 2008.

59. J. W. Choi, A. C. Singer, J Lee and N. I. Cho, "Improved linear soft-input soft-output detection via soft feedback successive interference cancellation," *IEEE Transactions on Communications*, vol.58, no.3, pp.986-996, March 2010.

60. M. J. Wainwright, T. S. Jaakkola, and A. S. Willsky, "A new class of upper bounds on the log partition function," *IEEE Transactions on Information Theory*, vol. 51, no. 7, pp. 2313 - 2335, July 2005.

61. H. Wymeersch, F. Penna, and V. Savic, "Uniformly Reweighted Belief Propagation for Estimation and Detection in Wireless Networks," *IEEE Transactions on Wireless Communications*, vol. 11, no. 4, April 2012.

62. J. Liu, R. C. de Lamare, "Low-Latency Reweighted Belief Propagation Decoding for LDPC Codes," *IEEE Communications Letters*, vol. 16, no. 10, pp. 1660-1663, October 2012.

63. J. Jose, A. Ashikhmin, T. L. Marzetta, and S. Vishwanath, "Pilot Contamination and Precoding in Multi-Cell TDD Systems," *IEEE Transactions on Wireless Communications*, vol.10, no.8, pp. 2640-2651, August 2011.

64. A. Ashikhmin and T. L. Marzetta, "Pilot contamination precoding in multi-cell large scale antenna systems," *Proc. IEEE International Symposium on Information Theory (ISIT)*,

Cambridge, MA, July 2012.

65. R. Rogalin, O. Y. Bursalioglu, H. Papadopoulos, G. Caire, A. F. Molisch, A. Michaloli-akos, V. Balan, and K. Psounis, "Scalable Synchronization and Reciprocity Calibration for Distributed Multiuser MIMO", *IEEE Transactions on Wireless Communications*, vol.13, no.4, pp. 1815-1831, April 2014.

66. S. Haykin, *Adaptive Filter Theory*, 4th ed. Englewood Cliffs, NJ: Prentice- Hall, 2002.

67. M. Biguesh and A.B. Gershman, "Training-based MIMO channel estimation: a study of estimator tradeoffs and optimal training signals," *IEEE Transactions on Signal Processing*, vol. 54 no. 3, March 2006.

68. T. Wang, R. C. de Lamare, and P. D. Mitchell, "Low-Complexity Set-Membership Channel Estimation for Cooperative Wireless Sensor Networks," *IEEE Transactions on Vehicular Technology*, vol.60, no.6, pp. 2594,2607, July 2011.

69. R. C. de Lamare and P. S. R. Diniz, "Set-Membership Adaptive Algorithms Based on Time-Varying Error Bounds for CDMA Interference Suppression," *IEEE Transactions on Vehicular Technology*, vol.58, no.2, pp.644-654, February 2009.

70. H. Qian and S. N. Batalama, "Data-record-based criteria for the selection of an auxiliary vector estimator of the MMSE/MVDR filter," *IEEE Transactions on Communications*, vol. 51, no. 10, pp. 1700-1708, October 2003.

71. M. L. Honig and J. S. Goldstein, "Adaptive reduced-rank interference suppression based on the multistage Wiener filter," *IEEE Transactions on Communications*, vol. 50, no. 6, June 2002.

72. Y. Sun, V. Tripathi, and M. L. Honig, "Adaptive, iterative, reducedrank (turbo) equalization," *IEEE Transactions on Wireless Communications*, vol. 4, no. 6, pp. 2789-2800, November 2005.

73. R. C. de Lamare, M. Haardt, and R. Sampaio-Neto, "Blind Adaptive Constrained Reduced-Rank Parameter Estimation based on Constant Modulus Design for CDMA Interference Suppression," *IEEE Transactions on Signal Processing*, vol. 56, no. 6, June 2008.

74. N. Song, R. C. de Lamare, M. Haardt, and M. Wolf, "Adaptive Widely Linear Reduced-Rank Interference Suppression based on the Multi-Stage Wiener Filter," *IEEE Transactions on Signal Processing*, vol. 60, no. 8, August 2012.

75. R. C. de Lamare and R. Sampaio-Neto, "Adaptive reduced-rank MMSE filtering with interpolated FIR filters and adaptive interpolators," *IEEE Signal Processing Letters*, vol. 12, no. 3, March, 2005.

76. R. C. de Lamare and Raimundo Sampaio-Neto, "Reduced-rank Interference Suppression for DS-CDMA based on Interpolated FIR Filters," *IEEE Communications Letters*, vol. 9, no. 3, March 2005.

77. R. C. de Lamare and R. Sampaio-Neto, "Adaptive Interference Suppression for DS-CDMA Systems based on Interpolated FIR Filters with Adaptive Interpolators in Multipath Channels," *IEEE Trans. Vehicular Technology*, vol. 56, no. 6, September 2007.

78. R. C. de Lamare and R. Sampaio-Neto, "Adaptive Reduced-Rank MMSE Parameter Estimation based on an Adaptive Diversity Combined Decimation and Interpolation Scheme," *Proc. IEEE International Conference on Acoustics, Speech and Signal Processing*, April 15-20, 2007, vol. 3, pp. III-1317-III-1320.

79. R. C. de Lamare and R. Sampaio-Neto, "Reduced-Rank Adaptive Filtering Based on Joint Iterative Optimization of Adaptive Filters," *IEEE Signal Processing Letters*, vol. 14, no. 12, December 2007.

80. R. C. de Lamare and R. Sampaio-Neto, "Adaptive reduced-rank processing based on joint and iterative interpolation, decimation, and filtering," *IEEE Transactions on Signal Pro-

cessing, vol. 57, no. 7, July 2009, pp. 2503-2514.

81. R. C. de Lamare and R. Sampaio-Neto, "Reduced-Rank Space-Time Adaptive Interference Suppression With Joint Iterative Least Squares Algorithms for Spread-Spectrum Systems," *IEEE Transactions on Vehicular Technology*, vol. 59, no. 3, March 2010, pp.1217-1228.

82. R.C. de Lamare and R. Sampaio-Neto, "Adaptive reduced-rank equalization algorithms based on alternating optimization design techniques for MIMO systems," *IEEE Transactions on Vehicular Technology*, vol. 60, no. 6, pp. 2482-2494 , July 2011.

83. R. C. de Lamare, R. Sampaio-Neto, and M. Haardt, "Blind adaptive constrained constant-modulus reduced-rank interference suppression algorithms based on interpolation and switched decimation," *IEEE Transactions on Signal Processing*, vol. 59, no. 2, pp. 681-695, February 2011.

84. H. Dai, A. F. Molisch, and H. V. Poor, "Downlink capacity of interference- 537 limited MIMO systems with joint detection", *IEEE Transactions on Wireless Communications*, vol. 3, no. 2, pp. 442-453, March 2004.

85. P. Marsch and G. Fettweis, "Uplink CoMP under a constrained back- 534 haul and imperfect channel knowledge," *IEEE Transactions on Wireless Communications*, vol. 10, no. 6, pp. 1730-1742, June 2011.

86. P. Li and R. C. de Lamare, "Distributed Iterative Detection With Reduced Message 2 Passing for Networked MIMO Cellular Systems," *IEEE Transactions on Vehicular Technology*, vol.63, no.6, pp. 2947-2954, July 2014.

87. P. Chevalier and A. Blin, "Widely linear MVDR beamformers for the reception of an unknown signal corrupted by noncircular interferences," *IEEE Transactions on Signal Processing*, vol. 55, no. 11, pp. 5323-5336, November 2007.

88. N. Song, W. U. Alokozai, R. C. de Lamare and M. Haardt, " Adaptive Widely Linear Reduced-Rank Beamforming Based on Joint Iterative Optimization," *IEEE Signal Processing Letters*, vol. 21, no. 3, March 2014.

89. M. Honig, U. Madhow, and S. Verdu, "Blind adaptive multiuser detection," *IEEE Transactions on Information Theory*, vol. 41, no. 4, pp. 944–960, July 1995.

90. R. C. de Lamare and R. Sampaio-Neto, "Low-complexity variable step-size mechanisms for stochastic gradient algorithms in minimum variance CDMA receivers," *IEEE Transactions on Signal Processing*, vol. 54, no. 6, pp. 2302-2317, June 2006.

91. C. Xu, G. Feng, and K. S. Kwak, "A modified constrained constant modulus approach to blind adaptive multiuser detection," *IEEE Transactions on Communications*, vol. 49, no. 9, pp. 1642–1648, September 2001.

92. R. C. de Lamare and R. Sampaio-Neto, "Blind adaptive code-constrained constant modulus algorithms for CDMA interference suppression in multipath channels," *IEEE Communications Letters*, vol. 9, no. 4, pp. 334-336, April 2005.

93. R. C. de Lamare and R. Sampaio-Neto, "Blind adaptive MIMO receivers for space-time block-coded DS-CDMA systems in multipath channels using the constant modulus criterion," *IEEE Transactions on Communications*, vol. 58, no. 1, pp. 21-27, January 2010.

94. L. Wang and R. C. de Lamare, "Adaptive constrained constant modulus algorithm based on auxiliary vector filtering for beamforming," *IEEE Transactions on Signal Processing*, vol. 58, no. 10, pp. 5408-5413, October 2010.

14 Advances on Adaptive Sparse-Interpolated Filtering

Eduardo Luiz Ortiz Batista
Federal University of Santa Catarina (UFSC)

Rui Seara
Federal University of Santa Catarina (UFSC)

14.1 INTRODUCTION

Over the last few decades, many applications involving signal processing systems have played an important role in everyday life. Perhaps the most obvious examples of such applications can be found in modern smartphones, where several signal processing technologies are combined for supporting a ubiquitous access to telephony, messaging, video streaming, global positioning, among other things. The dissemination of these technologies has been supported not only by the emergence of powerful platforms for implementing signal processing systems, but mainly by a massive research effort towards the development of effective signal processing algorithms. A major concern for developing such algorithms is the computational complexity required for their implementation. In this context, several signal processing approaches have been developed with a special focus on reduced computational complexity, with applications ranging from intricate video encoding to more straightforward filtering operations.

The development of reduced-complexity algorithms to perform filtering operations is justified by the fact that the computational burden required in these operations may become a limiting factor when the memory size of the filter becomes large [1]. This problem is even more critical in cases involving nonlinear filters such as the Volterra filter [2], in which the computational burden grows exponentially with memory size. Moreover, both linear and nonlinear filters have been increasingly used in adaptive filtering applications that consider not only a filtering operation, but also the iterative update of the filter coefficients using an adaptive algorithm [3]. In these applications, the use of reduced-complexity algorithms may be mandatory for practical implementations.

The focus of this chapter is on a particular approach used for implementing standard and adaptive filters with reduced complexity, namely the sparse-interpolated approach. This approach is based on the use of sparseness, aiming to reduce the complexity required to implement a filter, along with interpolation to compensate for the performance loss arising from the use of a sparse filter [1], [4]. The sparse-interpolated approach was originally applied to linear finite-impulse-response (FIR) filters [1] and, thus, Section 14.2 of this chapter is dedicated to describing the foundations of the sparse-interpolated FIR (SIFIR) filters. As shown there, the SIFIR implementations are characterized by a reduced number of coefficients in comparison with the standard FIR filters. Consequently, such implementations are of great interest in adaptive filtering applications due to the smaller number of coefficients that need to be updated at each iteration of the adaptive algorithm [5], [6]. In this context, Section 14.3 is devoted to describing various aspects of the adaptive SIFIR filters, with a special focus on the use of the normalized least-mean-square (NLMS) algorithm to develop specific adaptive algorithms for such filters. Another type of filter that can greatly benefit from the use of the sparse-interpolated approach is the Volterra filter [2]. In this case, the complexity reduction provided by the sparse-interpolated approach potentially leads to an exponential reduction of computational complexity [4]. This characteristic is addressed in Section 14.4 of this chapter, where a detailed description of the different implementations of sparse-interpolated Volterra filters is introduced. Finishing the chapter, Section 14.5 discusses a case study involving the application of sparse-interpolated filters to a nonlinear network echo cancellation problem, and in Section 14.6 concluding remarks are presented.

14.2 FROM THE FIR FILTER TO THE SPARSE-INTERPOLATED FIR FILTER

The focus of this section is on both describing the basic notation used throughout this chapter and presenting the foundations of sparse-interpolated FIR filters. In this context, we start by giving some preliminary definitions of standard and sparse FIR filters. Afterwards, we describe a form of implementing FIR filters based on the singular value decomposition (SVD). Then, this section is finished with the description of the sparse-interpolated FIR filter and how it can be seen as a type of reduced-rank implementation of an FIR filter.

14.2.1 PRELIMINARY DEFINITIONS OF STANDARD AND SPARSE FIR FILTERS

The input-output relationship of a standard FIR filter with memory size N can be written as [3]

$$y(n) = \mathbf{w}^{\mathrm{T}}\mathbf{x}(n) \tag{14.1}$$

with the input vector given by

$$\mathbf{x}(n) = [x(n)\ x(n-1)\ x(n-2)\ \cdots\ x(n-N+1)]^{\mathrm{T}} \tag{14.2}$$

and the coefficient vector by

$$\mathbf{w} = [w_0\ w_1\ w_2\ \cdots\ w_{N-1}]^{\mathrm{T}}. \tag{14.3}$$

Now, let us consider an L-sparse FIR filter with memory size equal to N. Such a filter is characterized by an $N \times 1$ coefficient vector

$$\ddot{\mathbf{w}} = [w_0 \ 0 \ \cdots \ w_L \ 0 \ \cdots \ w_{2L} \ 0 \ \cdots \ w_{(N_s-1)L} \ 0 \ \cdots \ 0]^T \qquad (14.4)$$

having $L - 1$ of each L coefficients fixed to zero, which results in

$$N_s = \lfloor (N - 1)/L \rfloor + 1 \qquad (14.5)$$

nonzero elements with $\lfloor \cdot \rfloor$ representing the truncation operation. By considering (14.4), the input-output relationship of the L-sparse FIR filter can be written as

$$\ddot{y}(n) = \ddot{\mathbf{w}}^T \mathbf{x}(n). \qquad (14.6)$$

Now, using the Kronecker product (denoted here by $\otimes$) and padding $\ddot{\mathbf{w}}$ with zeros if N is not a multiple of L, one can express (14.4) as

$$\ddot{\mathbf{w}} = \dot{\mathbf{w}} \otimes \mathbf{n}_1 \qquad (14.7)$$

where

$$\dot{\mathbf{w}} = [w_0 \ w_L \ w_{2L} \ \cdots \ w_{(N_s-1)L}]^T \qquad (14.8)$$

denotes a compact representation of $\ddot{\mathbf{w}}$ composed only of its nonzero elements and $\mathbf{n}_1$ is an instance with $l = 1$ of a $L \times 1$ vector $\mathbf{n}_l$ that has one element equal to 1 in the lth position and remaining elements equal to zero elsewhere, i.e., $\mathbf{n}_1 = [1 \ 0 \ 0 \ \cdots \ 0]^T$. Then, taking into account (14.8), (14.6) can be rewritten as

$$\ddot{y}(n) = (\dot{\mathbf{w}} \otimes \mathbf{n}_1)^T \mathbf{x}(n) = (\dot{\mathbf{w}}^T \otimes \mathbf{n}_1^T) \mathbf{x}(n). \qquad (14.9)$$

In a general form, one can define

$$\ddot{y}(n - l + 1) = (\dot{\mathbf{w}}^T \otimes \mathbf{n}_l^T) \mathbf{x}(n) \qquad (14.10)$$

for $l = 1, 2, ..., L$. Thus, since the $L \times L$ identity matrix $\mathbf{I}_L$ can be written as

$$\mathbf{I}_L = [\mathbf{n}_1 \ \mathbf{n}_2 \ \mathbf{n}_3 \ \cdots \ \mathbf{n}_L] \qquad (14.11)$$

one has

$$\ddot{\mathbf{y}}(n) = (\dot{\mathbf{w}}^T \otimes \mathbf{I}_L) \mathbf{x}(n) \qquad (14.12)$$

with

$$\ddot{\mathbf{y}}(n) = [\ddot{y}(n) \ \ddot{y}(n - 1) \ \cdots \ \ddot{y}(n - L + 1)]^T \qquad (14.13)$$

denoting the output vector composed of the last L output samples of the sparse filter.

14.2.2 FIR IMPLEMENTATION USING SINGULAR VALUE DECOMPOSITION

Let us now consider two integers R and C with $R < C$ and $N = RC$. Thus, it is possible to define an $R \times C$ coefficient matrix for a standard FIR filter with memory size N as

$$
\mathbf{W} = \begin{bmatrix}
w_0 & w_R & \cdots & w_{(C-1)R} \\
w_1 & w_{R+1} & \cdots & w_{(C-1)R+1} \\
\vdots & \vdots & \ddots & \vdots \\
w_{R-1} & w_{2R-1} & \cdots & w_{N-1}
\end{bmatrix}. \tag{14.14}
$$

If, however, R and C are such that $N < RC$, a matrix similar to (14.14) can always be obtained by padding (14.3) with zeros. Therefore, such a definition of coefficient matrix is assumed to be general. The relationship between (14.3) and (14.14) is given by

$$
\mathbf{w} = \text{vec}(\mathbf{W}) \tag{14.15}
$$

where $\text{vec}(\cdot)$ is an operator that stacks the columns of a matrix to form a vector [7], [8]. Similarly to (14.14) and (14.15), one can define an input matrix as

$$
\mathbf{X}(n) = \begin{bmatrix}
x(n) & x(n-R) & \cdots & x(n-CR+R) \\
x(n-1) & x(n-R-1) & \cdots & x(n-CR+R-1) \\
\vdots & \vdots & \ddots & \vdots \\
x(n-R+1) & x(n-2R+1) & \cdots & x(n-N+1)
\end{bmatrix} \tag{14.16}
$$

such that

$$
\mathbf{x}(n) = \text{vec}[\mathbf{X}(n)]. \tag{14.17}
$$

As shown in [7],

$$
\text{tr}(\mathbf{A}^{\mathsf{T}}\mathbf{B}) = [\text{vec}(\mathbf{A})]^{\mathsf{T}} \text{vec}(\mathbf{B}) \tag{14.18}
$$

for a generic matrix $\mathbf{A}$ having dimensions similar to those of another generic matrix $\mathbf{B}$. Thus, considering (14.15), (14.17), and (14.18), the input-output relationship of the standard FIR filter [see (14.1)] can be rewritten as

$$
y(n) = \mathbf{w}^{\mathsf{T}}\mathbf{x}(n) = \mathbf{x}^{\mathsf{T}}(n)\mathbf{w} = \{\text{vec}[\mathbf{X}(n)]\}^{\mathsf{T}} \text{vec}(\mathbf{W}) = \text{tr}[\mathbf{X}^{\mathsf{T}}(n)\mathbf{W}]. \tag{14.19}
$$

Now, considering the singular value decomposition (SVD) [8], $\mathbf{W}$ can be represented as

$$
\mathbf{W} = \mathbf{U}\boldsymbol{\Lambda}\mathbf{V}^{\mathsf{T}} \tag{14.20}
$$

where $\boldsymbol{\Lambda}$ denotes an $R \times C$ diagonal matrix composed of the eigenvalues of $\mathbf{W}$, $\mathbf{U}$ is an $R \times R$ unitary matrix whose columns are the eigenvectors of $\mathbf{W}\mathbf{W}^{\mathsf{T}}$, and $\mathbf{V}$ is a $C \times C$ unitary matrix whose columns are the eigenvectors of $\mathbf{W}^{\mathsf{T}}\mathbf{W}$. Then, considering (14.18), (14.20), and the cyclic property of the trace [8], (14.19) can be rewritten as

$$
y(n) = \text{tr}[\mathbf{X}^{\mathsf{T}}(n)\mathbf{U}\boldsymbol{\Lambda}\mathbf{V}^{\mathsf{T}}] = \text{tr}[\mathbf{V}^{\mathsf{T}}\mathbf{X}^{\mathsf{T}}(n)\mathbf{U}\boldsymbol{\Lambda}] = \{\text{vec}[\mathbf{X}(n)\mathbf{V}]\}^{\mathsf{T}} \text{vec}(\mathbf{U}\boldsymbol{\Lambda}). \tag{14.21}
$$

Moreover, considering from [7] that

$$\text{vec}[\mathbf{X}(n)\mathbf{V}] = \text{vec}[\mathbf{I}_R\mathbf{X}(n)\mathbf{V}] = (\mathbf{V}^T \otimes \mathbf{I}_R)\,\text{vec}[\mathbf{X}(n)] \qquad (14.22)$$

(14.21) can be rewritten as

$$y(n) = [(\mathbf{V}^T \otimes \mathbf{I}_R)\mathbf{x}(n)]^T \text{vec}(\mathbf{U}\Lambda) \qquad (14.23)$$

with $\mathbf{I}_R$ representing an $R \times R$ identity matrix. Since we are assuming that $R < C$, Λ is a diagonal rectangular matrix with more columns than rows and, thus, the following $RC \times 1$ coefficient vector can be defined:

$$\tilde{\mathbf{u}} = \text{vec}(\mathbf{U}\Lambda) = [\lambda_0\mathbf{u}_0^T \ \ \lambda_1\mathbf{u}_1^T \ \ \cdots \ \ \lambda_{R-1}\mathbf{u}_{R-1}^T \ \ 0 \ \ \cdots \ \ 0]^T \qquad (14.24)$$

with λ_k and $\mathbf{u}_k$ denoting, respectively, the kth eigenvalue of $\mathbf{W}$ and the kth column of $\mathbf{U}$. Moreover, the following $RC \times 1$ input vector can be defined:

$$\hat{\mathbf{x}}(n) = (\mathbf{V}^T \otimes \mathbf{I}_R)\mathbf{x}(n) = \begin{bmatrix} (\mathbf{v}_0^T \otimes \mathbf{I}_R)\mathbf{x}(n) \\ (\mathbf{v}_1^T \otimes \mathbf{I}_R)\mathbf{x}(n) \\ \vdots \\ (\mathbf{v}_{C-1}^T \otimes \mathbf{I}_R)\mathbf{x}(n) \end{bmatrix} \qquad (14.25)$$

where $\mathbf{v}_k$ is the vector corresponding to the kth column of $\mathbf{V}$. Now, allowing for (14.24) and (14.25), (14.23) can be expressed as

$$y(n) = \tilde{\mathbf{u}}^T\hat{\mathbf{x}}(n) = \tilde{\mathbf{u}}_0^T\hat{\mathbf{x}}_0(n) + \tilde{\mathbf{u}}_1^T\hat{\mathbf{x}}_1(n) + \cdots + \tilde{\mathbf{u}}_{R-1}^T\hat{\mathbf{x}}_{R-1}(n) = \sum_{k=0}^{R-1} \tilde{\mathbf{u}}_k^T\hat{\mathbf{x}}_k(n) \qquad (14.26)$$

with

$$\tilde{\mathbf{u}}_k = \lambda_k\mathbf{u}_k \qquad (14.27)$$

and

$$\hat{\mathbf{x}}_k(n) = (\mathbf{v}_k^T \otimes \mathbf{I}_R)\mathbf{x}(n) = \ddot{\mathbf{v}}_k^T\mathbf{x}(n) \qquad (14.28)$$

for $k = 0, 1, \ldots, R - 1$. Then, similarly to (14.12) and (14.28), one notices that $\hat{\mathbf{x}}_k(n)$ corresponds to the $R \times 1$ output vector of an R-sparse FIR filter whose nonzero elements are given by the elements of $\mathbf{v}_k$, i.e., $\mathbf{v}_k$ is the compact coefficient vector [see (14.8)] of the R-sparse FIR filter. Thus, considering the input-output relationship of the standard FIR filter written as (14.26), one verifies that such a filter can be alternatively implemented by using the structure presented in Figure 14.1. This SVD-based structure, which was originally introduced in [9] and [10], is composed of R parallel branches, with each branch consisting of a cascade of an R-sparse FIR filter $\ddot{\mathbf{v}}_k$ with memory size $N = RC$ and a smaller standard FIR filter $\tilde{\mathbf{u}}_k$ whose memory size is equal to R. It is worth mentioning that, since $\ddot{\mathbf{v}}_k$ and $\tilde{\mathbf{u}}_k$ correspond to linear filters, their positions can be exchanged in each of the branches with no loss of generality.

Note that each of the branches from the SVD-based FIR implementation, shown in Figure 14.1, has $R+C$ nonzero coefficients and, as a result, the whole structure has

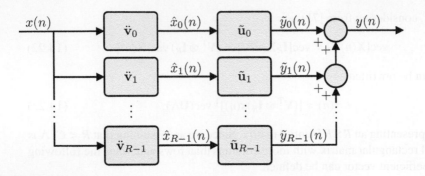

Figure 14.1   Block diagram of the SVD-based FIR implementation.

$R^2 + RC$ coefficients. This number is larger than the number of coefficients of the corresponding standard FIR filter ($N = RC$) and, thus, the direct use of the SVD-based implementation is in principle not interesting in practical applications. However, an appealing characteristic of the SVD-based structure arises from the fact that each $\tilde{\mathbf{u}}_k$ is obtained from the product of $\mathbf{u}_k$ (which is a unit vector since it is one of the eigenvectors of $\mathbf{W}\mathbf{W}^{\mathrm{T}}$) with one of the eigenvalues of $\mathbf{W}$ [see (14.27)]. As a consequence, if $\mathbf{W}$ is a rank-deficient matrix or if some eigenvalues of $\mathbf{W}$ have very small values, some of the branches of the SVD-based structure from Figure 14.1 may have a negligible contribution to the output $y(n)$ of the whole structure. Thus, in many practical applications, these branches can be neglected (i.e., to be removed), leading to an implementation structure with smaller computational complexity than that of the standard FIR filter [9], [11].

14.2.3 SPARSE-INTERPOLATED FIR FILTERS

As previously mentioned, the sparse-interpolated FIR (SIFIR) filter was originally introduced in [1] and since then has been attracting a considerable research interest. The idea behind the SIFIR filter is to use a sparse FIR filter for reducing the complexity along with an interpolator that seeks to compensate for the performance loss arising from the use of a sparseness-based implementation [1], [12]. Figure 14.2 shows the two forms of implementing SIFIR filters, one using the interpolator at the output and another using the interpolator at the input [6]. In this figure, $\ddot{\mathbf{h}}$ represents an L-sparse FIR filter with memory size N whose coefficient vector is

$$\ddot{\mathbf{h}} = [h_0 \; 0 \; \cdots \; h_L \; 0 \; \cdots \; h_{2L} \; 0 \; \cdots \; h_{(N_s-1)L} \; 0 \; \cdots \; 0]^{\mathrm{T}}. \qquad (14.29)$$

Vector $\mathbf{g}$ is the interpolator, which is nothing but an FIR filter with memory size M and coefficient vector given by

$$\mathbf{g} = [g_0 \; g_1 \; g_2 \; \cdots \; g_{M-1}]^{\mathrm{T}}. \qquad (14.30)$$

Also in Figure 14.2, $x(n)$ and $\tilde{y}(n)$ are, respectively, the input and output signals of the SIFIR filter, $\hat{x}(n)$ denotes the input signal filtered by the sparse filter $\ddot{\mathbf{h}}$, and $\tilde{x}(n)$

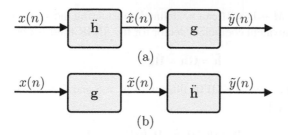

Figure 14.2 Block diagrams of SIFIR filters with (a) the interpolator at the output and (b) the interpolator at the input.

is the input signal filtered by the interpolator **g**.

The similarity between the SIFIR filter and one of the branches of the SVD-based FIR structure from Figure 14.1 is evident since both are composed of a (usually larger) sparse filter followed by a (smaller) standard FIR filter. As a consequence, considering that one branch in the SVD-based structure corresponds to a rank-one implementation of an FIR filter, one can infer that the SIFIR can be seen as a reduced-rank implementation of an FIR filter [11]. However, it is important to notice that the SIFIR filter is not exactly a rank-one implementation due to the fact that the memory size of the interpolator, which is typically given by

$$M = 1 + 2(L - 1) = 2L - 1 \qquad (14.31)$$

is larger than the memory size of the non-sparse filter from a rank-one implementation (i.e., an implementation carried out by using only one branch of the SVD-based FIR structure). This implies that the SIFIR filter is more general than a rank-one implementation, since the latter is in fact a special case of the former [11]. The main reason for choosing the memory size of the interpolator according to (14.31) is to obtain the interpolation characteristic that names the SIFIR filter. The idea is to compensate for the sparseness of **ḧ** by "recreating" the coefficients that are equal to zero as a function of their neighbors. To better understand this point, let us first consider that the equivalent impulse response of a SIFIR structure can be obtained by convolving the impulse responses of the interpolator **g** and the sparse filter **ḧ**. Thus, to obtain the equivalent coefficient vector, we define the following Toeplitz convolution matrix [13], [14]:

$$\mathbf{G} = \left\{ \begin{bmatrix} \mathbf{g} \\ 0 \\ 0 \\ \vdots \\ 0 \end{bmatrix} \begin{bmatrix} 0 \\ \mathbf{g} \\ 0 \\ \vdots \\ 0 \end{bmatrix} \begin{bmatrix} 0 \\ 0 \\ \mathbf{g} \\ \vdots \\ 0 \end{bmatrix} \begin{matrix} \cdots \\ \cdots \\ \cdots \\ \ddots \\ \cdots \end{matrix} \begin{bmatrix} 0 \\ 0 \\ 0 \\ \vdots \\ \mathbf{g} \end{bmatrix} \right\} \qquad (14.32)$$

with dimensions $(N + M - 1) \times N$. Similarly, another Toeplitz convolution matrix **Ḧ** with dimensions $(N + M - 1) \times M$ can be defined considering the sparse coefficient vector $\overset{..}{\mathbf{h}}$. Then, the equivalent coefficient vector for the SIFIR filter is obtained from

$$\bar{\mathbf{h}} = \mathbf{G}\overset{..}{\mathbf{h}} = \overset{..}{\mathbf{H}}\mathbf{g}. \tag{14.33}$$

Now, let us evaluate the case of a SIFIR filter with $N = 5$ and $L = 2$, which results in $M = 3$, $\mathbf{g} = [g_0 \ g_1 \ g_2]^T$,

$$\overset{..}{\mathbf{h}} = [h_0 \ 0 \ h_2 \ 0 \ h_4]^T \tag{14.34}$$

and

$$\bar{\mathbf{h}} = \left[\underline{g_0 h_0} \ \ \underline{g_1 h_0} \ \boxed{g_2 h_0 + g_0 h_2} \ \underline{g_1 h_2} \ \boxed{g_2 h_2 + g_0 h_4} \ \underline{g_1 h_4} \ \underline{g_2 h_4}\right]^T. \tag{14.35}$$

By comparing (14.34) with (14.35), one notices that the interpolator coefficients play distinct roles in the equivalent structure [15]. The central coefficient g_1, for instance, is a weight for the replication of the nonzero coefficients of $\overset{..}{\mathbf{h}}$ in the equivalent coefficient vector $\bar{\mathbf{h}}$, resulting in the coefficients that are underlined in (14.35). On the other hand, g_0 and g_2 are weights for the recreation of the zero coefficients of $\overset{..}{\mathbf{h}}$ as a function of their neighboring coefficients, which can be identified from the boxed coefficients of (14.35). In other words, g_0 and g_2 are the weights for the interpolation process that recreates the zero coefficients of $\overset{..}{\mathbf{h}}$ in $\bar{\mathbf{h}}$. Additionally, the remaining two coefficients in the boundaries of (14.35) result from an incomplete interpolation process that is denoted here as boundary effect [15], [16]. Now, considering the standard practice of using a triangular window for defining the coefficients of the interpolator aiming to obtain linear interpolation [1], [17], one has $\mathbf{g} = [0.5 \ 1 \ 0.5]^T$ and, thus, (14.35) becomes

$$\bar{\mathbf{h}} = \left[\underline{0.5h_0} \ \ \underline{h_0} \ \boxed{0.5h_0 + 0.5h_2} \ \underline{h_2} \ \boxed{0.5h_2 + 0.5h_4} \ \underline{h_4} \ \underline{0.5h_4}\right]^T. \tag{14.36}$$

From this equation, one can clearly see the aforementioned characteristics of the equivalent interpolated coefficient vector, i.e., the replication of the nonzero coefficients of $\overset{..}{\mathbf{h}}$ (underlined ones), the recreation of the zeroed coefficients of $\overset{..}{\mathbf{h}}$ (boxed ones), and the coefficients resulting from the boundary effect (remaining ones).

From (14.33), it is also easy to see that the equivalent coefficient vector of the SIFIR filter has a dimension $N + M - 1$. Thus, defining an extended input vector with $N + M - 1$ samples of the input signal as

$$\bar{\mathbf{x}}(n) = [x(n) \ x(n-1) \ x(n-2) \ \cdots \ x(n-N-M+2)]^T \tag{14.37}$$

one can write the input-output relationship of the SIFIR filter as

$$\tilde{y}(n) = \bar{\mathbf{h}}^T\bar{\mathbf{x}}(n) = \overset{..}{\mathbf{h}}^T\mathbf{G}^T\bar{\mathbf{x}}(n) = \mathbf{g}^T\overset{..}{\mathbf{H}}^T\bar{\mathbf{x}}(n). \tag{14.38}$$

Alternatively, one can define a vector of samples of the input signal filtered by the interpolator as

$$\tilde{\mathbf{x}}(n) = \mathbf{G}^T\bar{\mathbf{x}}(n) \tag{14.39}$$

and another vector of samples of the input signal filtered by the sparse filter as

$$\hat{\mathbf{x}}(n) = \ddot{\mathbf{H}}^T \bar{\mathbf{x}}(n) \tag{14.40}$$

and then rewrite (14.38) as

$$\tilde{y}(n) = \bar{\mathbf{h}}^T \bar{\mathbf{x}}(n) = \ddot{\mathbf{h}}^T \tilde{\mathbf{x}}(n) = \mathbf{g}^T \hat{\mathbf{x}}(n). \tag{14.41}$$

14.3 ADAPTIVE SPARSE-INTERPOLATED FIR FILTERS

In the previous section, the foundations of the SIFIR filters have been described, showing that such filters have a reduced number of coefficients in comparison with the standard FIR filters. This characteristic is particularly interesting in adaptive filtering applications in which the update of the values of all filter coefficients is usually required for each new sample of the input signal. In this context, this section is devoted to the description of different forms of implementation of adaptive filters based on the SIFIR structure. These filters are depicted here considering the normalized least-mean-square (NLMS) algorithm for updating their coefficients due to the very good tradeoff between complexity and performance presented by such an algorithm [3].

14.3.1 BACKGROUND ON THE NLMS ALGORITHM

Generally speaking, the goal of an adaptive filter is to minimize some convex function of an error signal, which for a standard FIR filter is defined as

$$e(n) = d(n) - y(n) = d(n) - \mathbf{w}^T(n)\mathbf{x}(n) \tag{14.42}$$

with $d(n)$ representing a desired or reference signal [3]. For system identification applications (such as echo cancellation), the desired signal $d(n)$ is the output of the plant (system to be modeled). Then, an arbitrarily accurate model for the plant can be obtained by somehow minimizing the energy of $e(n)$. In the case of the NLMS algorithm, such a modeling task is carried out in an iterative fashion by looking for a coefficient vector that makes $e(n)$ equal to zero considering the data available at each algorithm iteration, i.e., considering the instantaneous values of $d(n)$ and $\mathbf{x}(n)$. In other words, the NLMS searches for a coefficient vector $\mathbf{w}(n + 1)$ that makes the a posteriori error

$$\varepsilon(n) = d(n) - \mathbf{w}^T(n + 1)\mathbf{x}(n) \tag{14.43}$$

equal to zero at each algorithm iteration [3]. This search is carried out considering, as a reference, the current coefficient vector $\mathbf{w}(n)$. In this context, the goal is to obtain the closest coefficient vector (in terms of Euclidean distance) with respect to $\mathbf{w}(n)$ that leads to $\varepsilon(n) = 0$, which is achieved solving a minimization problem defined as

$$\begin{aligned} &\text{minimize } \|\mathbf{w}(n + 1) - \mathbf{w}(n)\|^2 \\ &\text{subject to } \varepsilon(n) = d(n) - \mathbf{w}^T(n + 1)\mathbf{x}(n) = 0. \end{aligned} \tag{14.44}$$

Then, considering the Lagrange multiplier method to solve such a problem, the following expression for updating the coefficients of a standard FIR filter using the NLMS algorithm can be obtained [3]:

$$\mathbf{w}(n+1) = \mathbf{w}(n) + \alpha \frac{e(n)}{\|\mathbf{x}(n)\|^2 + \psi} \mathbf{x}(n) \tag{14.45}$$

where α is the step-size control parameter and ψ represents a regularization parameter [used to avoid divisions by very small values in (14.45)].

14.3.2   STANDARD ADAPTIVE SPARSE-INTERPOLATED FIR FILTERS

The simplest adaptive implementations of SIFIR filters are carried out by using an adaptive sparse filter along with a fixed interpolator [6], [18]. In these implementations, called here adaptive SIFIR (ASIFIR) filters, the coefficients of the interpolator are set with arbitrary values (e.g., using a triangular window to obtain linear interpolation [17]) whereas the coefficients of the sparse filter are updated to iteratively minimize the energy of an error signal defined as

$$e(n) = d(n) - \tilde{y}(n) = d(n) - \ddot{\mathbf{h}}^{\mathrm{T}}(n)\mathbf{G}^{\mathrm{T}}\bar{\mathbf{x}}(n). \tag{14.46}$$

In this context, aiming to derive an update expression based on the NLMS algorithm, the idea is to look for a sparse coefficient vector $\ddot{\mathbf{h}}(n+1)$ that leads to an a posteriori error

$$\varepsilon(n) = d(n) - \ddot{\mathbf{h}}^{\mathrm{T}}(n+1)\mathbf{G}^{\mathrm{T}}\bar{\mathbf{x}}(n) \tag{14.47}$$

equal to zero at each algorithm iteration. Thus, the following minimization problem is defined:

$$\text{minimize } \|\ddot{\mathbf{h}}(n+1) - \ddot{\mathbf{h}}(n)\|^2$$

$$\text{subject to } \begin{cases} \varepsilon(n) = d(n) - \ddot{\mathbf{h}}^{\mathrm{T}}(n+1)\mathbf{G}^{\mathrm{T}}\bar{\mathbf{x}}(n) = 0 \\ \mathbf{C}^{\mathrm{T}}\ddot{\mathbf{h}}(n+1) = \mathbf{0}. \end{cases} \tag{14.48}$$

Note that this minimization problem has, in addition to the zero-error constraint, a second constraint given by $\mathbf{C}^{\mathrm{T}}\ddot{\mathbf{h}}(n+1) = \mathbf{0}$ that compels $\ddot{\mathbf{h}}(n+1)$ to be sparse, with $\mathbf{C}$ representing the constraint matrix and $\mathbf{0}$ a vector of zeros [17], [19]. Now, solving this problem in a similar way as (14.44), the following update equation for the NLMS ASIFIR filter is obtained:

$$\ddot{\mathbf{h}}(n+1) = \mathbf{P}\ddot{\mathbf{h}}(n) + \alpha_{\mathrm{h}} \frac{e(n)}{\|\mathbf{P}\mathbf{G}^{\mathrm{T}}\bar{\mathbf{x}}(n)\|^2 + \psi_{\mathrm{h}}} \mathbf{P}\mathbf{G}^{\mathrm{T}}\bar{\mathbf{x}}(n) \tag{14.49}$$

where α_{h} and ψ_{h} denote, respectively, the step size and the regularization parameter [similarly to α and ψ in (14.45)]. Moreover, $\mathbf{P} = \mathbf{I}_N - \mathbf{C}\mathbf{C}^{\mathrm{T}}$ is a diagonal matrix with dimension $N \times N$ and diagonal elements equal to either 0 or 1, whereas $\mathbf{I}_N$ represents the $N \times N$ identity matrix. Note that (14.49) depends on the evaluation of $\mathbf{G}^{\mathrm{T}}\bar{\mathbf{x}}(n)$, which can be rewritten as

$$\tilde{\mathbf{x}}(n) = \mathbf{G}^{\mathrm{T}}\bar{\mathbf{x}}(n) = [\mathbf{g}^{\mathrm{T}}\mathbf{x}_g(n) \quad \mathbf{g}^{\mathrm{T}}\mathbf{x}_g(n-1) \quad \cdots \quad \mathbf{g}^{\mathrm{T}}\mathbf{x}_g(n-N+1)]^{\mathrm{T}} \tag{14.50}$$

Table 14.1
NLMS ASIFIR Algorithm

Initialization

$$\ddot{\mathbf{h}}(0) = [0 \ \ 0 \ \ \cdots \ \ 0]^{\mathrm{T}}$$

$\mathbf{g} = \Delta(M)$, with $\Delta(M)$ representing a triangular window of size M

Do for $n \geq 0$

$$e(n) = d(n) - \ddot{\mathbf{h}}^{\mathrm{T}}(n)\tilde{\mathbf{x}}(n)$$

$$\ddot{\mathbf{h}}(n+1) = \mathbf{P}\ddot{\mathbf{h}}(n) + \alpha_{\mathrm{h}} \frac{e(n)}{\|\mathbf{P}\tilde{\mathbf{x}}(n)\|^2 + \psi_{\mathrm{h}}} \mathbf{P}\tilde{\mathbf{x}}(n)$$

with

$$\mathbf{x}_{\mathrm{g}}(n) = [x(n) \ \ x(n-1) \ \ \cdots \ \ x(n-M+1)]^{\mathrm{T}}. \tag{14.51}$$

From (14.50), it is easy to verify that the cost for evaluating $\tilde{\mathbf{x}}(n) = \mathbf{G}^{\mathrm{T}}\bar{\mathbf{x}}(n)$ at each iteration can be substantially reduced by exploiting data reuse, i.e., evaluating only $\mathbf{g}^{\mathrm{T}}\mathbf{x}_{\mathrm{g}}(n)$ at each iteration and reusing the result of such a product in the subsequent iterations to obtain the remaining elements of (14.50). In doing so, (14.49) can be simplified to

$$\ddot{\mathbf{h}}(n+1) = \mathbf{P}\ddot{\mathbf{h}}(n) + \alpha_{\mathrm{h}} \frac{e(n)}{\|\mathbf{P}\tilde{\mathbf{x}}(n)\|^2 + \psi_{\mathrm{h}}} \mathbf{P}\tilde{\mathbf{x}}(n). \tag{14.52}$$

It is worth mentioning that the matrix products that still need to be evaluated in (14.52) [i.e., $\mathbf{P}$ with either $\ddot{\mathbf{h}}(n)$ or $\tilde{\mathbf{x}}(n)$] do not result in significant computational cost due to the particular structure of $\mathbf{P}$ (diagonal matrix with elements equal to either 0 or 1). Thus, such products simply imply that some elements of the involved vectors are either set to zero or kept equal to zero during the adaptive process.

Note that the update expression given by (14.52) can be used to update the sparse filter in any of the two equivalent SIFIR implementations shown in Figure 14.2 (i.e., with the interpolator at the output or at the input). Yet, it is shown in [6] that the use of the SIFIR implementation with the interpolator at the input will result in a smaller computational burden. This is due to the fact that, with the interpolator at the output, $\hat{\mathbf{x}}(n)$ has to be evaluated to obtain $e(n)$, and $\tilde{\mathbf{x}}(n)$ also has to be evaluated for updating the coefficients using (14.52). In contrast, using the interpolator at the input, only $\tilde{\mathbf{x}}(n)$ has to be evaluated and then used for both evaluating $e(n)$ and updating the coefficients using (14.52), thus leading to smaller computational burden. Table 14.1 summarizes the steps for implementing the NLMS algorithm in the ASIFIR filter. Note that, in this case, the interpolator is initialized with a triangular window of size M, since this leads to a linear interpolation in the coefficient recreation process. Other types of interpolators can be used depending on the previous knowledge on the characteristics of the application considered.

14.3.3 FULLY ADAPTIVE SPARSE-INTERPOLATED FIR FILTERS

One of the main concerns in the implementation of ASIFIR filters is the choice of the coefficients of the interpolator. This is due to the fact that such coefficients are kept unchanged during the adaptive process and, as a consequence, a bad initial choice of their values will inevitably lead to poor performance [20], [21]. Fully adaptive SIFIR (FASIFIR) filters have been developed aiming to overcome this drawback [13], [20], [21], [22]. In these filters, both the interpolator and the sparse filter are adaptive, allowing an automatic adjustment of all coefficients of the SIFIR structure during the adaptive process [5].

The implementation of a FASIFIR filter involves the adaptation of two cascaded filters (interpolator and sparse filter). In this context, two optimization problems have to be defined for obtaining the update equations for the NLMS algorithm [21]. The first, for the sparse filter, is similar to (14.48) but with a time-varying version $\mathbf{G}(n)$ of matrix $\mathbf{G}$. Thus, one has

$$\text{minimize } \|\ddot{\mathbf{h}}(n + 1) - \ddot{\mathbf{h}}(n)\|^2$$

$$\text{subject to } \begin{cases} \varepsilon_h(n) = d(n) - \ddot{\mathbf{h}}^T(n + 1)\mathbf{G}^T(n)\bar{\mathbf{x}}(n) = 0 \\ \mathbf{C}^T\ddot{\mathbf{h}}(n + 1) = \mathbf{0}. \end{cases} \quad (14.53)$$

On the other hand, for the interpolator, one has

$$\text{minimize } \|\mathbf{g}(n + 1) - \mathbf{g}(n)\|^2$$

$$\text{subject to } \varepsilon_g(n) = d(n) - \mathbf{g}^T(n + 1)\ddot{\mathbf{H}}^T(n)\bar{\mathbf{x}}(n) = 0. \quad (14.54)$$

By solving (14.53), one obtains

$$\ddot{\mathbf{h}}(n + 1) = \mathbf{P}\ddot{\mathbf{h}}(n) + \alpha_h \frac{e(n)}{\|\mathbf{P}\mathbf{G}^T(n)\bar{\mathbf{x}}(n)\|^2 + \psi_h}\mathbf{P}\mathbf{G}^T(n)\bar{\mathbf{x}}(n) \quad (14.55)$$

whereas, solving (14.54), one gets

$$\mathbf{g}(n + 1) = \mathbf{g}(n) + \alpha_g \frac{e(n)}{\|\ddot{\mathbf{H}}^T(n)\bar{\mathbf{x}}(n)\|^2 + \psi_g}\ddot{\mathbf{H}}^T(n)\bar{\mathbf{x}}(n). \quad (14.56)$$

Note that the coefficient updating by using both (14.55) and (14.56) requires the evaluation of two matrix products, i.e., $\mathbf{G}^T(n)\bar{\mathbf{x}}(n)$ and $\ddot{\mathbf{H}}^T(n)\bar{\mathbf{x}}(n)$, thus resulting in a high computational cost for practical implementations. To circumvent this problem, a slow adaptation condition is assumed to allow the evaluation of such matrix products with a reduced computational load exploiting data reuse [21]. For instance, under slow adaptation condition, the product $\mathbf{G}^T(n)\bar{\mathbf{x}}(n)$, which is given by

$$\mathbf{G}^T(n)\bar{\mathbf{x}}(n) = [\mathbf{g}^T(n)\mathbf{x}_g(n) \ \mathbf{g}^T(n)\mathbf{x}_g(n - 1) \ \cdots \ \mathbf{g}^T(n)\mathbf{x}_g(n - N + 1)]^T \quad (14.57)$$

can be approximated to

$$\tilde{\mathbf{x}}'(n) = [\mathbf{g}^T(n)\mathbf{x}_g(n) \ \mathbf{g}^T(n - 1)\mathbf{x}_g(n - 1) \ \cdots$$
$$\mathbf{g}^T(n - N + 1)\mathbf{x}_g(n - N + 1)]^T \cong \mathbf{G}^T(n)\bar{\mathbf{x}}(n). \quad (14.58)$$

In doing so, only the first element of $\tilde{\mathbf{x}}'(n)$ needs to be evaluated at each iteration, whereas the remaining ones are obtained reusing the elements evaluated in previous iterations. A similar approximation can be carried out to evaluate $\ddot{\mathbf{H}}^{\mathrm{T}}(n)\bar{\mathbf{x}}(n)$, resulting in

$$\hat{\mathbf{x}}'(n) = [\ddot{\mathbf{h}}^{\mathrm{T}}(n)\mathbf{x}(n) \quad \ddot{\mathbf{h}}^{\mathrm{T}}(n-1)\mathbf{x}(n-1) \quad \cdots$$
$$\ddot{\mathbf{h}}^{\mathrm{T}}(n-N+1)\mathbf{x}(n-N+1)]^{\mathrm{T}} \cong \ddot{\mathbf{H}}^{\mathrm{T}}(n)\bar{\mathbf{x}}(n). \quad (14.59)$$

Thus, considering $\tilde{\mathbf{x}}'(n)$ and $\hat{\mathbf{x}}'(n)$ in lieu of $\mathbf{G}^{\mathrm{T}}(n)\bar{\mathbf{x}}(n)$ and $\ddot{\mathbf{H}}^{\mathrm{T}}(n)\bar{\mathbf{x}}(n)$, respectively, (14.55) is rewritten as

$$\ddot{\mathbf{h}}(n+1) = \mathbf{P}\ddot{\mathbf{h}}(n) + \alpha_{\mathrm{h}} \frac{e(n)}{\|\mathbf{P}\tilde{\mathbf{x}}'(n)\|^2 + \psi_{\mathrm{h}}} \mathbf{P}\tilde{\mathbf{x}}'(n) \quad (14.60)$$

and (14.56) as

$$\mathbf{g}(n+1) = \mathbf{g}(n) + \alpha_{\mathrm{g}} \frac{e(n)}{\|\hat{\mathbf{x}}'(n)\|^2 + \psi_{\mathrm{g}}} \hat{\mathbf{x}}'(n). \quad (14.61)$$

Besides the aforementioned approximations for the evaluation of $\mathbf{G}^{\mathrm{T}}(n)\bar{\mathbf{x}}(n)$ and $\ddot{\mathbf{H}}^{\mathrm{T}}(n)\bar{\mathbf{x}}(n)$, another important aspect to be considered, aiming to obtain FASIFIR filters with reduced computational burden, is the position of the interpolator in the structure. As mentioned in the previous section, implementations of the ASIFIR filter with the interpolator at the input present a smaller computational burden. In contrast, for the case of the FASIFIR filter, a smaller computational cost is obtained with the interpolator at the output [16]. This can be easily verified first considering that, regardless of the interpolator position, both $\hat{\mathbf{x}}'(n)$ and $\tilde{\mathbf{x}}'(n)$ have to be computed at each iteration in order to update the coefficients using (14.60) and (14.61). Moreover, the output signal $\tilde{y}(n)$ of the SIFIR structure also has to be evaluated at each iteration, resulting in $\tilde{y}(n) \cong \ddot{\mathbf{h}}^{\mathrm{T}}(n)\tilde{\mathbf{x}}'(n)$ for the SIFIR structure with the interpolator at the input and $\tilde{y}(n) \cong \mathbf{g}^{\mathrm{T}}(n)\hat{\mathbf{x}}'(n)$ for the structure with the interpolator at the output. Since $\mathbf{g}(n)$ usually has fewer coefficients than $\ddot{\mathbf{h}}(n)$, the evaluation of the output signal using the SIFIR structure with the interpolator at the output has a smaller computational load than that of the structure with the interpolator at the input. A comparison between the computational complexity required by ASIFIR and the FASIFIR filters implemented using the different SIFIR structures (with the interpolator either at the output or at the input) is shown in Figure 14.3 considering $L = 2$. From this figure, one observes that the FASIFIR filter with the interpolator at the output presents a computational burden that is very close to that required by the ASIFIR filter with the interpolator at the input. Table 14.2 summarizes the steps for implementing the NLMS FASIFIR filter with the interpolator at the output.

14.3.4 ADAPTIVE FIR-SIFIR FILTERS

In many adaptive filtering applications, the modeling of impulse responses having non-uniform characteristics is required. The echo response found in some digital

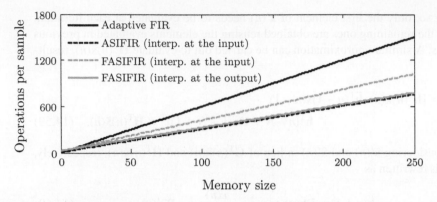

Figure 14.3 Computational complexity comparison between the adaptive FIR filter and different adaptive implementations of SIFIR filters with $L = 2$.

Table 14.2
NLMS FASIFIR Algorithm

Initialization

$$\ddot{\mathbf{h}}(0) = [0 \quad 0 \quad \cdots \quad 0]^{\mathrm{T}}$$

$$\mathbf{g}(0) \neq \mathbf{0}$$

Do for $n \geq 0$

$$e(n) = d(n) - \mathbf{g}^{\mathrm{T}}(n)\hat{\mathbf{x}}'(n)$$

$$\ddot{\mathbf{h}}(n+1) = \mathbf{P}\ddot{\mathbf{h}}(n) + \alpha_{\mathrm{h}}\frac{e(n)}{\|\mathbf{P}\tilde{\mathbf{x}}'(n)\|^2 + \psi_{\mathrm{h}}}\mathbf{P}\tilde{\mathbf{x}}'(n)$$

$$\mathbf{g}(n+1) = \mathbf{g}(n) + \alpha_{\mathrm{g}}\frac{e(n)}{\|\hat{\mathbf{x}}'(n)\|^2 + \psi_{\mathrm{g}}}\hat{\mathbf{x}}'(n)$$

---

subscriber line (DSL) systems, which is illustrated in Figure 14.4, is a very good example of this type of impulse response since it exhibits a short and rapidly-changing initial segment (the head) followed by a slowly-decaying final part (the tail) [23], [24]. An effective modeling of such a response can be obtained with a reduced computational cost by using a less-complex SIFIR filter (to model the smooth tail segment) in association with an FIR filter (to satisfactorily model the irregular head segment) [25]. The basic structure of the hybrid FIR-SIFIR filter resulting from such an association is illustrated in Figure 14.5. In this structure, by properly choosing the memory sizes of the filters and the value of the delay parameter, one can obtain an effective FIR-SIFIR filter that models non-uniform echo responses as the aforementioned one in a very accurate way. The FIR-SIFIR approach is extended in [26] by splitting the tail segment in smaller parts and using SIFIR filters with different sparseness levels to model each of these parts, which leads to an FIR-SIFIR-SIFIR structure with an even smaller computational complexity. Moreover, responses with characteristics similar to those of the response shown in Figure 14.4 are also found in the Ethernet systems described in [27]. There, FIR-SIFIR as well as three-stage SIFIR-FIR-SIFIR filters are considered for canceling both echo and near-end cross talk.

The FIR-SIFIR filter has been originally introduced in [23]. There, the focus is on the use of a SIFIR filter for canceling the echo generated by the tail part of the echo response in a DSL system. However, the problems arising from the association of a standard FIR filter and a SIFIR filter are overlooked in [23]. The most important of these problems is an outcome of the aforementioned boundary effect, which produces an undesirable ramp shape characteristic in the initial and in the final parts of the SIFIR impulse response [24], [28]. As a consequence, when FIR and SIFIR filters are associated to form an FIR-SIFIR structure, the boundary effect produces a distortion in the impulse response that often leads to a significant loss of performance [28]. Such a distortion can be clearly seen in the impulse response of an FIR-SIFIR filter illustrated in Figure 14.6, in which a mismatch between the echo response being modeled (gray line) and that obtained by using the FIR (dotted line) and SIFIR (dashed line) filters is noticeable. Aiming to offset such a mismatch, the approach introduced in [24] uses an FIR filter with larger memory size, which leads to an overlap between the FIR and SIFIR impulse responses and compensates for the mismatch caused by the boundary effect (see Figure 14.7). As a result, a more effective FIR-SIFIR implementation is obtained at the cost of an increase of computational burden. Another more effective FIR-SIFIR implementation is introduced in [28], in which a procedure for removing the boundary effect is used for the sake of performance improvement. Such implementation is discussed in the next section.

14.3.5 REMOVING THE BOUNDARY EFFECT

As described in [15] and [16], one important aspect that impairs the performance of sparse-interpolated filters is the boundary effect. Such an effect, whose impact on the performance of FIR-SIFIR filters has been described in detail in the previous section, is a characteristic that arises from the convolution of the impulse responses of the two components (interpolator and sparse filter) of a SIFIR filter. For instance, in the case of a SIFIR filter with $N = 5$, $L = 2$, and $\mathbf{g} = [g_0 \ g_1 \ g_2]^T$, the equivalent coefficient vector is given by

$$\bar{\mathbf{h}} = \left[g_0 h_0 \ \underline{g_1 h_0} \ \boxed{g_2 h_0 + g_0 h_2} \ \underline{g_1 h_2} \ \boxed{g_2 h_2 + g_0 h_4} \ \underline{g_1 h_4} \ g_2 h_4 \right]^T. \quad (14.62)$$

As described in Section 14.2.3, the underlined coefficients of (14.62) are those replicated from the sparse kernel, the boxed coefficients are the ones recreated by interpolation, and the remaining ones in the boundaries correspond to the boundary effect. Thus, it is easy to notice that a version with removed boundary effect of (14.62) is

$$\check{\mathbf{h}} = \left[\underline{g_1 h_1} \ \boxed{g_2 h_0 + g_0 h_2} \ \underline{g_1 h_2} \ \boxed{g_2 h_2 + g_0 h_4} \ \underline{g_1 h_4} \right]^T. \quad (14.63)$$

As discussed in [16], (14.63) can be obtained by pre-multiplying (14.62) by a transformation matrix $\mathbf{T}$ that eliminates the elements corresponding to the boundary effect. By considering that in general the first $L - 1$ and the last $L - 1$ elements of $\bar{\mathbf{h}}$

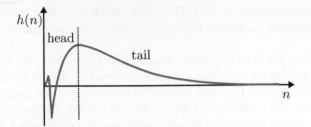

Figure 14.4 Illustration of an impulse response having a short rapidly-changing head segment followed by a slowly-decaying tail segment.

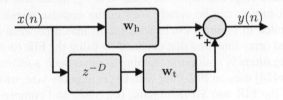

Figure 14.5 A typical block diagram of an FIR-SIFIR structure.

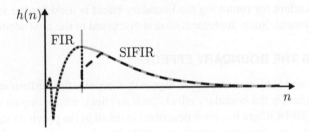

Figure 14.6 Impulse response of a conventional FIR-SIFIR structure.

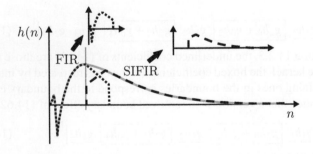

Figure 14.7 Impulse response of a filter based on the FIR-SIFIR structure from [24].

correspond to the boundary effect, such an $N \times (N + M - 1)$ transformation matrix is defined as

$$
\mathbf{T} = \begin{bmatrix}
0 & \cdots & 0 & 1 & 0 & \cdots & 0 & 0 & \cdots & 0 \\
0 & \cdots & 0 & 0 & 1 & \cdots & 0 & 0 & \cdots & 0 \\
0 & \cdots & 0 & \vdots & \vdots & \ddots & \vdots & 0 & \cdots & 0 \\
0 & \cdots & 0 & 0 & 0 & \cdots & 1 & 0 & \cdots & 0
\end{bmatrix}.
$$

$$\underbrace{}_{L-1 \text{ columns}} \quad \underbrace{}_{N \text{ columns}} \quad \underbrace{}_{L-1 \text{ columns}} \tag{14.64}$$

Thus, we have

$$\check{\mathbf{h}} = \mathbf{T}\bar{\mathbf{h}} = \mathbf{TG}\check{\mathbf{h}} = \mathbf{T}\ddot{\mathbf{H}}\mathbf{g} \tag{14.65}$$

and the input-output relationship of the SIFIR filter with removed boundary effect can be written as

$$y(n) = \check{\mathbf{h}}^{\mathrm{T}}\mathbf{x}(n) = \check{\mathbf{h}}^{\mathrm{T}}\mathbf{G}^{\mathrm{T}}\mathbf{T}^{\mathrm{T}}\mathbf{x}(n) = \mathbf{g}^{\mathrm{T}}\ddot{\mathbf{H}}^{\mathrm{T}}\mathbf{T}^{\mathrm{T}}\mathbf{x}(n). \tag{14.66}$$

Now, defining two transformed versions of (14.39) and (14.40) as

$$\check{\mathbf{x}}(n) = \mathbf{G}^{\mathrm{T}}\mathbf{T}^{\mathrm{T}}\mathbf{x}(n) \tag{14.67}$$

and

$$\acute{\mathbf{x}}(n) = \ddot{\mathbf{H}}^{\mathrm{T}}\mathbf{T}^{\mathrm{T}}\mathbf{x}(n) \tag{14.68}$$

respectively, (14.66) can be rewritten as

$$y(n) = \check{\mathbf{h}}^{\mathrm{T}}\mathbf{x}(n) = \ddot{\mathbf{h}}^{\mathrm{T}}\check{\mathbf{x}}(n) = \mathbf{g}^{\mathrm{T}}\acute{\mathbf{x}}(n). \tag{14.69}$$

By comparing (14.41) and (14.69), one can notice that the latter is obtained by replacing $\tilde{\mathbf{x}}(n)$ and $\hat{\mathbf{x}}(n)$ by their modified versions $\check{\mathbf{x}}(n)$ and $\acute{\mathbf{x}}(n)$ in the former. As shown in [16], such a replacement results in an increment of computational burden of $2L - 2$ operations per sample. Thus, one observes that the boundary effect removal can be carried out with relatively small computational cost since L is usually small, resulting in a SIFIR implementation with removed boundary effect denoted here by BSIFIR.

The derivation of adaptive algorithms for updating the coefficients of BSIFIR filters is carried out by using steps similar to those described in Sections 14.3.2 and 14.3.3 to derive the ASIFIR and FASIFIR algorithms. In doing so, one can obtain an update expression given by

$$\check{\mathbf{h}}(n + 1) = \mathbf{P}\check{\mathbf{h}}(n) + \alpha_{\mathrm{h}} \frac{e(n)}{\|\mathbf{P}\check{\mathbf{x}}(n)\|^2 + \psi_{\mathrm{h}}} \mathbf{P}\check{\mathbf{x}}(n) \tag{14.70}$$

for the sparse filter of an adaptive BSIFIR (ABSIFIR) implementation in which only such a filter is updated. Moreover, in a fully adaptive BSIFIR (FABISIFIR) implementation, along with (14.70), one has the following update equation for the interpolator:

$$\mathbf{g}(n + 1) = \mathbf{g}(n) + \alpha_{\mathrm{g}} \frac{e(n)}{\|\acute{\mathbf{x}}(n)\|^2 + \psi_{\mathrm{g}}} \acute{\mathbf{x}}(n). \tag{14.71}$$

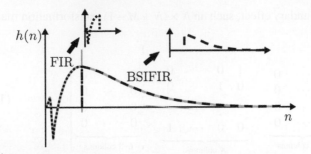

Figure 14.8   Impulse response of an FIR-BSIFIR structure.

It is important to mention that approximations for the assessment of the input vectors similar to those described for the FASIFIR filters must be considered for evaluating the input vectors $\breve{x}(n)$ and $\hat{x}(n)$ of the FABSIFIR filters. Otherwise, the computational burden becomes larger than that required by a standard FIR filter, hampering the use of the FABSIFIR implementations.

As shown in [28], the adaptive filters based on the BSIFIR structure are especially interesting for the implementation of hybrid FIR-SIFIR filters. This is due to the fact that, using the BSIFIR implementation, the memory size of the FIR filter does not need to be increased to offset the boundary effect as in the case of the FIR-SIFIR implementation from [24]. Moreover, since in the FIR-SIFIR implementations the removal of the boundary effect is usually only required in the head segment of the SIFIR impulse response, the cost of implementing the BSIFIR filter becomes equal to that of implementing a standard SIFIR filter [28]. As a result, one can obtain an effective FIR-BSIFIR filter presenting a better tradeoff between computational complexity and performance as compared with the other FIR-SIFIR implementations from the literature. The impulse response obtained by using such an FIR-BSIFIR filter is illustrated in Figure 14.8.

14.3.6   IMPROVED FULLY ADAPTIVE SPARSE-INTERPOLATED FIR FILTERS

From the discussion regarding adaptive SIFIR filters presented in this section, one can see that the ASIFIR-based implementations (ASIFIR and ABSIFIR) have a smaller computational complexity while the FASIFIR ones (FASIFIR and FABSIFIR) usually lead to superior performance. Since the difference of complexity between such implementations is usually small (see Figure 14.3), the use of. FASIFIR implementations is often preferred to the ASIFIR. However, the adaptation of FASIFIR filters involves a simultaneous update of two different cascaded structures, namely the interpolator and sparse filter. Such a kind of adaptation process exhibits certain drawbacks that are not common to simpler adaptive processes such as that of the ASIFIR filter [29], [30]. The first of these drawbacks is the freezing of the adaptive algorithm if both the interpolator and the sparse filter have all coefficients equal to zero. This can be easily seen in the case of the FABSIFIR filter by noticing that, if all coefficients are equal to zero, $\mathbf{G}$ and $\ddot{\mathbf{H}}$ become zero matrices and thus $\check{\mathbf{x}}(n)$ and $\acute{\mathbf{x}}(n)$ become zero vectors. Therefore, the coefficients remain unchanged since (14.70) and (14.71) result in $\ddot{\mathbf{h}}(n + 1) = \mathbf{P}\ddot{\mathbf{h}}(n)$ and $\mathbf{g}(n + 1) = \mathbf{g}(n)$, respectively. A second and more troublesome convergence issue of FASIFIR filters is related to the values attained by the coefficients of vectors $\mathbf{g}$ and $\ddot{\mathbf{h}}$ after the filter has converged. Note that the equivalent coefficient vectors $\bar{\mathbf{h}}$ (for the SIFIR structure) and $\check{\mathbf{h}}$ (for the BSIFIR structure) depend on $\mathbf{g}$ and $\ddot{\mathbf{h}}$ as described in (14.33) and (14.65), respectively. From such expressions, one notices that the use of coefficient vectors $\mathbf{g}_a = c\mathbf{g}$ and $\ddot{\mathbf{h}}_a = (1/c)\ddot{\mathbf{h}}$, with c denoting an arbitrary scalar value, results in the same equivalent coefficient vector. This can be easily observed by writing (14.65) as

$$\check{\mathbf{h}} = \mathbf{TG}\ddot{\mathbf{h}} = \mathbf{TG}_a\ddot{\mathbf{h}}_a = \mathbf{T}(c\mathbf{G})(\ddot{\mathbf{h}}/c). \tag{14.72}$$

Then, considering (14.72), one can notice that there are infinite combinations of $\mathbf{g}$ and $\ddot{\mathbf{h}}$ that result in the same $\check{\mathbf{h}}$. As a consequence, it is possible to obtain a given $\check{\mathbf{h}}$ from a vector $\mathbf{g}_a$ with small-valued elements and a vector $\ddot{\mathbf{h}}$ with large-valued ones (or vice versa). Such a characteristic is especially undesirable in implementations using fixed-point arithmetic. In these cases, small values may imply loss of precision and large values may exceed the maximum value supported by the adopted numerical representation (overflow). Thus, one has an important threat for both numerical stability and algorithm convergence. To overcome this problem, an alternative strategy for implementing FABSIFIR filters has been proposed in [29]. The main idea behind such a strategy is to exploit the role of the central coefficient of the interpolator in the FABSIFIR structure. From (14.62) and (14.63), one can notice that this coefficient has the specific task of reproducing the coefficients of the sparse filter in the equivalent structure. Thus, during the adaptive process, such a coefficient can be fixed to an arbitrary value k without loss of generality in the equivalent structure. Such a strategy prevents the previously mentioned stalling of the adaptive process, since the interpolator coefficient vector $\mathbf{g}$ always presents at least one nonzero coefficient. Additionally, the numerical stability issue is also overcome, since a given $\check{\mathbf{h}}$ can be obtained from two different combinations of $\ddot{\mathbf{h}}$ and the interpolator coefficient vector with central coefficient fixed to k (denoted here by $\mathbf{g}_{[k]}$), with both combinations having vectors with equal norm values. Such a feature is discussed in detail in [29].

In the improved FABSIFIR (IFABSIFIR) implementation obtained by setting the central coefficient of the interpolator to k, the update of the coefficients of the sparse filter is carried out in the same way as in the FABSIFIR implementation described in the previous section, i.e., using (14.70). On the other hand, for obtaining the update equation for the interpolator coefficients considering the NLMS algorithm, the following optimization problem is defined:

$$\text{minimize } \|\mathbf{g}_{[k]}(n + 1) - \mathbf{g}_{[k]}(n)\|^2$$

$$\text{subject to } \begin{cases} \varepsilon_g(n) = d(n) - \mathbf{g}_{[k]}^{\mathrm{T}}(n + 1)\ddot{\mathbf{H}}^{\mathrm{T}}(n)\mathbf{T}^{\mathrm{T}}\mathbf{x}(n) = 0 \\ \mathbf{c}_g^{\mathrm{T}}\mathbf{g}_{[k]} = k \end{cases} \tag{14.73}$$

where $\mathbf{c}_g$ is an $M \times 1$ constraint vector having the central element equal to 1 and the remaining ones equal to zero. Then, solving this problem using the Lagrange multiplier method, one gets

$$\mathbf{g}_{[k]}(n+1) = \mathbf{P}_g \mathbf{g}_{[k]}(n) + \frac{\alpha_g}{\left\| \mathbf{P}_g \acute{\mathbf{x}}(n) \right\|^2 + \psi_g} e(n) \mathbf{P}_g \acute{\mathbf{x}}(n) + k \mathbf{c}_g \qquad (14.74)$$

with $\mathbf{P}_g = \mathbf{I}_M - \mathbf{c}_g \mathbf{c}_g^{\mathrm{T}}$.

14.4 ADAPTIVE SPARSE-INTERPOLATED VOLTERRA FILTERS

The Volterra filter is one of the most appealing of the nonlinear structures used in filtering applications. This is due to the fact that such a filter is a universal approximator according to the Stone-Weierstrass approximation theorem [31]. Such a generality comes along with a large number of coefficients to cope with, which implies high computational complexity especially in adaptive filtering applications. As a consequence, practical implementations of adaptive Volterra filters often use specific strategies for reducing the computational complexity. One of these strategies, inspired by the aforementioned SIFIR structures, is based on the use of a sparse Volterra filter along with a linear interpolator. This approach gives rise to the sparse-interpolated Volterra (SIV) filters, which are described in this section with a special focus on their use in adaptive applications.

14.4.1 REVISITING THE VOLTERRA FILTER

A truncated Pth-order Volterra filter [2] is composed of a set of P kernels whose outputs are given by

$$y_p(n) = \sum_{m_1=0}^{N-1} \sum_{m_2=0}^{N-1} \cdots \sum_{m_p=0}^{N-1} h_{\widehat{p},m_1,m_2,\ldots,m_p} \times \prod_{k=1}^{p} x(n - m_k) \qquad (14.75)$$

with $p = 1, 2, \ldots, P$, $x(n)$ denoting the input signal, $h_{\widehat{p},m_1,m_2,\ldots,m_p}$ representing a pth-order coefficient, and N as the memory size. The input-output relationship of the Volterra filter is given by the sum of the outputs of all kernels, resulting in

$$y(n) = \sum_{p=1}^{P} y_p(n). \qquad (14.76)$$

From (14.75), it is easy to notice that the first-order kernel ($p = 1$) of the Volterra filter is equivalent to an FIR filter. Thus, one can write

$$y_1(n) = \mathbf{h}_1^{\mathrm{T}} \mathbf{x}_1(n) \qquad (14.77)$$

with

$$\mathbf{x}_1(n) = [x(n) \; x(n-1) \; \cdots \; x(n-N+1)]^{\mathrm{T}} \qquad (14.78)$$

and

$$\mathbf{h}_1 = [h_{\widehat{1},0} \ h_{\widehat{1},1} \ \cdots \ h_{\widehat{1},N-1}]^{\mathrm{T}}. \tag{14.79}$$

For the second-order kernel, defining a second-order input vector as

$$\mathbf{x}_2(n) = \mathbf{x}(n) \otimes \mathbf{x}(n) = [x^2(n) \ x(n)x(n-1) \ \cdots \ x(n)x(n-N+1)$$
$$x(n-1)x(n) \ x^2(n-1) \ \cdots \ x^2(n-N+1)]^{\mathrm{T}} \tag{14.80}$$

and a second-order coefficient vector as

$$\mathbf{h}_2 = [h_{\widehat{2},0,0} \ h_{\widehat{2},0,1} \ \cdots \ h_{\widehat{2},0,N-1} \ h_{\widehat{2},1,0} \ h_{\widehat{2},1,1} \ \cdots \ h_{\widehat{2},N-1,N-1}]^{\mathrm{T}} \tag{14.81}$$

one gets

$$y_2(n) = \mathbf{h}_2^{\mathrm{T}} \mathbf{x}_2(n). \tag{14.82}$$

Then, in general terms, the output of the pth-order kernel is given by

$$y_p(n) = \mathbf{h}_p^{\mathrm{T}} \mathbf{x}_p(n) \tag{14.83}$$

with a pth-order input vector given by

$$\mathbf{x}_p(n) = \mathbf{x}_1(n) \otimes \mathbf{x}_{p-1}(n) \tag{14.84}$$

and a coefficient vector

$$\mathbf{h}_p = [h_{\widehat{p},0,0,\ldots,0} \ h_{\widehat{p},0,0,\ldots,1} \ \cdots \ h_{\widehat{p},0,0,\ldots,N-1} \ \cdots \ h_{\widehat{p},N-1,N-1,\ldots,N-1}]^{\mathrm{T}}. \tag{14.85}$$

14.4.1.1 Spatial Representation of the Volterra Kernels

An alternative form to represent a pth-order kernel of a Volterra filter is through a spatial representation obtained by considering the indices m_1 to m_p of the coefficients as coordinates of the Euclidean space. In such representation, each kernel has a number of dimensions that is equal to the order of the kernel. Thus, the spatial representation of the first-order kernel is unidimensional, matching the vector representation given in (14.79). For the second-order kernel, one has a two-dimensional spatial representation, resulting in a coefficient matrix given by

$$\mathbf{H}_2 = \begin{bmatrix} h_{\widehat{2},0,0} & h_{\widehat{2},0,1} & \cdots & h_{\widehat{2},0,N-1} \\ h_{\widehat{2},1,0} & h_{\widehat{2},1,1} & \cdots & h_{\widehat{2},1,N-1} \\ \vdots & \vdots & \ddots & \vdots \\ h_{\widehat{2},N-1,0} & h_{\widehat{2},N-1,1} & \cdots & h_{\widehat{2},N-1,N-1} \end{bmatrix}. \tag{14.86}$$

For the third-order kernel, the spatial representation has three dimensions, resulting in a cube or a cubic tensor. In general, for kernels of higher orders (with $p \geq 4$), one has hypercubic structures that are not easy to visualize. Despite this, the idea behind a spatial representation is still very interesting from an analytical point of view. For instance, it gives rise to a notion of kernel diagonals that leads to an implementation based on diagonal coordinates [2].

14.4.1.2 Redundancy-Removed Implementation

In the standard form, Volterra filters have a considerable number of coefficients that are redundant for real-world applications [2], [4]. This characteristic can be observed considering, for instance, that the elements $x(n)x(n-1)$ and $x(n-1)x(n)$ of $\mathbf{x}_2(n)$ are multiplied by two different coefficients $h_{2,0,1}$ and $h_{2,1,0}$ of $\mathbf{h}_2$ for obtaining the output $y_2(n)$ of the second-order kernel. In this case, since $x(n)x(n-1) = x(n-1)x(n)$, it becomes clear that $h_{2,0,1}$ and $h_{2,1,0}$ can be replaced by a single coefficient $\underline{h}_{2,0,1} = h_{2,0,1} + h_{2,1,0}$ without modifying the input-output relationship of the filter. In this context, one can infer that the standard Volterra filter has a considerable number of redundant coefficients and, thus, these coefficients can be removed or neglected for the sake of simplicity. To illustrate this feature, let us consider a second-order kernel with $N = 3$ whose coefficient vector can be obtained from (14.81). By removing the redundant coefficients from such a vector, one obtains the following redundancy-removed second-order coefficient vector:

$$\underline{\mathbf{h}}_2 = [\underline{h}_{2,0,0} \ \ \underline{h}_{2,0,1} \ \ \underline{h}_{2,0,2} \ \ \underline{h}_{2,1,1} \ \ \underline{h}_{2,1,2} \ \ \underline{h}_{2,2,2}]^{\mathrm{T}}. \qquad (14.87)$$

Then, defining the corresponding redundancy-removed input vector as

$$\underline{\mathbf{x}}_2(n) = [x^2(n) \ \ x(n)x(n-1) \ \ x(n)x(n-2)$$
$$x^2(n-1) \ \ x(n-1)x(n-2) \ \ x^2(n-2)]^{\mathrm{T}} \quad (14.88)$$

one can rewrite the input-output relationship of the second-order kernel as

$$y_2(n) = \mathbf{h}_2^{\mathrm{T}}\mathbf{x}_2(n) = \underline{\mathbf{h}}_2^{\mathrm{T}}\underline{\mathbf{x}}_2(n). \qquad (14.89)$$

Moreover, using the spatial representation described in the previous section, (14.87) can be rearranged in the form of the following coefficient matrix:

$$\underline{\mathbf{H}}_2 = \begin{bmatrix} \underline{h}_{2,0,0} & \underline{h}_{2,0,1} & \underline{h}_{2,0,2} \\ 0 & \underline{h}_{2,1,1} & \underline{h}_{2,1,2} \\ 0 & 0 & \underline{h}_{2,2,2} \end{bmatrix}. \qquad (14.90)$$

Due to the upper-triangular characteristic of this matrix, the Volterra implementation with removed redundancy is also known in the literature as triangular Volterra implementation [2].

In general terms, the redundancy-removed (triangular) implementation of a pth-order Volterra kernel is obtained by replacing any set of coefficients of a standard pth-order kernel having permutated indices by a single coefficient whose value is given by the sum of the coefficients of the set [e.g., the set composed of $h_{3,0,1,2}$, $h_{3,0,2,1}$, $h_{3,1,0,2}$, $h_{3,1,2,0}$, $h_{3,2,0,1}$, and $h_{3,2,1,0}$ is replaced by a single coefficient $\underline{h}_{3,0,1,2}$]. By using this approach, one can define a pth-order redundancy-removed coefficient vector $\underline{\mathbf{h}}_p$ having a number of coefficients that is considerably smaller than the one of a pth-order standard coefficient vector $\mathbf{h}_p$ (except for the case of the first-order kernel, for which $\underline{\mathbf{h}}_1 = \mathbf{h}_1$). In this context, defining a corresponding pth-order redundancy-removed input vector $\underline{\mathbf{x}}_p(n)$ whose relationship with $\mathbf{x}_p(n)$ is described in [14], one can write the output of the pth-order kernel as

$$y_p(n) = \mathbf{h}_p^\mathrm{T}\mathbf{x}_p(n) = \underline{\mathbf{h}}_p^\mathrm{T}\underline{\mathbf{x}}_p(n). \tag{14.91}$$

Then, defining a redundancy-removed Volterra coefficient vector as

$$\underline{\mathbf{h}}_\mathrm{V} = [\underline{\mathbf{h}}_1^\mathrm{T} \ \underline{\mathbf{h}}_2^\mathrm{T} \ \cdots \ \underline{\mathbf{h}}_P^\mathrm{T}]^\mathrm{T} \tag{14.92}$$

and a corresponding input vector as

$$\underline{\mathbf{x}}_\mathrm{V}(n) = [\underline{\mathbf{x}}_1^\mathrm{T}(n) \ \underline{\mathbf{x}}_2^\mathrm{T}(n) \ \cdots \ \underline{\mathbf{x}}_P^\mathrm{T}(n)]^\mathrm{T} \tag{14.93}$$

the input-output relationship of the redundancy-removed Volterra filter can be rewritten as

$$y(n) = \underline{\mathbf{h}}_\mathrm{V}^\mathrm{T}\underline{\mathbf{x}}_\mathrm{V}(n). \tag{14.94}$$

14.4.2 SPARSE-INTERPOLATED VOLTERRA FILTERS

In Section 14.4.1.2, it has been shown that the number of coefficients of a standard Volterra filter can be reduced considerably by removing the redundant coefficients of the nonlinear kernels. Despite this, the number of coefficients of the resulting redundancy-removed Volterra filter is still very large [4] and, thus, additional complexity-reduction approaches are often considered for the practical implementation of Volterra filters [14]. One of these approaches, which is inspired by the SIFIR filters, has been discussed in the last few years. In such an approach, the complexity reduction is attained by using sparse Volterra kernels that present a homogeneous sparseness characteristic obtained by zeroing all coefficients having at least one index that is not a multiple of a given sparseness factor L [4], [14]. A sparse second-order kernel of this kind with $N = 3$ and $L = 2$ has a coefficient matrix given by

$$\ddot{\mathbf{H}}_2 = \begin{bmatrix} h_{\tilde{2},0,0} & 0 & h_{\tilde{2},0,2} \\ 0 & 0 & 0 \\ h_{\tilde{2},2,0} & 0 & h_{\tilde{2},2,2} \end{bmatrix}. \tag{14.95}$$

Note that all zeroed coefficients of (14.95) can be recreated as a function of their neighboring coefficients by means of a two-dimensional interpolation process. As shown in [4] and [32], this type of interpolation can be obtained by simply filtering the input signal using an FIR filter (called interpolator) with memory size $M = 2L-1$. As a result, if an interpolator with coefficient vector $\mathbf{g} = [0.5 \ 1 \ 0.5]^\mathrm{T}$ (linear interpolator [17]) is used, the following equivalent (interpolated) second-order coefficient matrix is obtained:

$$
\bar{\mathbf{H}}_2 =
\begin{bmatrix}
0.25h_{\widetilde{2},0,0} & 0.5h_{\widetilde{2},0,0} & \begin{array}{c}0.25h_{\widetilde{2},0,0}+\\0.25h_{\widetilde{2},0,2}\end{array} & 0.5h_{\widetilde{2},0,2} & 0.25h_{\widetilde{2},0,2} \\[2ex]
0.5h_{\widetilde{2},0,0} & h_{\widetilde{2},0,0} & \begin{array}{c}0.5h_{\widetilde{2},0,0}+\\0.5h_{\widetilde{2},0,2}\end{array} & h_{\widetilde{2},0,2} & 0.5h_{\widetilde{2},0,2} \\[2ex]
\begin{array}{c}0.25h_{\widetilde{2},0,0}+\\0.25h_{\widetilde{2},0,2}\end{array} & \begin{array}{c}0.5h_{\widetilde{2},0,0}+\\0.5h_{\widetilde{2},0,2}\end{array} & \begin{array}{c}0.25h_{\widetilde{2},0,0}+\\0.25h_{\widetilde{2},0,2}+\\0.25h_{\widetilde{2},2,2}+\\0.25h_{\widetilde{2},2,0}\end{array} & \begin{array}{c}0.5h_{\widetilde{2},0,2}+\\0.5h_{\widetilde{2},2,2}\end{array} & \begin{array}{c}0.25h_{\widetilde{2},0,2}+\\0.25h_{\widetilde{2},2,2}\end{array} \\[3ex]
0.5h_{\widetilde{2},2,0} & h_{\widetilde{2},2,0} & \begin{array}{c}0.5h_{\widetilde{2},2,0}+\\0.5h_{\widetilde{2},2,2}\end{array} & h_{\widetilde{2},2,2} & 0.5h_{\widetilde{2},2,2} \\[2ex]
0.25h_{\widetilde{2},2,0} & 0.5h_{\widetilde{2},2,0} & \begin{array}{c}0.25h_{\widetilde{2},2,0}+\\0.25h_{\widetilde{2},2,2}\end{array} & 0.5h_{\widetilde{2},2,2} & 0.25h_{\widetilde{2},2,2}
\end{bmatrix}.
\tag{14.96}
$$

In (14.96), the underlined coefficients are the ones replicated from the sparse kernel [see (14.95)] whereas the boxed coefficients are those recreated by interpolating the values of the underlined ones; the remaining coefficients in (14.96) correspond to the boundary effect. As described in [4], the use of such an approach (input FIR interpolator along with a sparse kernel) results in a multidimensional interpolation process for higher-order kernels. This process is illustrated in Figure 14.9 for the case involving an interpolator with $M = 3$ coefficients and a sparse third-order kernel with $N = 3$ and $L = 2$. In this figure, the black circles represent the coefficients replicated from the sparse kernel and the gray circles represent the coefficients recreated by interpolation. The boundary effect is omitted in Figure 14.9.

From the discussion above, one notices that a sparse-interpolated Volterra (SIV) filter with characteristics similar to those of the SIFIR filters can be obtained by using an input FIR interpolator cascaded with a sparse Volterra filter. Considering this fact along with the redundancy-removed Volterra implementation described in Section 14.4.1.2, one obtains the SIV filter structure depicted in Figure 14.10 [4]. In this figure, $\mathbf{g}$ represents the interpolator and $\ddot{\underline{\mathbf{h}}}_V$, a sparse Pth-order Volterra filter in its redundancy-removed form. Moreover, $x(n)$ and $y(n)$ are the input and output signals, respectively, and $\ddot{\underline{\mathbf{h}}}_p$ is a pth-order redundancy-removed kernel whose output is given by

$$
\hat{y}_p(n) = \ddot{\underline{\mathbf{h}}}_p^T \tilde{\underline{\mathbf{x}}}_p(n)
\tag{14.97}
$$

with $\tilde{\underline{\mathbf{x}}}_p(n)$ representing a pth-order redundancy-removed input vector obtained by using samples of the signal at the interpolator output. Then, defining a sparse Volterra coefficient vector as

$$
\ddot{\underline{\mathbf{h}}}_V = [\ddot{\underline{\mathbf{h}}}_1^T \ \ddot{\underline{\mathbf{h}}}_2^T \ \cdots \ \ddot{\underline{\mathbf{h}}}_P^T]^T
\tag{14.98}
$$

and a corresponding interpolated input vector as

$$
\tilde{\underline{\mathbf{x}}}_V(n) = [\tilde{\underline{\mathbf{x}}}_1^T(n) \ \tilde{\underline{\mathbf{x}}}_2^T(n) \ \cdots \ \tilde{\underline{\mathbf{x}}}_P^T(n)]^T
\tag{14.99}
$$

where $\underline{\check{\mathbf{x}}}_1(n) = \check{\mathbf{x}}(n)$ and $\underline{\check{\mathbf{x}}}_p(n)$ (with $p \geq 2$) is a pth-order redundancy-removed input vector obtained from $\underline{\check{\mathbf{x}}}_1(n)$. For the case of the PSIV filter, the boundary effect is removed by replacing $\dot{\tilde{\mathbf{x}}}_V(n)$ [see (14.103)] by

$$\dot{\tilde{\mathbf{x}}}_V(n) = [\check{\mathbf{x}}_1^T(n) \ \ \underline{\check{\mathbf{x}}}_2^T(n) \ \cdots \ \check{\mathbf{x}}_p^T(n)]^T. \tag{14.105}$$

14.4.3 ADAPTIVE IMPLEMENTATION OF SIV FILTERS

The SIV and PSIV filters described in the previous section are especially interesting for adaptive filtering applications involving Volterra filters. In such applications, one typically has both a filtering operation and the update of all values of filter coefficients for each new sample of the input signal. Thus, the use of Volterra implementations with reduced number of coefficients, such as the SIV and PSIV, potentially leads to a considerable reduction in the required number of operations per sample. In this section, the main approaches for implementing adaptive versions of SIV and PSIV filters are briefly described.

14.4.3.1 Standard Approach

The first adaptive implementations of SIV filters were carried out by adapting only the sparse Volterra filter while the interpolator is kept fixed [4]. This type of implementation, called adaptive SIV (ASIV) filter, can thus be seen as a generalization of the ASIFIR implementation described in Section 14.3.2. As a consequence, the update rule for the coefficients of the ASIV filter using the NLMS algorithm can be obtained from an optimization problem similar to that defined in (14.48) [4]. Therefore, one has

$$\text{minimize } \|\ddot{\underline{\mathbf{h}}}_V(n + 1) - \ddot{\underline{\mathbf{h}}}_V(n)\|^2$$

$$\text{subject to } \begin{cases} \varepsilon(n) = d(n) - \ddot{\underline{\mathbf{h}}}_V^T(n + 1)\tilde{\underline{\mathbf{x}}}_V(n) = 0 \\ \mathbf{C}_V^T\ddot{\underline{\mathbf{h}}}_V(n + 1) = \mathbf{0} \end{cases} \tag{14.106}$$

where $\mathbf{C}_V$ is a generalized version of the constraint matrix $\mathbf{C}$ described in Section 14.3.2 [14]. By solving this problem, one obtains the following update equation for the NLMS ASIV filter:

$$\ddot{\underline{\mathbf{h}}}_V(n + 1) = \mathbf{P}_V\ddot{\underline{\mathbf{h}}}_V(n) + \alpha_h \frac{e(n)}{\|\mathbf{P}_V\tilde{\underline{\mathbf{x}}}_V(n)\|^2 + \psi_h}\mathbf{P}_V\tilde{\underline{\mathbf{x}}}_V(n) \tag{14.107}$$

where α_h denotes the step size, ψ_h is the regularization parameter, and $\mathbf{P}_V = \mathbf{I} - \mathbf{C}_V\mathbf{C}_V^T$.

An adaptive implementation of the PSIV filter (called APSIV implementation) can be obtained from (14.107) simply by replacing, respectively, $\ddot{\underline{\mathbf{h}}}_V(n)$ and $\tilde{\underline{\mathbf{x}}}_V(n)$ by $\underline{\mathbf{h}}_V(n)$ and $\underline{\tilde{\mathbf{x}}}_V(n)$ in such an expression. Moreover, versions with removed boundary effect (BASIV and BAPSIV implementations) can be easily obtained from the ASIV and APSIV implementations simply by replacing either $\tilde{\underline{\mathbf{x}}}_V(n)$ by $\check{\underline{\mathbf{x}}}_V(n)$ or $\dot{\underline{\mathbf{x}}}_V(n)$ by $\dot{\check{\mathbf{x}}}_V(n)$ in the corresponding update expressions.

14.4.3.2   Fully LMS/NLMS Adaptive Approach

As in the case of the ASIFIR filter, the ASIV and APSIV filters may also present poor performance if the fixed coefficients of the interpolator are not chosen properly. Thus, fully adaptive implementations of SIV filters, in which the interpolator is also adaptive, are of great practical interest. However, due to the nonlinear structure of the SIV filters, the development of such fully adaptive implementations is not a straightforward task. To overcome this drawback, the approach proposed in [33] uses a combination of the LMS and NLMS algorithms. The faster one (NLMS) is used for updating the coefficients of a sparse Volterra filter with removed boundary effect, resulting in the following update equation:

$$\underline{\ddot{\mathbf{h}}}_V(n+1) = \mathbf{P}_V\underline{\ddot{\mathbf{h}}}_V(n) + \alpha_h \frac{e(n)}{\|\mathbf{P}_V\underline{\check{\mathbf{x}}}_V(n)\|^2 + \psi_h}\mathbf{P}_V\underline{\check{\mathbf{x}}}_V(n). \tag{14.108}$$

On the other hand, the simpler LMS algorithm is applied to the interpolator due to the small number of coefficients involved and also for the sake of mathematical simplicity. Thus, the coefficients of the interpolator are updated according to

$$\mathbf{g}(n+1) = \mathbf{g}(n) + \mu_g\nabla_{\mathbf{g}}e^2(n) \tag{14.109}$$

with μ_g denoting the step-size parameter and $\nabla_{\mathbf{g}}e^2(n)$, the gradient of the squared error with respect to $\mathbf{g}(n)$. By evaluating $\nabla_{\mathbf{g}}e^2(n)$ as described in [33] and substituting the obtained expression into (14.109), one can get the following update equation for the interpolator of the fully LMS/NLMS adaptive SIV filter with removed boundary effect:

$$\mathbf{g}(n+1) = \mathbf{g}(n) - 2\mu_g e(n)\sum_{p=1}^{P} p[\check{\mathbf{X}}(n) \otimes \check{\mathbf{x}}_{p-1}^{\mathrm{T}}(n)]\ddot{\mathbf{h}}_p(n) \tag{14.110}$$

where $\check{\mathbf{X}}(n)$ represents an input matrix with removed boundary effect whose structure is described in detail in [33]. In addition, the update expression for the interpolator of the fully LMS/NLMS adaptive PSIV filter with removed boundary effect is similar to (14.110), differing only by the range of the summation (which for the PSIV filter is from $p = 2$ to $p = P$ [33]).

14.4.4   DIAGONALLY INTERPOLATED VOLTERRA FILTERS

As discussed in the previous sections, the idea behind the SIV and PSIV implementations of Volterra filters is similar to that of the SIFIR filters, i.e., to use a sparse filter with homogeneous sparseness for complexity reduction along with an interpolator to compensate for performance loss. As mentioned in Section 14.4.2, the p-dimensional interpolation required for recreating the coefficients of a pth-order sparse Volterra kernel is obtained by placing the interpolator at the input of the kernel. On the other hand, if the interpolator is placed at the output, a different type of interpolation process takes place [34]. In this case, the interpolation is carried out along the diagonals of the sparse kernel and, as a result, the equivalent kernel will still be sparse [34]. This characteristic becomes evident considering that some of the diagonals of (14.95) (coefficient matrix of the second-order sparse kernel) are composed solely of zeros and, thus, any interpolation along such diagonals will not modify their all-zero elements.

The interpolation along diagonals described above can be exploited to obtain a different type of sparse-interpolated implementation that also results in a non-sparse equivalent structure. As shown in [34], such implementation, termed diagonally-interpolated Volterra (DIV) filter, uses sparse Volterra kernels obtained by zeroing all coefficients whose first index is not a multiple of the sparseness factor L, which results in the following coefficient matrix for a second-order kernel with $N = 3$ and $L = 2$:

$$\dddot{\mathbf{H}}_2 = \begin{bmatrix} h_{\widetilde{2},0,0} & h_{\widetilde{2},0,1} & h_{\widetilde{2},0,2} \\ 0 & 0 & 0 \\ h_{\widetilde{2},2,0} & h_{\widetilde{2},2,1} & h_{\widetilde{2},2,2} \end{bmatrix}. \tag{14.111}$$

This type of sparse kernel clearly has a number of coefficients that is larger than that of the sparse kernels considered in SIV structures [compare with (14.95)]. As a consequence, one can notice that the reduction of complexity obtained by using a DIV filter is smaller than that obtained by a SIV filter with the same values of N and L. However, the larger number of coefficients of the DIV filter usually leads to superior performance in comparison with the SIV filter (for same values of N and L). Moreover, as shown in [34], DIV filters can benefit from the use of different interpolators for each kernel diagonal, which may lead to considerable performance improvement. In addition, certain diagonal-pruning strategies used for obtaining other reduced-complexity Volterra implementations [14] can be easily applied to a DIV filter [34], thus leading to hybrid implementations with a changeable trade-off between complexity and performance.

The block diagram of a typical second-order DIV filter is illustrated in Figure 14.12. In this figure, $\ddot{\mathbf{h}}_1$ is the first-order sparse kernel and $\mathbf{g}_1$ is the corresponding interpolator. Moreover, $\ddot{\mathbf{h}}_{2,n}$ represents the nth diagonal of the second-order redundancy-removed sparse kernel (the main diagonal is the one with $n = 0$), whereas the associated interpolator is represented by $\mathbf{g}_{2,n}$. The aforementioned pruning of diagonals can be carried out simply by removing the diagonals that are far from the main diagonal, i.e., removing the branches of the structure from Figure 14.12 for which $n \geq D$, with D representing the number of diagonals considered in the implementation.

14.5 A CASE STUDY OF NETWORK ECHO CANCELLATION

In this section, numerical simulation results are presented aiming to illustrate the potential of the sparse-interpolated approach to provide effective solutions to adaptive filtering problems. In this context, different sparse-interpolated nonlinear filters are applied in a network echo cancellation problem with signals obtained from an analog telephone adapter (ATA) used in VoIP systems. For comparison purposes, both an FIR filter and a Volterra filter are also applied to such a problem. Thus, the following filters are considered in the simulations: i) standard FIR filter with $N = 51$;

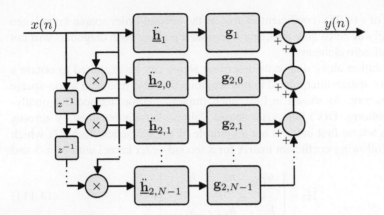

Figure 14.12   Block diagram of a second-order diagonally-interpolated Volterra filter.

ii) redundancy-removed Volterra filter with a first-order kernel having memory size $N_1 = 51$ and a second-order kernel with memory size $N_2 = 21$; iii) APSIV filter (see Section 14.4.3.1) with $N_1 = 51$, $N_2 = 21$, $L = 2$, and $\mathbf{g} = [0.5 \ 1 \ 0.5]^T$; iv) LMS/NLMS fully adaptive PSIV (FAPSIV) filter (see Section 14.4.3.2) with $N_1 = 51$, $N_2 = 21$, $L = 2$, and $\mathbf{g} = [0.5 \ 1 \ 0.5]^T$; and v) a fully adaptive DIV (FADIV) implementation (see Section 14.4.4) composed of a non-sparse first-order kernel with $N_1 = 51$ and a second-order DIV kernel having $N_2 = 21$, $D = 3$, $L = 2$, and with the interpolators of each branch initially set to $\mathbf{g} = [0.5 \ 1 \ 0.5]^T$. The parameters of the NLMS algorithm used to update all filters considered here are $\alpha = 0.5$ and $\psi = 0.1$. The considerable complexity reduction provided by the sparse-interpolated implementations can be observed in Table 14.3, where the number of operations per sample required by each of the considered filters is shown.

The curves of echo return-loss enhancement (ERLE) [3] obtained in simulations considering signals of around 6 seconds acquired from the ATA are shown in Figure 14.13. From this figure and also allowing for Table 14.3, one observes the very good tradeoff between performance and complexity obtained by either the FAPSIV or the FADIV filter. Moreover, as expected, intermediate results are obtained with the APSIV filter, which is due to the use of a fixed interpolator.

14.6   CONCLUDING REMARKS

This chapter was devoted to the sparse-interpolated approach used for implementing standard and adaptive filters with reduced computational complexity. In this context, the first part of the chapter was dedicated to describe the foundations of the first filters implemented using the sparse-interpolated approach, namely the sparse-interpolated FIR filters. The second part of the chapter was focused on the adaptive implementations of the sparse-interpolated FIR filters, describing the different strategies used to carry out such implementations. In the third part of this chapter, the application of the sparse-interpolated approach to Volterra filters was discussed, showing the differ-

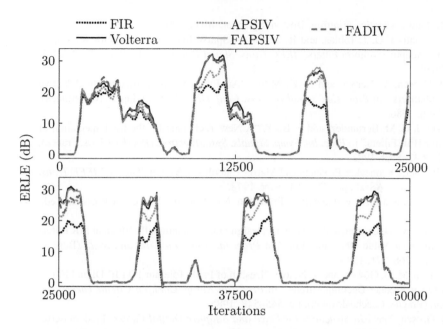

Figure 14.13 ERLE curves obtained from simulations using real-world signals of a network echo cancellation application.

ent adaptive Volterra implementations that can be obtained from such an application. The effectiveness of these implementations is verified in the last part of this chapter, where a case study of nonlinear network echo cancellation involving real-world signals is discussed.

Table 14.3

Computational Complexity of the Filters Considered in Section 14.5

Filter	Operations per sample	Relative percentage
Volterra	1718	**100%**
APSIV	728	42.4%
FAPSIV	907	52.8%
FADIV	579	33.7%
FIR	309	18.0%

REFERENCES

1. Y. Neuvo, D. Cheng-Yu, and S. K. Mitra, "Interpolated Finite Impulse Response Filters," *IEEE Trans. Acoust., Speech, Signal Process.*, vol. 32, pp. 563–570, July 1984.
2. V. J. Mathews and G. L. Sicuranza, *Polynomial Signal Processing*. New York: Wiley, 2000.

3. P. S. R. Diniz, *Adaptive Filtering*. Boston: Springer, 4th ed., 2013.

4. E. L. O. Batista, O. J. Tobias, and R. Seara, "A Sparse-Interpolated Scheme for Implementing Adaptive Volterra Filters," *IEEE Trans. Signal Process.*, vol. 58, pp. 2022–2035, Apr. 2010.

5. M. D. Grosen, Y. Neuvo, and S. K. Mitra, "An Efficient FIR Structure for Adaptive Line Enhancers," in *Proc. Eur. Signal Process. Conf. (EUSIPCO)*, (Hague, Netherlands), pp. 93–96, 1986.

6. R. Seara, J. C. M. Bermudez, and E. Beck, "A New Technique for the Implementation of Adaptive IFIR Filters," in *Proc. Int. Symp. Signals, Syst., Electron. (ISSSE)*, (Paris, France), pp. 644–647, Sept. 1992.

7. J. W. Brewer, "Kronecker Products and Matrix Calculus in System Theory," *IEEE Trans. Circ. Syst.*, vol. CAS-25, pp. 772–781, Sept. 1978.

8. D. S. Bernstein, *Matrix Mathematics*. Princeton, NJ: Princeton University Press, 2nd ed., 2009.

9. S. K. Mitra, M. D. Grosen, and Y. Neuvo, "Design of Computationally Efficient FIR Filters Using Singular Value Decomposition," in *Proc. Int. Conf. on Circ. and Syst.*, (Beijing, China), pp. 268–271, 1985.

10. S. K. Mitra, M. D. Grosen, and Y. Neuvo, "Design of Fault-Tolerant 1D FIR Digital Filters using Outer Product Decompositions," in *Proc. 7th Europ. Conf. on Circuit Theory and Design*, (Prague, Czechoslovakia), pp. 545–548, 1985.

11. M. D. Grosen, *New FIR Structures for Fixed and Adaptive Digital Filters*. Ph.d. dissertation, University of California Santa Barbara, 1987.

12. T. Saramaki, T. Neuvo, and S. K. Mitra, "Design of Computationally Efficient Interpolated FIR Filters," *IEEE Trans. Circ. Syst.*, vol. 35, no. 1, pp. 70–88, 1988.

13. R. C. de Lamare and R. Sampaio-Neto, "Adaptive Reduced-Rank Processing Based on Joint and Iterative Interpolation, Decimation, and Filtering," *IEEE Trans. Signal Process.*, vol. 57, pp. 2503–2514, July 2009.

14. E. L. O. Batista and R. Seara, "On the Performance of Adaptive Pruned Volterra Filters," *Signal Processing*, vol. 93, pp. 1909–1920, July 2013.

15. E. L. O. Batista, O. J. Tobias, and R. Seara, "Border Effect Removal for IFIR and Interpolated Volterra Filters," in *Proc. IEEE Int. Conf. Acoust., Speech, Signal Process. (ICASSP)*, vol. 3, (Honolulu, HW), pp. 1329–1332, Apr. 2007.

16. E. L. O. Batista, O. J. Tobias, and R. Seara, "A Fully Adaptive IFIR Filter with Removed Border Effect," in *Proc. IEEE Int. Conf. Acoust., Speech, Signal Process. (ICASSP)*, (Las Vegas, NV), pp. 3821–3824, Mar. 2008.

17. O. J. Tobias and R. Seara, "Analytical Model for the First and Second Moments of an Adaptive Interpolated FIR Filter Using the Constrained Filtered-X LMS Algorithm," *IEE Proc.-Vis. Image Signal Process.*, vol. 148, pp. 337–347, Oct. 2001.

18. L. S. Resende, C. A. F. Rocha, and M. G. Bellanger, "A New Structure for Adaptive IFIR Filtering," in *Proc. IEEE Int. Conf. Acoust., Speech, Signal Process. (ICASSP)*, vol. 1, (Istanbul, Turkey), pp. 400–403, June 2000.

19. O. L. Frost, "An Algorithm for Linearly Constrained Adaptive Array Processing," *Proc. IEEE*, vol. 60, pp. 926–935, Aug. 1972.

20. R. C. Bilcu, P. Kuosmanen, and K. Egiazarian, "On Adaptive Interpolated FIR Filters," in *Proc. IEEE Int. Conf. Acoust., Speech, Signal Process. (ICASSP)*, vol. 2, (Montreal, Canada), pp. 665–668, 2004.

21. E. L. O. Batista, O. J. Tobias, and R. Seara, "New Insights in Adaptive Cascaded FIR Structure: Application to Fully Adaptive Interpolated FIR Structures," in *Proc. Eur. Signal*

Process. Conf. (EUSIPCO), (Poznan, Poland), pp. 370–374, Sept. 2007.

22. R. C. de Lamare and R. Sampaio-Neto, "Adaptive Interference Suppression for DS-CDMA Systems Based on Interpolated FIR Filters With Adaptive Interpolators in Multipath Channels," *IEEE Trans. Veh. Technol.*, vol. 56, pp. 2457–2474, Sept. 2007.

23. A. Abousaada, T. Aboulnasr, and W. Steenaart, "An Echo Tail Canceller Based on Adaptive Interpolated FIR Filtering," *IEEE Trans. Circuits Syst. II, Analog Digit. Signal Process.(1993-2003)*, vol. 39, pp. 409–416, July 1992.

24. S.-S. Lin and W.-R. Wu, "A Low-Complexity Adaptive Echo Canceller for xDSL Applications," *IEEE Trans. Signal Process.*, vol. 52, pp. 1461–1465, May 2004.

25. V. da Maia, "FIR-IFIR Hybrid Echo Canceller Degradation Analysis," in *Proc. IEEE Int. Conf. on Comm. Systems*, (Singapore), pp. 1–5, Oct. 2006.

26. C.-S. Wu and A.-Y. Wu, "A Novel Cost-Effective Multi-Path Adaptive Interpolated FIR (IFIR)-Based Echo Canceller," in *Proc. IEEE Int. Symp. Circuits Syst. (ISCAS)*, vol. 5, (Phoenix, AZ), pp. 453–456, May 2002.

27. Y.-L. Chen, C.-Z. Zhan, and A.-Y. Wu, "Cost-Effective Echo and NEXT Canceller Designs for 10GBASE-T Ethernet System," in *Proc. IEEE Int. Symp. Circuits Syst. (ISCAS)*, (Seattle, WA), pp. 3150–3153, May 2008.

28. E. L. O. Batista, O. J. Tobias, and R. Seara, "Efficient Implementation of Adaptive FIR-IFIR Filters Using the LMS Algorithm," in *Proc. Eur. Signal Process. Conf. (EUSIPCO)*, (Aalborg, Denmark), pp. 1713–1717, Aug. 2010.

29. E. L. O. Batista and R. Seara, "On the Implementation of Fully Adaptive Interpolated FIR Filters," in *Proc. Eur. Signal Process. Conf. (EUSIPCO)*, (Bucharest, Romania), pp. 2173–2177, Aug. 2012.

30. S. Wahls and H. Boche, "A Projection Approach to Adaptive IFIR Filtering," in *Proc. IEEE Int. Conf. Acoust., Speech, Signal Process. (ICASSP)*, (Kyoto, Japan), pp. 3741–3744, Mar. 2012.

31. A. Carini and G. L. Sicuranza, "Fourier Nonlinear Filters," *Signal Processing*, vol. 94, pp. 183–194, Jan. 2014.

32. E. L. O. Batista, O. J. Tobias, and R. Seara, "A Mathematical Framework to Describe Interpolated Adaptive Volterra Filters," in *Proc. IEEE Int. Telecommun. Symp. (ITS)*, (Fortaleza, Brazil), pp. 906–911, Sept. 2006.

33. E. L. O. Batista and R. Seara, "A Fully LMS/NLMS Adaptive Scheme Applied to Sparse-Interpolated Volterra Filters with Removed Boundary Effect," *Signal Processing*, vol. 92, pp. 2381–2393, Oct. 2012.

34. E. L. O. Batista and R. Seara, "Adaptive NLMS Diagonally-Interpolated Volterra Filters for Network Echo Cancellation," in *Proc. Eur. Signal Process. Conf. (EUSIPCO)*, (Barcelona, Spain), pp. 1430–1434, Aug. 2011.

15 Cognitive Power Line Communication

Alam Silva Menezes
Petrobras Transporte S.A.

Laryssa Ramos Amado
Federal University of Juiz de Fora (UFJF)

Weiler Alves Finamore
Federal University of Juiz de Fora (UFJF)

Moisés Vidal Ribeiro
Federal University of Juiz de Fora (UFJF)

15.1 INTRODUCTION

Nowadays, there is an increasing interest towards efficient, reliable, two-way, and low-cost communication technologies to fulfill the demands of electric utilities [1]-[6] and, mainly, to introduce smart grid technologies into electric power systems. Worldwide efforts have been made to raise the needs for smart grid communications (SG-Comm), to assess the the SG-Comm infrastructure and, to specify the constraints and the requirements for designing the system and deploying the SG-Comm service [7]-[13]. Another worldwide issue challenging the communication industry is to attend to the demands posed by the need to lessen the digital division numbers—today around 70% of the world population does not have Internet access. Several attempts at addressing this issue with the use of current technologies, such as satellite communication, wireless communication (RF, Wimax, and LTE), optic communication, and DSL are being investigated and deployed. All attempts to use such technologies fail to deliver proper Internet access to this group of the population correctly due to the constraints related to over 2/3 of the worldwide population.

The use of electric power grids as the communication infrastructure is a promising alternative for SGComm and as the solution to the digital division worries [14]. Indeed, the expenditure to deploy PLC systems can be shared between both electric and telecom utilities, considerably reducing the cost of this deployment, operation, and maintenance of the SGComm infrastructure. This synergy surely enhances the motivation for a business plan to put the communication infrastructure in service of also the 2/3 of the worldwide population eager for the benefit of digital communications.

The electric power grids, however, were not conceived for high-frequency data communication—power cables are not shielded, for instance, which increases the possibilities of interference from electromagnetic radiations into radio communication services. The potential disturbances, originated by PLC signals, which may cause interference in other communication systems operating in the same frequency band near and far from the power cables is the subject of debates among researchers in academia, regulatory authorities, and companies, because PLC systems are considered as non-preferential users of the frequency band between 0 and 500 MHz, being allowed to use this frequency range opportunistically only. PLC systems, are then, classified as secondary users of the specified spectrum.

The spectrum scarcity as well as the the incessant search for efficiency has intensified the research activities to quantify how occasionally the spectrum is used and to design spectrum sensing techniques aimed at systems with opportunistic operation in the frequency band allocated to the primary users. The need for cognitive PLC systems, such as cognitive radio, cannot be overlooked. In the recent years, the PLC industry focused on increasing the performance and throughput by launching the standards IEEE Std 1901-2010 [15]-[16] and ITU-T G.9960/G.hn [17] for broadband PLC systems. Also, both IEEE and ITU-T introduced new standards for narrowband PLC systems. Nevertheless, cognitive issues are not addressed in these new efforts to standardize the PLC technologies [11]. Actually, only techniques to avoid interference have been included. Therefore, designing cognitive PLC systems is a new and challenging issue for the industry.

The potential disturbances brought by the use of PLC technology, such as interference with other communication systems operating in the same frequency band, near and far from power cables, were discussed, investigated, and measured in [18]-[26]. Regulations, by several national and international telecom agencies, have been imposed to limit the interference from signals emitted by PLC systems with and thus protect, among others, systems with sensitive monitoring services, both governmental and military.

In this direction, several mitigation techniques have been proposed and adopted in some regulation and standardization documents. These regulation standards aim at protecting vital and sensitive radio communication from suffering unacceptable performance degradation. Services such as life safety and vital governmental frequencies should, in general, operate free from harmful interference from the radiation and, in particular, by emissions from the PLC technologies [27]. Techniques to reduce electromagnetic interference (EMI) emissions have been addressed in the literature [28]-[39]. The European Telecommunication Standardization Institute (ETSI), approved in 2008 [39], requires the detection of broadcasting signals by sensing the "noise" (including radio broadcast picked up on the power cable) at an electrical outlet. Frequencies where short wave radio broadcasting signals are identified must be excluded from the transmitted signal by inserting a notch into the transmitting spectrum. Here, the cognitive idea is applied in the time, frequency, and the location domain. We recognize that the concept of cognitive communication [45] has been somehow adopted in [40]. In this new standard provisions to adaptively manage the transmission power is the included cognitive feature.

Also, techniques based on fixed, dynamic, and pre-fixed notching approaches have been discussed [38]-[41]. These techniques are based on the assumption that frequency bands are only allocated to primary users and, for these users, the interference problem can be considerably eliminated if these cognitive techniques are applied. On the other hand, the usage of the spectrum by primary users being time and space dependent surely result in an inefficient usage of the spectrum by the primary users. In addition, the scarcity of the spectrum for data communication, mainly for SGComm and access network, brings attention to the fact that it may be advantageous to introduce techniques that allow PLC systems to operate and coexist, opportunistically, with primary users in the same frequency band conducting an efficiently and harmonious exploitation of the frequency spectrum [42]-[44].

The severity of interference over the frequency spectrum will definitely endanger smart grid communications and access network based on PLC technologies. Similar to wireless communication, a possible solution for the aforementioned problems is the introduction of cognitive PLC systems. Endowing the PLC systems with spectrum sensing capability will introduce the possibility to, opportunistically, use it for data communication. In this regard, it is a challenging issue to design PLC data communication systems to operate on the frequency range that goes from 0 to 500 MHz, perturbed by colored additive noise, and coexisting, besides, with a large number of primary users, such as amateur radio, army radio signals, AM stations, FM stations, police department communications, navy radio signals, TV stations, and so forth. It should be noted that the frequency range from 0 up to 500 kHz is already allocated to SGComm (narrowband data communication) and the range starting at 1.7 MHz up to 100 MHz is being investigated for network access and in-home systems (broadband data communication). The frequencies between 0.5 MHz and 1.7 MHz, allocated to AM radio stations, could be used, opportunistically, by PLC technologies for SGComm if cognitive concepts are taken into account.

15.2   COGNITIVE PLC

15.2.1   DEFINITION AND SPECTRUM CONSIDERATIONS

Cognition is derived from the Latin *cognitione* that means "knowledge acquisition by perception." The use of knowledge to deal with, and to adapt to, action towards predetermined goals is the foundation of the cognitive process or cognitive systems. The cognitive process is based on two fundamental skills: perception and memory. Perception is the process of detecting changes in the surrounding environment, through sensing its inputs. Memory is used to store and recover information. Therefore, cognition is the capacity of associating new information, acquired by the perception of the existing ones, and directly adapt action towards some goals.

The term cognitive radio was introduced for the first time in Mitola's dissertation [45], where, cognitive radio (CR) was defined as an evolution of software-defined radio (SDR). SDR is basically composed of radio frequency front-end circuits, analog-to-digital and digital-to-analog converters, and software for demodulation/modulation [45]. In an SDR based communication system the users are allowed to demand the switching from one carrier frequency to another as well as to demand for the switching from one radio standard to another. Also, cognitive radio should be able to transparently reconfigure itself, switching to the most effective configuration, without notifying the user. This means that the CR is capable of operating automatically in multi-band, even sharing bands with other users (using different swaths of frequency) and employing the most suitable radio standard available, in software, for its use.

In order to do this and to improve the flexibility of personal wireless services, Mitola proposed the Radio Knowledge Representation Language (RKRL) and introduced the idea of cognitive cycle [45]. As proposed in [46], the cognitive cycle, the central idea of CR, can be understood as the interaction of the three interconnected processes: i) radio scene analysis; ii) channel-state estimation and; iii) transmit power control and spectrum management.

In general, CR denotes a wireless communication system that has the ability to discover an underutilized band of the frequency spectrum and occupy this free band, opportunistically, without causing harmful interferences. Similar to CR, Cognitive PLC (CogPLC) is a system in which the ability to sense the spectrum and opportunistically transmit data through the electric power grids using these free bands. In a CogPLC system the frequency spectrum occupancy is periodically checked with a spectrum sensing technique. In addition, the system makes use of resource allocation techniques to allow the CogPLC system to transmit data through those spectrum bands, in which the licensed incumbent (primary users) are absent or, if transmission in this band is enabled, the level of interference generated by such transmission is acceptable, increasing usage efficiency of the available resources.

The CogPLC awareness of its communication environment and its ability to dynamically adapt to the changes in the environment helps to increase spectrum usage efficiency allowing more data to be transmitted while keeping the interference with the primary users below the required levels. CogPLC must then provide protection and restrictions while transmitting its signal. In other words, the CogPLC is able to transmit data trough the electric power grid, and, at the same time, sense the frequency spectrum to identify the bands occupied by the primary users.

With its spectrum sensing capabilities, the CogPLC can incorporate restrictions avoiding harmful interference in detected primary or other secondary users. Restrictions violation can be prevented by applying notching filters to avoid emission in some frequency bands or by using power control techniques.

The concept of CogPLC for network access, in low voltage electric power grids, was discussed in 2007 [47]. In that work, the POWERNET project addressed the problem of preventing interference by electromagnetic radiation from PLC technology. According to this project, the CogPLC master modem senses the spectrum and identifies free channels which can be allocated to CogPLC users, which are named slaves, yielding communication without interfering with other PLC users plugged into the electric circuit. This contribution does not offer, however, information on how to sense the spectrum, or, in other words, on the digital signal processing procedures applied to sense the environment.

The CogPLC based on the combination of orthogonal frequency division multiple accesses (OFDMA) and CSMA has been introduced in [43] allowing multiple users to access the spectrum at the same time without introducing interference. The cognitive feature of this system is based on a map of multiple channels and on the use of the diversity of multiple channels in the frequency domain. Recently, some techniques based on the cognitive concept have been proposed for using spectrum access and allocation in indoor PLC systems without interfering with other data communication systems, such as VDSL2 [48]-[49]. A proposal for fair cognitive sharing of PLC channel resources among non-coexisting PLC modems is presented in [38]— the PLC systems detect interferences from originating on other, non-coexisting systems, by comparing the interfering spectrum with the ground noise. Frequency bands where the SNR is high are kept for communication and the frequency bands where SNR is low are not used for PLC transmissions, so that other systems can use them. After multiple cognitive steps, each PLC modem can find which optimal communication resources are reserved for them.

Figure 15.1 illustrates the environment where the CogPLC system will be deployed. In this scenario, PLC cells [50], with distinct geometries, are established according to the topologies of electric grids, and, complying to the already deployed electric grid infrastructures. The majority of electric distribution systems apply radial topology with the medium voltage/low voltage (MV/LV) transformers feeding blocks composed of a few up to over forty customers. Overall, each MV/LV transformer defines PLC cells with distinct geometries mainly due to LV power system infrastructure. One possible consideration is to divide the LV distribution grids into smaller PLC cells, following models adopted in wireless communications. This is, however, questionable for practical reasons. It may happen that, in the near future, when electric grids are designed for both efficient energy delivery and data communication, the PLC cell geometry will naturally look for a compromise between energy delivery and data communication.

PLC cells, which can be indoor or outdoor and cover home, neighborhood, or wide areas due to their proximities, can result in neighboring networks (NN) interference [51]. The NN interference is a typical problem that CogPLC can effectively handle. In this context, PLC cells demand intelligent devices/terminals to manage communication within each PLC, communications between different PLC cells and, the coexistence with the primary users. One can state thus that PLC cells will compete or cooperate with the others, operating in the vicinity, as well as with several primary users, such as aeronautics mobiles, police stations, broadcasting stations, amateur radios, and so forth.

The success of CogPLC relies heavily on the design of effective and powerful techniques to monitor and to share the available resources to bring harmonious coexistence among PLC cells. The first attempts to investigate cognitive concepts in PLC systems are discussed in [52]-[55]. The frequency band that has been regulated for PLC systems in Brazil ranges from 1.7 MHz up to 50 MHz [56]. In the future, this range will be extended to go up to 500 MHz.

Currently telecommunications regulatory authorities as well as standardization institutions are focused on the frequency range from 1.7 to 30 MHz as well as 1.7 up to 100 MHz, which implies sharing the entire HF and part of the very high frequency (VHF) bands [57] Also, new contributions forecast that the frequency band up to 500 MHz will be, in the future, used for data communication through electric power grids [58]-[59]. Although the spectrum for possible use by PLC technologies has considerably increased during the past two decades, regulations restrictions to protect primary users from interferences originated in PLC systems resulted, in the worst-case scenario, an availability of less than 40% of the 1.7 up to 30 MHz band allocated to PLC systems (see Tables A.1 and A.2 in [37]). Conjecturing that only 40% of the bandwidth is allocated to PLC systems leads to the conclusion that if the spectral efficiency is, on average, equal to 10 bits/s/Hz, then the maximum bit rate achievable by the system can be, thus, as high as 113 Mbps. This indicates that cognitive concepts have to be introduced to offer theoretical foundations for the design of tools that can maximize the spectrum sharing in order to advance the applicability and usefulness of PLC technologies.

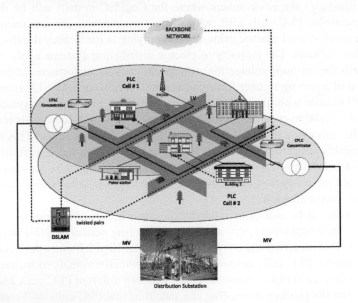

Figure 15.1   CogPLC environment.

15.2.2   DYNAMIC NOTCHING AND POWER CONTROL

To comply with the current regulation limits on interference, originated in PLC systems, on licensed incumbent users, static notching of pre-determined radio frequencies frequency bands are imposed, to prevent non-licensed emission at these radio frequencies. These frequencies are listed in a database (into which the PLC system connects prior to data transmission). Dynamic notching is also imposed to prevent emission on frequencies in which operational radio services are detected, so that the occupied frequency sub-band docs not receive interference. In order to avoid harmful interference, PLC modems apply a mask with static notching, in a set of frequency subbands, which are used by primary users. Emission on these frequencies are forbiden even if primary users are not operating which frequently occurs. A notching of pre-determined radio frequencies could eventually improve the performance of PLC systems, in terms of bit rate. The use of this strategy demands that a database, to be accessed by PLC systems, must be available. This issue raises then the following questions: i) How will the PLC modem communicate with the database? ii) Where will the database be located? iii) How can the database be updated? iv) What is the procedure to include frequencies in the database? A solution built around these questions is not so attractive due to the complexity and amount of resources that should be available. However, dynamic notching combined with power control is being considered as a good approach for CogPLC.

Essentially, the dynamic notching is applied only at frequencies that are being used by primary users. The set of frequencies occupied by primary users should previously be obtained by the PLC system itself. This system has, to achieve this task, to perform a spectrum sensing as, for instance, standardized by ETSI [39].

The power control is responsible for controlling the interference by reducing the transmitted power. While the use of dynamic notching can easily be implemented in PLC cells, the adoption of power control, to guarantee the harmonic coexistence among the cells, can be a challenging issue because the sharing of information among all entities involved (e.g., PLC cells and primary users or, at least, among PLC cells) is required.

The CogPLC can sense the environment by employing, basically, two distinct type of sensors: (a) a PLC coupler and/or (b) an antenna. If the sensors are used simultaneously, and cooperative spectrum sensing techniques are considered [42], a better performance in detecting the primary users can be attained. In fact, the diversity offered by the antenna and PLC coupler can be exploited to design effective and robust detection techniques. Cognitive PLC Systems have been demonstrated by ETSI plug tests [60]-[61]. The simultaneous use of both sensors for spectrum sensing for indoor and outdoor electric power grids was evaluated in [62]-[63].

Figures 15.2 and 15.3 depict the spectrogram of the frequency band between 1.7 and 100 MHz in residences and in the outdoor electric grid, respectively, in the city of Juiz de Fora, Brazil. To show these plots, a measurement campaign was carried out with a setup constituted by an industrial personnel computer, a 200 Msps (mega samples per second) and a 14 bits signal acquisition card with a Peripheral Component Interconnect (PCI) interface, PLC coupler, and an omnidirectional antenna. Both signals were acquired at the same time period. The ESD plots show that these sensors present diversity that can eventually be exploited by CogPLC systems.

The spectrum awareness brought by spectrum sensing techniques raises the following and main questions:

- Should the spectrum sensing technique be designed to statistically characterize the spectrum usage or should it also be used in a CogPLC system?

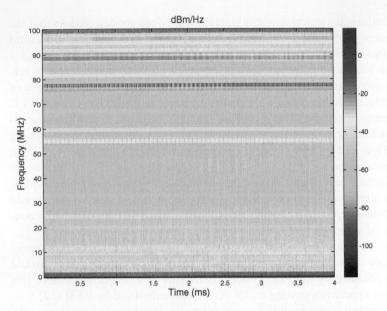

Figure 15.2 Spectrogram for indoor signals acquired with PLC coupler connected to a power cable outlet.

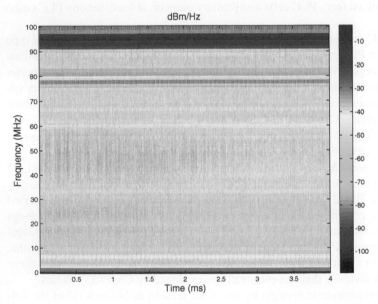

Figure 15.3 Spectrogram for outdoor signals acquired with antenna close to a power cable outlet.

- Should the design of spectrum sensing techniques comply with the current PLC standards or should their use be considered to make changes or updates in such standards?
- How would one design spectrum sensing techniques without models? Should their use be considered to make changes or updates in such standards?
- Which frequency-bands should be addressed?
- Which mechanism should be considered to allow information exchange between devices from distinct neighboring PLC cells so that resource allocation can be efficiently applied?
- What kind of signal processing techniques are more suitable for spectrum sensing purposes in the frequency band between 9 kHz and 100/200/300/500 MHz?
- Should they be the same signal processing techniques that have been investigated for wireless communication?
- What is the most appropriate sensor to monitor the spectrum?

The list of issues to be addressed to advance the theory for the design CogPLC systems can include many other subjects. However, in this chapter we try to formulate spectrum sensing problems for PLC systems, to review spectrum sensing techniques applied to these systems, and to present some results related to a new spectrum sensing technique that can be applied to statistically characterize the spectrum as well as to monitor the spectrum by the devices in PLC cells.

15.3 SPECTRUM SENSING TECHNIQUES

The most common spectrum sensing signal processing techniques [64]-[67] and properties used in cognitive wireless communications, such as matched filtering, energy detector, cyclostationarity, signal identification, waveform based sensing, wavelets, multi-taper spectrum estimation, higher-order statistics, compressive sensing, can be applied in CogPLC context. A brief explanation of some spectrum sensing techniques used in the cognitive radio context, considered as adequate for Cog-PLC, will be presented below.

15.3.1 MATCHED FILTERING

The optimum spectrum sensing technique to detect the presence of primary users in the scenario, where the CogPLC knows all the characteristics of the signals emitted by such users [68], is called matched filtering. The primary user characteristics of interest are carrier frequency, frequency bandwidth, modulation type, pulse shaping, signaling rate, and so forth. When the number of primary users is small, this technique could be applied if computational complexity is regarded as not an important constraint. The highly probable fact that some primary users apply telecommunication technologies unknown to the public, reduces drastically the chances of this kind of technique to be successful in CogPLC. It may be an interesting choice when the aim is to detect the presence of AM and FM radio stations.

15.3.2   ENERGY DETECTION

Energy detection is the most investigated spectrum sensing technique, for detecting the presence of primary users on the network, mainly because of the reduced computational complexity required for its use [69] since, for this kind of technique dispenses with the knowledge of the specific characteristic of the signals emitted by the primary users. The primary users' presence is detected by comparing the level of energy of the signals arriving at the CogPLC sensor with a threshold. This threshold is dependent upon the noise floor and the knowledge of the power losses in the links among the primary users and the CogPLC. The choice of a reliable threshold can be an intricate task to be accomplished, since it depends on the kind of transducer in use — which can be either a PLC coupler or an antenna — and, in addition, depends on the estimation of the corresponding power losses. Furthermore, the presence of impulsive noise can reduce the effectiveness of the energy detector method if the primary users occupy lower frequency bands.

15.3.3   CYCLOSTATIONARITY DETECTION

Cyclostationarity detection is a technique that exploits the cyclostationarity behavior inherent to the emitted signals, brought by the need to periodically transmit information such as header, preamble, cyclic prefix, and so forth. The periodicity present on these signals can be used for detection purposes [69]. The transmitted signal $y(t)$ can be modeled as a stochastic process $Y(t)$ with Cyclic Spectrum Density (CSD) that can be expressed as

$$S(f, \alpha) = \sum_{-\infty}^{\infty} R_Y^{\alpha}(\tau) e^{-j2\pi\tau}, \tag{15.1}$$

where

$$R_Y^{\alpha}(\tau) = \mathbb{E}\left\{ Y(n+\tau)Y^*(n-\tau)e^{j2\pi n} \right\} \tag{15.2}$$

is the Cyclic Autocorrelation Function (CAF) of process and α is the cyclic frequency. Peaks will appear in the CSD. These peaks inform that the cyclic frequency, α, is equal to the fundamental frequency of the signal transmitted by a primary user. The cyclostationarity allows differentiating noise from the signal transmitted by primary users.

15.3.4   RADIO IDENTIFICATION-BASED SENSING

The radio identification based sensing allows a CogPLC device to identify the telecommunication standards that are being used by primary users [70]. It is very similar to the functionalities of a high-quality spectrum analyzer. With this technique, the CogPLC device is capable of demodulating and extracting information from the signals. For the frequency band between 500 kHz and 100 MHz, a CogPLC device could demodulate signals transmitted by AM and FM stations in order to obtain the geolocation. This is a kind of technique that is more appropriate for a spectrum monitoring device than for a CogPLC device.

15.3.5 WAVEFORM BASED-SENSING

Waveform based sensing can be used when the CogPLC knows a priori some characteristics of the primary users, such as preamble, pilot pattern, spreading sequences, and so forth [71]. A CogPLC device could use this information to obtain a correlation with incoming signals and, as a consequence, detect the presence of primary users.

15.3.6 WAVELET

Wavelet sensing, which uses wavelet transform, has been deployed to identify edge in the spectrum density [72]. This edge corresponds to the transition between sub-bands that are in use and spectrum holes or vice versa and, therefore, it could be used to make a map of spectrum usage.

15.3.7 MULTITAPER SPECTRUM ESTIMATIONS

The Multitaper Spectrum Estimation (MTM) was proposed as a spectrum sensing technique for cognitive radio to detect broadband signals [64]. The MTM result in power spectrum estimation close to a maximum likelihood approximation of PSD. However, it can be applied to detect narrowband signals.

15.3.8 SPECTRUM SENSING FOR COGPLC

A few spectrum sensing techniques for CogPLC have been proposed in the literature [28], [41], [48], [54], [73], and [74]. In [54], a CogPLC system based on distributed Orthogonal Frequency Division Multiple Access (DOFDMA) has been introduced with sub-carriers sensing using energy technique selected to detect primary users. In [41] and [74] spectrum sensing techniques that rely on signal to noise ratio (SNR) were used for detecting the emission of radio systems with frequencies on the short wave (1.6-30 MHz) range. The investigation of the use of a PLC coupler and of an antenna for sensing the spectrum was carried out in [48]. A spectrum sensing technique based on cross-correlation techniques was addressed in [28] and [73]. Pattern recognition based techniques for spectrum sensing were initially discussed in [38] and [63]. In [38] it was shown that the energy detection technique worked for both detections of analog modulation (AM) and Digital Radio Mondiale (DRM) signals.

15.4 SPECTRUM SENSING TECHNIQUES FOR COGPLC

Although spectrum sensing techniques could take into account time versus frequency representations to reveal the presence of primary users, only the frequency representation of signals to highlight spectrum occupancy will be addressed. By considering this representation, the power spectral density (PSD) or energy spectral density (ESD) can be considered. The use of PSD for spectrum analysis is well established in the literature. However, little attention has been given to the use of ESD for spectrum sensing purpose. In this section, we present some results regarding the use of ESD as well as frequency representations of monitored signals together with pattern recognition based techniques for providing high detection rate of primary users. In this context, let the monitored signal be modeled as

$$x(t) = s(t) + \upsilon(t)$$
$$= \sum_{k=1}^{K} s_k(t) + \upsilon(t), \tag{15.3}$$

where, for a given t, $s_k(t) \in \mathbb{R}$ is the signal generated by the k-th data communication system (analog or digital), K is the number of data communication system or spectrum users, $\upsilon(t) \in \mathbb{R}$ is the additive noise, and $x(t) \in \mathbb{R}$ is a time domain representation of the monitored signal, which is corrupted by the presence of additive noise. The acquisition of the signal $x(t)$ for spectrum sensing purpose can be made with different sensors (transducers together with Analog-to-Digital (ADC) boards). In general there are M_s different types of sensors, characterized by the vector function $\mathcal{P} = (P_1, P_2, \ldots, P_{M_s})$, then the use of the m_s-th sensor with input $x(t)$ and output signal $x_{m_s}(t)$ yields the vector $\mathbf{x}_{m_s}$ which is then segmented in ℓ sub-vectors. Let $\mathbf{x}_m$ be m-th sub-vector produced by the segmenting process. We have then $\mathbf{x}_{m_s} = [\mathbf{x}_1, \ldots, \mathbf{x}_m, \ldots, \mathbf{x}_\ell]$. It should be noted that

$$\mathbf{x}_m = \mathbf{s}_m + \mathbf{v}_m, \tag{15.4}$$

where $\mathbf{x}_m \in \mathbb{R}^{N \times 1}$, $\mathbf{s}_m \in \mathbb{R}^{N \times 1}$, and $\mathbf{v}_m \in \mathbb{R}^{N \times 1}$ are vectors with N samples in time domain, with components obtained from the discrete time representations of signals $x_{m_s}(t)$, $s_{m_s}(t)$, and $v_{m_s}(t)$, respectively. Since we will be considering that the frequency band of $x(t)$ extends from 0 up to B Hz, the sampling frequency is $2B$. In the following equation, (15.5), $\check{\mathbf{x}}_m$ is the resulting vector when a linear transformation A_p (DFT, the most common) is applied to the vector $\check{\mathbf{x}}_m$. In other words,

$$\check{\mathbf{x}}_m = \mathbf{A}_p \mathbf{x}_m$$
$$= \check{\mathbf{s}}_m + \check{\mathbf{v}}_m$$
$$= \sum_{u=1}^{U} \check{\mathbf{s}}_{u,m} + \check{\mathbf{v}}_m, \tag{15.5}$$

where $\check{\mathbf{x}}_{m,p} \in \mathbb{C}^{N \times 1}$, $\check{\mathbf{s}}_{u,m} = \mathbf{A}_p \mathbf{s}_{u,m} \in \mathbb{C}^{N \times 1}$, and $\check{\mathbf{v}}_{u,m} = \mathbf{A}_p \mathbf{v}_{k,m} \in \mathbb{C}^{N \times 1}$. The set $\mathcal{A} = \{\mathbf{A}_1, \mathbf{A}_2, \ldots, \mathbf{A}_P\}$ denotes a set of P transform techniques. It should be noticed that the transform in (15.5) exhibit explicitly the spectral content of $\mathbf{x}_m$. The signal $\check{\mathbf{s}}_u$ corresponding to the u-th primary user can be expressed as

$$\check{\mathbf{s}}_{u,m} = \left[\mathbf{0}_{k,a}^T \; [\check{s}_{u,m}(L_{a,u,m} + 1) \ldots \check{s}_{u,m}(L_{a,u,m} + L_{u,m})] \; \mathbf{0}_{u,m,b}^T \right]^T \tag{15.6}$$

in which $\mathbf{0}_{a,u,m}$ and $\mathbf{0}_{b,u,m}$ refer to $L_{a,u,m} \times 1$ and $L_{b,u,m} \times 1$ vectors constituted by zeros and

$$L_{a,u,m} + L_{b,u,m} + L_{u,m} = N, \quad \forall k \in \mathbb{Z}. \tag{15.7}$$

The processing requires that vector $\check{\mathbf{x}}_m$ be segmented into Q segments, each one with L_q samples. The q-th segment of $\check{\mathbf{x}}_m$ is represented by

$$\check{\mathbf{x}}_{m,q} = \check{\mathbf{s}}_{m,q} + \check{\mathbf{v}}_{m,q}$$

$$= \sum_{u=1}^{K} \check{\mathbf{s}}_{u,m,p,q} + \check{\mathbf{v}}_{m,q}, \qquad (15.8)$$

where $q = 1, 2, \ldots, Q$, and

$$\check{\mathbf{x}}_{m,q} = \left[\check{x}_{m,q}((d-1)L_q) \ \ldots \ \check{x}_{m,q}(dL_q - 1) \right]^T \qquad (15.9)$$

is a vector constituted by L_q consecutive elements of the vector $\check{\mathbf{x}}_m$. Similarly, we define

$$\check{\mathbf{s}}_{m,q} = \left[\check{s}_{m,q}((d-1)L_q) \ \ldots \ \check{s}_{m,q}(dL_q - 1) \right]^T, \qquad (15.10)$$

and

$$\check{\mathbf{v}}_{m,q} = \left[\check{v}_{m,q}((d-1)L_q) \ \ldots \ \check{v}_{m,q}(dL_q) - 1) \right]^T. \qquad (15.11)$$

Finally, take the set $\mathcal{L} = \{L_1, L_2, \ldots, L_Q\}$ to be the set of the Q different lengths that arise when segmenting the vector $\check{\mathbf{x}}_m$.

After the segmentation, the spectrum sensing for CogPLC systems can be formulated as the simple hypotheses test problem

$$\check{\mathbf{x}}_{m,q} = \begin{cases} \check{\mathbf{v}}_{m,q}, & H_0 \\ \check{\mathbf{s}}_{m,q} + \check{\mathbf{v}}_{m,q}, & H_1 \end{cases}, \qquad (15.12)$$

where H_0 and H_1 are the hypothesis associated with the absence and presence, respectively, of a primary user in the frequency band covered by the vector. If a length R feature vector,

$$\check{\mathbf{r}}_{m,q} \in \mathbb{R}^{R \times 1} \qquad (15.13)$$

is extracted from $\check{\mathbf{x}}_{m,q}$ the problem dimensionality $R \leqslant L_q$ is reduced. In this case distinct feature extraction and selection techniques can be applied to obtain the vector $\check{\mathbf{r}}_{m,q}$. Let the set of G techniques, which combines feature extraction and selection, that can be applied be represented by $\Psi = \{\Psi_1, \Psi_2, \ldots, \Psi_G\}$.

The probabilities P_D and P_F, the probability of detection, and the false alarm probability respectively, are expressed by

$$P_D = P_r\left(H_0 \mid \check{\mathbf{r}}_{m,q}\right)\Big|_{\check{x}_{m,q}=\check{v}_{m,q}} + P_r\left(H_1 \mid \check{\mathbf{r}}_{m,q}\right)\Big|_{\check{x}_{m,q}=\check{s}_{m,q}+\check{v}_{m,q}} \qquad (15.14)$$

and

$$P_F = P_r\left(H_1 \mid \check{\mathbf{r}}_{m,q}\right)\Big|_{\check{x}_{m,q}=\check{v}_{m,q}} + P_r\left(H_0 \mid \check{\mathbf{r}}_{m,q}\right)\Big|_{\check{x}_{m,q}=\check{s}_{m,q}+\check{v}_{m,q}} \qquad (15.15)$$

In (15.15), P_r is the probability measure. Probabilities P_D and P_F can, in principle, be obtained by applying a set of M distinct detection techniques that can be selected from the set $\mathcal{D} = \{D_1, D_2, \ldots, D_M\}$. Finally, each solution results in a set of J distinct computational complexity represented by the set $C = \{C_1, C_2, \ldots, C_{F|}\}$.

By considering all possible choices of transforms among those belonging to $\mathcal{A}$, of detection techniques in $\mathcal{D}$, of sensors in $\mathcal{P}$, of selection techniques in Ψ, and all choices of segmentation sets, $\mathcal{L}$, the problem is to find the best choice, namely, the choice which gives the maximum value of P_D such that the complexity does not exceed a maximum value of complexity $C_{\max}$. In other words, the problem can be formulated as follows:

Find the choice P^*, A^*, D^*, L^*, such that

$$P_D^* = \max_{q \in \mathcal{L}} P_D(q) \tag{15.16}$$

subject to $C_j \leq C_{\max} \ \forall \ j \in \{1, 2, \ldots, F\}$.

A solution to this problem has to consider too many choices which are dependent on the characteristics of primary users. It is important to emphasize that the sets $\mathcal{A}$, $\mathcal{L}$, $\mathcal{D}$, $\mathcal{P}$, and Ψ play a crucial role in the choice of a spectrum sensing technique for a CogPLC. Solving this optimization problem is a very difficult task to be accomplished. One way to overcome this difficulty is to find a suboptimal solution by applying well-known pattern recognition techniques. In [75] and [76] results have been presented revealing a promising spectrum sensing technique for CogPLC. This technique, which was initially introduced in [38] and [62]-[63], can be summarized by the diagram depicted by the block diagram in Figure 15.4. The first block in this figure applies a time to frequency transformation G to the discrete-time signal, producing, at its output, the vector $\check{x}$ which is then processed in the second block, output a reduced set of features, vector $\check{r}_{m,q}$, which were a priori selected using a feature selection technique. The features are extracted so that the primary user detection can be performed by separately processing each segment of vector. The third and last block implements a technique which, by observing the set of features $\check{r}_{m,q}$, decides for the presence or absence of the primary user in the monitored frequency band.

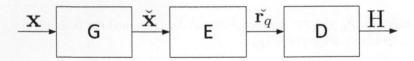

Figure 15.4 Block diagram of the spectrum sensing technique as proposed in [63].

A detailed description of each block is presented next.

Processor G is in charge of applying the transform operator to the vector **x**. The most common transforms used in the implementation of this block are the DFT

(Discrete Fourier Transform), the DHT (Discrete Hartley Transform), and the MCLT (Modulated Complex Lapped Transform). These transforms are formalized in Section 4.1.

Processor E produces vectors $\check{\mathbf{x}}_{m,q}$ by appropriately selecting components from vector $\check{\mathbf{x}}$ extracting also, a feature vector $\check{\mathbf{r}}_{m,q} \in \mathbb{R}^{L_R \times 1}$ constituted by features that were a priori selected, such that $L_R \leqslant L_q$. The features selected to compose vector $\check{\mathbf{r}}_{m,q}$ and their selections are addressed in Section 4.2.

Processor D implements the algorithm that selects the most probable hypothesis, H_0, or H_1, deciding that either there is not a primary user transmitting, i.e., $\check{\mathbf{x}}_{m,q} = \check{\mathbf{v}}_{m,q}$ or, $\check{\mathbf{x}}_{m,q} = \check{\mathbf{v}}_{m,q} + \check{\mathbf{v}}_{m,q}$ or, on the contrary, there is. There are many ways to solve this problem; however, in this contribution, it was chosen to design a Multilayer Perceptron Neural Network (MLPNN), to show the gains that can be obtained when a particular technique, a computational intelligence technique, has been used. The chosen MLPNN is succinctly described in Section 15.4.4.

15.4.1 FREQUENCY DOMAIN TRANSFORMS

Spectrum sensing applications require the choice of a transform to make apparent the spectral content of the signal. Any transforms that provide the time versus frequency representation of the signal should, therefore, be included in the set A. The current section describes the DFT, DHT, and MCLT, each of them, chosen for their deficiency in revealing spectrum information of the signal.

Following some of the indices m, p, and q, will be dropped becoming implicit then that the notation $\check{\mathbf{x}}_{m,q}$, $\mathbf{x}_{m,q}$, etc, will be replaced by the simpler notation $\check{\mathbf{x}}$, $\mathbf{x}$, etc.

Let us consider that the vector to be transformed, defined in (15.4) is $\mathbf{x}_m$, given by the deterministic finite discrete time sequence of N values

$$\{x[n]\}_{n=0}^{N-1}, \quad x[n] \in \mathbb{R}.$$

- The DFT can be expressed by

$$\check{x}[k] = \sum_{n=0}^{N-1} x[n]\exp(-2\pi kn/N), \quad 0 \leqslant k \leqslant N-1, \tag{15.17}$$

- The DHT [66] can be given by

$$\check{x}[k] = \sum_{n=0}^{N-1} x[n]\left(\cos\left(\frac{2\pi kn}{N}\right) + \sin\left(\frac{2\pi kn}{N}\right)\right), \quad 0 \leqslant k \leqslant N-1. \tag{15.18}$$

- The MCLT [73]-[74] is given by

$$\check{x}[k] = \sum_{n=0}^{2N-1} x[n]\,p[n,k], \quad 0 \leqslant k \leqslant N-1, \tag{15.19}$$

in which $p[n, k] \triangleq p_C [n, k] + j p_S [n, k]$ is a basic analysis complex function that, with $j = \sqrt{-1}$, is specified by the functions

$$
\begin{aligned}
p_C[n, k] &= \sqrt{\frac{2}{N}} \, h[n] \cos\left(\left(n + \frac{N+1}{2}\right)\left(k + \frac{1}{2}\right)\frac{\pi}{N}\right), \\
p_S[n, k] &= \sqrt{\frac{2}{N}} \, h[n] \sin\left(\left(n + \frac{N+1}{2}\right)\left(k + \frac{1}{2}\right)\frac{\pi}{N}\right),
\end{aligned}
\tag{15.20}
$$

in which $h[n]$ is the modulated impulse response of the filter.

15.4.2 FEATURE EXTRACTION TECHNIQUES

The feature extraction techniques provide a large set of features related to a vector. A premium is placed on extracting relevant features that are somewhat invariant to the change of the vector $\check{\mathbf{x}}_{m,q}$. There are several kinds of feature extraction techniques for both real and complex sequences. In this contribution we focus on feature extraction techniques based on HOS (higher order statistics), skewness, energy, and kurtosis-based features due to their effectiveness when used in conjunction with other detection techniques [79]. Let us now consider that $\{\check{x}[k]\}_{k=0}^{N-1}$, an observation that corresponds to the N-length sequence of (deterministic) values, is modeled as the random vector $\check{\mathbf{X}}$. We will consider $\{\check{X}[k]\}$ to be a sequence of independent and identically distributed random variables with $\mathbb{E}\{\check{X}[k]\} = 0$, in which $\mathbb{E}\{\cdot\}$ denotes the expectation operator. The feature detection techniques to be considered are described next.

15.4.2.1 Higher Order Statistics

HOS is a widely used technique in non-Gaussian processes and non-linear systems. The use of HOS as a feature can result in detection, classification, and identification with high performance [79]. Based on the fact that the monitored signal is a non-Gaussian process, then HOS can be a powerful candidate to constitute the feature vector $\check{\mathbf{r}}_{m,q}$. The second-, third-, and fourth-order cumulants are given by [75]

$$
C_{2,\check{X}_{m,q}}[i] = \mathbb{E}\left\{\check{X}_{m,q}[k] \, \check{X}_{m,q}[k + i]\right\},
\tag{15.21}
$$

$$
C_{3,\check{X}_{m,q}}[i] = \mathbb{E}\left\{\check{X}_{m,q}[k] \, \check{X}_{m,q}^2[k + i]\right\},
\tag{15.22}
$$

$$
C_{4,\check{X}_{m,q}}[i] = \mathbb{E}\left\{\check{X}_{m,q}[k] \, \check{X}_{m,q}^3[k + i]\right\} - 3 C_{2,\check{X}_{m,q}}[i] C_{2,\check{X}_{m,q}}[0],
\tag{15.23}
$$

respectively, where i corresponds to the i-th lag. An estimate for the HOS can be obtained by

$$
\hat{C}_{2,\check{X}_{m,q}}[i] \cong \frac{2}{N} \sum_{k=0}^{N/2-1} \check{x}_{m,q}[k] \, \check{x}_{m,q}[k + i],
\tag{15.24}
$$

$$
\hat{C}_{3,\check{X}_{m,q}}[i] \cong \frac{2}{N} \sum_{k=0}^{N/2-1} \check{x}_{m,q}[k] \, \check{x}_{m,q}^2[k + i],
\tag{15.25}
$$

$$\hat{C}_{4,\check{X}_{m,q}}[i] \cong \frac{2}{N} \sum_{k=0}^{N/2-1} \check{x}_{m,q}[k] \, \check{x}_{m,q}^2[k+i]$$

$$-\frac{12}{L^2} \sum_{k=0}^{N/2-1} \check{x}_{m,q}[k] \, \check{x}_{m,q}[k+i] \sum_{k=0}^{N/2-1} \check{x}_{m,q}^2[k], \qquad (15.26)$$

15.4.2.2 Skewness

Skewness measures the degree of symmetry on a probability density function (pdf) of a random variable. For the random variable $\check{X}[k]$ it would be defined as

$$\gamma_3(\check{X}_{m,q}) = \frac{\mathbb{E}\left\{ \left(\check{X}[k] - \mu_{\check{X}} \right)^3 \right\}}{\left(\mathbb{E}\left\{ \left(\check{X}[k] - \mu_{\check{X}} \right)^2 \right\} \right)^{3/2}}, \qquad (15.27)$$

where $\mathbb{E}\left\{ \check{X}[k] \right\} = \mu_{\check{X}}$. If the random variables $\check{X}[k]$ are equidistributed $\hat{\gamma}_3(\check{X})$ is the estimate of $\mathbb{E}\left\{ \gamma_3 \left[\check{X} \right] \right\}$.

15.4.2.3 Kurtosis

Kurtosis measures the peakedness of the pdf of a random process. It is evaluated by

$$\kappa_3(\check{X}_{m,q}) = \frac{\mathbb{E}\left\{ \left(\check{X}[k] - \mu_{\check{X}} \right)^4 \right\}}{\left(\mathbb{E}\left\{ \left(\check{X}[k] - \mu_{\check{X}} \right)^2 \right\} \right)^{1/2}} - 3 . \qquad (15.28)$$

Again, $\hat{\kappa}_3(\check{X}_{m,q})$ is the estimate of $\kappa_3(\check{X}_{m,q})$.

15.4.2.4 Signal Energy

It is the most common feature applied for spectrum sensing due to its low computational complexity and the lack of necessity of previous information about the transmitted signal. For a given value of m, L_q-length deterministic sequence, the expression

$$\hat{E}_{\check{X}_{m,q}} = \sum_{k=0}^{L_q-1} |\check{x}[k]|^2 , \qquad (15.29)$$

where $|\cdot|$ stands for the absolute operator, gives the estimate of the energy of the random sequence $\check{X}_{m,q}$. Then, we end up with the vector of η features extracted from $\check{x}$ given by

$$\mathbf{z}_{\check{x}} = \left[\hat{E}_{(\check{x})}, \, \hat{\gamma}_{3(\check{x})}, \, \hat{\kappa}_{3(\check{x})}, \, \hat{\mathbf{C}}_{2,\check{x}}, \, \hat{\mathbf{C}}_{3,\check{x}}, \, \hat{\mathbf{C}}_{4,\check{x}} \right]^T \qquad (15.30)$$

where $\hat{\mathbf{C}}_{k,\check{x}} = \left[\hat{C}_{k,\check{x}}[0] \quad \hat{C}_{k,\check{x}}[1] \quad \cdots \quad \hat{C}_{k,\check{x}}[L_q - 1] \right]^T$, $k = 2, 3, 4$.

15.4.3 FEATURE SELECTION TECHNIQUE

Feature selection is a process that selects the most significant features of a set. This stage is crucial to the detection process, since reducing the number of features reduces, consequently, the dimensionality of the associated vector. One of the most used and simple feature extraction techniques that presents satisfactory results is the Fisher's Discriminant Ratio (FDR) [80]-[81].

All the features obtained (one for each segment m) are random, and we will model $z_{\check{x}}$ as a random vector $\mathbf{Z}$ with coefficients that are independent, Gaussian distributed, random variables. The two important events to be considered are the two hypothesis $H_0 = \{$signal is present$\}$, and $H_1 = \{$signal is absent$\}$.

We will use the compact notation $\mathcal{N}(\mathbf{0}, \mathbf{K_Z})$ to represent the joint distribution, $f_{\mathbf{Z}|H_0}(\alpha)$. Notice that the auto-covariance matrix of $\mathbf{Z}$ is a diagonal matrix $\mathbf{K_Z}$ with the vector $[\sigma_1^2, \ldots, \sigma_\eta^2]$. We have then

$$f_{\mathbf{Z}|H_0}(\alpha) \cong \mathcal{N}(\mathbb{E}\{\mathbf{Z} \mid H_0\}, \mathbf{K_Z}). \tag{15.31}$$

Similarly, we have, when H_1 is true,

$$f_{\mathbf{Z}|H_1}(\alpha) \cong \mathcal{N}(\mathbb{E}\{\mathbf{Z} \mid H_1\}, \mathbf{K_Z}). \tag{15.32}$$

Then, the FDR vector which leads to a reduced dimensional space of feature vectors, can be defined by the following expression

$$\mathbf{J}_c = (\mathbb{E}\{\mathbf{Z} \mid H_0\} - \mathbb{E}\{\mathbf{Z} \mid H_1\}) \odot \left[\frac{1}{2\sigma_1^2}, \ldots, \frac{1}{2\sigma_\eta^2} \right] \tag{15.33}$$

where $\mathbf{J}_c = \begin{bmatrix} J_1 & J_2 & \cdots J_\eta \end{bmatrix}^T$ and, η is the total number of features. The symbol $\odot$ refers to the Hadamard's product, defined by $\mathbf{a} \odot \mathbf{b} = [a_0 b_0 \; a_1 b_1 \ldots a_l b_l]^T$. The vector $\mathbf{J}_R$, of length $L_R \ll \eta$, formed with the elements of the FDR vector $\mathbf{J}_c$ having the highest values, define the best feature vector for detecting the primary users in the feature space whose dimension is L_R.

15.4.4 DETECTION TECHNIQUE

The reduced dimensionality vector $\check{x}_R$ obtained by selecting the best coefficients from $\check{x}_{m,q}$ needs to be afterwards tested to find out among the detectors which is the detector most suitable. There are several detection techniques based on detection theory to do it, and some of these quite complex since finding a proper probability density function to match the problem randomness is a hard task. The Multi-Layer Perceptron Neural Network (MLPNN) was chosen and the training procedure highlighted in [82] as the detection technique, a technique which rendered good performance in detection and classification problems related to other applications.

A matricial representation of a MLPNN with one hidden layer can be expressed by

$$\check{\mathbf{x}}_{m,q} = \mathbf{B}^T \begin{bmatrix} \check{\mathbf{r}}_{m,q} \\ 1 \end{bmatrix}, \tag{15.34}$$

$$y_{nn} = \mathbf{c}^T \begin{bmatrix} \tanh(\check{\mathbf{x}}_{m,q}) \\ 1 \end{bmatrix}, \tag{15.35}$$

where $\check{\mathbf{r}}_{m,q}$ is a vector constituted of L_R features, which are extracted according to the feature extraction technique described in Section 4.2; L_{nn} is the number of neurons at the hidden layer; $\mathbf{B} \in \mathbb{R}^{(L_R+1)\times L_{nn}}$ is the synaptic weight matrix between the input and the hidden layers; and $\mathbf{c} \in \mathbb{R}^{(L_{nn}+1)\times 1}$ is the matrix of the synaptic weights between the hidden and the output layers. For an in-depth description of the optimum synaptic weight matrices $\mathbf{A}^*$ and $\mathbf{b}^*$ the reader is referred to [38] where the training procedure is described.

15.4.5 PERFORMANCE ASSESSMENT

The results, obtained by simulating the discussed technique, applied to measured data are exhibited in Sections 15.4.5.1 and 15.4.5.2, respectively. Additionally, in both sections, answers to the questions formulated in Section 15.2.2 are presented.

The performance results obtained, when MLPNN is the detection technique used, are reported in terms of detection rate and false alarm rate.

15.4.5.1 Simulated Data

Four signals, occupying different frequency bands, constitute the data used in our simulations to assess the performance of the spectrum sensing technique discussed. These signals represent primary users occupying the frequency bandwidths 40 kHz, 125 kHz, 250 kHz, and 1 MHz. The frequency band occupied by these signals are arbitrary allocated at non-overlapping bands. All primary users make use of a Quadrature Phase-Shift Keying (QPSK) digital modulation scheme. The passband signal representation related to the of k-th user is modeled as the stochastic process characterized by

$$S_{k,m}(t) = \sum_{\ell=-\infty}^{\infty} A_{c,k,m} p_k(t - \ell T_k) \cos(2\pi k f_{c,k} t) \tag{15.36}$$
$$-A_{s,k,m} p_k(t - \ell T_k) \sin(2\pi k f_{c,k} t),$$

in which $A_{c,k,m}$, $A_{s,k,m} \in \left\{ -\frac{1}{\sqrt{2}}, +\frac{1}{\sqrt{2}} \right\}$ are i.i.d binary, equiprobable random variables, $f_{c,k}$ is the carrier frequency, T_k is the symbol period for the k-th user, and $p_k(t)$ is k-th user raised cosine pulse, having roll-off factor α, defined by

$$p_k(t) = \frac{\sin(1/T_k)}{1/T_k} \left(\frac{\cos(\frac{\alpha\pi t}{T_k})}{1 - \left(\frac{2\alpha\pi t}{T_k}\right)^2} \right). \tag{15.37}$$

Table 15.1

$f_{c,k}$ **and** T_k **values for each carrier**

Carrier (k)	$f_{c,k}$ (MHz)	T_k (μs)
1	9.96	50
2	30.0	16
3	54.88	8
4	88.87	2

In our simulation $\alpha = 0,1988$. Table 15.1 lists the values chosen for the carrier frequencies and the symbol periods of signals $s_k(t)$, transmitted by users $k = 1, 2, 3,$ and 4. The noise is additive white Gaussian noise (AWGN) and the frequency band varies from 1.7 to 100 MHz, the sampling frequency used in our simulations is thus $f_s = 200$ mega samples per second (Msps); $N = 8192$, which is the length of one block. The set of transforms $\mathcal{A} = \{$DFT, DHT, MCLT$\}$ was used with the set of segments $\mathcal{L} = \{16, 32, 64, 128, 256\}$; $L_R = 3$ and $L_{nn} = 3$; and the number of blocks of simulated data is 3600.

In our first analysis only one detector was used to detect the presence of signals with distinct frequency bands. The best features are selected based on all the signals in the database used for designing the cognitive procedure. The same applies to find the combination of transforms and window sizes L_q of the monitored signal to give the best performance results in terms of detection and false alarm rates.

Figures 15.5 to 15.10 display the performance, in terms of the Correct Detection Rate (P_D) and of False Alarm Rate (P_F), of systems using several spectrum sensing techniques, with SNR ranging from -15 to 20 dB. The detection schemes based in DFT, DHT, and MCLT, as well as those based on an ANN (Artificial Neural Network) and on Bayesian detection rules (based on the Gaussianity assumption and maximum likelihood criterion) are compared.

It should be noticed, from examined systems, irrespective of the value of the parameter L_q, that the ANN based schemes of the detection error rate, P_D, exceeds that of the system based on a Bayes scheme and the other three transform based schemes, noticeably for values of SNR below zero. On the other hand, the system based on Bayesian detection rules, has better performance in terms of P_F, the False Alarm Rate.

The examination of Figures 15.7 and 15.9 and, Figures 15.6 and 15.10, shows that the performance of the system, in terms of P_D, the three techniques DFT, DHT, and MCLT are used, improves as the values of L_q grows larger. On the other hand, when confronting the systems in terms of P_F, this trend no long holds, or, in other words, this performance when $L_q = 16$ is smaller $L_q = 4$, as can be seen on Figures 15.9 and 15.5, respectively.

A comparison of the schemes based on the transforms, for the three values of L_q

investigated, reveals that DFT exhibits better performance, and the MCLT and DHT performances are, in this order, smaller.

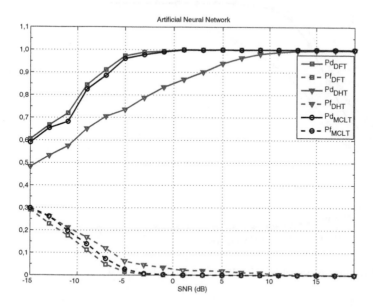

Figure 15.5 Performance evaluation using ANN and $L_q = 4$.

15.4.5.2 Spectrum Measured Data

In order to evaluate the spectrum sensing technique introduced in this chapter using real data, measurements of the signal at frequency band from 1.7 up to 100 MHz were made. The measured data were captured with a Gage Razor CompuScope 1642 acquisition board [83], which makes use of analog-to-digital converters of 16 bits of vertical resolution and a maximum sampling rate of 200 MS/s by channel.

Following this procedure, a specialist can use the spectrogram as presented in Figure 15.2 to map those frequency bands being used by primary users. Two kinds of sensors were used to measure the spectrum occupancy: i) an omnidirectional antenna with the frequency band ranging from 1 MHz up to 1000 MHz and impedance equals to 50 Ω [84], and ii) a capacitive coupler (PREMO P-1240-021) [85], which suffers from attenuation that keeps increasing as the frequency goes beyond 30 MHz and is equipped with a high-pass filter with cutoff frequency equal to 1.7 MHz. Our measurements indicate that the increasing attenuation behavior introduced by the capacitive coupler can be neglected, if it is not high, for two following reasons: i) since the PLC coupler is connected to the electric cable, by means of the power cable outlet, it can be regarded a part of background noise in power cables and, as a consequence, its attenuation profile cannot be verified; and ii) the increasing attenuation mitigates the radio as well as the noise signal and if this attenuation is

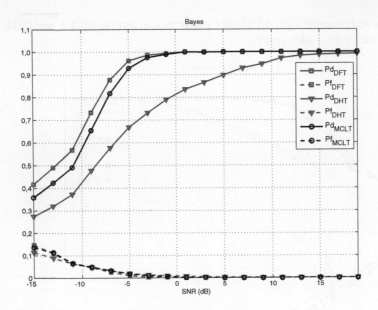

Figure 15.6 Performance evaluation using Bayes' detector for $L_q = 4$.

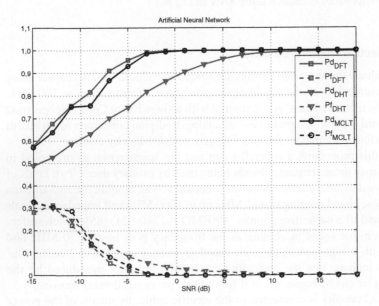

Figure 15.7 Performance evaluation using ANN and $L_q = 8$.

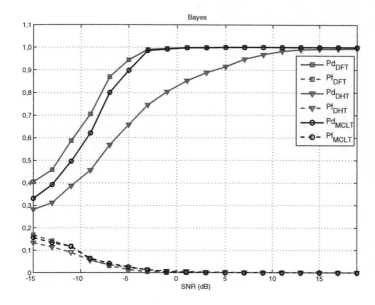

Figure 15.8 Performance evaluation using Bayes' detector for $L_q = 8$.

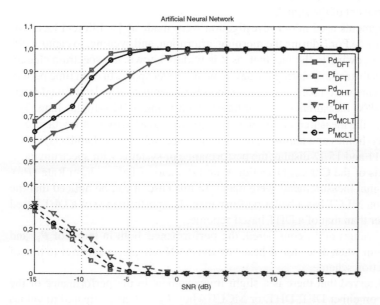

Figure 15.9 Performance evaluation using ANN and $L_q = 16$.

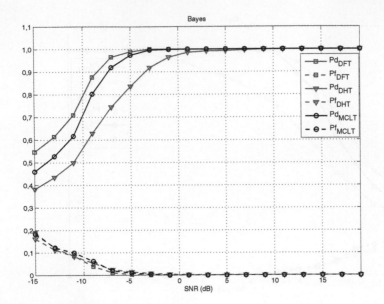

Figure 15.10   Performance evaluation using Bayes' detector for $L_q = 16$.

expressive, then the output of the A/D converter will offer the same SNR for each spectrum component of the signal.

The measurements used in our investigation were made at 16 different locations in the city of Juiz de Fora, during the morning, afternoon, and evening, to capture different influences, owing to the fact that, as it is widely known, the spectrum characteristics varies substantially during the day as well as over space. In each time period, 100 measurements were taken at every 6 seconds (an empirically determined interval). Overall, 4800 measurements were taken, each being 8192 samples long. The dwellings where the measurements took place are close to an airport, army facilities, and urban areas (homes and buildings of low- and medium-income habitants).

Figures 15.11 and 15.12 display the performance of systems with spectrum monitoring, in terms of the Correct Detection Rate (P_d) and of False Alarm Rate (P_F), when actual signal measurements were made. In this case, it can be noticed that the scheme based on MLCT has a better performance. The performance of a DFT based scheme is larger than that of a DHT based scheme.

For the two values of L_q considered, the performance, both in terms of P_D and P_F, of the system when detection is based on the ANN scheme, is superior to that of Bayesian detection scheme.

It is also observed that there is a slight improvement in the performance of the systems based on either DFT, DHT, or MCLT when $L_q = 8$, as compared to values obtained for $L_q = 4$.

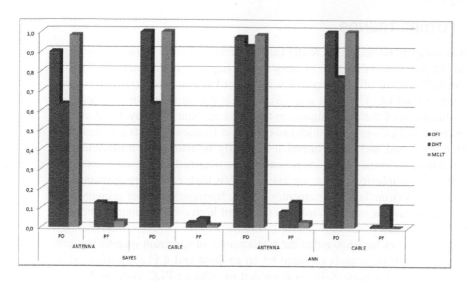

Figure 15.11 Performance evaluation for $L_q = 4$.

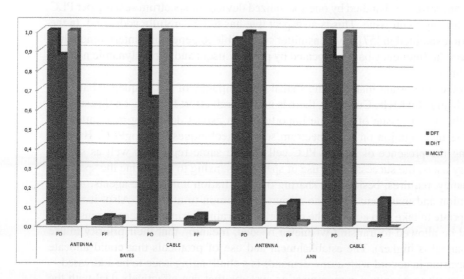

Figure 15.12 Performance evaluation for $L_q = 8$.

15.5 FUTURE TRENDS

An initial comparison between CR and CogPLC that make use of spectral sensing, leads to the conclusion that a lot of work is still required to advance the CogPLC technology. Of course, all techniques developed for CR can, with some adjustment, be applied to CogPLC. However, it is apparent that the introduction of adaptations and modifications of these techniques, to take into account, directly, the particular characteristics of PLC systems, which can operate in frequency bands between 0 and 500 MHz, or beyond, and are perturbed, intrinsically, by man-made noise, can bring substantial improvements. These improvements are obtained at the cost of increased computational complexity. Notice, for instance, that high-resolution monitoring of the frequency spectrum is a requirement, since the frequency band of primary users signals being as narrow as 5 kHz, asks for the use of high computational complexity detection techniques.

Some recent work has shown that the performance of spectrum monitoring techniques, which examine signals captured by an antenna and a PLC coupler, on indoor PLC systems are quite different from that of outdoor PLC systems. It is not clear yet if exploiting the diversity offered by both sensors can improve the detection performance. The same question arises when cooperative detection techniques are considered, since the presence of primary users, which occupy the frequency band between 0 and 500 MHz, can be easily detected by antenna based monitoring equipment. Investigating if spectrum monitoring techniques should be part of all PLC transceivers or handled by one specialized device for spectrum sensing per PLC cell is an important issue. This is an aspect that has to be carefully addressed. In the scenario examined in [57] any transmitter susceptible to generating interferences has to detect the frequency bands allocated by primary users and apply dynamic notches accordingly.

Another important matter is the statistical occupancy of the frequency band between 0 and 500 MHz. Currently, there is no research effort focusing on this subject. Actually, it is an issue of ultimate importance, an aspect deserving more attention than the investigation on new spectrum sensing techniques for CogPLC. By considering the presence of several PLC cells in the same region, as well as several primary users, the success of the use of spectrum sensing to determine the spectrum occupancy, requires, besides the sharing of information among the agents, the coordination and management of the resources common to all users. Also, it will be appropriate to take into account the network constraints (primary users, for instance, should be allowed to transmit whenever needed, meaning that their priority to use the channel is higher). The establishment and use of protocols that could allocate the spectrum in accord with users' demands is a challenging issue to be addressed— it will result in cognitive communication systems that can effectively deal with the scarcity of resources and regulatory constraints.

Another important issue, that has been addressed in ETSI TS 102 578 [56]-[57], is the restriction to put on the time interval required for the CogPLC to sense the occupancy of a frequency band, and subsequently leave or enter the network. A minimum interference level should be set, to avoid interference in the primary users as well as

PLC cells. These recommendations, in order to protect the HF radio communication have limited this time interval to 15 seconds.

The real time monitoring of the spectrum, based on signal processing techniques (cooperative or not) that use time versus frequency representations of the monitored signals, which demand fast and powerful processing devices make room for improvements. The bulk of results on spectrum sensing techniques are based on SNR and energy. Several good results using this point of view have been obtained; however, improved performance can be achieved if more advanced techniques are investigated.

Finally, an issue of paramount importance is finding spectrum sensing approaches of reduced complexity—this relevant aspect will drive the R&D research directions since the next generation of PLC technologies have to bring into their conception the current society demands, being summarized by the words green, flexibility, and sustainability.

15.6   CONCLUSIONS

This chapter discussed CogPLC as applied to smart grid communications and cheap network access—to lessen the digital divide gap. These are challenging topics also in the Internet of Things context. According to the provided information, the success of PLC technologies for SGComm and digital division can be considerably increased if cognitive aspects can be taken into account. The chapter has highlighted the fact that CogPLC is a subject that has been addressed little in the literature and that there are several challenging and interesting research questions deserving attention.

Some results related to the use of pattern recognition based detection techniques were presented. Some relevant aspects associated with spectrum sensing technique for powering PLC systems for SGComm applications were discussed and investigated. Also, a spectrum sensing technique has been discussed in detail under several assumptions and constraints. Its performance has been investigated based on simulation, synthetic generated data as well as on measured data.

Our simulation results with synthetic data has shown that the DFT based sensing presents higher detection rates when compared to sensing based on the other transforms. Simulation results with measured ones led to a different conclusion—as a mater of fact, the DFT based sensing achieved the worst performance. Performance with MCLT based sensing, on the other hand, provided the highest detection rate.

The window size, L_q, selection which exhibited the best detection rate when synthetic data was used was $L_q = 16$. When simulating with measured data, the best selection, yielding higher detection rates, is $L_q = 32$. The discrepancies between the data obtained with synthetic and measured data reveals, in fact, that the theoretical model is far from an ideal representation of the measured data due to the difficulty to capture, with the model, the myriad signals observed in the real spectrum. Hence, the results obtained can be taken as an indication that the proposed technique is effective for spectrum sensing and gives some directions for the design of cognitive PLC systems.

Finally, it can be concluded that the detection of spectrum usage by PLC systems, aiming at SGComm applications, can be enhanced by applying spectrum sensing techniques, such as the one discussed in this chapter. The attained results reveal that there is room for improvement and more research efforts are needed in this respect. The techniques and considerations made by this chapter emphasize the issues that are important for the development of CogPLC.

REFERENCES

1. A. Ipakchi and F. Albuyeh, "Grid of the future," *IEEE Power Energy Magazine*, vol. 7, no. 4, pp. 52-62, Mar. 2009.
2. G. W. Arnold, "Challenges and opportunities in smart grids: A position article," *Proc. IEEE*, vol. 99, no. 6, pp. 922-927, June 2011.
3. V. C. Gungor et al., "Opportunities and challenges of wireless sensor networks in smart grid," *IEEE Trans. Ind. Elec.*, vol. 57, no. 57, pp. 3557-3564, Oct. 2010.
4. N. Ghasemi and S. M. Hosseini, "Comparison of smart grid with cognitive radio: Solutions to spectrum scarcity," in *Proc. 12th Int. Conf. Advanced Communication Technology*, Phoenix Park, 2010, pp. 898–903.
5. J. Gaoa et al., "A survey of communication/networking in Smart Grids," *Future Generation Computer Systems*, vol. 28, no. 5, pp. 755-768, May 2012.
6. V. C. Gungor and F. C. Lambert, "A survey on communication networks for electric system automation," *Computer Networks*, vol. 50, pp. 877–897, Feb. 2006.
7. Y. Zhang et al., "Mitigating blackouts via smart relays: A machine learning approach," *Proc. IEEE*, vol. 99, no. 1, pp. 94–118, Jan. 2011.
8. M. Erol-Kantarci and H. T. Mouftah, "Wireless sensor networks for cost-efficient residential energy management in the smart grid," *IEEE Trans. Smart Grid*, vol. 2, no. 2, pp. 314–325, June 2011.
9. Z. M. Fadlullah et al., "Toward intelligent machine-to-machine communications in smart grid," *IEEE Commun. Mag.*, vol. 49, no. 4, pp. 60–65, Apr. 2011.
10. T. Sauter and M. Lobashov, "End-to-end communication architecture for smart grids," *IEEE Trans. Ind. Elec.*, vol. 58, no. 4, pp. 1218–1228, Apr. 2011.
11. S. Galli et al., "Industrial and international standards on PLC base networking technologies," in *Power Line Communications: Theory and Applications for Narrowband and Broadband Communications over Power Lines*, 1st ed., H. C. Ferreira et al., eds. United Kingdom: John Wiley and Sons Ltd. 2010, pp. 377–426.
12. S. Galli et al., "For the grid and through the grid: The role of power line communications in the smart grid," *Proc. IEEE*, vol. 99, no. 6 , pp. 998–1027, June 2011.
13. R. Yu et al., "Cognitive radio based hierarchical communications infrastructure for smart grid," *IEEE Network*, vol. 25, no. 5, pp. 6–14, Sept. 2011.
14. M. V. Ribeiro. "Power line communications: A promising communication systems paradigm for last miles and last meters applications," in *Telecommunications: Advances and Trends in Transmission, Networking and Applications*, 1st ed., C. C. Cavalcante et al., eds. Fortaleza: UNIFOR, 2006, pp. 133–156.
15. S. Goldfisher and S. Tanabe, "IEEE 1901 access system: An overview of its uniqueness and motivation," *IEEE Commun. Mag.*, vol. 48, no. 10, pp. 150–157, Oct. 2010.
16. *IEEE Standard for Broadband Over Power Line Networks: Medium Access Control and Physical Layer Specifications*, IEEE Std. 1901-2010, Sept. 2010.
17. V. Oksman and S. Galli, "G.hn: The new ITU-T home networking standard," *IEEE Commun. Mag.*, vol. 47, no. 10, pp. 138–145, Oct. 2009.

18. P. S. Henry, "Interference characteristics of broadband power line communication systems using aerial medium voltage wires," *IEEE Commun. Mag.*, vol. 43, no. 4, pp. 92–98, Apr. 2005.
19. A. Chubukjian et al., "Potential effects of broadband wireline telecommunications on the HF spectrum," *IEEE Commun. Mag.*, vol. 46, no. 11, pp. 49–54, Nov. 2008.
20. M. Gebhardt *et al.*, "Physical and regulatory constraints for communication over the power supply grid," *IEEE Commun. Mag.*, vol. 41, no. 5, pp. 84–90, May 2003.
21. W. Liu et al., "Broadband PLC access systems and field deployment in European power line networks," *IEEE Commun. Mag.*, vol. 41, no. 5, pp. 114–118, May 2003.
22. M. D'Amore and M. S. Sarto, "Electromagnetic field radiated from broadband signal transmission on power line carrier channels," *IEEE Trans. Power Del.*, vol. 12, no. 2, pp. 624–631, Apr. 1997.
23. Y. R. Ferreira, "Inspection method on measuring unwanted emissions from broadband power line using mobile monitoring station," in *Proc. IEEE Int. Sym. on Power Line Communications and Its Applications*, 2010, Rio de Janeiro, pp. 231–235.
24. J. H. Stott and J. Salter, "The effects of power line telecommunications on broadcast reception: Brief trial in Crieff," White Paper WHP067, BBC Research and Development, Sep. 2003.
25. J. H. Stott, "PLT and broadcasting - can they co-exist?" White Paper WHP099, BBC Research and Development, 2004.
26. J. H. Stott, "Co-existence of PLT and Radio Services - a possibility?" White Paper WHP114, BBC Research and Development, 2005.
27. P. Pagani et al., "Electromagnetic compatibility for power line communications: Regulatory issues and countermeasures," in *Proc. IEEE 21st Int. Sym. on Personal Indoor and Mobile Radio Communications*, Istanbul, Sept. 2010, pp. 2799–2804.
28. I. S. Areni et al., "A study of radiation detection methods for cognitive PLC system," in *Proc. IEEE Int. Sym. on Power Line Communications and Its Applications*, 2011, Udine, pp. 430–433.
29. D. Hansen, "Review of EMC main aspects in fast PLC including some history," in *Proc. IEEE Int. Sym. on Electromagnetic Compatibility*, Istanbul, 2003, pp. 184–192.
30. K. Salehian et al., "Field measurements of EM radiation from in-house power line telecommunications (PLT) devices," *IEEE Trans. Broadcasting*, vol. 57, no. 1, pp. 57–65, Marc. 2011.
31. J. G. Rhee et al., "Electromagnetic interferences caused by power line communications in the HF bands," in *Proc. IEEE Int. Sym. on Power Line Communications and Its Applications*, Jeju City, Jeju Island, 2008, pp. 249–252.
32. K.S. Kho, "Protection requirements for military HF radio services based on a generic sharing criteria for HF systems," in *Proc. IEEE Int. Sym. on Electromagnetic Compatibility*, Istanbul, 2003, pp. 180–183.
33. N. Oka et al., "Reduction of radiated emission from PLC systems by studying electrical unbalance of the PLC device and T-ISN," in *Proc IEEE Int. Sym. on Electromagnetic Compatibility*, Honolulu, HI, 2007, pp. 1–5.
34. T. S. Pang et al., "Feasibility Study of a New Injection Method for EMI Reduction in Indoor Broadband PLC Networks," *IEEE Trans. on Power Del.*, vol. 25, no. 4, pp. 2392–2398, Oct. 2012.
35. P. Favre, "Radiation and disturbance mitigation in PLC networks," in *Proc. IEEE Int. Sym. on Electromagnetic Compatibility*, Zurich, 2009, pp. 5–8.
36. A. Schwager and H. Hirsch, "Smart Notching – New concepts for EMC coordination,"

in *Proc. EMV 2008 - Internationale Fachmesse und Kongress für Elektromagnetische Verträglichkeit*, Düsseldorf, 2008, pp. 1–8.

37. *Power line communication apparatus used in low-voltage installations - Radio disturbance characteristics - Limits and methods of measurement - Part 1: Apparatus for in-home use*, FprEN 50561-1 FINAL DRAFT CENELEC, Aug. 2012.

38. L. R. Amado et al., "Spectrum sensing for powering power line communications," in *Proc. 30th Brasilian Telecommunication Simp.*, Brasilia, Brazil, 2012.

39. *PowerLine Telecommunications (PLT): Coexistence between PLT Modems and Short Wave Radio broadcasting services*, ETSI TS 102 578 V1.2.1, Aug. 2008.

40. *Technical Guidance Note TGN on Assessment of Powerline Telecommunications (PLT) Equipment*, ECANB TGN17 V2.0, July 2009.

41. N. Weling, "Feasibility study on detecting short wave radio stations on the powerlines for dynamic psd reduction as method for cognitive PLC," in *Proc. IEEE Int. Sym. on Power Line Communications and Its Applications*, Udine, 2011, pp. 311–316.

42. B. Praho et al., "Cognitive detection method of radio frequencies on power line networks," in *Proc. IEEE Int. Sym. on Power Line Communications and Its Applications*, Rio de Janeiro, 2010, pp. 225–230.

43. S. W. Oh et al., "Cognitive power line communication system for multiple channel access," in *Proc. IEEE Int. Sym. on Power Line Communications and Its Applications*, Dresden, 2009, pp. 47–52.

44. B. Praho et al., "A cognitive method for ensuring the coexistence between PLC and VDSL2 technologies in home network," in *Proc. IEEE Int. Conf. on Microwaves, Communications, Antennas and Electronics Systems*, 2011, pp. 1–4.

45. J. Mitola, "Cognitive radio: an integrated agent architecture for software defined radio," Ph.D. dissertation, Royal Inst. Technol. (KTH), Stockholm, Sweden, 2000.

46. S. Haykin, "Cognitive radio: brain-empowered wireless communications," *IEEE J. Sel. Areas on Commun.*, vol. 23, no. 2, pp. 201–220, Feb. 2005.

47. S. Rao et al., "Powerline communication system with alternative technology based on new modulation schemes," in *Proc. of BBEurope*, Antwerp, 2007, pp. 1-4.

48. B. Praho et al., "PLC coupling effect on VDSL2," in *Proc. IEEE Int. Sym. on Power Line Communications and Its Applications*, Udine, 2011, pp. 317–322.

49. B. Praho et al.,"Study of the coexistence of VDSL2 and PLC by analyzing the coupling between power line and telecommunications cable in the home network," in *Proc. 30th General Assembly and Scientific Sym.*, Istanbul, 2011, pp. 1-4.

50. L. P. Do and R. Lehnert, "Distributed dynamic resource allocation for multi-cell PLC networks," in *Proc. IEEE Int. Sym. on Power Line Communications and Its Applications*, Dresden, 2009, pp. 29–31.

51. J. Egan (Jan. 31, 2015). *PLC Neighboring Networks* [Online], Available: http://www.marvell.com

52. R. Dong et al., "A cognitive cross-layer resource allocation scheme for in-home power line communications," in *Proc. IEEE Int. Conf. on Communications*, Cape Town, 2010, pp. 1–5.

53. W. A. Finamore et al., "Advancing power line communication: cognitive, cooperative, and MIMO communication," in *Proc. 30th Brasilian Telecommunication Sym.*, Brasilia, Brazil, 2012.

54. Y. Zeng et al., "Subcarrier sensing for distributed OFDMA in powerline communication," in *Proc. IEEE Int. Conf. on Communications*, Dresden, 2010, pp. 1–5.

55. K. Kastner et al., "Dynamic spectrum allocation in low-bandwidth power line communica-

tions," in *Proc. The 2nd Int. Conf. on Advances in Cognitive Radio*, Chamonix, 2012, pp. 28–33.

56. E. A. Teixeira et al., "Regulation issues about broadband PLC: A Brazilian experience and perspective," in *Proc. IEEE Int. Sym. on Power Line Communications and Its Applications*, Pisa, 2007, pp. 523–527.

57. *IEEE Standard for broadband over power line networks: Medium access control and physical layer specifications*, IEEE Standard 1901-2010, 2010.

58. F. Versolatto and A. M. Tonello, "PLC channel characterization up to 300 MHz: Frequency response and access impedance," *Proc. of IEEE Global Telecommunications Conf.*, Anaheim, CA, 2012, pp. 3525–3530.

59. M. U. Rehman et al., "Achieving high data rate in multiband-OFDM UWB over power-line communication system," *IEEE Trans. Power Del.*, vol. 27, no. 3, pp. 1172–1177, Jul. 2012.

60. *Power Line Telecommunications*, ETSI TR 102 616 V1.1.1, 2008.

61. A. Schwager, "Powerline Communications: significant technologies to become ready for integration," Ph.D. dissertation, University Duisburg, Essen, 2010.

62. L. R. Amado, "A contribution to techniques analysis for spectrum sensing in PLC systems," M.S. thesis in Portuguese, Federal University of Juiz de Fora, Brazil, 2011.

63. L. R. Amado et al., "A contribution for spectrum sensing in power line communication systems," in *Proc. The 7th Int. Telecommunications Sym.*, Manaus, AM, 2010, pp. 1–6.

64. S. Haykin et al., "Spectrum sensing for cognitive radio," *Proc. of the IEEE*, vol. 97, no. 5, pp. 849–877, May 2009.

65. J. Ma et al., "Signal processing in cognitive radio," *Proc. of the IEEE*, vol. 97, no. 5, pp. 805–823, May 2009.

66. T. Yucek and H. Arslan, "A Survey of spectrum sensing algorithms for cognitive radio applications," *IEEE Commun. Surveys & Tutorials*, vol. 11, no. 1, pp. 116–130, May 2009.

67. B. Farhang-Boroujeny, "Filter bank spectrum sensing for cognitive radios," *IEEE Trans. on Signal Process.*, vol. 56, no. 5, pp. 1801–1811, May 2008.

68. J. G. Proakis, *Digital Communications*. New York: McGraw-Hill, 2008.

69. P. Wang et al., "Optimization of detection time for channel efficiency in cognitive radio systems," in *Proc. IEEE Wireless Communication and Networking Conference*, Kowloon, 2007, pp. 111–115.

70. S. Shankar et al., "Spectrum agile radios: utilization and sensing architectures," in *Proc. IEEE Int. Sym. on New Frontiers in Dynamic Spectrum Access Networks*, Baltimore, MD, 2005, pp. 160–169.

71. T. Yucek and H. Arslan, "Spectrum characterization for opportunistic cognitive radio systems," in *Proc. IEEE Military Communication Conf.*, Washington, DC, 2006, pp. 1–6.

72. Z. Tian and G. B. Giannakis, "A wavelet approach to wideband spectrum sensing for cognitive radios," in *Proc. IEEE Int. Conf. Cognitive Radio Oriented Wireless Networks and Communication*, Mykonos Island, 2006, pp. 1–5.

73. S. Tsuzuki et al., "Radiation detection and mode selection for a cognitive PLC system," in *Proc. IEEE Int. Sym. on Power Line Communications and Its Applications*, Udine, 2011, pp. 323–328.

74. N. Weling, "SNR-based detection of broadcast radio stations on powerlines as mitigation method toward a cognitive PLC solution," in *Proc. IEEE Int. Sym. on Power Line Communications and Its Applications*, Beijing, 2012, pp. 52–59.

75. M. V. Ribeiro et al., "Detection of disturbances in voltage signals for power quality analysis using HOS," *EURASIP Journal on Advances in Signal Processing*, vol. 2007, no. 1, pp. 36–54, Feb. 2007.

76. M. V. Ribeiro and J. L. R. Pereira, "Classification of single and multiple disturbances in electric signals," *EURASIP Journal on Advances in Signal Processing*, vol. 2007, no. 1, pp. 74–87, Feb. 2007.

77. H. S. Malvar. *Signal processing with Lapped transforms*. Ann Arbor, MI: Artech House, 1992.

78. H. S. Malvar, "Fast algorithm for the modulated complex lapped transform," *IEEE Signal Process. Letters*, vol. 10, no. 1, pp. 8–10, Jan. 2003.

79. J. M. Mendel, "Tutorial on higher-order statistics (spectra) in signal processing and system theory: theoretical results and some applications," *Proc. of the IEEE*, vol. 79, no. 3, pp. 278–305, Mar. 1991.

80. A. K. Jain et al., "Statistical pattern recognition: A review," *IEEE Trans. on Pattern Analysis and Machine Intelligence*, vol. 22, no. 1, pp. 4–37, Jan. 2000.

81. S. Theodoridis and K. Koutroumbas, *Pattern Recognition*. New York: Academic, 2006.

82. S. Haykin, *Neural Networks and Learnig Machines*. Upper Saddle River, NJ: Pearson Education, 2009.

83. Gage (Jan. 31, 2015). *Razor compuscope 16XX* [Online]. Available: www.gage-applied.com

84. Fractumrf (Jan. 31, 2015). *Antennas and calbles* [Online]. Available: http://www.fractumrf.com

85. PREMO (Jan. 31, 2015). *Capacitive Coupler P-1240-021* [Online]. http://www.grupopremo.com

Selected Topics in Signal Processing

João Marcos Travassos Romano
Charles Casimiro Cavalcante

16 Information Geometry: An Introduction to New Models for Signal Processing

RUI FACUNDO VIGELIS
Federal University of Ceará (UFC)

CHARLES CASIMIRO CAVALCANTE
Federal University of Ceará (UFC)

16.1 INTRODUCTION

Statistical models are an intrinsic part of signal processing methods for performing tasks such as estimation, filtering, classification, clustering, among others. The nature of the model usually defines the performance of the methods in terms of limits and attainable performance [1, 2].

As a typical example we have the Viterbi detector, in digital communication systems, that estimates the most probable sequence transmitted observing the received signal and minimizing the Euclidean distance of the received sequence with respect to all possible transmitted sequences. This is performed based on the fact that, if the noise is white and has a Gaussian distribution, the metric will rely on the Euclidean space [3]. For colored noise, a Mahalanobis distance is used, but the problem still can be viewed as a simple distance minimization problem [4].

However, those models that are based on Euclidean measures or projections of data in a flat space where such metric can be used are limited and do not serve some applications which have, for example, nonlinearities or high-dimensional parameters. In those cases, curved spaces are then required to provide a more realistic model for the case at hand and concepts of differential geometry are also needed in order to make the calculus over manifolds. A differentiable manifold is a set endowed with a differentiable structure, where each point has a neighborhood that is homeomorphic to the Euclidean space. Roughly speaking, we can say that the vicinity of each point of a manifold can be treated as a Euclidean space. Hence, in order to profit from such mathematical notions different models for signals are required.

Whenever evaluating statistical manifolds, characterized by the geometry of their probability density functions, which since the seminal work from Shannon [5] define the information, we rely on the area of *information geometry*. This area uses tools from differential geometry in order to study and evaluate information and its applications such as estimation. Although the rudiments of information geometry can be traced back to 1975 and the work from Efron [6], it is credited to Amari the concepts of its fundamental results [7, 8].

The application of manifolds in signal processing applications has begun in some optimization problems for adaptive filtering [9, 10]. From that work, many have tried to insert the notion of curvature and/or geometric notion about the space in their problems and solutions [11, 12, 13, 14, 15, 16, 17]. Other works have devoted their attention to the geometric aspects of more specific problems, as independent component analysis, with important insights from their approaches [18, 19, 20, 21], and some researchers have also put a lot of effort toward providing advances in the optimization procedures when working on manifolds, see for instance [22, 23] and references therein, trying to go further than the well-known natural gradient learning method, which takes into account the curvature of the space [24, 25]

More recently, a growth in the interest on related topics, from both theoretical and applied streams, has boosted the development of new methods and synergy among different areas such as mathematics, information theory, signal processing, and econometrics, among others. Focusing in signal processing, we can cite [26] and the publications originating from the first conferences in the area of information geometry [27, 28].

This chapter aims to introduce the main concepts of information geometry, based on some results from infinite-dimensional modeling [29], in order to provide the required tools for those interested in such an elegant way of treating information. The work is far from being complete due to the extensive amount of results and concepts provided in the last 40 years. For more comprehensive and detailed works the reader is pointed to the excellent texts [30, 7, 31].

The rest of the chapter is organized as follows: Section 16.2 describes the modeling of statistical manifolds and its properties. A generalization of such statistical manifolds for a more general probability model is provided in Section 16.3 and Section 16.4 discusses the geometry of the generalized statistical manifold model. A summary of the chapter and research directions are provided in Section 16.5 and, for the sake of completeness, a brief introduction to key results of Riemannian geometry, which are need to the developments throughout the chapter, is given in Appendix 16.A.

16.2 STATISTICAL MANIFOLDS

Let (T, Σ, μ) be a measure space, and let L^0 denote the set of all real-valued, measurable functions on T, with equality μ-a.e. A statistical manifold is a family of probability distributions, endowed with some geometric entities (for example, a metric), which is defined to be contained in

$$\mathcal{P}_\mu = \left\{ p \in L^0 : p > 0 \text{ and } \int_T p d\mu = 1 \right\}.$$

In other words, $\mathcal{P}_\mu$ represents the set of all probability measures on T which are equivalent to μ. As usual, in the case that $T = \mathbb{R}$ (or T is a discrete set), the measure

μ is identified as the Lebesgue measure (or the counting measure). We will consider families of probability distributions of the form

$$\mathcal{P} = \{p(t; \theta) : \theta \in \Theta\},$$

where each $p_\theta(t) := p(t; \theta)$ is given in terms of parameters $\theta = (\theta^1, \ldots, \theta^n) \in \Theta$ by one-to-one mapping. The family $\mathcal{P}$ is assumed to satisfy the following conditions:

(P1) Θ is a domain (an open and connected set) in $\mathbb{R}^n$.
(P2) $p(t; \theta)$ is a differentiable function with respect to θ.
(P3)   The operations of integration with respect to μ and differentiation with respect to θ^i are commutative.
(P4)   The *Fisher information matrix* (or *Fisher metric*) $g = (g_{ij})$, which is defined by

$$g_{ij} = -E_\theta\Big[\frac{\partial^2 l_\theta}{\partial\theta^i \partial\theta^j}\Big], \tag{16.1}$$

is positive definite at each $\theta \in \Theta$, where $l_\theta(t) = l(t; \theta) = \log p(t; \theta)$ and $E_\theta[\cdot]$ denotes the expectation with respect to $p_\theta \cdot \mu$.

Under condition (P4), expression (16.1) defines a Riemannian metric in $\mathcal{P}$. We will use (P3) to derive another expression for (g_{ij}). Thanks to this expression, we will show that (g_{ij}) is always positive semi-definite. We will also find an identification for the tangent space $T_{p_\theta}\mathcal{P}$.

The Fisher information matrix (g_{ij}) may also be written as

$$g_{ij} = E_\theta\Big[\frac{\partial l_\theta}{\partial\theta^i} \frac{\partial l_\theta}{\partial\theta^j}\Big]. \tag{16.2}$$

Using condition (P3), we get

$$0 = \frac{\partial}{\partial\theta^i} \int_T p(t; \theta) d\mu = \int_T \frac{\partial}{\partial\theta^i} p(t; \theta) d\mu = \int_T \frac{\partial}{\partial\theta^i} l(t; \theta) p(t; \theta) d\mu. \tag{16.3}$$

As a result, the *score function* $\partial l_\theta / \partial\theta^i$ has zero expectation:

$$E_\theta\Big[\frac{\partial l_\theta}{\partial\theta^i}\Big] = 0. \tag{16.4}$$

Now, differentiating expression (16.3) with respect to θ^j, we have

$$0 = \int_T \frac{\partial^2 l(t; \theta)}{\partial\theta^i \partial\theta^j} p(t; \theta) d\mu + \int_T \frac{\partial l(t; \theta)}{\partial\theta^i} \frac{\partial l(t; \theta)}{\partial\theta^j} p(t; \theta) d\mu,$$

which shows expression (16.2).

Because the inner product of vectors

$$X = \sum_i a^i \frac{\partial}{\partial\theta^i} \qquad \text{and} \qquad Y = \sum_i b^j \frac{\partial}{\partial\theta^j}$$

can be expressed as

$$g(X, Y) = \sum_{i,j} g_{ij} a^i b^j = \sum_{i,j} E_\theta \Big[\frac{\partial l_\theta}{\partial \theta^i} \frac{\partial l_\theta}{\partial \theta^j} \Big] a^i b^j = E_\theta[\widetilde{X}\widetilde{Y}],$$

where

$$\widetilde{X} = \sum_i a^i \frac{\partial l_\theta}{\partial \theta^i} \qquad \text{and} \qquad \widetilde{Y} = \sum_i b^j \frac{\partial l_\theta}{\partial \theta^j}.$$

We see that the tangent space $T_{p_\theta}\mathcal{P}$ can be identified with $\widetilde{T}_{p_\theta}\mathcal{P}$, the vector space spanned by the functions $\partial l_\theta / \partial \theta^i$, equipped with the inner product $\langle \widetilde{X}, \widetilde{Y} \rangle_\theta = E_\theta[\widetilde{X}\widetilde{Y}]$. By (16.4) we see that if a vector $\widetilde{X}$ belongs to $\widetilde{T}_{p_\theta}\mathcal{P}$ then $E_\theta[\widetilde{X}] = 0$. Expression (16.2) also implies that (g_{ij}) is always positive semi-definite, since

$$\sum_{i,j} g_{ij} a^i a^j = E_\theta \Big[\Big(\sum_i a^i \frac{\partial l_\theta}{\partial \theta^i} \Big)^2 \Big] \geq 0.$$

The α-connection, which we denote by $D^{(\alpha)}$, is defined by the Christoffel symbols

$$\Gamma_{ij}^{(\alpha)k} = \sum_m g^{mk} \Gamma_{ijm}^{(\alpha)},$$

where

$$\Gamma_{ijk}^{(\alpha)} = E_\theta \Big[\frac{\partial^2 l_\theta}{\partial \theta^i \partial \theta^j} \frac{\partial l_\theta}{\partial \theta^k} \Big] + \frac{1-\alpha}{2} E_\theta \Big[\frac{\partial l_\theta}{\partial \theta^i} \frac{\partial l_\theta}{\partial \theta^j} \frac{\partial l_\theta}{\partial \theta^k} \Big]. \tag{16.5}$$

Note that this expression defines a torsion-free connection (i.e., $\Gamma_{ij}^{(\alpha)k} = \Gamma_{ji}^{(\alpha)k}$). The cases $\alpha = 0$, $\alpha = 1$, and $\alpha = -1$ are of particular interest. The connection $D^{(\alpha)}$ reduces to the Levi–Civita connection when $\alpha = 0$. If Γ_{ij}^k is the Christoffel symbol of the Levi–Civita connection ∇ associated with g, then

$$\Gamma_{ijk} := \sum_m \Gamma_{ij}^m g_{mk} = \frac{1}{2} \Big(\frac{\partial g_{ki}}{\partial \theta^j} + \frac{\partial g_{kj}}{\partial \theta^i} - \frac{\partial g_{ij}}{\partial \theta^k} \Big).$$

Making use of (16.2), we obtain

$$\frac{\partial g_{ij}}{\partial \theta^k} = E_\theta \Big[\frac{\partial^2 l_\theta}{\partial \theta^i \partial \theta^k} \frac{\partial l_\theta}{\partial \theta^j} \Big] + E_\theta \Big[\frac{\partial l_\theta}{\partial \theta^i} \frac{\partial^2 l_\theta}{\partial \theta^j \partial \theta^k} \Big] + E_\theta \Big[\frac{\partial l_\theta}{\partial \theta^i} \frac{\partial l_\theta}{\partial \theta^j} \frac{\partial l_\theta}{\partial \theta^k} \Big],$$

and so

$$\Gamma_{ijk} = E_\theta \Big[\frac{\partial^2 l_\theta}{\partial \theta^i \partial \theta^j} \frac{\partial l_\theta}{\partial \theta^k} \Big] + \frac{1}{2} E_\theta \Big[\frac{\partial l_\theta}{\partial \theta^i} \frac{\partial l_\theta}{\partial \theta^j} \frac{\partial l_\theta}{\partial \theta^k} \Big], \tag{16.6}$$

showing that $\Gamma_{ijk}^{(0)} = \Gamma_{ijk}$. As a result, we may write

$$\Gamma_{ijk}^{(\alpha)} = \Gamma_{ijk} - \alpha T_{ijk}, \tag{16.7}$$

where

$$T_{ijk} = \frac{1}{2} E_\theta \Big[\frac{\partial l_\theta}{\partial \theta^i} \frac{\partial l_\theta}{\partial \theta^j} \frac{\partial l_\theta}{\partial \theta^k} \Big].$$

The connection $D^{(e)} := D^{(1)}$, respectively, $D^{(m)} := D^{(-1)}$, is called the *exponential connection* or *e-connection*, respectively, the *mixture connection* or *m-connection*. The connections $D^{(\alpha)}$ can be expressed in terms of $D^{(e)}$ and $D^{(m)}$:

$$D^{(\alpha)} = \frac{1+\alpha}{2} D^{(e)} + \frac{1-\alpha}{2} D^{(m)}.$$

The exponential and mixture connections are intrinsically associated with the exponential and mixture families of probabilities distributions. In these families they vanish identically. This statement is verified in Examples 1 and 4, which are presented below.

Example 1 (Exponential Family). A family of probability distributions $\mathcal{E}_p = \{p(t;\theta) : \theta \in \Theta\} \subseteq \mathcal{P}_\mu$ is said to be an *exponential family centered at* $p \in \mathcal{P}_\mu$ if its elements are expressed by

$$p(t;\theta) = \exp\left(\sum_{i=1}^{n} \theta^i u_i(t) - \psi(\theta)\right) p(t), \tag{16.8}$$

where $u_1, \ldots, u_n : T \to \mathbb{R}$ are measurable functions satisfying some properties, which are stated below; and $\psi : \Theta \to [0, \infty)$ is a function which is introduced so that expression (16.18) results in a probability distribution in $\mathcal{P}_\mu$. The function ψ is called the *cumulant-generating function*.

The set $\Theta \subset \mathbb{R}^n$ is considered to be maximal. In other words, we define Θ as the set of all vectors $\theta = (\theta^i)$ for which

$$E_p\left[\exp\left(\lambda \sum_{i=1}^{n} \theta^i u_i\right)\right] < \infty, \qquad \text{for some } \lambda > 1,$$

where $E_p[\cdot]$ denotes the expectation with respect to $p \cdot \mu$. Moreover, each function u_k is assumed to satisfy:

(i) $E_p[u_k] = 0$, and
(ii) there exists $\varepsilon > 0$ such that

$$E_p[\exp(\lambda u_k)] < \infty, \qquad \text{for all } \lambda \in (-\varepsilon, \varepsilon).$$

These conditions are motivated by technical reasons. Under (i)–(ii), the exponential family $\mathcal{E}_p$ is a submanifold of a non-parametric exponential family [32, 33, 34]. Item (i) implies that the function ψ is always non-negative. From condition (ii), one can infer, for example, that the set Θ is open. It is shown in [35] that the function ψ is strictly convex (the set Θ is convex). If the set T is finite, condition (ii) is always satisfied.

In $\mathcal{E}_p$, the Fisher information matrix (g_{ij}) is the Hessian of ψ, and the exponential connection $D^{(e)}$ vanishes identically. Expressions

$$\frac{\partial l_\theta}{\partial \theta^i} = u_i(t) - \frac{\partial \psi}{\partial \theta^i}, \qquad -\frac{\partial^2 l_\theta}{\partial \theta^i \partial \theta^j} = -\frac{\partial^2 \psi}{\partial \theta^i \partial \theta^j},$$

imply that

$$g_{ij} = \frac{\partial^2 \psi}{\partial \theta^i \partial \theta^j}.$$

In addition, the exponential connection $D^{(e)}$ vanishes identically, since

$$\Gamma_{ijk}^{(e)} = E_\theta\left[\frac{\partial^2 l_\theta}{\partial \theta^i \partial \theta^j}\frac{\partial l_\theta}{\partial \theta^k}\right] = -\frac{\partial^2 \psi}{\partial \theta^i \partial \theta^j}E_\theta\left[\frac{\partial l_\theta}{\partial \theta^k}\right] = 0.$$

In Section 16.4, where we assume that T is a finite set, we will show that $\mathcal{E}_p$ admits a parametrization in which the Christoffel symbols satisfy $\Gamma_{ij}^{(m)k} = 0$.

The cumulant-generating function is intimately related with the Kullback–Leibler divergence. A simple manipulation leads to

$$\psi(\theta) = \sum_{i=1}^{n} \theta^i u_i(t) + \log\frac{p(t)}{p(t;\theta)}.$$

From (i), it follows that

$$\psi(\theta) = \int_T \log\left(\frac{p}{p_\theta}\right)pd\mu =: \mathcal{D}_{\text{KL}}(p \parallel p_\theta),$$

i.e., $\psi(\theta)$ is the Kullback–Leibler divergence $\mathcal{D}_{\text{KL}}(p \parallel p_\theta)$. □

Many families of probability distributions are exponential families.

Example 2. Let us consider the discrete set $T = \{0, 1, \ldots, n\}$ and the domain

$$\Lambda = \left\{\lambda = (\lambda_i) \in \mathbb{R}^{n+1} : \sum_{i=0}^{n} \lambda_i = 1 \text{ and } \lambda_i > 0\right\}.$$

The family $\mathcal{P} = \{p(i; \lambda) : \lambda \in \Lambda\}$, whose members are given by $p(i; \lambda) = \lambda_i$, is an exponential family. Fix any probability distribution $p(i)$ in $\mathcal{P}$. Let $u_j(i) = \delta_{ij} - \delta_{i0}p(j)/p(0)$, where δ_{ij} is the Kronecker's delta function. It is easy to verify that $E_p[u_j] = 0$. Then

$$p(i; \lambda) = \exp\left\{\sum_{j=1}^{n}\left(\log\frac{\lambda_j}{p(j)} - \log\frac{\lambda_0}{p(0)}\right)u_j(i) + \log\frac{\lambda_0}{p(0)}\right\}p(i). \tag{16.9}$$

Denoting

$$\theta^j = \log\frac{\lambda_j}{p(j)} - \log\frac{\lambda_0}{p(0)},$$

expression (16.9) results in

$$p(i; \theta) = \exp\left(\sum_{j=1}^{n} \theta^j u_j(i) - \psi(\theta)\right)p(i),$$

where

$$\psi(\theta) = \log\left(p(0) + \sum_{j=1}^{n} p(j) \exp \theta^j\right).$$

Thus the family $\mathcal{P}$ is an exponential family. □

Example 3. The Poisson distribution with parameter $\lambda > 0$ is given by

$$p(i; \lambda) = \frac{\lambda^i}{i!} e^{-\lambda}, \qquad i \in \{0, 1, 2, \dots\}.$$

The family of Poisson distributions is also an exponential family. Notice that

$$\frac{\lambda^i}{i!} e^{-\lambda} = \exp[(i-1)\log(\lambda) - (\lambda - \log(\lambda) - 1)]\frac{e^{-1}}{i!}.$$

Denoting

$$u(i) = i - 1, \qquad p(i) = \frac{e^{-1}}{i!},$$

and

$$\psi(\theta) = e^\theta - \theta - 1, \qquad \theta = \log \lambda,$$

we have

$$p(i; \theta) = \exp(\theta u(i) - \psi(\theta)) p(i), \qquad \theta \in \mathbb{R}.$$

Some calculation shows that $E_p[u] = 0$ and $E_p[\exp(\lambda u_k)] < \infty$, for all $\lambda > 0$. □

Example 4 (Mixture Family). A family of probability distributions $\mathcal{M} = \{p(t; \eta) : \eta \in H\} \subseteq \mathcal{P}_\mu$ is called a *mixture family* if

$$p(t; \eta) = \sum_{i=1}^{n} \eta_i p_i(t) + \left(1 - \sum_{i=1}^{n} \eta_i\right) p_0(t), \qquad (16.10)$$

where p_0 and p_i are probability distributions, and $\eta = (\eta_i)$ is taken so that the right-hand side of (16.10) defines a probability distribution in $\mathcal{P}_\mu$. Clearly, the set $\{\eta = (\eta_i) : \eta_i \in (0, 1) \text{ and } \sum_{i=1}^{n} \eta_i < 1\}$ is contained in H. Differentiating $l(t; \eta) = \log p(t; \eta)$ with respect to η_i, we get

$$\frac{\partial l(t; \eta)}{\partial \eta_i} = \frac{p_i(t) - p_0(t)}{p(t; \eta)},$$

and then

$$g_{ij} = E_\eta\left[\frac{p_i - p_0}{p_\eta} \frac{p_j - p_0}{p_\eta}\right].$$

The mixture connection $D^{(m)}$ vanishes identically in $\mathcal{M}$. Since

$$\frac{\partial^2 l_\eta}{\partial \eta_i \partial \eta_j} = -\frac{\partial l_\eta}{\partial \eta_i} \frac{\partial l_\eta}{\partial \eta_j},$$

it follows that

$$\Gamma_{ijk}^{(m)} = \Gamma_{ijk} + T_{ijk}$$

$$= E_\theta\left[\frac{\partial^2 l_\eta}{\partial\eta_i\partial\eta_j}\frac{\partial l_\eta}{\partial\eta_k}\right] + E_\theta\left[\frac{\partial l_\eta}{\partial\eta_i}\frac{\partial l_\eta}{\partial\eta_j}\frac{\partial l_\eta}{\partial\eta_k}\right]$$

$$= 0.$$

In Example 17, we show (at least in the case that T is a finite set) how the functions u_i in an exponential family $\mathcal{E}_p$, and the distributions p_i constituting a mixture family $\mathcal{M}$, can be selected so that (θ^i) are (η_i) are dual coordinates systems.   □

On a Riemannian manifold (M, g), the gradient of a smooth function $f: M \to \mathbb{R}$ is the vector field ∇f for which

$$g(\nabla f, X) = f_*(X),$$

for any vector field X. In a coordinate system (θ^i), the gradient is expressed as

$$\nabla f = \sum_{i,j} g^{ij}\frac{\partial f}{\partial\theta^j}\frac{\partial}{\partial\theta^i}.$$

In practical situations where the gradient appears, e.g., the gradient descent method [25, 36, 37], its use is limited by the computation of g^{ij}. The next example shows a statistical manifold whose inverse g^{ij} is given in an explicit form.

Example 5. The family presented in Example 2 is a mixture family. Let $p_0, p_1, \ldots, p_n$ be linearly independent probability distributions in $\mathcal{P}$. Then

$$p(j; \eta) = \sum_{i=1}^{n} \eta_i p_i(j) + \left(1 - \sum_{i=1}^{n} \eta_i\right)p_0(j)$$

is a parametrization for $\mathcal{P}$. The case $p_i(j) = \delta_{ij}$ is of special interest. For these distributions, we have

$$p(j, \eta) = \eta_j,$$

where $\eta_0 = 1 - \sum_{i=1}^{n} \eta_i$. Thus the Fisher information matrix is given by

$$g_{ij} = E_\eta\left[\frac{p_i - p_0}{p_\eta}\frac{p_j - p_0}{p_\eta}\right]$$

$$= \sum_{k=0}^{n} \frac{p_i(k) - p_0(k)}{p(k; \eta)}\frac{p_j(k) - p_0(k)}{p(k; \eta)}p(k; \eta)$$

$$= \sum_{k=0}^{n} \frac{(\delta_{ik} - \delta_{0k})(\delta_{jk} - \delta_{0k})}{\eta_k}$$

$$= \frac{\delta_{ij}}{\eta_i} + \frac{1}{\eta_0},$$

whose inverse is

$$g^{ij} = \eta_i \delta_{ij} - \eta_i \eta_j.$$

This simple form of g^{ij} is particularly useful in frameworks involving the gradient of some cost function on manifolds. $\square$

16.3   GENERALIZED STATISTICAL MANIFOLDS

The metric and α-connection introduced in (16.1) and (16.7) are given in terms of the derivative of $l(t; \theta) = \log p(t; \theta)$. A metric and a connection can be defined by replacing the logarithm by the inverse of a φ-function. Instead of $l(t; \theta) = \log p(t; \theta)$, we can consider $f(t; \theta) = \varphi^{-1}(p(t; \theta))$. The manifold equipped with these metric and connection is called a *generalized statistical manifolds*.

A function $\varphi \colon \mathbb{R} \to (0, \infty)$ is said to be a φ-*function* if the following conditions are satisfied:

(a1) $\varphi(\cdot)$ is convex,
(a2) $\lim_{u \to -\infty} \varphi(u) = 0$ and $\lim_{u \to \infty} \varphi(u) = \infty$.

Moreover, we assume that a measurable function $u_0 \colon T \to (0, \infty)$ can be found such that, for every measurable function $c \colon T \to \mathbb{R}$ for which $\varphi(c(t))$ is a probability density in $\mathcal{P}_\mu$, we have that

(a3) $\displaystyle\int_T \varphi(c(t) + \lambda u_0(t))d\mu < \infty$, for all $\lambda > 0$.

Condition (a3) appears in the framework of non-parametric families of probability distributions [35]. Notice that if the set T is finite, condition (a3) is always satisfied. The exponential function is an example of φ-function, since $\varphi(u) = \exp(u)$ satisfies conditions (a1)–(a3) with $u_0 = \mathbf{1}_T$, where $\mathbf{1}_A$ is the indicator function of a subset $A \subseteq T$. Another example of φ-function is the Kaniadakis' κ-exponential [38, 35].

Let $\phi \colon (0, \infty) \to (0, \infty)$ be any increasing function. The *deformed logarithm function* (or ϕ-*logarithm function*) $\log_\phi \colon (0, \infty) \to (0, \infty)$ is defined by

$$\log_\phi(v) = \int_1^v \frac{1}{\phi(y)} dy, \qquad \text{for each } v > 0.$$

The deformed logarithm function $\log_\phi(\cdot)$ is concave and satisfies $\log_\phi(1) = 0$. In the range $(-m, M)$, where

$$m = \int_0^1 \frac{1}{\phi(y)} dy, \qquad M = \int_1^\infty \frac{1}{\phi(y)} dy,$$

the function $\log_\phi(\cdot)$ has an inverse. The *deformed exponential function* (or ϕ-*exponential function*) $\exp_\phi \colon \mathbb{R} \to [0, \infty]$ is defined by

$$\exp_\phi(u) = \begin{cases} 0, & \text{if } u \in (-\infty, -m], \\ \log_\phi^{-1}(u), & \text{if } u \in (-m, M), \\ \infty, & \text{if } u \in [M, \infty). \end{cases}$$

The deformed exponential function $\exp_\phi(\cdot)$ can also be expressed in terms of an integral:

$$\exp_\phi(u) = \int_{-\infty}^{u} \lambda(x)dx, \qquad \text{for all } u \in \mathbb{R},$$

where

$$\lambda(u) = \begin{cases} 0, & \text{if } u \in (-\infty, -m], \\ \phi(\exp_\phi(u)), & \text{if } u \in (-m, M), \\ \infty, & \text{if } u \in [M, \infty). \end{cases}$$

It is clear that $\phi(v) = \lambda(\log_\phi(v))$ for all $v > 0$. In the case $\phi(u) = u$, the functions $\log_\phi(\cdot)$ and $\exp_\phi(\cdot)$ coincide with the natural logarithm and the exponential function. Deformed exponential functions and φ-functions do not coincide. If a φ-function $\varphi(\cdot)$ satisfies $\varphi(0) = 1$, then $\varphi(\cdot)$ is a deformed exponential function. Deformed exponential functions whose image is not contained in $(0, \infty)$ are not φ-functions. For $q \neq 1$, the q-exponential function $\exp_q(\cdot)$ [39] is an example of deformed exponential function whose image is not contained in $(0, \infty)$.

Let $\mathcal{P} = \{p(t; \theta) : \theta \in \Theta\}$ be a family of probability distributions satisfying conditions (P1)–(P3). We also assume that condition (P4) is satisfied for the metric g_{ij} given below. Denote $f_\theta(t) = f(t; \theta) = \varphi^{-1}(p(t; \theta))$. We equip the family $\mathcal{P}$ with the metric

$$g_{ij} = -E_\theta'\left[\frac{\partial^2 f_\theta}{\partial\theta^i \partial\theta^j}\right], \tag{16.11}$$

where

$$E_\theta'[\cdot] = \frac{\int_T (\cdot)\varphi'(f_\theta)d\mu}{\int_T u_0 \varphi'(f_\theta)d\mu}.$$

Notice that when φ is the exponential function and $u_0 = 1$, expression (16.11) reduces to the Fisher information matrix.

The matrix (g_{ij}) can also be expressed as

$$g_{ij} = E_\theta''\left[\frac{\partial f_\theta}{\partial\theta^i}\frac{\partial f_\theta}{\partial\theta^j}\right], \tag{16.12}$$

where

$$E_\theta''[\cdot] = \frac{\int_T (\cdot)\varphi''(f_\theta)d\mu}{\int_T u_0 \varphi'(f_\theta)d\mu}.$$

Because the operations of integration with respect to μ and differentiation with respect to θ^i are commutative, we have

$$0 = \frac{\partial}{\partial\theta^i}\int_T p_\theta d\mu = \int_T \frac{\partial}{\partial\theta^i}\varphi(f_\theta)d\mu = \int_T \frac{\partial f_\theta}{\partial\theta^i}\varphi'(f_\theta)d\mu, \tag{16.13}$$

and

$$0 = \int_T \frac{\partial^2 f_\theta}{\partial\theta^i \partial\theta^j}\varphi'(f_\theta)d\mu + \int_T \frac{\partial f_\theta}{\partial\theta^i}\frac{\partial f_\theta}{\partial\theta^j}\varphi''(f_\theta)d\mu. \tag{16.14}$$

Equation (16.13) implies

$$E'_\theta\Big[\frac{\partial f_\theta}{\partial \theta^i}\Big] = 0. \tag{16.15}$$

Clearly, expression (16.12) is a consequence of (16.14). Using (16.12), we can infer that the matrix (g_{ij}) is positive semi-definitive, since

$$\sum_{i,j} g_{ij}a^i a^j = E''_\theta\Big[\Big(\sum_i a^i \frac{\partial f_\theta}{\partial \theta^i}\Big)^2\Big] \geq 0.$$

In view (16.12), the tangent space $T_{p_\theta}\mathcal{P}$ can be identified with $\widetilde{T}_{p_\theta}\mathcal{P}$, the vector space spanned by $\partial f_\theta/\partial \theta^i$, equipped with the inner product $\langle \widetilde{X}, \widetilde{Y}\rangle_\theta = E''_\theta[\widetilde{X}\widetilde{Y}]$. By (16.15), if a vector $\widetilde{X}$ belongs to $\widetilde{T}_{p_\theta}\mathcal{P}$, then $E'_\theta[\widetilde{X}] = 0$.

Let Γ^k_{ij} be the Christoffel symbols of the Levi–Civita connection ∇ associated with (g_{ij}). Expression

$$\Gamma_{ijk} := \sum_m \Gamma^m_{ij}g_{mk} = \frac{1}{2}\Big(\frac{\partial g_{ki}}{\partial \theta^j} + \frac{\partial g_{kj}}{\partial \theta^i} - \frac{\partial g_{ij}}{\partial \theta^k}\Big) \tag{16.16}$$

gives Γ_{ijk} in terms of g_{ij}. Differentiating with respect to θ^k the expression for the metric given in (16.12), we find

$$\frac{\partial g_{ij}}{\partial \theta^k} = E''_\theta\Big[\frac{\partial^2 f_\theta}{\partial \theta^i \partial \theta^k}\frac{\partial f_\theta}{\partial \theta^j}\Big] + E''_\theta\Big[\frac{\partial f_\theta}{\partial \theta^i}\frac{\partial^2 f_\theta}{\partial \theta^j \partial \theta^k}\Big]$$
$$+ E'''_\theta\Big[\frac{\partial f_\theta}{\partial \theta^i}\frac{\partial f_\theta}{\partial \theta^j}\frac{\partial f_\theta}{\partial \theta^k}\Big] - E''_\theta\Big[\frac{\partial f_\theta}{\partial \theta^i}\frac{\partial f_\theta}{\partial \theta^j}\Big]E''_\theta\Big[u_0\frac{\partial f_\theta}{\partial \theta^k}\Big],$$

where

$$E'''_\theta[\cdot] = \frac{\int_T (\cdot)\varphi'''(f_\theta)d\mu}{\int_T u_0\varphi'(f_\theta)d\mu}.$$

Then we get

$$\Gamma_{ijk} = E''_\theta\Big[\frac{\partial^2 f_\theta}{\partial \theta^i \partial \theta^j}\frac{\partial f_\theta}{\partial \theta^k}\Big] + \frac{1}{2}E'''_\theta\Big[\frac{\partial f_\theta}{\partial \theta^i}\frac{\partial f_\theta}{\partial \theta^j}\frac{\partial f_\theta}{\partial \theta^k}\Big]$$
$$- \frac{1}{2}E''_\theta\Big[\frac{\partial f_\theta}{\partial \theta^k}\frac{\partial f_\theta}{\partial \theta^i}\Big]E''_\theta\Big[u_0\frac{\partial f_\theta}{\partial \theta^j}\Big] - \frac{1}{2}E''_\theta\Big[\frac{\partial f_\theta}{\partial \theta^k}\frac{\partial f_\theta}{\partial \theta^j}\Big]E''_\theta\Big[u_0\frac{\partial f_\theta}{\partial \theta^i}\Big]$$
$$+ \frac{1}{2}E''_\theta\Big[\frac{\partial f_\theta}{\partial \theta^i}\frac{\partial f_\theta}{\partial \theta^j}\Big]E''_\theta\Big[u_0\frac{\partial f_\theta}{\partial \theta^k}\Big]. \tag{16.17}$$

Again this expression reduces to (16.6) in the case that φ is the exponential function and $u_0 = 1$. In Example 19, a family of α-connections $D^{(\alpha)}$, which generalizes the α-connections given in (16.5), is defined in terms of the φ-divergence $\mathcal{D}_\varphi(\cdot \| \cdot)$.

Example 6 (φ-Family). Fixed any probability density $p \in \mathcal{P}_\mu$, let $c: T \to \mathbb{R}$ be a measurable function for which $p = \varphi(c)$. We select any measurable functions $u_1, \dots u_n: T \to \mathbb{R}$, which are assumed to satisfy the following conditions:

(i) $\int_T u_i \varphi'(c) d\mu = 0$, and
(ii)   there exists $\varepsilon > 0$ such that

$$\int_T \varphi(c + \lambda u_i) d\mu < \infty, \qquad \text{for all } \lambda \in (-\varepsilon, \varepsilon).$$

Let $\Theta \subseteq \mathbb{R}^n$ be the set of all vectors $\theta = (\theta^i)$ for which

$$\int_T \varphi\left(c + \lambda \sum_{k=1}^n \theta^i u_i\right) d\mu < \infty, \qquad \text{for some } \lambda > 1.$$

The members of the *(parametric)* φ-*family* $\mathcal{F}_p = \{p(t; \theta) : \theta \in \Theta\}$ centered at $p = \varphi(c)$ are given by one-to-one mapping

$$p(t; \theta) := \varphi\left(c(t) + \sum_{i=1}^n \theta^i u_i(t) - \psi(\theta) u_0(t)\right), \qquad \text{for each } \theta = (\theta^i) \in \Theta. \qquad (16.18)$$

where the *normalizing function* $\psi \colon \Theta \to [0, \infty)$ is introduced so that expression (16.18) defines a probability distribution in $\mathcal{P}_\mu$.

Notice that condition (ii) is always satisfied if the set T is finite. One can show that Θ is open and convex. The normalizing function ψ is also convex. Conditions (i)–(ii) are also used in the definition of non-parametric φ-families. For details, see [35, 40].

Observing that

$$\frac{\partial f_\theta}{\partial \theta^i} = u_i(t) - \frac{\partial \psi}{\partial \theta^i}, \qquad -\frac{\partial^2 f_\theta}{\partial \theta^i \partial \theta^j} = -\frac{\partial^2 \psi}{\partial \theta^i \partial \theta^j},$$

we have

$$g_{ij} = \frac{\partial^2 \psi}{\partial \theta^i \partial \theta^j}.$$

Thus in $\mathcal{F}_p$ the matrix (g_{ij}) is the Hessian of ψ. As a result, expression (16.16) implies

$$\Gamma_{ijk} = \frac{1}{2} \frac{\partial g_{ij}}{\partial \theta^k} = \frac{1}{2} \frac{\partial^2 \psi}{\partial \theta^i \partial \theta^j \partial \theta^j}.$$

Some manipulation involving (16.18) yields

$$\psi(\theta) u_0(t) = \sum_{i=1}^n \theta^i u_i(t) + \varphi^{-1}(p(t)) - \varphi^{-1}(p(t; \theta)).$$

From condition (i), we can write

$$\psi(\theta) \int_T u_0 \varphi'(c) d\mu = \int_T [\varphi^{-1}(p) - \varphi^{-1}(p_\theta)] \varphi'(c) d\mu.$$

Then, using $\varphi'(c) = 1/(\varphi^{-1})'(p)$, we get

$$\psi(\theta) = \frac{\displaystyle\int_T \frac{\varphi^{-1}(p) - \varphi^{-1}(p_\theta)}{(\varphi^{-1})'(p)} d\mu}{\displaystyle\int_T \frac{u_0}{(\varphi^{-1})'(p)} d\mu} =: \mathcal{D}_\varphi(p \parallel p_\theta),$$

which defines the φ-divergence $\mathcal{D}_\varphi(p \parallel p_\theta)$. □

Definition 7 (Deformed Exponential Family). A deformed exponential family is a φ-family where the underlying measure is of the form $p \cdot \mu$ for some $p \in \mathcal{P}_\mu$, and the φ-function is a deformed exponential function $\exp_\phi \colon \mathbb{R} \to (0, \infty)$, and $u_0 = 1$. For these choices, the members of the *(parametric) deformed exponential family* $\mathcal{F}_p^\phi = \{p(t; \theta) : \theta \in \Theta\}$ centered at p are given by one-to-one mapping

$$p(t; \theta) := \exp_\phi\left(\sum_{i=1}^n \theta^i u_i(t) - \psi(\theta)\right) p(t), \qquad \text{for each } \theta = (\theta^k) \in \Theta. \tag{16.19}$$

where the *normalizing function* $\psi \colon \Theta \to [0, \infty)$ is introduced so that expression (16.19) is a probability density in $\mathcal{P}_\mu$, and $u_1, \dots u_n \colon T \to \mathbb{R}$ are measurable functions satisfying conditions (i)–(ii) in the definition of φ-fmilies.

16.4 GEOMETRY OF (GENERALIZED) STATISTICAL MANIFOLDS

In this section we present the notions of dual connections and dual coordinate systems. These concepts appear naturally in the framework of (generalized) statistical manifolds. For example, the connections $D^{(\alpha)}$ and $D^{(-\alpha)}$ presented in (16.5) are dual. We also show that a metric and a pair of dual connections can be defined in terms of a divergence. Using this result, we introduce a family of α-connections $D^{(\alpha)}$ in generalized statistical manifolds.

16.4.1 DUAL CONNECTIONS

Let (M, g) be a Riemannian manifold. Two connections D and D^* are said to be *(mutually) dual* if the relation

$$Xg(Y, Z) = g(D_X Y, Z) + g(Y, D_X^* Z) \tag{16.20}$$

holds for any vector fields X, Y, and Z. A connection compatible with g (see 29) is also called *self-dual*, since clearly a connection compatible with g is dual to itself.

Theorem 8. *The dual of any connection always exists and is unique. Moreover, the dual of the dual is the connection itself.*

Proof: Given a connection D, we can define its dual connection D^* by the expression

$$\frac{\partial g_{jk}}{\partial x^i} = \Gamma_{ijk} + \Gamma_{ikj}^*, \tag{16.21}$$

where Γ_{ijk} and Γ^*_{ijk} are the Christoffel symbols of D and D^*, respectively. Clearly, the connection D^* is unique. $\qquad\qquad\qquad\qquad\qquad\qquad\qquad\qquad\qquad\qquad\qquad\qquad\qquad$ $\square$

Example 9. The connections $D^{(\alpha)}$ and $D^{(-\alpha)}$ introduced in (16.7) are mutually dual. From expression (16.7), we have

$$\Gamma^{(\alpha)}_{ijk} + \Gamma^{(-\alpha)}_{ikj} = \Gamma_{ijk} - \alpha T_{ijk} + \Gamma_{ikj} + \alpha T_{ikj}$$
$$= \Gamma_{ijk} + \Gamma_{ikj}$$
$$= \frac{\partial g_{jk}}{\partial \theta^i}. \qquad (16.22)$$

Thus, according to (16.21), the connections $D^{(\alpha)}$ and $D^{(-\alpha)}$ are mutually dual. $\qquad$ $\square$

Consider two vector fields X and Y along a curve $\gamma\colon I \to M$ as the parallel transport of vectors $X(t_0)$ and $Y(t_0)$, for $t_0 \in I$, relative to some connection D. In general, parallel transport does not preserve inner product. In other words,

$$g(X(t), Y(t)) \neq g(X(t_0), Y(t_0)), \qquad \text{for some } t \in I.$$

If the inner product between any pair of vectors is preserved across parallel transport, the connection D is said to be compatible with g. Given any connection D, we can find another connection, which we denote by D^*, such that

$$g(X(t), Y^*(t)) = g(X(t_0), Y(t_0)), \qquad \text{for all } t \in I, \qquad (16.23)$$

where Y^* is a vector field along γ, defined as the parallel transport of $Y(t_0)$ with respect to D^*. Two connections D and D^* are said to be *(mutually) dual* if the inner product in (16.23) is preserved for any two vectors $X(t_0)$ and $Y(t_0)$. Below we show that the notions of duality presented in (16.20) and (16.23) coincide.

Theorem 10. *The definitions of duality given by expressions (16.20) and (16.23) are equivalent.*

Proof: Assume that (16.23) is satisfied. Consider any curve $\gamma\colon (-\varepsilon, \varepsilon) \to M$ with $\gamma(0) = p \in M$, which we express locally as $(\gamma^1(t), \ldots, \gamma^n(t))$ in some coordinate system $(x^1, \ldots, x^n)$. Let X and Y^* be the parallel transport of vectors $X(0)$ and $Y(0)$ along γ, relative to D and D^*, respectively. We express these vectors in $(x^1, \ldots, x^n)$ as

$$X = \sum_i a^i \frac{\partial}{\partial x^i}, \qquad Y^* = \sum_j b^j \frac{\partial}{\partial x^j}.$$

Denoting by Γ^k_{ij} and Γ^{*k}_{ij} the Christoffel symbols associated with D and D^*, respectively, we get

$$0 = D_{d\gamma/dt}X = \sum_m \left(\frac{da^m}{dt} + \sum_{i,j} \frac{d\gamma^i}{dt} a^j \Gamma^m_{ij} \right) \frac{\partial}{\partial x^m},$$

$$0 = D^*_{d\gamma/dt}Y^* = \sum_m \left(\frac{db^m}{dt} + \sum_{i,k} \frac{d\gamma^i}{dt} b^k \Gamma^{*m}_{ik} \right) \frac{\partial}{\partial x^m},$$

which provide the expressions

$$\frac{da^m}{dt} = -\sum_{i,j} \frac{d\gamma^i}{dt} a^j \Gamma_{ij}^m,$$

$$\frac{db^m}{dt} = -\sum_{i,k} \frac{d\gamma^i}{dt} b^k \Gamma_{ik}^{*m},$$

for each m. Using these formulas, we obtain

$$0 = \frac{d}{dt} g(X, Y^*) = \frac{d}{dt} \sum_{j,k} g_{jk} a^j b^k$$

$$= \sum_{j,k} \frac{dg_{jk}}{dt} a^j b^k + \sum_{m,k} g_{mk} \frac{da^m}{dt} b^k + \sum_{j,m} g_{jm} a^j \frac{db^m}{dt}$$

$$= \sum_{j,k} \frac{dg_{jk}}{dt} a^j b^k - \sum_{m,k} g_{mk} \left(\sum_{i,j} \frac{d\gamma^i}{dt} a^j \Gamma_{ij}^m \right) b^k - \sum_{j,m} g_{jm} a^j \left(\sum_{i,k} \frac{d\gamma^i}{dt} b^k \Gamma_{ik}^{*m} \right)$$

$$= \sum_{i,j,k} \frac{\partial g_{jk}}{\partial x^i} \frac{d\gamma^i}{dt} a^j b^k - \sum_{i,j,k} \Gamma_{ijk} \frac{d\gamma^i}{dt} a^j b^k - \sum_{i,j,k} \Gamma_{ikj}^* \frac{d\gamma^i}{dt} a^j b^k$$

$$= \sum_{i,j,k} \left(\frac{\partial g_{jk}}{\partial x^i} - \Gamma_{ijk} - \Gamma_{ikj}^* \right) \frac{d\gamma^i}{dt} a^j b^k.$$

Because the vectors $X(0)$, $Y(0)$, and $(d\gamma/dt)(0)$ may assume any value, we conclude that $\partial g_{jk}/\partial x^i = \Gamma_{ijk} + \Gamma_{ikj}^*$, which is equivalent to (16.20).

Now, if (16.20) is satisfied, we can write

$$\frac{d}{dt} g(X, Y^*) = g(D_{d\gamma/dt} X, Y^*) + g(X, D_{d\gamma/dt}^* Y^*)$$

$$= g(0, Y^*) + g(X, 0) = 0,$$

which implies (16.23). $\qquad\qquad\qquad\qquad\qquad\qquad\qquad\qquad\qquad\qquad\qquad\square$

A connection D is said to be *flat* if its torsion tensor T and curvature tensor R vanish identically. The following theorem provides an equivalent definition for flat connections.

Theorem 11. (a) *If M admits a flat connection D, then there exist local coordinate systems on M such that $\Gamma_{ij}^k = 0$. The transition functions between these coordinates systems are affine transformations.*

(b) *Suppose that M admits local coordinate systems whose transition functions are affine transformations. Then there exists a flat connection D whose Christoffel symbols satisfy $\Gamma_{ij}^k = 0$ for all such local coordinate systems.*

For a proof of Theorem 11, we refer to [41]. A coordinate system (x^i) for which $\Gamma_{ij}^k = 0$ is called an *affine coordinate system* with respect to D.

Theorem 12. *Let M be a manifold that is flat with respect to a torsion-free connection D. Then M is also flat with respect to the dual connection D^*.*

This theorem is a consequence of the following lemma.

Lemma 13. *The symbols R_{ijkl} and R^*_{ijkl} of curvature tensors of dual connections D and D^* are related by*

$$R_{ijkl} = -R^*_{ijlk}.$$

Proof: Using expression (16.20), we can write

$$
g\left(D_{\partial/\partial x^i} D_{\partial/\partial x^j} \frac{\partial}{\partial x^k}, \frac{\partial}{\partial x^l}\right) = \frac{\partial}{\partial x^i} g\left(D_{\partial/\partial x^j} \frac{\partial}{\partial x^k}, \frac{\partial}{\partial x^l}\right)
$$
$$
- g\left(D_{\partial/\partial x^j} \frac{\partial}{\partial x^k}, D^*_{\partial/\partial x^i} \frac{\partial}{\partial x^l}\right)
$$
$$
= \frac{\partial^2}{\partial x^i \partial x^j} g\left(\frac{\partial}{\partial x^k}, \frac{\partial}{\partial x^l}\right) - \frac{\partial}{\partial x^i} g\left(\frac{\partial}{\partial x^k}, D^*_{\partial/\partial x^j} \frac{\partial}{\partial x^l}\right)
$$
$$
- \frac{\partial}{\partial x^j} g\left(\frac{\partial}{\partial x^k}, D^*_{\partial/\partial x^i} \frac{\partial}{\partial x^l}\right)
$$
$$
+ g\left(\frac{\partial}{\partial x^k}, D^*_{\partial/\partial x^j} D^*_{\partial/\partial x^i} \frac{\partial}{\partial x^l}\right)
$$

and

$$
g\left(D_{\partial/\partial x^j} D_{\partial/\partial x^i} \frac{\partial}{\partial x^k}, \frac{\partial}{\partial x^l}\right) = \frac{\partial}{\partial x^j} g\left(D_{\partial/\partial x^i} \frac{\partial}{\partial x^k}, \frac{\partial}{\partial x^l}\right)
$$
$$
- g\left(D_{\partial/\partial x^i} \frac{\partial}{\partial x^k}, D^*_{\partial/\partial x^j} \frac{\partial}{\partial x^l}\right)
$$
$$
= \frac{\partial^2}{\partial x^i \partial x^j} g\left(\frac{\partial}{\partial x^k}, \frac{\partial}{\partial x^l}\right) - \frac{\partial}{\partial x^j} g\left(\frac{\partial}{\partial x^k}, D^*_{\partial/\partial x^i} \frac{\partial}{\partial x^l}\right)
$$
$$
- \frac{\partial}{\partial x^i} g\left(\frac{\partial}{\partial x^k}, D^*_{\partial/\partial x^j} \frac{\partial}{\partial x^l}\right)
$$
$$
+ g\left(\frac{\partial}{\partial x^k}, D^*_{\partial/\partial x^i} D^*_{\partial/\partial x^j} \frac{\partial}{\partial x^l}\right).
$$

Hence it follows that

$$
R_{ijkl} = g\left(D_{\partial/\partial x^i} D_{\partial/\partial x^j} \frac{\partial}{\partial x^k}, \frac{\partial}{\partial x^l}\right) - g\left(D_{\partial/\partial x^j} D_{\partial/\partial x^i} \frac{\partial}{\partial x^k}, \frac{\partial}{\partial x^l}\right)
$$
$$
= g\left(\frac{\partial}{\partial x^k}, D^*_{\partial/\partial x^j} D^*_{\partial/\partial x^i} \frac{\partial}{\partial x^l}\right) - g\left(\frac{\partial}{\partial x^k}, D^*_{\partial/\partial x^i} D^*_{\partial/\partial x^j} \frac{\partial}{\partial x^l}\right)
$$
$$
= -R^*_{ijlk},
$$

which is the desired expression. $\square$

A manifold that is flat with respect to a dual pair of connections is said to be *dually flat*.

Example 14. In Example 1, we verified that the symbols $\Gamma_{ij}^{(e)k}$ of the exponential connection $D^{(e)}$ vanish identically. Because the connections $D^{(m)}$ and $D^{(e)}$ are mutually dual, the Christoffel symbols of the mixture connection, in some system of coordinates (η_i) for the exponential family, satisfy $\Gamma_{ij}^{(m)k} = 0$. For the mixture family introduced in Example 4, the Christoffel symbols $\Gamma_{ij}^{(m)k}$ associated with the mixture connection $D^{(m)}$ vanish identically. Theorem 12 implies the existence of a system of coordinates (θ^i) for the mixture family, which satisfies $\Gamma_{ij}^{(e)k} = 0$.   □

16.4.2 DUAL COORDINATE SYSTEMS

In a dually flat manifold, the flat coordinate systems with respect to each of the connections are related by a duality relation. They are dual coordinate systems:

Definition 15. Two coordinate systems (θ^i) and (η_j) are said to be *(mutually) dual* if their coordinate basis vectors satisfy the relation

$$g\left(\frac{\partial}{\partial\theta^i}, \frac{\partial}{\partial\eta_j}\right) = \delta_i^j.$$

There is no guarantee that dual coordinate systems exist for general manifolds. Dual coordinate systems (θ^i) and (η_j) are not necessarily affine coordinate systems. However, for a dually flat manifold, dual coordinate systems always exist. As we will show in what follows, they are the affine coordinate systems with respect to each of the dual connections.

Let (θ^i) and (η_j) be dual coordinate systems. Expressing (θ^i) and (η_j) according to

$$\theta^i = \theta^i(\eta_1, \ldots, \eta_n), \qquad \eta_j = \eta_j(\theta^1, \ldots, \theta^n),$$

we find that the basis vectors are related by

$$\frac{\partial}{\partial\theta^i} = \sum_j \frac{\partial\eta_j}{\partial\theta^i} \frac{\partial}{\partial\eta_j}, \qquad \frac{\partial}{\partial\eta_j} = \sum_i \frac{\partial\theta^i}{\partial\eta_j} \frac{\partial}{\partial\theta^i}. \qquad (16.24)$$

The matrices $g_{ij} = g(\frac{\partial}{\partial\theta^i}, \frac{\partial}{\partial\theta^j})$ and $h_{ji} = g(\frac{\partial}{\partial\eta_i}, \frac{\partial}{\partial\eta_j})$ induced by the coordinate basis vectors $\frac{\partial}{\partial\theta^i}$ and $\frac{\partial}{\partial\eta_i}$ are inverses of each other. In other words, $h_{ji} = g^{ij}$. This result is a consequence of the relations

$$g_{ij} = g\left(\frac{\partial}{\partial\theta^i}, \frac{\partial}{\partial\theta^j}\right) = \sum_k \frac{\partial\eta_k}{\partial\theta^i} g\left(\frac{\partial}{\partial\eta_k}, \frac{\partial}{\partial\theta^j}\right) = \frac{\partial\eta_j}{\partial\theta^i}, \qquad (16.25)$$

and

$$h_{ji} = g\left(\frac{\partial}{\partial\eta_i}, \frac{\partial}{\partial\eta_j}\right) = \sum_k \frac{\partial\theta^k}{\partial\eta_j} g\left(\frac{\partial}{\partial\eta_i}, \frac{\partial}{\partial\theta^k}\right) = \frac{\partial\theta^i}{\partial\eta_j}, \qquad (16.26)$$

since the Jacobians $(\frac{\partial \eta_j}{\partial \theta^i})$ and $(\frac{\partial \theta^i}{\partial \eta_j})$ are each other's matrix inverse. Expression (16.24) can be rewritten as

$$\frac{\partial}{\partial \theta^i} = \sum_j g_{ij} \frac{\partial}{\partial \eta_j}, \qquad \frac{\partial}{\partial \eta_j} = \sum_i g^{ij} \frac{\partial}{\partial \theta^i}. \tag{16.27}$$

The next theorem shows that dual coordinate systems can be given in terms of potential functions.

Theorem 16. *If (θ^i) and (η_j) are dual coordinate systems, then there exist potential functions ψ and ψ^* on M such that*

$$\theta^i = \frac{\partial \psi^*}{\partial \eta_i}, \qquad \eta_j = \frac{\partial \psi}{\partial \theta^j}, \tag{16.28}$$

and

$$\psi(p) + \psi^*(p) = \sum_i \theta^i(p)\eta_i(p). \tag{16.29}$$

In addition,

$$g_{ij} = \frac{\partial^2 \psi}{\partial \theta^i \partial \theta^j}, \qquad g^{ij} = \frac{\partial^2 \psi^*}{\partial \eta_i \partial \eta_j}. \tag{16.30}$$

Conversely, if there exists a potential function ψ on M such that $g_{ij} = \frac{\partial^2 \psi}{\partial \theta^i \partial \theta^j}$, then (16.28) provides a coordinate system (η_j) that is dual to (θ^i), and the other potential function may be derived from (16.29).

Proof: Let (θ^i) and (η_j) be dual coordinate systems. In view of expression (16.25) we already know that $g_{ij} = \frac{\partial \eta_j}{\partial \theta^i}$. By the symmetry of g_{ij}, we get $\frac{\partial \eta_j}{\partial \theta^i} = \frac{\partial \eta_i}{\partial \theta^j}$, from which we conclude that (η_j) is the gradient of some potential function. In other words, there exists a function ψ such that $\eta_j = \frac{\partial \psi}{\partial \theta^j}$. Analogously, expression (16.26) in conjunction with the symmetry of $h_{ji} = g^{ij}$ implies the existence of a potential function ψ^* such that $\theta^i = \frac{\partial \psi^*}{\partial \eta_i}$. Expression (16.30) is a consequence of inserting $\eta_j = \frac{\partial \psi}{\partial \theta^j}$ and $\theta^i = \frac{\partial \psi^*}{\partial \eta_i}$ into (16.25) and (16.26), respectively. To show (16.29), we begin with the calculation

$$\frac{\partial^2}{\partial \eta_i \partial \eta_j}\left(\sum_l \theta^l \eta_l\right) = \frac{\partial}{\partial \eta_i}\left(\sum_l \frac{\partial \theta^l}{\partial \eta_j} \eta_l + \theta^j\right)$$

$$= \sum_l \frac{\partial^2 \theta^l}{\partial \eta_i \partial \eta_j} \eta_l + \frac{\partial \theta^i}{\partial \eta_j} + \frac{\partial \theta^j}{\partial \eta_i}$$

$$= \left(\sum_l \frac{\partial^2 \theta^l}{\partial \eta_i \partial \eta_j} \eta_l + \frac{\partial \psi^*}{\partial \eta_i \partial \eta_j}\right) + \frac{\partial \psi^*}{\partial \eta_i \partial \eta_j}. \tag{16.31}$$

Noting that

$$\frac{\partial^2 \psi^*}{\partial \eta_i \partial \eta_j} = g\left(\frac{\partial}{\partial \eta_i}, \frac{\partial}{\partial \eta_j}\right)$$

$$= \sum_{k,l} g\left(\frac{\partial}{\partial \theta^k}, \frac{\partial}{\partial \theta^l}\right) \frac{\partial \theta^k}{\partial \eta_i} \frac{\partial \theta^l}{\partial \eta_j}$$

$$= \sum_{k,l} g_{kl} \frac{\partial \theta^k}{\partial \eta_i} \frac{\partial \theta^l}{\partial \eta_j}$$

$$= \sum_{k,l} \frac{\partial^2 \psi}{\partial \theta^k \partial \theta^l} \frac{\partial \theta^k}{\partial \eta_i} \frac{\partial \theta^l}{\partial \eta_j},$$

we can write

$$\frac{\partial^2 \psi^*}{\partial \eta_i \partial \eta_j} = \sum_{l} \frac{\partial}{\partial \eta_i}\left(\frac{\partial \psi}{\partial \theta^l}\right) \frac{\partial \theta^l}{\partial \eta_j}$$

$$= \frac{\partial}{\partial \eta_i} \sum_{l}\left(\frac{\partial \psi}{\partial \theta^l} \frac{\partial \theta^l}{\partial \eta_j}\right) - \sum_{l} \frac{\partial \psi}{\partial \theta^l} \frac{\partial^2 \theta^l}{\partial \eta_i \partial \eta_j}$$

$$= \frac{\partial^2 \psi}{\partial \eta_i \partial \eta_j} - \sum_{l} \frac{\partial^2 \theta^l}{\partial \eta_i \partial \eta_j} \eta_l. \tag{16.32}$$

Using (16.32) in (16.31), we get

$$\frac{\partial^2 \psi}{\partial \eta_i \partial \eta_j} + \frac{\partial^2 \psi^*}{\partial \eta_i \partial \eta_j} = \frac{\partial^2}{\partial \eta_i \partial \eta_j}\left(\sum_{l} \theta^l \eta_l\right),$$

which results in

$$\psi + \psi^* = \sum_{i} \theta^i \eta_i + \sum_{i} a^i \eta_i + a, \tag{16.33}$$

for some constants a^i and a. Differentiating both sides of (16.33) with respect to η_i, we obtain $a^i = 0$. Clearly we may assume $a = 0$.

Conversely, assume that ψ is a potential function on M such that $g_{ij} = \frac{\partial^2 \psi}{\partial \theta^i \partial \theta^j}$. The systems of coordinates $(\eta_j) = \left(\frac{\partial \psi}{\partial \theta^j}\right)$ satisfies

$$g\left(\frac{\partial}{\partial \theta^i}, \frac{\partial}{\partial \eta_j}\right) = g\left(\frac{\partial}{\partial \theta^i}, \sum_{k} \frac{\partial \theta^k}{\partial \eta_j} \frac{\partial}{\partial \theta^k}\right)$$

$$= \sum_{k} g\left(\frac{\partial}{\partial \theta^i}, \frac{\partial}{\partial \theta^k}\right) \frac{\partial \theta^k}{\partial \eta_j}$$

$$= \sum_{k} g_{ik} \frac{\partial \theta^k}{\partial \eta_j}.$$

Notice that $\left(\frac{\partial \theta^k}{\partial \eta_j}\right)$ is the inverse of $(g_{kj}) = \left(\frac{\partial \eta_j}{\partial \theta^k}\right)$. Then we conclude that $g\left(\frac{\partial}{\partial \theta^i}, \frac{\partial}{\partial \eta_j}\right) = \delta_i^j$, showing that (θ^i) and (η_j) are dual coordinate systems. $\qquad\square$

Example 17. To avoid some technicalities, we assume that T is a finite set. Fix any probability distribution $p \in \mathcal{P}_\mu$. Let $\mathcal{E}_p = \{p(t; \theta) : \theta \in \Theta\}$ be an exponential family centered at p, whose members are given by

$$p(t; \theta) = \exp\left(\sum_{i=1}^{n} \theta^i u_i(t) - \psi(\theta)\right) p(t),$$

for measurable functions $u_1, \ldots, u_n : T \to \mathbb{R}$ satisfying conditions (i) and (ii) in the definition of exponential families. Given probability distributions $p_1, \ldots, p_n$, we consider the mixture family $\mathcal{M}_p = \{q(t; \eta) : \eta \in H\} \subseteq \mathcal{P}_\mu$ composed by probability distributions of the form

$$q(t; \eta) = \sum_{i=1}^{n} \eta_i p_i(t) + \left(1 - \sum_{i=1}^{n} \eta_i\right) p(t).$$

Because T is a finite set, we have $\mathcal{P}_\mu = \mathcal{E}_p = \mathcal{M}_p$. We will impose conditions on the functions $u_1, \ldots, u_n$ and on the probability distributions $p_1, \ldots, p_n$ so that the coordinate systems (θ^i) and (η_j) are mutually dual.

Suppose that (θ^i) and (η_j) are mutually dual. Differentiating $p(t; \theta) = q(t; \eta)$ with respect to θ^j, we get

$$\sum_{i=1}^{n} \frac{\partial \eta_i}{\partial \theta^j}(p_i - p) = p_\theta\left(u_j - \frac{\partial \psi}{\partial \theta^j}\right).$$

Since $\frac{\partial \psi}{\partial \theta^j}(0) = 0$ and $g_{ij} = \frac{\partial \eta_i}{\partial \theta^j}$, it follows that

$$\sum_{i=1}^{n} \frac{p_i - p}{p} g_{ij}(0) = u_j. \tag{16.34}$$

Conversely, assume that $u_1, \ldots, u_n$ and $p_1, \ldots, p_n$ are related by the expression (16.34). Let $\theta = (\theta^i)$ and $\eta = (\eta_i)$ be such that $p(t; \theta) = q(t; \eta)$. By the property $E_p[u_i] = 0$, we can write

$$E_\theta[u_j] = \sum_{i=1}^{n} \eta_i E_{p_i}[u_j]. \tag{16.35}$$

Taking the derivative of $\int_T p(t; \theta) = 1$ with respect to θ^i, we have

$$0 = \int_T \frac{\partial p_\theta}{\partial \theta^i} d\mu = \int_T p_\theta\left(u_i(t) - \frac{\partial \psi}{\partial \theta^i}\right) d\mu,$$

which implies

$$\frac{\partial \psi}{\partial \theta^i} = E_\theta[u_i]. \tag{16.36}$$

If we multiply both sides of (16.34) by u_k, and apply the expectation $E_p[\cdot]$, we will get

$$\sum_{i=1}^{n} E_{p_i}[u_k] g_{ij}(0) = E_p[u_k u_j] = g_{kj}(0).$$

Because the matrix (g_{ij}) is positive definite, it follows that $E_{p_i}[u_k] = \delta_i^k$. This result together with expressions (16.35) and (16.36) implies $\frac{\partial \psi}{\partial \theta^i} = \eta_i$. By Theorem 16, the coordinate systems (θ^i) and (η_j) are mutually dual.

Thus, in order that (θ^i) and (η_j) be dual coordinate systems, it is necessary and sufficient that $u_1, \ldots, u_n$ and $p_1, \ldots, p_n$ satisfy relation (16.34) (or $E_{p_i}[u_j] = \delta_i^j$). □

The next result states that, in a dually flat manifold, dual coordinate systems always exist.

Theorem 18. *If a manifold M is flat with respect to a dual pair of torsion-free connections D and D^*, then there exist dual, affine coordinate systems (θ^i) and (η_j) with respect to D and D^*, respectively.*

Proof: Let (θ^i) be a coordinate system for which $\Gamma_{ijk} = 0$. Thus expression (16.21) implies $\Gamma^*_{ijk} = \frac{\partial g_{jk}}{\partial \theta^i}$. Because the connection D^* is torsion-free, we get $\frac{\partial g_{jk}}{\partial \theta^i} = \Gamma^*_{ijk} = \Gamma^*_{jik} = \frac{\partial g_{ik}}{\partial \theta^j}$, showing the existence of a function η_k such that $g_{jk} = \frac{\partial \eta_k}{\partial \theta^j}$. By the symmetry $g_{ij} = g_{ji}$, it follow that $\frac{\partial \eta_i}{\partial \theta^j} = \frac{\partial \eta_j}{\partial \theta^i}$. As a result, there exists a potential function ψ on M such that $g_{ij} = \frac{\partial^2 \psi}{\partial \theta^i \partial \theta^j}$. According to Theorem 16, there exists a coordinate system (η_j) that is dual to (θ^i). Next we verify that (η_j) is an affine coordinate system with respect to D^*. By the definition of duality,

$$\frac{\partial}{\partial \theta^i} g\left(\frac{\partial}{\partial \theta^j}, \frac{\partial}{\partial \eta_k}\right) = \frac{\partial}{\partial \theta^i} \delta_j^k = 0.$$

Using the relations in (16.27), we can write

$$\frac{\partial}{\partial \theta^i} g\left(\frac{\partial}{\partial \theta^j}, \frac{\partial}{\partial \eta_k}\right) = g\left(D_{\partial/\partial \theta^i} \frac{\partial}{\partial \theta^j}, \frac{\partial}{\partial \eta_k}\right) + g\left(\frac{\partial}{\partial \theta^j}, D^*_{\partial/\partial \theta^i} \frac{\partial}{\partial \eta_k}\right)$$

$$= \sum_l g^{lk} g\left(D_{\partial/\partial \theta^i} \frac{\partial}{\partial \theta^j}, \frac{\partial}{\partial \theta^l}\right) + \sum_{l,m} g_{jl} g_{im} g\left(\frac{\partial}{\partial \eta_l}, D^*_{\partial/\partial \eta_m} \frac{\partial}{\partial \eta_k}\right)$$

$$= \sum_{l,m} g_{jl} g_{im} g\left(\frac{\partial}{\partial \eta_l}, D^*_{\partial/\partial \eta_m} \frac{\partial}{\partial \eta_k}\right),$$

which shows that the Christoffel symbols of D^* with respect to (η_j) vanish identically. Thus (η_j) is an affine coordinate system with respect to D^*. □

16.4.3 DIVERGENCE

Let M be a smooth manifold. We consider on M a non-negative, differentiable function $\mathcal{D}: M \times M \to [0, \infty)$ satisfying

$$\mathcal{D}(p \| q) = 0 \quad \text{if and only if} \quad p = q. \tag{16.37}$$

The function $\mathcal{D}(\cdot \| \cdot)$ is called a *divergence* if the matrix (g_{ij}), whose entries are given by

$$g_{ij}(p) = -\left[\left(\frac{\partial}{\partial \theta^i}\right)_p \left(\frac{\partial}{\partial \theta^j}\right)_q \mathcal{D}(p \| q)\right]_{q=p}, \tag{16.38}$$

is positive definite for each $p \in M$. In view of (16.37), the matrix (g_{ij}) is always positive semi-definite. Condition (16.37) also implies

$$0 = \left[\left(\frac{\partial}{\partial \theta^i}\right)_p \mathcal{D}(p \| q)\right]_{q=p} = \left[\left(\frac{\partial}{\partial \theta^i}\right)_q \mathcal{D}(p \| q)\right]_{q=p}$$

and

$$g_{ij}(p) = \left[\left(\frac{\partial^2}{\partial \theta^i \partial \theta^j}\right)_p \mathcal{D}(p \| q)\right]_{q=p} = \left[\left(\frac{\partial^2}{\partial \theta^i \partial \theta^j}\right)_q \mathcal{D}(p \| q)\right]_{q=p}. \tag{16.39}$$

Thus, associated with a divergence $\mathcal{D}(\cdot \| \cdot)$, we have a metric $g = (g_{ij})$, which is defined by (16.38) or (16.39).

A divergence $\mathcal{D}(\cdot \| \cdot)$ also induces a connection D, whose Christoffel symbols are given by

$$\Gamma_{ijk} = -\left[\left(\frac{\partial^2}{\partial \theta^i \partial \theta^j}\right)_p \left(\frac{\partial}{\partial \theta^k}\right)_q \mathcal{D}(p \| q)\right]_{q=p}. \tag{16.40}$$

Let $\mathcal{D}^*(p \| q) = \mathcal{D}(q \| p)$ define the *dual divergence* of $\mathcal{D}(\cdot \| \cdot)$. The connection D^* induced from $\mathcal{D}^*(\cdot \| \cdot)$, whose Christoffel symbols are

$$\Gamma_{ijk}^* = -\left[\left(\frac{\partial^2}{\partial \theta^i \partial \theta^j}\right)_p \left(\frac{\partial}{\partial \theta^k}\right)_q \mathcal{D}^*(p \| q)\right]_{q=p} \tag{16.41}$$

$$= -\left[\left(\frac{\partial}{\partial \theta^k}\right)_p \left(\frac{\partial^2}{\partial \theta^i \partial \theta^j}\right)_q \mathcal{D}(p \| q)\right]_{q=p}, \tag{16.42}$$

is dual to D. A simple computation shows that

$$\frac{\partial g_{jk}}{\partial \theta^i} = \Gamma_{ijk} + \Gamma_{ikj}^*.$$

Example 19 (α-Connections in generalized statistical manifolds). In Example 6, we introduced the φ-divergence between two probability distributions p and q in $\mathcal{P}_\mu$ as

$$\mathcal{D}_\varphi(p \| q) := \frac{\displaystyle\int_T \frac{\varphi^{-1}(p) - \varphi^{-1}(q)}{(\varphi^{-1})'(p)} d\mu}{\displaystyle\int_T \frac{u_0}{(\varphi^{-1})'(p)} d\mu}.$$

Clearly, $\mathcal{D}_\varphi(p \| q) \geq 0$ for all $p, q \in \mathcal{P}_\mu$. Moreover, assuming that $\varphi(\cdot)$ is strictly convex, then $\mathcal{D}_\varphi(p \| q) = 0$ if and only if $p = q$.

In a generalized statistical manifold $\mathcal{P} = \{p(t; \theta) : \theta \in \Theta\}$, the metric defined in (16.11) coincides with the metric derived from the divergence $\mathcal{D}(q \| p) := \mathcal{D}_\varphi(p \| q)$. To see this, we express the divergence $\mathcal{D}(\cdot \| \cdot)$ between p_θ and $p_{\hat{\theta}}$ as

$$\mathcal{D}(p_\theta \| p_{\hat{\theta}}) = E_{\hat{\theta}}'[(f_{\hat{\theta}} - f_\theta)].$$

Then we get

$$g_{ij} = -\left[\left(\frac{\partial}{\partial \theta^i}\right)_p \left(\frac{\partial}{\partial \theta^j}\right)_q \mathcal{D}(p \| q)\right]_{q=p}$$

$$= -E_{\hat{\theta}}'\left[\frac{\partial^2 f_\theta}{\partial \theta^i \partial \theta^j}\right].$$

The pair of dual connections derived from the divergence $\mathcal{D}(\cdot \parallel \cdot)$ can be used to define a family of α-connections $D^{(\alpha)}$ in generalized statistical manifolds. Expressions (16.40) and (16.41) provide a pair of dual connections $D^{(1)}$ and $D^{(-1)}$, whose symbols $\Gamma_{ijk}^{(1)}$ and $\Gamma_{ijk}^{(-1)}$ are given by

$$
\begin{aligned}
\Gamma_{ijk}^{(1)} &= -\left[\left(\frac{\partial^2}{\partial\theta^i\partial\theta^j}\right)_p \left(\frac{\partial}{\partial\theta^k}\right)_q \mathcal{D}(p \parallel q)\right]_{q=p} \\
&= E_\theta''\left[\frac{\partial^2 f_\theta}{\partial\theta^i\partial\theta^j}\frac{\partial f_\theta}{\partial\theta^k}\right] - E_\theta'\left[\frac{\partial^2 f_\theta}{\partial\theta^i\partial\theta^j}\right]E_\theta''\left[u_0\frac{\partial f_\theta}{\partial\theta^k}\right]
\end{aligned}
$$

and

$$
\begin{aligned}
\Gamma_{ijk}^{(-1)} &= -\left[\left(\frac{\partial}{\partial\theta^k}\right)_p \left(\frac{\partial^2}{\partial\theta^i\partial\theta^j}\right)_q \mathcal{D}(p \parallel q)\right]_{q=p} \\
&= E_\theta''\left[\frac{\partial^2 f_\theta}{\partial\theta^i\partial\theta^j}\frac{\partial f_\theta}{\partial\theta^k}\right] + E_\theta'''\left[\frac{\partial f_\theta}{\partial\theta^i}\frac{\partial f_\theta}{\partial\theta^j}\frac{\partial f_\theta}{\partial\theta^k}\right] \\
&\quad - E_\theta''\left[\frac{\partial f_\theta}{\partial\theta^j}\frac{\partial f_\theta}{\partial\theta^k}\right]E_\theta''\left[u_0\frac{\partial f_\theta}{\partial\theta^i}\right] - E_\theta''\left[\frac{\partial f_\theta}{\partial\theta^i}\frac{\partial f_\theta}{\partial\theta^k}\right]E_\theta''\left[u_0\frac{\partial f_\theta}{\partial\theta^j}\right].
\end{aligned}
$$

We define the α-connection $D^{(\alpha)}$ as

$$
\Gamma_{ijk}^{(\alpha)} = \frac{1+\alpha}{2}\Gamma_{ijk}^{(1)} + \frac{1-\alpha}{2}\Gamma_{ijk}^{(-1)}. \tag{16.43}
$$

Notice that $\Gamma_{ijk}^{(0)}$ corresponds to the Levi–Civita connection given in (16.17). Denoting

$$
\begin{aligned}
T_{ijk} &= \frac{1}{2}E_\theta'''\left[\frac{\partial f_\theta}{\partial\theta^i}\frac{\partial f_\theta}{\partial\theta^j}\frac{\partial f_\theta}{\partial\theta^k}\right] - \frac{1}{2}E_\theta''\left[\frac{\partial f_\theta}{\partial\theta^k}\frac{\partial f_\theta}{\partial\theta^i}\right]E_\theta''\left[u_0\frac{\partial f_\theta}{\partial\theta^j}\right] \\
&\quad - \frac{1}{2}E_\theta''\left[\frac{\partial f_\theta}{\partial\theta^k}\frac{\partial f_\theta}{\partial\theta^j}\right]E_\theta''\left[u_0\frac{\partial f_\theta}{\partial\theta^i}\right] - \frac{1}{2}E_\theta''\left[\frac{\partial f_\theta}{\partial\theta^i}\frac{\partial f_\theta}{\partial\theta^j}\right]E_\theta''\left[u_0\frac{\partial f_\theta}{\partial\theta^k}\right],
\end{aligned}
$$

then we can write

$$
\Gamma_{ijk}^{(\alpha)} = \Gamma_{ijk}^{(0)} - \alpha T_{ijk}.
$$

By (16.22), we see that the pair of connections $D^{(\alpha)}$ and $D^{(-\alpha)}$ is mutually dual.   □

Example 20 (Parallel transport). Assume that T is a finite set. Let $\mathcal{P} = \{p(t;\theta) : \theta \in \Theta\}$ be a generalized statistical manifold such that $\mathcal{P} = \mathcal{P}_\mu$. We saw in Section 16.3 that the tangent space $T_p\mathcal{P}$ can be identified with $\widetilde{T}_p\mathcal{P}$, the vector space spanned by the functions $\partial f_\theta/\partial\theta^i$, equipped with the inner product $\langle \widetilde{X}, \widetilde{Y} \rangle = E_\theta''[\widetilde{X}\widetilde{Y}]$, where $p = p_\theta$. We will use this identification to define the connection $D^{(1)}$, introduced in Example 19, in terms of the parallel transport

$$
P_{q,p} : \widetilde{T}_q\mathcal{P} \to \widetilde{T}_p\mathcal{P}
$$

given by

$$
\widetilde{X} \mapsto \widetilde{X} - E_\theta'[\widetilde{X}]u_0,
$$

where $p = p_\theta$. Notice that the parallel transport $P_{q,p}$ does not depend on the curve joining q and p. This result is consistent with the fact that $D^{(1)}$ is a flat connection. Let $\gamma(t)$ denote the coordinate curve given locally by $\theta(t) = (\theta^1, \ldots, \theta^i + t, \ldots, \theta^n)$. Clearly, the inverse $P^{-1}_{\gamma(0),\gamma(t)}$ maps the vector $\frac{\partial f_\theta}{\partial \theta^j}(t)$ to

$$\frac{\partial f_\theta}{\partial \theta^j}(t) - E'_{\theta(0)}\Big[\frac{\partial f_\theta}{\partial \theta^j}(t)\Big]u_0.$$

Thus, using expression (16.52), we define the connection

$$\begin{aligned}\widetilde{D}_{\partial f_\theta/\partial\theta_i}\frac{\partial f_\theta}{\partial\theta_j} &= \frac{d}{dt}P^{-1}_{\gamma(0),\gamma(t)}\Big(\frac{\partial f_\theta}{\partial\theta_j}(\gamma(t))\Big)\Big|_{t=0}\\ &= \frac{d}{dt}\Big(\frac{\partial f_{\theta(t)}}{\partial\theta^j} - E'_{\theta(0)}\Big[\frac{\partial f_{\theta(t)}}{\partial\theta^j}\Big]u_0\Big)\Big|_{t=0}\\ &= \frac{\partial^2 f_\theta}{\partial\theta^i\partial\theta^j} - E'_\theta\Big[\frac{\partial^2 f_\theta}{\partial\theta^i\partial\theta^j}\Big]u_0.\end{aligned}$$

Denoting by D the connection acting on $T_p\mathcal{P}\times T_q\mathcal{P}$, that is equivalent to $\widetilde{D}$, it follows that

$$\begin{aligned}g\Big(D_{\partial/\partial\theta_i}\frac{\partial}{\partial\theta_j},\frac{\partial}{\partial\theta_k}\Big) &= \Big\langle\widetilde{D}_{\partial f_\theta/\partial\theta_i}\frac{\partial f_\theta}{\partial\theta_j},\frac{\partial f_\theta}{\partial\theta_k}\Big\rangle\\ &= E''_\theta\Big[\frac{\partial^2 f_\theta}{\partial\theta^i\partial\theta^j}\frac{\partial f_\theta}{\partial\theta^k}\Big] - E'_\theta\Big[\frac{\partial^2 f_\theta}{\partial\theta^i\partial\theta^j}\Big]E''_\theta\Big[u_0\frac{\partial f_\theta}{\partial\theta^k}\Big]\\ &= \Gamma^{(1)}_{ijk},\end{aligned}$$

which shows that $D = D^{(1)}$.   □

Let M be a manifold that is flat with respect to a dual pair of torsion-free connections D and D^*. Let (θ^i) and (η_j) be affine coordinate systems with respect to D and D^*, whose potential functions are denoted by ψ and ψ^*. In a dually flat manifold M, the divergence between points p and q is defined as

$$\mathcal{D}(p \| q) = \psi(p) + \psi^*(q) - \sum_i \theta^i(p)\eta_i(q). \tag{16.44}$$

Clearly, the metric (g_{ij}) can be recovered from this divergence by using expression (16.38). In the coordinate system (θ^i), the Christoffel symbols of D and D^*, which are derived from $\mathcal{D}(\cdot \| \cdot)$ and $\mathcal{D}^*(\cdot \| \cdot)$, are given by

$$\Gamma_{ijk} = 0$$

and

$$\Gamma^*_{ijk} = \frac{\partial^2 \eta_k}{\partial\theta^i\partial\theta^j} = \frac{\partial^3 \psi}{\partial\theta^i\partial\theta^j\partial\theta^k}.$$

Using expressions (16.40) and (16.41) with respect to the coordinate system (η_j), we get

$$\Gamma^{ijk} = \frac{\partial^2 \theta^k}{\partial\eta_i\partial\eta_j} = \frac{\partial^3 \psi^*}{\partial\eta_i\partial\eta_j\partial\eta_k}$$

and

$$\Gamma^{*ijk} = 0.$$

The divergence given by (16.44) satisfies the triangular relation

$$\mathcal{D}(p \| q) + \mathcal{D}(q \| r) = \mathcal{D}(p \| r) + \sum_i [\theta^i(p) - \theta^i(q)][\eta_i(r) - \eta_i(q)] \qquad (16.45)$$

for any points p, q, and r in M. This relation is the result of some manipulations. By the definition of $\mathcal{D}(\cdot \| \cdot)$, we have

$$\mathcal{D}(p \| q) = \psi(p) + \psi^*(q) - \sum_i \theta^i(p)\eta_i(q)$$

and

$$\mathcal{D}(q \| r) = \psi(q) + \psi^*(r) - \sum_i \theta^i(q)\eta_i(r).$$

Using

$$\psi(q) + \psi^*(q) = \sum_i \theta^i(q)\eta_i(q)$$

and

$$\psi(p) + \psi^*(r) = \mathcal{D}(p \| r) + \sum_i \theta^i(p)\eta_i(r),$$

we get

$$\begin{aligned}\mathcal{D}(p \| q) + \mathcal{D}(q \| r) &= \mathcal{D}(p \| r) + \sum_i \theta^i(p)\eta_i(r) + \sum_i \theta^i(q)\eta_i(q) \\ &\quad - \sum_i \theta^i(p)\eta_i(q) - \sum_i \theta^i(q)\eta_i(r) \\ &= \mathcal{D}(p \| r) + \sum_i [\theta^i(p) - \theta^i(q)][\eta_i(r) - \eta_i(q)].\end{aligned}$$

An immediate consequence of (16.45) is the *Pythagorean relation*.

Theorem 21 (Pythagorean relation). *Let p, q, and r be three points in M. Assume that the D-geodesic γ_1 connecting q and p, and the D*-geodesic γ_2 connecting q and r, are orthogonal at q (see Figure 16.1). Then p, q, and r satisfy the* Pythagorean relation

$$\mathcal{D}(p \| q) + \mathcal{D}(q \| r) = \mathcal{D}(p \| r). \qquad (16.46)$$

Let (θ^i) and (η_j) be affine coordinate systems with respect to D and D^*. In these coordinate systems, the D- and D^*-geodesics γ_1 and γ_2 are given by $t\theta(p) + (1-t)\theta(q)$ and $t\eta(r) + (1-t)\eta(q)$. Thus the tangent vector to these geodesics are expressed as

$$\frac{d\gamma_1}{dt} = \sum_i [\theta^i(p) - \theta^i(q)]\frac{\partial}{\partial\theta^i}$$

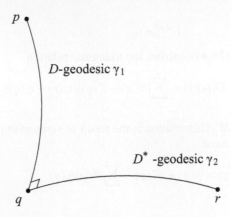

Figure 16.1 Pythagorean relation.

and

$$\frac{d\gamma_2}{dt} = \sum_j [\eta_i(r) - \eta_i(q)] \frac{\partial}{\partial \eta_i}.$$

Because (θ^i) and (η_j) are dual, we have

$$g\left(\frac{d\gamma_1}{dt}, \frac{d\gamma_2}{dt}\right) = \sum_i [\theta^i(p) - \theta^i(q)][\eta_i(r) - \eta_i(q)] = 0.$$

Thus, in the case that γ_1 and γ_2 are orthogonal, expression (16.45) reduces to the Pythagorean relation (16.46).

The following result is a consequence of the Pythagorean relation (Theorem 21):

Corollary 22 (Projection Theorem). *Let M be a dually flat manifold, and let N be a submanifold of M which is D^*-autoparallel (i.e., $D_X^* Y$ is also a vector field on N, for any vectors fields X and Y on N). Fix $p \in M \setminus N$. Then the minimum of $r \mapsto \mathcal{D}(p \| r)$ is attained at q if, and only if, the D-geodesic joining p and q is orthogonal to N at q (see Figure 16.2).*

Let (ζ^k) be a coordinate system of N around $r \in N$. Let γ denote the D-geodesic joining p and r, whose local expression is given by $t\theta(p) + (1-t)\theta(r)$. From (16.44), it follows that

$$\frac{\partial}{\partial \eta_j} \mathcal{D}(p \| r) = \frac{\partial \psi^*}{\partial \eta_j}(r) - \theta^j(p) = \theta^j(r) - \theta^j(p).$$

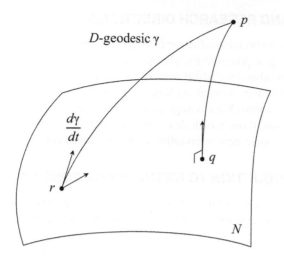

Figure 16.2 Projection Theorem.

Using this expression, we can write

$$\frac{\partial}{\partial\zeta^k}\mathcal{D}(p\,\|\,r) = \sum_{i,j}\frac{\partial\theta^i}{\partial\zeta^k}\frac{\partial\eta_j}{\partial\theta^i}\frac{\partial}{\partial\eta_j}\mathcal{D}(p\,\|\,r) = -\sum_{i,j}[\theta^j(p) - \theta^j(r)]\frac{\partial\theta^i}{\partial\zeta^k}g_{ij}$$

$$= -g\Big(\sum_j[\theta^j(p) - \theta^j(r)]\frac{\partial}{\partial\theta^j}, \sum_i\frac{\partial\theta^i}{\partial\zeta^k}\frac{\partial}{\partial\theta^i}\Big)$$

$$= -g\Big(\frac{d\gamma}{dt}, \frac{\partial}{\partial\zeta^k}\Big).$$

Thus $q \in N$ is a stationary point of $r \mapsto \mathcal{D}(p\,\|\,r)$ if, and only if, the D-geodesic joining p and q is orthogonal to N at q.

Let (h_{jk}) be the metric on N induced from (M, g), i.e.,

$$h_{jk} = g\Big(\frac{\partial}{\partial\zeta^j}, \frac{\partial}{\partial\zeta^k}\Big).$$

Then we can express the gradient of $r \mapsto \mathcal{D}(p\,\|\,r)$ on N as

$$\nabla\mathcal{D}(p\,\|\,r) = \sum_{j,k}h^{jk}\frac{\partial}{\partial\zeta^k}\mathcal{D}(p\,\|\,r)\frac{\partial}{\partial\zeta^j}$$

$$= -\sum_{j,k}h^{jk}g\Big(\frac{d\gamma}{dt}, \frac{\partial}{\partial\zeta^k}\Big)\frac{\partial}{\partial\zeta^j}.$$

In other words, the gradient of $r \mapsto \mathcal{D}(p\,\|\,r)$ is the negative of the projection of $d\gamma/dt$ onto T_rN.

16.5   SUMMARY AND RESEARCH DIRECTIONS

This chapter has provided a brief introduction to information geometry models which can be very helpful for signal processing applications, such as optimization and information estimation. Generalized statistical manifolds and their geometric properties are also provided and allow us to work with problems with a more broad range of statistical probability functions. We envisage as interesting and important directions to be followed in this area, those which deal with the definition of new divergence functions that can reveal some new information about processes and signals.

16.A   BRIEF INTRODUCTION TO RIEMANNIAN GEOMETRY

In this section we provide a brief introduction to Riemannian geometry so that the text can be understood with a minimum of prerequisites. A more detailed exposition is found in [42, 43, 44].

16.A.1   SMOOTH MANIFOLDS

A smooth manifold of dimension n is a set M and a collection of pairs $(U_i, \phi_i)_{i \in I}$, composed by sets $U_i \subseteq M$, and injective mappings $\phi_i \colon U_i \to \mathbb{R}^n$, such that

(i) $\bigcup_{i \in I} U_i = M$;
(ii)   for any pair i, j with $U_i \cap U_j = W \neq \emptyset$, the sets $\phi_i^{-1}(W)$ and $\phi_j^{-1}(W)$ are open in $\mathbb{R}^n$ and the mapping $\phi_j \circ \phi_i^{-1}$ is smooth.
(iii) The collection $(U_i, \phi_i)_{i \in I}$ is maximal with respect to conditions (i) and (ii).

The pair (U_i, ϕ_i) with $p \in U_i$ is called a *chart* (or *system of coordinates*) for M around p. The set U_i is called a *coordinate neighborhood* around p. A collection $(U_i, \phi_i)_{i \in I}$ satisfying (i) and (ii) is called a *smooth atlas* on M.

Let M and N be smooth manifolds. A mapping $f \colon M \to N$ is *smooth* at $p \in M$ if given any chart (V, ψ) around $f(p)$ there exists a chart (U, ϕ) around p such that $f(U) \subseteq V$ and the mapping $\psi^{-1} \circ f \circ \phi$ from $\phi^{-1}(U)$ to $\psi(V)$ is smooth at $\phi(p)$. The mapping f is smooth on an open set of M if it is smooth at all points of this open set. We denote by $C^\infty(M)$ (respectively, $C_p^\infty(M)$) the set of all functions $f \colon M \to \mathbb{R}$ that are smooth on M (respectively, smooth at p).

16.A.2   TANGENT VECTORS AND VECTOR FIELDS

Definition 23. Let M be a smooth manifold. Given $p \in M$, let $\gamma \colon (-\varepsilon, \varepsilon) \to M$ be a smooth curve (a smooth function) on M such that $\gamma(0) = p$. The *tangent vector* to the curve γ at $t = 0$ is a function $\frac{d\gamma}{dt}(0) \colon C_p^\infty(M) \to \mathbb{R}$ given by

$$\left(\frac{d\gamma}{dt}(0) \right) f = \frac{d(f \circ \gamma)}{dt} \Big|_{t=0}, \qquad \text{for each } f \in C_p^\infty(M).$$

A tangent vector at p is the tangent vector at $t = 0$ of some curve $\gamma \colon (-\varepsilon, \varepsilon) \to M$ with $\gamma(0) = p$. The set of all tangent vectors to M at p is denoted by $T_p M$.

In a chart (U, ϕ) around $p = \phi^{-1}(0)$, a function f and a curve γ are expressed locally as

$$f \circ \phi^{-1}(x) = f(x^1, \ldots, x^n), \qquad \text{for } x = (x^1, \ldots, x^n) \in \phi(U),$$

and

$$\phi \circ \gamma(t) = (\gamma^1(t), \ldots, \gamma^n(t)).$$

Using this representation, we get

$$\gamma'(0)f = \frac{d(f \circ \gamma)}{dt}\Big|_{t=0}$$

$$= \frac{d}{dt} f(\gamma^1(t), \ldots, \gamma^n(t))\Big|_{t=0}$$

$$= \sum_{i=1}^{n} \frac{d\gamma^i}{dt}(0) \frac{\partial f}{\partial x^i}(0)$$

$$= \left[\sum_{i=1}^{n} \frac{d\gamma^i}{dt}(0) \left(\frac{\partial}{\partial x^i}\right)_{\phi^{-1}(0)}\right] f.$$

Then, locally, the vector $\gamma'(0)$ can be written as

$$\gamma'(0) = \sum_{i=1}^{n} \frac{d\gamma^i}{dt}(0) \left(\frac{\partial}{\partial x^i}\right)_p.$$

Notice that $(\partial/\partial x^i)_p$ is the tangent vector to the coordinate curve $x^i \mapsto \phi^{-1}(0, \ldots, 0, x^i, 0, \ldots, 0)$. The set $T_p M$ is a vector space of dimension n, and the set $\{(\partial/\partial x^1)_p, \ldots, (\partial/\partial x^n)_p\}$ associated with the chart (U, ϕ) constitutes a base for $T_p M$. The vector space $T_p M$ is called the *tangent space* of M at p.

Theorem 24. *Let $F: M \to N$ be a smooth mapping between smooth manifolds M and N. Fixed $p \in M$ and $v \in T_p M$, take a smooth curve $\gamma: (-\varepsilon, \varepsilon) \to M$ with $\gamma(0) = p$ and $\gamma'(0) = v$. Let β be the curve in N defined by the composition $\beta = F \circ \gamma$. The mapping $F_*: T_p M \to T_{F(p)} N$ given by $F_*(v) = \beta'(0)$ is a linear mapping that does not depend on the choice of γ. The linear mapping F_* is called the* pushforward *of F at p.*

Definition 25. A vector field X on a smooth manifold M is a correspondence that associates to each point $p \in M$ a vector $X_p \in T_p M$. A vector field X is called *smooth* if the mapping Xf, defined by $f(p) \mapsto Xf(p)$, belongs to $C^\infty(M)$, for each $f \in C^\infty(M)$. The set of all smooth vector fields on M is denoted by $X(M)$.

If X and Y are smooth vector fields on M, and $f: M \to \mathbb{R}$ is a smooth function, we can consider the compositions $X(Yf)$ and $Y(Xf)$. In general, a composition of smooth vector fields does not result in another vector field. However, the following result can be stated.

Lemma 26. *Let X and Y be smooth vector fields on a smooth manifold M. Then there exists a unique vector field $[X, Y]$, called the* Lie bracket *of X and Y, such that*

$$[X, Y]f = X(Yf) - Y(Xf), \qquad \text{for all } f \in C^\infty(M).$$

In any coordinate system (x^i), the vectors $\partial/\partial x^i$ and $\partial/\partial x^j$ satisfy $[\partial/\partial x^i, \partial/\partial x^j] = 0$.

16.A.3 CONNECTIONS

Definition 27. A *connection D* on a smooth manifold M is a mapping $D: X(M) \times X(M) \to X(M)$, which associates to any pair of vector fields X and Y, another vector field $D_X Y$, satisfying the following properties:

(i) $D_{fX+gY}Z = fD_X Z + gD_Y Z$,
(ii) $D_X(Y + Z) = D_X Y + D_X Z$,
(iii) $D_X(fY) = fD_X Y + X(f)Y$,

for any vector fields $X, Y, Z \in X(M)$ and functions $f, g \in C^\infty(M)$.

Let X and Y be smooth vector fields, which we express locally in some coordinate system (x^i) as

$$X = \sum_i a^i \frac{\partial}{\partial x^i}, \qquad \text{and} \qquad Y = \sum_j b^j \frac{\partial}{\partial x^j}.$$

By the properties of the connection D, we write

$$D_X Y = \sum_i a^i D_{\partial/\partial x^i}\left(\sum_j b^j \frac{\partial}{\partial x^j}\right)$$

$$= \sum_i \sum_j a^i \frac{\partial b^j}{\partial x^i} \frac{\partial}{\partial x^j} + \sum_i \sum_j a^i b^j D_{\partial/\partial x^i}\left(\frac{\partial}{\partial x^j}\right). \qquad (16.47)$$

Introducing the *Christoffel symbols (of the second kind)* Γ^k_{ij} by $D_{\partial/\partial x^i}\partial/\partial x^j = \sum_k \Gamma^k_{ij}\partial/\partial x^k$, equation (16.47) results in

$$D_X Y = \sum_k \left(\sum_i a^i \frac{\partial b^k}{\partial x^i} + \sum_{i,j} a^i b^j \Gamma^k_{ij}\right)\frac{\partial}{\partial x^k}, \qquad (16.48)$$

which is the local expression of $D_X Y$ in some coordinate system (x^i).

Definition 28. A *Riemannian metric g* on a smooth manifold M is a correspondence which associates to each point $p \in M$ an inner product $g_p(\cdot, \cdot)$ on the tangent space $T_p M$, which varies smoothly in the following sense: If (U, ϕ) is a chart around p, then

$$g_{ij}(x^1, \ldots, x^n) = g_q\left(\left(\frac{\partial}{\partial x^i}\right)_q, \left(\frac{\partial}{\partial x^j}\right)_q\right), \qquad \text{with } \phi^{-1}(x^1, \ldots, x^n) = q \in U,$$

is a smooth function on $\phi^{-1}(U)$.

A *Riemannian manifold* is a pair (M, g), where M is a smooth manifold and g is a Riemannian metric on M.

Definition 29. A connection D on a Riemannian manifold (M, g) is said to be *compatible* with the metric if

$$Xg(Y, Z) = g(D_X Y, Z) + g(Y, D_X Z), \qquad \text{for all } X, Y, Z \in X(M).$$

Definition 30. A connection D on a smooth manifold M is called *torsion-free* (or *symmetric*) if

$$D_X Y - D_Y X = [X, Y], \qquad \text{for all } X, Y \in X(M). \tag{16.49}$$

With $X = \partial/\partial x^i$ and $Y = \partial/\partial x^j$, equation (16.49) reduces to

$$D_{\partial/\partial x^i}\left(\frac{\partial}{\partial x^j}\right) - D_{\partial/\partial x^j}\left(\frac{\partial}{\partial x^i}\right) = \left[\frac{\partial}{\partial x^i}, \frac{\partial}{\partial x^j}\right] = 0.$$

As a result, the Christoffel symbols satisfy $\Gamma_{ij}^k = \Gamma_{ji}^k$. This justifies the terminology for a connection satisfying (16.49).

Theorem 31 (Levi–Civita). *Given a Riemannian manifold (M, g), there exists a unique torsion-free connection D on M that is compatible with the Riemannian metric.*

The connection given in the theorem above is called the *Levi–Civita connection* on M. We will use the symbol ∇ to represent a Levi–Civita connection.

The Christoffel symbols Γ_{ij}^k of a Levi–Civita ∇ can be expressed in terms of g_{ij}. Because ∇ is compatible with g, we get

$$\frac{\partial g_{jk}}{\partial x^i} = g\left(\nabla_{\frac{\partial}{\partial x^i}} \frac{\partial}{\partial x^j}, \frac{\partial}{\partial x^k}\right) + g\left(\frac{\partial}{\partial x^j}, \nabla_{\frac{\partial}{\partial x^i}} \frac{\partial}{\partial x^k}\right),$$

$$\frac{\partial g_{ki}}{\partial x^j} = g\left(\nabla_{\frac{\partial}{\partial x^j}} \frac{\partial}{\partial x^k}, \frac{\partial}{\partial x^i}\right) + g\left(\frac{\partial}{\partial x^k}, \nabla_{\frac{\partial}{\partial x^j}} \frac{\partial}{\partial x^i}\right),$$

$$\frac{\partial g_{ij}}{\partial x^k} = g\left(\nabla_{\frac{\partial}{\partial x^k}} \frac{\partial}{\partial x^i}, \frac{\partial}{\partial x^j}\right) + g\left(\frac{\partial}{\partial x^i}, \nabla_{\frac{\partial}{\partial x^k}} \frac{\partial}{\partial x^j}\right).$$

By the symmetry of ∇, these expressions result in

$$\frac{\partial g_{jk}}{\partial x^i} + \frac{\partial g_{ki}}{\partial x^j} - \frac{\partial g_{ij}}{\partial x^k} = 2g\left(\frac{\partial}{\partial x^k}, \nabla_{\frac{\partial}{\partial x^j}} \frac{\partial}{\partial x^i}\right).$$

From the definition of the Christoffel symbols of the second kind, it follows that

$$\Gamma_{ijk} := \sum_m \Gamma_{ij}^m g_{mk} = \frac{1}{2}\left(\frac{\partial g_{jk}}{\partial x^i} + \frac{\partial g_{ki}}{\partial x^j} - \frac{\partial g_{ij}}{\partial x^k}\right),$$

where Γ_{ijk} is the *Christoffel symbols (of the first kind)*. Denoting by (g^{km}) the inverse of (g_{km}), we obtain

$$\Gamma_{ij}^m = \frac{1}{2} \sum_k \left(\frac{\partial g_{jk}}{\partial x^i} + \frac{\partial g_{ki}}{\partial x^j} - \frac{\partial g_{ij}}{\partial x^k} \right) g^{km}. \tag{16.50}$$

Equation (16.50) is the classical expression for the Christoffel symbols of the Levi–Civita in terms of g_{ij}.

A *vector field* V *along a curve* $\gamma \colon I \to M$ is a mapping that associates to each $t \in I$ a tangent vector $V(t) \in T_{\gamma(t)}M$. A vector field along γ is smooth if, for any smooth function f on M, the function $t \mapsto Vf(t)$ is smooth on I. The vector field $\gamma_*(d/dt)$, denoted by $d\gamma/dt$, is called the *tangent vector field* (or *velocity field*) of γ.

A connection gives rise to derivatives of vector fields along curves, as shown by the following proposition.

Proposition 32. *Let M be a smooth manifold with a connection D. There exists a unique correspondence $\frac{D}{dt}$, which associates to a vector field V along the smooth curve $\gamma \colon I \to M$ another vector field $\frac{DV}{dt}$ along γ, called the* covariant derivative *of V along γ, such that*

(a) $\frac{D}{dt}(V + W) = \frac{DV}{dt} + \frac{DW}{dt}$, *for any vector fields V and W along γ;*

(b) $\frac{D}{dt}(fV) = \frac{df}{dt}V + f\frac{DV}{dt}$, *for any smooth function $f \colon I \to \mathbb{R}$, and any vector field V along γ; and*

(c) *if V is induced by a smooth vector field X, i.e., $V(t) = X(\gamma(t))$, then $\frac{DV}{dt} = D_{d\gamma/dt}X$.*

Let $(\gamma^1(t), \dots \gamma^n(t))$ be the local expression of γ in some coordinate system (x^i). Then, in this coordinate system, we express the vector fields $d\gamma/dt$ and V along γ as

$$\frac{d\gamma}{dt} = \sum_i \frac{d\gamma^i}{dt} \frac{\partial}{\partial x^i}, \quad \text{and} \quad V = \sum_j v_j \frac{\partial}{\partial x^j}.$$

Item (c) and (16.48) imply

$$\frac{DV}{dt} = \sum_k \left(\frac{dv_k}{dt} + \sum_{i,j} \frac{d\gamma^i}{dt} v_j \Gamma_{ij}^k \right) \frac{\partial}{\partial x^k}. \tag{16.51}$$

Definition 33. Let M be a smooth manifold with a connection D. A vector field V along a curve $\gamma \colon I \to M$ is called *parallel* if $\frac{DV}{dt} = 0$ for all $t \in I$.

Theorem 34. *Let M be a smooth manifold with a connection D. Let $\gamma \colon I \to M$ be a smooth curve in M. Fix any tangent vector V_0 at $\gamma(t_0)$, for some $t_0 \in I$. Then there exists a unique parallel vector field V along γ such that $V(t_0) = V_0$. The vector field $V(t)$ is called the* parallel transport *of V_0 along γ.*

A connection D can be recovered from the parallel transport as follows. Let X and Y be vector fields on M. Let $p \in M$ and $\gamma \colon I \to M$ be an integral curve of X passing through p, i.e., $\gamma(t_0) = p$ and $\frac{d\gamma}{dt} = X(\gamma(t))$. Consider the mapping

$$P_{\gamma, t_0, t} \colon T_{\gamma(t_0)}M \to T_{\gamma(t)}M,$$

which is defined as the parallel transport of a vector along γ from t_0 to t. Then we have

$$(D_X Y)(p) = \frac{d}{dt} P_{\gamma, t_0, t}^{-1}(Y(c(t)))\Big|_{t=t_0}. \tag{16.52}$$

Theorem 35. *Let (M, g) be a Riemannian manifold. A connection D is compatible with the metric if, and only if, $g(V, W) = $ constant for any pair of parallel vector fields V and W along any smooth curve γ.*

Theorem 36. *Let (M, g) be a Riemannian manifold. A connection D on M is compatible with the metric if, and only if,*

$$\frac{d}{dt} g(V, W) = g\Big(\frac{DV}{dt}, W\Big) + g\Big(V, \frac{DW}{dt}\Big),$$

for any vector fields V and W along any smooth curve $\gamma \colon I \to M$.

16.A.4   GEODESICS

In this section we assume that (M, g) is a Riemannian manifold endowed with its Levi–Civita connection ∇.

Definition 37. A curve $\gamma \colon I \to M$ is a *geodesic* if $\frac{D}{dt}(\frac{d\gamma}{dt}) = 0$ for all $t \in I$.

The *length* of a curve $\gamma \colon I = (a, b) \to M$ is defined by

$$L(\gamma) = \int_a^b \Big|\frac{d\gamma}{dt}\Big|_g \, dt,$$

where $|\frac{d\gamma}{dt}|_g = g(\frac{d\gamma}{dt}, \frac{d\gamma}{dt})^{1/2}$. If γ is a geodesic then

$$\frac{d}{dt} g\Big(\frac{d\gamma}{dt}, \frac{d\gamma}{dt}\Big) = g\Big(\frac{D}{dt}\frac{d\gamma}{dt}, \frac{d\gamma}{dt}\Big) + g\Big(\frac{d\gamma}{dt}, \frac{D}{dt}\frac{d\gamma}{dt}\Big) = 0.$$

Thus the length of the tangent vector $d\gamma/dt$ is constant. From now on, we will assume that $|\frac{d\gamma}{dt}|_g = c \neq 0$. In this case, the arc length of γ from a fixed point $t_0 \in I$ is given by

$$s(t) = \int_{t_0}^t \Big|\frac{d\gamma}{dt}\Big|_g \, dt = c(t - t_0).$$

Denoting by $(\gamma^1(t), \ldots, \gamma^n(t))$ the local expression of γ in some coordinate system (x^i), expression (16.51) implies

$$\frac{D}{dt}\Big(\frac{d\gamma}{dt}\Big) = \sum_k \Big(\frac{d^2\gamma^k}{dt^2} + \sum_{i,j} \frac{d\gamma^i}{dt}\frac{d\gamma^j}{dt}\Gamma_{ij}^k\Big)\frac{\partial}{\partial x^k}.$$

Thus γ is a geodesic if, and only if, the coordinates $(\gamma^1(t), \ldots, \gamma^n(t))$ satisfy the system of second-order differential equations

$$\frac{d^2\gamma^k}{dt^2} + \sum_{i,j} \frac{d\gamma^i}{dt} \frac{d\gamma^j}{dt} \Gamma^k_{ij} = 0, \qquad \text{for each } k.$$

Theorem 38. *For any point $p \in M$ and for any tangent vector X_p at p, there exists locally a unique geodesic $\gamma: (-\varepsilon, \varepsilon) \to M$ satisfying the initial conditions*

$$\gamma(0) = p, \qquad \frac{d\gamma}{dt}(0) = X_p. \tag{16.53}$$

A geodesic satisfying the initial conditions (16.53) is denoted by $\exp(tX_p)$.

Theorem 39. *For a tangent space T_pM at any point $p \in M$ there exists a neighborhood N_p of 0 such that, for any $X_p \in N_p$, the mapping $t \mapsto \exp(tX_p)$ is defined on an open interval containing $[-1, 1]$.*

The mapping on N_p given by $X_p \mapsto \exp(X_p)$ is called the *exponential mapping* at p.

16.A.5 CURVATURE

The *torsion tensor T* of a connection D is defined by

$$T(X, Y) = D_X Y - D_Y X - [X, Y].$$

The components T^k_{ij} of the torsion tensor T, which are given by

$$T\left(\frac{\partial}{\partial x^i}, \frac{\partial}{\partial x^j}\right) = \sum_k T^k_{ij} \frac{\partial}{\partial x^k},$$

satisfy the relation

$$T^k_{ij} = \Gamma^k_{ij} - \Gamma^k_{ji}.$$

Clearly, a connection D is torsion free if, and only if, its torsion tensor T vanishes identically.

The *curvature tensor R* of a connection D is defined as

$$R(X, Y)Z = D_X D_Y Z - D_Y D_X Z - D_{[X,Y]}Z,$$

whose components R^i_{jkl} are given by

$$R\left(\frac{\partial}{\partial x^k}, \frac{\partial}{\partial x^l}\right)\frac{\partial}{\partial x^j} = \sum_i R^i_{jkl} \frac{\partial}{\partial x^i}.$$

The components R^i_{jkl} are expressed in terms of the Christoffel symbols Γ^k_{ij} by

$$R^i_{jkl} = \frac{\partial \Gamma^i_{lj}}{\partial x^k} - \frac{\partial \Gamma^i_{kj}}{\partial x^l} + \sum_m (\Gamma^m_{lj}\Gamma^i_{km} - \Gamma^m_{kj}\Gamma^i_{lm}).$$

A connection D is said to be *flat* if the torsion tensor T and the curvature tensor R vanish identically. A manifold M endowed with a flat connection D is called a *flat manifold*. Some properties of flat manifolds are presented in Section 16.4.

REFERENCES

1. H. L. Van Trees, K. L. Bell, and Z. Tian, *Detection, Estimation and Modulation Theory*, 2nd ed.   Wiley, 2013, vol. 1.
2. S. M. Kay, *Fundamentals of Statistical Signal Processing: Estimation Theory*, ser. Prentice Hall Signal Processing Series.   Prentice-Hall, 1993, vol. 1.
3. A. J. Viterbi and J. K. Omura, *Principles of Digital Communication and Coding*.   Dover Publications, 2009.
4. C. M. Bishop, *Pattern Recognition and Machine Learning*, ser. Information Science and Statistics.   Springer, 2007.
5. C. E. Shannon, "A Mathematical Theory of Communication," *Bell Systems Technical Journal*, vol. Vol. 27, pp. 379–423, July 1948.
6. B. Efron, "Defining the Curvature of a Statistical Problem (with Application to Second Order Efficiency)," *Annals of Statistics*, vol. 3, no. 6, pp. 1189–1242, 1975.
7. S.-I. Amari and H. Nagaoka, *Methods of information geometry*, ser. Translations of Mathematical Monographs.   American Mathematical Society, Providence, RI; Oxford University Press, Oxford, 2000, vol. 191.
8. S.-I Amari, "Information Geometry on Hierarchy of Probability Distributions," *IEEE Transactions on Information Theory*, vol. 47, no. 5, pp. 1701–1711, July 2001.
9. S. T. Smith, "Geometric Optimization Methods for Adaptive Filtering," Ph.D. dissertation, Harvard University, Cambridge, Massachussetts, 1993.
10. S. T. Smith, "Optimization Techniques on Riemannian Manifolds," *Fields Institute Communications*, vol. 3, pp. 113–146, 1994.
11. A. Manikas, A. Sleiman, and I. Dacos, "Manifold Studies of Nonlinear Antenna Array Geometries," *IEEE Transactions on Signal Processing*, vol. 49, no. 3, pp. 497–506, March 2001.
12. J. H. Manton, "Optimization Algorithms Exploiting Unitary Constraints," *IEEE Transactions on Signal Processing*, vol. 50, no. 3, pp. 635–650, March 2002.
13. J. H. Manton, "On the Role of Differential Geometry in Signal Processing," in *Proceedings of IEEE International Conference on Acoustics, Speech, and Signal Processing, 2005 (ICASSP '05)*, vol. 5, 2005, pp. v/1021–v/1024.
14. S. T. Smith, "Covariance, Subspace, and Intrinsic Crámer-Rao Bounds," *IEEE Transactions on Signal Processing*, vol. 53, no. 5, pp. 1610–1630, May 2005.
15. T. Abrudan, J. Eriksson, and V. Koivunen, "Efficient Riemannian algorithms for optimization under unitary matrix constraint," in *IEEE International Conference on Acoustics, Speech and Signal Processing (ICASSP 2008)*, April 2008, pp. 2353–2356.
16. T. E. Abrudan, J. Eriksson, and V. Koivunen, "Steepest Descent Algorithms for Optimization Under Unitary Matrix Constraint," *IEEE Transactions on Signal Processing*, vol. 56, no. 3, pp. 1134–1147, March 2008.
17. T. E. Abrudan, J. Eriksson, and V. Koivunen, "Conjugate gradient algorithm for optimization under unitary matrix constraint," *Signal Processing*, vol. 89, no. 9, pp. 1704–1714, September 2009.
18. J.-F. Cardoso, *Entropic Contrasts for Source Separation: Geometry and Stability*.   John Wiley & Sons, 2000, vol. 1, pp. 139–190.

19. J.-F. Cardoso, "The Three Easy Routes to Independent Component Analysis; Contrasts and Geometry," in *Proceedings of 2001 Independent Component Analysis (ICA) Workshop*, San Diego, CA, December 2001.

20. M. D. Plumbley, "Geometrical Methods for Non-Negative ICA: Manifolds, Lie Groups and Toral Subalgebras," *Neurocomputing*, vol. 67, pp. 161–197, August 2005.

21. M. D. Plumbley, "Geometry and Manifolds for Independent Component Analysis," in *Proceedings of the IEEE International Conference on Acoustics, Speech and Signal Processing (ICASSP2007)*, vol. 4, Honolulu, Hawaii, April 2007, pp. IV–1397–IV–1400.

22. P.-A. Absil, R. Mahony, and R. Sepulchre, *Optimization Algorithms on Matrix Manifolds*. Princeton University Press, 2008.

23. N. Boumal, B. Mishra, P.-A. Absil, and R. Sepulchre, "Manopt, a Matlab toolbox for optimization on manifolds," *Journal of Machine Learning Research*, vol. 15, pp. 1455–1459, 2014.

24. S.-I Amari and J.-F. Cardoso, "Blind Source Separation - Semiparametric Statistical Approach," *IEEE Transactions on Signal Processing*, vol. 45, no. 11, pp. 2692–2700, November 1997.

25. S.-I. Amari, "Natural gradient works efficiently in learning," *Neural Computation*, vol. 10, no. 2, pp. 251–276, 1998.

26. J. H. Manton, D. Applebaum, S. Ikeda, and N. L. Bihan, "Introduction to the Issue on Differential Geometry in Signal Processing," *IEEE Journal of Selected Topics in Signal Processing*, vol. 7, no. 4, pp. 573–575, August 2013.

27. F. Nielsen and R. Bhatia, Eds., *Matrix Information Geometry*.   Springer, 2013.

28. F. Nielsen, Ed., *Geometric Theory of Information*, ser. Signals and Communication Technology.   Springer, 2014.

29. G. Pistone, "Nonparametric information geometry," in *Geometric science of information*, ser. Lecture Notes in Comput. Sci.   Springer, Heidelberg, 2013, vol. 8085, pp. 5–36.

30. M. K. Murray and J. W. Rice, *Differential Geometry and Statistics*, ser. (Monographs on Statistics and Applied Probability).   Chapman & Hall, 1993, no. 48.

31. O. Calin and C. Udrişte, *Geometric Modeling in Probability and Statistics*.   Springer, 2104.

32. G. Pistone and C. Sempi, "An infinite-dimensional geometric structure on the space of all the probability measures equivalent to a given one," *Ann. Statist.*, vol. 23, no. 5, pp. 1543–1561, 1995.

33. A. Cena and G. Pistone, "Exponential statistical manifold," *Ann. Inst. Statist. Math.*, vol. 59, no. 1, pp. 27–56, 2007.

34. M. R. Grasselli, "Dual connections in nonparametric classical information geometry," *Ann. Inst. Statist. Math.*, vol. 62, no. 5, pp. 873–896, 2010.

35. R. F. Vigelis and C. C. Cavalcante, "On φ-families of probability distributions," *J. Theoret. Probab.*, vol. 26, no. 3, pp. 870–884, 2013.

36. M. J. S. Barão, "Métodos de controlo probabilístico," Ph.D. dissertation, Universidade Técnica de Lisboa, Instituto Superior Técnico, Sep. 2008.

37. L. Malagò and G. Pistone, "Combinatorial optimization with information geometry: the Newton method," *Entropy*, vol. 16, no. 8, pp. 4260–4289, 2014.

38. G. Kaniadakis, "Statistical mechanics in the context of special relativity," *Phys. Rev. E (3)*, vol. 66, no. 5, pp. 056 125, 17, 2002.

39. C. Tsallis, "What are the numbers that experiments provide?" *Quimica Nova*, vol. 17, no. 6, pp. 468–471, 1994.

40. R. F. Vigelis and C. C. Cavalcante, "The Δ_2-condition and φ-families of probability

distributions," in *Geometric science of information*, ser. Lecture Notes in Comput. Sci. Springer, Heidelberg, 2013, vol. 8085, pp. 729–736.

41. H. Shima, *The geometry of Hessian structures*.   World Scientific Publishing Co. Pte. Ltd., Hackensack, NJ, 2007.

42. M. P. do Carmo, *Riemannian geometry*.   Boston: Birkhäuser, 1992.

43. S. Gallot, D. Hulin, and J. Lafontaine, *Riemannian geometry*, 3rd ed., ser. Universitext. Springer-Verlag, Berlin, 2004.

44. J. M. Lee, *Introduction to smooth manifolds*, ser. Graduate Texts in Mathematics. Springer-Verlag, New York, 2003, vol. 218.

17 Bio-Inspired and Information-Theoretic Signal Processing

ROMIS RIBEIRO DE FAISSOL ATTUX
University of Campinas (UNICAMP)

LEVY BOCCATO
University of Campinas (UNICAMP)

DENIS GUSTAVO FANTINATO
University of Campinas (UNICAMP)

JUGURTA ROSA MONTALVÃO FILHO
Federal University of Sergipe (UFS)

ALINE DE OLIVEIRA NEVES PANAZIO
Federal University of ABC (UFABC)

RICARDO SUYAMA
Federal University of ABC (UFABC)

KENJI NOSE-FILHO
University of Campinas (UNICAMP)

DANIEL GUERREIRO E SILVA
University of Brasilia (UnB)

17.1 INTRODUCTION

Signal processing is a broader field than it may seem *prima facie*. Essentially, it is related to the notion of using a specially designed system (which will be termed filter) to mold an input signal to an output signal possessing some sort of desirable quality. Classically, one thinks of a filter either as an analog device built from resistors, capacitors, and inductors or, perhaps, a software-defined structure running on a digital computer, but this is not mandatory: a filter can be understood in terms of coding theory or may obey the rather intangible structure of a statistical classification machine. In Figure 17.1, we present a signal processing task in a very general form.

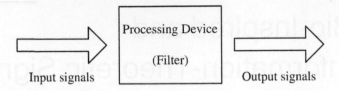

Figure 17.1 Scheme of a general signal processing task.

Basically, the design of a filter capable of properly carrying out a given task must encompass three aspects:

- The filtering structure, i.e., the mathematical model that relates the filter output to the filter input. This structure can be dynamically adapted in any level, but, in general, the performed mapping is controlled by a set of free parameters.
- The adaptation criterion, which expresses in mathematical terms and according to the available statistical information the aims of the problem to be solved, allowing that a relationship be established with the aforementioned free parameters.
- The adaptation algorithm, which is responsible from actualizing the potential engendered in the previous steps by effectively finding good quality solutions either in batch or online modes.

In general, after the application of interest is defined, it is possible to form a relatively clear view regarding structural demands, and one is also able to delimit the scope of the criterion described in the second entry of the list. Finally, given the characteristics of both structure and criterion, it is possible to choose a suitable search method.

In order to illustrate each point, we chose a specific paradigm characterized by: (i) belonging to the state of the art of its respective domain; (ii) having been consistently applied in filtering tasks and (iii) being sufficiently flexible to deal with the corresponding aspect in a general way. Our choices fill the paradigms of artificial neural networks [65], evolutionary computation [35], and information theoretic learning [113].

Structural aspects are addressed in Section 17.2. Basically, the main points to be taken into account are the linear / nonlinear and feedforward / recurrent "dilemma"s. Section 17.3 gravitates around the second item, being devoted to an analysis of the statistical description power associated with different paradigms. Finally, advanced topics regarding the last subject are addressed in Section 17.4.

As the reader will notice, the entire design cycle was based on a computational intelligence framework. This is essential, as those methods are sufficiently powerful and flexible to allow that robust filters be designed to cope with a broad range of environments and application demands. If the reader forms a clearer view of the motivations underlying such a daring statement, it is our belief that the chapter will have fulfilled its aim.

17.2   ARTIFICIAL NEURAL NETWORKS

Artificial neural networks (ANNs) are nonlinear adaptive signal processing devices whose *modus operandi* is inspired in theories regarding the nervous systems of living beings. They are, in general, structured in layers composed of nonlinear processing units (neurons), the behavior of which is defined by the values assumed by a set of synaptic weights and by a nonlinear memoryless activation function. A basic scheme is given in Figure 17.2.

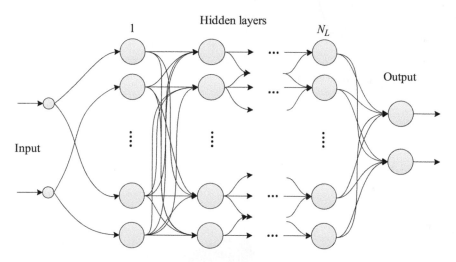

Figure 17.2   Generic structure of an artificial neural network.

The structure of an ANN is essentially defined by the following factors: (i) the neuron model (or models, if multiple responses are adopted); (ii) the number of neuron layers; (iii) the number of neurons in each layer and (iv) the existence or absence of feedback.

Let us discuss each of these points. In terms of a neuron model, there are two classical options: McCulloch-Pitts-like models [103], which give rise to approximation models built according to ridge functions, and radial-basis functions (RBF) [21], which have an activation pattern concentrated on a certain region of the input space.

The basic McCulloch-Pitts model is described in Figure 17.3, which, in mathematical terms, expresses an input-output relationship of the kind:

$$y(n) = f(\upsilon(n)), \tag{17.1}$$

being $\upsilon(n) = \mathbf{w}^T \mathbf{u}(n)$ the neuron activation and $f(\cdot)$ the activation function, which is typically a sigmoid (e.g., a hyperbolic tangent) [65]. The synaptic weights are contained in vector $\mathbf{w} = [w_0, \cdots, w_K]^T$, and $\mathbf{u}(n) = [1, u_1(n), \cdots, u_2(n)]^T$ is composed of all neuron inputs. The relationship described in (17.1) leads to a map of the general shape shown in Figure 17.4.

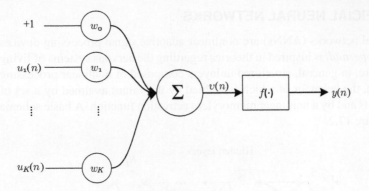

Figure 17.3 McCulloch-Pitts neuron model.

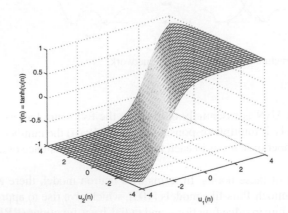

Figure 17.4 Example of a mapping generated by a McCulloch-Pitts neuron considering two input signals, $\mathbf{w} = [0.8, -0.5]^T$, and a sigmoidal activation function.

On the other hand, a RBF obeys the following pattern: it assumes a maximum value at a point termed center and has a value that decreases (or increases) radially with the distance thereto. The most employed RBF model is the Gaussian function, which engenders an input-output response of the form shown in Figure 17.5.

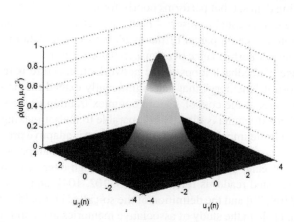

Figure 17.5 Example of a mapping generated by an RBF neuron considering two input signals, $\mu = [0, 0]^T$ and $\sigma^2 = 1$.

Mathematically, we may describe the neuron behavior, in the Gaussian case, as:

$$y(n) = \rho(\mathbf{u}(n); \mu; \sigma^2) = \exp\left(-\frac{\|\mathbf{u}(n) - \mu\|^2}{\sigma^2}\right), \qquad (17.2)$$

where $\mu \in \mathbb{R}^K$ is the center and σ^2 i the variance, which can be seen as a dispersion factor. Notice that the produced mapping has a local character, in contrast with the McCulloch-Pitts model, which, due to the mutual influence of a linear combiner and a nonlinear function, generates a non-local mapping. The choice for a neuron model depends on the nature of the problem to be solved, and may have a huge impact on the overall network performance [65].

The number of network layers can be, in theory, arbitrarily large. The modern research trend of deep neural networks [65] indicates this direction as effective in certain applications, but, nonetheless, an important result usually associated with the name of Cybenko [32] shows that a neural network with a single intermediate layer of sigmoidal McCulloch-Pitts neurons and a linear output layer is enough to endow the network with the potential of universal approximation [72, 32]. RBF networks, in contrast, are almost exclusively thought of in terms of a single hidden layer followed by a linear combiner, and these networks are also universal approximators under certain conditions [111]. Therefore, in this text, as a rule, we will consider no more than two neuron layers in a neural signal processing device.

The number of neurons in each layer is a fundamental parameter, since it defines, in simple terms, the approximation potential of a given network. For instance, it is roughly comparable to the number of terms used in a truncated Taylor series. As a consequence, it may seem that the usage of an elevated number of neurons is always desirable. However, this is not the case: given the limited amount of available data to guide the parameter adaptation process, it is possible that the model experiences the phenomenon known as overfitting – the generated mapping becomes extremely well suited to the training data set, but performs poorly for unseen data [65]. In order to deal with this problem, it is usual to resort – both for McCulloch-Pitts and RBF models – to a validation set, i.e., a parcel of the available training data that is put apart to monitor the network performance over unseen patterns [65].

Finally, the existence of feedback loops in the network architecture is responsible for generating a sort of "memory" that can significantly extend the processing capabilities of the structure, at the cost of a more complex training process and of the risk of instability. This dilemma is directly comparable to that of making an option for either finite or infinite impulse response filters in linear adaptive processing. In this chapter, the concept of recurrent neural networks (RNNs) shall be addressed later, in the discussion of echo state networks. For a thorough overview about the topic of RNNs, the interested reader is referred to [65, 102, 104], and, also, to the work of John Hopfield [69, 70] and the definition of the so-called Hopfield networks, which play an important role in the study of associative memories and combinatorial optimization problems [71].

17.2.1 THE MULTILAYER PERCEPTRON

The neural network known as multilayer perceptron (MLP) is probably the most emblematic nonlinear bio-inspired filter. Following the thread of our line of reasoning, we may define an MLP as a neural network composed of McCulloch-Pitts neurons, and with, at least, one hidden layer. The term "perceptron" refers to Rosenblatt's work on single layer networks [117], with the MLP being an extension of this pioneering notion. An example of an MLP is given in Figure. 17.6.

A difficulty that arises when one uses an MLP within a supervised training framework is that there is no direct dependence between the error and the synaptic weights of layers aside from the output layer. This demands the use of a mathematical procedure known as error backpropagation (BP) [136, 118]. This procedure is based on the sequential application of the chain rule to obtain the derivatives of a cost function, which is usually given by the mean squared error (MSE):

$$J_{\text{MSE}} = \frac{1}{N_s} \sum_{n=1}^{N_s} (d(n) - y(n))^2 , \qquad (17.3)$$

where N_s is the number of samples with the respective desired values, which form the training set, and $d(n)$ represents the desired output.

As it is not possible to obtain the minima of this function, in a closed form, for all the parameters, one is forced to resort to iterative methods. The standard option is to use the information brought by the derivatives of the cost function. The BP algorithm plays exactly this role: it allows that any derivative of the cost function, with respect to any synaptic weight, be obtained on a layer-by-layer basis [65].

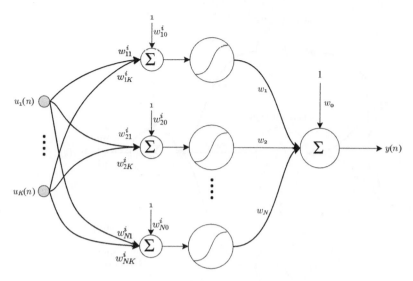

Figure 17.6 General architecture of an MLP network containing a single hidden layer with N neurons and receiving K inputs, where w_{jk}^i represents the synaptic weight associated with the connection between the j-th neuron and the k-th input and w_l gives the output weight corresponding to the l-th hidden neuron.

17.2.2 RADIAL-BASIS FUNCTIONS NETWORKS

RBF networks are, as a rule, composed of two layers: one layer of neurons that follows the model exposed in (17.2) and a linear output layer, which combines the activation of each hidden element. Figure 17.7 describes the model.

The input-output relationship of a network of this kind can be expressed as follows:

$$y(n) = \mathbf{w}^T \rho, \tag{17.4}$$

where $\mathbf{w} = [w_0, \cdots, w_N]^T$ contains the weights of the output linear combination and $\rho = [1, \rho(\mathbf{u}(n); \boldsymbol{\mu}_1; \sigma_1^2), \cdots, \rho(\mathbf{u}(n); \boldsymbol{\mu}_N; \sigma_N^2)]^T$ specifies the responses of the hidden neurons, including a constant value that acts as a bias.

As shown in equation (17.4), there are three classes of parameters that influence the network: (i) the neuron centers $\boldsymbol{\mu}$, (ii) the neuron dispersions σ^2, and (iii) the synaptic weights of the output layer $\mathbf{w}$. These parameters are of very different natures: the first two are arguments of a nonlinear (e.g., Gaussian) function, while the remaining are elements of a linear combiner, in a McCulloch-Pitts-like fashion. Although all of these parameters can be jointly adapted using a sort of backpropagation scheme, it is more usual to obtain the centers and dispersions using unsupervised methods (e.g., clustering) and, afterwards, to treat the problem of obtaining the synaptic weights as one of least squares (LS) [65]. The LS solution is obtained, given a training dataset, using the Moore-Penrose pseudoinverse:

$$\mathbf{w}_{\text{opt}} = \mathbf{H}^\dagger \mathbf{d} \tag{17.5}$$

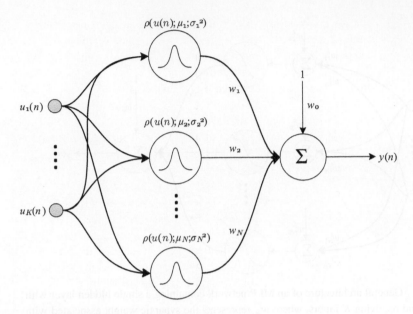

Figure 17.7 Architecture of an RBF network.

where the entries of $\mathbf{H} \in \mathbb{R}^{N_s \times M}$ are the values of the radial-basis functions evaluated at the points $\mathbf{u}(i), i = 1, \ldots, N_s$ according to (17.4). N_s is the number of available samples, M the number of neurons in the hidden layer plus a bias term ($M = N + 1$), and $\mathbf{d} = [d(1), \cdots, d(N_s)]^T$ specifies the desired responses of the network for each input vector $\mathbf{u}(i)$. The form of the pseudoinverse, denoted by $(^\dagger)$, in (17.5) depends on whether $M < N_s$ or $M > N_s$.

17.2.3 GENERAL REGRESSION NEURAL NETWORK (GRNN)

The general regression neural network (GRNN) can be seen as a special case of the RBF network [21] and was first presented by Specht in 1991 [130].

Similarly to the RBF, the GRNN is basically composed of two layers: an input layer composed by radial-basis functions and a linear output layer. The output of the GRNN is generally expressed as

$$y(n) = \frac{\sum_{i=1}^{N} w_i \rho(\mathbf{u}(n); \mu_i; \sigma^2)}{\sum_{i=1}^{N} \rho(\mathbf{u}(n); \mu_i; \sigma^2)}, \qquad (17.6)$$

where $\mathbf{u}(n) \in \mathbb{R}^K$ represent the input vector, N is the number of neurons, $\mu \in \mathbb{R}^K$ is the center vector of the i-th neuron, and w_i is the weight associated with the i-th neuron in the linear output layer. The radial basis function ρ is usually given by the Gaussian function, as expressed in Eq. (17.2).

The main difference between these two ANNs lies in the way that the parameters $\mathbf{w}$ and μ are determined. The GRNN is based on the estimation of the conditional expectation of a scalar random variable (RV) y given a particular measured value $\mathbf{X}$ of a RV $\mathbf{x}$. For the nonparametric estimate of the joint PDF, Specht employed the Parzen estimator [112], resulting in a nonlinear similarity measure known, in the field of information theoretic learning, as correntropy function [98].

The center vectors $\boldsymbol{\mu}$ of the GRNN are given by the inputs $\mathbf{u}(i), i = 1, \ldots, N_s$ of the training set and the weights w_i are given by the desired outputs $d(i)$. The only parameter to be set is σ, which controls the level of selectivity of the ANN. The higher its value, the smoother the fitness. The output is, then, a linear combination of all desired outputs weighted by the correntropy between the actual input and the recorded patterns. Its applicability comprises several applications in digital signal processing, including prediction and noise filtering [110, 109].

Despite the fact that this network is a one-pass learning algorithm with a highly parallel structure, the estimation time of the GRNN is proportional to the number of samples of the training set.

In order to reduce this time and make it practical in real-time applications, the authors in [110] proposed a modification of this network, that reduces the number of calculations performed by the GRNN. In this case, the output is given by

$$y(n) = \frac{\sum_{i \in K} w_i \rho(\mathbf{u}(n); \boldsymbol{\mu}_i; \sigma^2)}{\sum_{i \in k} \rho(\mathbf{u}(n); \boldsymbol{\mu}_i; \sigma^2)}, \qquad (17.7)$$

where K is the set of index i that produces the N_K smallest Euclidean distances $\|\mathbf{u}(n) - \boldsymbol{\mu}_i\|$, being N_K a parameter set by the user.

17.2.4 UNORGANIZED MACHINES

The ANNs we have discussed so far — MLPs and RBF-like networks — represent general nonlinear structures capable of modeling and generating nonlinear input-output mappings. The additional flexibility of these models can be useful in many relevant applications, such as channel equalization and source separation [3, 105, 24, 131, 43], in spite of the higher computational complexity, in comparison with a classical linear system [65].

Interestingly, recent works brought a different perspective to the network training process [78, 76, 99]: the hidden layers of neurons can be set in advance and remain fixed during the training, whereas the parameters of the output layer are effectively adjusted using information concerning the particular task that the network should perform. Hence, in accordance with the spirit of these approaches, the intermediate portion of the network is responsible for generating a vast repertoire of signals in response to input stimuli, which is adequately combined by means of an adjustable output layer so as to approximate a target signal. Additionally, by choosing a linear combiner (which is analogous to a finite impulse response (FIR) filter), the overall complexity of the adaptation process is reduced to that of solving a linear regression problem [99, 74].

These models, which can be collectively called unorganized machines [19], represent promising alternatives, especially from a signal processing standpoint, in view of the possibility of allying the simplicity that usually characterizes linear systems to the extended processing capability of nonlinear and/or dynamic models. In this section, we briefly present the main aspects of two modern unorganized machines, viz., extreme learning machines (ELMs) and echo state networks (ESNs).

17.2.4.1 Extreme Learning Machines

Extreme learning machines, proposed in [75, 76], are feedforward neural networks with a single hidden layer composed of N neurons, as depicted in Figure 17.8. The main feature of ELMs, which clearly distinguishes them from other feedforward architectures, like MLPs, is associated with the training process: instead of fully adapting the synaptic weights, only the coefficients of the linear combiner at the output are adjusted aiming at minimizing an error function between the ELM outputs and the target signals, while the weights of the connections between the inputs and the intermediate neurons can be randomly selected according to some predefined probability density function.

Hence, the ELM approach circumvents the necessity and the corresponding computational cost of backpropagating an error signal through the network architecture. In fact, the optimum solution, in the least-squares sense, for the output weights, can be obtained both in offline and online scenarios, with the aid of the generalized Moore-Penrose pseudoinverse operation [76] or iterative algorithms, such as the least-mean-squared (LMS) and the recursive least-squares (RLS) [64]. Moreover, the possibility of setting the intermediate layer without taking into account any information about the particular task is encouraged by theoretical analysis associated with the universal approximation capability of ELMs [76, 73].

Let $\mathbf{u}(n) = [u_1(n), \cdots, u_K(n)]^T$ be the vector containing the set of K input signals. The activations of the intermediate neurons, represented by $\mathbf{x}(n) = [x_1(n), \cdots, x_N(n)]^T$ are determined as follows:

$$\mathbf{x}(n) = \mathbf{f}\left(\mathbf{W}^i \mathbf{u}(n)\right), \tag{17.8}$$

where $\mathbf{W}^i \in \mathbb{R}^{N \times K}$ specifies the input weights of the network and $\mathbf{f}(\cdot) = (f_1(\cdot), \ldots, f_N(\cdot))$ denotes the activation functions of the internal neurons. Then, these activations are linearly combined according to the coefficients given in $\mathbf{W}^o \in \mathbb{R}^{L \times N}$ so as to produce the network outputs, represented by $\mathbf{y}(n) = [y_1(n), \cdots, y_L(n)]^T$, as shown in the following expression:

$$\mathbf{y}(n) = \mathbf{W}^o \mathbf{x}(n). \tag{17.9}$$

17.2.4.2 Echo State Networks

Echo state networks, proposed in [78], can also be characterized by the neural architecture shown in Figure 17.8, with the difference that feedback connections between the neurons within the hidden layer, known as dynamical reservoir, are allowed. Hence, the activations of the intermediate neurons are determined not only by the

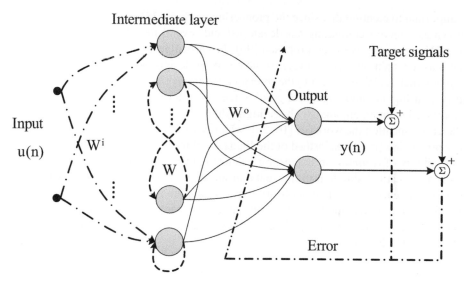

Figure 17.8 Standard architecture of ELMs and ESNs. The dashed lines (- -) represent the feedback connections, which exist only in ESNs.

current input stimuli, but also by the past activations of the reservoir, as indicated by the following expression:

$$\mathbf{x}(n) = \mathbf{f}\left(\mathbf{W}^i\mathbf{u}(n) + \mathbf{W}\mathbf{x}(n-1)\right), \tag{17.10}$$

where $\mathbf{W} \in \mathbb{R}^{N \times N}$ contains the synaptic weights of the recurrent connections between the reservoir neurons. The ESN output signals are generated through linear combinations of the reservoir signals contained in $\mathbf{x}(n)$ and can be determined by Eq. (17.9).

As previously mentioned, the training process of ESNs involves solely the parameters of the linear combiner at the output, while the weights of the input and recurrent connections of the reservoir can be predefined and, thus, remain unchanged. This means that it is possible to explore the benefits of the nonlinear and recurrent structure of ESNs and, at the same time, avoid the complexity and well-known difficulties that usually accompany conventional training strategies of recurrent neural networks [65, 78].

The idea of keeping the parameters of the reservoir fixed is supported by the so-called echo state property (ESP) [78, 141], which ensures that the activation of each reservoir neuron becomes a nonlinear transformation of the recent history of the input signals as long as the recurrent weight matrix $\mathbf{W}$ satisfies particular spectral conditions. Hence, by designing a weight matrix $\mathbf{W}$ that meets the requirements for the ESP, the intermediate layer shall produce a repertoire of signals that carry, to a certain extent, information about the recent history of the input signal, which can be useful in order that the output layer achieves a better approximation of the target signal.

It is important to mention that since the pioneering work of [78], a vivid research area known as reservoir computing has developed and several modifications to the original ESN approach have been proposed [99]. In particular, two aspects can be considered as the main objects of study: (i) the reservoir design method and (ii) the output layer structure. With respect to the reservoir, recent works have addressed the possibility of using the input signal for guiding the construction of the reservoir in an unsupervised manner, among which we highlight [122, 14]. With respect to the output layer, we mention the works of [16, 23], which investigated the use of special nonlinear processing structures instead of the linear combiner.

The possibility of reaching an interesting balance between approximation capability and mathematical simplicity represents an attractive feature of unorganized machines and has encouraged their application to several relevant tasks, such as pattern classification, function approximation, and time series prediction, as well as to important signal processing problems, like channel equalization and source separation [79, 80, 96, 16, 20, 18, 14].

17.2.4.3 General Remarks

We have provided the reader, in this section, with an overview on artificial neural networks, both the canonical MLP and RBF structures and the more recent unorganized machines known as extreme learning machines and echo state networks. The latter, in particular, seem to be most promising options for real-time operation in signal processing, as they have a supervised training effort analogous to that of a finite impulse response filter — this line of investigation is certainly most attractive.

Neural networks have been regularly used in adaptive filtering applications since the 1990s, and have definitely allowed the field to increase its applicability, as references like [83, 25, 26] attest. One point that must be highlighted is, undoubtedly, the connection between RBFs and fuzzy systems and the optimal Bayesian equalizer, which was instrumental in justifying the use of these nonlinear structures in the context of equalization [26, 54]. More recently, applications related to the compensation of nonlinear distortions in optical communications deserve to be mentioned [81].

17.3 INFORMATION THEORETIC LEARNING

In the previous section, the focus was on the possibilities of structures to be employed in signal processing problems, which constitute crucial means of achieving the desired objective, mathematically specified by the adoption of an optimization criterion. This last element must reliably translate the rationale of the entire framework according to the problem at hand. For this, it is desired that the criterion be able to extract and evaluate, in a more extensive manner, the statistical information about the signals of interest. In particular, this is the motivation underlying the field of information theoretic learning, which provides a rich set of promising optimization criteria for filtering problems. In that sense, our focus in this section will be that of providing the reader with the most useful criteria from this field, for both supervised and unsupervised approaches.

17.3.1 FUNDAMENTALS

The seminal works of Wiener and Kolmogorov [137, 89] established the canonical statistic approach when dealing with information retrieval problems. By relying on second-order statistical moments, the classical mean squared error (MSE) method found a vast palette of applications in scenarios based on the premises of linearity and Gaussianity. However, with respect to non-classical scenarios, e.g., for nonlinear processing structures or even for non-Gaussian distributed data, the use of second-order statistics does not suffice. In these cases, it becomes interesting to employ concepts and entities able to extract a richer statistical content beyond second-order. In this context, the research field known as information theoretic learning (ITL) provides a very attractive set of alternative adaptation criteria capable of reaching a more effective extraction of the statistical information available, by resorting to fundamental concepts borrowed from information theory [113].

The origins of the ITL field can be traced back to the exceptional work of Shannon [124], in which a remarkable quantity capable of measuring uncertainty, termed *entropy*, was presented:

$$h_S(E) = -\int f_E(e) \ln(f_E(e)) \, de = -\mathrm{E}\left[\ln(f_E(e))\right], \qquad (17.11)$$

where $f_E(e)$ is the PDF of the random variable E and $\mathrm{E}\left[\cdot\right]$ is the statistical expectation operator. The importance of this measure relies on its capability of dealing with all statistical moments present on the PDF. Measures like mutual information and the Kullback-Leibler divergence also contributed to enlarge this repertoire. However, in order to form the *corpus* of the ITL framework in regards to signal processing structures like adaptive filtering and machine learning, the approach based on Rényi's entropy provided by Prof. Príncipe's group was an essential contribution and also responsible for unifying the concepts behind the ITL paradigm. Presently, the scope of ITL encompasses the set of criteria capable of making use of a large amount of the statistical content present in data, like Rényi's divergence, quadratic PDF matching, and correntropy [113].

17.3.2 ESTIMATORS

The fundamental basis for any ITL approach is probability and entropy estimation methods, of which a myriad competing techniques can be found in literature. The necessity of probability estimators emerges directly from Shannon's and Rényi's definitions of information, which require knowledge of the probability mass function (PMF) – for discrete distributions – or of the PDF – for continuous distributions. On the other hand, direct entropy estimation methods are also attractive possibilities due to their capability of providing alternative straight estimates of entropy without handling the distributions. In this brief section, we arbitrarily present some methods that we believe to be amongst the most useful.

17.3.2.1 Probability Mass Function

Given a finite set of symbols, $\mathcal{X} = \{x_1, x_2, \ldots, x_K\}$, and an observed signal formed by a sequence of symbols randomly drawn from $\mathcal{X}$, one could be tempted to use a simple histogram as a naive PMF estimator for the underlying random variable X, such as $\hat{p}_X(x_i) = <$ *number of times x_i was observed $> / <$ total number of observations $>$*. For instance, if $K = 4$ and one observes the sequence $\mathbf{x} = [x_4, x_4, x_2, x_3, x_2, x_4, x_3, x_4]$, the estimated PMF would be $\hat{p}_X(x_1) = 0$, $\hat{p}_X(x_2) = 0.25$, $\hat{p}_X(x_3) = 0.25$, $\hat{p}_X(x_4) = 0.5$.

This PMF is naive because it assumes that the probability of symbol x_1 is zero. In other words, it assumes that x_1 will *never* be observed. A better assumption is that x_1 may still appear as the number of observations grow. This conservative perspective corresponds to the Laplace's rule [101], which assumes that every symbol has, *a priori*, an equal probability of being observed and that this *a priori* uniform PMF is slightly adapted (Bayesian adaptation) after every new observation.

This perspective allows a straightforward adaptation of this method to the estimation of transition probabilities of Lth order Markov processes [114], by simply limiting the observation past to the L most recent ones, $p_{X(n+1)}(x|X(n), X(n-1), \ldots \ldots, X(n-L+1))$. For independent and identically distributed stochastic processes, the dependence on $n+1$ can be dropped, thus yielding a PMF estimator of a single random variable.

17.3.2.2 Probability Density Function

PDF estimation demands stronger assumptions than PMF, either through the choice of a parametric model (e.g., Gaussian, Gamma) or through an arbitrary scheme for density computation from observations (e.g., Parzen, KNN). In domains such as pattern recognition and signal or image processing, a common choice is the parametric model based on Gaussian Mixture Models (GMM) [44], which are typically adjusted with the Expectation-Maximization (EM) method [40]. Notwithstanding, in recent years, works by José C. Príncipe and collaborators on information theoretic learning and Rényi's entropy [113] emphasized the nonparametric Parzen model [44, 135], a kernel-based approach.

The objective of Parzen's work [112] was to provide PDF estimates from a finite set of data, and the crucial element for achieving this goal was the use of kernel methods. The derivation follows a statistical standpoint by the interpretation of data as outcomes from stochastic processes. Following this line of reasoning, an information signal $x(n)$ is associated with a continuous RV called as X. Assuming that there are L independent and identically distributed (i.i.d.) samples $\{x_1, \ldots, x_L\}$ of X at disposal, the Parzen window estimate [112] of the PDF associated with X is:

$$\hat{f}_X(x) = \frac{1}{L} \sum_{i=1}^{L} \kappa_\sigma \left(\frac{x - x_i}{\sigma} \right), \tag{17.12}$$

where $\kappa(\cdot)$ is an arbitrary kernel function, which must obey the Mercer conditions [113], and σ is the kernel size.

Undoubtedly, the most used kernel is the Gaussian, which can be defined as

$$G_\sigma(x-x_i) = \frac{1}{\sqrt{2\pi}\,\sigma} \exp\left(\frac{-|x-x_i|^2}{2\sigma^2}\right), \tag{17.13}$$

where the only free parameter is σ.

17.3.2.3 Direct Entropy Estimation

A natural approach for entropy estimation is to take as many samples as possible to adjust PMF or PDF models and then to use these models directly in Shannon's or Rényi's formula. These approaches are known as plug-in methods [8]. However, an interesting question can be formulated: can we get rid of PMF or PDF estimates in entropy estimation? Fortunately, the answer is yes. And this answer brings together a series of interesting points of view. Indeed, entropy estimation directly through coincidence counting was proposed in [100] as a method to be used in statistical mechanics, but that can be easily adapted for problems belonging to a variety of domains. We chose the very simple but powerful Ma's method to be explained in this section in terms of our notation. For either a discrete or a symbolic signal with N samples, the method corresponds to the following steps:

1. Compare symbols in every $N_t = N(N - 1)/2$ distinct pair of samples, and count the N_c detected coincidences.
2. Compute the estimated set cardinality (i.e., the hypothetical finite set from which samples were independently drawn) as $\hat{K} = N_t/N_c$, for $N_c > 0$ (cardinality diverges to infinity if no coincidence is detected).
3. Estimate the entropy in bits as $\hat{H} = \log_2(\hat{K})$.

On the other hand, the entropy of continuous random variables diverges to infinity, and the usual approach to see how it diverges [31] is that of partitioning the variable domain into regular cells of size Δ and to reduce the size of these cells while calculating the corresponding entropy, $H(\Delta)$, under the requirement that the PDF is continuous inside the cells. Thus, if Δ is sufficiently small, the probability density inside each cell tends to be uniform, so that the corresponding entropy increases by 1 bit whenever the cell size is divided by 2. Therefore, entropy tends to be a linear function of $\delta = \log_2(1/\Delta)$ for sufficiently small values of Δ. Notice that Δ may stand for bin width for 1D variables, bin area for 2D variables, bin volume for 3D variables, and so forth.

Accordingly, the differential entropy, h, can be defined as the difference between $H(\delta)$ and the reference $R(\delta) = \delta$ that linearly increases 1 bit per unitary increment of δ. Consequently, h remains almost constant for small enough values of Δ, as illustrated in Figure 17.9.

By assuming that Δ plays the role of a sample neighborhood, coincidence can be (re)defined for continuously valued variables as being a signal sample falling

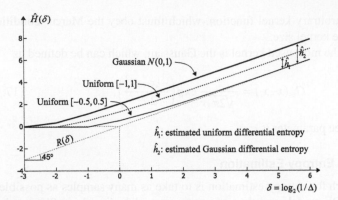

Figure 17.9 Graphic explanation of differential entropy.

inside the neighborhood of another sample. Consequently, for small values of Δ, Ma's method can be easily extended to differential entropy estimation, as follows:

1. Arbitrarily set a small sample neighborhood Δ.
2. Compare signals in every $N_t = N(N - 1)/2$ distinct pair of samples, and count the N_c detected *coincidences*.
3. Compute the estimated (hypothetical) set cardinality as $\hat{K} = N_t/N_c$.
4. Estimate the differential entropy in bits as $\hat{h} = \hat{H}(\Delta) + \log_2 \Delta$.

17.3.3 SUPERVISED CRITERIA

The use of estimators allied to the concepts and metrics from information theory greatly contributed to the flourishing of a rich set of criteria that belong to the ITL field. In the following, we present some of them which were conceived from a supervised perspective.

Rényi's Entropy of the Error

The proposal that vigorously contributed to establish and unify the ITL methods was introduced in a supervised task by combining the use of the quadratic Rényi's entropy and the Parzen window method for PDF estimation [119, 113], an arrangement that presents the advantage of the differentiability property, necessary for online adaptive algorithms based on the stochastic gradient. The resulting criterion which aims to minimize the error entropy (MEE) can be defined as:

$$\hat{h}_{R_2}(E) = -\log\left(\frac{1}{L^2} \sum_{i=n-L}^{n-1} \sum_{j=n-L}^{n-1} G_{\sigma\sqrt{2}}(e_j - e_i)\right) = -\log \hat{V}_2(E), \qquad (17.14)$$

where e_n denotes the error signal, e_i is the error sample at time instant i, $G_\sigma(\cdot)$ is the Gaussian kernel function, σ denotes the kernel size, L is the number of samples (time window) used to estimate the error PDF, and $\hat{V}_2(\cdot)$ is the term called information potential (IP). As $\hat{h}_{R_2}(\cdot)$ is a monotonic function of the negative of $\hat{V}_2(\cdot)$, the MEE is equivalent to the maximization of the IP, which is simpler to compute. Its associated gradient-based method is referred as stochastic information gradient for MEE (MEE-SIG) [113].

For linear adaptive structures, the MEE criterion shows itself to be more adequate when dealing with non-Gaussian, fat-tail distributions and with the occurrence of outliers, like the impulsive noise; while in the classical Gaussian scenario, the performance of the MEE is similar to that of the MSE [119, 17]. The approach is also an interesting option in nonlinear scenarios [46, 47], in which we detach its importance in the face of the increasing possibility of the use of nonlinear processing structures like neural networks [119] and fuzzy filters.

Shannon's Entropy of the Error

Although the MEE criterion occupies a prominent position in the ITL-based framework, the reminiscence of Shannon's work contributed to preceding formulations including kernel density estimators [2]. However, its employment in solving engineering problems is still recent [128]. This is particularly due to the inherent difficulty on the estimation of Shannon's entropy, which demands further approximations to achieve a computational cost similar to the Rényi's approach. In that sense, the estimation of Shannon's entropy is generally made by assuming mean value over a sample instead of the expectation operator allied to the use of the Parzen window method [2], which results in:

$$\hat{h}_S(E) = -\frac{1}{L} \sum_{i=1}^{L} \log \left(\frac{1}{L} \sum_{j=1}^{L} G_{\sigma\sqrt{2}}(e_i - e_j) \right). \tag{17.15}$$

This criterion, also called Shannon's Error Entropy (SEE), allows a gradient-based method to be employed, which, just like the MEE, can be used to train nonlinear structures. In addition, the SEE is also recommended for classification problems [128].

Correntropy

It is also worth it to mention the ITL-based criterion known as correntropy, whose main features are its simplicity and lower sensitivity to the kernel size σ parameter in comparison to other ITL approaches. It also presents interesting relations to the correlation function and the estimated Rényi's quadratic entropy (17.14)[113], hence the name. This criterion is defined as

$$\hat{v}(e) = \frac{1}{L} \sum_{i=1}^{L} \kappa_\sigma(e_i), \tag{17.16}$$

where $\kappa(\cdot)$ is a kernel function. Depending on the choice of the kernel, correntropy is able to encompass different statistical moments. The Gaussian kernel, for example, allows correntropy to take into account all even moments [113]. Correntropy has been used successfully in a variety of applications, in which we cite its use in system identification and noise cancellation for non-Gaussian disturbances, like the impulsive noise [129]; and in nonlinear dynamic systems identification [61], since correntropy is capable of quantifying nonlinear statistical relationships.

17.3.3.1 ℓ_p-Norms and Connections with ITL

In supervised approaches, the employment of distinct ℓ_p-norms of the error is not a recent issue [115, 7, 59]. In classical scenarios, the use of the ℓ_2-norm criterion, in light of its correspondences to the classical MSE, shows some advantages. However, for cases in which the error is non-Gaussian, this assertion may not hold. In such cases, in order to find the more adequate ℓ_p-norm, we illustrate some relations between the ℓ_p-norm criterion and a specific ITL criterion, more precisely, the error entropy criterion given by (17.11).

Let us consider that the error vector, given by

$$\mathbf{e} = \{e_1, \dots e_L\} \subset \mathbb{R}, \tag{17.17}$$

is composed of i.i.d. samples of a generalized Gaussian random variable E of zero mean and parameters α and β with PDF [106]

$$f_E(e) = \frac{\beta}{2\alpha\Gamma\left(\frac{1}{\beta}\right)} \exp\left(-\left(\frac{|e|}{\alpha}\right)^\beta\right), \tag{17.18}$$

where $\Gamma(\cdot)$ is the gamma function and

$$\alpha^\beta = \beta \mathrm{E}\left[|E|^\beta\right]. \tag{17.19}$$

The parameters α and β are concerned, respectively, to the scale and the shape of the generalized Gaussian distribution. This distribution covers a wide gamma of known distributions for different values of β, e.g., the Laplacian distribution is given for $\beta = 1$, the Gaussian for $\beta = 2$, and in the limit $\beta \to +\infty$ it converges to the PDF of the uniform distribution $[-\alpha, \alpha]$.

The differential entropy of the generalized Gaussian random variable is given by [107]

$$h_S(E) = -\ln\left(\frac{\beta}{2\alpha\Gamma\left(\frac{1}{\beta}\right)}\right) + \mathrm{E}\left[\left(\frac{|E|}{\alpha}\right)^\beta\right]. \tag{17.20}$$

Substituting (17.19) in (17.20) we obtain

$$h_S(E) = \frac{1}{\beta}(1 + \ln(\beta)) - \ln\left(\frac{\beta}{2\Gamma\left(\frac{1}{\beta}\right)}\right) + \ln\left(\left(\mathrm{E}\left[|E|^\beta\right]\right)^{\frac{1}{\beta}}\right). \tag{17.21}$$

Thus, if there is a sufficient number of samples L of the vector $\mathbf{e}(n)$, the absolute moment $\mathrm{E}\left[|E|^\beta\right]$ can be estimated by $\frac{1}{L}\|\mathbf{e}(n)\|_\beta^\beta$ and the differential entropy is

$$h_S(E) \approx \frac{1}{\beta}(1 + \ln(\beta)) - \ln\left(\frac{\beta}{2\Gamma\left(\frac{1}{\beta}\right)}\right) + \ln\left(\left(\frac{1}{L}\right)^{\frac{1}{\beta}}\|\mathbf{e}(n)\|_\beta\right), \tag{17.22}$$

which is proportional to the ℓ_p-norm of the vector $\mathbf{e}(n)$, with $p = \beta$

$$h_S(E) \propto \|\mathbf{e}(n)\|_{p=\beta}. \tag{17.23}$$

In summary, from an ITL point of view, to minimize $\|\mathbf{e}(n)\|_p$, where $\{e_1, \ldots e_L\} \subset \mathbb{R}$ are i.i.d. samples of a generalized Gaussian random variable of zero mean and parameters α and $\beta = p$, corresponds to minimize its entropy.

17.3.3.2 Entropy-Based Intersymbol Interference

So far, it has been stated that the possibility of dealing with entropy-based criteria alternative to the MSE can lead to relevant performance improvements. However, when it comes to classical performance evaluation metrics, like signal-to-noise ratio (SNR), MSE, and intersymbol intereference (ISI), they are strongly related to a quadratic framework.

In order to circumvent these problems, in [108], the authors proposed a performance metric for the deconvolution problem, namely entropy-based intersymbol interference (HISI). In that work, it has been shown that the use of a criterion based on the ℓ_p-norm, in determined situations, led to significant improvements in terms of the HISI.

In deconvolution, the usual performance evaluation measure is the ISI, which makes use of the complete knowledge of the channel:

$$IS I_{dB} = 10 \log_{10} \frac{\left(\sum_{i=0}^{M} |c_i|^2\right) - \max_i |c_i|^2}{\max_i |c_i|^2}, \tag{17.24}$$

where $\mathbf{c}$ is the combined channel + equalizer impulse response, which is assumed to be causal and M is the maximum length of $\mathbf{c}$. Ideally, in this problem, the desired combined channel + equalizer impulse response is given by a delayed and scaled impulse. The ISI metric basically measures the ratio between the interference energy and the signal energy. On the other hand, HISI makes use of Shannon's entropy in order to measure the level of uncertainty associated with the combined channel + equalizer response, treating it as a discrete probability distribution,

$$HIS I_{dB} = 10 \log_{10} H_S (\alpha|\mathbf{c}|),$$

$$= 10 \log_{10} \left(-\sum_{i=0}^{M} \alpha|c_i| log_2 (\alpha|c_i|) \right), \tag{17.25}$$

where $H_S(\cdot)$ denotes Shannon's entropy and $\alpha = 1/\sum_i |c_i|$ is a unit-norm correction term. A desired result is to keep the HISI as low as possible in order to reduce the presence of interference components.

17.3.4 UNSUPERVISED CRITERIA

Beyond the favorable achievements found in the supervised approach, the use of the ITL framework in unsupervised scenarios is quite useful due to its inherent capability of using the higher-order statistical information present in data. As in the supervised case, there is a set of interesting criteria, among which we present some of them.

Blind Rényi's Entropy

We start with the criterion that suggests the conciliation of α-Rényi's entropy and the idea behind the well-known blind constant modulus (CM) criterion [58], which results in $J_{SFA} = h_{R_2}\left(|y(n)|^2 - R_p\right) = h_{R_2}\left(|y(n)|^2\right)$ [121], where $p \in \mathcal{Z}$, $y(n)$ is the filter output signal and $R_p = \frac{E[|s(n)|^{2p}]}{E[|s(n)|^p]}$, being $s(n)$ the the original signal. Using the minimization correspondence of the Rényi's entropy to the maximization of the IP descriptor, the cost function reduces to:

$$\hat{J}_{SFA} = \frac{1}{L^2} \sum_{i=0}^{L-1} \sum_{j=0}^{L-1} G_{\sigma\sqrt{2}}(|y_{n-j}|^2 - |y_{n-i}|^2). \tag{17.26}$$

The associated steepest descent algorithm was named stochastic fast algorithm (SFA) [121], which must consider a restriction to the filter coefficients, such as unitary norm, in order to avoid the trivial solution. It is important to remark that, like the CM criterion, the SFA also assumes that the signal $s(n)$ presents constant modulus.

PDF Matching via Quadratic Distance

A parallel approach in blind ITL criteria considers the premise firmed by the Benveniste-Goursat-Ruget theorem [10], which proposes the matching of the PDF of the equalizer output to that of the transmitted signal. By using a quadratic distance (QD), the criterion can be writen as $J_{QD} = \int \left(f_y(v) - f_s(v)\right)^2 dv$ [94, 95, 93], where f_y and f_s are the PDFs associated with the filter output and with the reference signal, respectively. Estimating the PDFs through the use of the Parzen window method and Gaussian kernels, the cost function results in:

$$\hat{J}_{QD} = \frac{1}{L^2} \sum_{i=0}^{L-1} \sum_{j=0}^{L-1} G_{\sigma\sqrt{2}}(|y_{n-j}|^2 - |y_{n-i}|^2) - \frac{2}{L_s L} \sum_{i=1}^{L_s} \sum_{j=0}^{L-1} G_{\sigma\sqrt{2}}(|y_{n-j}|^2 - |s_i|^2) \tag{17.27}$$

where L_s is the cardinality of the transmitted symbols alphabet and s_i its ith symbol. Note that we considered only the terms dependent on the filter output. The corresponding gradient-based algorithm was named stochastic QD, or simply, SQD. Some variants of this approach consider the evaluation of the PDFs in some specific target values [95], or even the use of only the last term of (17.27), as shown in [94]. We refer to the cost of this last approach as $\hat{J}_{MQD}$.

These blind criteria establish interesting relationships to each other, but differ in aspects like robustness and computational cost. At one end, $\hat{J}_{QD}$ is shown to be more complex and robust. At the other end, $\hat{J}_{SFA}$ is simpler, but tends to be more susceptible to local convergence, while the $\hat{J}_{MQD}$ provides a good trade-off.

Other similar ITL blind methods can be found in [49],[45], and [48].

Blind Correntropy

A common framework of the methods discussed above is the idea of working directly with the PDF, to explore more information about the signals involved than using just second-order or a few higher-order moments. Another step in this direction was achieved by the unsupervised version of correntropy [120, 113], which is able to encompass not only the statistical distribution of signals but also its time structure. Instead of using the error signal, the blind operation of correntropy considers differences between delayed samples of the RV in focus. For stationary stochastic processes, the simple relation is valid: $\hat{v}(x_n - x_{n-m}) = \hat{v}_x(m)$. In that sense, an interesting criterion is to match the correntropy of the filter output with that of the transmitted signal:

$$J_{corr} = \sum_{m=1}^{P} \left(\hat{v}_s(m) - \hat{v}_y(m) \right)^2, \qquad (17.28)$$

where $\hat{v}_s$ is the correntropy of the source and $\hat{v}_y$ is the correntropy of the filter output. The steepest descent algorithm derived from (17.28) has shown to perform better than CMA in scenarios with correlated sources and impulsive noise [120]. If compared to the SQD and MQD algorithms previously discussed, it performs better only when the correlation of the sources increases. Correntropy has been applied with good results in several problems such as time series analysis, robust regression, PCA, and blind source separation [97, 61, 113]. An alternative approach when dealing with correlated sources is to employ a multivariate PDF matching similar to (17.27), which has shown additional potentiality for extracting the statistical dependence information in certain scenarios [50].

17.3.5　INDEPENDENT COMPONENT ANALYSIS

Unsupervised ITL methods also play an important role as efficient tools to the so-called independent component analysis (ICA) [30], most generally applied to blind source separation (BSS) problems. Basically, ICA techniques are built to explore the assumption of statistical independence of the sources. In linear BSS problems, this is a recurrent case, where N independent sources $\mathbf{s}$ are linearly combined to give rise to M dependent mixtures, i.e., $\mathbf{x} = \mathbf{As}$ – being $\mathbf{A} \in \mathbb{R}^{M \times N}$ the mixing matrix. Without *a priori* knowledge of $\mathbf{A}$ and $\mathbf{s}$, a demixing matrix $\mathbf{W}$ with output $\mathbf{y} = \mathbf{Wx}$ is adjusted in order to recover the statistical independence and, consequently, the original information of the sources $\mathbf{s}$.

The ICA main concept lies in the achievement of independence by means of the mutual information (MI) minimization between the outputs $\mathbf{y}$, which, in the linear BSS case, will be equivalent to the minimization of the sum of marginal output entropies:

$$J_{ICA} = \sum_{i=1}^{N} h_S(Y_i) - \ln |\det(\mathbf{W})|, \qquad (17.29)$$

where Y_i is the RV associated with the i-th output signal y_i. In this case, Shannon's entropy can be approximated by different estimators.

Alternative approaches to perform ICA are (i) negentropy [77], a criterion to maximize non-Gaussianity; (ii) the Infomax principle [9], which aims at maximizing the information flow in the demixing system; (iii) cumulants and (iv) kurtosis. We also highlight the ITL-based Minimum Rényi's mutual information (MRMI) algorithm [66], which replaces Shannon's differential entropy in (17.29) by Rényi's definition with $\alpha = 2$ allied to the Parzen window method. However, a careful analysis of the scenario at hand is recommended [67].

17.3.6 SIGNAL PROCESSING OVER FINITE FIELDS

So far, the study in the area of signal processing has been mainly developed in terms of formulations that consider systems and signals models defined over the fields of the real and complex numbers. Undoubtedly, these formulations suit very well in many practical applications, such as wireless communications, geophysical signal processing, and chemical sensor analysis [116], however, they are not representative of inherently discrete/symbolic domains, like those of digital and genetic data. In addition, in the face of the existence of enormous binary and genomic databases, as well as of inherently discrete problems related to bio-inspired computing [1], the relevance of this research area should not be underestimated.

The pioneering efforts in signal processing over finite fields can be attributed to Yeredor, Gutch, Gruber, and Theis [63], which established a theory of blind source separation over finite fields that can be applied to linear and instantaneous mixing models. The most extensively studied case is the extension of ICA to the finite domain [138, 62], which gave rise to algorithms like AMERICA [139] and MEXICO [63]. In all cases, the techniques comprise parameter adaptation based on ITL cost functions for performing the process of extraction and deflation [39], mainly through the minimization of entropy. Extensions of these works also consider the use of immune-inspired algorithms [126, 127] and the problem of convolutive mixtures [52]. Other problems in finite fields are linear equalization and prediction [51, 53, 140], where the minimization of the error entropy is also considered. From a general perspective, it is possible to affirm that the research area of signal processing over finite fields has shown to be a very promising element inside ITL, as it opens plenty of possibilities to directly employ the metrics and ideas of Shannon's work.

17.4 BIO-INSPIRED OPTIMIZATION

As emphasized in Section 17.1, a general signal processing framework involves (1) the selection of a structure, (2) an optimality criterion, which establishes the desired objective, and (3) a search methodology, which is responsible for efficiently adapting the parameters of the chosen structure, in order to hopefully determine the optimum solution with respect to the adopted criterion.

Interestingly, biological systems/phenomena not only serve as the basis for the development of structural options, like the artificial neural networks discussed in Section 17.2, but also inspire the design of search strategies with wide applicability to hard optimization problems, such as those that usually arise in complex signal processing tasks. In this section, we briefly present the main aspects of bio-inspired optimization algorithms belonging to three different classes [35]: evolutionary computation, artificial immune systems and swarm intelligence.

17.4.1 EVOLUTIONARY COMPUTATION

Evolution is a most prolific notion from a philosophical standpoint, as attested by the ideas contained in works as diverse as those written by Darwin [33], Bergson [11], Teilhard de Chardin [132], Gould [60], Dennett [41] and Dawkins [34]. Let not the reader then be disappointed with the relative structural simplicity of the evolutionary algorithms we will discuss in this chapter:

1. Our aim is not to study a biological phenomenon using computational tools, but to use these elements as means to a clear end: to solve complex optimization tasks.
2. The computational framework on which the algorithms we will discuss is based on a bottom-up underlying structure, which means that a relatively simple code can engender, when time unfolds, rather complex behaviors.

Evolutionary algorithms (EAs) can be understood as being stochastic optimization methods whose operation is, to some extent, inspired by concepts and/or theories concerning evolution. Typically, due to the synergy existing between both fields, EAs also include elements of modern genetics [4].

In very simple terms, we may begin to give the reader an idea of the *modus operandi* of EAs considering the following general properties:

- EAs are populational methods, although, in some special cases, this property may not hold (e.g. when there is very slight available computational power).
- The individuals of the population affect each other, in the search process, either in a direct or indirect way.
- New solutions can be generated by combining properties of existing solutions (which gives rise to the notion of the crossover operator).
- New solutions can also be generated by spurious modification of existing solutions (which models the concept of mutation).
- The cost function (or even cost functions, in the multiobjective case) to be optimized provides a measure of the fitness of each solution (hence the expression fitness function), which is used in one or more selection stages. The influence of this (these) stage (s) is an "invisible hand" that guides the search process.
- The crossover and mutation operators evoke, with a varying degree of fidelity, the biological view on reproduction.
- EAs, in general, do not offer any guarantee of global (or even local) convergence over a finite period of computational time. They, nonetheless, as a rule, include mechanisms that potentially allow the scape from local optima.
- EAs do not require the manipulation of the cost function at hand, since this function is used only to provide a fitness measure. This makes them very attractive options to cope with mathematically complex, dynamical and uncertain environments.

17.4.1.1 Basic Structure of an EA

An EA operates in successive cycles — called generations —, which includes the potential application of all solution-generating and selection operators. Fig. 17.10 illustrates the basic structure of a generation.

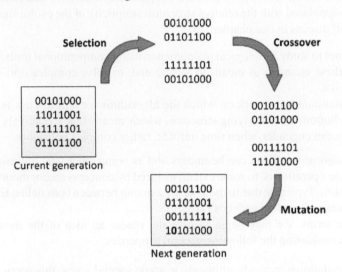

Figure 17.10 Basic structure of an EA generation.

The concatenation of multiple generations is ruled by a stopping criterion that can be based, for instance, on the fitness of the best individual obtained so far or simply on a maximal number of iterations. As for the initial population, its choice is up to the user. Two classical possibilities — which may apropos be combined — are:

- Random generation of the parameters of the initial individuals.
- To use *a priori* information about the problem and/or about features of desirable solutions in this process.

In summary, in metaphorical terms, each individual of the population is the analogous of the genotype of a living organism. The operators are responsible for generating variability due to reproduction and the cost function polarizes, to some degree, the selection of individuals that will form the subsequent generation. Therefore, the search process takes place in a genotypical space, albeit the final product is undoubtedly a phenotype, which is the interpretation of the parameter vector that is optimized. For instance, one may optimize vectors containing real values (genotype) that are used as coefficients of the transfer function of a digital filter (phenotype).

Before using an EA, it is necessary to make a number of choices, most of which are generally far from trivial:

- What will be the adopted coding? This demands concrete answers to questions like: what kind of data structure will be used? Will the free parameters be Boolean, integer or real values?

- What will be the structures of the mutation and crossover operators? Will both be used together?
- What selection scheme will be employed? Will it be used to control the action of the operators or only as a final stage in the process of defining the next generation?
- Will the best individual of the population always allowed to survive (elitism) or not?

It is important to remark that the performance of an EA can vary immensely according to the adequacy of the answers given to each of the above questions (and also to more specific questions that are not in the list).

17.4.1.2 Canonical Divisions

For historical reasons, EAs are divided into four classes that, presently, have a remarkable degree of intersection, to the point that this division is gradually losing relevance from a strictly pragmatic standpoint.

Genetic Algorithms (GAs)

GAs are algorithms based on the ideas put forward by John Holland and collaborators [68]. In its original form, they employ binary coding, roulette wheel selection and allow the possibility of using the so-called Michigan population structure, for which the solution to the problem being addressed is interpreted as being the entire population, and not a single "best individual". The canonical GA operators are the one-point crossover and the binary mutation. The first has the following dynamic: two parents are selected, as well as a random position of the parameter vector; two descendants are created — they are identical copies of the two parents up to the random point of cut, and have the remaining parameters taken from the "opposite parent" for the subsequent position. Fig. 17.11 illustrates the process.

Mutation is very simple: there is a given probability (p_m) that a bit of a given individual be swapped from 0 to 1 or from 1 to 0.

Roulette wheel selection means that the probability of selecting an individual is directly proportional to its fitness, which is assumed to be a nonnegative value that is larger for better solutions. This means that, in more intuitive terms, each application of the selection process is like the operation of a roulette with different areas.

Evolution Strategies (ESs)

ESs are methods whose origin is strongly application-oriented. This explains why they use real coding (which bears no resemblance with the discrete character of the alphabet of the biological genome), thus requiring mutation operators that depend on probability density functions, the Gaussian being the most popular choice. Originally, these methods did not consider the use of crossover operators, but the latter have become, in due time, also canonical. The use of real parameters allows that arithmetic crossovers be employed, based, for instance, on a convex combination of

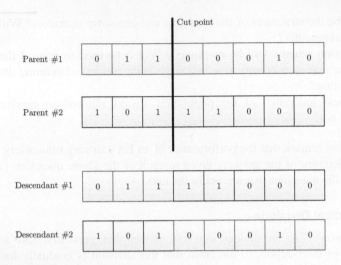

Figure 17.11   One-point crossover operator.

the parents' parameter vector [4]. ESs also became intimately associated with auto-adaptation i.e. with the evolution of the parameters of the very operators that form the core of the optimization process. In other words, aside from optimizing the parameters of the cost function, ESs can attempt to optimize also the parameters of its own operators with the purpose of potentially increasing the adequacy of the algorithm to the problem at hand.

ESs are also characterized by a lack of emphasis on the use of selection operators before the creation of new solutions. What happens to these solutions depends on the chosen population structure — the most widespread options are the $(\mu + \lambda)$ and (μ, λ) strategies [123, 13]. In the first case, the μ individuals of the population compete for the right of forming the next generation with their λ descendants in a completely greedy scheme (the μ best solutions survive); in the second case, the μ best individuals among the λ descendants are automatically chosen ($\mu < \lambda$). The second option might seem to be "weak" in terms of selective pressure, but this can be of great value, for instance, in dynamical environments.

Evolutionary Programming (EP)

The conceptual framework of EP was built in an appealing context: to evolve information-processing structures — more specifically finite state machines (FSMs) — in order that computational problems be emergently solved. Therefore, each individual, in a classical EP method, corresponds to an FSM that is used to process a Boolean input in order to generate an output as close as possible to a desired Boolean signal [57]. The fitness function is built to reflect the matching between these two sequences.

A classical EP approach uses only mutation operators, which can modify any aspect of an FSM: state structure, transitions and input/output aspects. No crossover operator was originally devised [56]. Later, David Fogel [55, 4] used concepts of the original formulation of EP to create an evolutionary optimization algorithm used to handle real-valued tasks.

Genetic Programming (GP)

Genetic programming [90] can be seen as an extension of the conception of Fogel [57] in two senses: handling a broader class of problem-solving computational frameworks and using the structure of genetic algorithms as a basic roadmap.

The canonical data structures of the method were LISP statements or high-level language parse trees, and it follows a conventional GA *modus operandi*, with suitable crossover and mutation operators. However, the concept of GP has been substantially increased, and, presently, it can be evoked in the context of the evolutionary design of a number of hardware and software structures [90].

17.4.2 ARTIFICIAL IMMUNE SYSTEMS

The immune system is, in simple words, the entity which protects the animal against the incessant attack of foreign agents [133]. Its key importance to the being's survival requires that it works efficiently; it is a complex system redundantly built, with a series of components and mechanisms that are adapted to neutralize a single invader or to protect it against a collection of infectious agents. Two fundamental aspects to the operation of this system are noteworthy: the memory and learning mechanisms. The system can capture information about the infectious agents and store them for a later use, in the case of a new invasion by the same (or a similar) agent.

Hence, the system recognizes and eliminates hazardous organisms, but also develops and improves its defense mechanisms in order to obtain an even more effective performance against future attacks. Such properties have naturally instigated many researchers to replicate the basic principles of the immune system in the development of intelligent computational systems.

Consequently, in the beginning of the 90's the development of artificial immune systems (AIS) became a fact. This concept comprises bio-inspired algorithms with operators inspired in the natural immune system of vertebrates, which can carry out several tasks, such as pattern recognition, clustering and optimization. Immune-inspired algorithms have a desirable feature in comparison with other bio-inspired proposals: to present a higher diversity maintenance capability [38], which is interesting to avoid premature convergence and to locate multiple optima [36, 38].

There are several algorithms and techniques, each one employing different principles and mechanisms of the immune system, as the reader can see in [36]. In particular, two immune theories, which are strongly related to the adaptive response of the immune system against invaders, have received a considerable attention to support the design of immune-inspired algorithms: the clonal selection principle [22] and the immune network theory [82].

The clonal selection principle [22] basically states that, when the antigens (molecular signatures) of a foreign pathogen are identified by immune cells, these cells proliferate through a cloning process, followed by a controlled mutation in the presence of selective pressure, so that the recognition capability of those cells can be improved and the organism can properly respond against the infection. On the other hand, the immune network theory [82] states that the immune cells are also capable of recognizing other immune cells of the organism, which may lead to two distinct responses: a positive response, characterized by the activation of the recognizing cell, followed by cloning and mutation; and a negative response, characterized by tolerance and even the suppression (elimination) of the recognized cell (i.e., another antibody that is sufficiently similar). Thus, the immune network theory states that the main aspects of the natural immune system are emergent properties, derived from regulatory mechanisms which intend to maintain the network in a dynamic steady state [12].

To illustrate how a immune-inspired algorithm works, in the following we present the CLONALG, an important algorithm that was built based on the clonal selection principle.

17.4.2.1 The CLONALG Algorithm

The CLONALG algorithm is an emblematic immune-inspired technique, with a number of successful applications since its proposal, in 2002 [37]. Such success may be explained due to its simple implementation and the flexibility of application to pattern recognition or optimization problems.

The method, as the name suggests, is based on the clonal selection principle. To fully understand its proposal, it is necessary to associate some agents of the immune system with elements of the optimization problem[1]. There is a population of elements over the search space, Ab, each one (i.e. each candidate solution) being an abstraction of a B-cell structure with an specific antibody. The objective function associated with this individual in Ab represents its affinity or "matching" with respect to a certain antigen (Ag) — the foreign agent / substance that activates the immune response. Then, the main steps of the algorithm are introduced in Alg. 1.

The algorithm starts with a randomly-defined population Ab of antibodies or cells[2]. Each Ab_i represents an element in the parameter space to be optimized. Inside the main loop, the affinity evaluation corresponds to calculating the objective function value for each Ab_i, which is followed by the cloning step (nCl copies per individual). Then, the clones are submitted to a mutation process whose ratio is inversely proportional to their parents' affinity. Generally, this step is implemented by the definition of a mutation probability according to

$$p_m = \exp\left(-\rho f^{Ag}(Ab_i)\right), \tag{17.30}$$

where $f^{Ag}(Ab_i)$ is the (parent) cell's affinity, normalized within $[0, 1]$, and ρ is a parameter that controls the exponential function decay. When this step is finished, the affinities of C^* — a new set that corresponds to the clones after the mutation

---

[1]The pattern recognition case is not considered in this chapter [37].

[2]We assume, in the context of immune-inspired algorithms, that both names are equivalent.

Algorithm 1 Main steps of CLONALG algorithm for optimization.

Parameters:
- nAB: initial number of cells;
- nCl: number of clones per cell;
- $maxIT$: maximum number of iterations;
- b: percent value of new cells inserted;

1- Generate an initial population with nAB individuals;
while (iteration $\leq$ max$_{it}$) **do**
    2- Evaluate the affinity (*fitness*) of cells in Ab;
    3- Generate nCl clones for each cell i in Ab;
    4- Apply the mutation operator in each clone;
    5- Evaluate the affinity of the mutated clones;
    6- Choose, between the parent cell and its clones, who is kept in the new population;
    7- Substitute the b worst cells by new ones, randomly generated;
end while

---

process — are evaluated in order to select the best candidate between the parent and its clones. The loop ends with an insertion of $b\%$ new randomly generated cells that replace the worst affinity individuals of Ab. The process is repeated for a maximum number of max$_{it}$ iterations, when the element Ab_i with highest affinity is the solution of the optimization problem.

There is a similarity between this algorithm and evolutionary methods, such as evolution strategies: both methods are population-based and adopt selection, fitness and mutation operators. However, besides the distinct biological inspiration, CLONALG differs due to the cloning process, the mutation scheme which is inversely proportional to the fitness value and the regular insertion of new candidate solutions in the population. Differently from a classical ES, those aspects provide an intrinsic capacity of maintaining diverse solutions during the algorithm execution [36], which may be relevant to increment the probability of locating a global (or a sufficiently good) optimal point. This particular feature encouraged the application of AIS to a number of signal processing problems, such as blind source separation and blind deconvolution [42, 134, 127].

17.4.3 PARTICLE SWARM OPTIMIZATION

Swarm systems are characterized by the presence of a collection of individuals whose behavior can be described by a set of relatively simple rules and by the existence of interactions among the individuals and with the environment [87, 35]. An interesting phenomenon that can be observed in swarm systems is the emergence of intelligent behavior that could not be directly expressed by isolated individuals. This property, known as swarm intelligence, highlights the potential benefits of living within a community of social individuals and has motivated the development of computational models that explore the principles of swarm systems to solve challenging problems.

In this section, we concentrate our attention on a search methodology that exploits the notion of swarm intelligence to adapt a repertoire of candidate solutions as they interact with each other and with the environment, which is represented by a given cost function that provides information about the problem to be solved. The classical implementation of the so-called particle swarm optimization (PSO) algorithm, proposed by [86, 84], updates each candidate solution, named particle, by taking into account two aspects: (1) the information about the individual experience of the particle, given by the best position it has occupied so far, and (2) the information about the experience of the neighbors of such particle, given by the best position visited by one of these neighbors.

Let $\mathbf{x}_i(t) \in \mathbb{R}^M$ represent the current position of the i-th particle of the swarm. The particle shall move according to the direction given by the so-called velocity vector v_i, as shown in the following expression:

$$\mathbf{x}_i(t+1) = \mathbf{x}_i(t) + v_i(t+1). \tag{17.31}$$

Consider that $\mathbf{p}_i(t) \in \mathbb{R}^M$ denotes the best position visited by the i-th particle and $\mathbf{g}_i(t)$ the best position occupied by the neighbors of the i-th particle until iteration t. The velocity vector consists of a combination of the two aforementioned positions, which are associated with the individual experience and the global knowledge, respectively. More specifically,

$$v_i(t+1) = v_i(t) + \psi_1 \otimes (\mathbf{p}_i(t) - \mathbf{x}_i(t)) + \psi_2 \otimes (\mathbf{g}_i(t) - \mathbf{x}_i(t)), \tag{17.32}$$

in which $\otimes$ denotes an element-wise multiplication between vectors, ψ_1 and ψ_2 correspond to random vectors whose elements are uniformly distributed in the intervals $(0, L_1)$ and $(0, L_2)$, respectively, where L_1 and L_2, known as acceleration constants, are parameters of the algorithm [3]. In order to limit the extent of the changes in the position of each particle, it is recommended to define lower and upper bounds to the elements of the velocity vector v, denoted as v_{min} and v_{max}, ensuring that the particle movement is restricted to small steps at each iteration.

It is important to recognize in Eq. (17.32) that the term $(\mathbf{g}_i(t) - \mathbf{x}_i(t))$ represents the social interaction between the particle and its neighbors, whereas the term $(\mathbf{p}_i(t) - \mathbf{x}_i(t))$ corresponds to the cognitive portion of the behavior of the particle. Additionally, [125] discussed the role of each of the terms in Eq. (17.32) from the standpoint of global and local searches: while the first two terms aim at performing local exploitation in the region where the particle is located, the latter contributes to the exploration of new regions of the search space. These desirable features of PSO, however, must be adequately balanced in order that the algorithm be successful in the task of determining the global optimum of the given cost function.

The social interaction between particles is determined by a predefined topological structure, so that the particles have access to the positions and performances of

---

[3] According to [85], the sum of the acceleration constants should be equal to $L_1 + L_2 = 4.1$, and the usual choice is $L_1 = L_2 = 2.05$.

the units directly connected to them. There are several strategies for choosing the neighborhood, among which we mention the ring topology (each particle has two neighbors – right and left) and the globally connected topology (all particles are interconnected) [87, 35].

The main steps of PSO are summarized in Alg. 2, where we consider that $f(\mathbf{x}) : \mathbb{R}^M \longmapsto \mathbb{R}$ yields the quality of the particles with respect to the cost function associated with the problem at hand.

Algorithm 2 PSO algorithm.

Parameters:
- N: number of particles;
- M: number of dimensions of the search space;
- $v_{\min}, v_{\max}$: lower and upper bounds of the components of the velocity vector;
- L_1, L_2: acceleration constants;
- $\max_{it}$: maximum number of iterations;

1- Generate an initial population with N particles;
2- Randomly generate initial velocity vectors for each particle;
while (iteration $\leq \max_{it}$) **do**
    3- Compute $f(\mathbf{x}_i), i = 1, \ldots, N$;
    4- Update the best position of the neighbors ($\mathbf{g}_i, i = 1, \ldots, N$);
    5- Update the best position each particle has visited ($\mathbf{p}_i, i = 1, \ldots, N$);
    6- Update the velocity vector of each particle using Eq. (17.32);
    7- Move each particle according Eq. (17.31);
end while

The PSO algorithm represents an interesting alternative to classical optimization methods due to: 1) its conceptual simplicity, 2) the capability of dealing with complex cost functions, e.g., nonlinear and non-differentiable functions, and 3) the possibility of escaping local optima by combining local and global information in the adaptation of the particles. Moreover, when compared with other competing bio-inspired metaheuristics, like genetic algorithms, PSO presents a reduced number of parameters to be adjusted. These attractive features encouraged the application of PSO in the context of signal processing tasks, such as channel equalization, source separation and direction of arrival estimation [142, 88, 15].

Nevertheless, several variants of the classical PSO have been proposed in recent years, which attempt to rigorously analyze and improve its performance, as well as its robustness [125, 28, 5], and also to extend the applicability of PSO to more challenging scenarios, e.g., multiobjective [29, 6] and discrete optimization [92, 6].

17.4.4 GENERAL REMARKS

The use of evolutionary computation in signal processing came, on the one hand, as a natural consequence of the more complex filtering criteria arising from the development of blind methods and also of statistical decision and information-theoretic formulations [26, 113], and, on the other, from the increasing adoption of nonlinear and recurrent filtering structures [116, 64, 65].

The most widespread methods in signal processing are, historically, evolutionary algorithms [27], but particle swarm optimization has also received a significant deal of attention [91]. The use of artificial immune systems is more restricted, albeit some groups have used these tools with remarkable success [116].

17.5 CONCLUDING REMARKS

Most of the traditional signal processing techniques found in the literature have been developed based on two key assumptions: that the signals involved may be modeled as a Gaussian random process and that the systems producing or distorting the signals of interest are linear. When these two hypotheses hold, it is natural to consider linear filters and criteria based on second-order statistics to process the signals. Moreover, in this context, simple optimization algorithms based on the gradient of the cost function are able to obtain the correct filter parameters.

However, when either the systems are inherently nonlinear or the signals present a non-Gaussian distribution, considering nonlinear structures and optimization criteria that can handle, in a more extensive manner, the statistical information about the signals of interest, may have advantages over traditional signal processing techniques. Nevertheless, since the optimization problem, in this scenario, becomes more intricate, the search of alternative optimization methods to improve the global search potential is necessary.

In this chapter we discussed methods and algorithms that fit into this second scenario, covering the three basic components of a general signal processing framework: structure, optimization criteria, and search methodology.

Artificial neural networks represent a very general and flexible nonlinear structure that has been extensively used for signal processing. The development of new approaches in the field, specifically those related to unorganized machines, have attracted the attention of an increasing number of researchers, mainly due to the fact that such machines have a much simpler training process when compared to MLP and RBF networks.

As far as optimization criteria are concerned, information theoretic learning provides a very attractive set of optimization criteria capable of extracting more efficiently the statistical content of the signals of interest. These criteria are essential to deal with unsupervised signal processing problems, such as the blind source separation and blind channel equalization problems, but may also improve the performance achieved in supervised filtering applications.

Finally, the use of bio-inspired optimization is justified due to the nature of the optimization task, which may be significantly different from that found in more classical frameworks. The bio-inspired methods discussed in the chapter present a very good compromise between local and global search capabilities, being particularly useful in problems where the cost function is multimodal.

It is important to mention that it was not our intention to provide a complete view of all these different approaches. Our main goal was to present and analyze essential features of these tools, indicating their potential in different signal processing applications, and to instigate researchers to consider these methods as valuable additions to their repertoire of signal processing techniques.

ACKNOWLEDGMENT

The authors would like to thank FAPESP (2013/14185-2 and 2013/06322-0), CAPES, and CNPq for the financial support.

REFERENCES

1. L. M. Adleman. Molecular computation of solutions to combinatorial problems. *Science*, 266(5187):1021–1024, 1994.
2. I. A. Ahmad and P. E. Lin. A nonparametric estimation of the entropy for absolutely continuous distributions. *IEEE Transactions on Information Theory*, 22:372–375, 1976.
3. S.-I. Amari and A. Cichocki. Adaptive blind signal processing – neural network approaches. *Proceedings of the IEEE*, 86(10):2026–2048, 1998.
4. T. Back, D. B. Fogel, and Z. Michalewicz, Eds. *Evolutionary computation 1: basic algorithms and operators*. Taylor & Francis, 2000.
5. A. Banks, J. Vincent, and C. Anyakoha. A review of particle swarm optimization. Part I: background and development. *Natural Computing*, 6(4):467–484, 2007.
6. A. Banks, J. Vincent, and C. Anyakoha. A review of particle swarm optimization. Part II: hybridisation, combinatorial, multicriteria and constrained optimization, and indicative applications. *Natural Computing*, 7(1):109–124, 2008.
7. J. Bednar, R. Yarlagadda, and T. Watt. l_1 deconvolution and its application to seismic signal processing. *IEEE Transactions on Acoustics, Speech and Signal Processing*, 34(6):1655–1658, 1986.
8. J. Beirlant, E. J. Dudewicz, L. Gyorfi, and E. C. Van Der Meulen. Nonparametric entropy estimation: an overview. *International Journal of Mathematical and Statistics Sciences*, (6):17–39, 1997.
9. A.J. Bell and T.J. Sejnowski. An information-maximization approach to blind separation and blind deconvolution. *Neural computation*, 7(6):1129–1159, 1995.
10. A. Benveniste, M. Goursat, and G. Ruget. Robust identification of a nonminimum phase system: blind adjustment of a linear equalizer in data communications. *IEEE Transactions on Automatic Control*, AC-25(3):385–399, 1980.
11. H. Bergson. *Creative evolution*. Henry Holt and Company, 1911.
12. H. Bersini. Revisiting idiotypic immune networks. In *Proceedings of the 2003 European Conference on Advances in Artificial Life (ECAL)*, 164–174, 2003.
13. H.-G. Beyer and H.-P. Schwefel. Evolution strategies: a comprehensive introduction. *Natural Computing*, (1):3–52, 2002.
14. L. Boccato, R. Attux, and F. J. Von Zuben. Self-organization and lateral interaction in echo state network reservoirs. *Neurocomputing*, 138:297–309, 2014.
15. L. Boccato, R. Krummenauer, R. Attux, and A. Lopes. Application of natural computing algorithms to maximum likelihood estimation of direction of arrival. *Signal Processing*, 92(5):1338–1352, 2012.
16. L. Boccato, A. Lopes, R. Attux, and F. J. Von Zuben. An extended echo state network using Volterra filtering and principal component analysis. *Neural Networks*, 32:292–302, 2012.

17. L. Boccato, D. G. Silva, D. Fantinato, R. Ferrari, and R. Attux. A comparative study of non-MSE criteria in nonlinear equalization. In *Proceedings of the 2014 International Telecommunications Symposium (ITS)*, 2014.

18. L. Boccato, D. G. Silva, D. Fantinato, K. Nose-Filho, R. Ferrari, R. R. F. Attux, A. Neves, J. Montalvão, and J. M. T. Romano. Error entropy criterion in echo state network training. In *Proceedings of the 2013 European Symposium on Artificial Neural Networks, Computational Intelligence and Machine Learning (ESANN)*, 35–40, 2013.

19. L. Boccato, E. S. Soares, M. M. L. P. Fernandes, D. C. Soriano, and R. Attux. Unorganized machines: from Turing's ideas to modern connectionist approaches. *International Journal of Natural Computing Research*, 2(4):1–16, 2011.

20. L. Boccato, D. C. Soriano, R. Attux, and F. J. Von Zuben. Performance analysis of nonlinear echo state network readouts in signal processing tasks. In *Proceedings of the 2012 International Joint Conference on Neural Networks (IJCNN)*, 1–8, 2012.

21. D. S. Broomhead and D. Lowe. Multivariate functional interpolation and adaptive networks. *Complex Systems*, 2:321–355, 1988.

22. F. M. Burnet. *The clonal selection theory of acquired immunity*. Cambridge University Press, 1959.

23. J. B. Butcher, D. Verstraeten, B. Schrauwen, C. R. Day, and P. W. Haycock. Reservoir computing and extreme learning machines for non-linear time-series data analysis. *Neural Networks*, 38:76–89, 2013.

24. C. C. Cavalcante, J. R. Montalvão, B. Dorizzi, and J. C. M. Mota. A neural predictor for blind equalization in digital communications. In *Proceedings of 2000 Adaptive Systems for Signal Processing, Communications and Control (AS-SPCC)*, 347–351, 2000.

25. S. Chen, G. J. Gibson, C. F. N. Cowan, and P. M. Grant. Adaptive equalization of nonlinear channels using multilayer perceptrons. *Signal Processing*, 20:107–119, 1990.

26. S. Chen, B. Mulgrew, and P. M. Grant. A clustering technique for digital communications channel equalization using radial-basis networks. *IEEE Transactions on Neural Networks*, 4:570–590, 1993.

27. S. Chen and Y. Wu. Maximum likelihood joint channel and data estimation using genetic algorithms. *IEEE Transactions on Signal Processing*, 46:1469–1473, 1998.

28. M. Clerc and J. Kennedy. The particle swarm explosion, stability, and convergence in a multidimensional complex space. *IEEE Transactions on Evolutionary Computation*, 6(1):58–73, 2002.

29. C. A. Coello Coello, G. T. Pulido, and M. S. Lechuga. Handling multiple objectives with particle swarm optimization. *IEEE Transactions on Evolutionary Computation*, 8(3):256–279, 2004.

30. P. Comon. Independent component analysis, a new concept? *Signal Processing*, 36(3):287–314, 1994.

31. T. M. Cover and J. A. Thomas. *Elements of information theory*. Wiley-Interscience, 2006.

32. G. Cybenko. Approximation by superpositions of a sigmoidal function. *Mathematics of Control Signals and Systems*, 2:303–314, 1989.

33. C. Darwin. *On the origin of species by means of natural selection, or the preservation of favoured races in the struggle for life*. John Murray, 1859.

34. R. Dawkins. *The selfish gene*. Oxford University Press, 1976.

35. L. N. De Castro. *Fundamentals of natural computing: basic concepts, algorithms and applications*. Chapman & Hall/CRC, 2006.

36. L. N. De Castro and J. Timmis. *Artificial immune systems: a new computational intelligence approach*. Springer Verlag, London, UK, 2002.

37. L. N. De Castro and F. J. Von Zuben. Learning and optimization using the clonal selection principle. *IEEE Transactions on Evolutionary Computation*, 6(3):239–251, June 2002.
38. F. O. De França, G. P. Coelho, and F. J. Von Zuben. On the diversity mechanisms of optainet: a comparative study with fitness sharing. In *Proceedings of the 2010 IEEE Congress on Evolutionary Computation (CEC)*, 1–8, July 2010.
39. N. Delfosse and P. Loubaton. Adaptive blind separation of independent sources: a deflation approach. *Signal processing*, 45(1):59–83, 1995.
40. A. Dempster, N. Laird, and D. Rubin. Maximum likelihood estimation from incomplete data using the EM algorithm. *Journal of the Royal Statistics Society*, 39:1–38, 1977.
41. D. C. Dennet. *Darwin's dangerous idea*. Simon & Schuster, 1995.
42. T. M. Dias, R. Attux, J. M. T. Romano, and R. Suyama. Blind source separation of post-nonlinear mixtures using evolutionary computation and gaussianization. *Proceedings of the 2009 Independent Component Analysis and Signal Separation (ICA), LNCS 5441*, 5441:235–242, 2009.
43. P. S. R. Diniz. *Adaptive filtering: algorithms and practical implementation*. Springer, 4th edition, 2013.
44. R. O. Duda, P. E. Hart, and D. G. Stork. *Pattern Classification*. Wiley-Interscience, 2001.
45. D. Erdogmus, K. E. Hild II, and J. C. Príncipe. Online entropy manipulation: Stochastic information gradient. *IEEE Signal Processing Letters*, 10(8):242–245, 2003.
46. D. Erdogmus and J. C. Príncipe. An error-entropy minimization algorithm for supervised training of nonlinear adaptive systems. *IEEE Transactions on Signal Processing*, 50(7):1780 – 1786, 2002.
47. D. Erdogmus and J. C. Príncipe. Generalized information potential criterion for adaptive system training. *IEEE Transactions on Neural Networks*, 13(5):1035–1044, 2002.
48. D. Erdogmus and J. C. Príncipe. Adaptive blind deconvolution of linear channels using Renyi's entropy with Parzen window estimation. *IEEE Transactions on Signal Processing*, 52(6):1489–1498, 2004.
49. D. Erdogmus, J. C. Príncipe, and L. Vielva. Blind deconvolution with minimum Renyi's entropy. In *Proceedings of the 2002 European Signal processing Conference (EUSIPCO)*, 2:71–74, 2002.
50. D. G Fantinato, L. Boccato, A. Neves, and R. Attux. Multivariate PDF matching via kernel density estimation. In *Proceedings ot the 2014 IEEE Symposium Series on Computational Intelligence (SSCI)*, 2014.
51. D. G. Fantinato, D. G. Silva, R. Attux, R. Ferrari, L. T. Duarte, R. Suyama, J. R. Montalvão, A. Neves, , and J. M. T. Romano. Optimal linear filtering over a Galois field: equalization and prediction. In *Anais do V Encontro dos Alunos e Docentes do Departamento de Engenharia de Computação e Automação Industrial (EADCA)*, 2012.
52. D. G. Fantinato, D. G. Silva, E. Nadalin, A. Neves, J. R. Montalvão, R. Attux, and J. M. T. Romano. Blind separation of convolutive mixtures over Galois fields. In *Proceedings of the 2013 IEEE International Workshop on Machine Learning for Signal Processing (MLSP)*, 1–6, 2013.
53. D. G Fantinato, Daniel G. Silva, R. Attux, R. Ferrari, L. T. Duarte, R. Suyama, A. Neves, J. R. Montalvão, and J. M. T. Romano. Optimal time-series prediction over Galois fields. In *Anais do III Simpósio de Processamento de Sinais da UNICAMP*, 2012.
54. R. Ferrari, R. Suyama, R. R. Lopes, R. Attux, and J. M. T. Romano. An optimal MMSE fuzzy predictor for SISO and MIMO blind equalization. In *Proceedings of the 2008 IAPR Workshop on Cognitive Information Processing (CIP)*, 2008.
55. G. B. Fogel and D. B. Fogel. Continuous evolutionary programming: analysis and experi-

ments. *Cybernetics and Systems*, 26:79–90, 1993.

56. L. J. Fogel. *Intelligence through simulated evolution: forty years of evolutionary programming*. John Wiley & Sons, 1999.

57. L. J. Fogel, A. J. Owens, and M. J. Walsh. *Artificial intelligence through simulated evolution*. John Wiley & Sons, 1966.

58. D. Godard. Self-recovering equalization and carrier tracking in two-dimensional data communication systems. *IEEE Transactions on Communications*, COM-28(11):1867–1875, 1980.

59. R. Gonin and A. H. Money. *Nonlinear L_p-norm estimation*. CRC Press, 1989.

60. S. J. Gould. *Ontogeny and Phylogeny*. Belknap Press of Harvard University Press, 1977.

61. A. Gunduz and J. Príncipe. Correntropy as a novel measure for nonlinearity tests. *Signal Processing*, 89:14–23, 2009.

62. H. Gutch, P. Gruber, and F. Theis. ICA over finite fields. In *Proceedings of the 2010 Latent Variable Analysis and Signal Separation (LVA)*, 645–652. Springer, 2010.

63. H. W. Gutch, P. Gruber, A. Yeredor, and F. J. Theis. ICA over finite fields – separability and algorithms. *Signal Processing*, 92(8):1796–1808, 2012.

64. S. Haykin. *Adaptive filter theory*. NJ: Prentice Hall, 3rd edition, 1996.

65. S. Haykin. *Neural networks: a comprehensive foundation*. Prentice Hall, 2nd edition, 1998.

66. K. E. Hild II, D. Erdogmus, and J. C. Príncipe. Blind source separation using Renyi's mutual information. *IEEE Signal Processing Letters*, 8(6):174–176, 2001.

67. K. E. Hild II, D. Erdogmus, and J. C. Príncipe. An analysis of entropy estimators for blind source separation. *Signal processing*, 86(1):182–194, 2006.

68. J. H. Holland. Outline for a logical theory of adaptive systems. *Journal of the ACM*, 9(3):297–314, 1962.

69. J. J. Hopfield. Neural networks and physical systems with emergent collective computational properties. In *Proceedings of the 1982 National Academy of Sciences of the USA*, 79(8):2554–2558, 1982.

70. J. J. Hopfield. Neurons with graded response have collective computational properties like those of two-state neurons. In *Proceedings of the 1984 National Academy of Sciences of the USA*, 81:3088–3092, 1984.

71. J. J. Hopfield and D. W. Tank. Neural computation of decisions in optimization problems. *Biological Cybernetics*, 52:141–152, 1985.

72. K. Hornik, M. Stinchcombe, and H. White. Multilayer feedforward neural networks are universal approximators. *Neural Networks*, 2(5):359–366, 1989.

73. G.-B. Huang, L. Chen, and C.-K. Siew. Universal approximation using incremental constructive feedforward networks with random hidden nodes. *IEEE Transactions on Neural Networks*, 17(4):879–892, 2006.

74. G.-B. Huang, D. H. Wang, and Y. Lan. Extreme learning machines: a survey. *International Journal of Machine Learning and Cybernetics*, 2(2):107–122, 2011.

75. G.-B. Huang, Q.-Y. Zhu, and C.-K. Siew. Extreme learning machine: a new learning scheme of feedforward neural networks. In *Proceedings of the 2004 IEEE International Joint Conference on Neural Networks (IJCNN)*, 985–990, 2004.

76. G.-B. Huang, Q.-Y. Zhu, and C.-K. Siew. Extreme learning machine: theory and applications. *Neurocomputing*, 70:489–501, 2006.

77. A. Hyvárinen, J. Karhunen, and E. Oja. *Independent component analysis*. Wiley-Interscience, 2001.

78. H. Jaeger. The echo state approach to analyzing and training recurrent neural networks.

Technical Report 148, German National Research Center for Information Technology, 2001.

79. H. Jaeger. Adaptive nonlinear system identification with echo state networks. In *Advances in Neural Information Processing Systems*, 593–600, 2003.

80. H. Jaeger and H. Hass. Harnessing nonlinearity: predicting chaotic systems and saving energy in wireless communication. *Science*, 304(5667):78–80, 2004.

81. M. A. Jarajreh, S. Rajbhandari, E. Giacoumidis, and N. J. Doran. Fibre impairment compensation using artificial neural network equalizer for high-capacity coherent optical OFDM systems. In *Proceedings of the 2014 International Symposium on Communication Systems, Networks & Digital Signal Processing (CSNDSP)*, 2014.

82. N. K. Jerne. Towards a network theory of the immune system. In *1974 Annales d'Immunologie*, 125(1-2):373–389, 1974.

83. G. Kechriotis, E. Zervas, and E. S. Manolakos. Using recurrent neural networks for adaptive communication channel equalization. *IEEE Transactions on Neural Networks*, 5:267–278, 1994.

84. J. Kennedy. The particle swarm: social adaptation of knowledge. In *Proceedings of the 1997 IEEE International Conference on Evolutionary Computation (CEC)*, 303–308, 1997.

85. J. Kennedy. Particle swarms: optimization based on sociocognition. In L. N. de Castro and F. J. Von Zuben, eds., *Recent Developments in Biologically Inspired Computing*, chapter X, 235–269. Idea Group Publishing, 2004.

86. J. Kennedy and R. Eberhart. Particle swarm optimization. In *Proceedings of the 1995 IEEE International Conference on Neural Networks (IJCNN)*, 4:1942–1948, 1995.

87. J. Kennedy, R. Eberhart, and Y. Shi. *Swarm intelligence*. Morgan Kaufmann Publishers, 2001.

88. V. Khanagha, A. Khanagha, and V. T. Vakili. Modified particle swarm optimization for blind deconvolution and identification of multichannel FIR filters. *EURASIP Journal on Advances in Signal Processing*, 2010(1):716–862, 2010.

89. A. Kolmogorov. *Interpolation and extrapolation of stationary random processes*. Rand Co. (Russian Translation), Santa Monica, 1962.

90. J. R. Koza. *Genetic programming: on the programming of computers by means of natural selection*. MIT Press, Cambridge, 1992.

91. D. J. Krusienski and W. K. Jenkins. Particle swarm optimization for adaptive IIR structures. In *Proceedings of the 2004 IEEE Congress on Evolutionary Computation (CEC)*, 2004.

92. E. C. Laskari, K. E. Parsopoulos, and M. N. Vrahatis. Particle swarm optimization for integer programming. In *Proceedings of the 2002 IEEE Congress on Evolutionary Computation (CEC)*, 1576–1581, 2002.

93. M. Lazaro, I. Santamaria, D. Erdogmus, K. Hild II, C. Pantaleon, and J. C. Príncipe. Stochastic blind equalization based on PDF fitting using parzen estimator. *IEEE Transactions on Signal Processing*, 53(2):696–704, 2005.

94. M. Lazaro, I. Santamaria, C. Pantaleon, D. Erdogmus, and J. C. Príncipe. Matched PDF-based blind equalization. In *Proceedings of the 2003 IEEE International Conference on Acoustics, Speech, and Signal Processing (ICASSP)*, 4(IV):297–300, 2003.

95. M. Lazaro, I. Santamaria, C. Pantaleon, D. Erdogmus, K. E. Hild II, and J. C. Príncipe. Blind equalization by sampled PDF fitting. In *Proceedings of the 2003 International Conference on Independent Component Analysis (ICA)*, 1041–1046, 2003.

96. M.-B. Li., G.-B. Huang, P. Saratchandran, and N. Sundararajan. Channel equalization using complex extreme learning machine with RBF kernels. In *Lecture Notes in Computer*

Science, 3973:114–119. 2006.

97. R. Li, W. Liu, and J. C. Príncipe. A unifying criterion for instantaneous blind source separation based on correntropy. *Signal Processing*, 87:1872–1881, 2007.

98. W. Liu, P. P. Pokharel, and J. C. Príncipe. Correntropy: properties and application in non-Gaussian signal processing. *IEEE Transactions on Signal Processing*, 55:5286–5298, 2007.

99. M. Lukosevicius and H. Jaeger. Reservoir computing approaches to recurrent neural network training. *Computer Science Review*, 3:127–149, 2009.

100. S.-K. Ma. Calculation of entropy from data of motion. *Journal of Statistical Physics*, 26(2):221–240, 1980.

101. D. J. C. MacKay. *Information theory, inference, and learning algorithms*. Cambridge University Press, 2003.

102. D. P. Mandic and J. A. Chambers. *Recurrent neural networks for prediction*. Wiley-Interscience, 2001.

103. W. McCulloch and W. Pitts. A logical calculus of the ideas immanent in nervous activity. *Bulletin of Mathematical Biophysics*, 5(4):115–133, 1943.

104. L. Medsker and L. C. Jain, Eds. *Recurrent neural networks: design and applications*. CRC Press, 1999.

105. J. Montalvão, B. Dorizzi, and J. C. M. Mota. Some theoretical limits of efficiency of linear and nonlinear equalizers. *Journal of Communications and Information Systems*, 14:85–92, 1999.

106. S. Nadarajah. A generalized normal distribution. *Journal of Applied Statistics*, 32:685–694, 2005.

107. K. Nose-Filho, L. T. Duarte, R. Attux, E. Nadalin, R. Ferrari, and J. M. T. Romano. Sobre filtragem l_p. In *Anais do XXX Simpósio Brasileiro de Telecomunicações (SBrT)*, 2012.

108. K. Nose-Filho, D. G. Fantinato, R. Attux, A. Neves, and J. M. T. Romano. A novel entropy-based equalization performance measure and relations to l_p-norm deconvolution. In *Anais do XXXI Simpósio Brasileiro de Telecomunicações (SBrT)*, 2013.

109. K. Nose-Filho, A. D. P. Lotufo, and C.R. Minussi. Preprocessing data for short-term load forecasting with a general regression neural network and a moving average filter. In *Proceedings of the 2011 IEEE PowerTech*, 2011.

110. K. Nose-Filho, A. D. P. Lotufo, and C.R. Minussi. Short-term multinodal load forecasting using a modified general regression neural network. *IEEE Transactions on Power Delivery*, 26(4):2862–2869, 2011.

111. J. Park and I. W. Sandberg. Universal approximation using radial-basis-function networks. *Neural Computation*, 3(2):246–257, 1991.

112. E. Parzen. On estimation of a probability density function and mode. *The annals of mathematical statistics*, 33(3):1065–1076, 1962.

113. J. C. Príncipe. *Information theoretic learning: Renyi's entropy and Kernel perspectives*. Springer Publishing Company, Incorporated, 1st edition, 2010.

114. V. Rajagopalan. *Symbolic dynamic filtering of complex systems*. PhD thesis, Pennsylvania State University, University Park, PA, USA, 2007. AAI3380627.

115. J. R. Rice and J. S. White. Norms for smoothing and estimation. *SIAM Review*, 6(3):243–256, 1964.

116. J. M. T. Romano, R. R. F. Attux, C. C. Cavalcante, and R. Suyama. *Unsupervised signal processing: channel equalization and source separation*. CRC Press, 2010.

117. F. Rosenblatt. The perceptron: a probabilistic model for information storage and organization in the brain. *Psychological Review*, 65(6):386–408, 1958.

118. D. E. Rumelhart, G. E. Hinton, and R. J. Williams. Learning representations by back-propagating errors. *Nature*, 323:533–536, 1986.

119. I. Santamaria, D. Erdogmus, and J. C. Príncipe. Entropy minimization for supervised digital communications channel equalization. *IEEE Transactions on Signal Processing*, 50(5):1184–1192, 2002.

120. I. Santamaria, P. Pokharel, and J. C. Príncipe. Generalized correlation function: definition, properties and application to blind equalization. *IEEE Transactions on Signal Processing*, 54(6):2187–2197, 2006.

121. I. Santamaria, C. Vielva, and J. C. Príncipe. A fast algorithm for adaptive blind equalization using order-α Renyi's entropy. In *Proceedings of the 2002 International Conference on Acoustics, Speech and Signal Processing (ICASSP)*, 2(1):2657–2660, 2002.

122. B. Schrauwen, M. Wardermann, D. Verstraeten, J. J. Steil, and D. Stroobandt. Improving reservoirs using intrinsic plasticity. *Neurocomputing*, 71:1159–1171, 2008.

123. H.-P. Schwefel. *Numerical optimization of computer models*. John Wiley & Sons, Inc., New York, NY, 1981.

124. C. Shannon. A mathematical theory of communication. *The Bell System Technical Journal*, 27:379–423, 623–656, 1948.

125. Y. Shi and R. C. Eberhart. Empirical study of particle swarm optimization. In *Proceedings of the 1999 IEEE Congress on Evolutionary Computation (CEC)*, 1945–1950, 1999.

126. D. G. Silva, R. Attux, E. Z. Nadalin, L. T. Duarte, and R. Suyama. An immune-inspired information-theoretic approach to the problem of ICA over a Galois field. In *Information Theory Workshop (ITW), 2011 IEEE*, 618–622, oct. 2011.

127. D. G. Silva, E. Z. Nadalin, G. P. Coelho, L. T. Duarte, R. Suyama, R. Attux, F. J. Von Zuben, and J. Montalvão. A Michigan-like immune-inspired framework for performing independent component analysis over Galois fields of prime order. *Signal Processing*, 96:153–163, March 2014.

128. L. M. Silva, J. M. de Sá, and L. A. Alexandre. Neural network classification using Shannon's entropy. In *Proceedings of the 2005 European Symposium on Artificial Neural Networks, Computational Intelligence and Machine Learning (ESANN)*, 217–222, 2005.

129. A. Singh and J. Príncipe. Using correntropy as a cost function in linear adaptive filters. In *Proceedings of the 2009 International Joint Conference on Neural Networks (IJCNN)*, 2950–2955, 2009.

130. D. F. Specht. A general regression neural network. *IEEE Transactions on Neural Networks*, 2:568–576, 1991.

131. Y. Tan, J. Wang, and J. M. Zurada. Nonlinear blind source separation using a radial basis function network. *IEEE Transactions on Neural Networks*, 12(1):124–134, 2001.

132. P. Teilhard de Chardin. *Le phénomène humain*. Éditions du Seuil, 1955.

133. I. R. Tizard. *Immunology an introduction*. Saunders College Publishing, 4th edition, 1995.

134. C. Wada, D. M. Consolaro, R. Ferrari, R. Suyama, R. Attux, and F. J. Von Zuben. Nonlinear blind source deconvolution using recurrent prediction-error filters and an artificial immune system. In *Proceedings of the 2009 International Conference on Independent Component Analysis (ICA), LNCS 5441*, 5441:371–378, 2009.

135. A. Webb. *Statistical pattern recognition*. John Wiley & Sons, 2002.

136. P. J. Werbos. *Beyond regression: new tools for prediction and analysis in the behavioral sciences*. PhD thesis, Harvard University, 1974.

137. N. Wiener. *Nonlinear problems in random theory*, volume 1. MIT, 1958.

138. A. Yeredor. ICA in boolean XOR mixtures. In *Proceedings of the 2007 Independent Component Analysis and Signal Separation (ICA)*, 827–835. Springer, 2007.

139. A. Yeredor. Independent component analysis over Galois fields of prime order. *IEEE Transactions on Information Theory*, 57(8):5342–5359, 2011.

140. A. Yeredor. On blind channel identification and equalization over Galois fields. In *Proceedings of the 2014 IEEE International Conference on Acoustics, Speech and Signal Processing (ICASSP)*, 4239–4243, 2014.

141. I. B. Yildiz, H. Jaeger, and S. J. Kiebel. Re-visiting the echo state property. *Neural Networks*, 35:1–9, 2012.

142. Y. Zhao and J. Zheng. Particle swarm optimization algorithm in signal detection and blind extraction. In *Proceedings of the 2004 International Symposium on Parallel Architectures, Algorithms and Networks*, 37–41, 2004.

18 High-Resolution Techniques for Seismic Signal Prospecting

RAFAEL KRUMMENAUER
DSPGeo

ANDRÉ KAZUO TAKAHATA
University of Campinas (UNICAMP)

TIAGO TAVARES LEITE BARROS
University of Campinas (UNICAMP)

MARCOS RICARDO COVRE
University of Campinas (UNICAMP)

RENATO DA ROCHA LOPES
University of Campinas (UNICAMP)

18.1 INTRODUCTION

In this chapter, we describe a very important application of digital signal processing: the processing of seismic signals, which are the main tool for the discovery of new reserves of oil and gas, and for the proper management and exploration of the corresponding fields. For instance, the decision of whether or not a well should be drilled, and where it should be drilled, is based in great part on information extracted from seismic signals. Clearly, this application is essential for the whole oil and gas industry.

The main goal of seismic signal processing is to estimate the structure of the subsurface of a region of interest, such as the location of rock interfaces, as well as some properties of these rocks. This information helps a geologist to assess, for instance, the potential for oil in the region. The seismic signal itself is generated by shooting a seismic source, such as dynamite, at the surface. This shot generates a seismic wave that propagates downward, through rock layers with different acoustic impedances. At the interface between different rocks, the impedance mismatch causes some of the wave energy to be reflected back to the surface.

At the same time that a shot is fired, several sensors located at the surface start measuring the response of the subsurface to the shot.[1] This measurement captures

---

[1] The signal recorded at a given receiver in response to a shot is called a *trace*.

the aforementioned reflected waves, which contain information on the location of the interfaces, as well as noise originating from several sources. After some measuring time, usually around ten seconds, the seismic sources and receivers are shifted by a few meters, usually around 25 m, and the firing and recording process is repeated, until the area of interest has been covered.

After the seismic data is acquired in the field, it goes through several processing stages until an image of the subsurface is created. This processing has several goals: eliminating coherent noise such as surface waves that carry no information on the subsurface; enhancing the signal-to-noise ratio of the signals; placing the reflections at the correct position, etc. It is interesting to note that many of these tasks form the essence of digital signal processing in general, and there has been a strong link between these two areas of research. In fact, many of the tools in signal processing were originally developed for geophysical applications. Examples include the prediction error filter and minimum-phase digital systems [29], and the modern and general form of the wavelet transform [12].

In this chapter, we discuss signal processing methods that aim at improving the resolution of seismic processing. The word resolution is a bit subjective and difficult to quantify but it is common sense that it is related to the ability to distinguish the presence of energy in two or more closely spaced points, frequencies, or parameters. For instance, if a signal contains energy at two nearby frequencies, we should be able to see this, as opposed to seeing a single frequency range with energy. Here, we focus on the resolution of three problems: the estimation of seismic velocities, the detection of seismic events, and the placement of the seismic reflections at their proper position. In the first two problems, the definition of resolution is somewhat obvious: we should be able to resolve the velocities of reflections arriving at approximately the same time, and we should be able to distinguish two close events. For these problems, we improve the resolution by employing methods that originate from the literature in array signal processing. For the third problem, known in seismic literature as migration, we define resolution as the ability to resolve, for instance, two diffractions that occur at points that are close in space. The migration resolution is improved using techniques from the image processing literature.

High resolution methods have a long and rich history in digital signal processing. The first high-resolution techniques were purely based on Fourier analysis, being Schuster a pioneer with the introduction of the periodogram approach [33]. A refinement of the periodogram-based methods, introduced by Blackman and Tuckey in 1959 [6], reduces the variance of the spectrum estimates at the cost of reduction in resolution [35]. Some years later, the computationally efficient Fast Fourier Transform (FFT) algorithm was developed by Cooley and Tukey [7], renewing the interest in the periodogram approach for spectral estimation [16]. These methods are usually classified as nonparametric and make no assumption upon the signals, except for stationarity.

The resolution of spectral analysis can be improved if parametric descriptions of the signals are incorporated into the design of the estimator. One of the pioneers in the parametric approach was Burg in 1967, who introduced the maximum entropy method with application to geophysical data processing, and suggested the use of a maximum entropy spectrum for certain signals. Almost in parallel, autoregressive (AR) based estimators were developed, with strong contributions given by the studies of Parzen [25, 26].

The application of spectral estimators is not constrained to time series analysis, where determination of frequency content is the main objective. For instance, in an array of sensors spatially distributed over a field, the spectral information to be extracted may represent the angle, phase, or velocity of an impinging wave. For this particular application, in 1969 Capon proposed a high resolution method for frequency-wavenumber spectrum analysis. This method is known as Capon's beamformer, and has been used tremendously over the last decades in sonar, radar, and biomedical applications.

The greatest step towards super-resolution came in 1973, when Pisarenko proposed a method of harmonic decomposition of signals immersed in white noise [27]. This work marks the beginning of a new era in spectral analysis, being one of the first to explore the eigendecomposition of the covariance matrix in all signal processing fields. Some years later, two independent publications, [4, 31], came up with the most famous subspace-based approach for spectral analysis until now, the so-called MUSIC (Multiple Signal Classification) algorithm. Then, from the '80s on, a vast repertoire of advances has been devised. These are, to a large extent, variants of the former techniques mainly concerning more specific signal models, incorporating any information that can help in improving resolution and taking into account different criteria, like robustness and numerical efficiency. For a detailed survey on some of these methods and also the history of high-resolution methods from the spectral analysis point of view, we refer the reader to [16, 19, 35].

High resolution methods also find widespread application in image processing, especially in areas such as medical and astronomical images. The challenges here are either to build a high-resolution image from several low-resolution images, or to produce a high-resolution image by restoring a noisy and blurred image [24]. The techniques used to achieve these goal are somewhat different and more varied than those used in radar and array signal processing, but the final goal is the same: if a certain part of the image of interest corresponds to a single pixel in the digital image, then this information should not leak to adjacent pixels. For a great review on this topic, we refer the reader to [28].

One concept from digital image processing that is of particular interest to this work is the point spread function [11]. This function models some distortions that affect images in general, especially the distortions associated with measurement devices. For instance, in microscopy, the PSF may refer to the distortion introduced by the lenses. The PSF appears traditionally in the image processing literature, where many methods have been proposed to mitigate some of its effects. As we will see, PSFs can also model some of the distortions introduced by migration, which, as mentioned before, is the process that places the reflections at the appropriate position. This interpretation of the migration distortions as a PSF allows us to apply well-known PSF-mitigating filters to improve the resolution of seismic images.

18.2 SUMMARY

This chapter is organized as follows. In Section 18.3, we discuss the problem of velocity analysis, and describe how the traditional methods used in the literature can be linked to high-resolution methods such as MUSIC. We propose a simple interpretation of MUSIC in the context of seismic signals, and we show that the resulting method indeed provides a much better resolution when compared to the most widely used method of velocity analysis. Then, in Section 18.4, we discuss the importance of detecting seismic events. We formulate this problem as a hypothesis test, which enables us to use the huge body of literature in this traditional area of statistics to detect these events. To enable this formulation, we conduct some interesting studies on the statistics of seismic signals. Finally, in Section 18.5, we show how the final image produced by seismic signal processing can be, at least locally, seen as a convolution of the desired image with a point spread function (PSF). We show how this PSF can be estimated with ray tracing, and how some methods from image restoration can be effectively used to improve seismic images.

18.3 HIGH-RESOLUTION VELOCITY SPECTRA

One of the most important geological parameters that can be estimated from the seismic data is the velocity of the seismic wave at different points of the subsurface. This velocity can help to determine what rock is present at a given region, as different rocks have different velocities. It is also crucial to transform the information recorded at the receivers, which is the *time* it took a reflection to return to the surface, into information about the *depth* of the interface that generated this reflection. In fact, estimating the velocity is so important, and so tricky, that it is not done in a single step. Instead, the velocity model is continuously improved throughout the processing of the seismic data. In this section, we describe the first step of this estimation, called velocity analysis. We begin by describing the problem in general, and then we discuss its most widely used solution. Then we show how methods from array signal processing can be used to produce estimates with higher resolution.

18.3.1 THE VELOCITY ANALYSIS PROBLEM

To describe velocity analysis, we need first to describe the acquisition process in more detail, and define some variables. To that end, consider Figure 18.1(a), which shows the recording geometry and ray paths associated with a single shot and a horizontal reflector. Note that source and receiver are placed along a single line. This is typical of 2D acquisitions, which are assumed here. In this case, the position of a source or a receiver can be described by a scalar, s or g, respectively, that denotes its position along the recording line [43]. Recall that a trace is the recording, at a given receiver at g, of the subsurface response to a shot fired at a position s. Thus, each trace is associated with the source-receiver pair at coordinates (s, g).

After the acquisition, we can sort the seismic traces in different ways. In Figure 18.1(b) a common-midpoint (CMP) geometry is illustrated, also with the ray

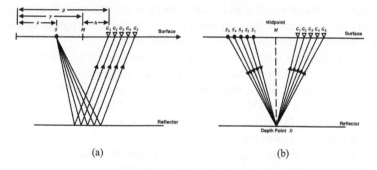

Figure 18.1 (a) Common-shot and (b) common-midpoint geometries.

paths associated with a horizontal reflector. The CMP gather is generated by grouping the traces associated to source-receiver pairs with the same midpoint, y [43]. In this case, the coordinates of the traces in the CMP gather can be expressed in terms of the midpoint, y, and half offset, h, as $y = (g + s)/2$ and $h = (g - s)/2$. Note that the half offset h is half the distance between source and receiver.

The importance of the CMP family is that, as seen in Figure 18.1(b), the reflections recorded in the traces come from the same point in the subsurface. However, since the reflections in the traces in a CMP gather were recorded at different locations and time instants, they are hopefully corrupted by independent noise samples. Thus, processing seismic data sorted in CMP gathers allows us to benefit from this diversity to improve their signal-to-noise ratio.

To properly combine the reflections in the CMP gather, so that they add coherently, we need to take into account the different traveltimes of the reflections in each trace. To that end, consider a zero-offset trace, for which both source and receiver are collocated at y. This trace cannot be generated in practice, since the receiver would be damaged if it were too close to the source. Still, if the velocity were known, it would be possible to relate the traveltime of the reflection in the zero-offset trace to that of the other traces. Indeed, for a CMP gather configuration, a well-accepted traveltime is the normal moveout (NMO) [41, 43], given by

$$t^2(t_0, h) = t_0^2 + \frac{4h^2}{v^2}, \qquad (18.1)$$

where t_0 denotes the two-way zero-offset traveltime and v can be viewed as an effective velocity. The NMO equation can be obtained by an application of the Pythagoras theorem to the rays in Figure 18.1(b). Note that for each value of t_0 we would have a different velocity v, so that we should write $v = v(t_0)$. To keep the notation simple, we will write simply v.

Obviously the velocity in (18.1) is unknown, but it can be estimated by fitting the CMP data to the traveltime model (18.1). The reasoning is that, if we actually had a reflection at t_0 with a velocity v, then the data in the CMP gather, at the time instants given by (18.1), would all correspond to the same reflection, and would thus be related, or coherent. Thus, for each t_0 and a test velocity v, we take a small window of data around $t(t_0, h)$ and measure the coherence of the data in this window. In the remainder of this section, we discuss this process in more detail.

In Figure 18.2, we illustrate how the windowing operation is applied to seismic data. The data here is a CMP gather of a real offshore survey, acquired in the Jequitinhonha Basin, Brazil. For each t_0 and v, the seismic data is windowed with windows of L samples before and after the NMO traveltime given by (18.1) with the corresponding values of t_0 and v. In particular, in Figure18.2(a), we show the limits of two windows, both with the same t_0 but with different velocities, while the resulting windowed data are shown in Figures 18.2(b) and 18.2(c). Note how the use of a good velocity leads to a window with horizontal events in Figure 18.2(c).

For each window, resulting from a choice of t_0 and v, we may define an $N_r \times N_t$ matrix $\mathbf{D}$, where $N_t = 2L + 1$ is the number of time samples in the window and N_r is the number of traces in the CMP gather. The entries of the matrix $\mathbf{D}$ are the amplitudes of the seismic signal in the corresponding point in the window. In keeping with the notation in the literature, each row of $\mathbf{D}$ corresponds to the amplitudes on the window for a given receiver. In other words, the rows of $\mathbf{D}$ appear as vertical lines in Figures 18.2(b) and 18.2(c). Furthermore, note that different values of t_0 and v lead to different matrices $\mathbf{D}$; however, for notational simplicity, we do not make this dependence explicit.

To determine if the data window contains an event, we usually seek to measure whether the rows of $\mathbf{D}$ are coherent. The most employed coherence measure is the *semblance* [39], given by

$$S = \frac{\mathbf{1}^T \hat{\mathbf{R}} \, \mathbf{1}}{N_t \text{Tr}\{\hat{\mathbf{R}}\}}, \tag{18.2}$$

where $\hat{\mathbf{R}}$ is the spatial sample covariance matrix (SCM) of $\mathbf{D}$, $\mathbf{1}$ is a column vector of ones and $\text{Tr}\{\cdot\}$ denotes the trace operator. The value of S falls within the interval $[0, 1]$. This spectral-based criterion evaluates coherence of the reflection events in terms of spatial alignment within the data window.

The semblance is computed for several values of t_0 and v, resulting in a velocity spectrum called a *coherence map* that conveys information about the existence of reflection energy and *the most likely* effective velocity value for each t_0. Soon, we will see that the correct use of the term "the most likely" depends on the adopted hypothesis for the existence of events. In Figure 18.3 we show the semblance coherence map of the CMP gather shown in Figure 18.2(a). This figure suggests, for instance, that there was probably a reflection at 2.5 s with a velocity of 1500 m/s. In practice, an interpreter would pick the velocities of the strongest events, and a velocity function $v(t_0)$ would be generated by the interpolation of these picked values.

18.3.2 THE STRUCTURE OF THE WINDOWED DATA

As mentioned before, the velocity analysis procedure consists in, for each t_0, designing several traveltime curves using equation (18.1) with different velocities, v, and

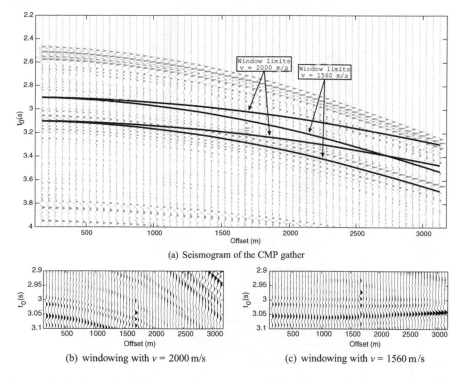

(a) Seismogram of the CMP gather

(b) windowing with $v = 2000$ m/s   (c) windowing with $v = 1560$ m/s

Figure 18.2 Data windowing. (a) seismogram of a CMP gather highlighting the windowing procedure at $t_0 = 3.0$ s for two different velocities. The windows are shown by the solid lines. The traces are in gray only to highlight the window limits; normally, they are in black. (b) and (c) highlight the interpolated samples that form the corresponding data windows.

to choosing the velocity that results in a maximum value of a coherence function, usually the semblance. However, [5] and [17] showed that eigenstructure methods originating from the radar literature can lead to spectra with higher resolution than semblance. One of most commonly used high-resolution methods is the Multiple Signal Classification (MUSIC), introduced in [4, 31, 32], which is based in some properties of the eigendecomposition of the data. Recently, a new version of MUSIC, based on the temporal correlation of the seismic data, has been presented [3, 2]. In this section, we describe the structure of the windowed data matrix, and its impact on the spatial and temporal correlation matrices, which are the basis of MUSIC.

In a way, all we need to understand about how MUSIC works in this application is the following powerful approximation, which states that, when a window with correct values of t_0 and v is applied, the windowed data matrix $\mathbf{D}$ contains several repetitions of the reflection, all arriving at the same time instant at all the receivers, plus noise terms. In this case, $\mathbf{D}$ can be written as

$$\mathbf{D} = \mathbf{1s}^T + \mathbf{N}, \tag{18.3}$$

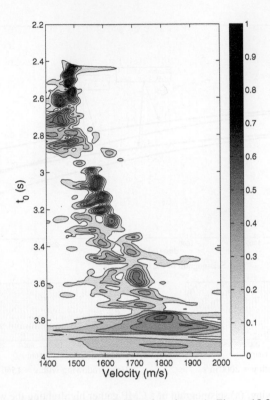

Figure 18.3   Velocity spectrum for the CMP gather shown in Figure 18.2(a).

where $\mathbf{s}$ is an $N_t \times 1$ vector that contains the samples from the reflected wavelet, $\mathbf{1}$ is a $N_r \times 1$ vector of ones, $\mathbf{N}$ is an $N_r \times N_t$ noise matrix independent of $\mathbf{s}$, which may also contain interfering reflections, and the superscript T refers to the transpose operation. Note that this expression itself does not contain any approximation. However, we will assume that the noise term is zero-mean and uncorrelated with the signal, which is clearly not exact.

18.3.3   S-MUSIC

The MUSIC high-resolution coherence function based in the spatial correlation matrix, also known as S-MUSIC, has been widely used to produce seismic velocity spectra [5, 17, 18]. This method is based on the spatial correlation matrix of the windowed data, which can be estimated by employing time averaging, resulting in

$$\hat{\mathbf{R}} = \frac{1}{N_t} \mathbf{D} \mathbf{D}^T. \tag{18.4}$$

The matrix $\hat{\mathbf{R}}$ has dimensions $N_r \times N_r$, and its elements contain the estimated correlation between the data at different receivers. This explains why it is called a *spatial* correlation matrix. Finally, using (18.3), and assuming that the noise matrix $\mathbf{N}$ is independent of the data, we may write

$$\hat{\mathbf{R}} \approx \frac{\|\mathbf{s}\|^2}{N_t} \mathbf{1}\mathbf{1}^T + \sigma_n^2 \mathbf{I}, \tag{18.5}$$

where σ_n^2 is the noise variance and $\mathbf{I}$ is an $N_r \times N_r$ identity matrix.

The approximation of $\hat{\mathbf{R}}$ in (18.5) has a very interesting eigenstructure. Indeed, it is not hard to show that its largest eigenvalue is

$$\frac{\|\mathbf{s}\|^2 N_r}{N_t} + \sigma_n^2, \tag{18.6}$$

with $\mathbf{1}$ as the associated eigenvector. We will refer to this as the largest eigenvector of $\hat{\mathbf{R}}$. The other eigenvalues are σ_n^2, with eigenvectors orthogonal to $\mathbf{1}$. This observation can be used to interpret the role of MUSIC in this problem. Essentially, it tries to determine whether $\mathbf{1}$ is the largest eigenvector of $\hat{\mathbf{R}}$. If the answer is positive, then it is likely that the values of t_0 and v that originated the windowed data $\mathbf{D}$ and the associated correlation matrix $\hat{\mathbf{R}}$ correspond to an event. The S-MUSIC coherence measure is given by

$$P_S = \frac{N_r}{N_r - |\mathbf{1}^T\mathbf{v}_1|^2}, \tag{18.7}$$

where $\mathbf{v}_1$ is the unit-norm eigenvector of $\hat{\mathbf{R}}$, with dimension $N_r \times 1$, associated to its largest eigenvalue. If the window is perfectly matched to a single event, then $\mathbf{v}_1$ and $\mathbf{1}$ will be perfectly aligned and P_S will be infinity.

18.3.4 T-MUSIC

In this section, we show how the MUSIC coherency measure can be obtained from a matrix smaller than $\hat{\mathbf{R}}$, namely the temporal correlation matrix. The resulting high-resolution coherence measure was proposed in [2] and its main goal is to reduce the computational cost of the eigendecomposition required by S-MUSIC. To that end, define the estimated temporal correlation matrix

$$\hat{\mathbf{r}} = \frac{1}{N_r} \mathbf{D}^T \mathbf{D}. \tag{18.8}$$

The dimension of $\hat{\mathbf{r}}$ is $N_t \times N_t$, and it contains the correlation between different time samples of the windowed data. Recall that N_t is the number of samples in the window, which is usually smaller than the number of receivers, N_r. This fact justifies the abuse of notation of using a lower-case letter to denote the matrix $\hat{\mathbf{r}}$: this notation calls attention to the (usually) smaller dimension of $\hat{\mathbf{r}}$ when compared to the $N_r \times N_r$ matrix $\hat{\mathbf{R}}$. This fact also ensures that the computation of the largest eigenvector of $\hat{\mathbf{r}}$ is simpler.

As was done with S-MUSIC, using the approximation in (18.3) it can be shown [2] that the largest eigenvalue of $\hat{\mathbf{r}}$ is $\lambda_1 \approx \|\mathbf{s}\|^2 + \sigma^2$, associated with the eigenvector $\mathbf{u}_1 \approx \mathbf{s}$. Therefore, instead of testing whether the all-ones vector, $\mathbf{1}$, is the largest eigenvector of $\hat{\mathbf{R}}$, as is done in S-MUSIC, we may test whether $\mathbf{s}$ is the largest eigenvector of $\hat{\mathbf{r}}$. Note, however, that $\mathbf{s}$ is the waveform of the seismic reflection, which is unknown. To compute a coherence measure from $\hat{\mathbf{r}}$ without knowledge of $\mathbf{s}$, we use the fact that the eigenstructures of $\hat{\mathbf{R}}$ and $\hat{\mathbf{r}}$ are related. This allows us to show that [2] if $\mathbf{1}$ is the largest eigenvector of $\hat{\mathbf{R}}$, then $\mathbf{D}^T \mathbf{1}$ is the largest eigenvector of $\hat{\mathbf{r}}$.

The term $\mathbf{D}^T \mathbf{1}$ has an interesting interpretation. Indeed, let

$$\hat{\mathbf{s}} = \frac{1}{N_r} \mathbf{D}^T \mathbf{1}. \tag{18.9}$$

This operation computes the average of the columns of $\mathbf{D}^T$. Using again the approximation in (18.3), we see that this average should give us an estimate of the seismic wavelet $\mathbf{s}$. Now, recall that, if t_0 and v are correct, then all the traces in the window contain repetitions of $\mathbf{s}$. In consequence, we may interpret T-MUSIC as trying to verify whether the largest eigenvector of $\hat{\mathbf{r}}$ is proportional to the estimated wavelet $\hat{\mathbf{s}}$. The resulting T-MUSIC coherence measure is given by

$$P_T = \frac{\|\hat{\mathbf{s}}\|^2}{\|\hat{\mathbf{s}}\|^2 - |\hat{\mathbf{s}}^T \mathbf{u}_1|^2}, \tag{18.10}$$

where $\mathbf{u}_1$ is the largest eigenvector of $\hat{\mathbf{r}}$, with dimension $N_t \times 1$. If $\hat{\mathbf{s}}$ is the largest eigenvector of $\hat{\mathbf{r}}$, then P_T will be infinity.

18.3.5 NUMERICAL EXAMPLES

To illustrate the high resolution of the eigendecomposition-based coherence measures, we compute the semblance, S- and T-MUSIC velocity spectra for the marine CMP shown in Figure 18.2. The results are shown in Figure 18.4. For better visualization of the S- and T-MUSIC velocity spectra, we used the *semblance weighting* normalization function, presented in [2]. The number of traces in that CMP is $N_r = 30$ and in all the spectra from Figure 18.4 we used windows of $N_t = 15$ samples. The coherence functions were computed for each time sample at a sampling period of 4 ms, with velocities going from 1000 m/s to 3000 m/s in increments of 20 m/s. We can see that, for this data, both S- and T-MUSIC present higher resolution than semblance. Although the complexity of MUSIC is higher than the one from semblance, there are algorithms to decrease the implementations of MUSIC complexity [2]. In the computation of Figure 18.4, the complexity of T-MUSIC was approximately three times the one from semblance.

18.4 EVENT DETECTION SCHEMES FOR SEISMIC DATA ANALYSIS

Seismic processing involves many procedures that require the identification of seismic events. An example is the picking events in the velocity spectrum to establish the

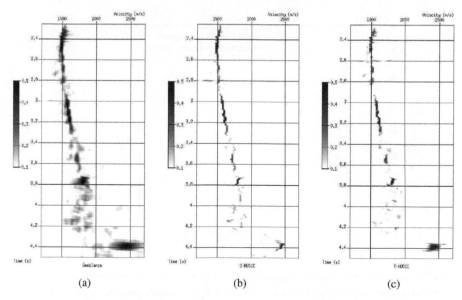

Figure 18.4 Velocity spectra for the marine CMP, computed with semblance (a), S-MUSIC (b), and T-MUSIC (c) coherence functions.

velocity profile in CMP data gathers. Another is the picking of reflections for inversion problems, such as seismic tomography [20, 42]. These are normally performed manually, by a human expert that examines each separate region of the data. However, in processing large data sets, manual picking may become a challenging and unworkable task. Thus, there is a large practical interest in methods for automatic picking.

A natural way for modeling and detecting seismic reflection events is to consider the coherence measure, usually represented by the semblance: large values of semblance are judged to represent an event, or at least a portion of data likely to contain one. The detection would then involve establishing a threshold, such that values of semblance above this threshold would be considered to be events. An interesting approach in order to define this threshold for event detection would be to use the probability density function (PDF) of the semblance. This would allow us to use the great body of literature in detecting radar reflections [15, 30]. The PDF of the semblance can be estimated via histogram evaluation. Figure 18.5 shows an example of a histogram obtained from real seismic data.

Interpreting Figure 18.5 as a PDF is equivalent to saying that the semblance S is a random variable. Observe in this figure that the events, represented by large values of S, are rare. In this example, 2.83% of the outcomes of $S(t_0, v)$ are greater than 0.3 and only 0.26% of the outcomes are greater than 0.5. In other words, the reflections can be characterized as sparse and, consequently, detection schemes really make sense in seismic data analysis. To quantify the level of sparsity, we use the Gini Index [14], which measures the inequality among values of a distribution. For non-negative

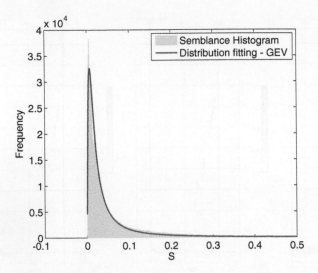

Figure 18.5 Statistical distribution of semblance. This figure shows the histogram of the semblance. We also show a generalized extreme value (GEV) pdf, which provides a good fit to the data.

samples, exactly the case of S, the Gini Index range is $[0, 1]$. Specifically for this example, the Gini Index is 0.622, indicating a strong inequality and, consequently, significant sparsity. In Figure 18.5, we also show the PDF of a generalized extreme value (GEV) distribution. As indicated in the figure, and in other experiments, the semblance seems to follow this distribution.

Nowadays, the semblance is widely used to detect events. However, the complication of this approach is the need to define a threshold, which depends upon a previous knowledge of the data set. Besides, in the deepest regions of the data, the reflections tend to present smaller alignment energies, so the event detection via coherence measure should consider a variable threshold value along the time axis.

One way to overcome the difficulty of defining a consistent threshold is to treat the problem as a hypothesis test. In this case, the threshold can be uniformly defined for the whole data volume through a probability index, such as a desired probability of detection. Furthermore, its definition becomes not just intuitive but also independent from the data quality. It is important to note that the hypothesis model analyzed in the sequel can be used to represent other classes of seismic events in any data gather configuration and, consequently, the detection scheme proposed here can be used to solve a vast list of seismic imaging problems.

The proposed detector is inspired by the data windowing discussed in Section 18.3. We propose the following hypotheses model for event detection in a data window:

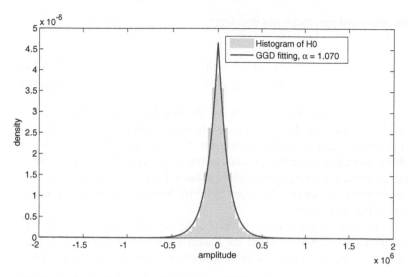

Figure 18.6 Analysis of the noise distribution. Notice that in this example, based on real data, the noise has a generalized Gaussian distribution very close to a Laplacian.

$$\mathcal{H}_0 \quad : \quad \mathbf{x}_k = w_0 \mathbf{n}_k \tag{18.11}$$

$$\mathcal{H}_1 \quad : \quad \mathbf{x}_k = A\,\mathbf{s}_k + w_1 \mathbf{n}_k \tag{18.12}$$

where $\mathcal{H}_0$ and $\mathcal{H}_1$ denote the null and alternative hypotheses, respectively, $\mathbf{x}_k \in \mathbb{R}^{N_r}$ represents a snapshot vector with data samples from the N_r sensors at instant k, for $k = 1, \ldots, N_t$ (i.e., $\mathbf{x}_k$ is a row in the windowed data matrix $\mathbf{D}^T$ in Figure 18.2(b) or, equivalently, $\mathbf{D} = [\mathbf{x}_1, \mathbf{x}_2, \ldots, \mathbf{x}_{N_t}]$), $\mathbf{n}_k$ represents the noise and interferers, and the factors w_0 and w_1 represent the noise scale for the null and alternative hypotheses, respectively. The vectors $\mathbf{s}_k \in \mathbb{R}^{N_r}$ represent the waveform of the signal/event along the space in the selected window. We assume these vectors are known, which is usually reasonable, since we expect that the amplitudes of the events in different receivers are similar. On the other hand, the amplitude of the event, parameter A, is inevitably unknown. Consequently, the PDF of $\mathcal{H}_1$ cannot be determined, and the Neymam-Pearson detector is unfeasible. Nevertheless, as the signal component in $\mathcal{H}_1$ is deterministic and the class of noise PDF may be obtained via distribution estimation, we can design suboptimal detectors through a prior analysis of the data set. In this text we focus on *Generalized Likelihood Ratio Test* (GLRT) detectors. It is important to highlight that the noise in the data sets we have analyzed tend to have a heavy-tailed distribution, usually fitting the Laplacian PDF. One example of this match is shown in Figure 18.6. This fact led us to adopt this PDF class to represent the noise model. This, in turn, results in the use of the l_1 norm in our detector, making it robust to outliers.

Adopting the linear model for the hypotheses in (18.11) and (18.12), assuming $\mathbf{s}_k = \mathbf{1}$ and incorporating the Laplacian distribution for the noise $\mathbf{n}_k$, we can design a

GLRT for the two-sided parameter test problem:

$$\mathcal{H}_0 \quad : \quad A = 0; \tag{18.13}$$

$$\mathcal{H}_1 \quad : \quad A \neq 0; \tag{18.14}$$

with $-\infty \leq A \leq \infty$ and data set given by the windowed data matrix $\mathbf{D} = [\mathbf{x}_1, \mathbf{x}_2, \ldots, \mathbf{x}_{N_t}]$. To make the detector completely independent from the data, we suppose that the scale factor w_0 of $\mathcal{H}_0$ is unknown. Note that in marine data this parameter could be estimated, since the existence of the water layer allows the isolation of noise samples (there is a time period with no reflected energy due to the two-way traveltime between the surface and the ocean bottom); however, we do not consider this possibility in our derivations.

The GLRTs are defined by the ratio of the probability density functions of $\mathcal{H}_1$ and $\mathcal{H}_0$, and these PDFs are computed using the maximum likelihood (ML) estimates (finite number of samples) of their unknown parameters. Thus, the ratio is given by

$$L_G(\mathbf{y}) = \frac{p(\mathbf{y}; \hat{A}, \hat{\sigma}_1^2, \mathcal{H}_1)}{p(\mathbf{y}; \hat{\sigma}_0^2, \mathcal{H}_0)} \tag{18.15}$$

where $\mathbf{y} \in \mathbb{R}^N$, with $N = N_r N_t$, is a vector with all N samples of the data window $\{x_k[m]\}$ for $m = 0, 1, \ldots, N_s - 1$ and $k = 1, 2, \ldots, N_t$.

The ML estimates (MLE) of A and σ_1^2 are obtained by maximizing $p(\mathbf{y}; A, \sigma_1^2, \mathcal{H}_1)$ and the MLE of σ_0^2 is obtained by maximizing $p(\mathbf{y}; \sigma_0^2, \mathcal{H}_0)$. Distinguishing the variance estimate in $\mathcal{H}_1$ with respect to $\mathcal{H}_0$ is needed because there is a bias in the former due to the presence of the signal with amplitude $A \neq 0$. It can be shown that, under the several assumptions made in this chapter, such as a Laplacian noise, the ML estimators are given by

$$\hat{A} \quad = \quad \mathrm{median}(\mathbf{y}) \triangleq y_{\mathrm{med}} \tag{18.16}$$

$$\hat{\omega}_1 \quad = \quad \frac{1}{N} \sum_{n=0}^{N-1} |y[n] - y_{\mathrm{med}}| \tag{18.17}$$

$$\hat{\omega}_0 \quad = \quad \frac{1}{N} \sum_{n=0}^{N-1} |y[n]| \tag{18.18}$$

where $\omega_1 = \sigma_1 / \sqrt{2}$ and $\omega_0 = \sigma_0 / \sqrt{2}$. The PDF in $\mathcal{H}_1$ is described by

$$p(\mathbf{y}; A, \omega_1, \mathcal{H}_1) \quad = \quad \left(\frac{1}{2\sigma_1^2} \right)^{\frac{N}{2}} \exp\left(-\frac{\sum_n |y[n] - A|}{\sqrt{\sigma_1^2/2}} \right) \tag{18.19}$$

$$= \quad \left(\frac{1}{2\omega_1} \right)^N \exp\left(-\frac{\sum_n |y[n] - A|}{\omega_1} \right). \tag{18.20}$$

Finally, after some further manipulations, the likelihood ratio for the hypotheses test takes the form

$$L_G(\mathbf{y}) = \left(\frac{\hat{\omega}_0}{\hat{\omega}_1}\right)^N. \tag{18.21}$$

A remarkable characteristic of this particular detector is that it has a constant false alarm ratio (CFAR). This makes it suitable to process any portion of the data, from the shallow to the deep reflection times, without suffering the problem of nonuniform alignment of reflected energy. The proof of CFAR is simple. Consider that $w[n] = \omega u[n]$ where $u[n]$ is Laplacian zero mean with unitary variance. Under $\mathcal{H}_0$ we have that $y[n] = \omega u[n]$, with $\omega = \sigma/\sqrt{2} > 0$, so that the likelihood ratio becomes

$$L_G^{\frac{1}{N}}(\mathbf{y}) = \frac{\frac{1}{N}\sum_n |\sigma u[n]|}{\frac{1}{N}\sum_n |\sigma u[n] - \sigma u_{\text{med}}|} \tag{18.22}$$

$$= \frac{\frac{1}{N}\sum_n |u[n]|}{\frac{1}{N}\sum_n |u[n] - u_{\text{med}}|} \tag{18.23}$$

whose PDF does not depend on σ^2. This detector is CFAR because its detection threshold is independent from the energy of the noise.

The asymptotic detection performance of the GLRT is given by a χ_1^2 (central Chi-squared PDF with 1 degree of freedom) under $\mathcal{H}_0$ and a $\chi_1'^2(\lambda)$ (noncentral Chi-squared PDF with 1 degree of freedom and noncentrality parameter λ) under $\mathcal{H}_1$, with $\lambda = 2NA^2/\sigma^2$ (see, e.g., [30, 15]). As the number of samples $N = N_r N_t$ is generally large for seismic data, the asymptotic PDFs for the test statistics are really representative. This approximation allows us to compute several performance metrics for the proposed detector, and to establish the threshold.

In the following we show the performance of the proposed GLRT detector when applied to the velocity analysis problem in real seismic data surveys. We considered data sets from two Brazilian Basins: i) offshore data from Jequitinhonha Basin; and ii) onshore data from the Amazon Tacutu Basin. Figure 18.7 shows the results of velocity analysis using the conventional procedure (Figure 18.7(a)) and the GLRT based procedure (Figure 18.7(b)), for a given CMP gather from the Jequitinhonha data set. Both measures range from 0 to 1, and suggest a reliability index about the presence of events in each time for each velocity being analyzed. The results for the GLRT are shown in terms of the probability of detection, which is given by the CDF (Cumulative Distribution Function) of the $\chi_1'^2(\lambda)$ random variable that approximates the statistics of the hypothesis test, when evaluated with the ML estimates $\hat{A}$ and $\hat{\sigma}_1$. To compute the probability of detection, this CDF was further masked by the generalized likelihood ratio L_G normalized by its maximum computed value. This mask is useful mainly for visual quality control because the estimated CDF becomes noisy in certain windows along the analysis where events are less likely to appear, and thus worse estimates of the amplitude A are obtained as the signal-to-noise ratio is decreased.

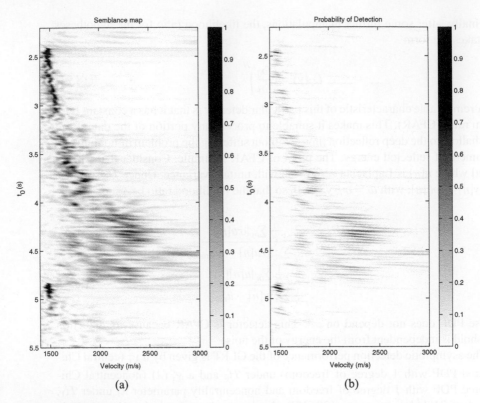

Figure 18.7 Coherence *versus* detection: (a) Semblance spectrum S; (b) Probability of detection.

Note that around 2.45 s and 4.9 s, the ocean bottom primary and multiple are precisely sensed in the case of the GLRT detector, while the semblance measure presents poor resolution in time and is unable to localize accurately the time of both reflections. Also note that the semblance and the GLRT detector present similar resolution in the velocity. This is because the detector uses no filtering process, like the subspace decomposition treated in Section 18.3.

As we process some kilometers of the seismic line we can verify the effect of the proposed detector. To see this, in Figure 18.8 we show a comparison of the conventional procedure and the GLRT detector for several neighboring CMP gathers of the same dataset. The time and midpoint pixels in the images consist, respectively, of the maximum value of the semblance for that time in the corresponding CMP gather, and the maximum probability of detection in the GLRT method. Notice the much cleaner image produced by the GLRT detector, which allows for a much better definition and detection of the events.

The results in Figure 18.8 represent an application to data sets of good quality, as commonly found in offshore surveys. A much more challenging scenario is found in onshore data. To show the efficacy of the detection based approach against the

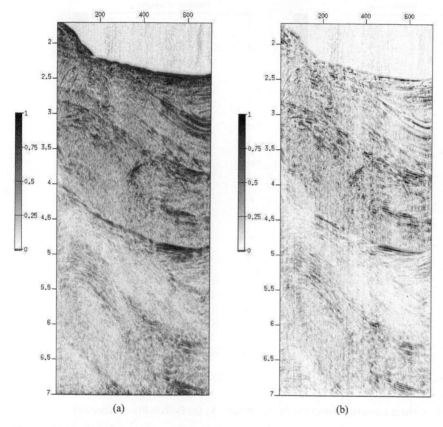

(a)   (b)

Figure 18.8 Coherence *versus* detection for about 8.8 km of the seismic line: (a) Semblance S; (b) Probability of detection.

conventional procedure in this case, we show in Figure 18.9 the velocity spectra for a given CMP gather of data from the Tacutu Basin, in Amazon, Brazil. The maximum fold is now 12 traces, as opposed to 60 in the previous results, which also had a much higher signal-to-noise ratio. Besides, the irregular acquisition geometry and the noise interference make this data set a challenging task to be solved with the conventional procedure.

Note in Figure 18.9 that the semblance is unable to assess adequately the available samples, failing to "see" several events. The difference lies just on how the information is exploited by the approaches since the traveltime model that defines the wavefront of an event within the data gather is exactly the same. More than the improved resolution in time, the GLRT detector really does a good job in sensing the presence of events, even in the deepest regions where the energy of reflection is quite weak. As shown in the figure, the GLRT detector again allows for much better detection.

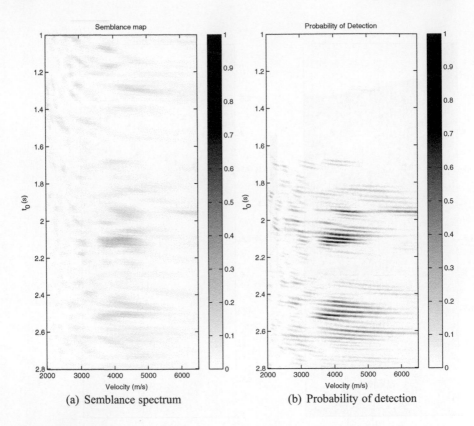

(a) Semblance spectrum   (b) Probability of detection

Figure 18.9 Coherence *versus* detection: (a) Semblance S; (b) Probability of detection.

One could say that the proposed detection scheme competes with the spectral-based high-resolution methods described in Section 18.3. This is not completely inaccurate, since both approaches aim at improving the resolution. However, their objectives are quite different. Consider the velocity analysis problem, where the data gather is an ensemble of traces with the same midpoint location. The spectral-based methods aim at estimating the best velocity at each t_0, while the detection scheme aims at obtaining a useful metric for the decision about the existence of an event at t_0. Although the velocity spectrum is a byproduct of the detector and a velocity profile could be extracted from this panel, note that the obtained metric is the result of a specific parameter test. The detector was designed to resolve the amplitude of events in time. No attention is given to improving the velocity estimation. On the contrary, attention is given to provide the best estimates of the amplitude of the events and also the noise power, which are needed to assess the likelihood ratio. If we could summarize into one question the job of each approach for the case of the velocity analysis problem, the closest ones would be: i) Spectral-based high-resolution methods: *Which is the best velocity at each recording time?* ii) Detector: *What is the probability of existence of an event at each recording time?*

18.5 2D DECONVOLUTION

We now turn to the third problem of interest in this chapter: using image processing methods to improve the resolution of migrated images. As mentioned in the introduction, the goal of reflection seismic techniques is to produce an image of the geological structures of the subsurface through the processing of a set of signals obtained from a sequence of seismic experiments [43]. Each seismic experiment produces traces that record seismic reflections that originated from geological formations in the subsurface [8]. However, the recorded seismic events generally do not correspond to waves that were reflected directly below to the receiver; instead, they are often reflected in geological formations located in different and previously unknown directions. The family of techniques that map these recorded seismic events to their true positions is known as seismic migration [8, 43]. In this section, we consider a particular migration technique called prestack depth migration (PSDM) using diffraction stacking [43]. For a formal introduction to this subject, we refer the reader to works such as [8, 43].

As with the rest of this chapter, our main concern here is with the resolution of the migrated image. It is well known that factors such as geologic complexities, limitations in acquisition geometry, and limits on the seismic signature bandwidth impose limits on the resolution of PSDM images. However, the distortions imposed by these limitations may be quantified by the so-called resolution function, as described in works such as [22, 9, 10, 34]. Under proper assumptions [9], the resolution function can be treated as a *point spread function* (PSF), so that the PSDM image can be modeled as the spatial convolution between the true reflectivity and the PSF. The interest in this interpretation is that PSFs have been widely used to model blurred images, and thus can be used to design filters that mitigate the blur. For a complete review on this subject, we refer the reader to works such as [1].

In this section, we exploit the parallel between the PSDM image and blurred images in general to propose methods to improve the resolution of migrated images. We initially present in Section 18.5.1, a brief review of the PSDM method with diffraction stacking. Next, in Section 18.5.2, we explain the concept of resolution functions, and how this results in the PSF and the 2D convolution model that describes the seismic image obtained by PSDM. In Section 18.5.3, we describe how the PSF can be used to perform 2D deconvolution of PSDM images. The goal is to use a filtering approach to mitigate the effects of the PSF, as in [37]. Finally, in Section 18.5.4, we present an example on synthetic data.

18.5.1 PSDM WITH DIFFRACTION STACKING

Let us consider first the imaging of a point scatterer embedded in a homogeneous model as in Figure 18.10. In this scenario, $\mathbf{r}_s = (s, 0)^T$, $\mathbf{r}_g = (g, 0)^T$ and $\mathbf{r} = (x, z)^T$ denote, respectively, the coordinates of the source (circle), the receiver (triangle), and the scatterer (asterisk), where we assume that the acquisition surface is at an elevation $z = 0$. If the medium velocity is v, then the traveltimes from the source to the image point and from the image point to the receiver are given respectively by

$$\tau_s(\mathbf{r}, \mathbf{r}_s) = \sqrt{(x_s - x)^2 + z^2}/v, \tag{18.24}$$

$$\tau_g(\mathbf{r}_g, \mathbf{r}) = \sqrt{(x_g - x)^2 + z^2}/v, \tag{18.25}$$

and the total traveltime is

$$\tau(\mathbf{r}_g, \mathbf{r}_s, \mathbf{r}) = \tau_s(\mathbf{r}, \mathbf{r}_s) + \tau_g(\mathbf{r}_g, \mathbf{r}) = \frac{\sqrt{(x_s - x)^2 + z^2} + \sqrt{(x_g - x)^2 + z^2}}{v}. \tag{18.26}$$

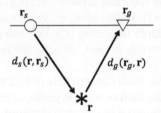

Figure 18.10 This figure shows a point scatterer, located in $\mathbf{r}$, embedded in a homogeneous model. The figure also shows the incident ray originating at the source in $\mathbf{r}_s$ and the scattered ray that reaches the receiver located in $\mathbf{r}_g$. $d_s(\mathbf{r}, \mathbf{r}_s)$ and $d_g(\mathbf{r}_g, \mathbf{r})$ represent the distances traveled by the respective rays.

Now, let us consider the acquisition geometry illustrated in Figure 18.11. In this scenario, the seismic pulse is generated by a single source, represented by a circle, and the response of the subsurface is recorded by a set of receivers represented by triangles. This family of traces is known as a common-shot (CS) gather. In this case, a point scatterer is located at $\mathbf{r} = (2000, 500)^T$ and the medium velocity is $v = 3000$ m/s. The resulting set of simulated traces is shown in Figure 18.12. The gray curve shows how the total traveltime of the diffracted wave varies as a function of the position of the receiver by using (18.26). Clearly, Figure 18.12 shows that the seismic events matches this traveltime.

In order to perform imaging, now we use the PSDM technique called diffraction stacking (for a more detailed description see e.g., [43]). As inputs for this procedure, there must be a seismic data with known acquisition geometry and a velocity model. In our example, the data is the CS gather presented in Figure 18.12, and the acquisition geometry is as shown in Figure 18.11 with $v = 3000$ m/s. Let us first consider the image point described by the cross in Figure 18.11, located at $\mathbf{r}' = (1800, 400)^T$. In diffraction stacking, each point in the migrated image is considered as a potential point scatterer. If there were a scatterer at $\mathbf{r}'$, then we would observe energies arriving at the receiver at times $\tau(\mathbf{r}_g, \mathbf{r}_s, \mathbf{r}')$ following (18.26). In Figure 18.13, the gray

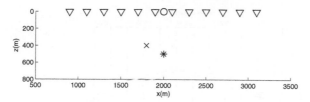

Figure 18.11 Acquisition geometry of a common-shot gather. The circle represents the source and the triangles the receivers. In this model, a point scatterer, represented by the asterisk, is embedded in a homogeneous model where $v = 3000$ m/s. The cross represents an image point.

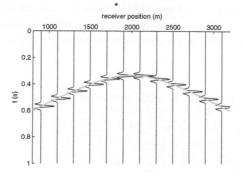

Figure 18.12 A common-shot (CS) gather resulting from the model described in Figure 18.11. The gray curve represents the total traveltime obtained by using (18.26) as a function of the receiver position. In this case, the source position is constant and we consider that the image point is coincident with the scatterer position and thus the curve follows the seismic events of the seismogram.

curve shows how $\tau(\mathbf{r}_g, \mathbf{r}_s, \mathbf{r}')$ varies as a function of the position of the receiver. Next, the amplitudes of the traces are summed or *stacked* along the curve and the result is assigned to the coordinate corresponding to $\mathbf{r}'$. If $\mathbf{r}'$ does correspond to a scatterer, then these amplitudes will add coherently, resulting in a large amplitude at that point in the migrated image. Otherwise, the result of stacking will have a small amplitude.

Given a region of interest and a spatial sampling rate, the diffraction stacking process is repeated for all image points $\mathbf{r}'$. The result for this process for the data in Figure 18.13 is shown in Figure 18.14. As expected, the amplitude resulting from the stacking at point $\mathbf{r}'$ in Figure 18.13 is not high, as shown in the position marked by a cross in Figure 18.14. In contrast to this, if $\mathbf{r}'$ is chosen to be at the scatterer position, then the respective $\tau(\mathbf{r}_g, \mathbf{r}_s, \mathbf{r}')$ curve will be as the one shown in Figure 18.12 and the sum of the amplitudes will be high. Thus, Figure 18.14 shows high amplitudes around the scatterer.

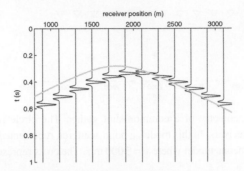

Figure 18.13 The same scenario as in Figure 18.11, but in this case the traveltime curve refers to the image point represented by a cross in Figure 18.11, which is not coincident with the scatterer position. Thus, as expected, the curve does not follow the seismic events of the seismogram.

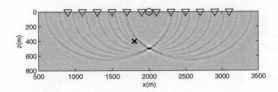

Figure 18.14 Result of diffraction stacking of the data acquired using the model displayed in Figure 18.11.

Observe that the result of PSDM in Figure 18.14 does not show a perfect point scatterer, but rather a blurred region with high amplitude around the scatterer. To show how the blur pattern depends on the acquisition geometry, we show in Figure 18.15 the result of diffraction stacking for the same subsurface model as in Figure 18.11, but with a different acquisition geometry, with all the receivers located to the right of the source. Comparing Figures 18.14 and 18.15, it is clear that the results depend on the acquistion geometry.

As a last observation, we note that reflections at the interfaces of geological layers are more common than point scatterers. Moreover, usually, the main aim of seismic signal processing is to accurately determine the structure of these interfaces in the subsurface. In spite of that, the diffraction stacking approach can still be used for this objective since, due to Huygens' principle [43], the response from the reflectors may be seen as the envelope of the response of an infinite number of point scatterers located along the reflector.

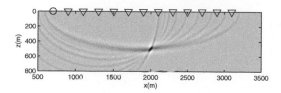

Figure 18.15 Result of diffraction stacking of CS gather data acquired using the model displayed in Figure 18.11, but with different positions of source (circle) and receivers (triangles), as indicated in this figure. Observe that the pattern formed around the scatterer position is different from the one obtained in Figure 18.14.

18.5.2 THE RESOLUTION FUNCTION, THE PSF AND THE 2D CONVOLUTION MODEL IN PSDM IMAGES

As seen in the examples in Section 18.5.1, PSDM images show a blurred pattern in response to a point scatterer, and thus offer only approximated information about the subsurface. These distortions occur due to limitations in the frequency bandwidth of the seismic wavelet and the acquisition geometry as well as complexities in the velocity field. As in [22, 9, 10], the effect of these limitations in the migrated image may be quantified with the use of the concept of resolution function. In this section, we briefly describe this concept, and how it allows us to obtain a PSF.

Let $s(\mathbf{r})$ be the reflectivity of the subsurface at a given point $\mathbf{r}$, and let $S(\mathbf{K})$ be its local Fourier transform, given by

$$S(\mathbf{K}) = \int_{\Omega} s(\mathbf{r}') \exp(-j\mathbf{K} \cdot \mathbf{r}') d\mathbf{r}', \qquad (18.27)$$

where $\mathbf{K}$ denotes the Fourier vector at the center point of interest $\mathbf{r}$, and $\mathbf{r}'$ are the spatial coordinates of points within a small region Ω around $\mathbf{r}$ [10]. The Fourier vector $\mathbf{K}$ is called *scattering number vector*. For each source-receiver pair, i.e., for each trace, $\mathbf{K}$ can be related to the acquisition geometry and the seismic wave propagation velocity model by

$$\mathbf{K}(\mathbf{r}_g, \mathbf{r}_s, \mathbf{r}) = \omega \mathbf{I}(\mathbf{r}_g, \mathbf{r}_s, \mathbf{r}), \qquad (18.28)$$

where ω correspond to the frequency of the source signature and $\mathbf{I}(\mathbf{r}_g, \mathbf{r}_s, \mathbf{r})$ is the *illumination vector* [21], illustrated in Figure 18.16 and given by

$$\mathbf{I}(\mathbf{r}_g, \mathbf{r}_s, \mathbf{r}) = \mathbf{p}_g(\mathbf{r}_g, \mathbf{r}) - \mathbf{p}_s(\mathbf{r}, \mathbf{r}_s) = \frac{\hat{\mathbf{u}}_g(\mathbf{r}_g, \mathbf{r}) - \hat{\mathbf{u}}_s(\mathbf{r}, \mathbf{r}_s)}{v(\mathbf{r})}, \qquad (18.29)$$

where $\mathbf{p}_s(\mathbf{r}, \mathbf{r}_s)$ and $\mathbf{p}_g(\mathbf{r}_g, \mathbf{r})$ are, respectively, the slowness vectors tangent to the rays connecting the source at $\mathbf{r}_s$ to the image point at $\mathbf{r}$ and the image point to the receiver located at $\mathbf{r}_g$. Also, $\hat{\mathbf{u}}_s(\mathbf{r}, \mathbf{r}_s)$ and $\hat{\mathbf{u}}_g(\mathbf{r}_g, \mathbf{r})$ are the corresponding unit-norm vectors and $v(\mathbf{r})$ is the velocity at $\mathbf{r}$. It is important to notice here that, given the velocity

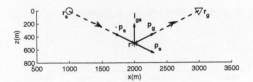

Figure 18.16 The illumination vector $\mathbf{I}_{gs}$ is calculated at the image point $\mathbf{r}$. The source and receiver are located at $\mathbf{r}_s$ and $\mathbf{r}_g$, respectively. $\mathbf{p}_s$ and $\mathbf{p}_g$ are the slowness vectors associated, respectively, with the rays connecting the source to the image point and the image point to the receiver.

model, the illumination vector can be computed for each source-receiver pair by ray tracing.

To illustrate the use of illumination vectors to obtain the PSF, Figure 18.17 shows the incident and scattered rays for the geometry used to generate Figure 18.15. The resulting illumination vectors are shown in Figure 18.18. Clearly, we cannot obtain a 360° coverage of the illumination vector $\mathbf{I}$ at the reflection point, and the same happens to the scattering wavenumber vector $\mathbf{K}$. This is because the positions of $\mathbf{r}_s$ and $\mathbf{r}_g$ are limited to a finite region of the surface. Also, because of the fact that $\mathbf{K}$ is proportional to ω, as shown in (18.28), the magnitude of $\mathbf{K}$ is also limited by the frequency band of the source signature.

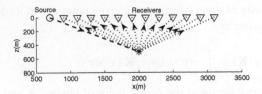

Figure 18.17 This figure shows the incident and scattered rays in a point scatterer using the same model used to generate Figure 18.15.

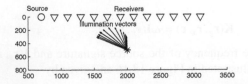

Figure 18.18 The illumination vectors calculated at the scatterer position corresponding to the traces obtained in Figure 18.17.

In general, a function $H(\mathbf{K})$ can be used to model the limitations in the coverage of $\mathbf{K}$. This function acts as a mask, so that $H(\mathbf{K})$ has non-zero values only in the directions of $\mathbf{K}$ allowed by the directions of the incident and scattered rays at $\mathbf{r}$ within a given acquisition and velocity model and in the magnitudes of $\mathbf{K}$ that are allowed by the bandwidth of the seismic wavelet. Under these assumptions, the PSDM image Fourier domain, $X(\mathbf{K})$, is given by [9, 10]:

$$X(\mathbf{K}) = S(\mathbf{K})H(\mathbf{K}). \tag{18.30}$$

Since a product in the Fourier domain corresponds to a convolution in space, the PSDM image can be locally described by

$$x(\mathbf{r}) = \int s(\mathbf{r}')h(\mathbf{r} - \mathbf{r}')d\mathbf{r}', \tag{18.31}$$

where $h(\mathbf{r})$ is the inverse Fourier transform of $H(\mathbf{K})$ given by

$$h(\mathbf{r}) = \int H(\mathbf{K})\exp(j\mathbf{K} \cdot \mathbf{r})d\mathbf{K}. \tag{18.32}$$

From (18.31), it is possible to verify that $x(\mathbf{r}) = s(\mathbf{r})$ if and only if $h(\mathbf{r}) = \delta(\mathbf{r})$, where $\delta(\mathbf{r})$ is a 2D Dirac delta, or $H(\mathbf{K}) = 1$ in the Fourier domain. However, this is not possible in practice because, as discussed previously, $H(\mathbf{K})$ is limited by the acquisition geometry and frequency bandwidth. Therefore, $h(\mathbf{r})$ is known as the *resolution function* [22, 9, 10], and it describes how the original model, $s(\mathbf{r})$, is distorted in the PSDM image. Also, in (18.31) it is assumed that PSDM acts locally as a linear and space invariant system around $\mathbf{r}$. Under this point of view, $h(\mathbf{r})$ can be seen as an impulse response of the PSDM around $\mathbf{r}$, i.e., it measures how a point of the original model is blurred or spread in the final image. Thus, under proper conditions [9], $h(\mathbf{r})$ becomes a point spread function (PSF).

In summary, to estimate the PSF using the concepts of illumination vectors and scattering wavenumber vectors, we use the following steps, having as inputs the seismic velocity model and the acquisition geometry, as required for PSDM, as well as an estimated spectrum of the seismic signature.

Step 1: Select an image point $\mathbf{r}$.
Step 2: For each trace (source-receiver pair) do:
  Step 2.1: Using ray tracing, estimate the ray from the source to the point $\mathbf{r}$ and the ray from the point $\mathbf{r}$ to the receiver. Then, estimate the illumination vector using (18.29). See Figures 18.16 or 18.18.
  Step 2.2: Map the magnitude spectrum of the wavelet into the wavenumber domain using (18.28).
Step 3: Since the PSDM image represents the superposition of the responses of individual traces (like the ones shown in Figures 18.14 and 18.15) the scattering wavenumber vectors corresponding to each trace must be summed. This result gives the PSF in the wavenumber domain, as in the example shown in Figure 18.19.
Step 4: Transform the PSF from the wavenumber domain to the space domain. If this is applied to the example in Figure 18.19, the result is displayed in Figure 18.20.

Comparing the result of this procedure, shown in Figure 18.20, to the pattern around the point scatterer observed in Figure 18.15, we verify that both patterns are similar in the vicinity of the scatterer.

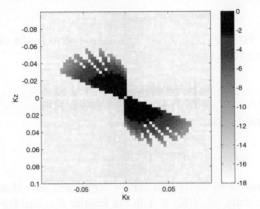

Figure 18.19 The 2D magnitude spectrum of the PSF resulting from the mapping of the magnitudes of the seismic signature spectrum into the illumination vectors in Figure 18.18 using (18.28).

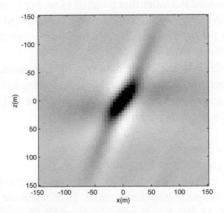

Figure 18.20 The PSF obtained in Figure 18.19 in the space domain. Note how this pattern is similar to the distortion around the scatterer in Figure 18.15.

From the point of view of the 2D convolutive model presented in (18.31), the examples shown in Figures 18.14 and 18.15 can be seen as the convolution between a reflectivity model with $s(\mathbf{r}) = \delta(\mathbf{r} - \mathbf{r}_0)$, where $\mathbf{r}_0$ is the coordinate of the point scatterer, and a PSF, which results in $x(\mathbf{r}) = h(\mathbf{r} - \mathbf{r}_0)$, i.e., a blurred version of the original model given by a PSF shifted to the scatterer position. However, the observed pattern is different for points more distant from the scatterer. This stems from the fact that the assumption that the PSF is spatially invariant is an approximation, as it is possible to infer from (18.29) and Figure 18.16 that the variation in position affects $\mathbf{p}_s$ and $\mathbf{p}_g$. However, this approximation is valid within a small region where these changes are negligible. If the changes in PSF become important, then (18.31) may be easily generalized to accommodate spatially variant PSFs.

18.5.3 2D DECONVOLUTION OF SEISMIC SECTIONS GENERATED BY PSDM

As shown in Sections 18.5.1 and 18.5.2, limitations on the acquisition geometry, seismic signature frequency band and geological complexities cause distortions in migrated images generated by PSDM. In this section, we introduce 2D deconvolution filters that aim to minimize the effects of these distortions. Techniques such as least-squares migration [23] and migration deconvolution [13] are based on the use of the wavefield modeling operator and the respective adjoint operator (migration). These approaches are very intensive in computer processing and often require simplifying procedures, such as using only the main diagonal of the so-called Hessian matrix, in order to make them usable in practice (for a more complete discussion see e.g., [40]). On the other hand, as described in Section 18.5.2, the distortions on sections obtained by PSDM can be modeled as a PSF [22, 9, 10, 21]. In works such as [10, 34], the concept of PSF is used for enhancing seismic sections obtained with PSDM, but only a small central vertical region of the PSF is considered. In this section, we describe an improvement of these works presented in [37, 36] in which a 2D linear filter is used to perform deconvolution. This approach allowed the use of the entire 2D PSF, in contrast to previous methods.

The 2D filtering approach is similar to the 1D case [43, 38], where a linear deconvolution filter is used to shape the original source signature into a narrower wavelet. In the 2D case, a wide PSF, $h(m,n)$, represented by $(2M + 1) \times (2N + 1)$ pixels is considered. Figure 18.21(a) shows an example of a PSF. To understand how the deconvolution filter can be designed, recall from (18.31) that the migrated image, $x(m,n)$ is the convolution between the desired reflectivity, $s(m,n)$ and the PSF. In the discrete form, $x(m,n)$ may be modeled as

$$x(m,n) = s(m,n) * *h(m,n) = \sum_{p=-M}^{M} \sum_{q=-N}^{N} s(m-p, n-q)h(p,q), \qquad (18.33)$$

where $**$ denotes a 2D convolution. If we apply a deconvolution filter $w(m,n)$ to $x(m,n)$, we obtain

$$y(m,n) = x(m,n) * *w(m,n) = s(m,n) * *h(m,n) * *w(m,n). \qquad (18.34)$$

Now, let the deconvolution filter $w(m,n)$ have size $(2P + 1) \times (2Q + 1)$, and let $c(m,n)$ be the combined response of the PSF and the filter. Then, $c(m,n)$ is given by the 2D convolution between $h(m,n)$ and $w(m,n)$:

$$c(m, n) = h(m, n) * *w(m, n) = \sum_{p=-P}^{P} \sum_{q=-Q}^{Q} h(m - p, n - q)w(p, q). \qquad (18.35)$$

Using (18.34), the response of the deconvolution filter may be designed so that

$$h(m, n) * *w(m, n) \approx \delta(m, n), \qquad (18.36)$$

where $\delta(m, n)$ is the 2D discrete delta function. In this case, we would have

$$y(m, n) = x(m, n) * *w(m, n) = s(m, n) * *h(m, n) * *w(m, n) \approx s(m, n). \qquad (18.37)$$

Figure 18.21(b) shows an example of $w(m, n)$ designed according to this criterion, while Figure 18.21(c) shows the resulting combined response $c(m, n)$. Note how $c(m, n)$ approaches an ideal impulse, shown in Figure 18.21(d). Note that the combined response can only be a perfect impulse if $w(m, n)$ had infinite size, analogous to the 1D case.

As presented in [37], $w(m, n)$ may be calculated using a least-squares approach. This can be done by rearranging the values of the pixels of 2D images into 1D vectors using the lexicographic ordering [34, 37], which allows us to express (18.35) as

$$\mathbf{c} = \mathbf{Hw}, \qquad (18.38)$$

where $\mathbf{w}$ and $\mathbf{c}$ are vectors representing $w(m, n)$ and $c(m, n)$ and $\mathbf{H}$ is the 2D convolution matrix corresponding to $h(m, n)$ [36]. If we consider δ as the vector representation of $\delta(m, n)$, then the optimal least-squares filter is obtained by solving

$$\min_{\mathbf{w}} \|\delta - \mathbf{Hw}\|^2 + \lambda\|\mathbf{w}\|^2, \qquad (18.39)$$

where $\lambda\|\mathbf{w}\|^2$ is a regularization term. Thus, the 2D deconvolution filter is given by

$$\mathbf{w} = (\mathbf{H}^T\mathbf{H} + \lambda\mathbf{I})^{-1}\mathbf{H}^T\delta. \qquad (18.40)$$

18.5.4   EXAMPLE ON SYNTHETIC DATA

In this section, we illustrate the performance of the proposed deconvolution method on synthetic data referring to a homogeneous model with two close scatterers, as shown in Figure 18.22. As seen in this figure, the acquisition geometry has the source located in the origin and the receivers ranging from $(-1600, 0)^T$ to $(1600, 0)^T$, with a spacing of 10 m. In this model, the velocity is $v = 2000$ m/s and the scatterers are located at $(-20, 2000)^T$ and $(30, 2000)^T$. The resulting CS gather is shown in Figure 18.23.

The result of PSDM on the CS gather is shown in Figure 18.24(a). As seen in this figure, the distance between the scatterers is small enough that they are diffraction

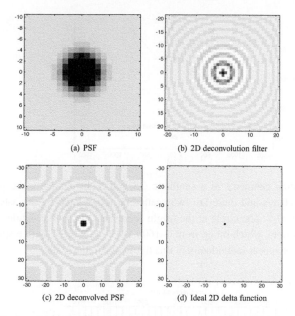

(a) PSF   (b) 2D deconvolution filter

(c) 2D deconvolved PSF   (d) Ideal 2D delta function

Figure 18.21   In an analogy to the 1D case, the 2D deconvolution filter is used to transform the wide original PSF into a narrower PSF, ideally a 2D delta function. This is not possible in practice, as this would require an infinite length filter.

limited, in the sense that only one lobe is displayed on the result, and it is not possible to identify two scatterers. Figure 18.24(b) shows the result of the 2D deconvolution method described in Section 18.5.3. The two scatterers are now separated, but high frequency artifacts are also present in the result. As shown in [37], the reason for this becomes clear if we analyze the PSF and the 2D deconvolution filter in the Fourier domain, as shown in Figures 18.25(a) and 18.25(b). As seen in (18.40), the 2D deconvolution filter is a least-squares approximation of the inverse filter of the PSF, so it has large magnitudes in frequencies that have low magnitude in the PSF. However, from (18.28) and (18.29), we observe that the components with small magnitudes in the Fourier domain represent frequencies outside of the seismic wavelet frequency band, or directions where there is no illumination. Thus, these components indicate data that do not contain information on the desired reflectivity function $s(m, n)$. It is the enhancement of these components by the deconvolution that leads to the artifacts. To mitigate this effect, we use a tapered mask in the Fourier domain, such as the one shown in Figure 18.25(c). This mask ensures that $w(m, n)$ does not try to enhance frequencies that have zero amplitude in the PSF. This is achieved by setting to zero the corresponding values of the Fourier transform of the deconvolution filter. The result of filtering the 2D deconvolved image with the mask is shown in Figure 18.24(c). It is possible to observe that the artifacts are attenuated, while the two scatterers are still differentiable.

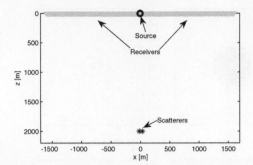

Figure 18.22 Acquisition geometry of a synthetic CS gather. In this case, the propagation velocity is $v = 2000$ m/s and there are two scatterers, represented by the asterisks at $(-20, 2000)^T$ and $(30, 2000)^T$. The circle at the origin represents the source, while the gray region represent the receivers, which range from $(-1600, 0)^T$ to $(1600, 0)^T$ with a spacing of 10 m. The receivers do not appear individually in this figure due to the scale.

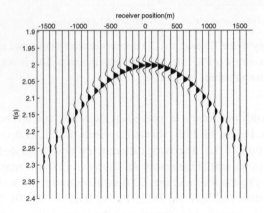

Figure 18.23 Synthetic CS section acquired with the geometry displayed in Figure 18.22.

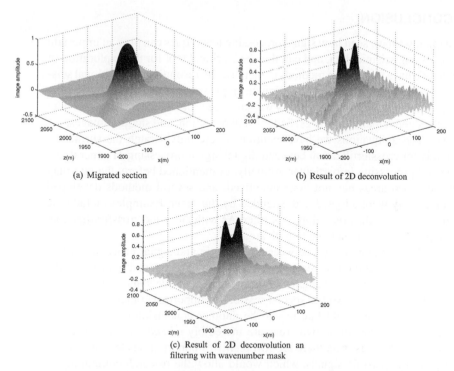

(a) Migrated section

(b) Result of 2D deconvolution

(c) Result of 2D deconvolution an filtering with wavenumber mask

Figure 18.24 PSDM results for the data in Figure 18.23. The migrated section shows only one lobe, so it is not possible to differentiate the two scatterers in this section. The result of 2D deconvolution shows the two scatterers separately, but high frequency artifacts are also produced. The use of the wavenumber mask displayed in Figure 18.25(c) attenuates these artifacts.

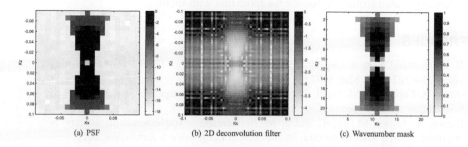

(a) PSF

(b) 2D deconvolution filter

(c) Wavenumber mask

Figure 18.25 2D magnitude spectra of the PSF, deconvolution filter, and wavenumber mask.

18.6 CONCLUSION

In this chapter, we have seen how signal processing techniques can play an important role in processing seismic data, and can thus be a critical tool for the exploration of oil and gas. We have focused on methods to increase the resolution of traditional seismic processing techniques, thereby showing the improvements that signal processing may bring to this important problem. As we have seen, the methods discussed here can provide better velocity estimates, better event detection, and sharper images.

However, the applications we have mentioned here only scratch the surface of the potentials for crossfertilization between digital signal processing in general and seismic signal processing in particular. Actually, as mentioned before, the interplay between these two areas has not gone unnoticed, and several methods developed by one community were adopted and extended by the other. Examples include the wavelet transform and the prediction error filter. As with any good crossfertilization, this exchange of ideas is beneficial to both areas.

In recent decades, the contact between the seismic and signal processing areas has faded somehow, but the last years have seen a rekindled interest in bringing the communities together again. Examples of these efforts include some special issues in journals of both areas, as well as special sessions in important conferences. Along these lines, perhaps we would like this to be another conclusion of this chapter: as our applications show, these two areas are incredibly related, and are based on similar tools and methods. And there are still plenty of opportunities to apply signal processing tools to seismic signals, which would allow the research community to advance the state of the art in both areas.

ACKNOWLEDGMENTS

The authors would like to thank the financial support of Petrobras, CNPq, CAPES, FAPESP, and INCT-GP, each of which funded different parts of this project. We would also like to acknowledge Professors Leiv Gelius and Isabelle Lecomte from University of Oslo (UiO) for sharing their knowledge with us. Many of the ideas in this chapter originated from discussions we have had with them.

REFERENCES

1. M. R. Banham and A. K. Katsaggelos. Digital image restoration. *IEEE Signal Processing Magazine*, 14(2):24 –41, Mar. 1997.
2. T. Barros, R. Lopes, and M. Tygel. Implementation aspects of eigendecomposition-based high-resolution velocity spectra. *Geophysical Prospecting*, 2014.
3. T. Barros, R. Lopes, M. Tygel, and J. M. T. Romano. Implementation aspects of eigenstructure-based velocity spectra. In *74th EAGE Conference & Exhibition*, 2012.
4. G. Bienvenu and L. Kopp. Adaptivity to background noise spatial coherence for high resolution passive methods. In *IEEE International Conference on Acoustics, Speech, and Signal Processing, ICASSP '80*, volume 5, pages 307–310, Apr. 1980.
5. B. L. Biondi and C. Kostov. High-resolution velocity spectra using eigenstructure methods. *Geophysics*, 54(7):832–842, 1989.

6. R. B. Blackman and J. W. Tukey. The measurement of power spectra, from the point of view of communications engineering. *Dover Publications.* New York. [1959, c1958].

7. J. W. Cooley and J. W. Tukey. An algorithm for machine calculation of complex Fourier series. *Mathematics of computing* 19:297–301, Apr. 1965.

8. J. Gazdag and P. Sguazzero. Migration of seismic data. *Proceedings of the IEEE,* 72(10):1302–1315, 1984.

9. L. J. Gelius and I. Lecomte. The resolution function in linearized born and kirchhoff inversion. In PerChristian Hansen, BoHolm Jacobsen, and Klaus Mosegaard, eds., *Methods and Applications of Inversion,* volume 92 of *Lecture Notes in Earth Sciences,* pages 129–141. Springer Berlin Heidelberg, 2000.

10. L. J. Gelius, I. Lecomte, and H. Tabti. Analysis of the resolution function in seismic prestack depth imaging. *Geophysical Prospecting,* 50(5):505–515, 2002.

11. R. C. Gonzalez and R. E. Woods. *Digital image processing.* Prentice Hall Upper Saddle River, NJ: 3rd edition, 2007.

12. P. Goupillaud, A. Grossman, and J. Morlet. Cycle-octave and related transforms in seismic signal analysis. *Geoexploration,* 23:85–102, 1984.

13. J. Hu, G. T. Schuster, and P. A. Valasek. Poststack migration deconvolution. *Geophysics,* 66(3):939–952, 2001.

14. N. Hurley and S. Rickard. Comparing measures of sparsity. *IEEE Transactions on Information Theory,* 55(10):4723–4741, 2009.

15. S. M. Kay. *Fundamentals of statistical signal processing, Vol. II: Detection Theory.* Prentice Hall Signal Processing Series. Upper Saddle River, NJ (Prentice-Hall PTR), 1998.

16. S. M. Kay and S. L. Marple Jr. Spectrum analysis-a modern perspective. *Proceedings of the IEEE,* 69(11):1380–1419, 1981.

17. R. L. Kirlin. The relationship between semblance and eigenstructure velocity estimators. *Geophysics,* 57(8):1027–1033, 1992.

18. R. L. Kirlin and W. J. Done. *Covariance Analysis for Seismic Signal Processing (Geophysical Development No. 8).* Society Of Exploration Geophysicists, 1999.

19. H. Krim and M. Viberg. Two decades of array signal processing research: the parametric approach. *IEEE Signal Processing Magazine,* 13(4):67–94, Jul. 1996.

20. G. Lambaré. Stereotomography. *Geophysics,* 73(5):VE25–VE34, 2008.

21. I. Lecomte. Resolution and illumination analyses in PSDM: A ray-based approach. *The Leading Edge,* 27(5):650–663, 2008.

22. I. Lecomte and L. J. Gelius. Have a look at the resolution of prestack depth migration for any model, survey and wavefields. *Society of Exploration Geophysicists,* pages 1112–1115, 1998.

23. T. Nemeth, C. Wu, and G. T. Schuster. Least-squares migration of incomplete reflection data. *Geophysics,* 64(1):208–221, 1999.

24. S. C. Park, M. K. Park, and M. G. Kang. Super-resolution image reconstruction: a technical overview. *IEEE Signal Processing Magazine,* 20(3):21–36, 2003.

25. E. Parzen. Mathematical considerations in the estimation of spectra. *Technometrics,* 3(2):167–190, 1961.

26. E. Parzen. Statistical spectral analysis (single channel case) in 1968. Technical report, Dep. Statistics, Stanford Univ., Stanford, CA, June, 1968.

27. V. F. Pisarenko. The retrieval of harmonics from a covariance function. *Geophysical Journal of the Royal Astronomical Society,* 33:347–366, 1973.

28. R. C. Puetter, T.R. Gosnell, and A. Yahil. Digital image reconstruction: deblurring and denoising. *Annual Review of Astronomy and Astrophysics,* 43:139–194, 2005.

29. E. A. Robinson. The MIT geophysical analysis group (GAG) from inception to 1954. *Geophysics*, 70(4):7JA–30, 2005.

30. L.L. Scharf and B. Friedlander. Matched subspace detectors. *IEEE Transactions on Signal Processing*, 42(8):2146–2157, 1994.

31. R. Schmidt. A signal subspace approach to multiple emitter location and spectral estimation. *Ph. D. Thesis, Stanford University*, Nov. 1981.

32. R. Schmidt. Multiple emitter location and signal parameter estimation. *Antennas and Propagation, IEEE Transactions on*, 34(3):276–280, 1986.

33. A. Schuster. The periodogram of magnetic declination as obtained from the records of the Greenwich Observatory during the years 1871-1895. *Cambridge, Philosophical Transactions of the Royal Society*, 18:107–135, 1889.

34. T. A. Sjoeberg, L.J. Gelius, and I. Lecomte. 2-D deconvolution of seismic image blur, *Society of Exploration Geophysicists*, chapter 269, pages 1055–1058. 2003.

35. P. Stoica and L. R. Moses. *Spectral analysis of signals*. Prentice Hall, 1st edition, 2005.

36. A. K. Takahata. *Unidimensional and Bidimensional Seismic Deconvolution*. Ph.d. thesis, University of Campinas, 2014.

37. A. K. Takahata, L.J. Gelius, R. R. Lopes, M. Tygel, and I. Lecomte. 2D spiking deconvolution approach to resolution enhancement of prestack depth migrated seismic images. In *75th EAGE Conference & Exhibition*, 2013.

38. A. K. Takahata, E. Z. Nadalin, R. Ferrari, L. T. Duarte, R. Suyama, R. R. Lopes, J. M. T. Romano, and M. Tygel. Unsupervised processing of geophysical signals: A review of some key aspects of blind deconvolution and blind source separation. *IEEE Signal Processing Magazine*, 29(4):27–35, 2012.

39. M. Turhan Taner and F. Koehler. Velocity spectra—digital computer derivation applications of velocity functions. *Geophysics*, 34(6):859–881, 1969.

40. Y. Tang. Target-oriented wave-equation least-squares migration/inversion with phase-encoded hessian. *Geophysics*, 74(6):WCA95–WCA107, 2009.

41. M. Tygel and L.T. Santos. Quadratic normal moveouts of symmetric reflections in elastic media: a quick tutorial. *Studia Geophysica et Geodaetica*, 51:185–206, 2007.

42. M. Woodward, D. Nichols, O. Zdraveva, P. Whitfield, and T. Johns. A decade of tomography. *Geophysics*, 73(5):VE5–VE11, 2008.

43. O. Yilmaz. *Seismic Data Analysis: Processing, Inversion, and Interpretation of Seismic Data, Volume 1*. Investigations in Geophysics. Society of Exploration Geophysicists, Tulsa, 2nd edition, 2001.

19 Synthetic Aperture Imaging for Ultrasonic Non-Destructive Testing

Cláudio Kitano
UNESP

David Romero-Laorden
CSIC-Madrid

Oscar Martínez-Graullera
CSIC-Madrid

Ricardo Tokio Higuti
UNESP

Silvio Cesar Garcia Granja
UNEMAT

Vander Teixeira Prado
UTFPR-CP

Non-destructive testing (NDT) consists of the inspection of materials/structures to check the possible existence of defects without producing changes to the object under inspection. Ultrasound is widely used in NDT in areas such as aerospace, petrochemical, and power generation, which require thorough inspection due to the high safety levels of operation. It has several advantages: reliability, ease of operation and speed of the test, high sensitivity, and it does not produce ionizing radiation and can propagate in solids, liquids, and gases. Ultrasonic NDT involves the use of bulk (longitudinal or shear), surface, or guided waves. They can be used to detect internal and external defects such as cracks, corrosion, delamination, and holes, and to measure material properties, such as elastic constants, for example.

Conventional ultrasonic NDT techniques generally employ a single element working in pulse-echo or a pair of elements operating in transmit-receive mode. In these cases, to inspect the entire structure, mechanical scans of the sensor across the region of interest are required, producing B-scan or C-scan images [16]. An array is a set of transducers geometrically arranged (linear or two dimensional, for example), whose acoustic beam can be electronically controlled [6]. Arrays are widely applied in NDT due to characteristics as beam control (beam focusing and deflection, lateral resolution, apodization) and speed of test, producing images of the structure and its defects [34, 5, 18, 31, 1, 25, 15, 38, 36, 28].

The quality of the image is limited by the array radiation pattern (main lobe width and side lobes levels), which in turn is dependent on the array characteristics, such as the number of elements and the inter-element spacing (pitch). Other factors that influence the quality of the image are the image beamforming technique, signal-to-noise ratio (SNR), and the existence of simultaneous propagation modes with different characteristics.

When all array elements, or part of them, are excited simultaneously, a transmit/receive channel dedicated to each element is required, which reduces data acquisition time and processing significantly. On the other hand, it increases the complexity and the cost of the system. This setup is called phased array (PA). PA systems allow the focusing at each point (x, z) of interest in transmission and/or reception.

An alternative to PA is the use of synthetic aperture (SA) techniques, in which one or a few transmit/receive channels are multiplexed, and the data is post-processed to obtain the image [37]. It is based on the technique of synthetic aperture radar (SAR), in which an aircraft captures several images, which are post-processed to increase the resolution, yielding a final image with a resolution equivalent to that of an antenna having the same dimensions of the aircraft displacement during the acquisitions [30]. In this technique the hardware complexity is reduced, without the necessity of reducing the number of elements of the aperture. Other characteristics are: possibility of dynamic focusing in transmission and reception, and the need of post-processing.

The focus of this chapter is to present the concept of SA beamforming and some techniques for improving defect detection with reduced hardware complexity. Section 19.1 introduces the beamforming process with PA and SA techniques, as well as the coarray concept (effective aperture) and the array radiation pattern. A coherence method for improving reflectors detectability and the quality of the images is presented in Section 19.2. In Section 19.3 two methods to reduce system complexity, with smaller number of signals to storage and processing are detailed. Section 19.4 presents the experimental setup, as well as the results and discussions. The final section presents general comments.

19.1 BEAMFORMING IN ULTRASONIC ARRAY IMAGING SYSTEMS

The beamforming process occurs when the action of several sources grouped in an array is compensated in time to change the acoustic beam at a determined location. Following a focusing law, this process can be used to focus or steer the acoustic beam in a region of interest by exciting the elements at different times. Figure 19.1 illustrates a linear array with eight elements. A subaperture of four active elements is used as an example. If all elements are excited at the same time, as presented in Figure 19.1(a), an approximate plane wave is generated in front of the aperture. In Figure 19.1(b), a focusing action at point (x_p, z_p) is performed. In this case, the elements that are more distant of the focus point are excited first, so that all signals arrive at the focus point in phase. If the elements are sequentially excited with a

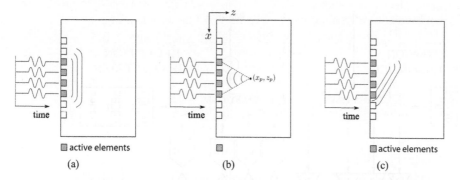

Figure 19.1 Different wave fronts generated from a linear array: (a) plane wave, (b) focused beam at (x_p, z_p), and (c) beam steering.

constant delay between neighbors, the acoustic beam can be steered, as observed in Figure 19.1(c).

19.1.1 BEAMFORMING WITH PHASED ARRAY SYSTEM

A generic PA imaging system is composed by a set of N transducers organized in a fixed distribution and two processing stages: the emission beamformer, that delays the transmitted pulse of each transducer to compensate the differences in the path from the transmitters to a determined location or emission focus point (x_{f_E}, z_{f_E}); and the reception beamformer, that introduces for each transducer a delay line to compensate the differences in the path from a determined location or reception focus point (x_{f_R}, z_{f_R}) to the receivers. A general picture of an electronic system that implements a PA beamformer is presented in Figure 19.2.

By considering that an ideal reflector exists at position (x_p, z_p), as illustrated in Figure 19.1(b), the signal generated by a PA imaging system of N elements organized in a linear array can be implemented by these two stages. The first stage of the beamforming is executed to compensate the propagation delay between elements in emission. This operation is called focusing in transmission. The signal in emission that reaches the focus can be written as:

$$S_E(t, x_{f_E}, z_{f_E}) = \sum_{e=0}^{N-1} s(t) * \delta(t - \tau_e(x_p, z_p)) * \delta(t - T_e(x_{f_E}, z_{f_E})), \qquad (19.1)$$

where $*$ is the convolution operator, $\delta(t)$ is the Dirac delta function, the index e is referred to the elements in emission of the aperture, $s(t)$ is the ultrasonic pulse shape, $\tau_e(x_p, z_p)$ is the propagation time from the eth element to the reflector position (x_p, z_p), and $T_e(x_{f_E}, z_{f_E})$ is the focus delay in emission to focus at (x_{f_E}, z_{f_E}) position, that is described by:

$$T_e(x, z) = \tau_e(x, z) - \max\{\tau_i(x, z)\}, \quad \text{for} \quad i = 0, 1, ..., N - 1. \qquad (19.2)$$

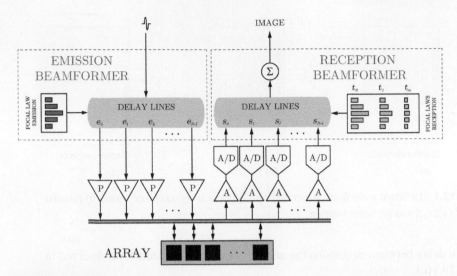

Figure 19.2 Generic description of a beamforming system. Two beamforming stages are identified: emission beamforming that introduces delays in the emitted pulses; and the reception beamforming where the data of each channel can be rearranged in real time to compensate the delays in reception.

Then, the received signal by the rth element from a reflector located at (x_p, z_p) is:

$$S_r(t) = S_E(t, x_{f_E}, z_{f_E}) * \delta(t - \tau_r(x_p, z_p)), \tag{19.3}$$

where $\tau_r(x_p, z_p)$ is the propagation time from the reflector position (x_p, z_p) to the rth element of the array.

By considering all received signals, in the second stage, the reception beamforming process can be applied as:

$$S_R(t, x_{f_E}, z_{f_E}, x_{f_R}, z_{f_R}) = \sum_{r=0}^{N-1} S_r(t) * \delta(t - T_r(x_{f_R}, z_{f_R})), \tag{19.4}$$

where the index r is referred to the transducers working in reception and $T_r(x_{f_R}, z_{f_R})$ is the reception focusing delay. The expression into the sum in (19.4) can be considered as the signal received by the rth channel. Although (x_{f_E}, z_{f_E}) and (x_{f_R}, z_{f_R}) can be different, it is a good practice to locate (x_{f_E}, z_{f_E}) at the same steering angle where the emitted energy is concentrated.

The array generates its maximum lateral resolution around the focusing point and its action is limited to the focal region, determined by [16]:

$$\Delta z = k_{pf} * \lambda \frac{|x_f|^2}{D^2}, \tag{19.5}$$

where Δz is the depth where the focus is maintained, k_{pf} is a constant that depends on the aperture geometry, λ is the wavelength, and D is the aperture size.

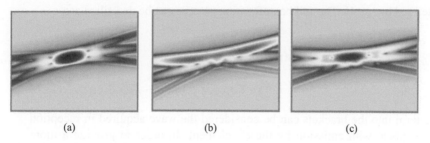

(a)   (b)   (c)

Figure 19.3 (a) $(x_{f_R}, z_{f_R}) = (x_{f_E}, z_{f_E}) = (x_p, z_p)$; (b) $(x_{f_R}, z_{f_R}) = (x_{f_E}, z_{f_E}) \neq (x_p, z_p)$; (c) $(x_{f_R}, z_{f_R}) = (x_p, z_p) \neq (x_{f_E}, z_{f_E})$.

When $(x_{f_R}, z_{f_R}) = (x_{f_E}, z_{f_E}) = (x_p, z_p)$ the system produces at this position its better performance (sonification level and lateral resolution) generating a high-quality image around that region.

An important simulation tool for analysis is the Point Spread Function (PSF), which is the image of a point reflector, i.e., it is observed how a point reflector at a certain position is visualized (contrast, resolution, artifacts) by the array with the beamforming technique. Figure 19.3 shows three different PSFs considering: (a) $(x_{f_R}, z_{f_R}) = (x_{f_E}, z_{f_E}) = (x_p, z_p)$, where the maximum quality is obtained; (b) $(x_{f_R}, z_{f_R}) = (x_{f_E}, z_{f_E}) \neq (x_p, z_p)$, that is an out-of-focus region where amplitude and resolution are highly reduced because of the elongation of the pulse caused by defocusing; (c) $(x_{f_R}, z_{f_R}) = (x_p, z_p) \neq (x_{f_E}, z_{f_E})$ where at least one of the focus regions contains the reflector point.

Both beamforming operations perform the same action, by compensating the propagation delays between elements. However, while the focus in emission is limited to a region determined by the focusing law, the reception beamforming operates over the acquired data and can be dynamically adapted, providing dynamic focusing capability and compensating part of the degradation process at least in the steering angle determined by the emission focus.

19.1.2 BEAMFORMING WITH SYNTHETIC APERTURE SYSTEM

The images of Figure 19.3 show that the image generation process in PA is mainly determined by the emission process, and the quality of the image far from the emission focus is degraded. To extend the efficiency of the focusing process it is necessary avoid the conventional emission focusing in the medium and find an alternative to compensate the propagation emission delays.

Equation (19.4) can be rearranged to decompose the emission beamforming process defined by (19.1). The full beamforming process can be described as:

$$S_R(t, x_{f_E}, z_{f_E}, x_{f_R}, z_{f_R}) = \sum_{e=0}^{N-1} \sum_{r=0}^{N-1} \left[s(t) * \delta(t - \tau_e(x_p, z_p)) * \delta(t - \tau_r(x_p, z_p)) \right]$$

$$* \delta(t - T_e(x_{f_E}, z_{f_E})) * \delta(t - T_r(x_{f_R}, z_{f_R})), \quad (19.6)$$

where the term into the brackets can be considered the wave acquired in reception by the rth element with emission by the eth element. In order to provide a more general expression, the reflected wave can be extended to an undetermined number of reflectors, each one with its own reflectivity coefficient R_i:

$$s_{er}(t) = s(t) * \sum_i R_i \delta(t - T_e(x_i, z_i)) * \delta(t - T_r(x_i, z_i)). \quad (19.7)$$

In essence this rearrangement means that all signals involved in the beamforming process can be acquired independently by pairs of emission-reception transducers. Once all data have been stored, the beamforming operation can be addressed to compensate at each point of the region of interest in emission and reception simultaneously, synthesizing the performance of a full aperture. This process is known as synthetic aperture imaging technique [7, 2, 14, 12] and has significant implications in the electronic system and signal processing.

By considering a linear array of N elements and all combinations of transmitter and receiver data $s_{er}(t)$, the emitted signal by element e and received by element r, the ultrasonic image at point (x, z) is obtained by:

$$I(x, z) = \frac{1}{N^2} \sum_{e=0}^{N-1} \sum_{r=0}^{N-1} s_{er}(t) * \delta\left(t - [\tau_e(x, z) + \tau_r(x, z)]\right), \quad (19.8)$$

which can also be written as:

$$I(x, z) = \frac{1}{N^2} \sum_{e=0}^{N-1} \sum_{r=0}^{N-1} s_{er}(\tau_e(x, z) + \tau_r(x, z)). \quad (19.9)$$

As the waveforms are delayed by $\tau_e(x, z) + \tau_r(x, z)$ and summed for every combination of emitter-receiver, this beamforming technique is also known as delay-and-sum (DAS).

When all N^2 possible combinations of emitter-receiver pairs are used in the image beamforming, a 3D data structure is required, where one axis is the emission channel, the second one is the reception channel, and the third one is the acquisition time. This data structure is known as Full Matrix Array (FMA). The process to obtain an image by beamforming over this FMA structure, described in (19.8) and (19.9), is known as Total Focusing Method (TFM) and it is able to produce an image where all image points are focused [9, 10, 11, 26, 6]. TFM produces a high-quality image but it needs a high degree of computational power to be implemented.

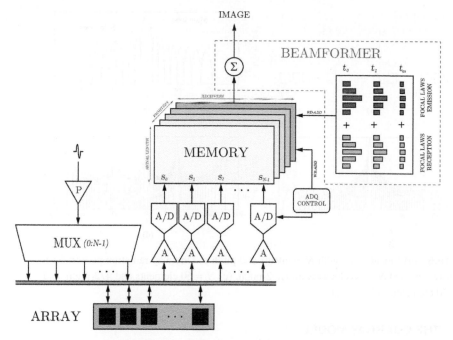

Figure 19.4 Generic description of a beamforming system for synthetic aperture image, with one pulser multiplexed over N channels and N receiver channels.

From the generic diagram proposed in Figure 19.2, for a phased array system, some simplifications can be made to obtain the generic diagram proposed for synthetic aperture system, to acquire the FMA data and obtain an image using the TFM method, as illustrated in Figure 19.4. First the net of emission pulsers and the circuits for the focusing synchronization are substituted by only one emission pulser and a multiplexer system. Second, although in the figure the receiver net has been maintained, it can be redesigned to operate with less channels at the cost of increasing the number of emissions needed to produce the image. In an extreme minimum configuration the acquisition process can be done by only one channel, multiplexing the transducers in reception, as it is done in transmission [13].

An alternative to reduce electronic complexity and signal processing related to PA and TFM methods is the use of the classic SAFT (Synthetic Aperture Focusing Technique) configuration [35, 14], where each element of the array operates only in pulse-echo mode. In this case (19.8) reduces to:

$$I(x, z) = \frac{1}{N} \sum_{e=0}^{N-1} s_{ee}(t) * \delta\left(t - 2\tau_e(x, z)\right). \qquad (19.10)$$

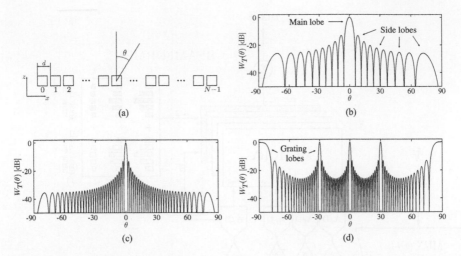

Figure 19.5 (a) Linear array with N elements and pitch d. One-way radiation pattern for a linear array with: (b) $N = 20$ elements and $\lambda/2$-pitch, (c) $N = 60$ elements and $\lambda/2$-pitch, and (d) $N = 20$ elements and 2λ-pitch.

19.1.3 THE COARRAY MODEL

By considering a linear array with N elements equally spaced by d, illustrated in Figure 19.5(a), the one-way radiation pattern of the array can be computed as the Discrete-Time Fourier Transform of the aperture [33]:

$$W_T(\theta) = \sum_{n=0}^{N-1} a[n]e^{-j\frac{2\pi}{\lambda}nd\sin(\theta)}, \qquad (19.11)$$

where $a[n]$ is the vibration amplitude of the nth array element, which is also called apodization function, and θ is the steering angle.

Figures 19.5(b), 19.5(c) and 19.5(d) illustrate the one-way radiation patterns ($W_T(\theta)$) for a linear array with different number of elements (N) and pitch (d) values, with uniform apodization ($a[n] = 1$ for $n \in [0, \quad N-1]$). By considering Figures 19.5(b) and 19.5(c), as N increases, the aperture size increases and the transmission is more concentrated around $\theta = 0^o$, i.e., the array is more directive, which is related to the main lobe width. In other directions other lobes can be observed, which are called side lobes. Side lobes are present in any array configuration, since the aperture has a finite size.

The main and side lobes characteristics are dependent on the number of elements, pitch and apodization function ($a[n]$). The use of different aperture sizes and apodizations results in different radiation patterns, in a similar way as windowing effects in spectral analysis [27]. The apodization can also be applied in the post-processing step, which means that different functions can be used for the same data set to extract the best characteristics from each one, with trade-off between main lobe width and side lobes levels [32].

Figure 19.5(d) illustrates a case when $d > \lambda/2$. In this case there is a spatial undersampling and high intensity lobes appear in directions different from 0°, called grating lobes. In this condition, artifacts appear in the generated image, leading to wrong interpretations of the structure integrity. Arrays whose elements are spaced by more than $\lambda/2$ are often called sparse arrays [17]. The use of sparse arrays is interesting as the number of elements is reduced, decreasing costs in hardware, data processing, and storage. Grating lobes are related to the discrete and periodic nature of the element distribution and are present in some sparse array configurations. Breaking the periodicity by randomly removing elements avoids the grating lobes [33]. Different apertures in transmission and reception [17], optimization techniques applied to elements distribution and weight [8] and coherence methods [3, 24] can be used to reduce the effects of grating lobes.

The pulse-echo radiation pattern is determined by the multiplication of the emission response ($W_T(\theta)$) by the reception response ($W_R(\theta)$). Assuming that $d = \lambda/2$, the pulse-echo response is:

$$W_{TR}(\theta) = \left[\sum_{n=0}^{N-1} a[n]e^{-j\pi n \sin(\theta)} \right] \left[\sum_{n=0}^{N-1} b[n]e^{-j\pi n \sin(\theta)} \right], \qquad (19.12)$$

where $a[n]$ and $b[n]$ are the apodization functions of the aperture in transmission and reception, respectively.

Using the Fourier transform and the convolution property, this response is equivalent to the one-way response produced by a linear array obtained by the convolution of emission and reception apertures:

$$c[n] = a[n] * b[n] = \sum_{m=0}^{2N-2} a[n-m]b[m], \qquad (19.13)$$

where $c[n]$ is called coarray distribution, or effective aperture.

The coarray concept is very useful in imaging systems in many steps, as in the design of the system, excitation/acquisition procedure, and post-processing. It is fast to compute and allows to evaluate the contribution of each signal to the beamforming process and image quality. For example, the jth element of $c[n]$, in (19.13), is composed by the sum of those elements in $a[k]$ and $b[l]$ that $k + l = j$. Although it is defined for narrowband operation, it is widely accepted that the same conclusions can be applied to wideband signals [17, 12, 19], which corresponds to the practical cases.

The coarray for a PA system is illustrated in Figure 19.6. Consequently, the two-way radiation pattern is equivalent to the one generated by a triangular apodization aperture with $2N - 1$ elements in transmission. If this process is applied to the FMA-SA scheme, as can be observed in Figure 19.7, the full coarray described for the SA case is progressively composed, introducing N new elements at each new transmission. Therefore, the two-way radiation pattern is identical to the one obtained with PA. However, the SNR is higher for PA, because all elements are used in transmission, while in SA only one element is used at a time.

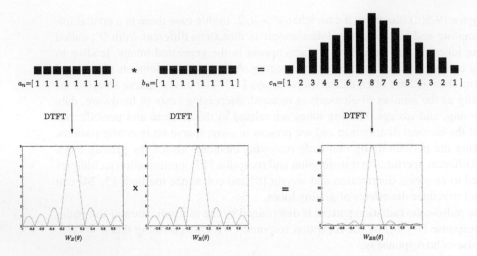

Figure 19.6 Aperture radiation pattern and coarray generation in PA for $N = 8$.

The beamformer is basically a spatial signal processor that exploits the redundancy in the signals to generate the image. Each coarray element position is considered a spatial frequency [13]. The coarray shows that in the beamforming process there is a certain degree of redundancy because several signals can fill the same spatial frequency.

SAFT uses only the pulse-echo signals and has reduced hardware complexity. Its coarray is illustrated in Figure 19.8. The resulting coarray has empty elements at the odd frequencies, which means that a $\lambda/2$-pitch array has a λ-pitch coarray. Consequently, there will be grating lobes in the two-way radiation pattern, at $\theta = \pm 90^o$. Then, to obtain a $\lambda/2$-pitch in the coarray, an array with $\lambda/4$-pitch is needed. In this case, to maintain the same aperture size, which is related to array directivity and lateral resolution, the number of elements must be doubled.

19.2   INSTANTANEOUS PHASE INFORMATION TO IMPROVE DE-FECT DETECTION

The use of SA techniques using the TFM method has the advantage that it generates the maximum lateral resolution available at each imaging point, producing high-quality images. However, because the individual signals can present low signal-to-noise ratio (SNR), there is a limitation in detecting reflectors or objects that are far from the array. For this reason, time-gain compensation is normally applied in conventional NDT techniques. Although the intensities of signals from the farther reflectors are increased, noise level is also amplified.

The coherence of the signals can be used to improve the quality of the images. At a defect position, all delayed signals should sum in phase, resulting in a high value (positive or negative) of the image at that point. At a position where there is not a defect, the amplitudes would sum randomly and lower values are expected.

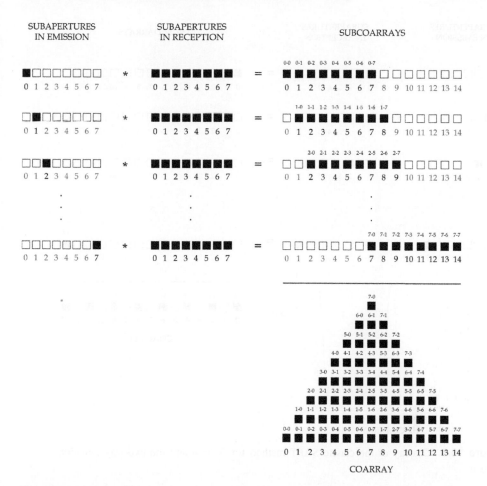

Figure 19.7 Transmission sequence of the elements for FMA process and its corresponding generated coarray for $N = 8$.

On the other hand, at this same point, for a given transmit-receive pair, an echo due to reflections from defects, or the boundaries of the structure coming from other directions, can arrive at the same time, increasing the magnitude of the sum and generating image artifacts at that point. A high noise level can also lead to similar results.

Methods that explore other characteristics of the signals involved in the beamforming process, such as the phase, have been proposed. Camacho, Parrilla, and Fritsch [3] presented a method based on the analysis of the phase diversity at the aperture data. Two coherence factors, the Phase Coherence Factor (PCF) and the Sign Coherence Factor (SCF), were proposed to weight the amplitude images. As a result, there is significant grating and side lobes suppression, as well as lateral resolution and SNR improvement.

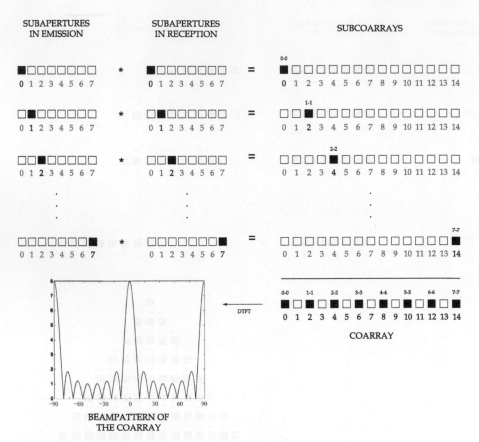

Figure 19.8 Coarray of the classic SAFT method for $N = 8$ and the two-way radiation pattern.

Martinez-Graullera et al. [24] proposed the Coherence Factor Map (CFM), which is a weighting factor obtained from a spectral analysis of the phase dispersion. The in-phase and quadrature components of the signals at the image point under consideration are used to obtain a descriptor that can be easily computed and introduced as a ponderation factor in the beamforming process. For a 64-element linear array operating with SA technique and dynamic focusing in emission and reception, the dynamic range improvement can reach up to 30 dB.

Traditionally, the coherence images have been used only as weighting factors, by directly multiplying them by the amplitude images, resulting in contrast and dynamic range improvements, and reduction in image artifacts due to side and grating lobes. However, they contain more information that can be used to further improve image quality and defects detection.

Following this idea, the use of the instantaneous phase (IP) of the signals has been proposed by Prado et al. [29] to generate a new image that complements the conventional beamformer process, providing more accurate information about the region of interest.

19.2.1 INSTANTANEOUS PHASE IMAGE

By considering a linear array of N elements and all combinations of transmitter and receiver data $s_{er}(t)$, the emitted signal by element e and received by element r, the IP image at point (x, z) is obtained by replacing the signal $s_{er}(t)$ by its instantaneous phase $\varphi_{er}(t)$ in (19.8):

$$I_{IP}(x, z) = \frac{1}{N^2} \sum_{e=0}^{N-1} \sum_{r=0}^{N-1} \varphi_{er}(t) * \delta(t - (\tau_e(x, z) + \tau_r(x, z))), \tag{19.14}$$

where $\varphi_{er}(t)$ can be calculated from [27]:

$$\varphi_{er}(t) = \arctan \left\{ \frac{\hat{s}_{er}(t)}{s_{er}(t)} \right\}, \tag{19.15}$$

and $\hat{s}_{er}(t)$ is the Hilbert transform of $s_{er}(t)$.

Equation (19.14) can also be written as:

$$I_{IP}(x, z) = \frac{1}{N^2} \sum_{e=0}^{N-1} \sum_{r=0}^{N-1} \varphi_{er}(\tau_e(x, z) + \tau_r(x, z)), \tag{19.16}$$

by replacing the amplitude signal by its instantaneous phase in (19.9).

A detailed analysis of the IP image is presented in [29], where a threshold parameter, to be applied to the IP image, is proposed:

$$\epsilon \triangleq \frac{1}{\sqrt{\log_{10} N_s}}, \tag{19.17}$$

where N_s is the number of signals used for imaging, which is equal to N^2 for (19.14) and (19.16).

As a result, a two-level image is obtained, which gives a statistical indication of whether the pixels of a region in the image are related to a reflector or noise/artifacts. The pixel value is considered due to, or part of, a reflector if $|I_{IP}(x, z)|$ (or its envelope, in practice) is above threshold, and noise if it is below. In the absence of a reflector there is a probability of false indication of a reflector, which is given by:

$$P_F \triangleq \frac{\sigma_0^2}{N_s} \log_{10} N_s, \tag{19.18}$$

where σ_0 is the standard deviation of the instantaneous phase of the noise, which is equal to $\pi/\sqrt{3}$ for an uniform distribution over 2π [4].

PSF images of a reflector 25 mm away from an $N = 32$ elements array with $\lambda/2$-pitch at 5 MHz in water ($\lambda = 0.3$ mm) are obtained by using (19.8) and (19.14).

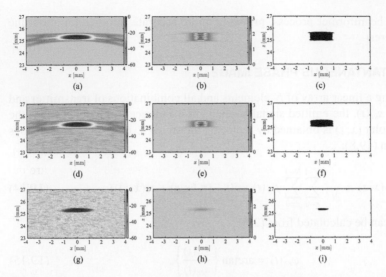

Figure 19.9 Amplitude (first column, in dB scale), IP (second column, in linear scale in radians), and two-level (zero or one) phase images (third column), respectively, for PSF scheme with different SNR values: (a-c): 20 dB; (d-f): 0 dB; (g-i): -30 dB.

Additive white Gaussian noise (AWGN) is added to the excitation signal, resulting in SNRs equal to 20, 0, and -30 dB. TFM and IP images are presented in Figure 19.9. The threshold in this case is $\epsilon = 0.5764$ rad, which is applied to the IP image, resulting in the two-level images presented in the third column in Figure 19.9, whose pixels assume 1 if they are above threshold and 0 otherwise.

For SNR equal or larger than -30 dB, the reflector is detected in the thresholded IP images, but for smaller SNR values there is no indication of the reflector. These results show that the IP method is very robust to SNR but has a limitation of a minimum SNR of operation, which is equal to -30 dB, for this simulation setup [29]. In practice, even if the SNR of each transmit-receive pair is low, the averaging effect of beamforming results in SNR improvement [9]. The SNR for each signal can be improved before beamforming by averaging in acquisition, resulting in an SNR improvement in proportion to the square root of the number of averages.

19.2.2 INSTANTANEOUS PHASE WEIGHTING FACTOR

As commented, coherence images can be used as weighting factors, by multiplying them by the amplitude images, resulting in contrast and dynamic range improvements, and reduction in image artifacts due to side and grating lobes. Camacho, Parrilla, and Fritsch [3] proposed the Sign Coherence Factor (SCF), which can be computed as:

$$I_{SCF}(x, z) = 1 - \sigma, \tag{19.19}$$

where

$$\sigma^2 = 1 - \left[\frac{1}{N^2} \sum_{e=0}^{N-1} \sum_{r=0}^{N-1} b_{er}(\tau_e(x,z) + \tau_r(x,z)) \right]^2, \qquad (19.20)$$

and $b_{er}(t)$ is the polarity signal of the aperture data:

$$b_{er}(t) = \begin{cases} -1, & \text{if} \quad s_{er}(t) < 0 \\ 1, & \text{if} \quad s_{er}(t) \geq 0 \end{cases}. \qquad (19.21)$$

σ is the standard deviation of the polarity $b_{er}(t)$ of the aperture data. The SCF measures the coincidence in algebraic sign of the received signals. Signals are defined as fully coherent if all of them have the same polarity ($I_{\text{SCF}}(x,z) = 1$). In other cases the value of $I_{\text{SCF}}(x,z)$ lies in the range $[0, 1[$.

The Coherence Factor Map (CFM) contains a measure of signal phase distribution combined at each pixel. The CFM is given by [24]:

$$I_{\text{CFM}}(x,z) = \frac{1}{N^2} \sum_{e=0}^{N-1} \sum_{r=0}^{N-1} \frac{s_{er}(\tau_e(x,z) + \tau_r(x,z))}{S_{er}(\tau_e(x,z) + \tau_r(x,z))}, \qquad (19.22)$$

where $S_{er}(t)$ is:

$$S_{er}(t) = \sqrt{s_{er}^2(t) + \hat{s}_{er}^2(t)}. \qquad (19.23)$$

In practice this method normalizes the signal, sample by sample, which is used to replace $s_{er}(t)$ in (19.9). The main difference between these methods is the signal that is summed in the beamforming process. If the signal of Figure 19.10(a) is used as the drive function, the IP image is related to the sum of the instantaneous phase of the signals, as illustrated in Figure 19.10(b). Likewise, the SCF is related to the sum of the polarity signals, as illustrated in Figure 19.10(c), and the CFM to the sample normalized signal, as illustrated in Figure 19.10(d). Direct (IP image) or indirectly (SCF and CFM), all methods are not using the amplitude information, but replacing it by the phase.

At a defect position, all delayed signals should sum in phase, resulting in a high value (positive or negative) in the coherence image. At a position where there is not a defect, the phases would sum randomly and lower values are expected. But the expected values and the distinction between reflector and noise have not been done by the authors of SCF and CFM methods. SCF is expected to be better than the CFM with respect to artifacts reduction, since the first is powered by two in (19.20), so attenuating the artifacts by a factor of two (in dB), and the last does not have any post-processing after beamforming.

As presented in the third column of Figure 19.9, a two-level image can be produced after thresholding the IP image using the factor ϵ. This thresholded image can then be used as reflector indication or as a weighting factor. However if a reflector is not detected by the thresholded IP image, due to its reflection coefficient or low SNR, for example, it would not be represented in the final weighted image. By assigning a value close to zero, in the range $]0, 1[$, the contrast is still improved, but weak reflectors can still be represented in the final weighted image.

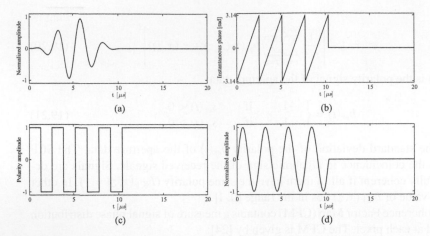

Figure 19.10 (a) Simulated 5 MHz four-cycles amplitude signal with a Gaussian envelope without noise and its: (b) instantaneous phase, (c) polarity, and (d) sample normalized signal.

Therefore the instantaneous phase weighting factor is defined as:

$$\text{IPWF}(x, z) = \begin{cases} 1, & \text{if} \quad |I_{IP}(x, z)| \geq \epsilon \\ P_F, & \text{if} \quad |I_{IP}(x, z)| < \epsilon \end{cases}, \qquad (19.24)$$

and the final weighted image is obtained by the multiplication of the amplitude image by the IPWF factor, pixel by pixel.

With simple simulations of PSFs with different array configurations and SNR conditions, the authors observed that the IPWF is very robust to side and grating lobes and noise, achieving better results when compared to other coherence factors, like SCF and CFM.

19.3 SIMPLIFIED PROCESS FOR SYNTHETIC APERTURE IMAGING

Although the efficiency of using FMA and TFM techniques to produce a high-quality image is well established, the huge volume of data which is necessary to acquire, transmit, and process is a serious drawback. Thus, it requires more storage resources and processing capability than any other synthetic aperture technique, which makes difficult its practical implementation with technology nowadays. To illustrate this, consider the following example: a 15 cm depth image, 40 MHz sampling rate, 64 channels, 1500 m/s medium velocity, and 2 bytes per sample. Each transmit-receive waveform demands approximately 1 MB of data, what means 64 MB of data to generate a single image frame when TFM is applied. For a frame rate of 20 images per second, it would be necessary to acquire and process 1.2 GB of data per second.

The bandwidth of most I/O standards available today put in evidence that any of current data protocols cannot deal with TFM requirements. Supposing a good efficiency and use of the resources, the USB 2.0 protocol would be able to transfer less than one image per second (48 MB/s). A similar situation occurs if USB 3.0

is employed, being the maximum transmission speed up to 480 MB/s, allowing us to transfer around 7 images per second, which is far beyond the necessary frame rate. Using recent standards as Thunderbolt, PCI Express, or DisplayPort, the transfer speed over copper wire is up to 960 MB/s, resulting in 14 images per second. Therefore, a reduction in data volume is desirable.

In order to reduce the number of signals involved in the process it is necessary to develop an acquisition strategy which is able to reduce the redundancy at each spatial frequency, but at the same time guaranteeing that the coarray has at least $\lambda/2$-pitch, in order to avoid grating lobes. The highest possible reduction is obtained when each coarray position is composed by just one signal (one emitter-receiver pair), which can be achieved by employing several acquisition strategies. These solutions are known as Minimum Redundancy Coarray (MRC) and allow a reduction from N^2 to $(2N-1)$ signals.

MRC solutions, in general, maintain the main characteristics of the original coarray aperture. Due to its flat shape (uniform apodization) it has better lateral resolution than the full aperture (triangular apodization) but higher side lobes. Nevertheless, the array radiation pattern properties can be changed in the post-processing stage, by the use of different apodization functions to produce an equivalent response to the TFM. In practice it has less dynamic range than the TFM solution because of the reduction in the number of signals in the average.

From the point of view of the system design and general performance of the imaging system, the reduction in the number of signals involved in the image process can improve velocity and reduce complexity. With this goal in mind, it is possible to establish several strategies which maintain a balance between the number of parallel channels and the number of transmissions during acquisition processes. Two methods are presented: 2R-SAFT and FAST-SAFT [20, 21, 22].

19.3.1   2R-SAFT

The 2R-SAFT technique [23] has some particular advantages that make it very useful for portable ultrasonic imaging systems. 2R-SAFT uses only one element in transmission and two elements in reception. As it is shown in Figure 19.11, all elements are sequentially activated as single emitters, without the use of any beamformer in emission. At each transmission, two consecutive channels are used as receivers, requiring to store two signals per emission.

Thus, when element i is used to emit a waveform, elements i and $i+1$ are used for receiving signals. For the last element of the array, only the pulse-echo signal is recorded. By employing only one emitter in each transmission, all received signals are completely uncorrelated, containing only information of a single transmitter-receiver pair. The coarray is fully populated, ensuring the suppression of grating lobes in the radiation pattern and producing good quality images.

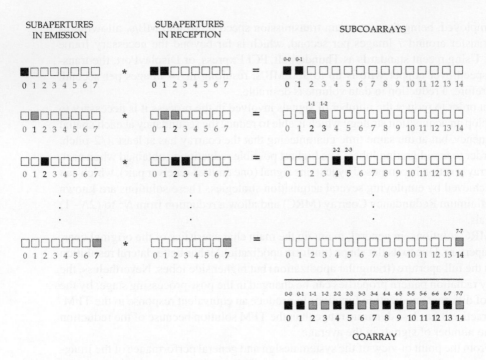

SUBAPERTURES
IN EMISSION

SUBAPERTURES
IN RECEPTION

SUBCOARRAYS

COARRAY

Figure 19.11   Coarray of the 2R-SAFT method.

19.3.2 FAST-SAFT

Another minimum-redundancy technique is denominated FAST-SAFT or, in its short form, kA-SAFT. The k index refers to the acceleration factor carried out during the acquisition stage, which can go from 2× to N×, depending on the number of channels used in reception. This strategy results in a slight increase in the cost involved in the acquisition system with respect to 2R-SAFT, but at the same time, reduces the number of transmissions by k times.

The kA-SAFT uses n_A consecutive elements to receive and a single element to emit, which is centered in the active subaperture (in the case of an even number of elements, the emitter is one of the two central elements of the subaperture). As shown in Figure 19.12, the elements in emission are sequentially activated with a shift of $n_A/2$ elements. At each transmission, n_A consecutive channels are used as receivers, and n_A signals are stored per emission, except for the first array element, where half of the signals are acquired, and the last array element, where one signal is acquired.

Figure 19.12 shows the coarray generated when kA-SAFT is employed for the case of $k = 2$ and $n_A = 4$. The number of signals to be recorded, as well as the coarray, are identical to that obtained with 2R-SAFT (Figure 19.11), preserving all its advantages, but the frame rate is doubled, in this case.

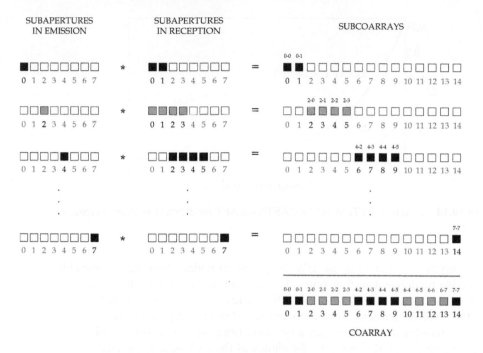

Figure 19.12 Coarray of the kA-SAFT method for $k = 2$ and $n_A = 4$.

19.3.3 CONSEQUENCES OF REDUNDANCY REDUCTION IN THE IP METHOD

As commented in Section 19.2.1, the IP image pixel is considered due to, or part of, a reflector if $|I_{IP}(x, z)|$ is above the threshold ϵ, and noise if it is below. In the absence of a reflector, the probability of false indication is described by P_F. As there are statistical issues in the definition of the threshold [29], redundancy is important for the IP defect detection methodology, because it improves certainty. Figure 19.13 presents the values of P_F for the three methods (TFM, 2R-SAFT, and kA-SAFT) for a different number of elements in an array, from 8 to 10000.

As expected, as the number of signals is reduced, the probability of false indication of defects is increased, since the method using the IP information comes from a statistical analysis of the number of elements. For example, for an 8-element array using the TFM, P_F is equal to 9.3%. By considering the same array using the 2R-SAFT or kA-SAFT, P_F is equal to 25.8%. These values are 0.3% and 5.4% for a 64-element array.

Although the probability of false indication of a defect is smaller using TFM than 2R-SAFT or kA-SAFT, it is not null, and spurious pixels can appear in the thresholded IP image for all cases. Furthermore, even if the minimum redundancy techniques are considered, which have higher probability of false indication, under high SNR conditions this may not be so significant. Furthermore, the IP approach has

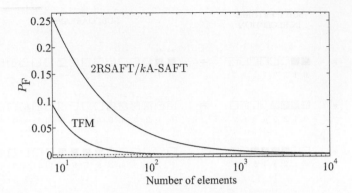

Figure 19.13 P_F values for TFM and 2R-SAFT/kA-SAFT for different number of elements in an array.

not the purpose of substituting the TFM image, but provides additional and valuable information for defect detection, improving conventional amplitude image analysis.

Due to the higher number of signals, TFM yields better results than 2R-SAFT or kA-SAFT, with better contrast and less artifacts. On the other hand, the cost with respect to number of transmissions, acquisition time, storage, and processing will also be higher for TFM. Therefore, the choice of the technique to be used will be related to the trade-off between the cost and image quality/defects detectability.

19.4   EXPERIMENTAL RESULTS

A commercial 128-element array (Imasonic, Besancon, France), 0.65 mm pitch, and central frequency of 5 MHz is used for imaging an aluminum block with defects, as illustrated in Figure 19.14. The defects are five pairs of drilled holes. The array is excited by a commercial 128-channel array system (SITAU-LF 32:128, Dasel SL, Madrid, Spain) with programmable digital pulses (bit width, excitation code) with -90 V amplitude, and the received digitized waveforms (12 bits resolution, up to 40 MHz sampling frequency) are transferred to the computer via USB interface for post-processing.

Figure 19.15 presents the images of the aluminum block obtained with 2R-SAFT, 4A-SAFT (kA-SAFT for k =4), and TFM. The first two methods required 255 signals in the beamformer, while the last one required 16384. For 2R-SAFT, 4A-SAFT, and TFM there were 128, 32, and 128 transmissions, respectively. In some cases, depending on the available hardware, the acquisition of the received signals is also multiplexed, using only one channel in transmission and one in reception. In that case, for TFM with a 128-element array, the number of required emissions would be 16384.

In all images it is possible to observe the multiple reflections between the reflectors, resulting in artifacts close to the true defects, with similar magnitudes. Other artifacts in the images are related to side lobes and noise.

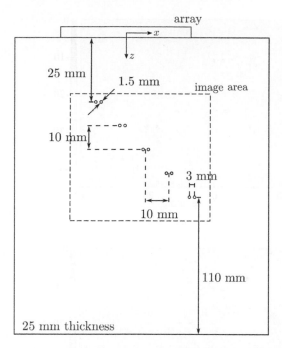

Figure 19.14   Aluminum block with five pairs of drilled holes as defects. The dashed rectangle limits the approximate area considered for imaging.

The images obtained with 2R-SAFT and 4A-SAFT are very similar to each other. In addition to the greater number of transmissions, the image obtained with 2R-SAFT results in slightly more artifacts, which can be observed in the region below the defects representation. In turn, TFM has better contrast and reduces the artifacts significantly, with the disadvantage of the higher number of signals used in beamforming. Even in the TFM image, artifacts related to the multiple reflections between the reflectors are represented with significant intensities.

Although the images in Figures 19.15(a) and 19.15(b) seem to be low-quality images, when compared to the one illustrated in Figure 19.15(c), all defects are represented with contrast higher than 15 dB, which allows the operator who is analyzing the structure to reach correct conclusions about its condition.

The IP method presented in this chapter was applied to the same dataset in order to discriminate defects from artifacts. The IP images as well as the thresholded IP images are illustrated in Figure 19.16. For 2R-SAFT (and 4A-SAFT) and TFM, the probability of false indication of defects is 3.10% and 0.08%, respectively, which is obtained by applying the thresholds of 0.644 rad and 0.487 rad, respectively.

As expected, as the number of signals decreases, the number of artifacts that are considered as defects increases. For the TFM method, which explores the redundancy, all artifacts are suppressed. Although the probability of false indication of defects for the minimum redundancy techniques is low (3.10%), there are spurious

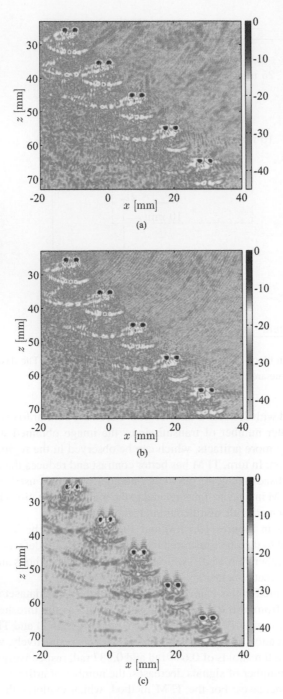

Figure 19.15 SA images of the aluminum block (image area described in Figure 19.14) obtained with (a) 2R-SAFT, (b) 4A-SAFT, and (c) TFM. Scales in dB.

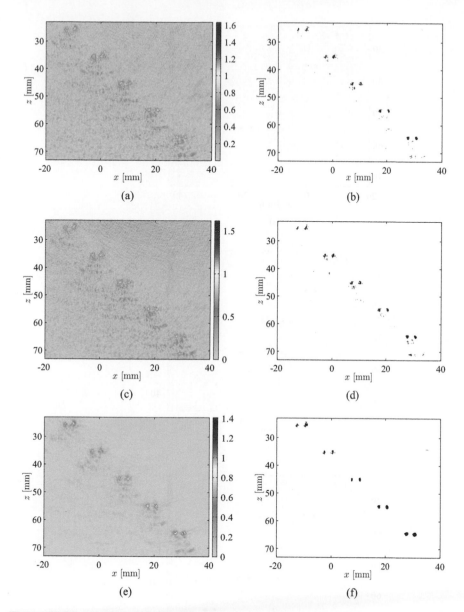

Figure 19.16 IP image (left, in radians) and thresholded IP image (right, zero or one) of the aluminum block (image area described in Figure 19.14) obtained with (a-b) 2R-SAFT, (c-d) 4A-SAFT, and (e-f) TFM.

pixels in the thresholded images. On the other hand, in general, artifacts due to high side lobes levels, noise, and due to multiple reflections between defects, are reduced and the thresholded IP images could be perfectly used as additional information for defect detection, improving conventional amplitude image analysis. Furthermore, if

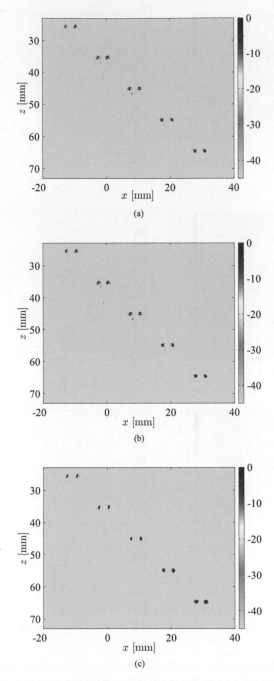

(a)

(b)

(c)

Figure 19.17 Weighted images of the aluminum block (image area described in Figure 19.14), obtained from the multiplication of the IPWF factor by (a) 2R-SAFT, (b) 4A-SAFT, and (c) TFM. Scales in dB.

it is considered that the minimum size of a defect that can be detected is, for example, a set of pixels with area greater than λ^2, a spatial filter could be applied to the thresholded images and the spurious artifacts related to false indications could be suppressed [29].

The amplitude images, illustrated in Figure 19.15, were multiplied by the IPWF factor and the results are presented in Figure 19.17. Despite some few (and small) false indications for the 2R-SAFT and 4A-SAFT, all defects are represented with improved contrast and resolution. Artifacts due to multiple reflections between the reflectors, side lobes, and noise are significantly reduced when compared to the original images, enhancing damage detection. Therefore, besides the use of the thresholded images for defect detection, the multiplication of the amplitude images by the IPWF factor is a good practice to improve the quality of images.

19.5   COMMENTS

Signal processing concepts are fundamental in the analysis, design, and implementation of ultrasonic NDT systems using arrays. This chapter described some basic concepts of image beamforming, such as phased array and synthetic aperture techniques. It also analyzed how redundancy is presented in sensor array systems, and how it can be exploited by signal processing techniques to improve the image quality or to establish a trade-off between hardware requirements and image quality in the design of the imaging systems.

The TFM conventional beamforming technique results in a high-quality image, but high data volume and processing is necessary. As an alternative, minimum redundancy coarray strategies can be used to reduce the cost of the system.

Coherence methods can be used to improve the conventional amplitude images, by using the phase information. The instantaneous phase information is a powerful tool to improve defects detection and image quality. From the same dataset used in TFM, the IP image is obtained, by replacing the amplitude data by its instantaneous phase in SA beamforming. The main contribution of the method proposed in [29] is the definition of a threshold level that is applied to the IP image, based on a statistical analysis of noise and the number of signals used in beamforming. The thresholded IP image can then be used to directly indicate the presence of a reflector, to create a weighting factor (IPWF) for amplitude images, or as a selection parameter in image compounding techniques.

When the coherence methods are used to weight MRC and TFM images, the results are very similar. As a consequence, real-time implementation may become possible by using these techniques.

REFERENCES

1. C. J. Brotherhood, B. W. Drinkwater, and R. J. Freemantle. An ultrasonic wheel-array sensor and its application to aerospace structures. *Insight*, 45(11):729–734, 2003.

2. C. B. Burckhardt, P. Grandchamp, and H. Hoffman. An experimental 2 MHz synthetic aperture sonar system intended for medical use. *IEEE Transactions on Sonics and Ultrasonics*, 21(1):1–6, 1974.

3. J. Camacho, M. Parrilla, and C. Fritsch. Phase coherence imaging. *IEEE Transactions on Ultrasonics, Ferroelectrics and Frequency Control*, 56(5):958–974, 2009.

4. A. B. Carlson and P. B. Crilly. *Communication systems*. McGraw-Hill Higher Education, 2009.

5. S. Chatillon, G. Cattiaux, M. Serre, and O. Roy. Ultrasonic non-destructive testing of pieces of complex geometry with a flexible array transducer. *Ultrasonics*, 38:131–134, 2000.

6. B. W. Drinkwater and P. D. Wilcox. Ultrasonic arrays for non-destructive evaluation: a review. *NDT&E International*, 39(7):525–541, 2006.

7. J. J. Flaherty, K. R. Erikson, and V. M. Lund. Synthetic aperture ultrasonic imaging systems, 1967.

8. S. Holm, B. Elgetun, and G. Dahl. Properties of the beampattern of weight- and layout-optimized sparse arrays. *IEEE Transactions on Ultrasonics, Ferroelectrics and Frequency Control*, New York, 44(5):983–991, 1997.

9. C. Holmes, B. W. Drinkwater, and P. D. Wilcox. Post-processing of the full matrix of ultrasonic transmit-receive array data for non-destructive evaluation. *NDT&E International*, 38(8):701–711, 2005.

10. C. Holmes, B. W. Drinkwater, and P. D. Wilcox. Advanced post-processing for scanned ultrasonic arrays: application to defect detection and classification in non-destructive evaluation. *Ultrasonics*, 48(6-7):636–42, November 2008.

11. A. J. Hunter, B. W. Drinkwater, and P. D. Wilcox. The wavenumber algorithm for full-matrix imaging using an ultrasonic array. *IEEE Transactions on Ultrasonics, Ferroelectrics and Frequency Control*, 55(11):2450–62, November 2008.

12. J. A. Jensen, S. I. Nikolov, K. L. Gammelmark, and M. H. Pedersen. Synthetic aperture ultrasound imaging. *Ultrasonics*, 44(1):e5–e15, 2006.

13. M. Karaman, H. S. Bilge, and M. O'Donnell. Adaptive multi-element synthetic aperture imaging with motion and phase aberration correction. *IEEE Transactions on Ultrasonics Ferroelectrics and Frequency Control*, 45(4):1077–87, 1998.

14. M. Karaman, P. Li, and M. O'Donnell. Synthetic aperture imaging for small scale systems. *IEEE Transactions on Ultrasonics, Ferroelectrics and Frequency Control*, 42(3):429–442, May 1995.

15. G. Konstantinidis, P. D. Wilcox, and B. W. Drinkwater. An investigation into the temperature stability of a guided wave Structural Health Monitoring system using permanently attached sensors. *IEEE Transactions on Ultrasonics, Ferroelectrics and Frequency Control*, 7(5):905–912, 2007.

16. J. Krautkramer and H. Krautkramer. *Ultrasonic testing of materials*. Springer-Verlag, 1983.

17. G. R. Lockwood, P. Li, M. O'Donnell, and F. S. Foster. Optimizing the radiation pattern of sparse periodic linear arrays. *IEEE Transactions on Ultrasonics, Ferroelectrics and Frequency Control*, 43(1):7–14, 1996.

18. S. Mahaut, O. Roy, C. Beroni, and B. Rotter. Development of phased array techniques to improve characterization of defect located in a component of complex geometry. *Ultrasonics*, 40:165–169, 2002.

19. C. J. Martin-Arguedas, O. Martinez-Graullera, G. Godoy, and L. G. Ullate. Coarray synthesis based on polynomial decomposition. *IEEE Transactions on Image Proccesing*,

19(4):1102–1107, 2010.

20. C. J. Martin-Arguedas, O. Martinez-Graullera, D. Romero-Laorden, R. T. Higuti, and L. G. Ullate. Linear scanning method based on the SAFT coarray. In *36th Annual Review of Progress in Quantitative Nondestructive Evaluation*, pages 2023–2030, Kingston (Rhode Island), 2009.

21. C. J. Martin-Arguedas, O. Martinez-Graullera, D. Romero-Laorden, M. Perez-Lopez, and L. G. Ullate. Improvement of synthetic aperture techniques by means of the coarray analysis. In *The International Congress on Ultrasonics*, pages 189–192, Gdansk, Poland, 2011.

22. C. J. Martin-Arguedas, O. Martinez-Graullera, D. Romero-Laorden, and L. G. Ullate. Method and architecture to accelerate multi-element synthetic aperture imaging. *Digital Signal Processing*, 23(4):1288–1295, February 2013.

23. C. J. Martin-Arguedas, O. Martinez-Graullera, L. G. Ullate, and G. Godoy. Reduction of grating lobes in SAFT images. In *IEEE International Ultrasonics Symposium*, number 1, pages 721–724, Beijing, China, 2008.

24. O. Martinez-Graullera, D. Romero-Laorden, C. J. Martin-Arguedas, A. Ibanez, and L. G. Ullate. A new beamforming process based on the phase dispersion analysis. In *The International Congress on Ultrasonics*, pages 185–188, Gdansk, Poland, 2011.

25. J. E. Michaels and T. E. Michaels. Guided wave signal processing and image fusion for in situ damage localization in plates. *Wave Motion*, 44(6):482–492, 2007.

26. L. Moreau, B. W. Drinkwater, and P. D. Wilcox. Transmission cycles for rapid nondestructive evaluation. *IEEE Transactions on Ultrasonics, Ferroelectrics and Frequency Control*, 56(9):1932–1944, 2009.

27. A. V. Oppenheim, R. W. Schafer, and J. R. Buck. *Discrete-Time Signal Processing*. Prentice-Hall, New Jersey, 1999.

28. W. Ostachowicz, P. Kudela, P. Malinowski, and T. Wandowski. Damage localisation in plate-like structures based on PZT sensors. *Mechanical Systems and Signal Processing*, 23(6):1805–1829, 2009.

29. V. T. Prado, R. T. Higuti, C. Kitano, and O. Martinez-Graullera. Instantaneous phase threshold for reflector detection in ultrasonic images. *IEEE Transactions on Ultrasonics, Ferroelectrics, and Frequency Control*, 61(7):1204–1215, 2014.

30. C. W. Sherwin, J. P. Ruina, and R. D. Rawcliffe. Some early developments in synthetic aperture radar systems. *IRE Transactions on Military Electronics*, MIL-6(2):111–115, April 1962.

31. R. A. Smith, J. M. Bending, L. D. Jones, and T. R. C. Jarman. Rapid ultrasonic inspection of ageing aircraft. *Insight*, 45(3):174–177, 2003.

32. H. C. Stankwitz, R. J. Dallaire, and J. R. Fienup. Nonlinear apodization for sidelobe control in SAR imagery. *IEEE Transactions on Aerospace and Electronic Systems*, New York, 31(1):267–279, 1995.

33. B. D. Steinberg. *Principles of aperture and array system design: including random and adaptive arrays*. A Wiley-Interscience publication. Wiley, 1976.

34. S. Sung-Jin, H. J. Shin, and Y. H. Jang. Development of an ultrasonic phased array system for non-destructive tests of nuclear power plant components. *Nuclear Engineering and Design*, 214:151–161, 2002.

35. R. N. Thomson. Transverse and longitudinal resolution of the synthetic aperture focusing technique. *Ultrasonics*, Surrey, 22(1):9–15, 1984.

36. A. Velichko and P. D. Wilcox. Guided wave arrays for high resolution inspection. *Journal of the Acoustical Society of America*, Melville, 123(1):186–196, 2008.

37. J. T. Ylitalo and H. Ermert. Ultrasound synthetic aperture imaging: monostatic approach.

IEEE Transactions on Ultrasonics, Ferroelectrics and Frequency Control, 41(3):333–339, 1994.

38. L. Yu and V. Giurgiutiu. In situ 2-D piezoelectric wafer active sensors arrays for guided wave damage detection. *Ultrasonics*, 48(2):117–134, 2008.

Index

Printed and bound by CPI Group (UK) Ltd, Croydon, CR0 4YY

23/10/2024

01777706-0003